Textbook on
Power Electronics and
Industrial Applications

Textbook on
Power Electronics and
Industrial Applications

Harish C Rai
PhD (Electrical Engg, IIT, Delhi), FIE (India), FIETE, MISTE, MAeSI

Former
Professor, Department of Electrical and Electronics Engineering
CR State College of Engineering
(Presently Deenbandhu Chhotu Ram University of Science and Technology)
Murthal, Haryana 131039

Controller of Examinations
Director, Academic Affairs
Director, Research Project Monitoring Cell
Director. Organization and Development
GGS Indraprastha University, Delhi
Advisor, All India Council of Technical Education
(Ministry of Human Resource and Development, Delhi)

CBS Publishers & Distributors Pvt Ltd

New Delhi • Bengaluru • Chennai • Kochi • Kolkata • Mumbai
Hyderabad • Jharkhand • Nagpur • Patna • Pune • Uttarakhand

Textbook on
Power Electronics and Industrial Applications

ISBN: 978-93-86827-86-9

Copyright © Author and Publisher

First Edition: 2018

Published by Satish Kumar Jain and produced by Varun Jain for
CBS Publishers & Distributors Pvt Ltd
4819/XI Prahlad Street, 24 Ansari Road, Daryaganj, New Delhi 110 002, India.
Ph: 23289259, 23266861, 23266867
Fax: 011-23243014
Website: www.cbspd.com
e-mail: delhi@cbspd.com; cbspubs@airtelmail.in.
Corporate Office: 204 FIE, Industrial Area, Patparganj, Delhi 110 092
Ph: 4934 4934   Fax: 4934 4935   e-mail: publishing@cbspd.com; publicity@cbspd.com

Branches

- **Bengaluru:** Seema House 2975, 17th Cross, K.R. Road,
  Banasankari 2nd Stage, Bengaluru 560 070, Karnataka
  Ph: +91-80-26771678/79   Fax: +91-80-26771680   e-mail: bangalore@cbspd.com
- **Chennai:** 7, Subbaraya Street, Shenoy Nagar, Chennai 600 030, Tamil Nadu
  Ph: +91-44-26680620, 26681266   Fax: +91-44-42032115   e-mail: chennai@cbspd.com
- **Kochi:** Ashana House, 39/1904, AM Thomas Road, Valanjambalam,
  Ernakulam 682 016, Kochi, Kerala
  Ph: +91-484-4059061-62-64-65   Fax: +91-484-4059065   e-mail: kochi@cbspd.com
- **Kolkata:** 6/B, Ground Floor, Rameswar Shaw Road, Kolkata-700 014, West Bengal
  Ph: +91-33-22891126, 22891127, 22891128   e-mail: kolkata@cbspd.com
- **Mumbai:** 83-C, Dr E Moses Road, Worli, Mumbai-400018, Maharashtra
  Ph: +91-22-24902340/41   Fax: +91-22-24902342   e-mail: mumbai@cbspd.com

Representatives

- **Hyderabad** 0-9885175004
- **Nagpur** 0-9021734563
- **Pune** 0-9623451994
- **Jharkhand** 0-9811541605
- **Patna** 0-9334159340
- **Uttarakhand** 0-9716462459

Printed at: Rashtriya Printers, Dilshad Garden, Delhi, India

to
_____

My respected parents

Late Sh Balraj

Late Smt Ram Devi

My lively children, Shivanshu and Himanshu

My loving wife, Sangeeta;

and Shipra and grandson Vivaan, whose

affection is always appreciated

Preface

Power electronics is an interdisciplinary area of applied science that functions using the members of thyristor family, control electronics to control the 'switch ON' and 'switch OFF' processes of the devices and principles of control theory. The invention of thyristor has revolutionized all fields of conversion and control, namely AC to DC, DC to DC, DC to AC and AC to AC. Hence, there is a necessity for understanding the thyristor and its family members, principles of power converters, for their efficient use in the industry. Power semiconductor technology is rapidly developing, after the advent of fast switching power devices with increasing voltage and current limits. Power switching devices such as power BJTs, power MOSFETs, SITs, IGBTs, MCTs, SITHs, SCRs, TRIACs, GTOs, and other semiconductor devices are finding increasing application in a wide range of products. With the availability of faster switching devices, the applications of modern microprocessors in synthesizing the control strategy for gating power devices to meet the conversion specifications are widening the scope of power electronics.

The book *Power Electronics and Its Industrial Applications* is a reflection of the entire subject, and explains all the aspects in detail with elaborate explanation and illustrations. The book begins with an introduction to thyristors, TRIACs, asymmetrical thyristors, GTOs, power transistors, power MOSFETs and hybrid devices. The entire content is divided into 18 chapters. The control circuits described for various applications using thyristors are not meant to be commercially feasible but provide the basic understanding necessary for synthesizing more sophisticated controllers.

Chapter 1 contains an overview of power electronics and its applications in brief.

Chapters 2–4 are devoted to power semiconductor devices, rectifying circuits, filters and Zenor diodes and thyristor and its family.

Chapters 5–7 deal with triggering devices, firing and commutation circuits. These chapters introduce different members of thyristor family and describe characteristics, working principles and application of DIACs, TRIACs, MOSFETs, IGBTs, UJT, PUTs etc. The triggering and commutation of SCRs have also been included. Controlled rectifiers, DC to DC converters, AC voltage controllers, AC to AC converters, cycloconverters and inverters are widely discussed topics in power electronics are reviewed in Chapters 8–12 with emphasis on the circuits and waveforms.

Thyristorised controlled DC and AC motors are covered extensively in Chapters 13–14.

Chapters 15 and 16 present with thyristor protection circuits and their industrial applications using solid state devices for the control of DC and AC motors and circuits used for various control purposes like heating, welding and many other industrial applications and also elaborate some microprocessors based applications.

Chapter 17 deals with regulated power supply; industrial drives applications using microprocessor are detailed in Chapter 18.

Each chapter is supplemented with elaborate illustrations, solved examples and multiple choice questions to provide an aid in comprehension of principles involved. At the end of the book, multiple choice questions (MCQs) from various examinations like Engineering Services Examination (IES), GATE and many other competitive examinations have been included.

The book will be useful to college and university students as well as engineers working in industry. I hope this book will be of immense use to the teachers and students of technical institutes. Suggestions from students and teachers for improvement in future editions of this book are welcome.

Harsh C Rai

Acknowledgements

I thank all my undergraduate students who suggested that I should write this book and indeed, all those who have encouraged me in this venture. I derive immense pleasure in expressing my sincere thanks to Prof Yogesh Singh, Vice Chancellor, Delhi Technological University; Prof Annu Singh Lather, Pro Vice Chancellor, Delhi Technological University, for the invaluable encouragement throughout this work. I am indebted to his guidance and invaluable suggestions.

I express my gratitude to Prof SS Murthy, former Vice Chancellor, Central University of Karnataka; Prof BP Singh; Prof ZH Zaidi, former Vice Chancellor, MJP Rohilkhand University, Bareilly, Prof Bhim Singh, Department of Electrical Engineering, IIT, Delhi, for sparing their valuable time and providing useful guidance on various chapters.

I thank my colleagues, Prof Alok Mittal, Member Secretary, AICTE, New Delhi; Prof JRP Gupta (NSIT, Delhi); Prof DR Bhaskar (DTU); Prof VK Sharma (NIT, Uttarakhand); Prof DR Rominder Randhwa, Director, Guru Tegh Bahadur Institute of Technology; Prof SS Tyagi, Director, BSA Institute of Technology, Haryana and Prof Lajpat Rai, IIT, Delhi with whom I have discussed power electronics while teaching courses on this subject.

I express my gratitude to my brother Dr Mahesh Popli (Income Tax Department), Rajasthan; Dr Vikas Gupta (DU); Mr Pankaj Munjal, Director, Training and Development, RVIT, Bijnore; Brig Pradeep Upmanu; Dr Nitin Malik, GGS Indraprastha University and Mr Ankit Popli for their immense help and constructive criticism on the manuscript.

My special thanks are due to Sh RC Taneja and late Sh KR Munjal for their moral support which has enabled me to complete this work. I am grateful to Sh Satish Kumar Jain (Mataji), CMD and Sh Varun Jain (Director), CBS Publishers & Distributors Pvt Ltd, New Delhi, for his patience, goodwill and cooperation. I express my gratitude to Mr YN Arjuna (Senior Vice President Publishing, Editorial and Publicity); Mrs Ritu Chawla (AGM Production); Mr Sumit Bhel; Ms Sanjubala Tripathy (Copy Editor) and Mrs Madhu Srivastva (Data Vision), for their skillful service and immense help in editing and figure work of the manuscript.

Finally, I appreciate the patience and solid support of my family—my wife Sangeeta Rai; children Shivanshu, Shipra and Himanshu.

Harsh C Rai

Contents

8. AC TO DC CONVERTER CONTROLLED RECTIFIER

Role of Power Electronics Applications

1.0 INTRODUCTION

The field of electrical engineering may be divided into three areas of specialisation, namely electronics, power and control. Electronics, basically deals with the study of semiconductor devices and circuits at low power levels. Power involves the generation, transmission and distribution of electrical energy, and rotating machines. The stability and response characteristics of closed loop system are the concern of control area of specialisation.

Power electronics deals with the use of electronics for the control of large power. Power electronics era started with high power tubes like thyratrons, ignitrons and mercury arc rectifiers. With the advent of power semiconductor devices like thyristor, TRIACS, power MOSFET and power transistors, power electronics has become very important in the control of large power. The major component of power electronics circuit is the thyristor, which is a fast switching semiconductor device whose function is to modulate the power in AC and DC systems. Modulation of power can vary from 10 W to 100 MW by turning the switch on and off in a particular mode.

1.1 APPLICATIONS OF POWER ELECTRONICS

Power electronics has already found an important role in modern technology. Examples of successful applications are as follows:

1. Rolling mill drives, in which the motor generator set of the conventional Ward–Leonard system is replaced by a controlled rectifier inverter
2. Electric drives for rapid-transit systems, in which wasteful control resistors are eliminated
3. Multiple drive systems for textile and paper mills
4. Machine tool control
5. Aircraft power supplies
6. Uninterruptible power supplies (UPS) for important loads such as computers and space applications
7. Illumination controls for lighting in trains, homes and theatres
8. High voltage direct current (HVDC) systems
9. Frequency converters for motor controls
10. Battery charging and welding systems

11. Vehicle propulsion systems
12. Excitation systems for alternators and synchronous condensers
13. Heat controls involving arc melting, induction heating and electrolysis

It is difficult to draw the boundaries for the applications of power electronics, specially with the present trends in the development of power devices and microprocessors, the upper limit is undefined.

1.2 ADVANTAGES AND DISADVANTAGES OF THYRISTOR CONVERTER SYSTEMS

1.2.1 Advantages

1. Fast response of the thyristor systems as compared to electromechanical converter systems
2. High efficiency due to less losses in thyristors
3. Long life and reliability due to the absence of moving parts
4. Less and easy maintenance
5. Less noise
6. Control flexibility
7. Small size and low weight, require less floor area

1.2.2 Disadvantages

1. Power semiconductor converters have a tendency to introduce current and voltage harmonics into supply systems and controlled systems
2. Harmonics in the supply system causes interference with communication systems and distortion of supply voltage
3. Thyristor controllers have low overload capacity
4. Regeneration of power is difficult
5. Some converters such as controlled rectifier, cycloconverter (AC to AC converters) and the AC voltage controller suffer from a poor power factor, particularly at low output voltages

1.3 POWER SEMICONDUCTOR CONVERTERS

The configuration of basic electric machines like DC machine, induction motors, and synchronous motors have remained essentially the same for the past several decades and will most likely remain so for many years in future. However, the techniques for controlling these machines has recently changed in a significant way. For example, a series of DC motors is used to propel subway cars. The speed of these cars has been controlled for many years by inserting resistances in series with the DC motors as shown in Fig. 1.1 (a).

In recent years, solid state choppers which can convert a fixed voltage DC into a variable voltage DC have been used for this purpose as shown in Fig. 1.1 (b). Solid state control provides smother control and higher efficiency. Other electric machines can be controlled by using appropriate converters which are discussed in this chapter. The following are various types of converters that are frequently used to control electric machines and conversion applications.

1. **Controlled rectifier (AC to DC converter):** A controlled rectifier converts a fixed voltage AC to a variable voltage DC. It is used primarily to control the speed of DC motors, such as those used in rolling mills. Figure 1.2 shows the block diagram of a closed-loop speed control scheme for DC motors using a thyristor phase controlled rectifier.

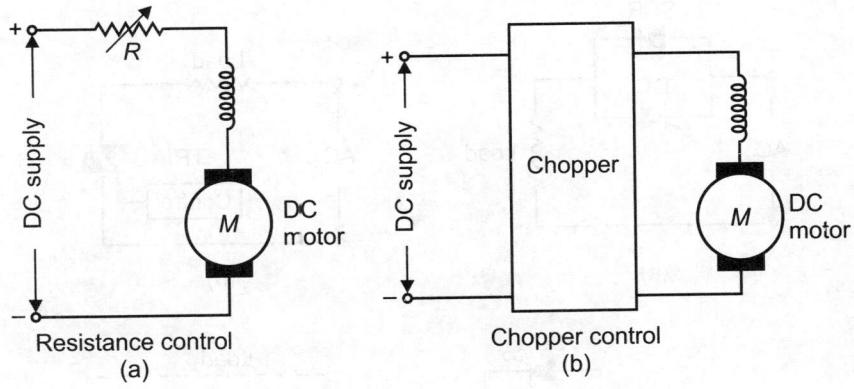

Fig. 1.1 Speed control of DC series motor

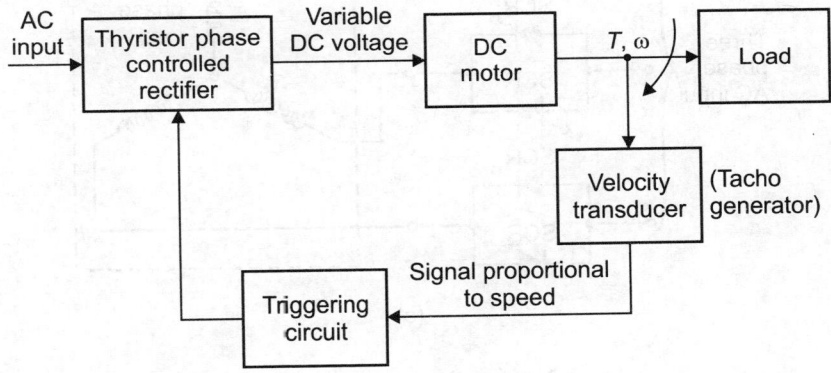

Fig. 1.2 Block diagram of thyristorised speed control DC motor

2. **AC voltage controllers (AC to AC):** An AC voltage controller converts a fixed voltage AC to a variable voltage AC. Variable voltage AC output is used for lighting control, speed control of fans, fractional horse power AC motors, three-phase induction motors, and for smooth induction motor starting. Three-phase AC voltage controller with Y-connected load is shown in Fig. 1.3.

3. **Chopper (DC to DC converter):** A DC chopper converts a fixed DC voltage to variable DC voltage. A DC chopper is used to control the speed to DC motor from a battery or DC supply. The block diagram of DC motor speed control scheme using the chopper is shown in Fig. 1.4.

4. **Inverter (DC to AC inverter):** An inverter converts a fixed voltage DC to a fixed (or variable) voltage AC with variable frequency. Inverters can be used to control the speed of three-phase induction and synchronous motors.

Some of the applications of inverters are:

(a) Generation of 50 Hz, fixed voltage AC from a DC source obtained from wind power, solar generation or batteries
(b) Uninterrupted power supply (UPS)
(c) Induction heating
(d) Standby power supply

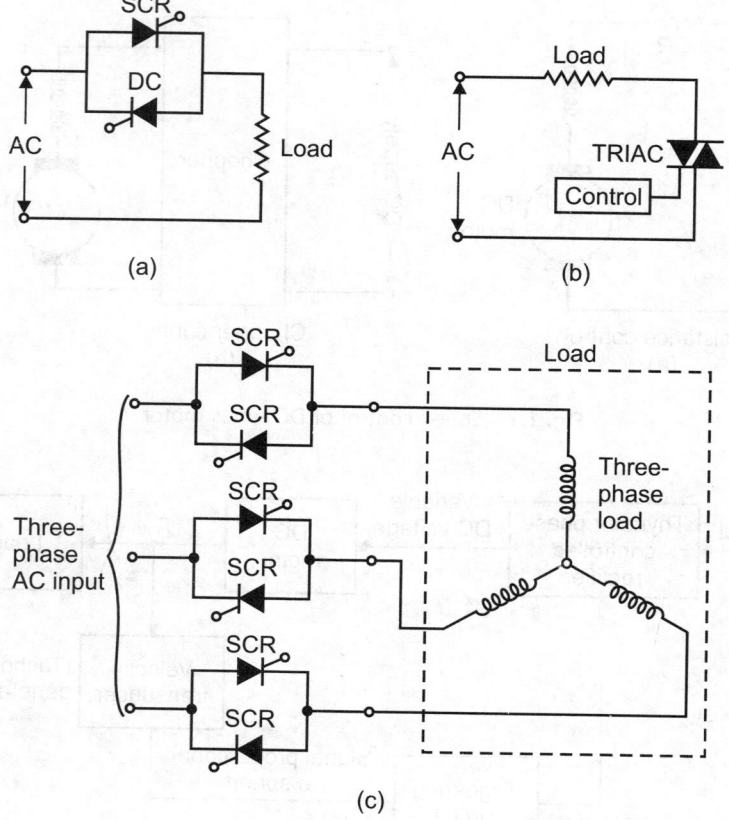

(a)

(b)

(c)

Fig. 1.3 AC voltage controller circuits

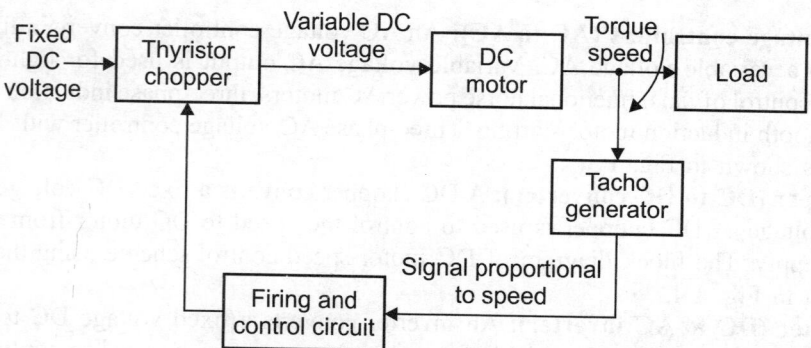

Fig. 1.4 Block diagram of a speed controlled DC motor using chopper

Figure 1.5 is a block diagram of the closed loop speed control system of a three-phase induction motor.

Figure 1.6 shows the scheme for an UPS application uninterrupted power supply when the power failure is of longer duration and unlimited protection time is required for domestic industrial applications.

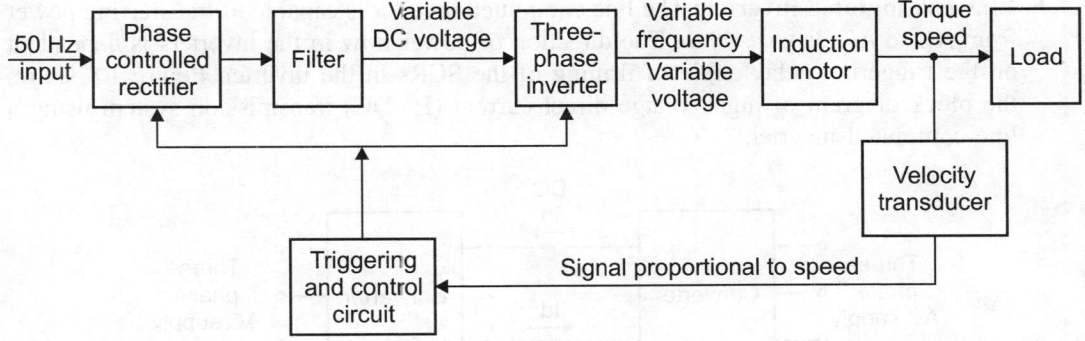

Fig. 1.5 Closed loop speed control system for a three-phase induction motor

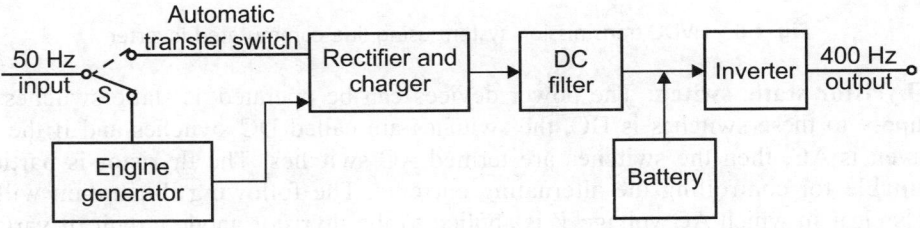

Fig. 1.6 Uninterrupted power supply system

5. **Cycloconverter (AC to AC converter):** A cycloconverter converts a fixed voltage and fixed frequency AC to a variable voltage and variable (lower) frequency AC. The output voltage is controlled by varying the firing delay angle. It can be used to generate a constant frequency, constant voltage from a variable frequency AC source. The cycloconverter principle is widely used in the following applications:

(a) Speed control of large AC drives like rotary kilns, etc.
(b) Static variable speed constant frequency (VSCF) generators for aircraft

The basic principle of operation of a static AC to AC frequency changer is shown in the block diagram of Fig. 1.7.

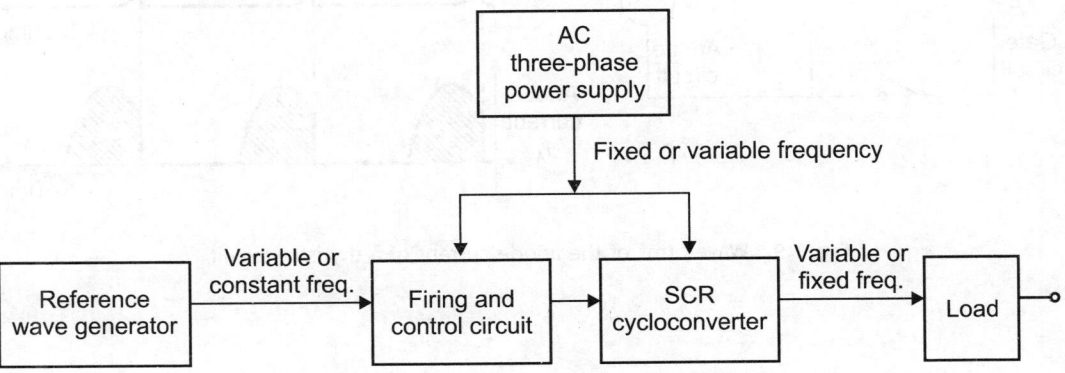

Fig. 1.7 AC to AC frequency changer system

6. Line commutated inverter: The line commuted inverter is capable of transferring power from AC to AC or *vice versa*. The direction of power flow in the inverters is dependent on the triggering delay angle of firming of the SCRs in the inverter. Figure 1.8 shows the block diagram of high voltage direct current (HVDC) transmission system using a line commuted inverter.

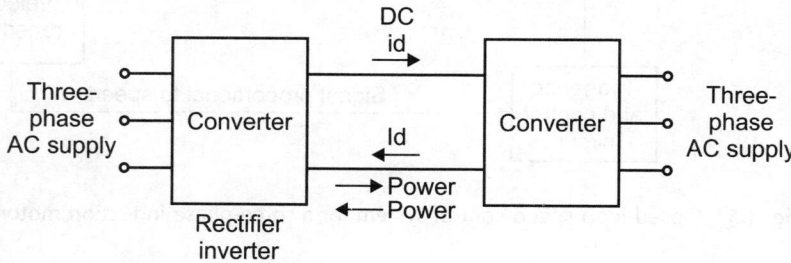

Fig. 1.8 HVDC transmission system using line commutated inverter

7. Thyristor static switch: The power devices can be operated as static switches. If the supply to these switches is DC, the switches are called DC switches and if the supply given is AC, then the switches are termed AC switches. The thyristor is particularly suitable for controlling the alternating currents. The following illustration will make this clear in which AC voltage V is applied to the thyristor anode circuit. It varies with time. By closing the switch S in the gate circuit momentarily, gate pulses against equal intervals of time are produced. Figure 1.9 shows the waveform of the anode current against time. Anode current flows from the instant when the triggering pulse is applied to the zero crossing point of the applied AC voltage.

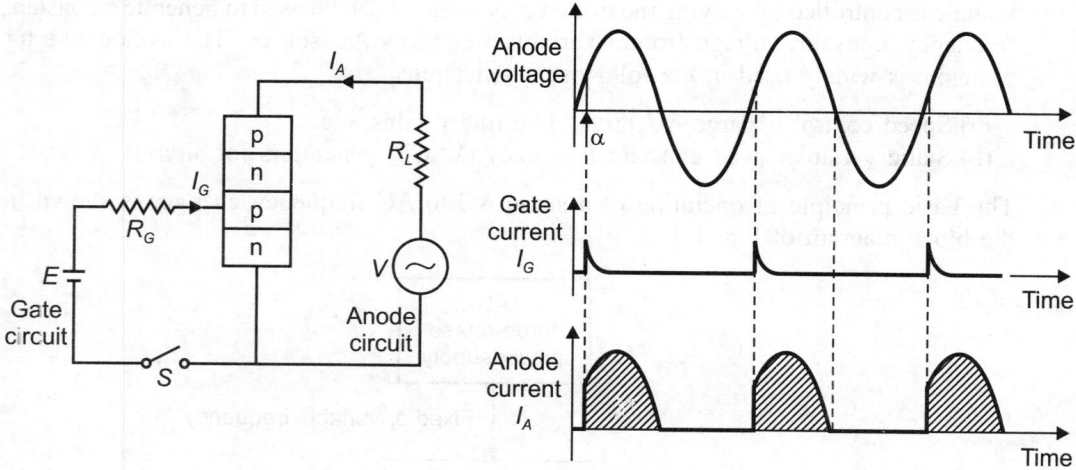

Fig. 1.9 Waveforms of the anode current of a thyristor circuit

Power Semiconductor Devices

2.0 INTRODUCTION

The electronics power conversion circuits that convert and control electrical power, use power semiconductor devices. The devices operate in the switching mode, which causes the losses to be reduced and therefore the conversion efficiency is to be improved. However, the disadvantages of switching-mode operation are the generation of harmonics and radio frequency interference (RFI). Electric power flow from a source to a load may be controlled either by varying the supply voltage or by inserting a regulator. The two common forms of series regulator are: (1) a switch and (2) an adjustable stepless impedance.

An ideal switch [Fig. 2.1(a)] changes between two states of operation, namely fully on which is a condition of zero impedance and zero voltage drop, and fully off, which is a condition of infinite impedance. An ideal switch changes between its 'on' and 'off' states instantaneously and hence there is no energy dissipated in the switching.

When power is regulated by controlled variation of series resistance, as shown in Fig. 2.1 (b) there occurs a waste of energy in the form of heat which reduces the system efficiency. For this reason the control of high power circuits is invariably approach via the use of semiconductor switching devices used as switches.

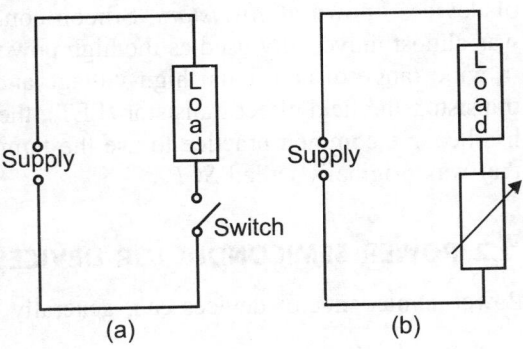

Fig. 2.1 Power regulated by (a) ideal switch, (b) variation of series resistance

A high power semiconductor switching device must possess two essential features:

1. It must be able to withstand large current flow in the ON state, I_{on} and high voltage in the OFF state, V_{off}.
2. It must possess on acceptability high power control gain ratio. A device that required most power to operate at the control electrode as the main channel power would be of a little use.

The output characteristics of liver diseases can be presented in graphical form by I_{on} and V_{on} across the device during conduction. The product of these variable represents device dissipation—largely heat loss and corresponds to inefficiency of control. The device can only withstand a certain fixed maximum amount of heat, P_{max} which corresponds to a parabolic I_{on} versus V_{on} locus on the output characteristic.

There are also maximum and minimum current limits. These boundaries enclose an area on the graph known as the safe operating area (SOA). Its size indicates the power handling capability of the device, the larger it is, the more rugged and useful the device.

2.1 LIMITATIONS OF PRACTICAL SEMICONDUCTOR SWITCHES

Practical semiconductor switches are imperfect. They possess a very low but finite on-state resistance which results in conductor voltage drop and device heating. The off-state resistance is very high but finite resulting in leakage current in either the forward or reverse direction depending on the polarity of the applied voltage. Further, switching on and switching off actions do not occur instantaneously. Each occupies a significant time span. Both switch-on and switch-off are accompanied by dissipation, which appears as heat and causes the device temperature rise.

In an application where the semiconductor device undergoes frequent switching, such as most AC load control situations, the switch-on and switch-off power losses may be added to the steady state conduction loss to form the incidental loss. The fact that semiconductor power switches are almost—but not quite ideal creates problems with regard to the design of semiconductor circuits and the ratings of their components. The maximum power handling capability of a device may be considered as the product of $V_{off\,max} \cdot I_{off\,max}$. The basic semiconductor switch is the solid-state diode which is a two-electrodes, anode voltage actuated, p-n junction device. The most comprehensive forms of silicon switches are found in a family of devices known as *thyristors*. Silicon controlled rectifier is a member of thyristor family was almost universally used as the high power semiconductor switch. There are now available in wide range of additional high voltage and high current rated devices such as the bipolar transistor, the field effect transistor (FET), the TRIAC and the gate turn off (GTO) thyristor. It has become common practice to use the family name thyristor to denote the particular device that was originally called *SCR*.

2.2 POWER SEMICONDUCTOR DEVICES

Power semiconductor devices can, generally be classified as follows:

1. Power diode
2. Power bipolar junction transistors (BJTs)
3. Power MOSFET
4. Insulated-gate bipolar junction transistors (IGBTs)
5. Thyristor

These devices are operated in the switching mode so that the losses are reduced and conversion efficiency is improved. The disadvantage of switching mode operation is the generation of harmonics and radio frequency interference (RFI). In this section, the external electrical characteristics of these devices are discussed briefly.

2.3 POWER DIODE

A power diode is a two-terminal p-n junction device and a p-n junction is normally formed by allowing diffusion, and epitaxial growth. The structure of a power diode and its symbol

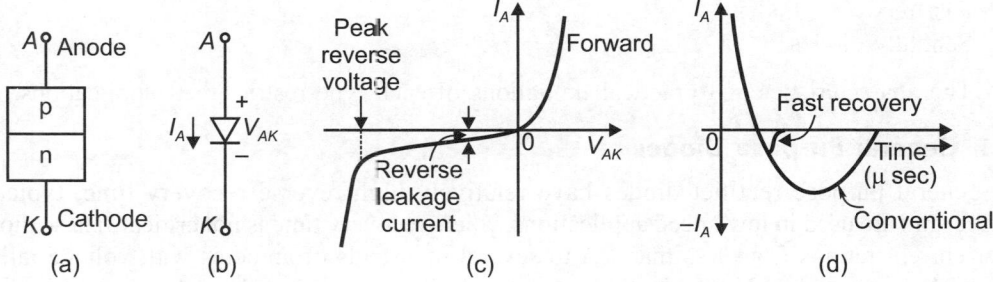

Fig. 2.2 Structure of a power diode and its symbol

are shown in Figs 2.2 (a) and (b). High power diodes are silicon rectifiers that can operate at high junction temperatures. Power diodes have larger power, voltage, and current handling capabilities than ordinary signal diodes. In addition, the switching frequencies of power diodes are low as compared to signal diodes.

The voltage–current characteristic of a diode is shown in Fig. 2.2 (c). When the anode potential is positive with respect to cathode, the diode is said to be forward biased, the diode conducts and behaves essentially as a closed switch. A conducting diode has a relatively small forward voltage drop across it, and the magnitude of this drop would depend on a manufacturing process and temperature. When the cathode potential is positive with respect to anode, the diode is said to be reverse biased, it behaves essentially as an open circuit. Under reverse biased condition, a small reverse current, known as leakage current in the range of micro- or milliampere, flows and this leakage current increases slowly in magnitude with the reverse voltage until the avalanche voltage is reached. The forward voltage drop when, it conducts current, is in the range of 0.8 to 1 V. Diodes with ratings as high as 4000 V and 2000 A are available.

Following the end of forward conduction in a diode, a reverse current flows for a short time. The device does not attain its full blocking capability until the reverse current ceases. The time interval during which reverse current flows is called rectifier recovery time. During this time, charge carriers stored in the diode at the end of forward conduction are removed. The recovery time is in the range of a few microseconds (1–5 μsec) in a conventional diode to several hundred nanoseconds in fast recovery diodes. This recovery time is of great significance in high frequency applications. The recovery characteristics of conventional and fast recovery diodes are shown in Fig. 2.2 (d).

The applications of these devices include electric traction, battery charging, electroplating, electromechanical processing, power supplies, welding, and uninterruptible power supply (UPS) systems.

2.3.1 Types of Power Diode

Ideally, a diode should have no reverse recovery time. However, manufacturing cost of such a diode would increase.

In many applications, the effects of reverse recovery time would not be significant, and inexpensive diodes can be used.

Depending on the recovery characteristics and manufacturing techniques, the power diodes can be classified into *three* categories:

1. General purpose diodes

2. Fast recovery diodes
3. Schottky diodes

The characteristics and practical limitations of each type restrict their applications.

2.3.2 General Purpose Diodes

The general purpose rectifier diodes have relatively high reverse recovery time, typically 25 μsec, and are used in low-speed applications, where recovery time is not critical. These diodes cover current ratings from less than 1 A to several thousands of amperes with voltage ratings from 50 V to around 5 kV. These diodes are generally manufactured by diffusion. However, alloyed types of rectifiers which are used in welding power supplies are most effective and rugged, and their ratings can go up to 300 A and 1000 V.

2.3.3 Fast Recovery Diodes

The fast recovery diodes have low recovery time, normally less than 5 μsec. They are used in chopper and inverter circuits, where the speed of recovery is often of critical importance. These diodes cover current ratings from less than 1 A to hundreds of amperes, with voltage ratings from 50 V to 3 kV.

For voltage ratings above 400 V, fast recovery diodes are generally made by diffusion and recovery time is controlled by platinum or gold diffusion. For voltage ratings below 400 V, epitaxial diodes provide faster switching speeds than that of diffused diodes. The epitaxial diodes have a narrow base width resulting in fast recovery time of as low as 50 nanosec.

2.3.4 Schottky Diodes

The charge storage problem of a p-n junction can be minimised in a Schottky diode. It is accomplished by setting up a 'barrier potential' with a contact between a metal and a semiconductor.

A layer of metal is deposited on a thin epitaxial layer of n-type silicon. The potential barrier simulates the behaviour of p-n junction. However, the rectifying action depends on the majority carriers only and as a result there is no excess minority carriers to recombine. The recovery effect is due to the self-capacitance of the semiconductor junction. The recovered charge of a Schottky diode is much less than that of an equivalent p-n junction diode. It has a relatively low forward voltage drop.

The leakage current of a Schottky diode is higher than that of a p-n junction diode. A Schottky diode with relatively low conduction voltage has relatively high leakage current. As a result, its maximum allowable voltage is limited to 100 V. The current ratings of Schottky diodes varies from 1 to 300 A. The Schottky diodes are ideal for high-current and low-voltage DC power supplies. However, they are used in low-current power supplies for increased efficiency.

2.4 PERFORMANCE PARAMETERS OF POWER DIODES

The semiconductor diodes are characterised by manufacturers in terms of certain performance parameters. The designer needs to determine the correct semiconductor diode to meet a given requirement. These parameters are non-linear and dependent on a number of factors. The manufacturers normally provide characteristic curves for important parameters in the form of data sheet having the following terms:

- *Junction temperature T_J*: T_J is the average temperature over the whole junction. $T_{J\,(max)}$ is the maximum junction temperature that a diode can withstand without failing due to thermal runway. This is a very important parameter and influences the ratings of other parameters. The range of T_J is typically –40 to 125°C, where $T_{J\,(max)}$ is 125°C. If a diode is operated below its minimum limit (–40°C), the crystal structure of the silicon wafer may be fractured.

- *Storage temperature T_{stg}*: T_{sig} defines the range of temperatures within which a non-conducting diode can be stored and transported. It is typically –40° to 150°C.
- *Ambient temperature T_A*: T_A is the temperature of the cooling medium and is measured with a thermometer close to the diode on the heat sink.
- *Case temperature T_C*: T_C is the temperature at the case, normally at the base of the diode and it can be measured with a thermocouple.
- *Sink temperature T_S*: T_S is the temperature of the hottest spot on the heat sink.
- *Junction to case thermal resistance, $R_{th\,JC}$*: $R_{th\,JC}$ is the effective thermal resistance between the junction and the outer case of the device. It is a measure of the heat transfer ability of the materials and mechanical construction of the diode and it is specified in °C/W and expressed as

$$R_{th\,JC} = T_J - T_C/P_D$$

where P_D is the power dissipated in the diode.

- *Case to sink thermal resistance $R_{th\,CS}$*: $R_{th\,CS}$ is the effective thermal resistance between the case and sink. Its value depends on the nature of the mounting surface. It is expressed as

$$R_{th\,CS} = \frac{T_C - T_S}{P_D}$$

- *Maximum average forward current $I_{F\,(AV)}$*: $I_{F\,(AV)}$ is the maximum allowable value of average forward current at a specified temperature. These data are normally quoted for a half-sine wave at a case temperature, $T_C = 85°C$.
- *Maximum rms forward current $I_{F(RMS)}$*: $I_{F(RMS)}$ is the maximum permissible rms value of forward current. This signifies the heating effect due to i^2R dissipation and is limited due to thermal stress on the device.
- *Maximum peak repetitive forward current I_{FRM}*: I_{FRM} is the maximum peak permissible current which can be applied on a repetitive basis. This is normally specified for a half-sinusoidal waveform.
- *Forward voltage drop V_F*: V_F is the instantaneous value of the forward voltage drop and is dependent on the junction temperature T_J.
- *Maximum peak forward voltage V_{FM}*: V_{FM} is the maximum forward voltage drop at a specified forward current and junction temperature.
- *Threshold voltage V_{TO}*: V_{TO} is the current-independent portion of forward voltage drop and depends on the temperature.
- *Reverse recovery time t_{rr}*: t_{rr} is defined as the time interval between the instant the current passes through zero during the changeover from the forward conduction to reverse blocking condition and the moment the reverse current has delayed to 25% of its peak reverse value i_{RR}.
- *i^2t for fusing*: The i^2t rating is a measure of the maximum forward non-recurring over current capability for a given period of 10 ms at a specified junction temperature. It is used to determine the thermal capability of fuses. For the protection of diode, i^2t rating of the fuse must be less than i^2t of the diode.

2.5 SERIES AND PARALLEL CONNECTED DIODES

Generally, the power handling capability of a single power diode is not sufficient. For example, in the converter station terminals of HVDC transmission lines and in rapid transit systems, the power requirements are such that a single diode will not be able to meet them. Diodes are connected in series to increase the reverse blocking capabilities. Steady state and transient voltages must be equally shared when diodes are connected in series.

In high-power applications, diodes are connected in parallel to increase the current carrying capability to meet the desired current requirements. When diodes are connected in parallel, each diode must share the total load current equally, both under transient and steady state conditions.

2.5.1 Series Operation of Diodes

Sometimes it is necessary to use semiconductor diodes which are outside the voltage rating range of commercially available diodes. In such cases, diodes may be connected in series to increase the voltage rating, to meet the requirements.

Assume two exactly similar diodes connected in series. The $V–I$ characteristics of the two diodes are seldom identical due to production spread. In the forward direction, both diodes conduct the same amount of current, and almost equal voltage appears across each diode. The current rating of the diodes in series is the same as the current rating of one of the diodes.

In the reverse direction, the same current flows in each diode, and each diode supports a different reverse voltage due to dissimilar characteristics. The ratio of the voltages across the diodes will depend on how dissimilar the diode characteristics are, as shown in Fig. 2.3. The voltage rating of the diode pair must be larger than the rating of one diode, because part of the total applied voltage will appear across the other diode.

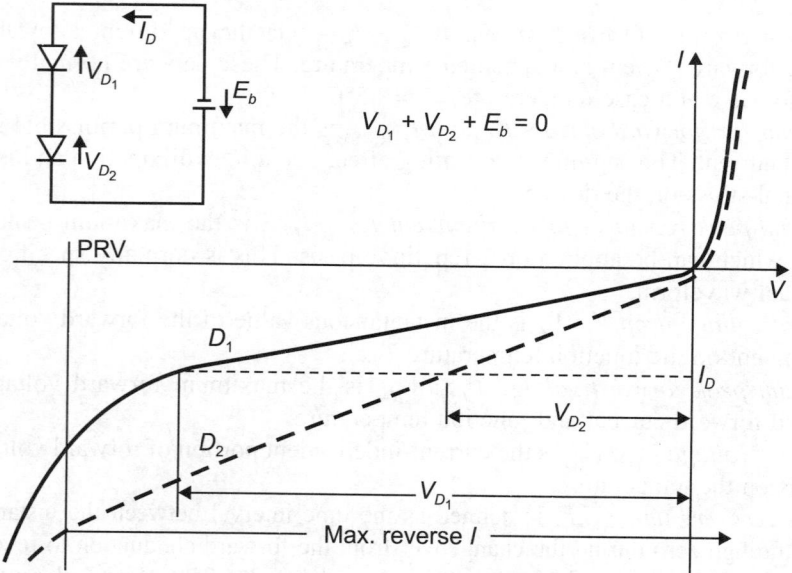

Fig. 2.3 Diode characteristics

Thus, the series connection of diodes requires either the selection of diodes with identical $V–I$ characteristics or the addition of circuitry to force equal sharing of voltage. This can be achieved by connecting a resistor across each diode. Although a different value of resistor could be placed across each diode to achieve some optimum voltage division, a more practical approach is to place the same value of resistor across each diode, which eliminates the problem of matching resistor values to individual diode characteristics, and making replacement of defective units simple. Figure 2.4 shows the effect of placing resistors 'R' across the diodes.

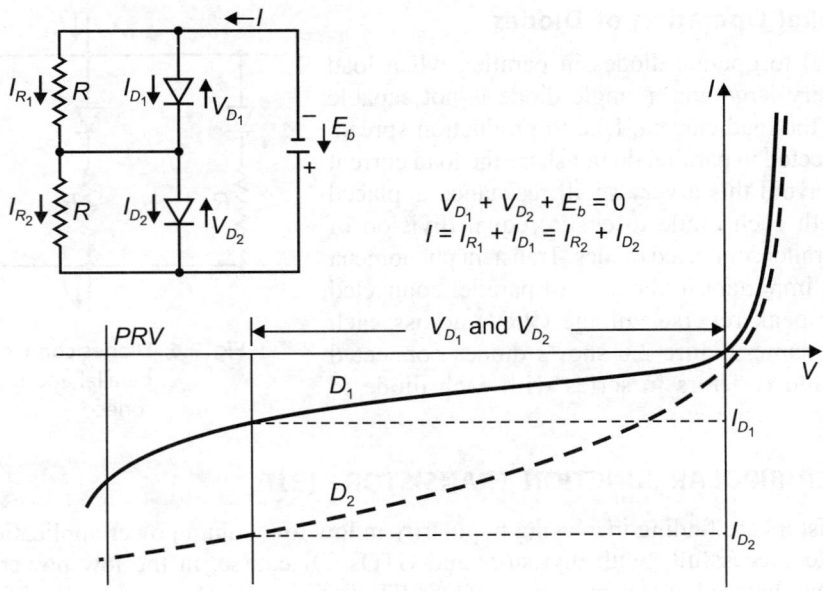

$$V_{D_1} + V_{D_2} + E_b = 0$$
$$I = I_{R_1} + I_{D_1} = I_{R_2} + I_{D_2}$$

Fig. 2.4 Effect of placing resistors

The steady state voltage distribution of a string of series connected diodes must be ensured by external circuitry. External circuitry must provide sharing of transient voltage changes caused by switching loads, lightning, or initial application of voltages to the diode circuit. If a voltage transient causes an increase in the reverse voltage applied to a series string of diodes, initially all of the diodes pass a changing current which depends upon the transient amplitude, the circuit load, and the diode characteristics. Due to production spread, some diodes will approach a new steady state distribution of carriers before other diodes. These faster acting diodes in series attempt to control and block the current associated with the voltage transient. Thus, the faster acting diodes will receive reverse voltage greater than their fair share (the total applied voltage divided by the number of diodes). The voltage across the faster acting diode may well exceed the peak reverse voltage rating of the diode and cause the diode to fail. If the diode fails by shorting out, the voltage on the other diodes of the string increases causing other diodes to fail.

Series connected diodes are protected from voltage transients by connecting a capacitor around each diode as shown in Fig. 2.5. The capacitors bypass the voltage transients around the diode string, dividing the voltage transients equally among the diodes by a capacitive divider action, and limiting the rate of change of voltage across the fast acting diodes.

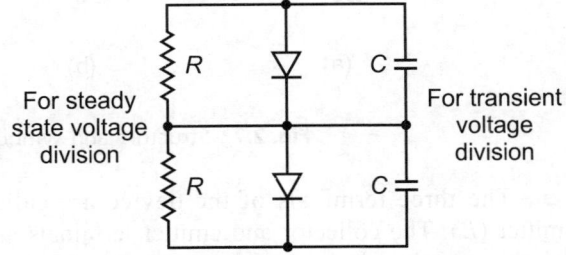

Fig. 2.5 Voltage transients by connecting a capacitor around each diode

The expected transient magnitude must be estimated in order to choose the capacitor voltage rating as well as the peak reverse voltage (PRV) of the diodes, and the entire circuit must be considered in choosing the value of capacitance necessary to limit the rate of change of voltage across the fastest acting diode.

2.5.2 Parallel Operation of Diodes

It is essential to operate diodes in parallel, when load current is very large and a single diode is not capable of handling the load current. Due to production spread, diodes connected in parallel do not share the load current equally. To avoid this a very small resistance is placed in series with each diode to ensure equal division of current in parallel connected diodes. Transient phenomena are not very important in the case of parallel connected diodes. The peak reverse voltage (PRV) across each diode is the same. Figure 2.6 shows diodes connected in parallel and resistors in series with each diode to force sharing of current.

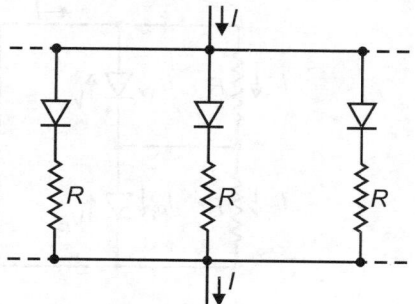

Fig. 2.6 Diodes connected in parallel and resistors in series

2.6 POWER BIPOLAR JUNCTION TRANSISTORS (BJTs)

Power transistors are finding increasing popularity in low to medium power applications, where they compete successfully with thyristors and GTOs. Of course, in the low power range, its popularity has been changed by power MOSFET devices, which are discussed in the next section. A power transistor has low current gain and requires continuous base drive during on-state conditions but does not require forced commutation circuitry. Power transistors can be used in high switching frequency, permitting size reduction of electromagnetic components and can provide current-limit protection by the base drive circuit. Power transistor cannot withstand reverse voltage and application is therefore limited to DC voltage fed inverters and choppers.

A transistor is a three-layered p-n-p or n-p-n semiconductor device having two junctions. This type of transistor is known as bipolar junction transistor (BJT). The structure and symbol of an n-p-n transistor are shown in Figs 2.7(a) and (b).

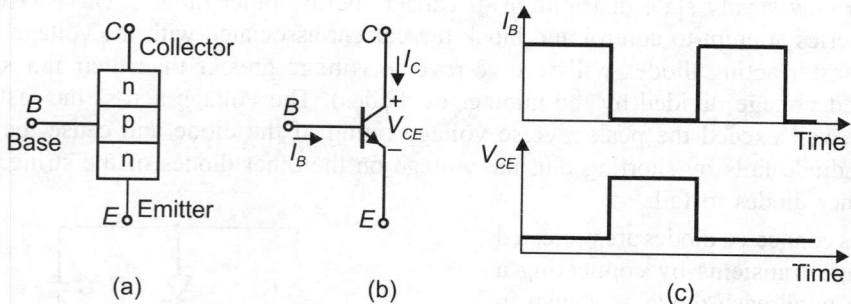

Fig. 2.7 Structure and symbol of an n-p-n transistor

The three terminals of the device are called the collector (C), the base (B) and the emitter (E). The collector and emitter terminals are connected to the main power circuit, and the base terminal is connected to control signal.

Like thyristors, transistors can also be operated in the switching mode. If the base current I_B is zero, the transistor is in an off-state and behaves as a switch. On the other hand, if the base is driven hard, that is, if the base current I_B is sufficient to drive the transistor into saturation, then the transistor behaves as a closed switch. This type of operation is illustrated in Fig. 2.7(c).

BJTs are current-controlled devices and base current must be supplied continuously to keep them in the on-state. The DC current gain life is usually only 5–10 in high-power transistors and so these devices are sometimes connected in a Darlington or Tripple Darlington configuration as shown in Figs 2.8(a) and (b) to achieve a larger current gain.

Some disadvantages accrue in this configuration including slightly higher overall $V_{CE(\text{sat})}$ values and slower switching speeds.

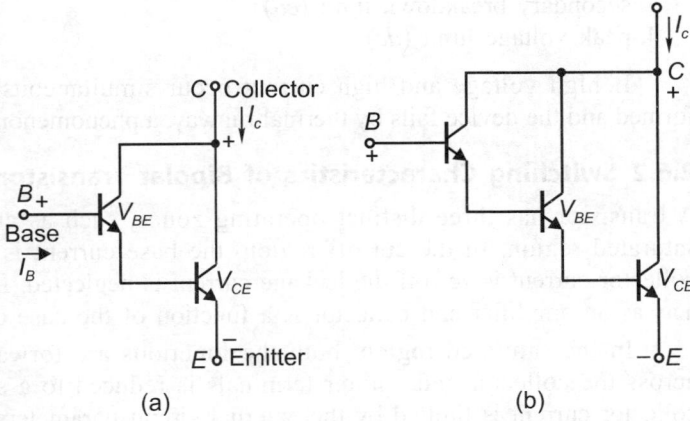

Fig. 2.8 Darlington configuration

BJTs, whether in single units or made as a Darlington configuration on a single chip [Monolittic Darlington (MD)] have significant storage time during the turn off transition. Typical switching times are in the range of a few hundred nonoseconds to a few microseconds.

BJTs including MDs are available in voltage ratings up to 1400 V and current ratings of a few hundred amperes. In spite of a negative temperature coefficient of on-state resistance, modern BJTs fabricated with good quality control can be paralleled provided that care is taken in the circuit layout and some extra current margin is provided, that is, where theoretically four transistors in parallel suffice based on equal current sharing, five may be used to tolerate a slight current imbalance.

The pair can be fabricated on one chip or two discrete transistors can be physically connected as a Darlington transistor. Power transistors switch on and switch off, much faster than thyristors. They may switch on in less than 1 μsec and turn off in less than 2 μsec. Therefore, power transistor can be used in applications where frequency is as high as 50 kHz. These devices are, however, very delicate. They fail under certain high voltage and high current conditions. They should be operated within the specified limits, known as safe operating area (SOA).

2.6.1 Safe Operating Area (SOA)

During switching, power transistors show a complex phenomenon known as the second breakdown effect. This may be the reason for frequent device failure if the circuit is not designed carefully. While avalanche breakdown is defined as the first breakdown, the second breakdown can be defined as breakdown of the junction due to localised heating effects.

The SOA is partitioned into four regions, as shown in Fig. 2.9 defined by the following limits:

1. peak current limit (*ab*)
2. power dissipation limit (*bc*)

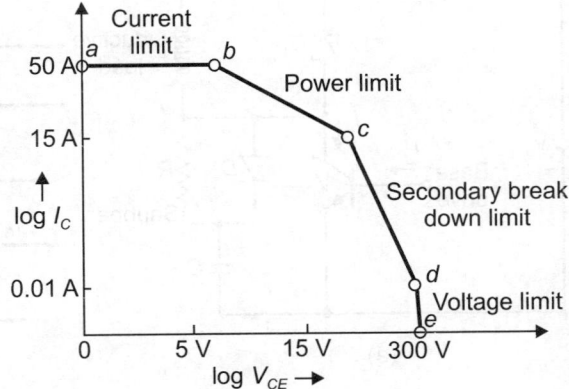

Fig. 2.9 Safe operating area curve

3. secondary breakdown limit (*cd*)
4. peak voltage limit (*de*)

If high voltage and high current occur simultaneously during turn off, a hot spot is formed and the device fails by thermal runway, a phenomenon known as *secondary breakdown*.

2.6.2 Switching Characteristics of Bipolar Transistor

A transistor has three distinct operating zones, such as cut off region, active region and saturated region. In the cut off region, the base current is either zero or negative and the collector current is zero if the leakage current is neglected. In the active region, the transistor acts as an amplifier and collector is a function of the base current.

In the saturated region, both the functions are forward biased and the voltage drop across the collector and emitter terminals is reduced to a small value about 0.7 V and the collector current is limited by the external circuit parameters. The transition from one region to other is effected by base current. As a result, the transistor can be operated as controlled switch, so that its operated point may assume two stable states only, one in the cut off region and the other in the saturated region. The cut off and the saturated states correspond to the 'off' and 'on' positions of the switch respectively. The transistor is operated as a switch in the common-emitter configuration because of maximum power gain available at this mode. The transistor cannot be called an ideal switch but it has many favourable points which enable it to operate as a switch. Polarised snubbers are used with power transistors to avoid simultaneous occurrence of peak voltage and peak current. Figure 2.10(a) shows the effects of snubber circuit on the turn off characteristics of a power transistor. A chopper circuit with an inductive load is considered.

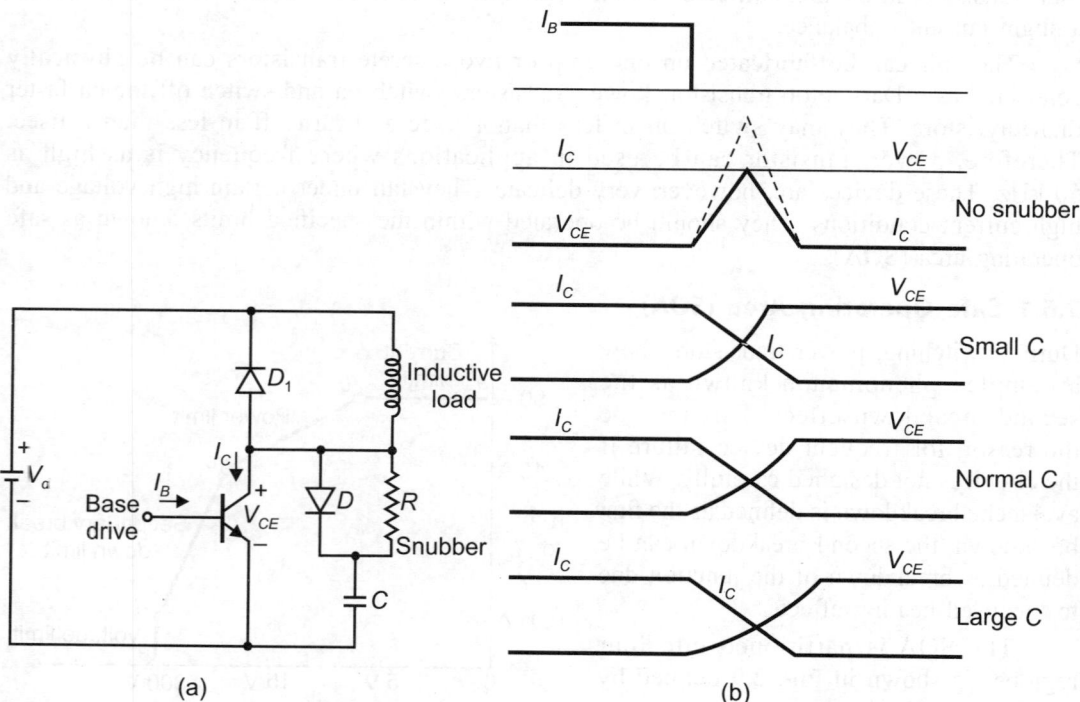

Fig. 2.10   Effects of snubber circuit

If no snubber circuit is used and the base current is removed to turn off the transistor, the voltage across the device V_{CE}, first rises, and when it reaches the DC supply voltage V_d, the collector current I_C falls. The power dissipation P during the turn off interval is shown in Fig. 2.10(b) by dashed line. In these idealised waveforms, the peaks of V_{CE} and I_C occur simultaneously, and this may lead to secondary breakdown failure.

If the snubber circuit is used and base current is removed to turn off the transistor, the collector current is diverted to the capacitor. The collector current I_C, therefore, decreases as the collector emitter voltage V_{CE} increases, avoiding the simultaneous occurrence of peak voltage and peak current. The effect of the size of snubber capacitor on the turn off characteristics is shown in Fig. 2.10(b).

Transistors do not have reverse blocking capability and they are shunted by antiparallel diodes if they are used in AC circuits. The power loss in the base drive circuit may be appreciable because the base current is required to keep a power transistor in the 'on' condition. Power transistors of ratings as high as 1000 V and 500 A are available.

2.6.3 Advantages of Bipolar Transistor as a Switch

1. High power gain, i.e. low control power requirement. In large collector current levels, the control power requirement is reduced by Darlington connection
2. Low switch-off and switch-on time
3. Fast switching and high frequency periodic switching
4. Long life
5. High reliability
6. High ratio of off-state to on-state resistance
7. Low power dissipation in both the states

2.6.4 Performance Parameters of BJT

The characteristics of transistors are normally specified by the manufacturers in terms of certain performance parameters for safe operation are given below:

- *DC gain h_{fe} or β*: It is the ratio of collector current I_C to base current I_B at a specified value of V_{CE}.
- *Crossover time t_c*: It is defined as the interval during which the collector voltage V_{CE} rises from 10% of its peak off-state value and collector current I_C falls to 10% of its on-state value.
- *Secondary breakdown*: The secondary breakdown (SB), which is a destructive phenomenon results from the current flow to a small portion of the base, producing localised hot spots. If the energy in these hot spots is sufficient, the excessive localised heating may damage the transistor. Thus, secondary breakdown is caused by a localised thermal runway, resulting from high current concentrations.
- *Forward-biased safe operating area FBSOA*: During turn on and on-state conditions, the average junction temperature and second breakdown, limit the power handling capability of a transistor. The manufacturers usually provide FBSOA curves under specified test conditions. FBSOA indicates the I_C–V_{CE} limits of the transistor and for reliable operation, the transistor must not be subjected to greater power dissipation than that shown by the FBSOA curve.
- *Breakdown voltages*: A breakdown voltage is defined as the absolute maximum voltage between two terminals with the third terminal open, shorted or biased in either forward or reverse direction. At breakdown the voltage remains relatively constant, where the current rises rapidly. The following breakdown voltages are quoted by the manufacturers.

- V_{EBO}: The maximum voltage between emitter terminal and base terminal with collector terminal open circuited.
- V_{CEV}: The maximum voltage between the collector terminal and emitter terminal at a specified negative charge applied between the base and emitter.
- V_{CEO}: The maximum sustaining voltage between the collector terminal and emitter terminal with the base open circuited. This rating is specified at the maximum collector current and voltage, appearing simultaneously across the device with a specified value of load inductance.

2.7 BASE DRIVE REQUIREMENTS (Design of Drive Circuits for BJTs)

Design considerations: The design of drive circuits for power transistor is considerably more complicated than for logic level devices for many reasons which are discussed here first, low beta of power transistor means that their base current will be large, sometimes tens of amperes, and consequently logic circuits cannot directly drive power transistors. An intermediate gain stage made up of transistors of moderate power and current capability must be used to provide the large base current needed to drive the high-power device. This leads to significant power dissipation in the drive circuits, which has to be considered as well as the dissipation in the main power transistor.

Secondly, it is mandatory that a negative base current be used in turning off the power BJT because otherwise the turn off time will be too long and will lead too much power dissipation during the turn off interval.

Thirdly, it is sometimes desirable to put several power BJTs in parallel to increase the total current capability of composite switch, and the drive circuit must ensure that all the parallel BJTs turn on and off simultaneously.

Consider the simple step-down converter circuit shown in Fig. 2.11, where the drive circuit is shown as a block. A single bar ground which is the BJT emitter is used as the voltage reference point for the base drive circuit. The power circuit may be safety grounded connected to utility system ground. The auxiliary power supplies needed for the base drive circuits must be referred to the emitter reference potential of the power transistor and must be supplied through isolation transformer.

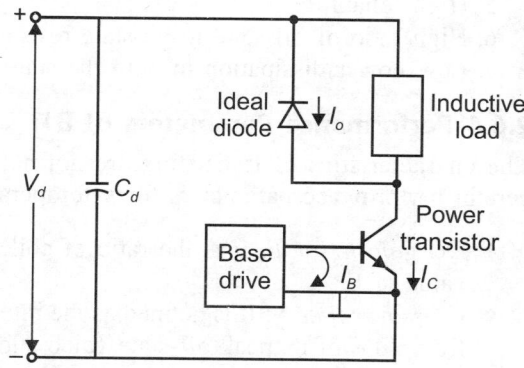

Fig. 2.11 Simple step-down converter circuit

2.7.1 Base Drive Circuits

A many simple base drive circuit suitable for converters with a single-switch topology is shown in Fig. 2.12 (a). At turn on, the p-n-p driver transistor is turned on by saturating one of the interval transistors in comparator (type LM31). This provides a base current for the main power BJT which is given by equation:

$$I_1 = I_{B(on)} + \frac{V_{BE(on)}}{R_2}$$

This base drive circuit should not be used in pulse width-modulated bridge converter circuits. Figure 2.12 (b) shows the associated current and voltage waveforms at turn off of base current drive circuit for a power BJT.

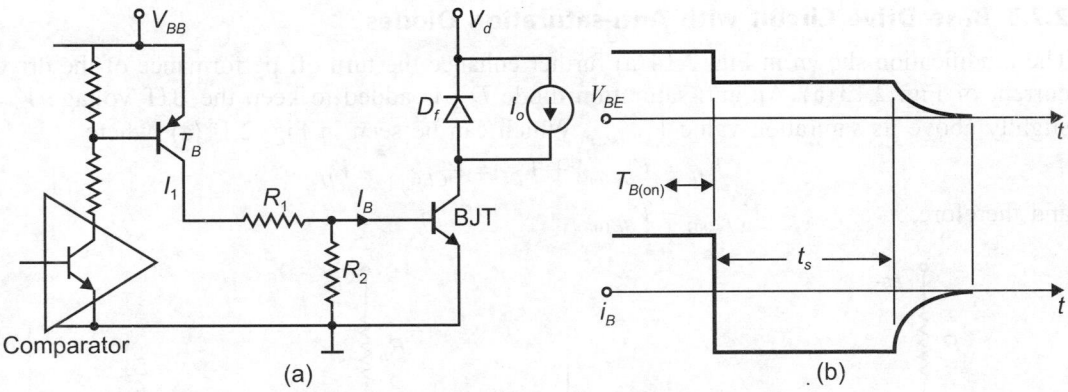

Fig. 2.12 (a) Simple base current drive circuit for a power BJT, and (b) the associated current and voltage waveforms at turn off

2.7.2 Base Current Drive Circuit with both Positive and Negative Voltages

A fast turn off can be provided by the base drive circuit shown in Fig. 2.13, where both positive and negative voltages supply with respect to the emitter one used. During the turn on interval of the BJT, the output transistor of the comparator is off, thus turning the transistor T_{B+} on, the on-state base current is given by

$$I_{B(on)} = \frac{V_{BB+} - V_{CE(sat)}(T_{B+}) - V_{BE(on)}}{R_B}.$$

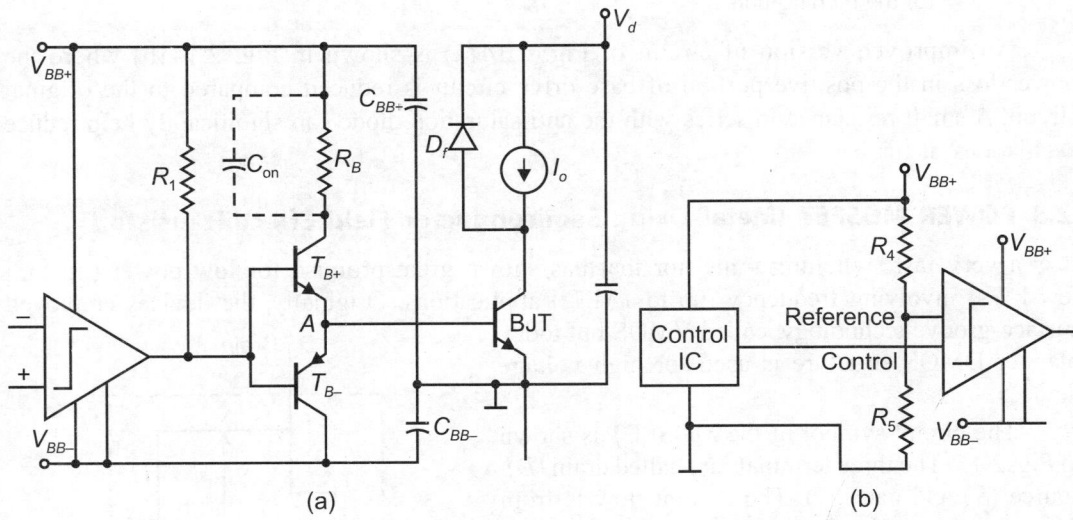

Fig. 2.13 (a) Base current drive circuit with both positive and negative voltages with respect to the BJT emitter for faster turn off of the power device, and (b) the BJT can be controlled by logic level circuits

Figure 2.13(b) shows a bipolar junction transistor (BJT), can be controlled by logic level circuits. If the control signal is supplied by a logic circuit connected between V_{BB+} and the emitter of BJT, then the reference input to the comparator should be at the mid potential between V_{BB+} and the BJT emitter terminal as in Fig. 2.13(b), where $R_4 = R_5$.

2.7.3 Base Drive Circuit with Anti-saturation Diodes

The modification shown in Fig. 2.14(a) further enhance the turn off performance of the drive current of Fig. 2.13(a). An anti-saturation diode D_{as} is added to keep the BJT voltage V_{CE} slightly above its saturation value $V_{CE(sat)}$. Which can be seen in Fig. 2.14(a), where

$$V_{AE} = V_{BE(on)} + V_{D_1} = V_{CE(on)} + V_{D_{as}}$$

and therefore, $$V_{CE(on)} = V_{BE(on)}.$$

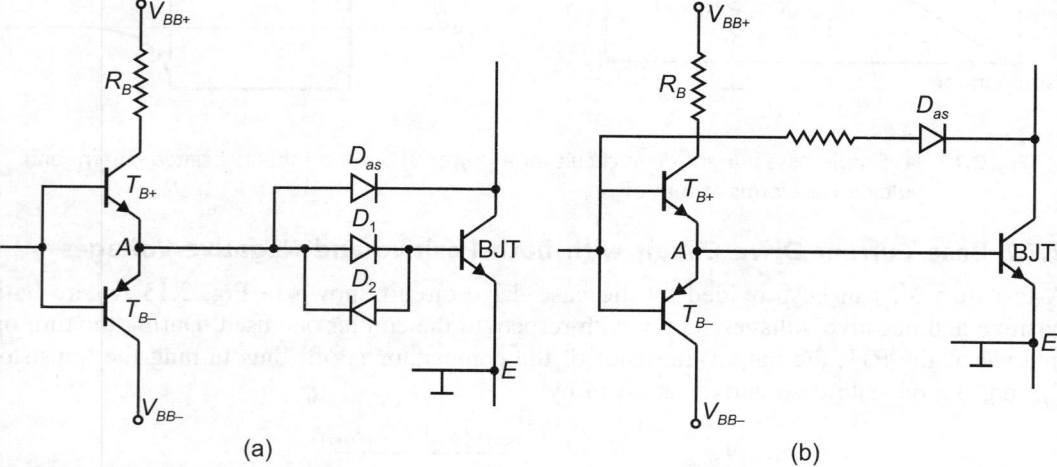

Fig. 2.14 (a) base drive circuit with anti-saturation diodes to minimize the storage time of the BJT
(b) the modification

An improved version of circuit of Fig. 2.14(a) is shown in Fig. 2.14(b) where the power loss in the positive portion of base drive circuit is reduced compared to the original circuit. A small resistance in series with the anti-saturation diode can significantly help reduce oscillations at turn-on.

2.8 POWER MOSFET (Metal Oxide Semiconductor Field Effect Transistor)

It is a very fast switching transistor that has shown great promise for low power (up to a few kWs) involving frequency (up to 1 MHz) applications. Originally, the devices employed surface-groove technology, called VMOS but today planner DMOS structure is used for high voltage devices.

The circuit symbol of the MOSFET is shown in Fig. 2.15. The three terminals are called drain (D), source (S) and gate (G). The current flow is from drain to source. The device has no reverse-voltage blocking capability and it always comes with an integrated reverse rectifier as shown in Fig. 2.15.

Unlike a bipolar transistor (which is a current driven device), a MOSFET is a voltage controlled majority carrier device. With positive voltage applied to the gate (i.e. V_{GS} positive), the transistor switches on. The gate is isolated by a silicon oxide SiO_2

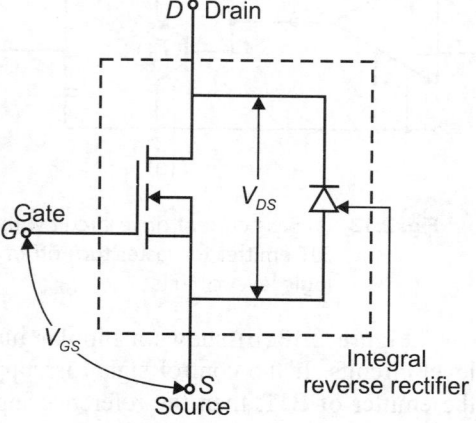

Fig. 2.15 Circuit symbol of the MOSFET

layer, and therefore the gate circuit input impedance is extremely high. This feature allows a MOSFET to be driven directly from CMOS or TTL logic. The gate drive current is therefore very low, it can be less than a milliampere.

The conduction loss of a high voltage MOSFET is very high, its switching loss is almost negligible. The device does not have minority-carrier delay time as in a bipolar device, and its switching times are determined essentially by the ability of the drive to charge and discharge a tiny input capacitance, C_{ISS} defined as $C_{GS} + C_{GD}$ with C_{DS} shorted, where C_{GS} is the gate to source capacitance C_{GD}, the gate to drain capacitance, and C_{DS} the drain to source capacitance. Although MOSFET can be controlled statistically by the voltage source, it is normal practice to drive it by a current source dynamically followed by voltage source to minimise switching delays.

2.8.1 Safe Operating Area of MOSFETs

MOSFET has a positive temperature coefficient of resistance and the possibility of secondary breakdown is almost non-existent. If local heating occurs, the effect of positive temperature coefficient of resistance forces the local concentrations of current to be distributed over the area, thereby avoiding the creation of local hot spots. The safe operating area of a MOSFET is shown in Fig. 2.16. It is bounded by three limits, namely:

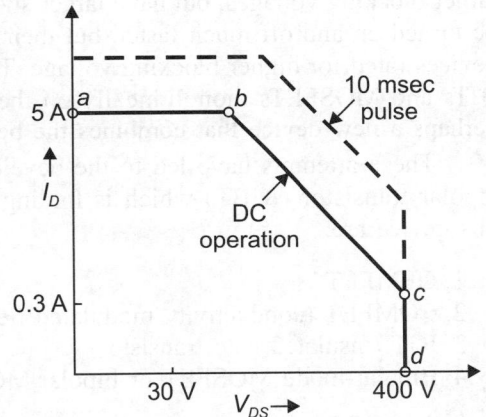

1. current limit (ab)
2. power dissipation limit (bc)
3. voltage limit (cd)

Fig. 2.16 Safe operating area of a MOSFET

The SOA can be increased for pulse operation of the device, shown dashed in Fig. 2.16. The switching characteristics of MOSFET are similar to those of BJT. However, MOSFETs switch on and off very fast, in less than 50 nanoseconds. Because MOSFETs can switch under high voltage and current conditions (practically no secondary breakdown), no current snubbing is required during turn off. However, these devices are very sensitive to voltage spikes appearing across them, and snubber circuits may be required to suppress voltage spikes. MOSFETs are still not available in high power ratings. MOSFETs with ratings of 500 V, 10 A, 50 nsec are available. These devices can be used in parallel for higher current ratings.

2.8.2 Advantages of Power MOSFETs Over Power Transistors

Since the power transistor is the power MOSFET's main competitor, it is good to know their relative advantages and disadvantages. The advantages of power MOSFET over the power transistor are:

1. no power switching losses
2. no second breakdown
3. much larger gain and hence simple and cheaper drive circuitry
4. better reliability, ruggedness, and thermal stability
5. higher peak current handling capability
6. easy to parallel due to the positive resistance coefficient
7. relatively more linear transfer characteristics and
8. much higher switching speed

2.8.3 Disadvantages

The only disadvantage is a higher conduction drop—typically 4.5 V compared to 1 V for the power transistor for ratings of 400 V and 10 A.

2.9 HYBRID DEVICES

A number of hybrid power semiconductor devices have recently received attention of which only two will be discussed briefly here namely (1) Insulated gate bipolar transistor (IGBT) and (2) FET controlled thyristors.

2.9.1 Insulated Gate Bipolar Transistor (IGBT)

Bipolar junction transistor (BJT) and MOSFET have characteristics that complement each other in some respects. BJTs have lower conduction losses in the on-state, especially in devices with larger blocking voltages, but have larger switching times, especially at turn off. MOSFETs can be turned on and off much faster, but their on-state conduction losses are larger especially in devices rated for higher blocking voltage. These observations have led to attempts to combine BJTs and MOSFETs monolithically on the same silicon wafer to achieve a circuit on even perhaps a new device that combines the best qualities of both types of device.

These attempts have led to the development of a new device called the insulated gate bipolar transistor (IGBT) which is finding increasingly wide applications. Other names for this device are:

1. GEMFET
2. COMFET (conductivity modulated field effect transistor)
3. IGT (insulated gate transistor)
4. Bipolar-mode MOSFET or bipolar-MOS transistor

An IGBT combines the advantages of BJTs and MOSFETs. An IGBT has high input impedance like MOSFETs and low on-state conduction losses like BJTs. But there is second breakdown problems like BJTs.

An IGBT is made of four alternative pnpn layers, and could latch like a thyristor given the necessary condition: $(\alpha_{\text{n-p-n}} + \alpha_{\text{p-n-p}}) > 1$. An IGBT is a voltage controlled device similar to a power MOSFET. It has lower switching and conducting losses while sharing many of the appealing features of power MOSFETs such as the case of gate drive, peak current, capability and ruggedness. An IGBT is inherently faster than a BJT. However, the switching speed of IGBT, is inferior to that of MOSFETs. The current rating of single IGBT can be up to 400 A, 1200 V and the switching frequency can be up to 20 kHz. IGBTs are finding increasing application in medium power application such as DC and AC motor drives, power supplies, solid-state relays and contactors.

This device has high input impedance of a MOSFET but a low on-state conduction drop similar to that of a bipolar transistor. The switching speed and safe operating area of bipolar transistor are retained. The storage time of the bipolar tends to be long because of its incapability to drive negative base current. However, the device has thyristor like reverse voltage blocking capability. The equivalent circuit of IGT is shown in Fig. 2.17 (a).

I–V characteristics: The *i–v* characteristic of an n-channel IGBT is shown in Fig. 2.17 (b). In the forward direction, they appear qualitatively similar to those of a logic level bipolar junction transistor except that the controlling parameter is an input voltage, the gate-source voltage, rather than an input current. The characteristics of a p-channel IGBT would be the same except that the polarities of the voltages and currents would be reversed.

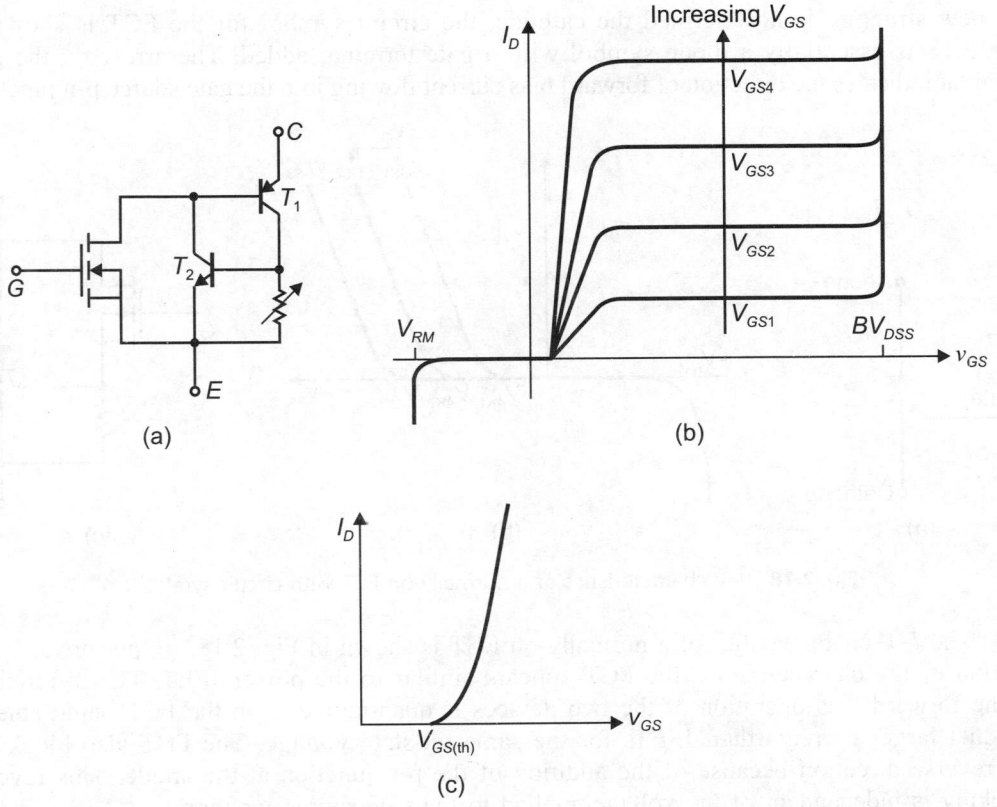

Fig. 2.17 IGBT circuit symbol, output and transfer characteristics

The transfer curve i_D–v_{GS} shown in Fig. 2.17(c) is identical to that of the power MOSFET. The curve is reasonably linear over most of the drain current range becoming nonlinear only at low drain currents where the gate-source voltage in approaching the threshold. If v_{GS} is less than the threshold voltage $V_{GS(th)}$, then the IGBT is in the off-state. The maximum voltage that should be applied to the gate-source terminals is usually limited by the maximum drain current that should be permitted to flow in the IGBT.

2.9.2 FET Controlled Thyristors

Research effects are made to improve the devices in respect to increase their blocking voltage capabilities, lowering their on-state losses, and increasing their switching speeds. A recent example of this is the development of the ring emitter BJT by Siemens. The concepts that have not yet found general commercial acceptance or are still in prototype stage, can be termed *emerging devices and circuits*. A list of such emerging devices would include power junction field effect transistors, also termed *static induction transistor*, controlled thyristors (bipolar static induction transistors), MOS-controlled thyristors (MCTs). Some of these devices may became widely used in the future in the industrial application.

If the drain of the power JFET structure is modified into an injecting contact by making it into a p-n junction, then a new device is produced. This new device, is termed a field controlled thyristor (FCT). The drain of an n-channel JFET is converted into a p-n junction and it becomes the anode of the device. The source of the JFET portion of the new structure is now termed the cathode, the circuit symbol for the FCT is shown in Fig. 2.18 is essentially a diode symbol with a gate terminal added. The arrow on the gate terminal indicates the direction of forward bias current flowing into the gate source p-n junction.

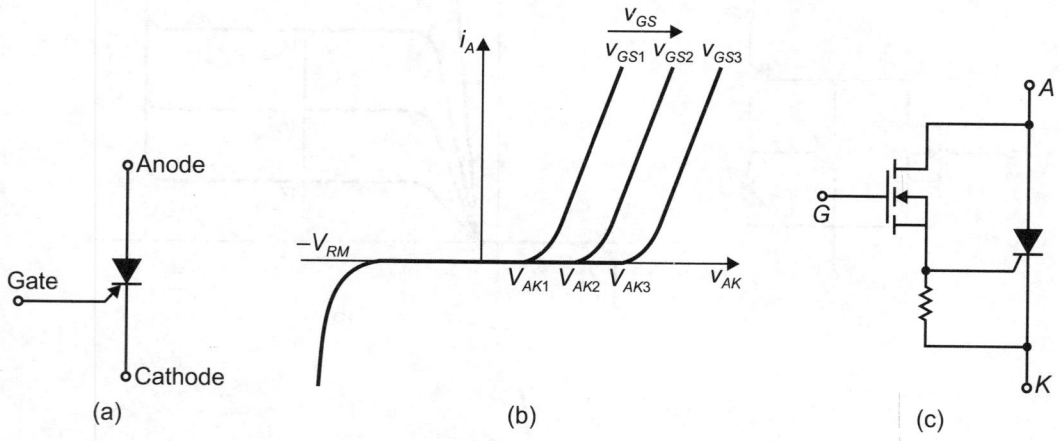

(a)   (b)   (c)

Fig. 2.18 *I–V* characteristics of a normally-on FCT with circuit symbol

The *I–V* characteristics of a normally-on FCT is shown in Fig. 2.18. In the forward bias portion of the characteristics, the FCT appears similar to the power JFET. The differences in the forward bias operation of the two devices is quantitative, with the FCT being able to conduct larger currents than JFETs for the same on-state voltage. The FCT also blocks in the reverse direction because of the addition of the p-n junction at the anode. This reverse blocking is independent of the voltage applied to the gate-source junction.

Even though the FCT is termed as field controlled thyristor, it is important to note that the device does not have any regenerative turn on or turn off as does a gate turn off thyristor. The FCT does not latch on or off. If the gate drive holding the normally-on FCT in the blocking state is removed, the FCT will turn on. Similarly removal of the gate drive holding the normally-off FCT in the on-state will cause the device to turn off. A FET controlled thyristor in which a MOSFET and a thyristor are connected in parallel is shown in Fig. 2.18(c). The device has the input impedance of a FET but the conduction drop of a thyristor, with the gate voltage exceeding the threshold voltage, current flows in the shunt resistance and switches on the thyristor. The device can be fired optically from an integrated circuit, providing electrical isolation between the control and power circuits.

2.10   THYRISTOR

The thyristor, also known as silicon controlled rectifier (SCR), is the main workhorse for bulk power conversion and control in industry. A thyristor has a p-n-p structure with three terminals and is fabricated by diffusion process. It is broadly discussed in Chapter 5. Light-fired thyristors have recently been developed for high voltage direct current (HVDC) applications. In this application, a number of devices are connected in series-parallel combination, and light firing provides the advantages of electrical isolation between the control and power circuits.

SOLVED EXAMPLES

Example 2.1   During turn on and turn off of a power transistor supplying a resistive load, the voltage and current variations are as shown in Fig. E2.1 as idealised linear changes with

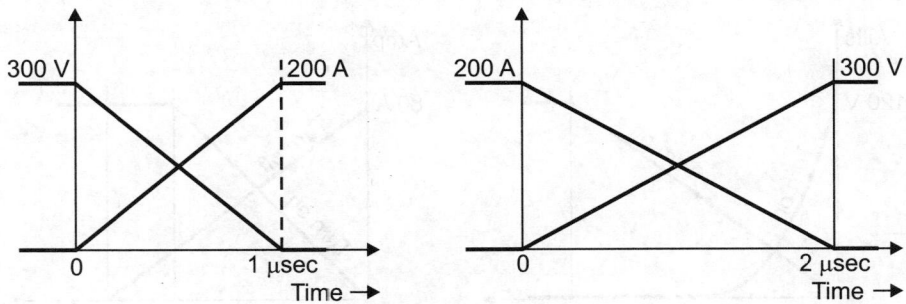

Fig. E2.1   The voltage and current variations

time. Show that, for each, the switching energy loss is given by $(VI/6)\,T$, where V is the off-state voltage, I the on-state current and T is the switching time. Determine the mean power loss due to switching at a frequency of 10 kHz. If the transistor is operating with a duty cycle of 0.5, that is equal to off and on times, determine the conduction loss if the collector-emitter voltage is 2.1 V at 200 A. Hence, determine the total device loss.

Solution   Referring to turn on waveforms, with V and I as on-state values, and taking time $t = 0$ at the start of switching, then at any time t,

$$i = \frac{I \cdot t}{T} \text{ and } v = V - \frac{V}{T}t$$

$$\text{Energy} = \int_{t=0}^{t=T} v \cdot i \, dt = \int_{0}^{T}\left(V - \frac{V \cdot t}{T}\right)\cdot\left(\frac{I \cdot t}{T}\right) dt$$

$$= VI\left[\frac{t^2}{2T} - \frac{t^3}{3T^2}\right] = \frac{VI}{6}T$$

For turn off condition, turn on loss

$$= \frac{300 \times 200}{6} \times 1 \times 10^{-6} = 0.01 \text{ J}$$

Turn off loss $$= \frac{300 \times 200}{6} \times 2 \times 10^{-6} = 0.02 \text{ J}$$

Mean switching power loss $= (0.01 + 0.02) \times 10 \times 10^3 = 300$ W

Conduction loss $= 200 \times 2.1 \times 0.5 = 210$ W

Total loss $= 300 + 210 = 510$ W.

Example 2.2   A transistor has the switching characteristic as shown in Fig. E2.2. If the mean power loss in the transistor is limited to 200 W, what is the maximum switching rate that can be achieved?

Solution   Energy loss in transistor $= \int_{0}^{i} i_c V_{CE} \, dt$

Turn on energy loss $$= \int_{0}^{40\times10^{-6}} 120\,(1 - 2.5 \times 10^4 t) \times 1.6 \times 10^6 t \, dt = 51 \text{ mJ}$$

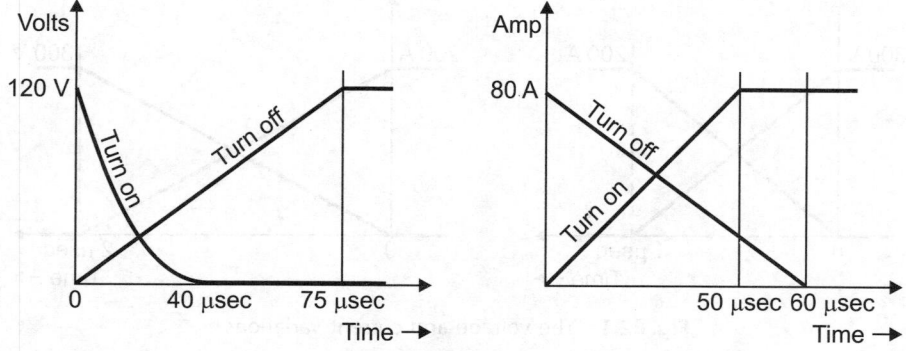

Fig. E2.2 A transistor with switching characteristic

Turn off energy loss $= \displaystyle\int_{0}^{60\times10^{-6}} 1.6 \times 10^{6} \cdot t \times 80\,(1 - 1.667 \times 10^{4}\,t)\,dt$

$= 76.8 \text{ mJ}$

Total loss in one cycle $= 51 + 76.8 = 127.8 \text{ mJ}$

Number of cycles in one sec $= \dfrac{200}{0.1278} = 1564.9$

EXERCISE

Review Questions

1. What are the types of power diodes?
2. What is a leakage current of diodes?
3. What is the cause of reverse recovery time in a p-n junction diode?
4. Why is it necessary to use fast recovery diodes for high-speed switching?
5. What are the main differences between p-n junction diodes and Schottky diodes?
6. What are the limitations of Schottky diodes?
7. What are the problems of series-connected diodes? What are the possible solutions?
8. What is the V–I characteristics of thyristors?
9. What is an off-state condition of thyristors?
10. What is a latching current of thyristors?
11. What is a holding current of thyristors?
12. What is a bipolar junction transistor (BJT)?
13. What is FBSOA of BJTs?
14. What is a secondary breakdown of BJTs?
15. What is a MOSFET?
16. What are the advantages and disadvantages of MOSFETs?
17. What is a threshold voltage of MOSFETs?
18. What are the problems of parallel operations of BJTs?

Rectifying Circuits, Filters and Zener Diodes

3.0 INTRODUCTION

Semiconductor diodes have found many applications in electronics and electrical engineering circuits. Rectification refers to the process of changing alternating current to direct current in applications. It is extensively used in charging batteries supplying DC motors, electrochemical processes and power supply sections of industrial equipment. Diodes are widely used in power electronics circuits for conversion of electric power. AC–DC converters provide a fixed DC output voltage. Most of the electronic devices require a DC voltage source for their operation. Whereas the electric supply available is almost always AC, the DC voltage can be obtained from batteries. Circuits which convert the alternating voltages to a direct voltage of suitable value are called power supplies. A regulated power supply consists of the following three systems:

1. Diode rectifier
2. Filter
3. Voltage regulator

Because of their unique ability to conduct current in one direction, diodes are used in rectifier circuits. Rectification is the process of converting the alternating voltages and currents to pulsating direct currents and the device is known as *rectifier*.

In most power supply applications, the standard 50 Hz AC power line voltage must be converted to a sufficiently constant DC voltage. The 50 Hz pulsating DC voltage must be filtered to virtually eliminate the large variations. The output of a rectifier contains AC and DC components. The AC component is undesirable and must be kept away from the load. Filter circuit is required which filters out the undesirable AC component and allows DC component to reach the loads, thus *filter* is a device which allows DC component to the load and blocks AC component of rectifier output. It should be installed between the rectifier output and the load.

In the superhetrodyne receiver, the regulated DC voltage is used to supply each of the circuits in the system. Frequently, design specification requires very stable and regulated output voltages from DC power supplies. In such situations, it is essential to add an additional device to the power supply to compensate for the regulation problems inherent in the power line. This is called a *voltage regulator*. Zener diodes are widely used for voltage regulation.

In Fig. 3.1, block diagram of a regulated power supply is shown.

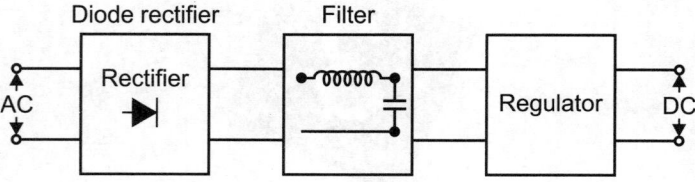

Fig. 3.1 Block diagram of a regulated power supply

The output from the rectifier is DC but pulsating. These pulsations are due to the AC component present in the rectifier output. The AC component is removed by the filter circuit and fed into the regulator to get DC output constant irrespective of changes in AC main or load.

3.1 HALF-WAVE RECTIFIER

One important use of a diode is rectification. A rectifier is a device which converts AC to unidirectional current. The half-wave rectifier with an ideal solid state diode is shown in Fig. 3.2. The diode begins to conduct as soon as its anode potential become greater than that of cathode as a result, the load current flows. In the negative half cycle, the diode is reverse-biased and only a small leakage current flows through the load resistance. This current is taken as zero for practical purposes. The diode behaves like an open switch. The voltage across R_L is zero over the time interval $T/2$ to T, where T is the time period or $\omega t = \pi$ to 2π. This means that the average value of the input AC voltage has been made non-zero. The waveforms across the load and diode are shown in Fig. 3.2.

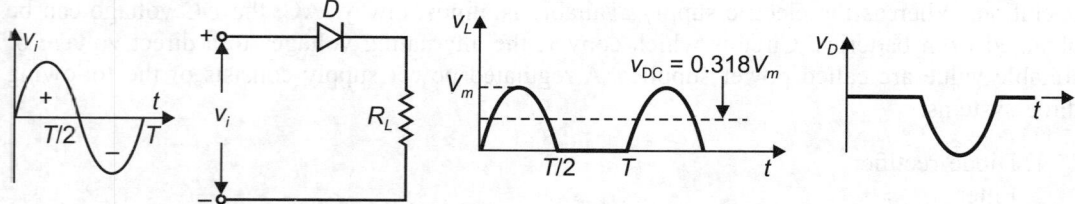

Fig. 3.2 Waveforms across the load and diode

Analysis of half-wave rectifier: Assuming voltage developed across load is alternate half cycle of the supply voltage. Hence, the current wave shape is exactly half for a sinusoidal supply voltage. Let the forward and load resistance be r_f and R_L respectively, and $v_i = \sqrt{2}V \sin \omega t$ $= V_m \sin \omega t$.

The instantaneous value of the current i_b is

$$i_b = \frac{V_m \sin \omega t}{r_f + R_L} \qquad 0 < \omega t \leq \pi,$$

where V = rms value of the input voltage

$$= I_m \sin \omega t, \qquad \text{where } I_m = \frac{V_m}{(r_f + R_L)}$$

$$i_b = 0, \qquad\qquad \pi < \omega t \leq 2\pi$$

DC value of current: The DC value of the current at output is given by

$$I_{DC} = \frac{1}{2\pi} \int_0^{2\pi} I_m \sin \omega t \, d(\omega t)$$

$$= \frac{1}{2\pi} \left[\int_0^{\pi} I_m \sin \omega t \, d(\omega t) + \int_{\pi}^{2\pi} 0 \cdot d(\omega t) \right]$$

$$= \frac{I_m}{2\pi} [-\cos \omega t]_0^{\pi} = \frac{I_m}{2\pi} [-\cos \pi + \cos 0]$$

$$= \frac{I_m}{\pi} = \frac{V_m}{\pi (r_f + R_L)} = \frac{0.318 V_m}{(r_f + R_L)} \qquad \text{...(3.1)}$$

Similarly DC value of voltage at output for ideal diode,

$$V_{DC} = 0.318 V_m.$$

When a silicon diode with diode voltage drop $V_T = 0.7$ V is used, the input will be 0.7 V, before the diode conducts, resulting in a zero output level until the transition occurs. When conducting, the difference between v_o and v_i is a fixed level of $V_T = 0.7$ V and $v_o = v_i - V_T$, which reduces the resulting DC voltage level. If V_m is closer to V_T then

$$V_{DC} = 0.318 (V_m - V_T) \qquad \text{...(3.2)}$$

RMS value of current: This is given by the expression

$$I_{rms} = \sqrt{\frac{1}{2\pi} \int_0^{2\pi} (I_m \sin \omega t)^2 \, d(\omega t)}$$

$$= \sqrt{\frac{1}{2\pi} \left[\int_0^{\pi} (I_m \sin \omega t)^2 \, d(\omega t) + \int_{\pi}^{2\pi} 0 \cdot d(\omega t) \right]}$$

$$I_{rms} = \frac{I_m}{2} = \frac{V_m}{2(r_f + R_L)} = \frac{0.5 V_m}{(r_f + R_L)} \qquad \text{...(3.3)}$$

DC voltage across the load,

$$V_{DC} = I_{DC} \cdot R_L = \frac{V_m}{\pi (r_f + R_L)} \cdot R_L \qquad \text{...(3.4)}$$

DC output power, P_{DC} is given by,

$$P_{DC} = I_{DC}^2 \cdot R_L = \left(\frac{I_m}{\pi} \right)^2 \cdot R_L = V_{DC} \cdot I_{DC}$$

$$= V_{DC} \cdot \frac{V_{DC}}{(R_L + r_f)} = \frac{(0.318 V_m)^2}{(R_L + r_f)} \qquad \text{...(3.5)}$$

AC power input to the rectifier,

$$P_{AC} = (I_{rms})^2 \cdot (r_f + R_L)$$

$$= \left(\frac{I_m}{2} \right)^2 (r_f + R_L) = (0.5 I_m)^2 (r_f + R_L) \qquad \text{...(3.6)}$$

***Ripple factor* γ:** The lack of smoothness of the waveform is given by ripple factor.

$$\gamma = \frac{\text{Effective value of AC component of current}}{\text{DC component of current}}$$

The total load current

$$I_{\text{rms}} = \sqrt{I_{\text{DC}}^2 + I_{\text{AC}}^2}$$

$$I_{\text{AC}} = \sqrt{I_{\text{rms}}^2 - I_{\text{DC}}^2}$$

$$\gamma = \frac{\sqrt{I_{\text{rms}}^2 - I_{\text{DC}}^2}}{I_{\text{DC}}} = \sqrt{\left(\frac{I_{\text{rms}}}{I_{\text{DC}}}\right)^2 - 1} = \sqrt{\left(\frac{I_m/2}{I_m/\pi}\right)^2 - 1} = 1.21 \qquad \qquad ...(3.7)$$

***Peak inverse voltage (PIV)*:** The peak inverse voltage is another factor in rectifier design. It is the maximum instantaneous voltage which the diode has to withstand without destroying the junction in the blocking state. In a half wave rectifier, the reverse voltage across the diode during nonconduction period equals the secondary voltage. Thus, PIV = $V_m = \sqrt{2}V = 1.414$ V.

***Rectifier efficiency*, η:** It is denoted by η

$$= \frac{\text{DC load power}}{\text{AC power input from transformer}}$$

$$\eta\% = \frac{P_{\text{DC}}}{P_{\text{AC}}} = \frac{(I_m/\pi)^2 \cdot R_L}{(I_m/2)^2 (r_f + R_L)} \times 100$$

$$\eta\% = \frac{0.406}{1 + r_f/R_L} \times 100 \qquad \qquad ...(3.8)$$

The efficiency will be maximum if r_f is negligible as compared to R_L,

i.e. $\qquad \qquad \eta_{\text{max}} \cong 40.6\%.$

Transformer utilisation factor

$$\text{TUF} = \frac{\text{DC load power}}{\text{AC rating of transformer secondary}} = \frac{P_{\text{DC}}}{V_s I_s}$$

$$= \frac{P_{\text{DC}}}{P_{\text{AC}}(\text{rated})} = \frac{(I_m/\pi)^2 \cdot R_L}{(V_m/\sqrt{2})(I_m/2)}$$

As $\qquad \qquad V_m = I_m (r_f + R_L)$

$\therefore \qquad \qquad \text{TUF} = \frac{0.286 R_L}{(r_f + R_L)} = \frac{0.286}{\left(1 + \dfrac{r_f}{R_L}\right)} \qquad \qquad ...(3.9)$

If $r_f \ll R_L$ then TUF = 0.286 $\approx$ 0.29.

***Displacement factor*:** The angle between the fundamental components of the input current and voltage is called the displacement angle. If α represents the displacement angle, then cos α is called the displacement factor, i.e.

$$DF = \cos\alpha \qquad \qquad ...(3.10)$$

Harmonic factor:

$$HF = \left[\frac{I_s^2 - I_{sf}^2}{I_{sf}} \right]^{1/2} = \left[\left(\frac{I_s}{I_{sf}} \right)^2 - 1 \right]^{1/2} \qquad \text{...(3.11)}$$

where I_{sf} is the fundamental component of input current I_s.

Input power factor: It is defined as

$$PF = \frac{V_s I_{sf}}{V_s I_s} \cos \alpha = \frac{I_{sf}}{I_s} \cdot \cos \alpha \qquad \text{...(3.12)}$$

Disadvantages of half-wave rectifier: Half-wave rectifier has the following disadvantages:

1. High ripple factor: The pulsating current in the load contains AC components, which make the ripple factor high.
2. Low rectification efficiency: The AC supply delivers power only half the time and hence output is low.
3. Low transformer utilization factor.
4. DC saturation of transformer secondary winding.

3.2 FULL-WAVE RECTIFIER

In full-wave rectification, current flows through load in the same direction for both half cycles of input AC voltage. There are two types of circuits used for full-wave rectifiers, namely (1) centre-tap circuit, and (2) bridge circuit.

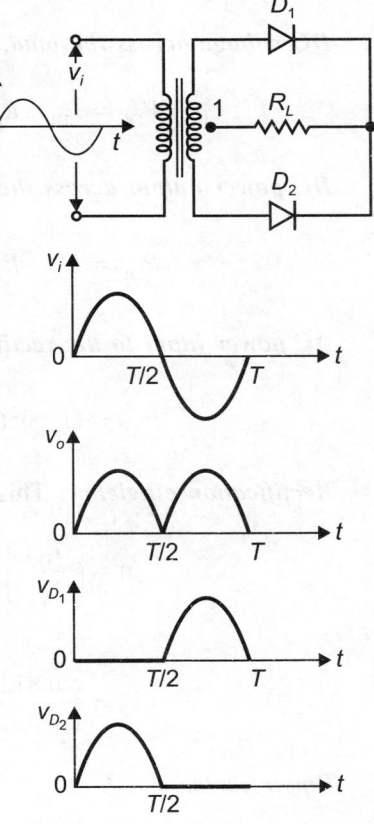

Centre-tap rectifier with resistive load: The circuit employs two-diodes D_1 and D_2 and a centre-tapped secondary winding. The diode D_1 is forward-biased and D_2 reverse-biased in the positive half cycle and current flows from point 2 to point 1 in the load R_L. Whereas in the negative half cycle D_1 is reverse-biased and D_2 is forward-biased and again current flows from point 2 to point 1. Hence, DC output is obtained across R_L, the corresponding waveforms are shown in Fig. 3.3 (a).

Analysis of full-wave rectifier: Let the forward and load resistance be r_f and R_L and the AC input voltage be

$$v_i = \sqrt{2} V \sin \omega t = V_m \sin \omega t$$

Current due to diodes D_1, D_2

$$\left. \begin{aligned} i_{d1} &= \frac{V_m}{r_f + R_L} \sin \omega t \\ i_{d2} &= 0 \end{aligned} \right\} \quad 0 \le \omega t \le \pi$$

$$\left. \begin{aligned} i_{d1} &= 0 \\ i_{d2} &= \frac{V_m}{r_f + R_L} \sin \omega t \\ &= I_m \sin \omega t \end{aligned} \right\} \quad \pi \le \omega t \le 2\pi$$

Fig. 3.3 (a) Centre-tap rectifier with resistive load and associated waveforms

where
$$I_m = \frac{V_m}{(r_f + R_L)}$$

Diode currents, i_{d1}, i_{d2} are however, the same as both diodes are similar.

DC value of current,

$$I_{DC} = \frac{1}{\pi} \int_0^\pi I_m \sin \omega t \, (\omega t)$$

$$I_{DC} = \frac{1}{\pi} \left[-I_m \cos \omega t \right]_0^\pi = \frac{2I_m}{\pi} = \frac{2V_m}{\pi(r_f + R_L)} = \frac{0.636V_m}{(r_f + R_L)} \qquad ...(3.13)$$

RMS value of current, I_{rms}

$$I_{rms} = \sqrt{\frac{1}{2\pi} \left[\int_0^\pi i_{d1}^2 d(\omega t) + \int_\pi^{2\pi} i_{d2}^2 d(\omega t) \right]}$$

$$I_{rms} = \frac{I_m}{\sqrt{2}} = \frac{V_m}{\sqrt{2}(r_f + R_L)} = \frac{0.707V_m}{(r_f + R_L)} \qquad ...(3.14)$$

DC voltage across the load,

$$V_{DC} = I_{DC} \cdot R_L = \frac{2I_m}{\pi} \cdot R_L = \frac{2}{\pi} \frac{V_m}{(r_f + R_L)} \cdot R_L \qquad ...(3.15)$$

DC power output across the load,

$$P_{DC} = I_{DC}^2 \cdot R_L = \left(\frac{2I_m}{\pi} \right)^2 \cdot R_L = \left[\frac{2}{\pi} \frac{V_m}{(r_f + R_L)} \right]^2 R_L \qquad ...(3.16)$$

AC power input to the rectifier,

$$P_{AC} = (I_{rms})^2 (r_f + R_L) = \left(\frac{I_m}{\sqrt{2}} \right)^2 \cdot (r_f + R_L) = 0.5 I_m^2 (r_f + R_L) \qquad ...(3.17)$$

Rectification efficiency: This is given by

$$\eta = \left(\frac{I_{DC}}{I_{rms}} \right)^2 \frac{1}{1 + \dfrac{r_f}{R_L}} = \left(\frac{2I_m/\pi}{I_m/\sqrt{2}} \right)^2 \cdot \frac{I}{1 + r_f/R_L}$$

$$\eta\% = \frac{0.812}{1 + r_f/R_L} \times 100 \qquad ...(3.18)$$

Ripple factor,

$$\gamma = \sqrt{\left(\frac{I_{rms}}{I_{DC}} \right)^2 - 1} = \sqrt{\left(\frac{I_m/\sqrt{2}}{2I_m/\pi} \right)^2 - 1} = 0.483 \qquad ...(3.19)$$

Peak inverse voltage, PIV: The peak inverse voltage of a full-wave rectifier, from Fig. 3.3 (b)

$$= V_{\text{sec}} + V_R = V_m + V_m = 2V_m \qquad \text{...(3.20)}$$

Voltage regulation: It is a measure of the ability of a rectifier to maintain a specified output voltage irrespective of the variation in load resistance. It is defined as

% voltage regulation

$$= \frac{\text{Output voltage at no load} - \text{voltage at full load}}{\text{Output voltage at full load}} \times 100 \quad \text{...(3.21)}$$

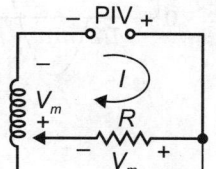

Fig. 3.3 (b) Peak inverse voltage of a full-wave rectifier circuit

Average value of the load voltage is given by

$$V_{\text{DC}} = \frac{2V_m}{\pi} \qquad \text{...(3.22)}$$

RMS value of the load voltage

$$V_{\text{rms}} = \frac{V_m}{\sqrt{2}} \qquad \text{...(3.23)}$$

Transformer utilization factor: For full-wave rectifier, it has a value equal to

$$\text{TUF} = 0.67. \qquad \text{...(3.24)}$$

3.3 BRIDGE RECTIFIER

The DC level obtained from a sinusoidal input can be improved 100% using a process called *full-wave rectification*. The most familiar network for performing such a function appears in Fig. 3.4 (a) with its four diodes in a bridge configuration. During the period $t = 0$ to $T/2$, the polarity of the input and across the ideal diodes are shown in Fig. 3.4 (b) to reveal that diodes D_2 and D_3 are conducting while D_1 and D_4 are in the off-state. The net result is the configuration of Fig. 3.4 (c) with its indicated current and polarity across R. Since the diodes are ideal, the load voltage $v_o = v_i$.

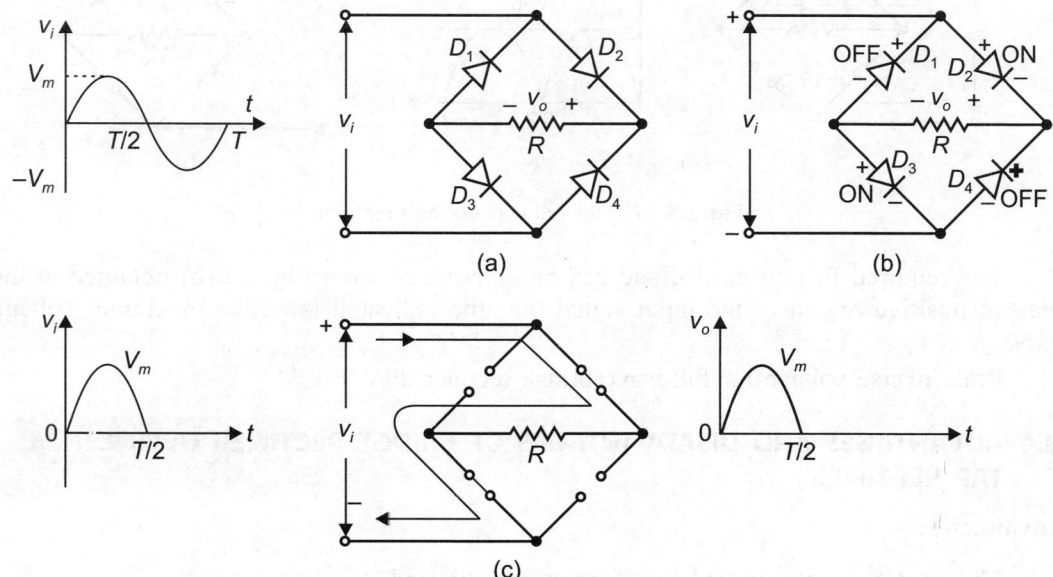

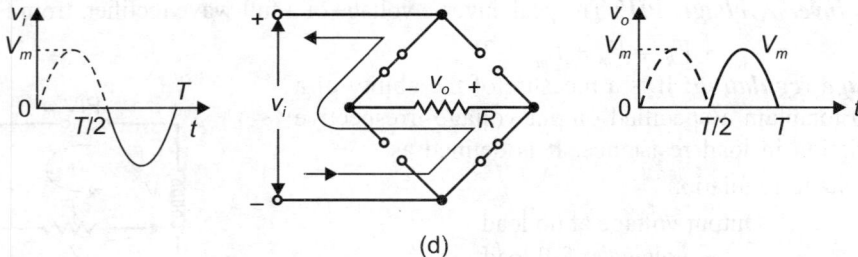

(d)

For the negative region of the input, the conducting diodes D_1 and D_4 resulting in the configuration of Fig. 3.4 (d), the important result is that the polarity across the load resistor R is same as shown in Fig. 3.4 (b) establishing a second positive pulse as shown in Fig. 3.4 (d). Over one full cycle, the input and output voltages will appear as shown in Fig. 3.4 (e).

Average DC value

$$V_{DC} = 0.636 V_m \qquad ...(3.25)$$

For silicon diodes during the positive conduction phase the effect of V_T has also doubled as shown in Fig. 3.5 (a). If $V_m \gg 2V_T$, then

$$V_{DC} \cong 0.636 V_m$$

If V_m is close to $2V_T$, then

$$V_{DC} \cong 0.636 (V_m - 2V_T) \qquad ...(3.26)$$

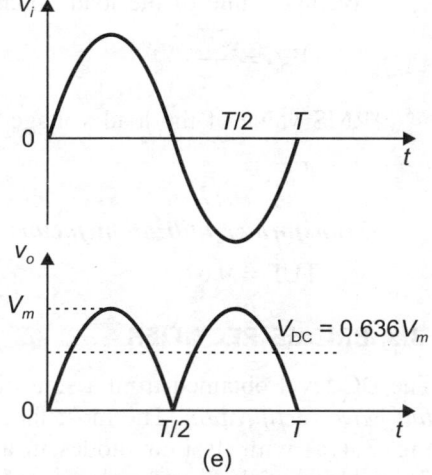

(e)

Fig. 3.4 Function of bridge rectifier and waveforms

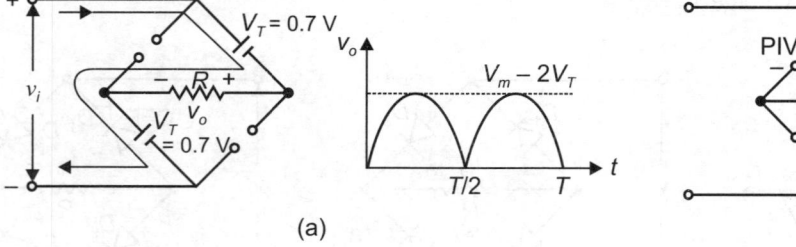

(a)   (b)

Fig. 3.5 PIV for full wave bridge rectifier

The required PIV of each diode can be determined from Fig. 3.5 (b) obtained at the peak of positive region of the input signal. For the indicated loop, the maximum voltage across R is V_m.

Peak inverse voltage for full-wave bridge rectifier, PIV = V_m.

3.4 ADVANTAGES AND DISADVANTAGES OF BRIDGE RECTIFIER OVER CENTRE-TAP RECTIFIER

Advantages:

1. The need for centre-tapped transformer is eliminated.

2. The output is twice that of the centre-tapped circuit for the same secondary voltage.
3. The peak inverse voltage is one-half of the centre-tap circuit.

Disadvantages:

1. It requires four diodes instead of two, in full-wave circuit.
2. There are always two-diodes in series that are conducting. Therefore, total voltage drop in the internal resistance of the rectifying unit and losses are increased. As a result, the rectification efficiency is poor

Comparison between half-wave and full-wave rectifiers

S. No.	Parameters	Half-wave rectifier	Full-wave rectifier	Bridge type
1.	Average value of current	I_m/π	$2I_m/\pi$	$2I_m/\pi$
2.	RMS value of current	$I_m/2$	$I_m/\sqrt{2}$	$I_m/\sqrt{2}$
3.	Ripple factor	1.21	0.483	0.483
4.	Fundamental frequency of ripple	Same as supply frequency, f	Double the supply frequency, $2f$	$2f$
5.	Peak inverse voltage	V_m	$2V_m$	V_m
6.	Maximum rectification efficiency	40.6%	81.2%	81.2%
7.	Voltage regulation	Good	Better	Better
8.	Transformer required	No	Yes	No
9.	Number of diodes	One	Two	Four
10.	Transformer utilization factor (TUF)	Low, TUF = (0.286)	High, TUF = (0.67)	High, TUF = (0.67)
11.	Form factor	1.57	1.11	1.11

3.5 MULTIPHASE RECTIFIERS

Single-phase full-wave rectifiers are used in applications up to a power level of 15 kW. For larger power output, three-phase and multiphase rectifiers are used. Such rectifiers use three, four or six diodes. The polyphase rectifiers operate from a three-phase power supply, yet they are called two-, three- or six-phase rectifiers depending on the way they operate. Considering the various polyphase transformer connections, it results in a large number of circuit configurations. Each such arrangement has its advantages and drawbacks. These circuits differ in the efficiency, transformer cost, ripple voltage and harmonic contents of line current. In practice, a filter is normally used to reduce the level of harmonics in the load. For high power rectification, power diodes or silicon controlled rectifiers can be employed. Some popular rectifier circuits are: (1) three-phase half-wave rectifier, (2) three-phase full-wave bridge rectifier, and (3) six-phase half-wave rectifier.

3.5.1 Three-phase Half-wave Rectifier with Resistive Load or Single Way

Most power rectifier circuits employ some modification of a star-connected transformer, secondary with a diode connected to each point of the start. Figure 3.6 shows a simple three-phase half-wave rectifier using $\Delta-Y$ transformer. The transformer primary is connected in delta, the secondary in Wye Y (star). The connection at the centre of Y is called the neutral which makes a fourth wire in this three-phase system. The voltage supplied between one winding

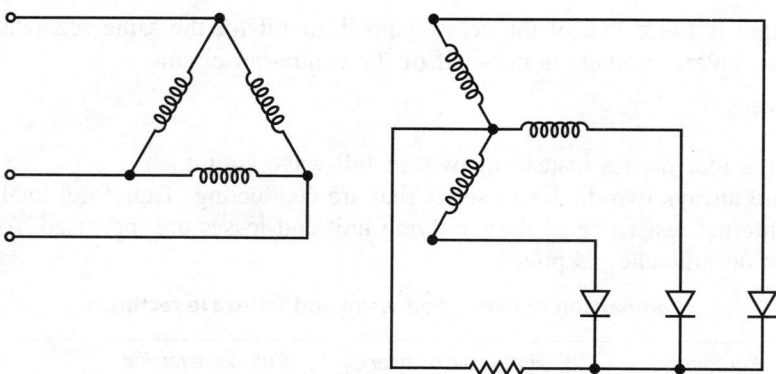

Fig. 3.6 Three-phase half-wave rectifier

and neutral is called the secondary phase voltage, the load resistance R is connected between the common cathode and the transformer neutral; it is customary to use the transformer neutral for voltage reference. The anode voltages are assumed to be sinusoidal and equal to primary voltages multiplied by appropriate transformation ratio.

Each of the three secondary winding voltages are displaced from each other in magnitude as well as phase angle by 120°. With a negligible diode drop, the potential of the common cathode connection is equal to that of the conducting anode. Assuming that only one rectifier conducts at a time. Thus, when v_1 is more positive than any other anode between $\pi/6$ and $5\pi/6$, diode D_1 will conduct. At $5\pi/6$ the voltage v_1 is dropping, diode D_1 ceases conduction and the current transfers or commutates to D_2. At $\omega t = 9\pi/6$, the current commutates from D_2 to D_3, each diode conducts for 120° of the cycle. The voltage sequence and current waveform are shown in Fig. 3.7.

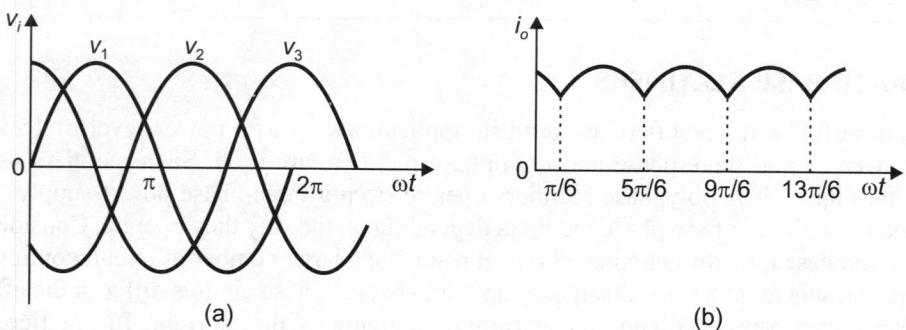

Fig. 3.7 Voltage sequence and current waveforms of a three-phase half-wave rectifier

When diode D_1 is conducting, then its cathode and the positive end of the load are at the potential v_1 [Fig. 3.7(a)]. The cathodes of diodes D_2 and D_3 will be positive with respect to their anodes, and therefore, cannot conduct. These potential relationships will change at time $\pi/6$, $5\pi/6$, $9\pi/6$,... These are the points of commutation as shown in Fig. 3.7(b). It is apparent that neither the voltage nor the current ever goes to zero in the load. The cathode current of diode D_1 can be written as

$$i_1 = \frac{V_m}{R}\sin \omega t, \qquad \frac{\pi}{6} < \omega t < \frac{5\pi}{6} \qquad \qquad ...(3.27)$$

The load current is three times the average current of one diode, or

$$I_{DC} = \frac{3}{2\pi} \int_{\pi/6}^{5\pi/6} \frac{V_m}{R} \sin \omega t \, (\omega t) = \frac{3\sqrt{3}}{2\pi R} V_m = 0.827 \frac{V_m}{R}$$

and
$$V_{DC} = 0.827V_m = 0.827 \times \sqrt{2}V_{rms} = 1.17V_{rms} \qquad ...(3.28)$$

The input power from the secondary of the transformer is

$$P_{AC} = \frac{3}{2\pi} \int_{\pi/6}^{5\pi/6} \frac{(V_m \sin \omega t)^2}{R} d(\omega t) = 0.706 \frac{V_m^2}{R}$$

% rectification efficiency

$$\eta = \frac{(0.827V_m)^2/R \times 100}{0.706V_m^2/R} = 96.5\%$$

The ripple have a fundamental frequency three times the supply frequency. The ripple magnitude may be found by computing the rms value of the load current as

$$I_{rms} = \left[\frac{3}{2\pi} \int_{\pi/6}^{5\pi/6} \left(\frac{V_m \sin \omega t}{R} \right)^2 d(\omega t) \right]^{1/2} = \frac{0.838V_m}{R} \qquad ...(3.29)$$

Then the ripple factor is given by

$$\gamma = 100\% \times \sqrt{\left(\frac{I_{rms}}{I_{DC}} \right)^2 - 1} = 100\% \sqrt{(1.014)^2 - 1} = 18.5\% \qquad ...(3.30)$$

The peak inverse voltage will occur when the maximum voltage appears across a diode while nonconducting. When diode D_1 is nonconducting, the voltage across it is $(v_1 - v_2)$ or $(v_1 - v_3)$, considering each as a rise from the neutral. Then

$$V_m \sin \omega t - V_m \sin (\omega t + \pi/3) = v_b \qquad ...(3.31)$$

The maximum inverse voltage occurs at

$$\omega t_1 = 60° + \frac{n\pi}{2}, \qquad n \text{ is an odd number}$$

Therefore, the peak inverse voltage occurs at 150° or $\frac{5\pi}{6}$ of the v_1 cycle.

The value of peak inverse voltage for the three-phase half-wave rectifier circuit is

$$PIV = \sqrt{3}V_m = 2.09V_{DC} \qquad ...(3.32)$$

The circuit suffers from DC saturation of the transformer core, since the DC component flows through each transformer is secondary.

3.5.2 Bridge Rectifier with Resistive Load (Double-way)

Another widely used three-phase rectifier is the three-phase bridge circuit shown in Fig. 3.8. It is popular in industrial applications because it can operate with or without a transformer and gives six pulse ripples on the output voltage. This is a full-wave rectifier. Most large rectifier circuit installations employ some variant of the double Y system of connections because it maintains conduction angles near the optimum value for high utility and reduces the number of commutations per cycle. Reference to Fig. 3.8 shows the cathodes of diodes D_1, D_2, D_3

are connected at a common point B. The anodes of D_4, D_5 and D_6 are tied together at point A. The load resistance R_L is connected between A and B. The waveforms of Fig. 3.9 shows the phase and amplitude relationships of phases 1, 2 and 3.

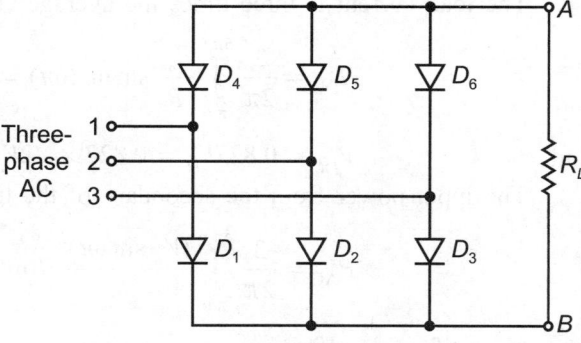

It is clear from the waveform that the sine wave of phase 2 lags that of phase 1 by 120° and the sine wave of phase 3 lags that of phase 2 by 120°. The horizontal axis marked in 30° units conventionally shows the phase

Fig. 3.8 Cathodes of diodes D_1, D_2, D_3

relationship and also shows the periods during which each waveform is most negative. Thus, waveform of phase 1 is more positive during interval P_1, P_2, P_3 and the most negative during the interval N_4, N_5, N_6. Waveform of phase 2 is the most positive during interval P_3, P_4, P_5 and the most negative during the interval N_6, N_7, N_8. Waveform of phase 3 is most positive during the interval P_5, P_6, P_7 and most negative during the interval N_8, N_9, N_{10} also N_2, N_3, N_4 of its preceding cycle.

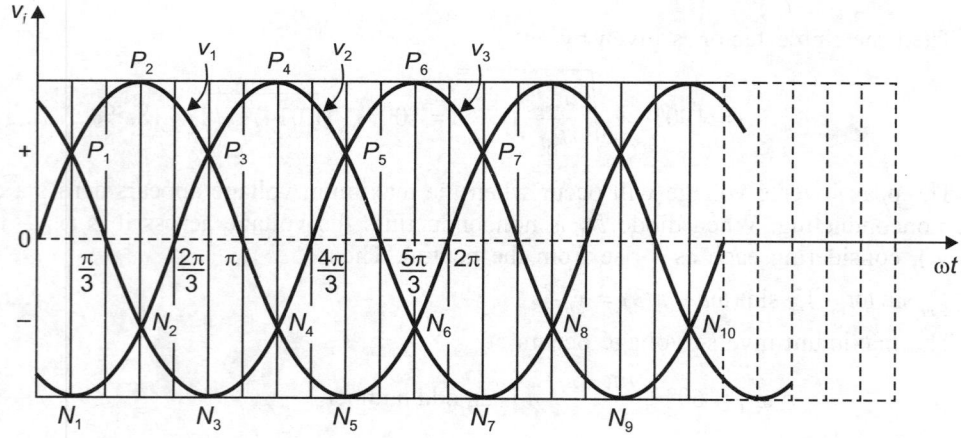

Fig. 3.9 Phase and amplitude relationships of phases 1, 2 and 3

The pair of diodes which are connected between that pair of supply lines having the highest amount of instantaneous line to line voltage will conduct. The circuit of Fig. 3.9 has been so designed that at any instant of time, the anode voltage of one of the diodes is more positive than that of others. At the same time, the cathode voltage of another six diodes is more negative than any of the others. These two-diodes will conduct at that instant of time. The circuit Fig. 3.8 is called the *bridge type rectifier* because current flows through diodes in a manner similar to current through diodes of a single phase bridge rectifier.

Referring to Figs 3.8 and 3.9, we note that during the interval P_1, P_2, P_3, waveform 1 is most positive. Therefore, anode of D_1 is more positive than the anodes of D_2 and D_3 during this interval. It is observed that during the interval N_1, N_2, corresponding to time P_1, P_2, waveform 2 is most negative. Hence, the cathodes of D_5 is more negative than the cathodes of D_4 and D_5 during the interval N_1, N_2. During the interval N_2, N_3 corresponding to P_2, P_3, waveform 3 is most negative. Hence, the cathode of D_6 is most negative during this interval. These facts indicate that during the interval P_1, P_2 or N_1, N_2, diodes D_1 and D_5 in series permit current to flow through the load from the phase 1 and 2 of the three-phase source. During the interval P_2, P_3 or N_2, N_3, diodes D_1 and D_6 in series permit current to flow through the load from phases 1 and 3 of the three-phase source.

The phase which has got the maximum value would conduct. The conduction of phase 1 and 3 means that the current starts flowing from phase 1 and returns to phase 3 through diode D_1, load resistance and diode D_6. Figure 3.10 is the waveform of load current in three-phase full-wave bridge rectifier, which is explained earlier. Each rectifier in the three-phase full-wave bridge circuit conducts for one-third of the full cycle. Due to the conduction period of 60°, the load current never falls below $\sqrt{3}I_m/2$. It is clear from Fig. 3.10 that the average and rms value of output voltage and current are obtained as given below.

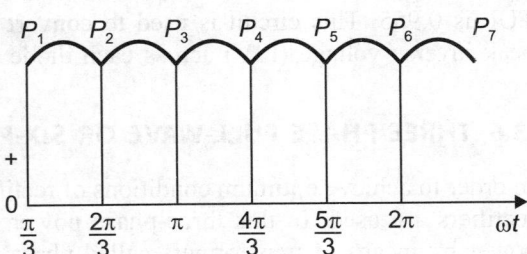

Fig. 3.10 Waveform of load current in three-phase full-wave bridge rectifier

The voltage and current factors may be calculated from the following equations.

The average DC voltage is given by:

$$V_{DC} = \frac{2}{2\pi/6} \int\limits_{0}^{\pi/6} \sqrt{3}V_m \cos \omega t \, d(\omega t) = \frac{3\sqrt{3}}{\pi} V_m$$

$$V_{DC} = 1.6542 V_m \qquad \qquad ...(3.33)$$

where V_m is the peak phase voltage.

rms output voltage; $\quad V_{rms} = \left[\dfrac{2}{2\pi/6} \int\limits_{0}^{\pi/6} 3V_m^2 \cos^2 \omega t \, d(\omega t) \right]^{1/2}$

$$= \left(\frac{3}{2} + \frac{9\sqrt{3}}{4\pi} \right)^{1/2} V_m = 1.6554 V_m \qquad \qquad ...(3.34)$$

If the load is purely resistive, the calculation of average DC current and rms current are identical. The peak current through the diode is $I_m = \sqrt{3}V_m/R$ and rms value of diode current I_r is

$$I_r = \left[\frac{4}{2\pi} \int\limits_{0}^{\pi/6} I_m^2 \cos^2 \omega t \, d(\omega t) \right]^{1/2}$$

$$= I_m \left[\frac{1}{\pi} \left(\frac{\pi}{6} + \frac{1}{2} \sin \frac{2\pi}{6} \right) \right]^{1/2} = 0.5518 I_m$$

and *the rms value of transformer secondary current,*

$$I_s = \left[\frac{8}{2\pi} \int\limits_{0}^{\pi/6} I_m^2 \cos^2 \omega t \, d(\omega t) \right]^{1/2}$$

$$= I_m \left[\frac{2}{\pi} \left(\frac{\pi}{6} + \frac{1}{2} \sin \frac{2\pi}{6} \right) \right]^{1/2} = 0.7804 I_m \qquad \qquad ...(3.35)$$

where I_m is the peak value of secondary line current.

Using these values, the ripple factor is 0.0374, the rectification efficiency is 99.5% and TUF is 0.955. This circuit is used to convert AC into DC in the automobile alternator. The peak inverse voltage (PIV) across each diode is only V_m.

3.6 THREE-PHASE FULL-WAVE OR SIX-PHASE HALF-WAVE RECTIFIER

In order to achieve optimum conditions of rectification, six-phase, twelve-phase and multiphase rectifiers are used. In this three-phase power is transformed into six-phase or twelve-phase power by means of transformers called phase transformers.

The operation is similar to three-phase rectifier except that in this case the current is commutated six times in each cycle instead of three times and each anode conducts for $2\pi/6$ radian in each cycle of the supply voltage. The principle of operation of the three-phase half-wave rectifier can be extended to explain the six-phase half-wave rectifier in Fig. 3.11. It consists of a transformer with primary windings P_1, P_2 and P_3. Q_1, Q_4, Q_2, Q_5 and Q_3, Q_6 are the centre-tapped secondaries of P_1, P_2 and P_3. N is common to which the centre tap of each of the secondary winding is connected. We know that the sine waves of voltage appearing across Q_1, Q_2 and Q_3 are 120° out of phase with each other. The voltage across Q_1 is 180° out of phase with that across Q_4. The voltage across Q_2 is 180° out of phase with that across Q_5. The voltage across Q_3 is 180° out of phase with that across Q_6. In Fig. 3.12, it is clear that the voltages across Q_1, Q_6, Q_2, Q_4, Q_3, and Q_5 are 60° out of phase with one another in the order shown. When one of the phases conducts, the other phases stop conduction because all the other diodes are reverse-biased, so only one-phase conducts at a time.

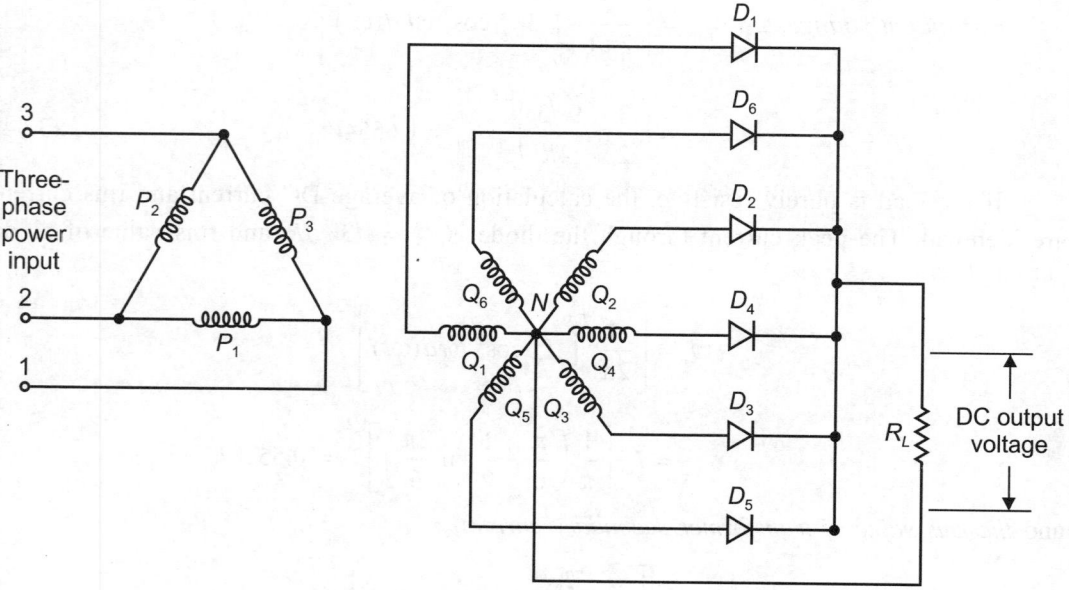

Fig. 3.11 Six-phase half-wave rectifier

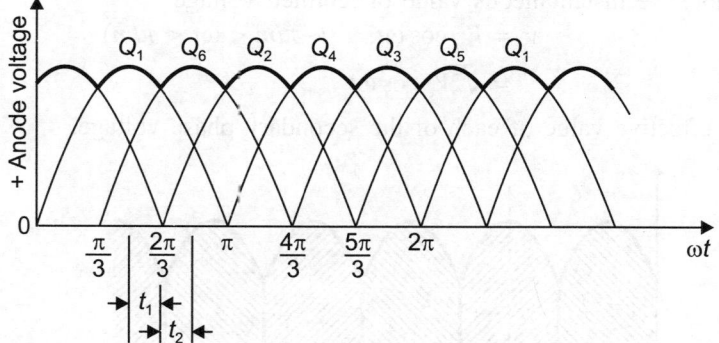

Fig. 3.12 Waveform of a six-phase half-wave rectifier

Operation of the six-phase half-wave rectifier is similar to that of three-phase half-wave rectifier. In six-phase half-wave rectifier only one of the six rectifiers will conduct at a time to supply current to the load. This is the diode whose positive anode voltage is highest during the interval. Diode D_1 will conduct during the interval t_1, when the voltage across Q_1 is most positive, D_6 will conduct during the interval t_2. Thus, in six-phase half-wave rectifier, we have just described each rectifier conducts during 60° of the entire circle and is cut off for 300°. When it is conducting, each diode supplies the full load current. Figure 3.13 is the waveform which appears across R_L, the load resistance. It is clear from the waveform that the current through the load is smoother. The three-phase full-wave circuit is often used instead of three-phase half-wave when less ripple is required and static magnetisation of the transformer core is avoided. The average DC voltage and current supplied to the load is higher than in the case of a three-phase half-wave rectifier. The zig-zag connection is used for eliminating DC saturation of the transformer core. As each primary phase carries a full cycle of current, there is no static magnetisation.

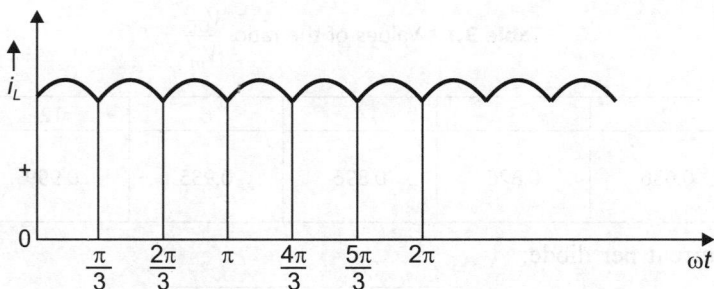

Fig. 3.13 Waveform across load resistance R_L

3.7 *m*-PHASE RECTIFIER CIRCUIT RELATIONS

It is frequently found that optimum rectification conditions or costs are obtained with 6, 12 or more phases. It is convenient to use star-connected transformers of *m*-phases, connected to *m* anodes, each anode conducting for an interval of $2\pi/m$ radians per cycle. For resistive load, current waveform will appear in Fig. 3.14 utilising a cosine wave because of the symmetry. Let us assume that: (1) the transformer resistances and reactances are negligible, and (2) anode voltages are sinusoidal.

Let v_d denote the instantaneous value of rectified voltage

$$v_d = V_m \cos \omega t, \quad (-\pi/m < \omega t < \pi/m) \qquad \qquad ...(3.36)$$
$$= \sqrt{2}V_s \cos \omega t$$

where V_s is the effective value of each of the secondary phase voltage.

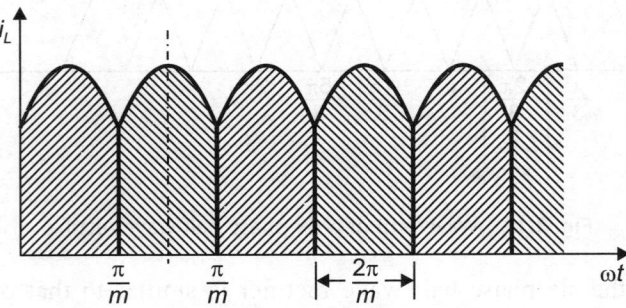

Fig. 3.14 Current waveform

A general expression for average value of rectified voltage V_{DC} is given by the following:

$$V_{DC} = \frac{m}{2\pi} \int_{-\pi/m}^{\pi/m} V_m \cos \omega t \, d(\omega t) = V_m \cdot \frac{m}{\pi} \sin \frac{\pi}{m} \qquad \qquad ...(3.37)$$

where V_m is the peak anode to neutral voltage of the transformer. Since $V_{DC} = I_{DC} R$, then

$$\frac{V_{DC}}{V_m} = \frac{m}{\pi} \sin \frac{\pi}{m} \qquad \qquad ...(3.38)$$

and values of this ratio can be calculated in terms of m showing that V_{DC} increases with the number of phases, although the increase is not large above $m = 6$ (Table 3.1).

Table 3.1 Values of the ratio $\dfrac{V_{DC}}{V_m}$

m	2	3	4	6	12	∞
$\dfrac{V_{DC}}{V_m}$	0.636	0.826	0.896	0.955	0.999	1.0

The rms current per diode:

$$I_{rms} = \sqrt{\frac{1}{2\pi} \int_{-\pi/m}^{\pi/m} \frac{V_m^2 \cos^2 \omega t}{R^2} \, d(\omega t)}$$

$$= \frac{V_m}{R} \sqrt{\frac{1}{2\pi}\left(\frac{\pi}{m} + \sin \frac{\pi}{m} \cos \frac{\pi}{m}\right)} \qquad \qquad ...(3.39)$$

The rms value of the load current having m pulses per cycle is $\sqrt{m}$ times the above value. The ripple in the load current can be calculated from the following equation:

$$\text{Ripple} = \sqrt{\left(\frac{I_{rms}}{I_{DC}}\right)^2 - 1} \times 100\% \qquad \qquad ...(3.40)$$

$$= 100\% \sqrt{\frac{(m/2\pi)\,[(\pi/m) + \sin(\pi/m)\cos(\pi/m)]}{[(m/\pi)\sin(\pi/m)]^2} - 1}$$

The ripple decreases rapidly with the number of phases, and this, with the higher ripple frequency, makes the output of the rectifier with six or more phases easy to filter. Distortion of the primary-current waveform is also less and this is very important, since harmonics introduced into the primary may cause interference in adjacent telephone lines. Rectifiers of 12, 18, 24 phases are generally employed to supply electrolytic processes.

3.8 TRANSFORMER UTILITY FACTOR (TUF)

Rectifier transformers do not have sinusoidal current through their primary and secondary windings. Secondary currents flow for the part of the cycle whereas primary current is continuous. Since the primary and secondary currents have different waveforms and rating of each winding is different, transformer design is based on volt-amperes (VA) rating and not on power delivered to the load. Transformers supplying rectifier circuits usually do not carry a sinusoidal current in their windings or may carry current in various coils only over a portions of a cycle. The harmonic components contribute to transformer heating but produce no useful DC output. To prevent these harmonic currents, a transformer rating in watts greater than contemplated DC power output is required, that is, the harmonic currents increase the volt-amperes without adding to the output DC watts.

The *utility factor* of a rectifier is given by the ratio of the DC power output of the winding to volt-amperes rating of the winding, the factor rates the efficiency with which a given winding is being used. The utility factor is a function of waveform and the ratio of rms to average voltage. Since primary and secondary windings have different VA ratings, a rectifier transformer has two primary and secondary utilization factors which may not be equal. The secondary utility factor of a single phase full-wave rectifier circuit is low because of the half-sinusoidal secondary current, it has the value 0.574 (Table 3.2).

In the single phase bridge circuit with sinusoidal secondary current, the secondary utility factor (SUF) has the value 0.813 with a resistance load. For the *m*-phase, general expressions for secondary and primary utility factors are expressed as follows. For a rectifier circuit such as the delta-star six-phase circuit, the volt-amperes per secondary phase are obtained by using the relation of I_{rms}:

$$\text{total SVA} = m\,\frac{V_m}{\sqrt{2}} \cdot \frac{V_m}{R}\,\sqrt{\frac{1}{2\pi}\left(\frac{\pi}{m} + \sin\frac{\pi}{m}\cos\frac{\pi}{m}\right)} \qquad \text{...(3.41)}$$

The DC power output, $P_{\text{DC}} = I_{\text{DC}}^2 R = \dfrac{V_m^2}{R}\left(\dfrac{m}{\pi}\right)^2 \sin^2\dfrac{\pi}{m}$ \qquad ...(3.42)

The ratio $(P_{\text{DC}}/\text{SVA})$ or is the secondary utility factor,

$$\text{SUF} = \frac{2\left[\left(\dfrac{m}{\pi}\right)\sin^2\left(\dfrac{\pi}{m}\right)\right]}{\sqrt{\pi\left[\dfrac{\pi}{m} + \sin\left(\dfrac{\pi}{m}\right)\cos\left(\dfrac{\pi}{m}\right)\right]}} \qquad \text{...(3.43)}$$

Calculation of secondary utility factor as function of *m* gives the value as in Table 3.2.

Table 3.2 Values of secondary utility factor (SUF)

m	2	3	4	6	12	24
$2\pi/m$, degree	180	120	90	60	30	15
SUF	0.57	0.675	0.636	0.551	0.399	0.286

The DC component is balanced out in star connected circuits shown in Fig. 3.11 by connecting the neutral to the common cathodes through the load of six-phase rectifier. The two secondary phases are supplied by a single primary phase. The rms current in a primary winding is the effective value of two current pulses.

$$\text{Primary,} \qquad I_{\text{rms}} = \sqrt{2}\,\frac{V_m}{2\pi}\sqrt{\frac{1}{2\pi}\left(\frac{\pi}{m} + \sin\frac{\pi}{m}\cos\frac{\pi}{m}\right)} \qquad \qquad ...(3.44)$$

Assuming 1: 1 voltage transformation. If the circuit employs p primary phases, the primary volt-amperes are:

$$PVA = p\,\frac{V_m^2}{R}\sqrt{\frac{1}{2\pi}\left(\frac{\pi}{m} + \sin\frac{\pi}{m}\cos\frac{\pi}{m}\right)} \qquad \qquad ...(3.45)$$

The primary utility factor (PUF)

$$= \frac{\sqrt{2}m}{p}\left\{\frac{\left(\dfrac{m}{\pi}\right)\sin^2\left(\dfrac{\pi}{m}\right)}{\sqrt{\pi\left[\dfrac{\pi}{m} + \sin\left(\dfrac{\pi}{m}\right)\cos\left(\dfrac{\pi}{m}\right)\right]}}\right\}$$

$$= \frac{m}{\sqrt{2}p}\cdot SUF \qquad \qquad ...(3.46)$$

In Equation (3.46), the factor $\dfrac{1}{\sqrt{2}}$ is present because of the fact that one primary phase feeds supply to two secondary phases at an angle of 180°.

3.9 RECTIFIER PERFORMANCE

It is seen that a polyphase rectifier supplies to the load, a voltage wave that never decreases to zero at any instant. For three-diode rectifier, the load voltage is half of its maximum value, the output voltage from three diodes has 17 percent ripple as compared with 48 percent ripple when the rectifier has two-diodes. In a two-diode single phase rectifier, current flows in each diode for a half cycle or 180°. The DC output voltage $V_{\text{DC}} = V_{\text{AC}} = 0.636 V_m$ while the AC applied input voltage $V_{\text{AC}} = V_{\text{rms}} = 0.707 V_m$.

$$\text{Therefore} \qquad \frac{V_{\text{DC}}}{V_{\text{AC}}} = \frac{0.636}{0.707} = 0.9 \qquad \qquad ...(3.47)$$

The DC output voltage is 0.9 times the AC input voltage. In three-diode rectifier, the current flows in each diode for one-third cycle or 120°. The load receives only the top 120° of each sine wave of voltage as shown in Fig. 3.15 (b). The average voltage during 120° is greater than the average during the entire 180°. For 120°,

$$V_{\text{DC}} = \frac{1}{2\pi/3}\int_{-\pi/3}^{\pi/3} \sqrt{2}V\cos\omega t\, d(\omega t) = 1.17\text{ V}$$

where V is the AC voltage per leg of the transformer secondary. The transformer secondary voltage per leg

$$\frac{1}{1.17} \cdot V_{DC} = 0.855 \text{ of the DC output voltage, } V_{DC}$$

Similarly for the six-phase half-wave rectifier, each diode conducts for the top 60° of its voltage wave. Figure 3.15(c) shows at the average height V_{DC} of this 60° section

$$= 1.35 \, V \qquad \qquad \qquad ...(3.48)$$

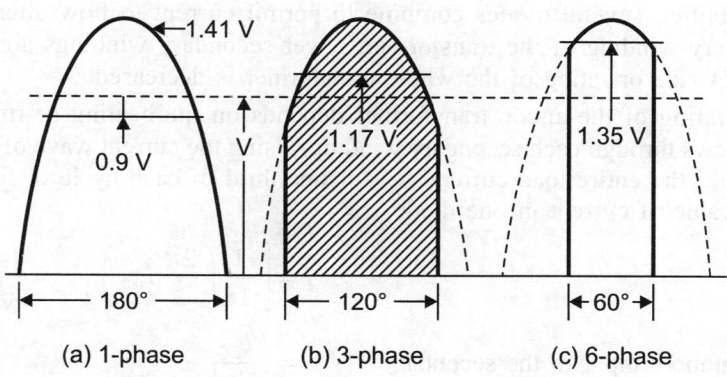

| ← 180° → | ← 120° → | ← 60° → |
| (a) 1-phase | (b) 3-phase | (c) 6-phase |

Fig. 3.15 Performance of a rectifier

The required V_{rms} per leg is

$$\frac{1}{1.35} V_{DC} = 0.74 V_{DC}$$

The important factors for this rectifier may be calculated as follows.

The DC value of the rectified output voltage

$$V_{DC} = \frac{6}{2\pi} \int_{\pi/3}^{2\pi/3} V_m \sin \omega t \, d(\omega t)$$

$$V_{DC} = \frac{3}{\pi} V_m = \frac{3(\sqrt{2}V_{rms})}{\pi} = 1.35 V_{rms}$$

The rms value of the load voltage

$$V_{o \, rms} = \left[\frac{6}{2\pi} \int_{\pi/3}^{2\pi/3} (\sqrt{2}V_{rms} \sin \omega t)^2 \, d\omega t \right]^{1/2}$$

$$V_{o \, rms} = 1.352 \cdot V_{rms}$$

$$\text{Foam factor} = \frac{\text{rms value of load voltage}}{\text{DC value of output voltage}} = \frac{1.352 V_{rms}}{1.35 V_{rms}} = 1.0015$$

$$\text{Ripple factor} = \frac{\left[V_{o\,\text{rms}}^2 - V_o^2\right]^{1/2}}{V_o} = \frac{\left[(1.352)^2 - (1.35)^2\right]^{1/2}}{1.35} = 0.06$$

In six-phase rectifier, the ripple factor is only 6% which is much less than that of the three-phase half-wave configuration and the ripple frequency is six times the fundamental frequency.

3.10 kVA RATING OF TRANSFORMER

The half-wave rectifier requires a larger transformer kVA rating than a full-wave rectifier. In two-diode rectifier, half of the transformer secondary carries current only half of the time because each diode permits current to flow in only one direction through each leg. In bridge or full-wave rectifier, several diodes combine to permit current to flow alternately in both directions in every winding of the transformer, fewer secondary windings are required; thus the required kVA size or rating of the whole transformer is decreased.

The kVA rating of the anode transformer depends on the heating or rms value of the current which flows through each secondary winding. Using the current wave of inductive load, each diode carries the entire load current I_{DC} for one-third of each cycle or for $2\pi/3$ radians therefore, rms value of current in one diode

$$I = \left[\frac{1}{2\pi}\left(\frac{2\pi}{3} I_{DC}\right)\right]^{1/2} = \frac{I_{DC}}{\sqrt{3}}$$

The volt-ampere input to the secondary $\qquad = \dfrac{VI_{DC}}{\sqrt{3}}$

Total volt-ampere rating of three secondaries $= \sqrt{3}VI_{DC}$

Resulting DC output of three diode rectifier $= V_{DC}\,I_{DC} = 1.17VI_{DC}$

Ratio of DC output to transformer total volt-ampere is known as secondary utilisation factor and is equal to

$$\frac{1.17VI_{DC}}{1.732VI_{DC}} = 0.675$$

When the rectifier has larger number of phases, the theoretical rectification efficiency nears 100 percent. At light load, diode losses are small since they vary in proportion to DC load current. The copper loss varies as the square of the DC load current. Transformer losses are small and the rectifier efficiency is good at light load. The efficiency drops at high load currents. The power factor of a rectifier is lagging, caused by transformer excitation, harmonic load components and transformer reactance.

3.11 EFFECT OF LEAKAGE REACTANCE

In the previous discussion, the transformer inductances/reactances are assumed to be negligible but finite inductances are always present in the practical rectifier circuit. In three-phase half-wave rectifier circuits, the load current I_{DC} flows in diode D_1 for 120° which seems to decrease instantly from I_{DC} to zero. Similarly, current in diode D_2 needs to rise instantly from zero to I_{DC}. These changes in current are prevented by inductance in each diode circuit. Due to leakage reactance L_s in Fig. 3.16, current actually starts to a second-diode before the first-diode current ceases, because of the emf generated in L_s, i.e. $(-L_s\,di/dt)$. The total current during the commutation interval will tend to remain constant, as current in one diode picks up, the other drops it. The current commutate between the two-diodes, the voltage across the

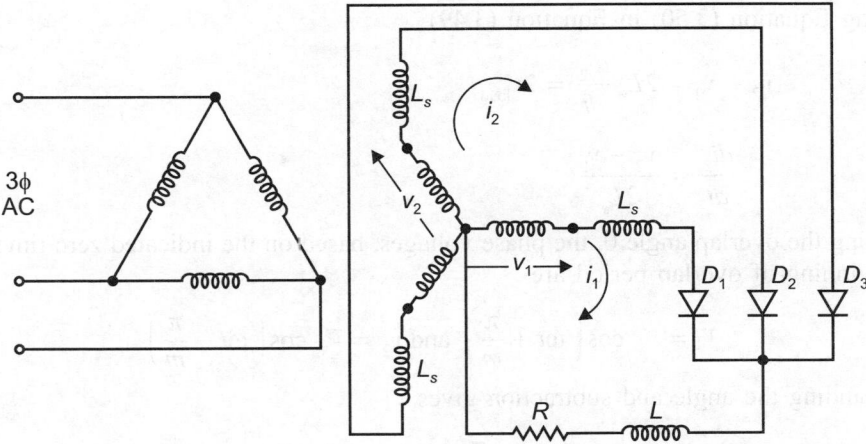

Fig. 3.16 Circuit of leakage reactance of a rectifier

leakage reactance subtracts from the transformer voltage and a lower average voltage is obtained. Figure 3.17 shows the current overlap and its effect on the voltage magnitude. During the overlap period, the load current is the addition of two-diode current, the assumption being made that the load is inductive enough to give a sensibly load current. The load voltage is the mean of two conducting phases, the effect of overlap being to reduce the mean level.

Let at any instant of cycle, the anode of diode D_2 has become positive with respect to cathode and is conducting before diode D_1 has stopped conducting, the paths from neutral to cathode through the two transformer windings are in parallel. Ignoring the diode volt-drops

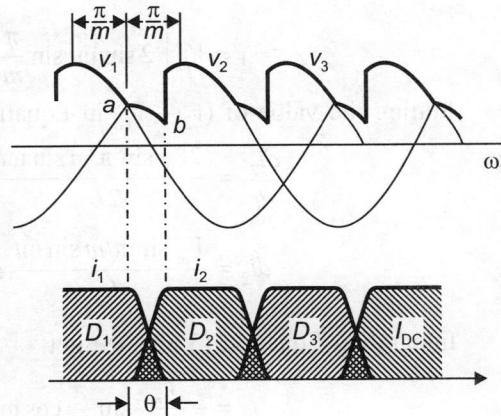

Fig. 3.17 Current overlap and its effect on the voltage magnitude

$$v_1 - L_s \frac{di_1}{dt} = v_2 - L_s \frac{di_2}{dt} = v_{\text{load}} \qquad \qquad ...(3.49)$$

As has been stated, during the commutation period, the total current remains approximately constant

$$i_1 + i_2 \cong I_{\text{DC}}$$

It follows from this statement:

$$\frac{di_1}{dt} + \frac{di_2}{dt} \cong 0$$

or

$$\frac{di_1}{dt} \cong -\frac{di_2}{dt} \qquad \qquad ...(3.50)$$

Using Equation (3.50) in Equation (3.49)

$$v_2 - v_1 = 2L_s \frac{di_2}{dt} = 2v_{\text{load}} \qquad \qquad ...(3.51)$$

or

$$\frac{di_2}{dt} = \frac{v_2 - v_1}{2L_s}$$

During the overlap angle θ, the phase voltages, based on the indicated zero time location at the beginning of overlap period are

$$v_1 = V_m \cos\left(\omega t + \frac{\pi}{m}\right) \text{ and } v_2 = V_m \cos\left(\omega t - \frac{\pi}{m}\right)$$

Expanding the angle and subtraction gives

$$v_2 - v_1 = V_m \left[\cos\left(\omega t - \frac{\pi}{m}\right) - \cos\left(\omega t + \frac{\pi}{m}\right)\right]$$

$$= V_m \left[\cos \omega t \cos \frac{\pi}{m} + \sin \omega t \sin \frac{\pi}{m} - \cos \omega t \cos \frac{\pi}{m} + \sin \omega t \sin \frac{\pi}{m}\right]$$

$$v_2 - v_1 = V_m \left[2 \sin \omega t \sin \frac{\pi}{m}\right]$$

Putting the value of $(v_2 - v_1)$ in Equation (3.51)

$$\frac{di_2}{dt} = \frac{2V_m \sin \pi/m \sin \omega t}{2L_s}$$

$$di_2 = \frac{V_m \sin \pi/m \sin \omega t}{L_s} dt \qquad \qquad ...(3.52)$$

Integrating Equation (3.52), we get

$$i_2 = -\frac{V_m}{\omega L_s} \sin \frac{\pi}{m} \cos \omega t + A$$

where ωL_s, is the leakage reactance of the transformer per phase at supply frequency.

Since $\qquad i_2 = 0$ at $t = 0$

$$0 = -\frac{V_m}{\omega L_s} \sin \frac{\pi}{m} + A$$

or

$$A = \frac{V_m}{\omega L_s} \sin \frac{\pi}{m}$$

The current in diode D_2 during commutation takes the form

$$i_2 = -\frac{V_m}{\omega L_s} \sin \frac{\pi}{m} [\cos \omega t - 1]$$

$$i_2 = \frac{V_m}{\omega L_s} \sin \frac{\pi}{m} [1 - \cos \omega t]$$

Since $\qquad i_1 \cong I_{\text{DC}} - i_2$

The current in diode D_1 can also be found during commutation, as:

$$i_1 \cong I_{DC} - \frac{V_m}{\omega L_s} \sin \frac{\pi}{m} [1 - \cos \omega t] \qquad ...(3.53)$$

The commutation angle θ may be found by noting that, $t = 0$, $i_1 = I_{DC}$

$$0 = I_{DC} - \frac{V_m}{\omega L_s} \sin \frac{\pi}{m} [1 - \cos \theta]$$

$$I_{DC} = \frac{V_m}{\omega L_s} \sin \frac{\pi}{m} [1 - \cos \theta]$$

$$\frac{\omega L_s I_{DC}}{V_m \sin \frac{\pi}{m}} = 1 - \cos \theta$$

$$\cos \theta = 1 - \frac{I_{DC} \omega L_s}{V_m \sin \frac{\pi}{m}} = 1 - \frac{I_{DC} \cdot X_L}{V_m \sin (\pi/m)}$$

$$\theta = \cos^{-1} \left[1 - \frac{I_{DC} \omega L_s}{V_m \sin \frac{\pi}{m}} \right] = \cos^{-1} \left[1 - \frac{I_{DC} X_L}{V_m \sin \pi/m} \right] \qquad ...(3.54)$$

The angle of overlap is seen to range from 10° to above 30°. For 12 phase rectifier, the angle of conduction is 30°.

3.12 DIODE LOAD AND CURRENT RATING

A rectifier diode is designed and rated to carry a certain amount of average current continuously. The maximum current, a diode can conduct safely at any instant, is limited by the electron emission from its cathode. If the load is assumed to be pure resistive, R in case of three-phase half-wave rectifier, the current wave flowing through R and each diode in turn, has the same variation or shape as the voltage. However, when load includes also a large inductive load L, this smooths out or removes the variation until the load direct current becomes steady at a value equal to the average value I_{DC} of the varying current wave.

If the load current is 21 A, diode A then carries 21 A for one-third of each cycle. Throughout the whole cycle, the average value of the current in one diode is 7 A. Since 21 A at 350 V equals 21 times 350 W. These three diodes can serve as the rectifier supply to operate 7.5 hp DC motor from three-phase AC supply.

3.13 PEAK REVERSE VOLTAGE IN POLYPHASE RECTIFIERS

Each diode in a rectifier circuit must withstand high reverse voltages during part of the cycle when that diode is not conducting. For any half-wave rectifier having an even number of legs and diodes, the peak inverse voltage is given by

$$PRV = \sqrt{2} V_{leg} \qquad ...(3.55)$$

For any bridge rectifier, $\qquad PRV = \sqrt{2} V_L$

Peak reverse voltage for three-phase half-wave rectifier is given by:

$$PRV = 2.45 V_{leg}. \qquad ...(3.56)$$

3.14 FILTERS

We have seen that the output of a rectifier is pulsating, i.e. it has AC and DC components known as ripples. The AC component is responsible for pulsations in the output of a rectifier. This component is undesirable and must be kept away from the load. This is achieved with the help of a device known as filter. Thus, a device that converts the pulsating output of a rectifier into a steady DC level is known as *filter*. This filters out the undesirable AC component and allows only constant DC component to reach to the load. A filter is installed between output of the rectifier and load. Various types of filter circuits are:

1. Inductor filter
3. L-C filter

2. Capacitor filter or C filter
4. R-C filter

3.14.1 Inductor Filter

An inductor in series with load represents a series impedance to the harmonic components in the rectifier circuit output and acts as a filter. The inductor starts storing energy due to flow of current as the output voltage of the rectifier rises when this voltage falls and eventually becomes zero at the end of the half cycle, the current does not become zero but continues to flow till the inductor has dissipated the magnetic energy stored by it (Fig. 3.18). This has the smoothening effect on the current waveform as shown in Fig. 3.18(b) and thus reducing the current variation or ripple.

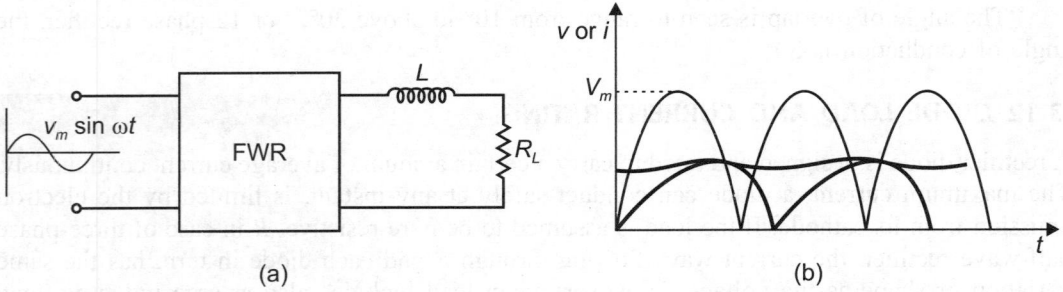

Fig. 3.18 Inductor filter circuit of FWR and waveforms

The Fourier series expression of the output voltage of the full-wave rectifier without filter is:

$$v_o = \frac{2V_m}{\pi} - \frac{4V_m}{3\pi} \cos 2\omega t - \frac{4}{15\pi} \cos 4\omega t \dots \frac{4V_m}{\pi} \left[\sum_{n=\text{even}} \frac{\cos n\omega t}{(n-1)(n+1)} \right]$$

Thus,
$$I_{\text{DC}} = \frac{2V_m}{\pi} \cdot \frac{1}{R_L}$$

Since the impedance of the inductor increases with frequency, hence there is a better filtering action for higher harmonic terms. For AC component, the impedance of L and R_L will be in series and is given by

$$Z = \sqrt{R_L^2 + (2\omega L)^2}$$

Frequency of AC component
$$= 2\omega$$

Assume that the AC component of v_o is represented only by first harmonic. Then

$$I_{AC1} = \frac{4V_m}{3\pi} \cdot \frac{\cos(2\omega t - \phi)}{\sqrt{R_L^2 + 4\omega^2 L^2}} \text{ and } \tan\phi = \frac{2\omega L}{R_L}$$

Therefore, $\quad I_{rms} = \dfrac{4V_m}{3 \cdot \sqrt{2} \cdot \pi \sqrt{R_L^2 + 4\omega^2 L^2}}$

Ripple factor is the ratio of rms value of AC component to the DC value of the wave. If $\omega L \gg R_L$ then ripple factor

$$\cong \frac{I_{rms}}{I_{DC}}$$

$$\gamma \cong \frac{1}{3\sqrt{2}} \cdot \frac{R_L}{\omega_L} \cong 0.236 \frac{R_L}{\omega L} \qquad \qquad ...(3.57)$$

Thus, the inductor filter should be used when R_L is consistently small, i.e. effective filtering takes place when load current is high. The effect is opposite to that of the capacitor filter, in which the ripple becomes greater with increase of load current. The inductor filter introduces no undue peak-current demands on the diodes, but it requires a high-input-voltage transformer for a given DC output and this will increase the cost of the unit. The inductor is also expensive and heavy, and the simple inductor filter is usually employed only as a building block for the L-C form of filter. This inductance introduces a time delay of the load current with respect to the input voltage, in case of single phase half-wave rectifier, a free wheeling diode is required to provide a path for this inductive current.

In L-C filter, the inductor serves to decrease the amplitude of the harmonic current components, and the capacitor bypasses them around the load, the result being a lower ripple than is possible with the same L or C alone.

Advantages and disadvantages of L-C filter

Advantages:

1. Smaller ripple factor than that obtained by multi-section L-C filter with the same total value of L and C
2. Higher DC output voltage

Disadvantages:

1. Poor voltage regulation
2. Increased peak anode current
3. Higher peak inverse voltage

3.14.2 Capacitor Filter or C-Filter

A capacitor connected in parallel with the load can reduce the ripple in the rectified output. If its reactance is very low at the input frequency, high frequency components will pass through it rather than the load. The capacitor acts as a filter under no load conditions with output open circuited, the ripple is zero.

When the instantaneous voltage v_s of a single phase bridge rectifier with capacitor filter [Fig. 3.19(a) circuit model] is higher than the instantaneous capacitor voltage v_c, the diodes (D_1 and D_4 or D_2 and D_3) conduct; and the capacitor is then charged from the supply. If the instantaneous supply voltage v_s falls below the instantaneous capacitor voltage v_c, the diodes

(D_1 and D_4 or D_2 and D_3) are reverse-biased and the capacitor C_e discharges through the load resistance R_L. The capacitor voltage v_c varies between a minimum $V_{C\,(min)}$ and maximum value $V_{C\,(max)}$. This is shown in Fig. 3.19(b).

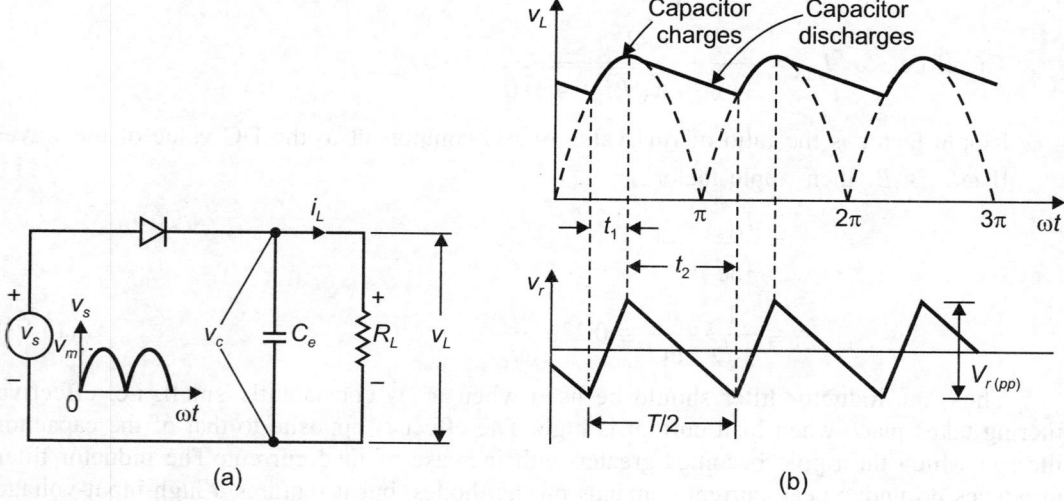

Fig. 3.19 Capacitor filter circuit and waveforms

Let us assume that t_1 is the charging time and t_2 is the discharging time of capacitor C_e. The capacitor charges almost instantaneous to the supply voltage V_s. The capacitor C_e will be charged to the peak supply voltage V_m, so that $v_c\,(t = t_1) = V_m$. The capacitor discharges exponentially through R_L

$$\frac{1}{C_e} \int i_L dt + v_c(t = 0) + Ri_L = 0 \qquad \ldots(3.58)$$

With an initial condition of $v_c\,(t = 0) = V_m$, gives the discharging current as

$$i_L = \frac{V_m}{R_L} e^{-t/R_L C_e} \qquad \ldots(3.59)$$

The output voltage (v_L) during the discharging period can be found from

$$v_L(t) = i_L \cdot R_L = \frac{V_m}{R_L} e^{-t/R_L C_e} \cdot R_L = V_m e^{-t/R_L C_e}$$

Peak to peak ripple voltage, $V_{r\,(pp)}$ can be found from

$$V_{r(pp)} = v_L(t_1) - v_L(t_2) = V_m - V_m e^{-t_2/R_L C_e}$$

$$= V_m(1 - e^{-t_2/R_L C_e}) \qquad \ldots(3.60)$$

Using the approximation, $e^{-x} \approx 1 - x$ in Equation (3.60)

$$V_{r(pp)} = V_m \left(1 - 1 + \frac{t_2}{R_L C_e}\right) = \frac{V_m \cdot t_2}{R_L C_e} = \frac{V_m}{2 f R_L C_e}$$

For full-wave light load ($f = 50$ Hz)

$$V_{r/rms} = \frac{I_{DC}}{4\sqrt{3}fC_e} = \frac{2.8I_{DC}}{C_e} = \frac{2.8V_{DC}}{R_LC} \qquad ...(3.61)$$

Average load voltage

$$V_{DC} = V_m - \frac{V_{r(pp)}}{2} = V_m - \frac{V_m}{2(2fR_LC_e)}$$

$$= V_m - \frac{V_m}{4fR_LC_e} \qquad ...(3.62)$$

rms value of output ripple voltage (approximately)

$$V_{AC} = \frac{V_{r(pp)}}{2\sqrt{2}} = \frac{V_m}{4\sqrt{2}fR_LC_e} \qquad ...(3.63)$$

Ripple factor $\quad\quad RF = \dfrac{V_{AC}}{V_{DC}} = \dfrac{V_m}{4\sqrt{2}fR_LC_e} \times \dfrac{4fR_LC_e}{V_m(4fR_LC_e - 1)} \qquad ...(3.64)$

$$RF = \frac{1}{\sqrt{2}(4fR_LC_e - 1)}$$

Since V_{DC} and I_{DC} relate to the filter load R_L,

$$r = \frac{2.8}{R_LC} \times 100\%$$

where I_{DC} is in mA, C is in microfarad and R_L in kΩ

or $\quad\quad\quad\quad\quad C_e = \dfrac{1}{4fR_L}\left(1 + \dfrac{1}{\sqrt{2}RF}\right). \qquad ...(3.65)$

3.14.3 L-C Filter

In the above section, a shunt capacitor C_e, used as a filter reduces the ripple voltage. But the current through the diode increases to charge the capacitor. This excessive current may damage the diode or cause increased heating of the power transformer resulting in decreased efficiency. On the other hand, a series inductance reduces both the peak and effective value of the rectified current. Thus, a combination of these is used to achieve the required amount of filtering without drawing excessive currents. The circuit and its equivalent for the harmonics is shown in Figs 3.20(a) and (b).

To make it easier for the nth harmonic ripple current to pass through the filter capacitor, the load impedance must be much greater than that of the capacitor, that is

$$\sqrt{R_L^2 + (n\omega L)^2} \gg \frac{1}{n\omega C_e} \qquad ...(3.66)$$

This condition is generally satisfied by the relation $\sqrt{R_L^2 + (n\omega L)^2} = \dfrac{10}{n\omega C_e}$ and under this condition the effect of load will be negligible. The rms value of the nth harmonic component appearing on the output can be found by using voltage divider rule.

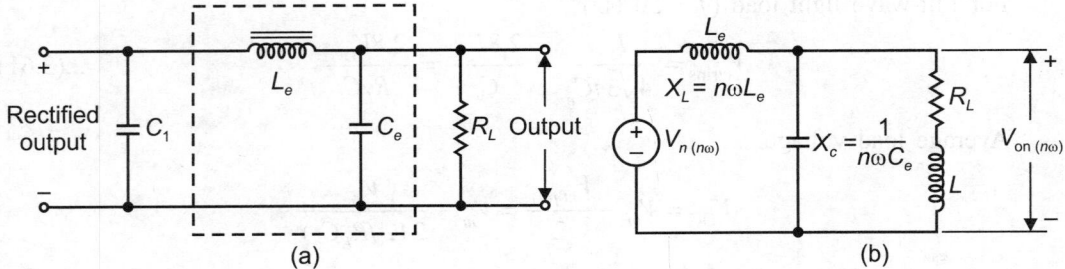

Fig. 3.20 L-C filter circuit

Therefore, $\qquad V_{02} = \left| \dfrac{-1/(n\omega C_e)}{(n\omega L_e - 1/(n\omega C_e))} \right| V_n = \left| \dfrac{-1}{(n\omega)^2 L_e C_e - 1} \right| V_n \qquad$ ...(3.67)

The total amount of ripple voltage due to all harmonics is

$$V_{AC} = \left(\sum_{n=2,4,6}^{\alpha} V_{on}^2 \right)^{1/2} \qquad \text{...(3.68)}$$

For a specified value of V_{AC} and with the value of C_e from Equation (3.60), the value of L_e can be computed. Considering only the dominant harmonic, second harmonic is the dominant one and its rms value is $V_2 = 4V_m/(3\sqrt{2}\pi)$ and its DC value is $V_{DC} = 2V_m/\pi$. For $n = 2$ above Equations (3.61) and (3.62) give

$$V_{AC} = V_{02} = \left| \frac{-1}{(2\omega)^2 L_e C_e - 1} \right| V_2$$

The value of the filter capacitor C_e is calculated from

$$\sqrt{R_L^2 + (2\omega L)^2} = \frac{10}{2\omega C_e}$$

or $\qquad C_e = \dfrac{10}{4\pi f \sqrt{R_L^2 + (4\pi f L)^2}}$

and ripple factor is $RF = \dfrac{V_{AC}}{V_{DC}} = \dfrac{V_2}{V_{DC}} \dfrac{1}{(4\pi f)^2 L_e C_e - 1} = \dfrac{\sqrt{2}}{3} \left| \dfrac{1}{[(4\pi f)^2 L_e C_e - 1]} \right| \qquad$ ...(3.69)

3.14.4 R-C Filter

It is possible to further reduce the amount of ripple across a filter capacitor while inducing the DC voltage by using an additional R-C filter section as shown in Fig. 3.21.

The purpose of the added R-C network is to pass as much of the DC component of the voltage developed across the first filter capacitor C_1 and to attenuate as much of the AC component of the ripple voltage developed across C_1 as possible. This action would reduce the amount of ripple in relation to the DC level, providing better filter operation than from the simple-capacitor filter. This filter includes a lower DC output voltage due to the DC voltage drop across the resistor and two additional components in the circuit will increase the cost.

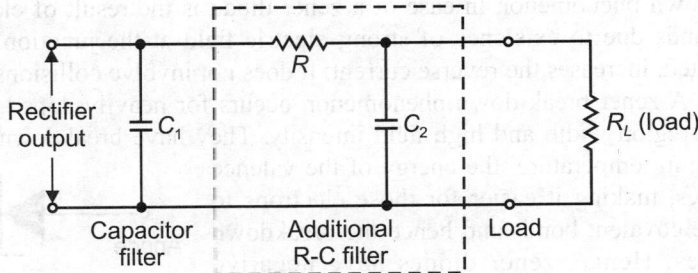

Fig. 3.21 An additional R-C filter section

Figure 3.22 shows the rectifier filter circuit for full-wave operation. Since the rectifier feeds directly into a capacitor, the peak currents through the diodes are many times the average current drawn from the supply. The voltage developed across capacitor C_1 is then further filtered by the resistor-capacitor section (R, C_2) providing an output voltage having less percent or ripple than that across C_1. The load, represented by resistor R_L, draws DC current through resistor R with an output DC voltage across the load being somewhat less than that across C_1 due to the voltage drop across R.

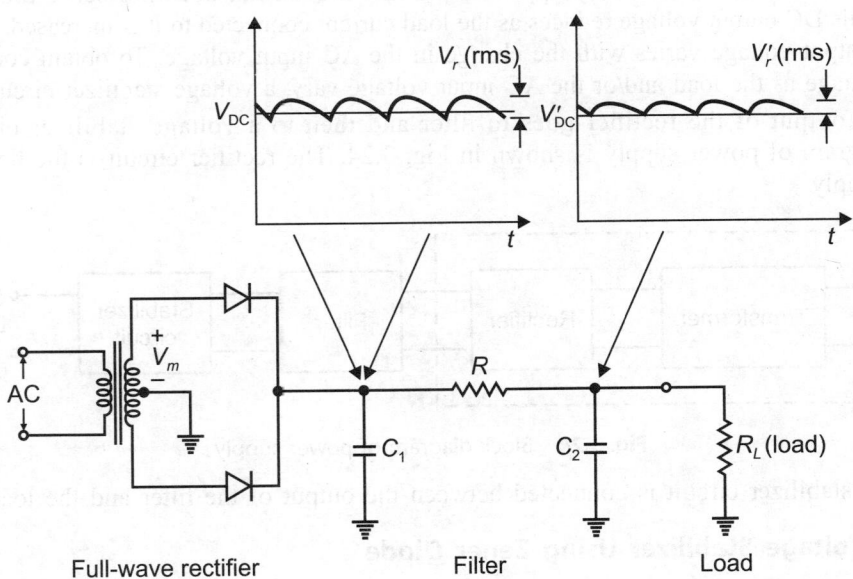

Fig. 3.22 R-C filter circuit for full-wave operation

This filter circuit, like the simple-capacitor filter circuit, provides best operation at light loads, with considerably poorer voltage regulation and higher percent of ripple at heavy loads.

3.15 ZENER DIODES

A zener diode is like an ordinary silicon p-n diode. The only difference between zener diode and ordinary crystal diode is that zener diode has a sharply defined knee in its reverse characteristics.

The breakdown phenomenon in case of a zener diode is the result of electrons breaking their covalent bonds due to existence of strong electric field at the junction. The new hole-electron pair created, increases the reverse current. It does not involve collisions of carriers with the lattice atoms. A zener break down phenomenon occurs for heavily doped diodes having a narrow depletion-region width and high field intensity. They have breakdown voltages below 6 V with increase in temperature, the energy of the valence electrons increases, making it easier for those electrons to break through the covalent bonds and hence the breakdown voltage decreases. Hence, zener diodes have negative temperature coefficient of breakdown voltage. If the applied reverse voltage exceeds the breakdown voltages, the zener diode acts like a constant voltage source.

Fig. 3.23 Circuit symbol of a zener diode

The circuit symbol of a zener diode is shown in Fig. 3.23.

A zener diode is specified by its breakdown voltage and the maximum power dissipation. The most common application of a zener diode is in the voltage stabilizing or regulator circuits.

3.16 VOLTAGE STABILIZER

After the ripples have been filtered from the rectifier output, a sufficiently steady state DC output voltage is obtained. But in many applications, this DC output does not serve the purposes because this DC output voltage reduces as the load current connected to it is increased. Secondly, the DC output voltage varies with the change in the AC input voltage. To obtain constant DC output voltage as the load and/or the AC input voltage vary, a voltage stabilizer circuit is used.

The output of the rectifier goes to filter and then to a voltage stabilizer circuit, the block diagram of power supply is shown in Fig. 3.24. The rectifier circuit is the heart of the power supply.

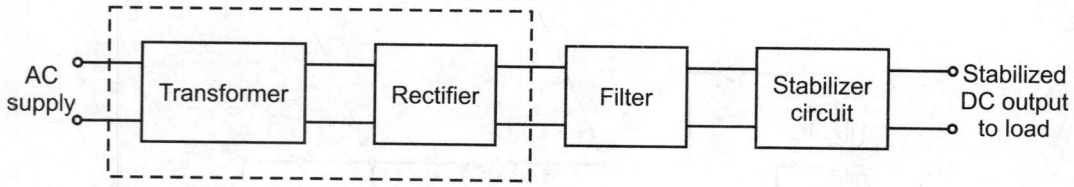

Fig. 3.24 Block diagram of power supply

The stabilizer circuit is connected between the output of the filter and the load.

3.16.1 Voltage Stabilizer Using Zener Diode

The simple stabilizer circuit consists of a resistor R connected in series with the input voltage, and a zener diode connected in parallel with the load as shown in Fig. 3.25.

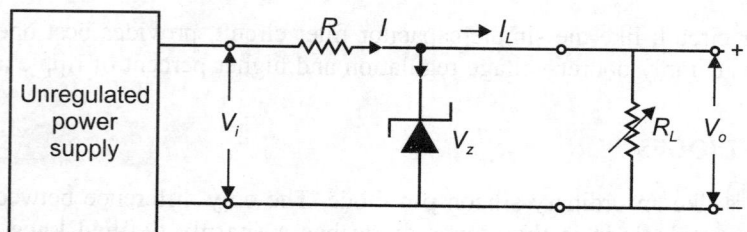

Fig. 3.25 A zener diode connected in parallel with the load

The voltage from an unregulated power supply is used as the input V_i to the stabilizer circuit as long as the voltage across R_L is less than the zener breakdown voltage V_z, the zener diode does not conduct. If the zener diode does not conduct the resistors R and R_L make a potential divider across V_i. At an increased V_i, the voltage across R_L becomes greater than the zener breakdown voltage. It then operates in its breakdown region. The registor R limits the zener current from exceeding its rated maximum $I_{z\,max}$.

The current from the unregulated power supply splits at the junction of zener diode and load resistor R_L

Therefore, $I = I_z + I_L$

When the zener diode operates in its breakdown region, the voltage V_z across it remains fairly constant even though the current I_z flowing through it may vary considerably. If the load current I_L increases because of reduction in load resistance, the current I_z through the zener diode falls by same percentage in order to maintain constant current I. This keeps the voltage drop across R constant. Hence, the output voltage V_o remains constant. If the load current decreases, the zener diode passes an extra current I_z such that the current I is kept constant. The output voltage is stabilized.

Let us examine the other cause of the output variation. If the input voltage V_i increases, the zener diode passes a large current so that extra voltage is dropped across R. Conversely, if input voltage V_i falls, the current I_z also falls and the voltage drop across R is reduced.

Because of the self-adjusting voltage drop across R, the output voltage V_0 fluctuates to a much lesser extent than does the input voltage V_i.

3.17 METAL RECTIFIERS

The metal rectifiers are nowadays preferred to valve rectifiers as they are mechanically strong, more reliable and do not require any voltage for filament heating. There are two types of metal rectifiers: (1) Copper oxide rectifiers, (2) Selenium rectifiers.

1. **Copper oxide rectifiers:** Such rectifiers are made by means of coating on one side of the disc with a layer of red cuprous oxide, which is very hard and is formed by heat treatment. Such a layer offers low resistance for the passage of current from oxide to copper plate but prevents their passage in the opposite direction.
2. **Selenium rectifier:** This is the most common type of rectifier used in the industries. It has an advantage that a layer of selenium can be formed over any type of metal, but generally nickel plated steel is used.

Construction of copper oxide rectifier and selenium rectifier: Copper oxide rectifiers are made from refined copper plates or discs which are usually 1 mm thick and 25 mm in diameter or smaller. These discs are heated in air to a temperature of about 1000°C until a layer of 1 mm of cuprous oxide is formed over the plates. Then the discs are annealed to modify the crystal structure of the oxide layer. Usually over the red cuprous oxide, a black layer of cupric oxide due to oxidization is also formed, which is removed chemically leaving the layer of cuprous oxide. Contact is made to the film through the copper on which it is formed and a soft metal disc pressed on the outer surface.

Over the red cuprous oxide is provided an aqueous suspension of colloidal graphite, over which comes the electrode of soft material such as lead. The construction of copper oxide is represented in Fig. 3.26. Each disc so formed is not generally used singly, since the maximum voltage which can be applied to it is about 6 to 8 volts and the maximum current in the forward direction should not exceed 0.1 to 0.15 A/cm^2 at an allowable voltage of

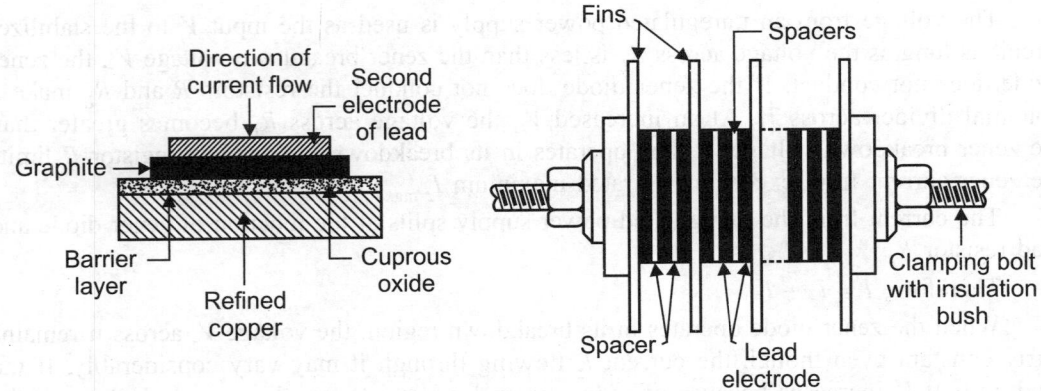

Fig. 3.26 Construction of a copper oxide rectifier

8 volts per element. Such units are connected in series and are put under the pressure of about 3 kg per square cm.

Selenium rectifier: It consists of a supporting plate made of nickel plated steel as shown in Fig. 3.27, about 1 mm thick layer of selenium is applied to the above base under a pressure at about 130°C. Then its temperature is raised to 180 to 215°C which changes the crystalline selenium into grey form. A low melting point alloy, such as tin alloy, is then sprayed over the selenium which acts as a second electrode. The cell is connected into circuit by means of tags which is in contact with both electrodes. The units can be connected in series or in parallel to make the rectifier suitable for various voltages and currents and are assembled in stacks.

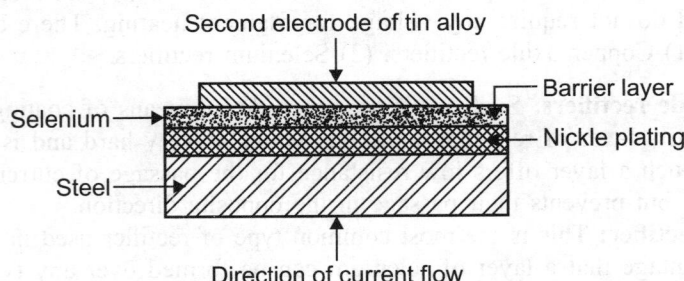

Fig. 3.27 A supporting plate made of nickel plated steel

The selenium cell has some advantages over the oxide cell. Changes in temperature have less effect on selenium cell then on the copper oxide unit. It can withstand large reverse voltages. It can be operated at temperaures as high as 75°C.

Characteristics of metal rectifiers: The rectifiers offer low resistance in the forward direction but offer high resistance in the reverse direction. Figure 3.28 represents the current relation of the copper and selenium rectifiers. For a reverse voltage of about 18 volts, the selenium rectifier is said to pass current in the reverse direction.

Applications: The rectifier instruments use metal rectifiers in connection with moving coil instruments for measurement of AC voltage and current at power and audio high frequency up to 100 kHz. They can be employed on polyphase circuits to supply large value of currents.

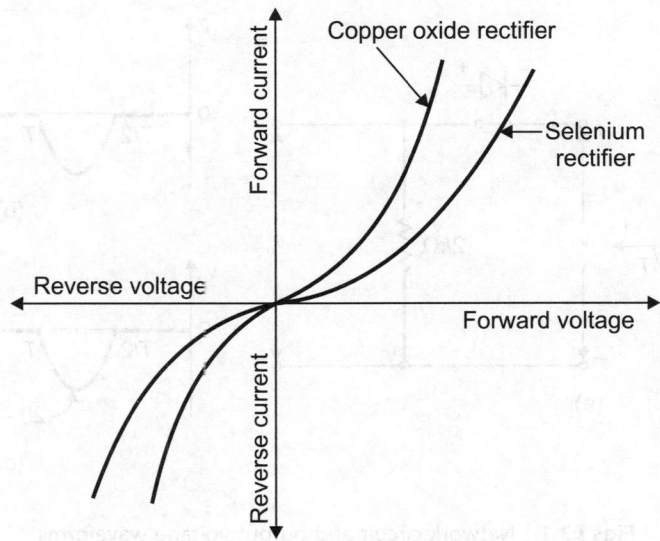

Fig. 3.28 Current relation of the copper and selenium rectifiers

SOLVED EXAMPLES

Example 3.1 Sketch the output v_o and determine DC level of the output for the network shown below, repeat if the ideal diode is replaced by silicon diode.

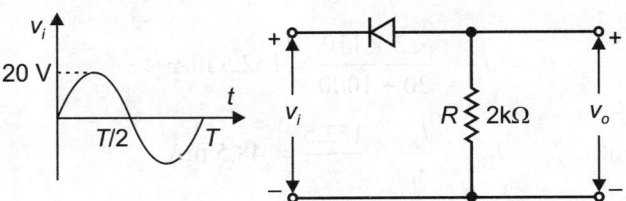

Fig. E3.1 DC level of the output for the network

Solution In this situation two-diodes will conduct during the negative part of the input and v_o will appear as shown in Fig. E3.1 (a). For the full period, the DC level

$$V_{DC} = -0.318 V_m$$
$$= -0.318 \times 20$$
$$= -6.36 \text{ V}$$

The negative sign indicates that the polarity of the output is opposite to the defined polarity of given example.

Using a silicon diode, the output has appearance of Fig. E3.1 (b).

$$V_{DC} \approx -0.318 \, (V_m - 0.7) \text{ V}$$
$$V_{DC} = -0.318 \, (19.3) \approx -6.14 \text{ V}$$

The resulting drop in DC level is 0.22 V of about 3.5%.

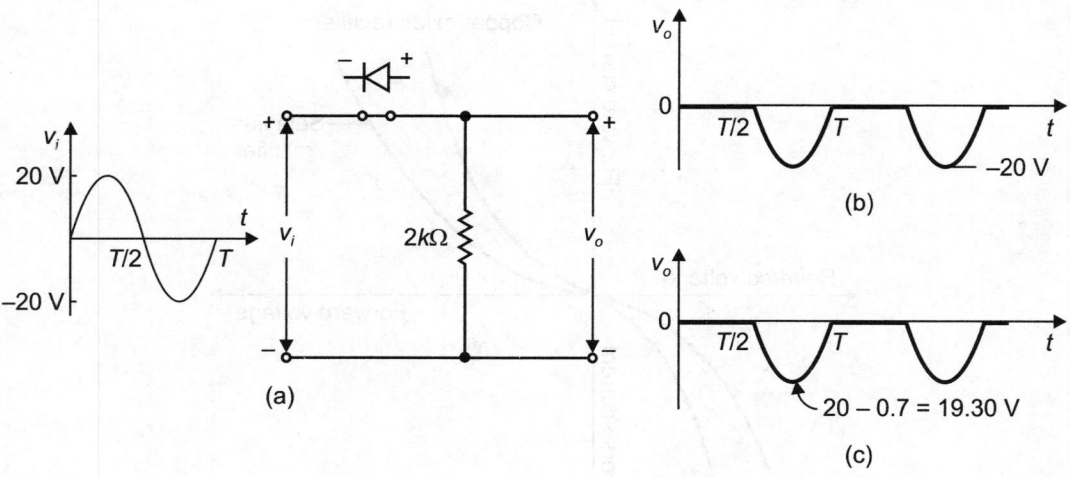

Figs E3.1 Network circuit and output voltage waveforms

Example 3.2 A diode whose internal resistance is $20\,\Omega$ is to supply power to a $1000\,\Omega$ load from a 110 V (rms) source of supply. Calculate: (a) peak load current (b) DC load current (c) AC load current (d) DC diode voltage (e) total input power to the circuit and (f) percentage regulation from no load to the given load.

Solution

(a) Peak load current

$$I_m = \frac{V_m}{r_f + R}$$

$$I_m = \frac{\sqrt{2} \times 110}{20 + 1000} = 152.5 \text{ mA}$$

(b) DC load current

$$I_{DC} = \frac{I_m}{\pi} = \frac{152.5}{\pi} = 48.5 \text{ mA}$$

(c) rms current

$$I_{rms} = \frac{I_m}{2} = \frac{152.5}{2} = 76.2 \text{ mA}$$

(d) DC diode voltage

$$V_{DC} = I_{DC} R_L = 48.5 \times 10^{-3} \times 10^3$$
$$= 48.5 \text{ V}$$

(e) Input power

$$P_i = I_{rms}^2 (r_f + R_L) = (76.2 \times 10^{-3})^2 \times 1020$$
$$= 5.92 \text{ W}$$

(f) % regulation

$$= \frac{V_{NL} - V_{FL}}{V_{FL}} \times 100 = \frac{(V_m/\pi) - I_{DC} \cdot R_L}{I_{DC} \cdot R_L} \times 100$$

$$= \frac{(\sqrt{2} \times 100/\pi) - 48.5}{48.5} = \frac{49.5 - 48.5}{48.5} \times 100$$

$$= 2.06\%.$$

Example 3.3 A half-wave rectifier has a forward resistance $500\,\Omega$. If the secondary voltage is $400 \sin 314t$, calculate: (a) I_m, I_{rms}, I_{DC} (b) DC output voltage (c) AC power input and DC power output for a load of 3.5 kΩ and (d) TUF.

Solution

(a)
$$V_m = 400 \text{ V}$$

$$I_m = \frac{V_m}{r_f + R_L} = \frac{400}{500 + 3500} = 100 \text{ mA}$$

$$I_{\text{rms}} = \frac{I_m}{2} = 50 \text{ mA}$$

$$I_{\text{DC}} = \frac{I_m}{\pi} = \frac{100}{\pi} = 31.8 \text{ mA}$$

(b) DC output voltage
$$V_L = I_{\text{DC}} \cdot R_L = 31.8 \times 10^{-3} \times 3500 = 111.3 \text{ V}$$

(c) AC power input
$$P_{\text{AC}} = \left(\frac{I_m}{2} \right)^2 (r_f + R_L)$$

$$= 50^2 \times 4000 \times 10^{-6} = 10 \text{ W}$$

Rectification efficiency for HWR is given by

$$\eta\% = \frac{40.6}{1 + \dfrac{r_f}{R_L}} = \frac{40.6}{1 + \dfrac{500}{3500}} = 35.5\%$$

DC power output
$$P_0 = \eta \times \text{AC power input}$$

$$= \frac{35.5}{100} \times 10 = 3.55 \text{ W}$$

(d)
$$\text{TUF} = \frac{0.286}{1 + \dfrac{r_f}{R_L}} = \frac{0.286}{1 + \dfrac{500}{3500}} = 0.249.$$

Example 3.4 A 40 kΩ load is fed by a supply obtained through a half-wave rectifier. The supply transformer is 120:120 V. Find: (a) V_m, V_{av} (b) I_m, I_{av} (c) PIV and (d) minimum manufacturer's PIV rating.

Solution

(a)
$$V_m = \sqrt{2} V_{\text{rms}}$$

$$= \sqrt{2} \times 120 = 170 \text{ V}$$

$$V_{av} = \frac{V_m}{\pi} = \frac{170}{\pi} = 54.1 \text{ V}$$

(b)
$$I_m = \frac{V_m}{R_L} = \frac{170}{40 \times 10^3} = 4.25 \text{ mA}$$

$$I_{av} = \frac{V_{av}}{R_L} = \frac{54.1}{40 \times 10^3} = 1.352 \text{ mA}$$

(c)
$$\text{PIV} = V_m = 170 \text{ V}$$

(d) Minimum manufacturer's PIV rating

$$= \frac{\text{Actual } PIV}{0.8} = \frac{170}{0.8} = 213 \text{ V}.$$

Example 3.5 A crystal diode having a forward resistance of $260\,\Omega$ is used as a half-wave rectifier. Find the rms value of voltage fed to it so as to act 100 V of DC across a load resistance of $2000\,\Omega$.

Solution DC output voltage across $2000\,\Omega$ load

$$= 100 \text{ V}$$

$$I_{\text{DC}} = \frac{100}{2000} = 0.05 \text{ A}$$

but

$$I_{\text{DC}} = \frac{I_m}{\pi}$$

or

$$I_m = \pi \cdot I_{\text{DC}} = \frac{22}{7} \times 0.05 = \frac{11}{70} \text{ A}$$

Again

$$I_m = \frac{V_m}{r_f + R_L}$$

$$V_m = I_m \cdot (r_f + R_L) = \frac{11}{70}(260 + 2000) = 355.1 \text{ V}$$

$$V_{\text{rms}} = \frac{V_m}{\sqrt{2}} = \frac{355.1}{\sqrt{2}} = 251.8 \text{ V}.$$

Example 3.6 In a full-wave rectifier the load resistance $R_L = 2$ kΩ. Each diode has an ideal characteristics having slope corresponding to a resistance of $400\,\Omega$, voltage applied to each diode is 240 sin 50t. Find: (a) the maximum value of current (b) DC value of current (c) rms value of current (d) rectifier efficiency and (e) ripple factor.

Solution

(a) Maximum value of current

$$I_m = \frac{V_m}{r_f + R_L} = \frac{240}{400 + 2000} = 100 \text{ mA}$$

(b) DC value of current

$$I_{\text{DC}} = \frac{2I_m}{\pi}$$

$$I_{\text{DC}} = \frac{2 \times 100}{\pi} = 63.84 \text{ mA}$$

(c) rms value of current $\quad I_{\text{rms}} = \frac{I_m}{\pi} = \frac{100}{\sqrt{2}} = 70.72 \text{ mA}$

(d) DC output power $\quad P_{\text{DC}} = I_{\text{DC}}^2 \cdot R_L = (63.84)^2 \times (10^{-3})^2 \times 2000 = 8.11 \text{ W}$

AC input power $\quad P_{\text{AC}} = I_{\text{rms}}^2 (r_f + R_L) = (70.72)^2 \times (10^{-3})^2 (400 + 2000) = 12 \text{ W}$

Hence, rectifier efficiency

$$\eta = \frac{P_{DC}}{P_{AC}} = \frac{8.11}{12} \times 100 = 67.6\%$$

(e) Ripple factor

$$\gamma = \sqrt{\left(\frac{I_{rms}}{I_{DC}^2}\right)^2 - 1} = \sqrt{\left(\frac{70.72}{63.84}\right)^2 - 1} = 0.482.$$

Example 3.7 A full-wave rectifier with 20 V rms sinusoidal input has a load resistor of 1 kΩ: (a) If silicon diodes are employed, what is the DC voltage available at the load? (b) determine the PIV rating of each diode (c) find the maximum current through the diode during conduction and (d) What is the required power rating of each diode?

Solution

(a) $$V_m = \sqrt{2}V = 1.414\,(20) = 28.28 \text{ V}$$

With silicon diodes, DC voltage available

$$V_{DC} = 0.636\,(V_m - 2V_T)$$
$$= 0.636\,(28.28 - 2 \times 0.7) = 17.096 \text{ V}$$

(b) $$\text{PIV} = V_m - V_T = 28.28 - 0.7 = 27.58 \text{ V}$$

(c) $$I_{max} = \frac{V_m - V_T}{R_L} = \frac{28.28 - 0.7}{1000} = 26.88 \text{ mA}$$

(d) Power $$VI = (0.7) \times (26.88 \times 10^{-3}) = 18.81 \text{ mW}.$$

Example 3.8 Show that the maximum DC output power $P_{DC} = V_{DC}\,I_{DC}$ in a half-wave single phase circuit occurs when the load resistance equals the diode resistance R_f.

Solution DC output power in a single phase half-wave rectifier

$$P_{DC} = I_{DC}^2 \cdot R_L$$

$$P_{DC} = \left(\frac{V_m}{\pi\,(R_f + R_L)}\right)^2 \times R_L$$

For maximum DC output power

$$\frac{dP_{DC}}{dR_L} = 0$$

$$\frac{dP_m}{dR_L} = \frac{V_m^2}{\pi^2}\left[\frac{(R_f + R_L)^2 - 2\,(R_f + R_L)\,R_L}{(R_f + R_L)^2}\right] = 0$$

or $$(R_f + R_L)^2 = 2\,(R_f + R_L)\,R_L$$
$$(R_f + R_L)^2 = 2R_L$$

∴ $$R_L = R_f.$$

Example 3.9 Prove that the regulation of both the full-wave and half-wave rectification is given by

$$\% \text{ regulation} = \frac{R_f}{R_L} \times 100.$$

Solution For a half-wave rectifier

$$V_{NL} = \frac{V_m}{\pi} \text{ and } V_{FL} = \frac{V_m}{\pi} \cdot \frac{R_L}{(R_f + R_L)}$$

$$\therefore \% \text{ regulation } = \frac{V_{NL} - V_{FL}}{V_{FL}} \times 100$$

$$= \left[\left(\frac{V_m}{\pi} \right) - \frac{(V_m/\pi) \cdot R_L}{(R_f + R_L)} \right] \bigg/ \frac{V_m}{\pi} (R_L/R_f + R_L)$$

$$= \frac{(V_m/\pi)[1 - R_L/(R_f + R_L)]}{(V_m/\pi)[R_L/(R_f + R_L)]} \times 100 = \frac{[1 - R_L/(R_f + R_L)]}{[R_L/(R_f + R_L)]}$$

$$= \left[\frac{R_f + R_L}{R_L} - 1 \right] \times 100 = \frac{R_f}{R_L} \times 100.$$

Example 3.10 In a full-wave rectifier the load resistance is 4.0 kΩ. Each diode has a forward resistance of 500 Ω and voltage applied to each diode is 1000 sin ωt. Calculate the reading of: (a) DC ammeter connected in series with load (b) AC ammeter connected in series with load.

Solution Here $V_m = 1000$ V

$$I_m = \frac{V_m}{r_f + R_L} = \frac{1000}{500 + 4000} = 0.222 \text{ A}$$

(a) The DC ammeter will read $I_{DC} = \frac{2I_m}{\pi} = \frac{2 \times 0.222}{\pi} = 0.1415$ A

(b) The AC ammeter will read $I_{rms} = \frac{I_m}{\sqrt{2}} = \frac{0.222}{\sqrt{2}} = 0.159$ A.

Example 3.11 A single phase bridge rectifier has a purely resistive load, $R = 10 \Omega$, the peak supply voltage $V_m = 170$ V, and the supply frequency $f = 60$ Hz. Determine the average output voltage of the rectifier if the source inductance is negligible.

Solution Peak voltage $V_m = 170$ V, load resistance, $R = 10 \Omega$

DC voltage $V_{DC} = 0.636 V_m = 0.636 \times 170 = 113.3$ V.

Example 3.12 The single phase bridge rectifier is required to supply an average voltage of $V_{DC} = 400$ V to a resistive load of $R_L = 10 \Omega$ through a transformer. Determine the voltage and current ratings of diodes and transformer.

Solution Given $V_{DC} = 400$ V, $R_L = 10 \Omega$

$$V_{DC} = 0.636 V_m$$

$\Rightarrow$ $400 = 0.636 V_m$

or $V_m = \frac{400}{0.636} = 628.34$ V

rms phase voltage, $V_{rms} = \frac{V_m}{\sqrt{2}} = \frac{628.34}{\sqrt{2}} = 444.3$ V

or $I_{DC} = \frac{V_{DC}}{R} = \frac{400}{10} = 40$ A

Diode:

Peak value of current $\qquad I_p = \dfrac{V_m}{R} = \dfrac{628.34}{10} = 628.34$ A

Average current $\qquad\qquad I_d = I_{DC}/2 = 40/2 = 20$ A

rms value of diode $\qquad\quad I_R = 62.834/2 = 31.417$ A

Transformer:

rms voltage $\qquad\qquad\quad V_s = \dfrac{V_m}{\sqrt{2}} = \dfrac{628.3}{\sqrt{2}} = 444.3$ V

rms value of current $\qquad I_s = I_m/\sqrt{2} = 62.834/\sqrt{2} = 44.43$ A

Volt-ampere $\qquad\qquad\quad VI = 444.3 \times 44.43 = 19.74$ kVA

$$P_{DC} = (0.6366V_m)^2/R \text{ and } P_{AC} = V_sI_s = V_m^2/2R$$
$$\text{TUF} = P_{DC}/P_{AC} = (0.6366)^2 \times 2 = 0.8105$$

and the derating factor of the transformer is

$$\frac{1}{\text{TUF}} = \frac{1}{0.8105} = 1.2338.$$

Example 3.13 Determine the output for the network given and calculate the output DC level and the required PIV of each diode.

Solution The network will appear as shown in Fig. E3.13 (a) for the positive region of the input voltage. Redrawing the network will result in the configuration of Fig. E3.13 (b), where $v_o = \dfrac{1}{2}v_i$ or

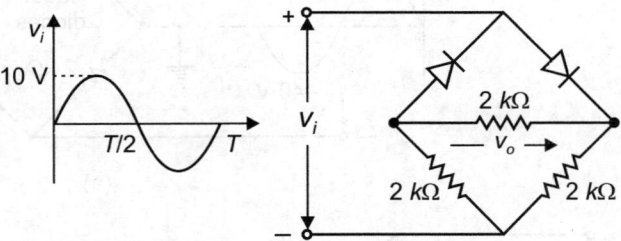

Fig. E3.13 Circuit of a network

$V_{o\,(max)} = \dfrac{1}{2} \cdot V_{i\,(max)} = \dfrac{1}{2}(10) = 5$ V as shown in Fig. E3.13 (b). For the negative part of the input, the roles of the diodes will be interchanged and v_o will appear as shown in Fig. E3.13 (c).

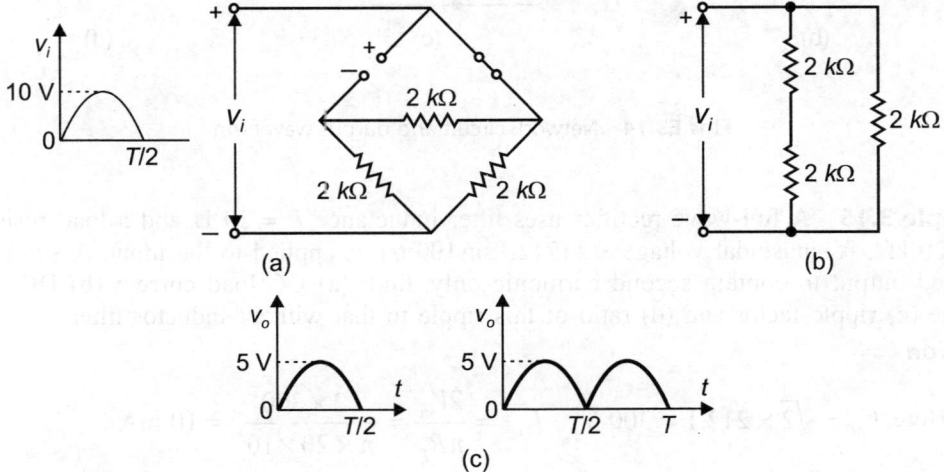

Figs E3.13 Functions of network

The effect of removing two-diodes from the bridge configuration was therefore to reduce the available DC level to the following:

$$V_{DC} = 0.636 \times 5 = 3.18 \text{ V}$$

This output is available from a half-wave rectifier with the same input. However, the PIV is equal to the maximum voltage across R, which is 5 V or half of that required for a half-wave rectifier with the same input.

Example 3.14 Sketch v_o for the network of Fig. E3.14(a) and determine the DC voltage available.

Solution The network will appear as shown in Fig. E3.14(a) for the positive region of the input voltage. Redrawing the network will result in the configuration of Fig. E3.14(b) where

$$v_{o\,(peak)} = \frac{40}{2} = 20 \text{ V as shown in the figure.}$$

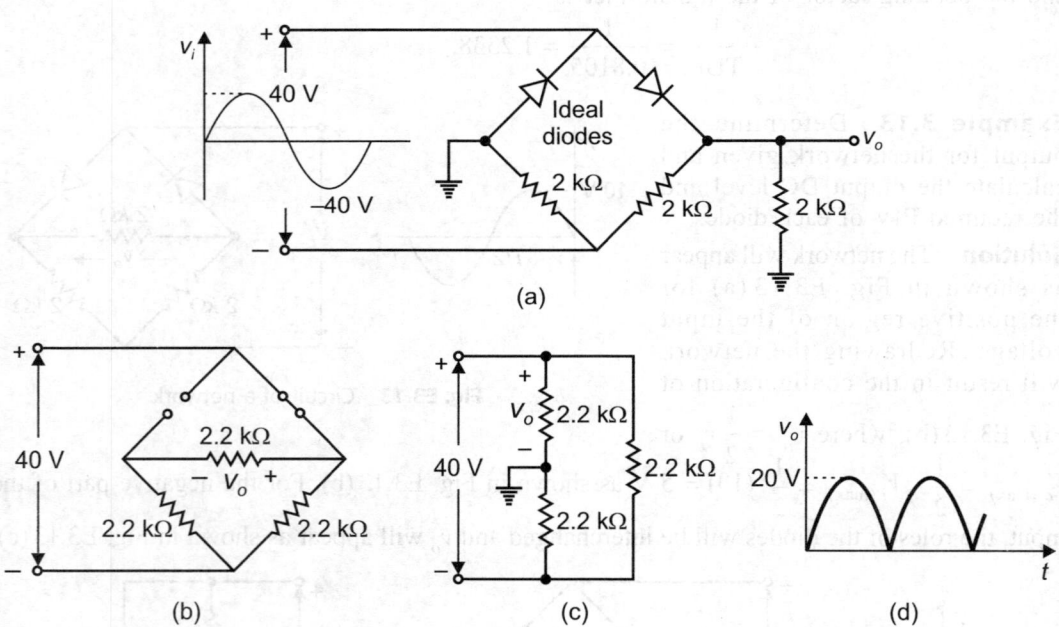

Figs E3.14 Network circuit and output waveform

Example 3.15 A full-wave rectifier uses filter inductance $L = 20$ H and a load resistance $R_L = 20$ kΩ. A sinusoidal voltage $\sqrt{2}\,(212.1\sin 100\pi t)$ is applied to the input. Assuming the rectified output to contain second harmonic only, find: (a) DC load current (b) DC output voltage (c) ripple factor and (d) ratio of this ripple to that without inductor filter.

Solution

(a) Here $V_m = \sqrt{2} \times 212.1 = 300$ V; $I_{DC} = \dfrac{2V_m}{\pi R_L} = \dfrac{2 \times 300}{\pi \times 20 \times 10^3} = 10$ mA

(b) $V_{DC} = I_{DC} \cdot R_L = 0.01 \times 20 \times 10^3 = 200$ V

(c) Ripple factor $\gamma = \dfrac{2}{3\sqrt{2}} \cdot \dfrac{1}{\sqrt{1 + (4\omega^2 L^2/R_L^2)}} = \dfrac{2}{3}\sqrt{2}\ \dfrac{1}{\sqrt{1 + \left(\dfrac{2 \times 2\pi \times 50 \times 20}{20 \times 10^3}\right)^2}} = 0.34$

Ripple factor for *FWR* without filter = 0.482.

(d) Ratio of this ripple factor to that without filter is $\dfrac{0.340}{0.482} = 0.7$.

Example 3.16   A full-wave rectifier uses choke input filter with $L = 20$ H and $C = 15\ \mu$F. The applied voltage is $v = 300 \sin 2\pi \times 50t$. Find the ripple factor.

Solution   The ripple factor for *LC* filter is given by

$$\gamma = \frac{1}{6\sqrt{2\omega^2 LC}}$$

$$= \frac{1}{6\sqrt{2 \times (2\pi \times 50)^2 \times 20 \times 15 \times 10^{-6}}}$$

$$= 0.002 \text{ or } 0.2\%.$$

Example 3.17   A three-phase full-wave rectifier has a purely resistive load of R. Determine: (a) the efficiency (b) form factor (c) ripple factor (d) transformer utilization factor (e) peak inverse voltage (PIV) of each diode and (f) peak current through a diode. The rectifier delivers $I_{DC} = 60$ A at an output voltage of $V_{DC} = 280.7$ V and the frequency of source is 60 Hz.

Solution

(a) From the relation, we have

$$V_{av} = 1.6542 V_m \text{ and } I_{av} = 1.6542 V_m/R$$
$$V_{rms} = 1.6554 V_m \text{ and } I_{rms} = 1.6554 V_m/R$$

Hence $P_{DC} = (1.6542 V_m)^2/R$ and

$\therefore$ $P_{AC} = (1.6554 V_m)^2/R$

Efficiency $= \dfrac{P_{DC}}{P_{AC}} = \dfrac{(1.6542 V_m)^2}{(1.6554 V_m)^2} = 99.86\%$

(b) The form factor $FF = \dfrac{1.6554}{1.6542} = 1.0007 = 100.07\%$

(c) Ripple factor $RF = \sqrt{1.0007^2 - 1} = 0.0374 = 3.74\%$

(d) The rms voltage of transformer secondary

$$V_s = 0.707 V_m$$

rms value of transformer secondary

$$I_s = 0.7804 I_m = 0.7804 \times \sqrt{3} V_m/R$$

Rating of transformer

$$= 3 V_s I_s = 3 \times 0.707 V_m \times 0.7804 \times \sqrt{3} V_m/R$$

$$TUF = \frac{1.6542^2}{3 \times \sqrt{3} \times 0.707 V_m \times 0.7804} = 0.9545$$

(e) The peak line neutral voltage $V_m = 280.7/1.6542 = 169.7$ V. The peak inverse voltage of each diode is equal to the peak value of the secondary line to line voltage.

$$\text{PIV} = \sqrt{3}V_m = \sqrt{3} \times 169.7 = 293.9 \text{ V}$$

(f) Average current through each diode is

$$I_{av} = \frac{4}{2\pi} \int_0^{\pi/6} I_m \cos \omega t \, d(\omega t) = I_m \frac{2}{\pi} \sin \frac{\pi}{6} = 0.3184 I_m$$

Average current through each diode $= 60/3 = 20$ A

Peak current, $\quad I_m = \dfrac{20}{0.3184} = 62.81$ A.

Example 3.18 The full-wave rectifier with centre tap transformer has an R-L load. (a) use the method of Fourier series to obtain expressions for output voltage $v_L(t)$ and load current $i_L(t)$ (b) if $V_m = 170$ V, $f = 60$ Hz and $R = 500\,\Omega$, determine the value of series inductance L to limit ripple current to 5% of I_{DC}.

Solution

(a) The rectifier output voltage may be described by a Fourier series as

$$v_L(t) = V_{DC} + \sum_{n=1,2...}^{\infty} (a_n \cos n\,\omega t + b_n \sin n\,\omega t)$$

where

$$V_{DC} = \frac{1}{2\pi} \int_0^{2\pi} v_L(t)\, d(\omega t) = \frac{2}{2\pi} \int_0^{\pi} V_m \sin \omega t \, d(\omega t) = \frac{2V_m}{\pi}$$

$$a_n = \frac{1}{\pi} \int_0^{2\pi} v_L \cos n\,\omega t \, d(\omega t) = \frac{2}{\pi} \int_0^{\pi} V_m \sin \omega t \cos n\,\omega t \, d(\omega t)$$

$$b_n = \frac{1}{\pi} \int_0^{2\pi} v_L \sin n\,\omega t \, d(\omega t) = \frac{2}{\pi} \int_0^{\pi} V_m \sin \omega t \sin n\,\omega t \, d(\omega t) = 0$$

$$= \frac{4V_m}{n} \sum_{n=2,4,...}^{\infty} \frac{1}{(n-1)(n+1)}$$

$$v_L(t) = \frac{2V_m}{\pi} - \frac{4V_m}{3\pi} \cos 2\omega t - \frac{4V_m}{15\pi} \cos 4\omega t - \frac{4V_m}{35\pi} \cos 6\omega t + ...$$

The load impedance

$$Z = R + f(n\omega L) = \sqrt{R^2 + (n\omega L)^2} \,\angle\theta_n$$

$$\theta_n = \tan^{-1} \frac{n\omega L}{R}$$

and the instantaneous current is

$$i_L(t) = I_{DC} - \frac{4V_m}{\pi\sqrt{R^2 + (n\omega L)^2}} \left[\frac{1}{3} \cos(2\omega t - \theta_2) - \frac{1}{15} \cos(4\omega t - \theta_4) ... \right]$$

where

$$I_{DC} = \frac{V_{DC}}{R} = \frac{2V_m}{\pi R}$$

(b) The rms value of ripple current is given by

$$I_{AC}^2 = \frac{(4V_m)^2}{2\pi^2[R^2 + (2\omega L)^2]}\left(\frac{1}{3}\right)^2 + \frac{(4V_m)^2}{2\pi^2[R^2 + (4\omega L)^2]}\left(\frac{1}{15}\right)^2$$

Considering only the lowest order harmonic ($n = 2$), we have

$$I_{AC} = \frac{4V_m \times 0.34}{\sqrt{2}\pi\sqrt{R^2 + (2\omega L)^2}}$$

Using the value of I_{DC} and after simplification, ripple factor

$$RF = \frac{I_{AC}}{I_{DC}} = \frac{0.481}{\sqrt{1 + (2\omega L/R)^2}} = 0.05$$

For $R = 500\,\Omega$, $f = 60$ Hz
or $L = 6.34$ H.

Example 3.19   (a) Express the output voltage of a m-phase rectifier in Fourier series (b) number of phases, $m = 6$, $V_m = 170$ V and supply frequency is 50 Hz, determine rms value of the dominant harmonic and its frequency.

Solution   The waveform for m phases are shown in Fig. 3.14 and the frequency of output is m times the fundamental component (mf). To obtain constants of the Fourier series, integrate from $-\pi/m$ to π/m and the constants are

$$b_n = 0$$

$$a_n = \frac{1}{\pi/m}\int_{-\pi/m}^{\pi/m} V_m \cos\omega t \cdot \cos n\,\omega t\, d(\omega t)$$

$$= \frac{mV_m}{\pi}\left\{\frac{\sin[(n-1)\,\pi/m]}{(n-1)} + \frac{\sin(n+1)\,\pi/m]}{(n+1)}\right\}$$

$$= \frac{mV_m}{\pi}\frac{(n+1)\sin[(n-1)\,\pi/m + (n-1)\sin(n+1)\,\pi/m]}{(n^2-1)}$$

On simplification and using the following results:

$$\sin(A + B) = \sin A \cos B + \cos A \sin B$$
$$\sin(A - B) = \sin A \cos B - \cos A \sin B$$

we get $$a_n = \frac{2mV_m}{\pi(n^2-1)}\left(n\sin\frac{n\pi}{m}\cos\frac{\pi}{m} - \cos\frac{n\pi}{m}\sin\frac{\pi}{m}\right)$$

For a rectifier with m pulses per cycle, the harmonics of the output voltage are qth, $2q$th, $3q$th, $4q$th and the above equation is valid for $n = 0, q, 2q, 3q$. The term $\sin(n\pi/m) = \sin\pi = 0$ and equation becomes

$$a_n = -\frac{2mV_m}{\pi(n^2-1)}\left(\cos\frac{n\pi_1}{m}\sin\frac{\pi}{m}\right)$$

The DC component is found by letting $n = 0$ and

$$V_{DC} = \frac{a_0}{2} = V_m\frac{m}{\pi}\sin\frac{\pi}{m}$$

The Fourier series of the output voltage is expressed as:

$$v_L(t) = \frac{a_0}{2} + \sum_{n=q,2q}^{\infty} a_n \cos n\,\omega t$$

Putting the value of a_n we have

$$v_L = V_m \frac{m}{\pi} \sin \frac{\pi}{m} \left(1 - \sum_{n=q,2q}^{\infty} \frac{2}{n^2-1} \cos \frac{n\pi}{m} \cos n\,\omega t \right)$$

(b) For $m = 6$ the output voltage is given by

$$v_L(t) = 0.9549V_m \left(1 + \frac{2}{35} \cos 6\omega t + \frac{2}{143} \cos 12\omega t + ... \right)$$

The sixth harmonic is dominant. The rms value of sinusoidal voltage is $1/\sqrt{2}$ times its peak value. The rms value of sixth harmonic is

$$V_6 = \frac{V_m \times 2 \times 0.9549}{35 \times \sqrt{2}} = 6.56 \text{ A}$$

and frequency of the sixth harmonic

$$f_6 = 6f = 300 \text{ Hz}.$$

Example 3.20   The single phase bridge rectifier is supplied from a 120 V, 60 Hz source. The load resistance is $R = 500\,\Omega$. Calculate the value of a series inductor 'L' that will limit, the rms ripple current I_{AC} to less than 5% of I_{DC}.

Solution   The load impedance

$$Z = R + j(n\omega L) = \sqrt{R^2 + (n\omega L)^2} \angle\theta_n$$

and

$$\theta_n = \tan^{-1} n\omega L/R$$

The instantaneous current is

$$i_L(t) = I_{DC} - \frac{4V_m}{\pi\sqrt{R^2 + (n\omega L)^2}} \left[\frac{1}{3} \cos(2\omega t - \theta_2) + \frac{1}{15} \cos(4\omega t - \theta_4)... \right]$$

where

$$I_{DC} = \frac{V_{DC}}{R} = \frac{2V_m}{\pi R}$$

The rms value of the ripple current

$$I_{AC}^2 = \frac{(4V_m)^2}{2\pi^2[R^2 + (2\omega L)^2]} \left(\frac{1}{3}\right)^2 + \frac{(4V_m)^2}{2\pi^2[R^2 + (4\omega L)^2]} \left(\frac{1}{15}\right)^2 + ...$$

Considering only the lowest order harmonic ($n = 2$), we have

$$I_{AC} = \frac{4V_m}{\sqrt{2}\pi\sqrt{R^2 + (2\omega L)^2}} \cdot \left(\frac{1}{3}\right)$$

Using the value of I_{DC} and after simplification the ripple factor

$$RF = \frac{I_{AC}}{I_{DC}} = \frac{0.4714}{\sqrt{1 + (2\omega L/R)^2}} = 0.05$$

For

$$R = 500\,\Omega \text{ and } f = 60 \text{ Hz}$$

Value of L is obtained from the equation

$$0.05 = \frac{0.4714}{\sqrt{1 + (2 \times 2\pi f\, L/500)^2}}$$

or $\qquad\qquad L = 6.22$ H.

Example 3.21 The single phase rectifier has an R-L load. If the peak input voltage is $V_m = 170$ V, the supply frequency $f = 60$ Hz, and load resistance $R = 15\,\Omega$, determine the load inductance L to limit the load current harmonic to 4% of average value I_{DC}.

Solution We know

$$V_m = 170 \text{ V}, \ f = 60 \text{ Hz}, \ R = 15\,\Omega \text{ and}$$
$$\omega = 2\pi f = 377 \text{ rad/sec}$$

Output voltage,

$$V_L(t) = \frac{2V_m}{\pi} - \frac{4V_m}{3\pi}\cos 2\omega t - \frac{4V_m}{15\pi}\cos 4\omega t - \frac{4V_m}{35\pi}\cos 6\omega t + \dots$$

Load impedance, $Z = R + j\,(n\omega L) = \sqrt{R^2 + (n\omega L)^2}\,\angle\theta_n$

and $\theta = \tan^{-1}(n\omega L/R)$ and load current is

$$i_L = I_{DC} - \frac{4V_m}{\sqrt{R^2 + (n\omega L)^2}}\left[\frac{1}{3}\cos(2\omega t - \theta_2) - \frac{1}{15}\cos(4\omega t - \theta_4)\dots\right]$$

where $\qquad\qquad I_{DC} = \dfrac{V_{DC}}{R} = \dfrac{2V_m}{\pi R}$

rms value of the ripple current is

$$I_{AC}^2 = \frac{(4V_m)^2}{2\pi^2[R^2 + (2\omega L)^2]}\left(\frac{1}{3}\right)^2 + \frac{(4V_m)^2}{2\pi^2[R^2 + (4\omega L)^2]}\left(\frac{1}{15}\right)^2 + \dots$$

Considering only the lowest order harmonic $(n = 2)$

$$I_{AC} = \frac{4V_m \times (1/3)}{\sqrt{2}\,\pi\sqrt{[R^2 + (2\omega L)^2]}}$$

Ripple factor, $RF = \dfrac{I_{AC}}{I_{DC}} = \dfrac{0.481}{\sqrt{1 + (2\omega L/R)^2}} = 0.04$

or $\qquad\qquad 0.481^2 = (0.4)^2\,[1 + (2 \times 377 \times L/15)^2$

or $\qquad\qquad L = 238.4$ mH.

Example 3.22 A single phase bridge rectifier is supplied from a 120 V, 60 Hz source. The load resistance is $R = 200\,\Omega$. (a) design a C-filter so that the ripple factor of the output voltage is less than 5% (b) with the value of capacitor C in part (a), calculate the average load voltage V_{DC}.

Solution Ripple factor $\qquad = 0.05, \ R = 200\,\Omega, \ f = 60$ Hz

Ripple factor $\qquad\qquad RF = \dfrac{1}{\sqrt{2}(4fRC_e - 1)}$ or $C_e = \dfrac{1}{4fR}\left(1 + \dfrac{1}{\sqrt{2}RF}\right)$

$$C_e = \frac{1}{4 \times 60 \times 200} \left(1 + \frac{1}{\sqrt{2} \times 0.05} \right) = 315.46\,\mu F$$

Average load voltage, $V_{DC} = V_m - \dfrac{V_m}{4fRC_e} = \sqrt{2} \times 120 - \dfrac{\sqrt{2} \times 120}{4 \times 60 \times 200 \times 315.46 \times 10^{-6}}$

$$= 169.7 - 11.21 = 158.49 \text{ V.}$$

Example 3.23 An L-C filter is used to reduce the ripple of the output voltage for a single phase full-wave rectifier. The load resistance $R_L = 40\,\Omega$, load inductance $L = 10$ mH and source frequency is 60 Hz. Determine the values of L_e and C_e so that the ripple factor of the output voltage is 10%.

Solution The value of filter capacitor C_e is given by

$$C_e = \frac{10}{4\pi f \sqrt{R_L^2 + (4\pi f L)^2}} = \frac{10}{4\pi \times 60\sqrt{40^2 + (4\pi \times 60 \times 10 \times 10^{-3})^2}}$$

$$= 326\,\mu F$$

$$RF = 0.1 = \frac{\sqrt{2}}{3} \left| \frac{1}{[(4\pi f)^2 L_e C_e - 1]} \right|$$

or $\qquad (4\pi f)^2 L_e C_e - 1 = 4.714$ or $(4\pi f)^2 L_e = 5.714/326 \times 10^{-6}$

or $\qquad\qquad L_e = 30.83$ mH.

Example 3.24 A six-phase star rectifier has a purely resistive load of $R_L = 10\,\Omega$, the peak supply voltage, $V_m = 170$ V and the supply frequency $f = 60$ Hz. Determine the average output voltage of the rectifier if the source inductance is negligible.

Solution Given: $R_L = 10\,\Omega$, $V_m = 170$ V, $f = 60$ Hz. For a six-phase star rectifier

$$V_{DC} = \frac{6V_m}{\pi} \sin(\pi/6)$$

or $\qquad\qquad V_{DC} = \frac{6}{\pi} \times 170 \sin(\pi/6) = 162.3$ V.

Example 3.25 A six-phase star connected rectifier has a resistive load of $10\,\Omega$, the peak supply voltage $V_m = 170$ V and the supply frequency 60 Hz. Determine the average output voltage of the rectifier if the source inductance per phase is $L_c = 0.5$ mH.

Solution Given load resistance $= 10\,\Omega$, $V_m = 170$ V, $f = 60$ Hz.

For a six-phase star rectifier, DC voltage, V_{DC}

$$V_{DC} = \frac{6}{\pi} V_m \sin(\pi/6) = \frac{6}{\pi} \times 170 \sin(\pi/6) = 162.3 \text{ V}$$

$$I_{DC} = \frac{V_{DC}}{R} = \frac{162.3}{10} = 16.23 \text{ A}$$

Average voltage reduction due to six commutations per cycle $V_x = 6fL\,I_{DC}$

or $\qquad\qquad V_x = 6 \times 60 \times 0.5 \times 10^{-3} \times 16.23 = 2.92$ V

Effective output voltage $\quad = 162.34 - 2.92$

$$= 159.42 \text{ V.}$$

Example 3.26 A three-phase bridge rectifier is supplied from a Wye-connected 208 V, 60 Hz supply. The average load current is 60 A and has negligible ripple. Calculate the percentage reduction of output voltage due to commutation if the line inductance per phase is 0.5 mH.

Solution Given

$$L = 0.5 \text{ mH}, \ V_s = 208/\sqrt{3} = 120 \text{ V}$$

$$f = 60 \text{ Hz}, \ I_{DC} = 60 \text{ A}, \ V_m = \sqrt{2} \times 120 = 169.7 \text{ V}$$

DC voltage $V_{DC} = 1.654 V_m = 1.654 \times 169.7 = 280.7 \text{ V}$

Average voltage reduction due to commutating inductance

$$V_x = 6fL \ I_{DC}$$

$$= 6 \times 60 \times 0.5 \times 10^{-3} \times 60 = 10.8 \text{ V}$$

Percentage reduction of output voltage

$$= \frac{10.8}{280.7} \times 100 = 3.85\%.$$

Example 3.27 A three-phase bridge rectifier is required to supply an average voltage of $V_{DC} = 750$ V at a ripple free current of $I_{DC} = 9000$ A. The primary and secondary of the transformer are connected in Wye. Determine the voltage and current ratings of diodes and transformers.

Solution Output DC voltage,

$$V_{DC} = 750 \text{ V}, \ I_{DC} = 9000 \text{ A or } V_{DC} = 1.6542 V_m$$

or

$$V_m = \frac{V_{DC}}{1.6542} = \frac{750}{1.6542} = 453.39 \text{ V}$$

rms phase voltage,

$$V_s = \frac{V_m}{\sqrt{2}} = \frac{453.39}{\sqrt{2}} = 320.59 \text{ V}$$

Diodes: Peak current $I_p = 9000$ A

Average diode current $I_d = \dfrac{I_{DC}}{2} = \dfrac{9000}{2} = 4500$ A

rms current $I_R = \dfrac{9000}{\sqrt{2}} = 6363.96$ A

Transformer: rms voltage

$$V_s = 320.59 \text{ V}$$

rms current $I_s = I_p = 9000$ A

VI per phase $= 320.59 \times 9000 = 2885.3$ kVA

$$\text{TUF} = \frac{P_{DC}}{P_{AC}} = \frac{750 \times 9000}{3 \times 2885.3} = 0.7798$$

Derating factor of transformer

$$= \frac{1}{\text{TUF}} = \frac{1}{0.7798} = 1.2824.$$

Example 3.28 A zener diode and a resistance are joined in series to provide 18 V stabilised output from 22 V supply. If load resistance is 150 Ω and the current through stabilizer is

12 mA, determine the values of series resistor and power dissipated in it.

If the load resistance is increased to $200\,\Omega$, determine the current which will flow through stabilizer diode. Assume ideal characteristics of zener.

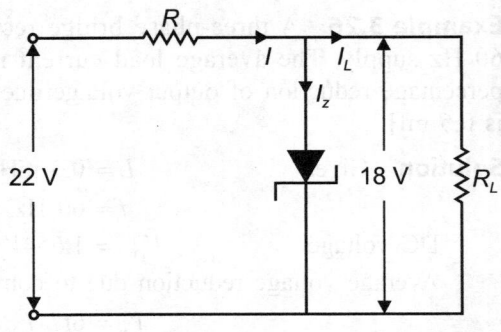

Fig. E3.28 Circuit diagram of stabilizer

Solution Current through zener,

$$I_z = 12 \text{ mA}$$

Current through $150\,\Omega$,

$$I_L = \frac{18 \text{ V}}{150} = 120 \text{ mA}$$

Total current

$$I_R = I_L + I_z$$
$$= 120 + 12 = 132 \text{ mA}$$

Voltage drop across R

$$V = 22 - 18 = 4 \text{ V}$$

Value of R

$$= \frac{4 \text{ V}}{132 \text{ mA}} = 30.3\,\Omega$$

Power dissipated in R

$$= \frac{V^2}{R} = \frac{4^2}{30.3} \approx 0.5 \text{ W}$$

When

$$R_L = 200\,\Omega$$

$$I_L = \frac{18}{200} = 90 \text{ mA}, \text{ as the voltage across } R \text{ must remain } 4\text{ V}$$

Current through $R = \dfrac{4}{30.3} = 132 \text{ mA}$

Current through zener diode,

$$I_z = 132 - 90 = 42 \text{ mA}.$$

Example 3.29 A zener diode of 30 V breakdown is connected in series with a resistance of $200\,\Omega$. If the load is $2000\,\Omega$ connected across zener diode, over what range of input voltage will the circuit operate? Maximum zener current is not to exceed 25 mA.

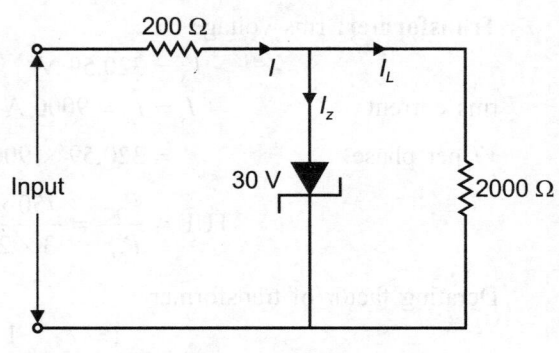

Fig. E3.29 Circuit diagram of a stabilizer

Solution Load resistance,

$$R_L = 2000\,\Omega$$

Zener breakdown voltage,

$$V_z = 30 \text{ V}$$

Load current $I_L = \dfrac{30}{2000} = 15 \text{ mA}$

If zener diode current, $I_z = 0$
before breakdown, minimum value of E_i

$$= I_L R + V_z$$

or
$$E_i = 15 \times 10^{-3} \times 200 + 30 = 33 \text{ V}$$

When zener diode current is 25 mA (max) current through series resistance R,

$$I = I_L + I_z$$
$$= 15 + 25 = 40 \text{ mA}$$

Minimum input voltage $E_i = I \times R + V_z$

$$E_i = 40 \times 10^{-3} \times 200 + 30 = 38 \text{ V}$$

If the input voltage falls below 33 V, the output will no longer remain 30 V but falls along with the input. On the other hand, if the input becomes more than 38 V, zener diode may destroy because of excessive current. Therefore, desired range of input voltage is 33 V to 38 V.

EXERCISES

Multiple Choice Questions

1. Which of the following circuits cannot be operated directly from the mains?
   (A) Half-wave rectifier
   (B) Centre-tap full-wave rectifier
   (C) Bridge rectifier
   (D) None of these

2. The ripple factor (RF) of a waveform is defined as $= \dfrac{\text{RMS of the ripple}}{\text{Average output level}}$. By the definition, the ripple factor of the half-wave rectifier is:
   (A) 0.246
   (B) 0.48
   (C) 1.21
   (D) none of these

3. In a rectifier circuit, the load connected is of low value. For proper filter operation, it is required that:
   (A) a capacitor is to be included in the circuit
   (B) a bleeder resistance is to be placed in the circuit
   (C) an inductor filter is to be included in the circuit
   (D) all of these

4. The function of bleeder resistance in filter circuit is:
   (A) to maintain minimum current necessary for optimum inductor filter operation
   (B) to work as voltage divider in order to provide variable output from the supply

   (C) to provide discharge path to capacitors so that output becomes zero when the circuit has been de-energised
   (D) all of these

5. The rectifier, which requires minimum amount of filtering is a:
   (A) half-wave rectifier
   (B) full-wave rectifier
   (C) voltage doubler circuit
   (D) none of these

6. A half-wave rectifier circuit with a capacitive filter is connected to 200 V, 50 Hz AC line, the output voltage across the capacitor should be approximately:
   (A) 100 V
   (B) 180 V
   (C) 280 V
   (D) 400 V

7. A zener diode:
   (A) has a high forward voltage rating
   (B) has a sharp breakdown at low reverse voltage
   (C) has a negative resistance
   (D) none of these

8. In a centre-tap full-wave rectifier, V_m is the peak voltage between the centre-tap and one end of the secondary. The maximum voltage across the reverse-biased diode is:
   (A) V_m
   (B) $2V_m$
   (C) $3V_m$
   (D) none of these

9. The important specification of a zener diode is:
   (A) its breakdown voltage and power dissipation
   (B) breakdown voltage, dynamic impedance and power dissipation
   (C) breakdown voltage and dynamic impedance
   (D) none of these

10. The major advantage of a bridge rectifier is that:
    (A) centre-tap transformer is not required
    (B) peak inverse voltage of each diode is double of that for a full-wave rectifier
    (C) peak inverse voltage of each diode is half of that for a full-wave rectifier
    (D) the output voltage is more smooth

11. If the ripple factor of the output wave of a rectifier is low, it means that:
    (A) output voltage will have less ripple
    (B) output voltage will be low
    (C) filter circuits may not be required
    (D) none of these

12. In a three-phase half-wave rectifier, the output voltage is equal to:
    (A) the most positive input phase voltage at any instant
    (B) the difference of most positive and most negative input phases at any instant
    (C) the average value of the three-phases
    (D) the difference of the two positive phase voltages

13. In a three-phase half-wave rectifier, if the input phase voltage is 200 V, the PIV required for each diode will be:
    (A) 440 V   (B) 346 V
    (C) 370 V   (D) 220 V

14. The effect of DC saturation in a rectifier transformer is to:
    (A) decrease the output
    (B) increase the efficiency

    (C) decrease the AC components of the output
    (D) increase the DC component

15. For a rectifier output of 100 kW at 400 V an average current rating of 40 A for each diode is sufficient in:
    (A) single phase full-wave rectifier
    (B) six-phase half-wave rectifier
    (C) three-phase half-wave rectifier
    (D) three-phase full-wave rectifier

16. As the number of phases in a multiphase rectifier are increased, the output will:
    (A) remains same
    (B) become more smooth
    (C) decrease
    (D) the diodes will require high peak inverse voltage

17. Ripple frequency of a six-phase half-wave rectifier for 220 V, 60 Hz input will be:
    (A) 150 Hz   (B) 60 Hz
    (C) 360 Hz   (D) 100 Hz

18. In a three-phase half-wave rectifier, each diode conducts for a duration of:
    (A) π (B) $\pi/6$
    (C) $\pi/3$ (D) $\pi/4$

19. A capacitor of 100 µF is charged to 10 V through a resistance of 10 kΩ. It will be fully charged in:
    (A) 5 sec   (B) 10 sec
    (C) 15 sec   (D) 0.5 sec

20. In the above question, in one second the capacitor will charge to:
    (A) 3 V   (B) 6.3 V
    (C) 12 V   (D) 15 V

21. A zener diode:
    (A) has a sharp breakdown at low reverse voltage
    (B) has a high forward voltage rating
    (C) has a negative resistance
    (D) none of these

ANSWER KEY

1. (B)	2. (C)	3. (C)	4. (D)	5. (B)	6. (C)	7. (B)	8. (B)
9. (B)	10. (C)	11. (A)	12. (A)	13. (C)	14. (A)	15. (B)	16. (B)
17. (C)	18. (C)	19. (A)	20. (B)	21. (A)			

Review Questions

1. What are the functions of: (a) rectifier (b) filter and (c) voltage regulator?

2. Define the following terms: (a) DC current I_{DC} (b) DC voltage V_{DC} (c) current I_{rms} (d) max voltage V_m (e) AC voltage V_{rms} in half-wave and full-wave rectifier.

3. Explain the following terms: (a) ripple factor γ (b) peak inverse voltage PIV (c) rectifier efficiency (d) transformer utilisation factor TUF, for FWR.

4. Explain the operation of a full-wave rectifier. Sketch different waveforms.

5. What are the advantages of a full-wave rectifier over a half-wave rectifier?

6. How will you find an AC voltmeter reading in a (a) half-wave rectifier (b) full-wave rectifier?

7. Draw the circuit of a bridge-rectifier. Give different waveforms in such circuit.

8. Give schematic representation of a: (a) three-phase half-wave rectifier circuit (b) three-phase full-wave rectifier circuit and (c) compare the rectification properties of the two.

9. For a m-phase rectifier, derive the relations for average and rms values of the output.

10. What would be the ripple frequency in six-phase full-wave rectifier?

11. Explain the filtering action of a simple capacitive filter.

12. Give a diagram of full-wave LC filtered rectifier. Explain how ripple factor is minimised?

13. Give a circuit of π-filter. What are its advantages and disadvantages?

Thyristor and Its Family

4.0 INTRODUCTION

In this chapter, a number of important devices not discussed in earlier chapters, are introduced. The two-layer semiconductor has led to three, four, and even five-layer devices. A family of four-layer pnpn devices will first be considered. Those four-layer devices with a control mechanism are commonly referred to as thyristors, although the term is most frequently applied to the silicon-controlled rectifier (SCR). It appears that the term thyristor is now becoming more common than the actual term SCR. The name thyristor derived from the word thyratron and transistor. This means that thyristor is a solid state device like a transistor and has the characteristics similar to the thyratron tube.

4.1 THYRISTORS AND ITS APPLICATIONS

Thyristor is the general name given to the family of semiconductor devices having four-layers and three p-n junctions. A special feature of the members of this family is that they exhibit the regenerative switching characteristic.

After semiconductor diodes and transistors were introduced in control applications, in addition to their compactness, thyristors had the advantages of being very fast in response, highly reliable and efficient. They require lesser power for operation and less maintenance.

Advancement in the fabrication techniques further led to the reduction in their cost and suitable for use in various control applications.

4.1.1 Applications

Thyristors, with large number of applications and tremendous control capabilities had a wide range of applications and it completely replaced electromagnetic control systems. Some of the applications are listed as follows:

1. DC to DC converters
2. AC to AC converters
3. AC to DC rectifiers
4. DC to AC inverters
5. Control of DC and AC motors
6. Temperature, level, position, and illumination controllers
7. HVDC transmission lines
8. Power switches

9. Improvement of power factor in transmission lines
10. Power switches and circuit breakers
11. Other controllers such as choppers, cycloconverter, dual controllers, etc.

These have been dealt in detail in the subsequent chapters.

4.2 THYRISTOR FAMILY

Thyristor is a general name given to a family of power semiconductor switching devices, having four-layers with three-junctions. Silicon controlled rectifier (SCR) is the most widely used and important member of thyristor family. The SCR is universally known by thyristor. Thyristor is a solid-state device like transistor and has characteristics similar to that of a thyratron tube. The other members of the thyristor family are:

1. DIAC (Bidirectional diode thyristor)
2. TRIAC (Bidirectional triode thyristor)
3. SUS (Silicon unilateral switch)
4. SCS (Silicon controlled switch)
5. SBS (Silicon bilateral switch)
6. LA SCR (Light activated SCR)
7. LA SCS (Light activated SCS)
8. PUT (Programmable unijunction transistor)
9. GTO (Gate turn off thyristor, etc.)

Since SCR and TRIAC are widely used in high power control circuits, they are termed as high power devices. All the above mentioned devices with the exception of TRIAC are generally called as low power devices as they are used in low power control circuits. Modern power devices which are finding wide use in power electronics field are asymmetrical thyristor (ASCRs), reverse conducting thyristor (RCT), power MOSFET, static induction thyristor (SIT) MOS-controlled thyristor (MCT) and so on.

4.3 THYRISTOR CONSTRUCTION

The thyristor structure consists of four-layers of semiconductor material of alternate conductivity type pnpn all formed in a single crystal. The material normally used is silicon wafer of diameter 30 mm and thickness 0.7 mm for a 2000 V, 300 A device. Electrical contacts are made to the end p-type and n-type regions. These are the contacts through which the main current flows. The contact to the p-region being known as anode and to the n-region as cathode. A third contact known as gate is a control electrode, is connected normally to the (inside) p-region. Figures 4.1 (a) and (b) show the schematic diagram of this device and its symbolic representation.

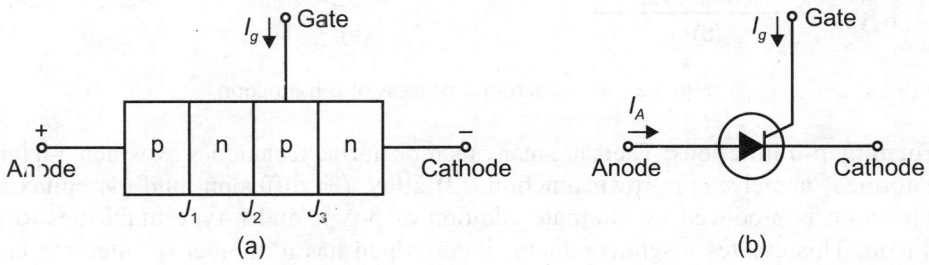

Fig. 4.1 Schematic diagram of thyristor

Constructional details: Thyristors are manufactured by a variety of highly specialized techniques involving different steps. Strict quality control and automated testing criterion are essential to ensure a good product during manufacturing.

Silicon from ordinary sand: The basic ingredient for manufacturing the semiconductor devices is derived from ordinary sand. The first step in producing any silicon semiconductor device is refining pure silicon from sand. Unfortunately, the refined silicon is in a polycrystalline form, each price of silicon material is composed of many tiny individual crystals aligned in random directions. Even when contaminated with n-type or p-type impurities, this silicon material would not be capable of conducting sufficient current for semiconductor operation.

Pure silicon can be converted to a monocrystalline form or single uniform crystalline structure using a growth process. During the process an n-type or p-type impurity may be added to produce a finished semiconductor material. Generally n-type silicon is produced.

The growth process starts with the heating of the refined silicon to molten state in a crucible. An n-type impurity is usually added to molten silicon carefully. The growing process starts with lowering a tiny piece or seed of monocrystalline silicon into hot molten silicon. The molten n-type silicon starts to crystallize on the solid silicon seed, and a larger mass of solid n-type silicon is formed. This silicon seed is gradually pulled out from the crucible of molten silicon shown in Figs 4.2 (a) and (b). At the same time the silicon seed and the crucible are rotated in opposite directions to ensure uniform structure of grown n-type silicon material. The cylindrical rod of n-type silicon material with a diameter of about 5 cm is formed shown in Fig. 4.2 (c). Each silicon rod is sliced into many thin discs or wafers.

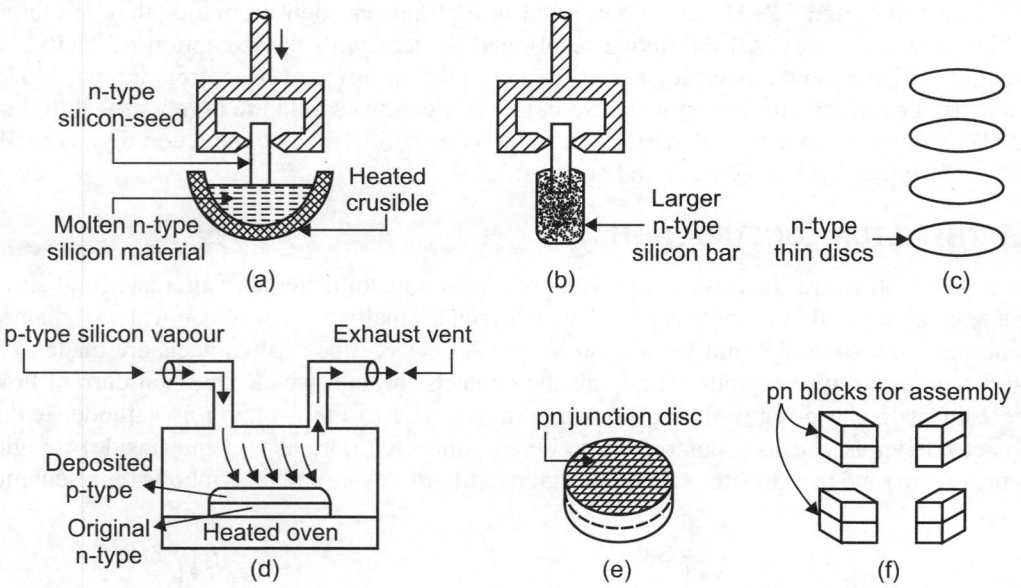

Fig. 4.2 Manufacturing process of p-n junction

Forming p-n junctions: There are many manufacturing techniques by which p-n junctions can be formed, namely: (1) grown junction, (2) alloy, (3) diffusion, and (4) epitaxial. The grown junction is produced by alternate addition of p-type and n-type impurities to molten pure silicon. This creates a semiconductor ingot which has a number of alternate layers of p-type and n-type materials. This ingot can be sliced into thin wafers of n-type material joined with p-type material to form p-n junction.

The alloy process involves placing small balls of a p-type impurity such as aluminium on the n-type silicon wafer. Then the two materials are heated in a furnace until aluminium melts and mixes with the silicon. The resulting alloy of aluminium and silicon forms a p-type silicon material and the desired p-n junction.

The diffusion process is similar to alloy process in that a p-type impurity is diffused into one side of n-type silicon wafer.

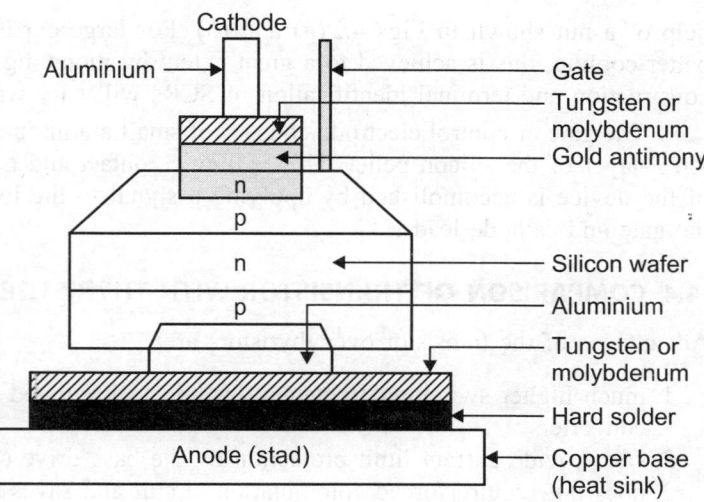

Fig. 4.2 (g)   Internal construction of thyristor

Another common method is the epitaxial or vapour depositing process to fabricate p-n junctions.

Figure 4.2 (d) illustrates the manufacturing process of p-n junction diodes by epitaxial deposition of p-type silicon material over n-type silicon discs. Finally, the wafer is sliced and lapped into the required shapes shown in Figs 4.2 (e) and (f).

Final assembly and packaging: The unmounted thyristor device is shaped to the required dimensions and cleaned to remove any impurities caused by manufacturing processes. Many thyristors, particularly the high current types are mounted on a metallic base such as molybdenum or tungsten. This base serves to protect the fragile silicon device from vibration or mechanical shock.

The molybdenum disc on the top and bottom are included to minimize the effects of thermal expansion and mechanical stresses on the complete assembly. One of these plates is hard-soldered to copper or an aluminium stud, which is threaded for attachment to a heat sink. This provides an efficient thermal path for conducting the internal losses to the surrounding medium. The use of hard solder between the pellet and backup plates minimises thermal fatigue when the SCRs are subjected to temperature induced stresses. The threaded portion is for the purpose of tightening the thyristor to the frame or heat sink with the

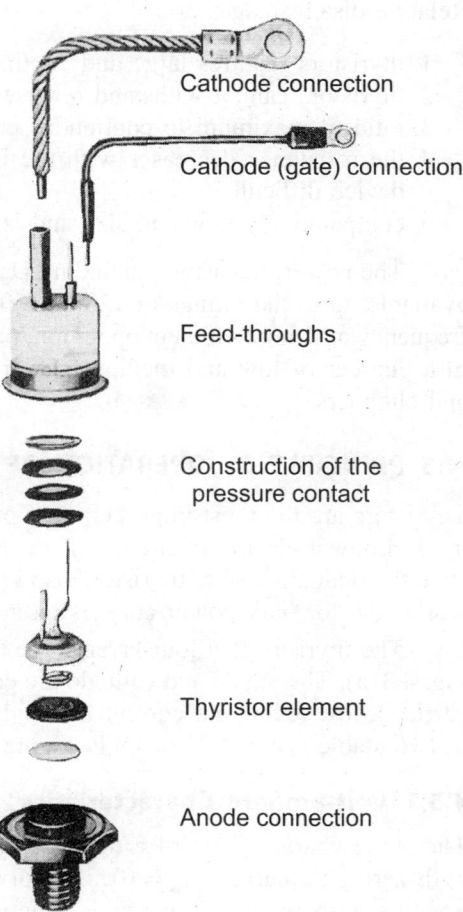

Fig. 4.2 (h)   Assembly of thyristor

help of a nut shown in Figs 4.2 (g) and (h). For large current applications, thyristors need better cooling, this is achieved to a great extent by mounting them onto heat sinks. The case construction and terminal identification of SCRs will vary with the application.

The gate or control electrode consists of small aluminium wire which is connected to the top P-layer of the silicon pellet. This is ohmic contact and not a rectifying junction. Control of the device is accomplished by applying a signal to the lower p-n junctions, i.e. between the gate and cathode leads.

4.4 COMPARISON OF TRANSISTOR WITH THYRISTOR

Advantages of the transistor over thyristors are:

1. much higher switching frequency leading to improved and more efficient operation of converters
2. can provide current limit protection by the base drive circuit
3. does not require forced commutation circuit and saves the associated switching losses, cost, weight and volume
4. low conduction drop.

Relative disadvantages are:

1. thyristors requires large and continuous base drive
2. thyristors cannot withstand reverse voltage
3. ratio of maximum to continuous current is low
4. the resistance decreases with the increasing temperature making the paralleling of the device difficult
5. comparatively larger in size and is costlier.

The power transistors, including Darlington transistors, suitable for drive applications are available up to the ratings of 120 V, 750 A and 1000 V, 60 A. Because of the high switching frequency and more efficient operation, power transistors have succeeded in replacing thyristors in a number of low and medium power (up to around 200 kW) drives employing inverters and choppers.

4.5 PRINCIPLE OF OPERATION OF THYRISTORS

Thyristors are the most important type of power semiconductor devices. They are extensively used in power electronic circuits. They are operated as bistable switches from non-conducting state to conducting state thyristor, also known as silicon controlled rectifier (SCR), is the main workhorse for bulk power conversion and control in industry.

The thyristor is a four-layer, three terminal, anode A, cathode K and gate G as shown in Fig. 4.3 (a). The anode and cathode are connected to the main power circuit. The gate terminal carries a low level gate current in the direction from gate to cathode. The thyristor operates in two stable states: ON or OFF. Figure 4.3 (b) shows the symbol of a thyristor.

4.5.1 Volt-ampere Characteristics

The static characteristic of a thyristor with no gate current applied is shown by Fig. 4.4 (a) with zero gate current ($i_g = 0$), if a forward voltage is applied across the device, i.e. anode positive with respect to cathode, junctions J_1 and J_3 are forward-biased while junction J_2 remains reverse biased, and therefore the anode current is a small leakage current. If the anode to cathode forward voltage reaches a critical limit, called the forward breakover voltage, the

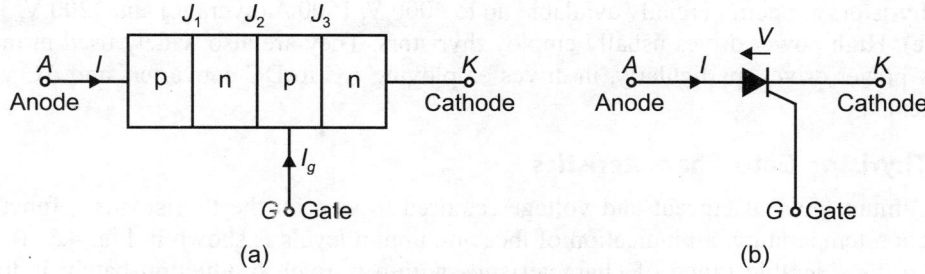

Fig. 4.3 Anode *A*, cathode *K* and gate *G*

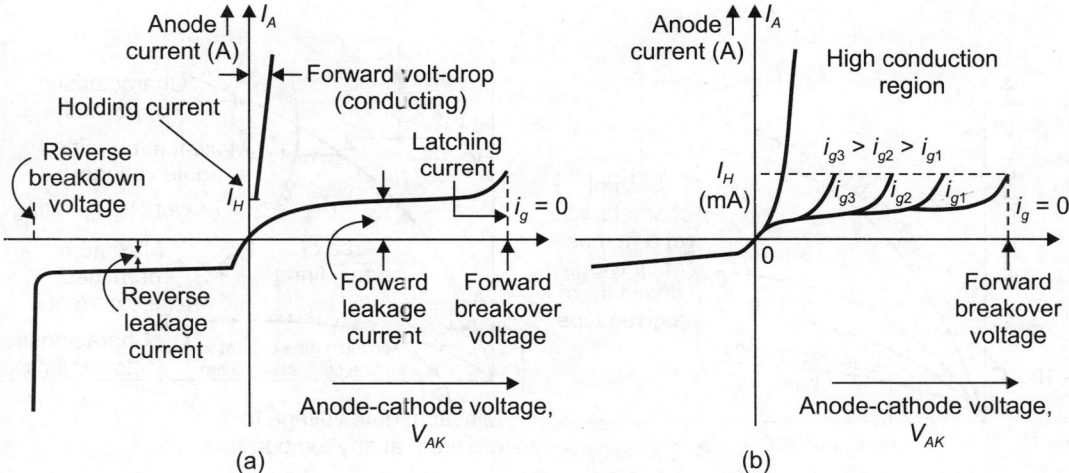

Fig. 4.4 *V-I* characteristics of thyristor with and without gate current

device switches into high conduction, thus forward biasing the junction J_2 to turn the thyristor ON. The forward voltage drop then falls to a value between 1 V and 2 V.

The thyristor can be switched to the ON or conducting state by injecting a current into the central p-type layer via the gate terminal. The injection of gate current provides additional holes in the central p-type layer, reducing the forward breakover voltage to a value less than the applied voltage and turning the thyristor ON. These conditions are shown in Fig. 4.4(b), showing the effect of increasing gate current on the level of forward breakover voltage. If the anode current falls below a critical limit, called the holding current, I_H, the device returns to its forward blocking state.

If a reverse voltage is applied across the device, i.e. anode negative with respect to cathode, the outer junctions J_1 and J_3 are reverse biased and the central junction J_2 is forward-biased. Therefore, only a small leakage current flows. If the reverse voltage is increased, then at the critical breakdown level known as reverse breakdown voltage, an avalanche will occur at J_1 and J_3 and the current will increase sharply. If this current is not limited to a safe value, power dissipation will increase to a dangerous level that will destroy the device.

The gate current is applied at the instant turn on is desired. The thyristor turns on provided $V_A > 0$. After turn on when I_A reaches a value known as latching current, the thyristor continues to conduct even after the gate signal has been removed. Hence, only a pulse of current is required for turn on.

Thyristors are commercially available up to 4000 V, 1500 A (average) and 1200 V, 3000 A (average). High power drives usually employ thyristors. They are also widely used in medium and low power drives, particularly, in drives employing AC to DC converters and AC voltage controllers.

4.5.2 Thyristor Gate Characteristics

The minimum level of current and voltage required to turn on the thyristor is a function of the junction temperature, an indication of these minimum levels is shown in Fig. 4.5 (a). There will be a considerable range of characteristics within a given production batch, individual thyristors having characteristics as shown in Fig. 4.5 (a). All the thyristors can be assumed to have a characteristic lying somewhere between the low and high resistance limits.

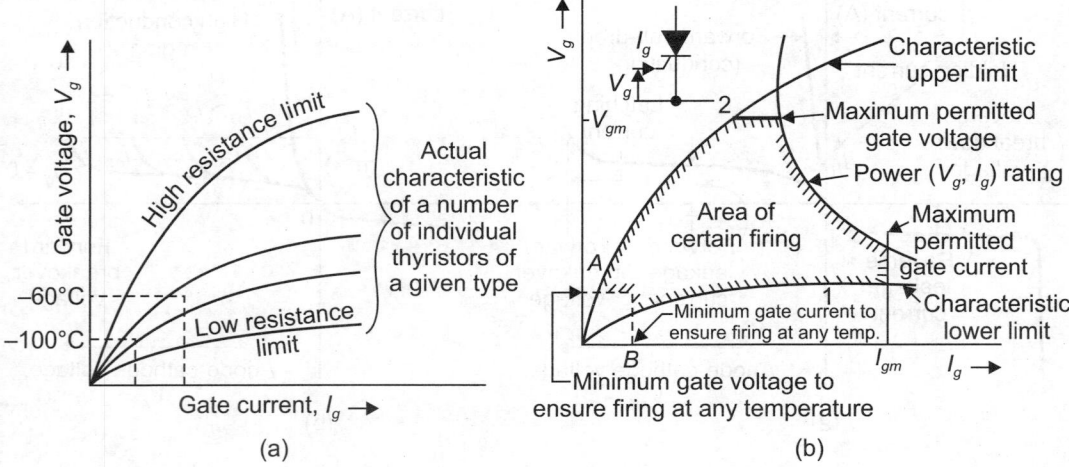

Fig. 4.5 Thyristor gate characteristics

The forward gate characteristics of a thyristor in the form of a graph between gate voltage and gate current are shown in Fig. 4.5 (b). Here positive gate to cathode voltage V_g and positive gate to cathode current I_g represent DC values. For a particular type of SCRs V_g–I_g characteristic has a spread between two curves 1 and 2 as shown in Fig. 4.5 (b). This spread is due to the difference in low dooping levels of p- and n-layers. The gate circuitry must be suitably designed to take care of this unavoidable scatter of characteristics. Curve 1 represents the lowest voltage values that must be applied to turn on the SCR. Curve 2 gives the highest possible voltage values that can be safely applied to gate circuit.

Each thyristor has maximum limits as V_{gm} and I_{gm} for gate voltage and gate current. There is also rated average gate power dissipation $(V_g I_g)$ specified for each SCR. OA and OB are the values of minimum gate voltage and minimum gate current to ensure firing at any temperature respectively. Figure 4.5 (b) shows these limits imposed on the gate cathode characteristic, giving the area into which the gate firing signal must be fitted for certain firing into the on-state to take place.

If the gate current and the gate voltage intersect within the boundaries of this curve the device will be successfully triggered. The gate current requirement increases as the temperature decreases. The gate of a thyristor is triggered by a pulse or a train of pulses to keep the gate power dissipation low. Typical trigger circuits are discussed in the next chapters.

4.6 TWO-TRANSISTOR MODEL OF THYRISTOR

A simple analogy that may be employed to explain many of the operating characteristics of a thyristor may be obtained by considering it to be made up of two interconnected transistors, one p-n-p transistor Q_1 and the other n-p-n transistor Q_2 shown in Fig. 4.6(a). The circuit corresponding to this arrangement is shown in Fig. 4.6(b).

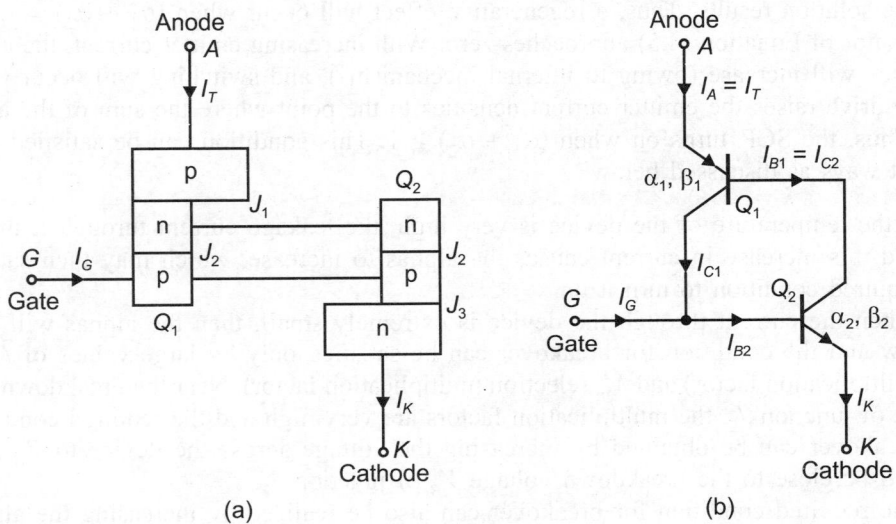

Fig. 4.6 Two-transistor analogy of pnpn switch

In general, the collector current I_C of a thyristor is related to the emitter current I_E and leakage current of the collector-base junction, I_{CBO}, as:

$$I_C = \alpha I_E + I_{CBO} \qquad ...(4.1)$$

and common base current gain is defined as $\alpha = I_C/I_E$.

For transistor Q_1: Emitter current is the anode current, I_A and the collector current, I_{C1} can be determined from Equation (4.1);

$$I_{C1} = \alpha_1 I_A + I_{CBO1} \qquad ...(4.2)$$

where α_1 is the current gain and I_{CBO1} is the leakage current for Q_1. Similarly, for transistor Q_2 (n-p-n), the collector current I_{C2}

$$= \alpha_2 I_K + I_{CBO2} \qquad ...(4.3)$$

where α_2 is the current gain and I_{CBO2} is the leakage current for transistor Q_2. By combining I_{C1} and I_{C2}, we get

$$I_A = I_{C1} + I_{C2} = \alpha_1 I_A + I_{CBO1} + \alpha_2 I_K + I_{CBO2} \qquad ...(4.4)$$

But for a gating current of I_G, $I_K = I_A + I_G$ and solving Equation (4.4) for I_A gives:

$$I_A = \alpha_1 I_A + \alpha_2 I_K + I_{CBO1} + I_{CBO2}$$

or

$$I_A = \alpha_1 I_A + \alpha_2 (I_A + I_G) + I_{CBO1} + I_{CBO2}$$

or

$$I_A = \alpha_1 I_A + \alpha_2 I_A + \alpha_2 I_G + I_{CBO1} + I_{CBO2}$$

or

$$I_A [1 - (\alpha_1 + \alpha_2)] = \alpha_2 I_G + I_{CBO1} + I_{CBO2}$$

or

$$I_A = \frac{\alpha_2 I_G + I_{CBO1} + I_{CBO2}}{[1 - (\alpha_1 + \alpha_2)]} \qquad ...(4.5)$$

Assuming the leakage currents of transistors Q_1 and Q_2 to be negligible, we have

$$I_A = \frac{\alpha_2 I_G}{(\alpha_1 + \alpha_2)}$$

The current gain α_1 varies with the emitter current $I_A = I_E$, and α_2 varies with $I_K = I_A + I_G$. If the base drive current $I_G + I_C$, is equated to the base current of the n-p-n transistor, exactly the same solution results. Thus, a regenerative effect will occur when $(\alpha_1 + \alpha_2) = 1$, i.e. the denominator of Equation (4.5) approaches zero. With increasing emitter current, the alphas of the device will increase (owing to internal mechanisms), and switching will occur provided the gate drive raises the emitter current densities to the point where the sum of the alphas is unity. Thus, the SCR turns on when $(\alpha_1 + \alpha_2) \geq 1$. This condition can be satisfied in three different ways as discussed below:

1. If the temperature of the device is very high, the leakage current through it increases, and this increase in current causes the alphas to increase, which may then satisfy the required condition to turn it on.
2. When the current through the device is extremely small, then the alphas will be very low and the condition for breakover can be satisfied only by large values of M_p (hole multiplication factor) and M_n (electron multiplication factor). Near the breakdown voltage V_B of junction J_2, the multiplication factors are very high and the required condition for breakover can be obtained by increasing the voltage across the device to V_{BO}, which will be close to the breakdown voltage V_B of junction J_2.
3. The required condition for breakover can also be realized by increasing the alphas. In Fig. 4.3, if a gate current I_g is injected into the base p_a in the same direction as the current I across J_2, the current gain of the n-p-n transistor can now be increased independently of V and I, because α_2 depends on $(I + I_g)$ and α_1 still, of course, depends on I. The total current gain will now depend on I_g and independent means of breaking over is obtained. The presence of gate current modifies the V–I characteristic, as shown in Fig. 4.4. This method of triggering is most frequently used in thyristor operation.

4.7 THYRISTOR TRANSIENT CHARACTERISTICS

When a forward voltage is applied across the thyristor, the two outer junctions J_1 and J_2 are forward-biased and the central junction J_2 is reverse biased. A reverse biased p-n junction has the characteristics of capacitor, therefore whenever forward voltage is applied, a charging current flows.

Under transient conditions, the capacitances of the p-n junctions will influence the characteristics of the thyristor. If a thyristor is in a blocking state, a rapidly rising voltage applied across the device would cause high current through capacitor C_{j2} can be expressed as

$$i_{j2} = \frac{d(q_{j2})}{dt} = \frac{d}{dt}(C_{j2} \cdot V_{j2}) = V_{j2}\frac{dC_{j2}}{dt} + C_{j2}\frac{dV_{j2}}{dt}$$

where C_{j2} and V_{j2} are the capacitance and voltage of junction J_2, respectively, q_{j2} is the charge in junction.

Figure 4.7 (a) shows an SCR, battery, and resistor connected in series. The SCR is triggered on by a step current into the gate. There is a finite time before the current through the SCR and the voltage across the SCR reach steady state. While the time required to reach steady state depends upon the SCR in the circuit. It is of the order of several microseconds.

The total time to reach steady state is subdivided into two shorter times t_d and t_r as shown in Fig. 4.7 (d).

The delay time, t_d is defined in the 10% point of the anode waveform, or 10% point of the anode current waveform. The rise time t_r, is the time required for the anode voltage to drop from 90% of its initial value to 10%. The turn on time is defined as $(t_d + t_r)$. Clearly the turn on time will vary with gate current 'drive' and from SCR to SCR. Turn on times are typically of the order of several microseconds.

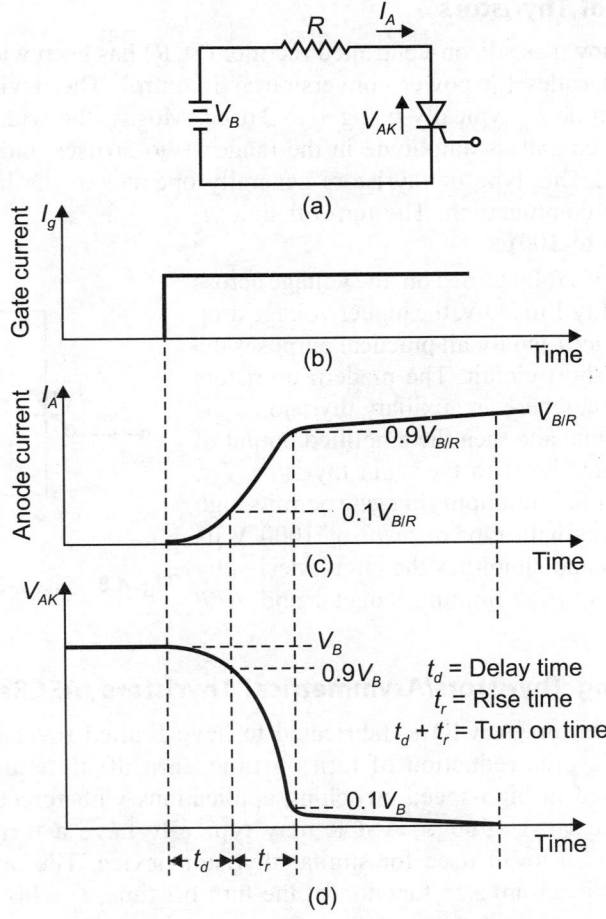

Fig. 4.7 Transient characteristics of thyristor

4.8 THYRISTOR TYPES

Thyristors are manufactured almost exclusively by diffusion. The anode current requires a finite time to propagate to the whole area of the junction, from the point near the gate when the gate signal is initiated for turning on the thyristor. The manufacturers use various gate structures to control the di/dt, turn on time, and turn off time. Depending on the physical construction, turn on and turn off behaviour, thyristors can be broadly classified into the following categories:

1. Phase-control thyristors (SCRs)
2. Fast-switching thyristors/asymmetrical thyristors (ASCRs)

3. Gate turn off thyristors (GTOs)
4. Bidirectional triode thyristors (TRIACs)
5. Reverse conducting thyristors (RCTs)
6. Static induction thyristors (SITHs)
7. Light-activated silicon-controlled rectifiers (LASCRs)
8. FET-controlled thyristors (FET-CTHs)
9. MOS-controlled thyristors (MCTs)

4.8.1 Phase-control Thyristors

The thyristor, also known as silicon-controlled rectifier (SCR) has been widely used in industry for more than two decades for power conversion and control. The device switches on very quickly, the turn on time t_{on} typically being 1 to 3 µsec. Mostly, the width of gate pulse is in the range 10 to 50 µsec and its amplitude in the range 10 to 50 µsec and its amplitude in the range 20 to 200 mA. This type of thyristors generally operates at the line frequency and is turned off by natural commutation. The turn off time, t_q is of the order of 50 to 100 µs.

When the device is fully turned on, the voltage across it is quite small typically 1 to 2.5 V, the higher voltage drop for higher current devices and for all practical purposes the device behaves as a short circuit. The modern thyristors use an amplifying gate, where an auxiliary thyristor, T_A is gated on by a gate signal and then the amplified output of T_A is applied as a gate signal to the main thyristor, T_M. This is shown in Fig. 4.8. The amplifying gate permits high dynamic characteristics with typical dv/dt of 1000 V/µs and di/dt of 500 A/µs and simplifies the circuit design by reducing or minimizing di/dt limiting inductor and dv/dt protection circuits.

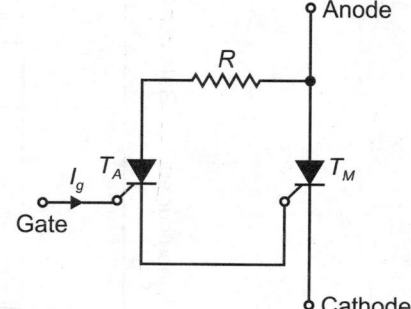

Fig. 4.8 Phase-control thyristor circuit

4.8.2 Fast-switching Thyristors/Asymmetrical Thyristors (ASCRs)

An asymmetrical thyristor (ASCR) is fabricated to have limited reverse voltage capability and, as a result, it permits reduction of turn on time, turn off time and conduction drop. These devices are used in high-speed switching applications with forced commutation (e.g. choppers and inverters). A 1500 V, ASCR may typically have a turn off time of 10 to 15 µsec compared to 20 to 30 µsec for similar thyristor device. The on-state forward drop varies, approximately as an inverse function of the turn off time, t_q. This type of thyristors is also known as an *inverter thyristor*.

These thyristors have high dv/dt of typically 1000 V/µs and di/dt of 1000 A/µs. The fast turn off and high di/dt are very important to minimize the size and weight of commutating and/or reactive circuit components. The on-state voltage of a 2200 A, 1800 V thyristor is typically 1.7 V. Inverter thyristors with a very limited reverse blocking capability, typically 10 V, and undesirable effect of stray inductance between the thyristor diode loop is also eliminated.

4.8.3 Gate Turn Off (GTO) Thyristor

The conventional thyristor as described earlier has over the years been developed such that two new devices of the thyristor family are now available, the asymmetrical thyristor and gate turn off (GTO) thyristor.

The conventional thyristor has two p-n junctions which can block high voltages in one or other direction, this being an essential requirement for applications in the rectifier circuits. However, for the inverter circuits the reverse blocking capability is not needed.

To reduce the time taken for the thyristor to recover its blocking state after turn off the silicon can be made thinner at the expense of it losing its ability to block a reverse voltage, this device now being known as *asymmetrical thyristor*.

The gate turn off thyristor, as its name implies, has a structure such that it can be turned off by removing current from the gate. Turn on is achieved by injecting current into the gate as in the conventional thyristor. The gate turn off thyristor has more complex structure than conventional thyristor. Referring to Fig. 4.9(a), the gate turn off thyristor has highly doped N spots in the P-layer at the anode, the plus sign indicating high doping levels. The gate-cathode structure is interdigitated, that is, each electrode is composed of a large number of narrow channels closely located.

With the gate turn off thyristor, in the absence of any gate current, a positive voltage at the anode with respect to the cathode is withstood at the centre N-P junction in a like manner to the conventional thyristor, but a reverse voltage with the cathode positive will breakdown the anode junction at a low level in a similar manner to the asymmetrical thyristor.

The turn on conditions for the gate turn off thyristor are similar to the conventional thyristor, but because of the differing structure the latching current is higher. The interdigitated nature of the gate results in a very rapid spread of conduction in the silicon, but it is necessary to maintain the gate current at a high level for a longer time to ensure that latching takes place. To minimise the anode-cathode voltage drop, it is advantageous to maintain a low level of gate current throughout conduction otherwise the on-state voltage and hence conduction losses will be slightly higher than necessary.

A GTO thyristor can be turned on by a single pulse of positive gate current like a thyristor, but in addition, it can be turned off by a pulse of negative gate current. Both on-state and off-state operation of the device are therefore controlled by the gate current. The symbols for the GTO thyristor frequently used are shown in Fig. 4.9(b).

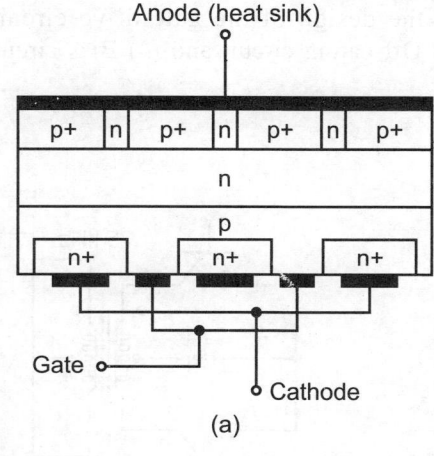

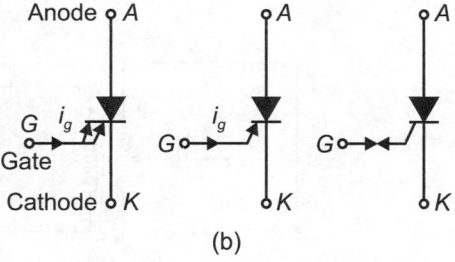

Fig. 4.9 Structures of GTO and symbols

4.8.3.1 Static V–I Characteristics

Typical static $V–I$ characteristics for GTO thyristor are shown in Fig. 4.10. It is observed from these characteristics that latching current for large power

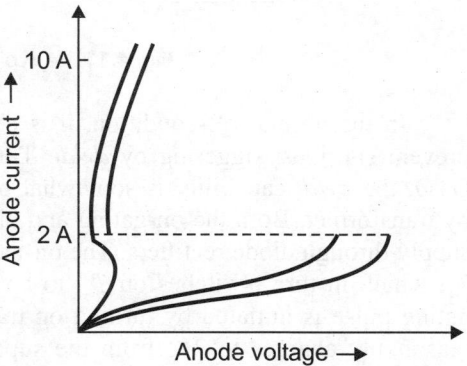

Fig. 4.10 Typical static $V–I$ characteristics for GTO thyristor

GTOs is several amperes as compared to conventional thyristors ($2A$) of the same rating. If the gate current is not able to turn on the GTO, it behaves like a high voltage, low gain transistor with considerable anode current. This leads to a noticeable power loss under such conditions. The on-state voltage of a typical 550 A, 1200 V, GTO is typically 3.4 V. GTOs are a serious competitor to thyristors and high powered transistors in inverter and chopper applications.

4.8.3.2 *Switching Characteristics*

Figure 4.11 shows a typical gate drive circuit of a GTO. GTO performance is greatly influenced by the design of the gate drive circuit. It consists of three parts: (1) On-gating circuit (2) Off-gating circuit and (3) Bias circuit.

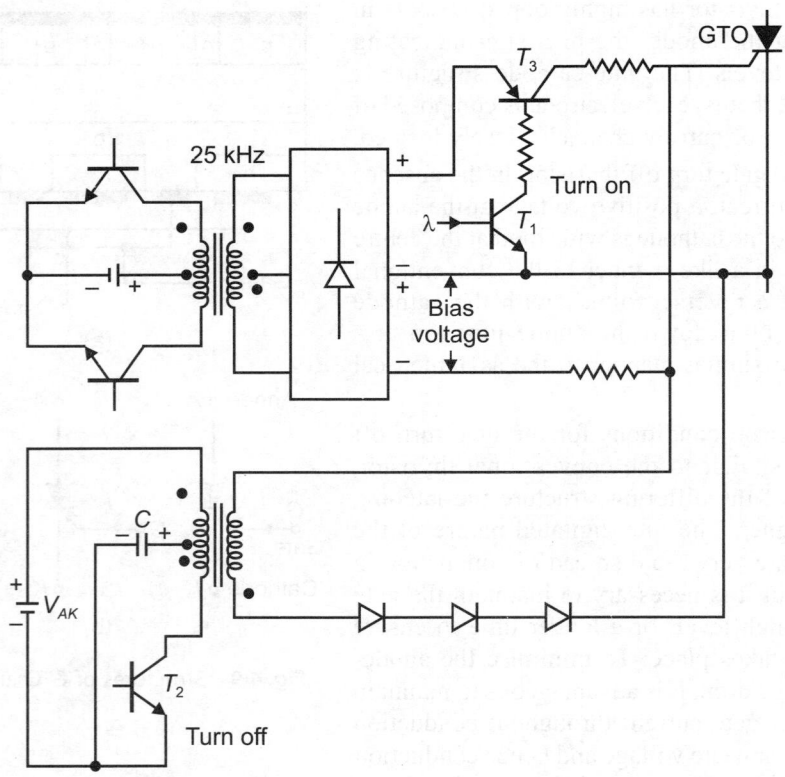

Fig. 4.11 A typical gate drive circuit of a GTO

In the normal off-condition, it is desirable to put a negative bias voltage in the gate to prevent spurious triggering by dv/dt. This is because with the unshorted emitter structure of GTO, the dv/dt capability is somewhat poor. The power supply for the gate drive is isolated by transformer. Both the on-gating and bias circuits are supplied from a high frequency power supply through diode rectifiers. The on-gating signal is optically isolated and turns on transistor T_1, which in turn, switches on T_3, to apply positive gate current pulse to the GTO. The off-gating pulse is initiated by turning on transistor T_2. Prior to switching on T_2, the capacitor C reasonably charges to $2V_d$ from the supply voltage V_d. When transistor T_2 is on, the stored energy of the capacitor creates a large negative gate current pulse. The series diodes and transistor T_3 prevent interaction between on-gating and off-gating circuits.

The turn on behaviour of a GTO is essentially similar to thyristors. Both the delay time and turn on time can be decreased by boosting the on-gating current. The turn off characteristics of a GTO are somewhat different and are shown in Fig. 4.12. When a negative voltage is applied across the gate and cathode terminals, the gate current i_g rises. When the gate current reaches its maximum value, I_{GR} the anode current begins to fall, and the voltage across the device, V_{AK} begins to rise. The fall time of I_A is abrupt, typically less than 1 μsec. Thereafter the anode current changes slowly, and this portion of the anode current is known as the *tail current*.

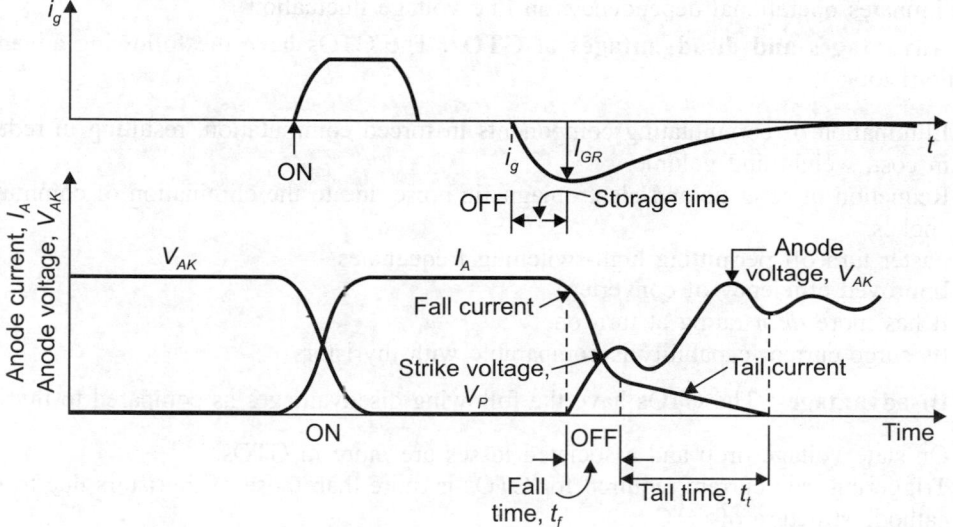

Fig. 4.12 The turn off characteristics of a GTO

The ratio (I_A/I_{GR}) of the anode current I_A (prior to turn off) to the maximum negative gate I_{GR} required for turn off is low, typically between 3 and 5. For example, a 2500 V, 1000 A, GTO typically requires a peak negative gate current of 250 A for turn off. Note that during turn off, both voltage and current are high. Therefore switching losses are somewhat higher in GTO thyristors. Consequently GTOs are restricted to operate at or below a 1 kHz switching frequency. If the spike voltage V_p is large, the device be destroyed. The power losses in the gate drive circuit are also somewhat higher than those of thyristors. Since no commutation circuits are required, the overall efficiency of the converter is improved. Elimination of commutation circuits also results in smaller and less expensive converter.

GTOs may have no reverse voltage blocking capability or else little −20% of the forward breakover voltage. New devices are developed having higher reverse voltage blocking capability. Therefore, an inverse diode must be used as shown in Fig. 4.13, if there is a possibility that appreciable reverse voltage may appear across the device. A polarized snubber consisting of a diode, capacitor, and resistor as shown in Fig. 4.13 is used for the following purposes.

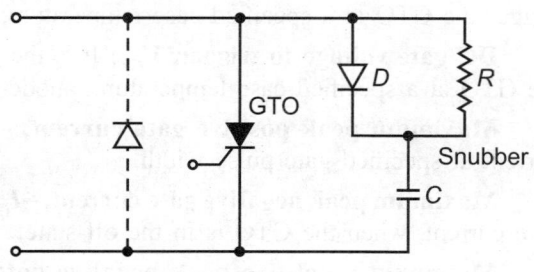

Fig. 4.13 An inverse diode

1. During the fall time of the turn off process the device current is diverted (known as *current snubbing*) to the snubber capacitor (charging it up).
2. The snubber limits the *dv/dt* across the device during turn off.

Recently these devices have been developed with large voltage and current ratings and improved performance (4500 V, 2000 A, 5–10 μsec GTOs are being used). They are becoming increasingly popular in power control equipment, and it is predicted that GTOs will replace thyristors where forced commutation is necessary as in choppers and inverters which reduces the overall efficiency of the converter. GTOs also result in small size and cost for the converter and eliminates operational dependency, on line voltage fluctuation.

Advantages and disadvantages of GTOs: The GTOs have the following advantages over thyristors:

1. Elimination of commutating components in forced commutation, resulting in reduction in cost, weight and volume.
2. Reduction in acoustic and electromagnetic noise due to the elimination of commutation chokes.
3. Faster turn off permitting high-switching frequencies.
4. Improved efficiency of converters.
5. It has more *di/dt* rating at turn on.
6. Its surge current capability is comparable with thyristors.

Disadvantages: The GTOs have the following disadvantages as compared to thyristors:

1. On-state voltage drop and associated losses are more in GTOs.
2. Triggering gate current required for GTOs is more than those of thyristors due to multi-cathode structure of GTO.
4. Magnitude of latching and holding currents is more in GTOs.
5. Gate drive circuit losses are more.
6. Its reverse voltage block capability is less than its forward blocking capability. But this is not a disadvantage so far as inverter circuits are concerned.

In view of the above advantages and disadvantages, GTO devices are used for adjustable-frequency inverter drives, traction purposes because of light weight and high performance drive systems such as field oriented control scheme used in rolling mills, machine tools, robotics control, etc.

GTO parameters: The on-state and off-state parameters are identical to that thyristor, only the turn off gate characteristics are different.

DC gate current to trigger, I_{GT}: It is the lowest value of the gate current which will trigger the GTO at a specified case temperature, anode-cathode voltage and on-state current.

DC gate voltage to trigger, V_{GT}: It is the lowest value of gate voltage that will trigger the GTO at a specified case temperature, anode-cathode voltage and on-state current.

Maximum peak positive gate current, $+I_{GM}$: It is the maximum peak positive gate current at specified gate pulse width.

Maximum peak negative gate current, $-I_{GM}$: It is the maximum allowable peak negative gate current, when the GTO is in the off-state.

Maximum repetitive peak negative gate voltage, $-V_{GRM}$: It is the maximum peak negative gate voltage that must be applied to the gate immediately after turn off to prevent retriggering of the GTO.

Controllable peak on-state current, I_{GTQ}: It is the peak value of on-state current which can be turned off by gate control. The off-state voltage is reapplied immediately after turn off and the reapplied dv/dt is only limited by the snubber capacitance. Once a GTO is turned off, the load current, I_L which is diverted through and charges the snubber capacitor, determines the reapplied dv/dt.

$$\frac{dv}{dt} = \frac{I_L}{C_s}$$

where C_s is the snubber capacitance.

Maximum gate-controlled turn off time, t_{gq}: It is the maximum value of time between the instant at which the gate current is 10% of its peak negative value and the instant at which the on-state current is reduced to 10% of the on-state current with resistive load. The turn off loss, depends on the off-state voltage and the snubber capacitance. An increase in the snubber capacitance increases the dv/dt of forward reapplied voltage as the anode current falls and as a result the turn off losses are reduced. However, the snubber loss is external to the GTO and does not affect the heating of GTO significantly. In practice, the discharging current of the snubber capacitor has to flow through the GTO and contribute to the power loss in the GTO.

Minimum off-time, t_{off}: It is the minimum value of time for which the GTO must be off before it may be retriggered. It is measured from the instant at which the turn off pulse is applied to the gate. A GTO requires this time to recombine the excess minority carriers in the cathode islands.

Turn on time, t_{qt}: It is the time required to complete the turn on process. It is measured from the instant of 10% of gate current to the instant at which the anode voltage reaches 10% of the supply voltage with a resistive load. This gate current must be maintained for a long enough time of t_{qt} to ensure that the turn on is complete. The turn on loss depends on the off-state voltage and becomes maximum with resistive load and minimum value of di/dt.

Minimum on-time, t_{on}: It is the minimum value of on-time to ensure that all cathode islands are fully turned on. This limit is also superimposed on the requirement that the GTO must be on for sufficient time to allow the snubber capacitor to discharge fully.

Maximum fall time, t_f: It is the minimum value of time for the on-state current to fall from 90% of on-state current to 10% of on-state current with resistive loads.

Minimum critical rate of rise of off-state voltage, dv/dt: It is the guaranted rate of rise of forward blocking voltage and depends on the negative gate bias. dv/dt is normally quoted for a gate bias of -2 V. Without a negative gate bias, dv/dt capability depends on the value of the bypass resistance between the gate and cathode.

4.8.4 Triac

The triac is a bidirectional single gate thyristor device that performs the circuit function of two thyristors connected in inverse parallel. It is used for the control of power in AC circuits. Alternating current power can be efficiently controlled today with a special device that switches conduction on and off during each alternation. Control of this type is accomplished with a special semiconductor device known as a triac. A *triac* is identified as a three-electrode or triode AC semiconductor switch. Conduction is triggered by a gate signal. A triac is the equivalent of two reverse-parallel connected SCRs with one common gate. Conduction can be achieved in either direction with an appropriate gate current. AC can be controlled efficiently and accurately with this device.

A *triac* is thus a bidirectional gate controlled thyristor with three terminals. Conduction is achieved by selecting an appropriate crystal combination. Selection depends on the polarity of the source. During one alternation, conduction is through a pnpn combination. Conduction for the next alternation is by an npnp combination. The crystal selection process is achieved automatically. P and N materials are jointly connected to each terminal.

Figure 4.14 (a) shows the cross-sectional view and its characteristics of the crystal structure, with schematic symbol of a triac. The terms anode and cathode are not applicable to triac. The crystal structure of this device forms an $N_1P_1N_2P_2$ structure between terminals T_1 and T_2. The gate is commonly connected to P_1 of the P_1N_4 junction. From terminals T_2 to T_1, the crystal structure is $N_3P_2N_2P_1$. The gate of this structure utilizes N_4 of the P_1N_4 junction. This shows that T_1, T_2 and G are in either an NPNP or PNPN combination depending on the voltage of T_1 and T_2. Selection of the crystal combination is based on the polarity of the applied voltage. When terminal T_1 is negative, N_1 is forward-biased and P_1 is reverse-biased. Current carriers will flow easily through the forward-biased material and not into the reverse-biased material. This means that the npnp or pnpn combination and the appropriate gate voltage for initiating conduction are selected by source voltage polarity. It operates in a conventional thyristor mode by means of a positive gate pulse with a voltage applied in the direction T_2, T_1. The other mode of operation is by means of a negative gate pulse when the applied voltage is in the direction of T_1, T_2. Triac is inferior to a thyristor in its peak ratings. This is because the same silicon wafer is required to operate as two thyristors.

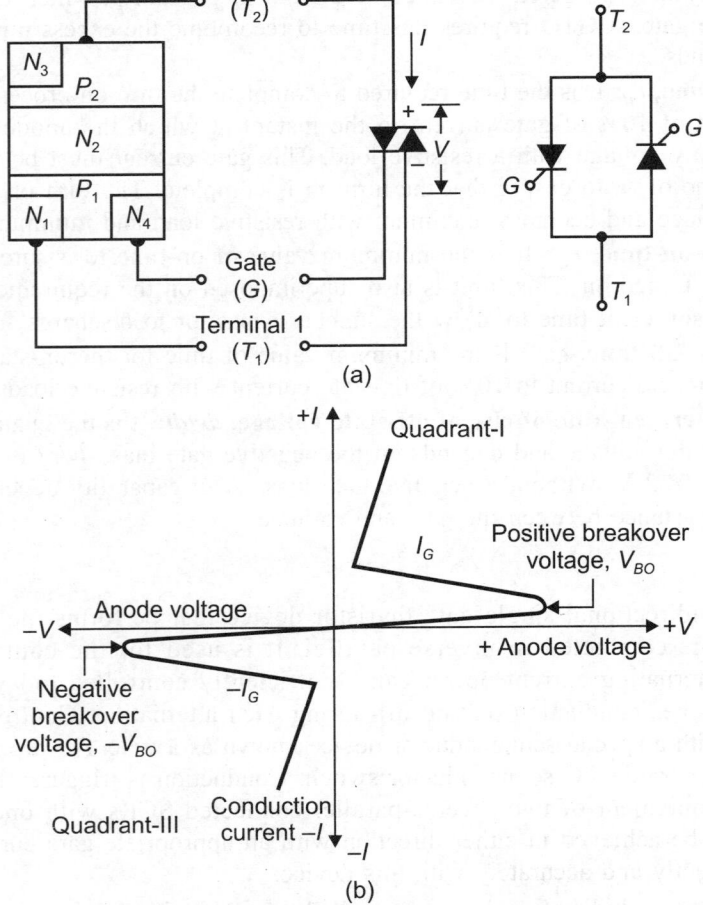

Fig. 4.14 Cross-sectional view of triac and its characteristics

Triac operation and I–V characteristics: The turn on process of triac can be explained as under. When AC is applied to a triac, terminal T_2 is positive and T_1 is negative for one alternation. For the next alternation, T_2 is negative and T_1 is positive. The current–voltage characteristics of a triac must therefore show how it will responds during both alternations of the AC source. Figure 4.14(b) shows a typical $I–V$ characteristic of a triac. Operation in quadrant-I occurs when terminal T_2 is positive and T_1 is negative. This represents the forward conduction mode of operation. Quadrant-III denotes reverse conduction. T_2 is made negative and T_1 positive for operation in quadrant-III. Hence, the first quadrant characteristic is just like an SCR but in third quadrant. The characteristic of triac is identical to its first quadrant, except that, as the polarities of the main terminals changes, the direction of current changes. The breakover voltage in quadrant-I or III is usually quite large when the gate current is zero. Generally, a triac is not designed for conduction when the gate current is zero.

Mode $I+$ is a conventional thyristor mode of operation when the applied voltage at terminal T_2 is positive with respect to T_1 and the device is fired by a positive gate current. It lowers the V_{BO}, which permits conduction at a lower voltage. In quadrant-I, either a positive or negative gate voltage produces gate current. Positive gate voltage tends to be more sensitive to triggering than negative values. When a suitable value of I_G occurs, the triac is triggered into conduction, current flows through T_1, the P_1 material, and the N_2, P_2 region, which is forward-biased by T_2. This current continues for the remainder of the alternation, when it moves into the holding current region of the $V–I$ characteristics.

Terminal T_1 is positive and terminal T_2 is negative during the negative cycle of the source. This condition of operation is shown in quadrant-III. With no gate current, the reverse voltage V_{BO} is quite large. No conduction occurs unless an extremely high value of V_{BO} is reached. This voltage value, like that of quadrant-I, is usually avoided in normal circuit operation. When the device is fired by a gate current, it lowers the reverse V_{BO} to a value that can cause conduction. In some triacs either a positive or negative gate voltage may be used to bias the gate. Ordinarily, a positive V_G is avoided in most applications. A negative V_G is therefore used to bias the gate and produce gate current I_G. Conduction is triggered and I_T flows from T_2, through N_3, P_2, N_2, P_1 and T_1. The N_2P_1 junction then sees a significant increase in current when I_G flows in the gate. This current added to that of the N_3P_2 junction lowers the reverse V_{BO} and causes full conduction to occur. Thus, triac can be triggered in four different modes given in Table 4.1.

Table 4.1 Firing modes of triac

Quadrant	Element polarities			Required gate signal, I_G
I +	T_2 +	T_1 –	I_G (+)	5–10 mA
I –	T_2 +	T_1 –	I_G (–)	10–20 mA
III –	T_2 –	T_1 +	I_G (–)	7–15 mA
III +	T_2 –	T_1 +	I_G (+)	>40 mA

Because of the dual polarity of the gate, terminals T_1 and T_2 of a triac have four possible trigger polarity combinations. Table 4.1 lists these combinations as triggering modes and representative gate signal (I_G) triggering values. Quadrant-I operation has T_1 negative and T_2 positive. Note that the positive gate voltage produces an I_G that is smaller than that of the negative value I_G. This indicates that the triac is more sensitive to the positive voltage value in quadrant-I. Quadrant-III operation occurs best when T_1 is positive and T_2 is negative. In this quadrant, the triac is more sensitive to a negative voltage polarity. Ordinarily, the positive V_G mode of operation is avoided in most circuit applications. Some manufacturers recommend that their triacs not be used in this triggering mode. A common practice used in selecting the most sensitive mode of operation is to match the polarity of the gate with the polarity of terminal T_2.

Sometimes special triacs are designed for operation in the III+ operational mode. This device generally has a special number designation or code to denote it unusual sensitivity. General Electric Company lists a standard triac as an SC141B. A device designed for +III operation is numbered SC141B13. The specialized device is usually more expensive than a standard unit. A triac is normally operated in the following modes:

Mode I+: This is a conventional thyristor mode of operation when the applied voltage at terminal T_2 is +ve with respect to T_1 and the device is fired by a positive gate current.

Mode III−: This is a remote thyristor operation when the terminal T_1 is positive with respect to T_2 and the device is fired by injecting a negative gate current.

In case of triac, the gate responds as a nonlinear low-impedance junction similar to that of a forward-biased diode. Therefore, triggering of the gate must be achieved by some type of low-impedance source. Most triacs, require a hundred or more milliamperes of I_G for a few microseconds to be triggered into conduction. Usually a capacitor is discharged into the gate to initiate the triggering process. As a rule, this necessitates some type of low-power on-off switching device to produce a suitable value of I_G. The device must respond equally well to both cycles of the AC input. Special AC diodes, known as *diacs*, have been developed for triac triggering. Now-a-days diac-triac pairs are being replaced by a single compact unit known as *quadrac*.

Applications: A triac is frequently used in many low power applications such as juice makers, blenders, vacuum cleaners, temperature control, illumination control, liquid level control and power switches, etc. It is economical and easy to control as compared to two SCRs connected antiparallel. However, a triac has a lower *dv/dt* capability and a longer turn off time. It is not available in high voltage and current ratings.

Limitations: Triac has the following limitations:

1. A triac has poor reapplied *dv/dt* capability, which makes it difficult to apply with inductive load.
2. The gate circuit sensitivity is somewhat poor.
3. Turn off time, t_q is longer.

4.8.5 Reverse-Conducting Thyristors (RCT)

In many choppers and inverter circuits, an antiparallel diode is connected across an SCR in order to allow a reverse current flow due to inductive load and to improve the turn off requirement of commutation circuit. The diode clamps the reverse blocking voltage of the SCR to 1 or 2 V under steady state conditions. However, under transient conditions, the reverse voltage may rise to 30 V due to induced voltage in the circuit stray inductance within the device.

An RCT is a compromise between the device characteristics and circuit requirement, and it may be considered as a thyristor with a built-in antiparallel diode as shown in Fig. 4.15.

An RCT is also called an asymmetrical thyristor (ASCR). The forward blocking voltage varies from 400 to 2000 V and the current rating goes up to 500 A. The reverse blocking voltage is

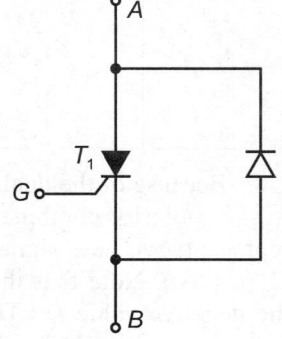

Fig. 4.15 Reverse conducting thyristor circuit

typically 30 to 40 V. Since the ratio of forward current through the thyristor to the reverse current of diode is fixed for a given device, their applications will be limited to specific circuit designs.

4.8.6 Static Induction Thyristor

The static induction thyristor (SITH) is in many respects similar to the gate turn off thyristor, but is a normally on device. Application of a reverse voltage to the gate with respect to the cathode will turn the SITH off. A SITH is a minority carrier device. As a result, SITH has low on-state resistance or voltage drop and it can be made with higher voltage and current ratings compared to other thyristor devices, the SITH is very fast switching and has lower switching losses. Normally off type SITH are also being developed. It has high dv/dt and di/dt capabilities. The switching time is of the order of 1 to 6 µs. The voltage rating can go up to 2500 V and current rating is limited to 500 A. This device is extremely process sensitive, and small perturbations in the manufacturing process would produce major changes in the device characteristics.

4.8.7 Light-Activated Silicon-Controlled Rectifiers (LASCRs)

This device is turned on by direct radiation on the silicon wafer with light. Electron-hole pairs which are created due to the radiation produce triggering current under the influence of electric field. The gate structure is designed to provide sufficient gate sensitivity for triggering from practical light sources.

The LASCRs are used in high voltage and high current applications like high voltage DC (HVDC) transmission and static reactive power or volt-ampere reactive (VAR) compensation LASCR offers complete electrical isolation between the light-triggering source and the switching device of a power converter, which floats at a potential of as high as a few hundred kilovolts. The voltage rating of a LASCR could be as high as 4 kV at 1500 A with light triggering power of less than 100 mW. The typical di/dt is 250 A/µsec and dv/dt could be as high as 2000 V/µsec.

4.8.8 FET-Controlled Thyristors (FET-CTHS)

A FET-controlled thyristor in which a MOSFET and a thyristor are connected in parallel is shown in Fig. 4.16. If a sufficient voltage is applied to the gate of the MOSFET, typically 3 V, a triggering current for the thyristor is generated internally. It has a high switching speed, high di/dt and high dv/dt. This device can be turned on like conventional thyristors, but it cannot be turned off by gate control. This would find applications where optical firing is to be used for providing electrical isolation between the input or control signal and the switching device of the power converter. The device has the input impedance of FET but the conduction drop of thyristor. With the gate voltage exceeding the threshold voltage, current flows in the shunt resistance and switches on the thyristor.

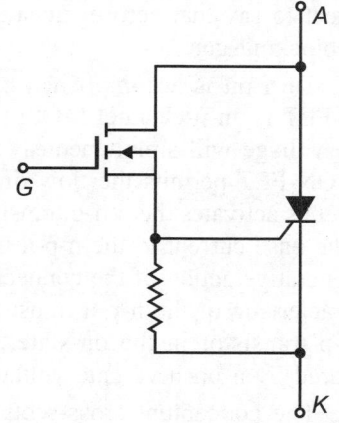

Fig. 4.16 MOSFET and a thyristor are connected in parallel

4.8.9 MOS-Controlled Thyristors (MCT)

The MOS-controlled thyristor or MCT is a new device that is in early stage of development. It is basically a thyristor with one or more MOSFETs built-into the gate structure. The equivalent circuit of the built-in MOSFETs turn on and turn off thyristor is shown in Fig. 4.17(a). Figure 4.17(b) shows the equivalent circuit of the device which is turned on by a gate current like a conventional thyristor and turned off by the built-in MOSFET. Both of these devices would have applications similar to those of GTO but the MCT would have simpler gating requirements, particularly in terms of the magnitude of the needed gating signals.

Figure 4.17(c) shows the symbol of MOS-controlled thyristor.

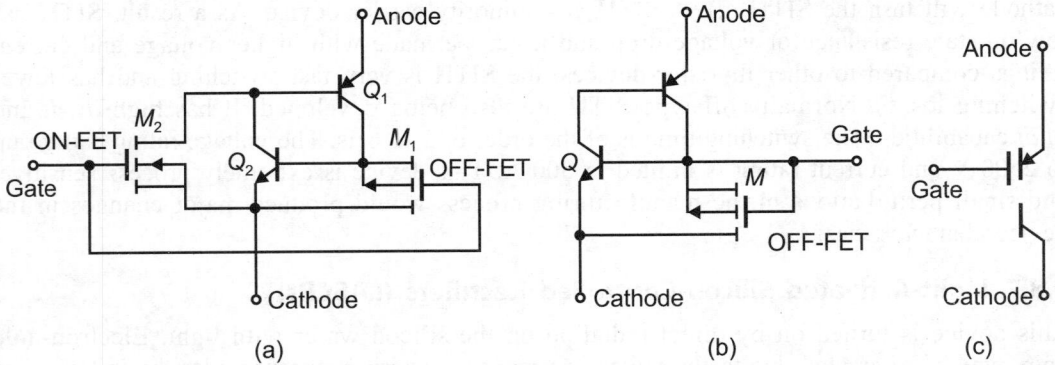

Fig. 4.17 Equivalent circuit of MOS-controlled thyristor (MCT), its symbol

4.8.9.1 *MOSFET-controlled Turn On and Turn Off*

Turn off is accomplished by turning on the OFF-FET, which in turn, shunts base current away from the n-p-n BJT in the thyristor pair. This causes the n-p-n transistor to begin to turn off as the stored charge in the p-base of the transistor disappears and its current gain falls to a low value where the latch-on condition of the thyristor is no longer satisfied. Once this occurs, the thyristor turns itself off by regenerative action. During the turn off of the thyristor, the other MOSFETs in the other circuit, the ON-FET, must be kept in the blocking state. It is correct to say that zero or negative gate-anode voltage will guarantee that the ON-FET is in the blocking gate.

Turn on is accomplished by driving the ON-FET into the conducting state. Since the ON-FET is an n-channel MOSFET and the OFF-FET is a p-channel device, a positive going gate voltage will simultaneously turn on the ON-FET and turn off the OFF-FET. Turn on of the ON-FET permits the flow of base current out of the p-n-p transistor in the thyristor pair that thus activates the p-n-p transistor. The collector current from the p-n-p transistor then flows as the base current to the n-p-n transistor and turn it on. Once both the transistors are on, the regenerative action of the connection will cause the thyristor to turn on. The better conduction characteristic of the n-p-n transistor ensures that it will carry most of the base current of the p-n-p transistor in the on-state, the OFF-FET must be kept in the blocking state which is ensured by a positive gate voltage.

The conceptual cross-section of the unit cell of this version of the MCT is shown in Fig. 4.18. A complete MCT would be composed of many of these cells connected in parallel in order to achieve the desired current capability. This complete device has not been fabricated. However, simpler derivatives of it, containing either just the ON-FET or the OFF-FET have been fabricated and tested. Structures with just ON-FET incorporated in the device have been termed FET-controlled thyristors or MOS gated SCRs. Experimental results achieved to date indicate that turn on and turn off for the MCT are faster than those realized with GTO.

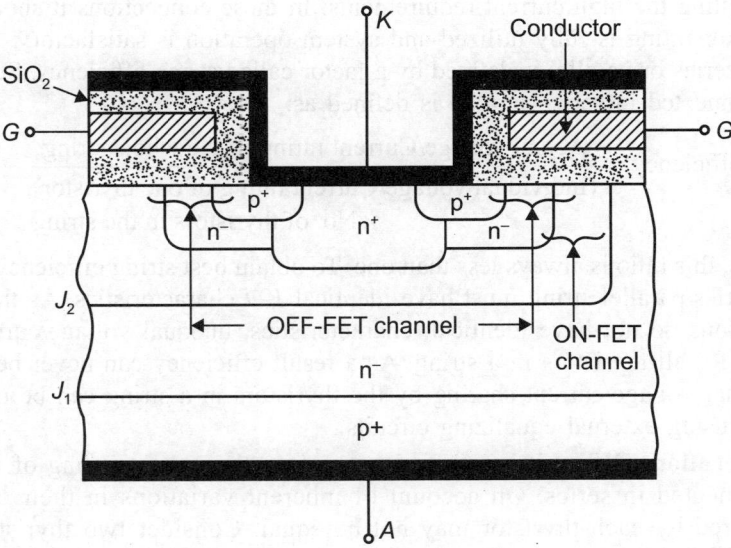

Fig. 4.18 The conceptual cross-section of the unit cell of this version of the MCT

4.8.9.2 *Conventional Gate Turn On and MOSFET Turn Off*

The unit cell of the MCT shown above indicates one significant disadvantage to the structure, the curvature of the n-p boundary which forms the J_2 junction of the thyristor. This curvature will lead to the reductions in the blocking voltage capabilities of the device. Removal of the ON-FET permits the np boundary to remain straight and thus the reduction in breakdown voltage because of field crowding is avoided. Turn off of this version of the MCT would be accomplished in the same manner as described previously for the other version of the MCT.

Both versions of the MCT will require more research and development to realise a practical device. If its full potential can be realised, it will become a desirable alternative for the GTO.

Advantages of MCT: MOS-controlled thyristor has:

1. a low forward voltage drop during conduction
2. a fast turn on time, typically 0.4 μs and fast turn off time typically 1.25 μsec for an MCT of 300 A, 500 V
3. low switching losses
4. a low reverse voltage blocking capability
5. a high gate input impedance which greatly simplifies the drive circuits

Disadvantages

1. It cannot easily be driven from a pulse transformer if a continuous bias is required to avoid state ambiguity.
2. Like GTO, the MCT cannot normally withstand a reverse voltage unlike the conventional thyristor which can block a high voltage in both forward and reverse directions.

4.9 SERIES AND PARALLEL OPERATIONS OF THYRISTORS

A single thyristor cannot fulfil the high voltage and high current requirements in any application. In such cases, thyristors are connected in series to meet the high voltage demand and in parallel for fulfilling the high current requirements. In these connections it should be ensured that each thyristor rating is fully utilized and system operation is satisfactory. The utilization of thyristors in series or parallel is defined by a factor called string efficiency. String efficiency of thyristors connected in series/parallel is defined as:

$$\text{String efficiency} = \frac{\text{Actual voltage/Current rating of the whole string}}{\substack{\text{(Individual voltage/Current rating of one thyristor)} \\ \times \text{No. of thyristors in the string}}}$$

In practice, this ratio is always less than one. To obtain best string efficiency, the thyristors connected in series/parallel string must have identical V–I characteristics. As thyristors of the same specifications do not have identical, characteristics, unequal voltage/current sharing is bound to occur for all thyristors in a string. As a result efficiency can never be equal to one. However, unequal voltage/current sharing by the thyristors in a string can be minimized to a great extent by using external equalizing circuits.

Series operations: When it is required to increase the voltage rating of a string, SCRs have to be connected in series. On account of inherent variations in their characteristics, the voltage shared by each thyristor may not be equal. Consider two thyristors with their static characteristics as shown in Fig. 4.19. For thyristor Th1 leakage resistance (V_1/I_o) is high whereas for thyristor Th2 it is low (V_2/I_o) for the same leakage current I_o in the series connected thyristors. Each thyristor is rated for a forward blocking voltage of V_1 volts which is always less than its forward breakover voltage. Here V_{BO1} and V_{BO2} are the forward breakover voltages for thyristors 1 and 2 respectively. The string efficiency for two series connected thyristors of Fig. 4.19 is

$$\frac{V_1 + V_2}{2V_1} = \frac{1}{2}\left(1 + \frac{V_2}{V_1}\right)$$

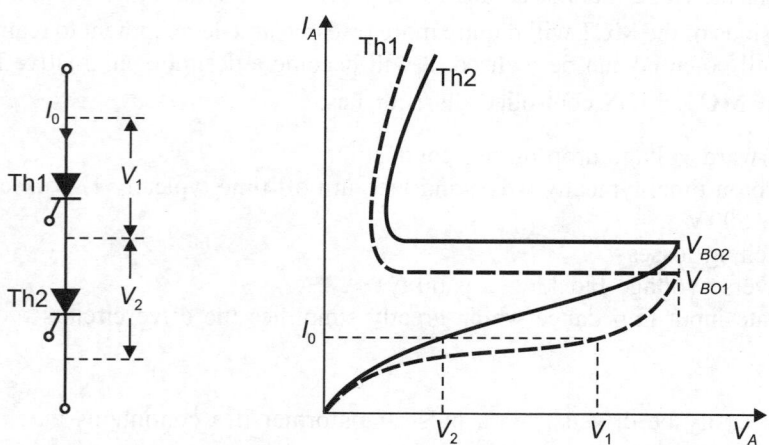

Fig. 4.19 Two thyristors with their static characteristics

It shows that even though thyristors have identical ratings, voltage shared by each thyristor is not the same and string efficiency is less than unity.

The voltage distribution can be made more uniform by connecting a shunt resistance across each thyristor as shown in Fig. 4.20. This is called *static equalising circuit*.

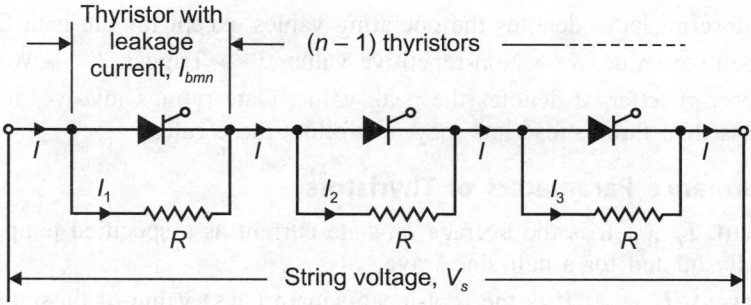

Fig. 4.20 Static equalising circuit

Let there are n thyristors connected in series as shown. Thyristor, Th1 has minimum leakage current, I_{bmn} and each of the remaining thyristors have the same leakage current, I_{bmx}, I_{bmn}. Thyristor with lower leakage current blocks more voltage. As thyristor Th1 has lower leakage current let it will block voltage, V_{bm}, which is more than that shared by each of the other $(n - 1)$ thyristors from Fig. 4.20.

$$I_1 = I - I_{bmn} \text{ and } I_2 = I - I_{bmx}, \text{ where } I \text{ is the total current.}$$

Voltage across thyristor Th1,

$$V_{bm} = I_1 R$$

Voltage across $(n - 1)$ thyristors

$$= (n - 1) I_2 R$$

String voltage,

$$V_s = I_1 R + (n - 1) I_2 R$$
$$V_s = V_{bm} + (n - 1) R (I - I_{bmx})$$
$$= V_{bm} + (n - 1) R [[(I_1 + I_{bmn}) - I_{bmx}]$$
$$= V_{bm} + (n - 1) R [I_1 - (I_{bmx} - I_{bmn})]$$
$$= V_{bm} + (n - 1) RI_1 - (n - 1) R \cdot \Delta I_b$$

where $\qquad \Delta I_b = I_{bmx} - I_{bmn}$ and $V_{bm} = I_1 R$

or $\qquad V_s = nV_{bm} - (n - 1) R \cdot \Delta I_b$

$$R = \frac{nV_{bm} - V_s}{(n - 1) \Delta I_b}$$

Once the value of R is determined, its power rating is given by

$$P_R = \frac{V_r^2}{R} \text{ where } V_r \text{ is rms voltage across } R.$$

4.10 RATINGS OF THYRISTORS

The ratings applicable to a particular device, in respect to steady state and transient values are numerous, relating to voltage, current, switching times dv/dt, di/dt, control parameters, losses, thermal and temperature values. If SCRs are operated under these specified conditions, no damage will be done to thyristors. For correct application of the device in thyristor circuits, a knowledge of these ratings is desirable.

Some subscripts are associated with voltage and current ratings for identifying them. First subscripts letter represents the state like

$D \rightarrow$ Off-state; $T \rightarrow$ On-state; $R \rightarrow$ Reverse, $F \rightarrow$ Forward

Second subscript letter denotes the operating values except for the gate G.

$R \rightarrow$ Repetitive value, $S \rightarrow$ Non-repetitive value, $T \rightarrow$ Trigger, $W \rightarrow$ Working value

Third subscript letter M denotes the peak value. Gate ratings involve the subscript G. Ratings with less than three subscripts may not follow these rules.

4.10.1 Performance Parameters of Thyristors

On-state current, $I_{T(AV)}$: It is the average on-state current as a specified temperature. These data are normally quoted for a half-sine wave.

RMS current, $I_{T(RMS)}$: It is the root-mean-square (rms) value of on-state current. This signifies the heating effect due to $i^2 R$ dissipation and is limited due to thermal stress on the device.

Non-repetitive rate of rise of on-state current, di/dt: It is the maximum value of the rate of rise of on-state current which the thyristor can withstand without destroying itself.

Maximum repetitive peak reverse voltage, V_{RRM}: It defines the maximum permissible instantaneous value of repetitive applied reverse voltage that a thyristor can block.

Maximum non-repetitive peak reverse voltage, V_{RSM}: It is the maximum instantaneous peak value of applied reverse voltage under transient conditions and for a specified time duration. V_{RSM} is typically 15% above V_{RRM}.

Maximum repetitive peak off-state voltage, V_{DRM}: It defines the maximum permissible instantaneous value of repetitive applied forward voltage that a thyristor can withstand.

Maximum non-repetitive peak off-state voltage, V_{DSM}: It is the maximum instantaneous peak value of applied forward voltage under transient conditions and for a specified time duration. V_{DSM} is typically 15% above V_{DRM}.

Finger voltage: The minimum voltage which is required between anode and cathode of the thyristor to trigger it to the conduction mode, is called its *finger voltage*. It is slightly more than the normal on-state voltage of the device.

On-state voltage drop, V_T: It is the instantaneous value of on-state voltage drop and is dependent on the junction temperature, T_j. V_T can be considered as being made up of (1) a value which is independent of the forward current, and (2) a value which is proportional to the instantaneous forward current.

Maximum peak on-state voltage, V_{TM}: It is the maximum on-state voltage drop at a specified on-state current and junction temperature.

Critical rate of rise of off-state voltage, dv/dt: It is the minimum value of the rate of rise of forward voltage which may cause switching from off-state to on-state.

Voltage safety factor: $V_f = \dfrac{\text{Peak inverse voltage}}{\sqrt{2} \times \text{rms value of operating voltage}}$

Its value normally lies between 2 and 2.7.

Latching current, I_L: It is the minimum anode current which is required to maintain the thyristor in the on-state immediately after a thyristor has been turned on and the gate signal has been removed.

Holding current, I_H: It is the minimum anode current to maintain the thyristor in the on-state. The holding current is less than the latching current.

Peak on-state current, I_P: It is the peak instantaneous value of on-state current. It depends on the di/dt and width of current pulse. The switching losses would depend on the value of I_P, pulse width, and di/dt.

Peak reverse current, I_{RM}: It is the peak value of reverse current at the maximum junction temperature and maximum repetitive peak reverse voltage with gate open. This current would cause junction heating.

Peak off-state current, I_{DM}: It is the peak value of off-state current at the maximum junction temperature and maximum repetitive peak reverse voltage with gate open. This current would also cause junction heating.

Reverse recovered charge, Q_{RR}: It is the amount of charge carriers which has to be recorered during the turn off process. Its value is determined from the area enclosed by the path of the reverse recovery current. The value of Q_{RR} depends on the rate of fall of on-state current and the peak value of on-state current before turn off. Q_{RR} causes corresponding energy loss within the device.

Junction-to-case thermal resistance, R_{thJC}: It is the effective thermal resistance between the junction and outer case of the device. It is a measure of the heat transfer ability of materials and mechanical construction of the thyristor.

DC gate current to trigger, I_{GT}: It is the recommended value of the gate current at a specified case temperature. The minimum and maximum values of I_{GT} are normally specified, namely 70 to 150 mA.

DC gate voltage to trigger, V_{GT}: It is the recommended value of gate voltage at a specified case temperature. The minimum and maximum values of V_{GT} are normally specified, namely 1.2 to 2.5 V.

DC gate voltage not to trigger, V_{GD}: It is the value of the gate voltage which does not cause the thyristor (with rated V_{DRM} between anode to cathode) to switch from off-state to on-state.

Maximum nonrepetitive surge current, I_{TSM}: It is the maximum permissible peak current of half-sine wave with a duration of normally 10 ms at a specified temperature. A repetition is permissible only after the expiration of a minimum interval to reduce the junction temperature to allowable range. This peak permissible value varies with the number of repetitions.

Turn off time, t_q: It is the minimum value of time interval between the instant when the on-state current has decreased to zero and the instant when the thyristor is capable of withstanding forward voltage without turning on. t_q depends on the peak value of on-state current and the instantaneous on-state voltage.

4.10.2 Surge Current Rating

When a thyristor is working under its repetitive voltage and current ratings, its permissible junction temperature is never exceeded. However, a thyristor may be subjected to abnormal operating conditions due to faults or short circuits. In order to withstand these unusual working conditions, surge current rating of thyristors is also specified. A surge rating indicates the maximum possible non-repetitive or surge current which the device can withstand. Surge current rating is inversely proportional to the duration of the surge. Surge currents are assured to be the sine waves with frequency of 50 or 60 Hz depending on the supply frequency. This rating is specified in terms of the number of surge cycles with corresponding surge current peak.

One cycle surge current rating is the peak value of the allowable non-recurrent half-sine wave of 10 msec duration for 50 Hz. For duration less than half-cycle, i.e. 10 msec; a subcycle surge current rating is also specified. This rating for 50 or 60 Hz supply is the peak value for a part of the half-sine wave. The subcycle surge current rating I_{sb} can be determined by equating the energies involved in one cycle surge and one subcycle surge as follows:

$$I_{sb}^2 t = I^2 T$$

or
$$I_{sb} = I\sqrt{\frac{T}{t}}$$

where I_{sb} is the subcycle surge current rating (A), I is one cycle surge current rating (A), t is the duration of subcycle surge (sec) and T is the time for one-half cycle of supply frequency (sec).

For 50 Hz supply, $T = 10$ msec

$$\therefore \qquad I_{sb} = I\sqrt{\frac{10 \times 10^{-3}}{t}} = \frac{I}{10}\sqrt{\frac{1}{t}}\,.$$

4.10.3 I^2t Rating

It is employed for the choice of a fuse or other protective equipment for thyristors. It specifies the energy that the device can absorb for a short time before the fault is cleared. I^2t rating is given by the relation:

[rms value of one cycle surge current]^2 × time for one cycle.

4.10.4 *di/dt* Rating

This rating of thyristor indicates the maximum rate of rise of current from anode to cathode. When a thyristor is turned on, conductor starts at a place near the gate. If the rate of rise of anode current (di/dt) is large as compared to the spreading velocity of carriers across the cathode junction, local hot spots will be formed near the gate connection on account of high current density. This causes the junction temperature to use above the safe limit and as a consequence, the SCR may be damaged permanently. Therefore a limit on the value of di/dt at turn on is specified in amperes per microsecond for all SCRs. Typical values are 60 to 800 A/μsec.

4.10.5 Junction Temperature

The operating junction temperature range of thyristors varies for the individual types. A low temperature limit may be required to limit thermal stress in the silicon crystal to safe values. This type of stress is due to the difference in the thermal coefficients of expansion of the materials used in fabricating the cell subassembly. The upper operating temperature limit is imposed because of the temperature dependence of the breakover voltage, turn off time and thermal stability considerations. The upper storage temperature limit in some cases may be higher than the operating limit. It is selected to achieve optimum reliability and stability of characteristics with time.

The rated maximum operating junction temperature can be used to determine steady state and recurrent overload capability for a given heat sink system and maximum ambient temperature. Conversely, the required heat sink system may be determined for a given loading of the semiconductor device by means of the classic thermal impedance approach.

Transiently the device may actually operate beyond its specified maximum operating junction temperature and still be applied within its ratings. An example of this type of operation occurs within the specified forward non-recurrent surge current rating. Another example is the local temperature rise of the junction due to the switching dissipation during the turn on of a thyristor under some conditions.

4.10.6 Power Dissipation

The power generated in the junction region in typical thyristor operation consists of the following five components of dissipation: (1) turn on switching, (2) conduction, (3) turn off or commutation, (4) blocking, and (5) triggering.

On-state conduction losses are the major source of junction heating for normal duty cycles and power frequencies. However, for very steep (high *di/dt*) current waveforms or high operating frequencies turn on switching losses may become the limiting consideration.

Figure 4.21 gives on-state conduction loss in average watts for the typical SCR as a function of average current in amperes for various conduction angles for operation up to 400 Hz. This type of information is given on the specification sheet for each type of SCR (with the exception of some inverter type SCRs). These curves are based on a current waveform which is the remainder of a half-sine wave which results when delayed angle triggering is used in a single phase resistive load circuit. Similar curves exist for rectangular current waveforms. These power curves are the integrated product of the instantaneous anode current and on-state voltage drop.

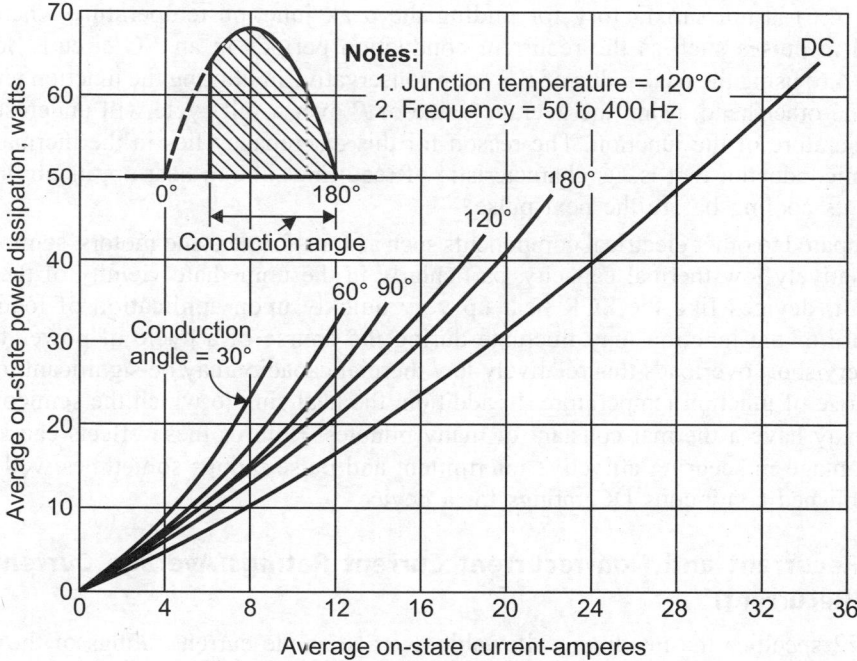

Fig. 4.21 On-state conduction loss in average watts for the typical SCR as a function of average current in amperes

Gate losses are negligible for pulse types of triggering signals. Losses may become more significant for gate signals with a high duty cycle, or for SCRs in a small package. *Highest gate dissipation will occur for an SCR whose gate characteristics intersect the gate circuit load line at its midpoint.*

4.10.7 Thermal Resistance

The heat developed at the junctions by the foregoing power losses flows into the case, then to the heat sink (if employed) and onto the surrounding ambient fluid. The junction temperature rises above the stud, or case, temperature in direct proportion to the amount of heat flowing from the junction and the thermal resistance of the device to the flow of heat. The following equation defines the relationship under steady state conditions:

$$T_J - T_C = PR_{\theta JC} \qquad \qquad ...(4.6)$$

where T_J = average junction temperature, °C

$$T_C = \text{case temperature, } °C$$

$$P = \text{average heat generation at junction, watts}$$

$$R_{\theta JC} = \text{steady state thermal resistance between junctions and bottom face of case, } °C/\text{watt}$$

The above Equation (4.6) can be used to determine the allowable power dissipation and thus the continuous pure DC on-state current rating of an SCR for a given case temperature through use of the on-state $V\text{–}I$ curves. For this purpose, T_J is the maximum allowable junction temperature for the specific device. The maximum values of $R_{\theta JC}$ and T_J are given in the specifications.

4.10.8 Transient Thermal Impedance

Equation (4.6) is not satisfactory for finding the peak junction temperature when the heat is applied in pulses such as the recurrent conduction periods in an AC circuit. Solution of Equation (4.6) using the peak value of P is over-conservative in limiting the junction temperature rise. On the other hand, using the average value of P over a full cycle will underestimate the peak temperature of the junction. The reason for this discrepancy lies in the thermal capacity of the semiconductor, that is, its characteristic of requiring time to heat up, its ability to store heat, and its cooling before the next pulse.

Compared to other electrical components such as transformers and motors, semiconductors have a relatively low thermal capacity, particularly in the immediate vicinity of the junction. As a result, devices like the SCR heat up very quickly upon application of load, and the temperature of the junction may fluctuate during the course of a cycle of power frequency. Yet, for very short overloads this relatively low thermal capacity may be significant in arresting the rapid rise of junction temperature. In addition, the heat sink to which the semiconductor is attached may have a thermal constant of many minutes. Both of these effects can be used to good advantage in securing attractive intermittent and pulse ratings sometimes well in excess of the published continuous DC ratings for a device.

4.10.9 Recurrent and Non-recurrent Current Ratings/Average Current Rating (Recurrent)

Figure 4.22 specifies the maximum allowable average anode current ratings of the SCR as a function of case temperature and conduction angle. Points on these curves are selected so that the junction temperature under the stated conditions does not exceed the maximum allowable value. The maximum rated junction temperature of the C380 SCR is 125°C.

The curves of Fig. 4.22 include the effects of the small contributionto total dissipation by reverse blocking, gate drive, and switching up to 400 Hz. For devices which are lead mounted or housed in small packages, the on-state current rating may be substantially affected by gate drive dissipation.

The slope of the curves shown in Fig. 4.22 is essentially dependent upon the $R_{\theta JC} \cdot P_D$ product. In some SCRs $R_{\theta JC}$ is not fixed but is a function of the method used to cool it.

If the C380 in a single phasi resistive load circuit is triggered as soon as its anode swings positive, the device will conduct for 180 electrical degrees. If the case temperature is maintained at 80°C, or less, the C380 is capable of handling 235 amperes average current as indicated in Fig. 4.22. If the triggering angle is retarded by 120° the C380 will conduct for only the 60 remaining degrees of the half cycle. Under these conditions of 60° conduction, the maximum rated average current at 80°C stud temperature (double side cooled) is 115 amperes, substantially less than for 180° conduction angle.

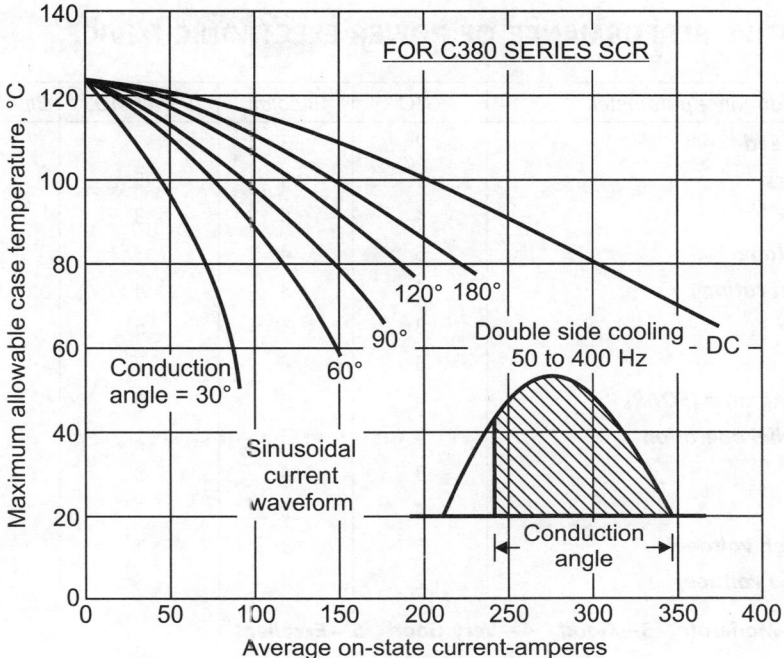

Fig. 4.22 Curves for various conduction waves

4.10.10 RMS Current (Recurrent)

It will be noted in Fig. 4.22 that the curves for the various conduction wave shapes have definite end points. These points represent identical rms values, and as such an rms rating is implicit in the curves of Fig. 4.22.

The rms current rating can be of importance when applying thyristors to high peak current, low duty cycle waveforms. Although the average value of the waveform may be well within the ratings, it may be that the allowable rms rating is being exceeded.

The average current values shown as phase control ratings in Fig. 4.22 fixed basic rms device current rating are for the resistive current waveform shown in the figure. Since the current form factor for the case of resistive loading is greatest, and since inductance in the path of the thyristor current will reduce its form factor, the average current ratings in Fig. 4.22 are conservative for inductive current waveforms.

For inductive waveforms in which the thyristor current waveform is essentially rectangular, such as may occur in a phase-controlled rectifier operating near full output, most specification sheets show separate rating curves to reflect the improvement in form factor. However, such current waveforms are, of course, subject to the restriction of the allowable turn on current rating of the device.

In other cases in which the current waveform may be half-sinusoidal in shape but of a base width less than half a period of the supply frequency as, for example, with discontinuous AC line current in an AC switch application at large phase retard, greater utilization of the thyristor in terms of its average current versus temperature ratings can be obtained by taking into account the improvement of form factor due to decreasing load power factor (greater inductance) when applying the device within its rms current rating.

4.11 RELATIVE PERFORMANCE OF POWER ELECTRONIC DEVICES

Performance parameter	GTO	Bipolar	Power MOS	Bipolar MOS hybrid
Switching speed	2	3	5	3
Switching loss	1	3	5	3
On-state loss	4	5	3	5
Blocking voltage	5	4	3	3
Surge current rating	5	3	4	4
Ease of drive	3	3	5	5
Cost	5	4	2	3
Safe operating area (SOAR)	4	3	5	3
Ease of parallel operation	3	3	5	3
$Q_{j\,max}$	3	4	3	4
Gain	4	3	5	4
Chip size (high voltage)	5	3	1	4
Chip size (low voltage)	2	4	5	3

1—Poor   2—Moderate   3—Good   4—Very Good   5—Excellent

Table 4.2   Symbols and V–I characteristics of important power devices

S. No.	Devices	Symbols	Characteristics (V–I)	No. of terminals
1.	Diode			2
2.	Thyristor: (SCR-silicon-controlled rectifier)			3
3.	TRIAC: (Bidirectional triode thyristor)			3

S. No.	Devices	Symbols	Characteristics (V–I)	No. of terminals
4.	DIAC: (Bidirectional trigger diode)			2
5.	GTO: (Gate turn off thyristor)			3
6	SCS: (Silicon-controlled switch)			4
7.	SUS: (Silicon unilateral switch)			3
8.	SBS: (Silicon bilateral switch)			3
9.	LASCR: (Light-activated silicon-controlled rectifier)			3

Contd...

Contd...

S. No.	Devices	Symbols	Characteristics (V–I)	No. of terminals
10.	N-channel MOSFET			3
11.	P-channel MOSFET			3
12.	PUT: (Programmable unijunction transistor)			3
13.	RCT: (Reverse conducting thyristor)			3

4.12 COMPARISON BETWEEN TRANSISTORS AND THYRISTORS

	Transistor	Thyristor
1.	Transistor is a three-layer, two-junction semiconductor device.	Thyristor is a four-layer, three-junction device.
2.	A continuous base current is needed to keep the transistor in conducting state.	Thyristor require gate pulse to make it conducting and thereafter it remain conducting.
3.	At present, voltage and current ratings of transistors are not as high as those of thyristors.	Thyristors with high voltage and current ratings are available.
4.	Power transistors can be used in very high-frequency applications.	Thyristors are used in comparatively low frequency applications.

Transistor	Thyristor
5. Circuits using power transistors will be in small in size and less costly.	Comparatively larger in size and is costlier.
6. When power transistors conduct appreciable current, the forward voltage drop is of the order of 0.3 to 0.8 V.	Forward voltage drop across the device is of the order 1.2 to 2 V.
7. Power transistors have no surge current capacity and can withstand only a low rate of change of current.	Thyristors have surge current rating and can withstand high rate of change of current compared to thyristors.

SOLVED EXAMPLES

Example 4.1 In two transistor analogy of a thyristor, the following data are given:

Gain of p-n-p transistor = 0.4

Gain of n-p-n transistor = 0.5

Gate current of device   = 60 mA

Calculate the value of anode current of the device.

Solution Anode current,

$$I_A = \frac{\alpha_2 I_G + I_{CBO1} + I_{CBO2}}{[1 - (\alpha_1 + \alpha_2)]}$$

where I_{CBO1} and I_{CBO2} are the leakage currents for transistors which are very small and can be neglected and α_1 and α_2 are the gains of transistors.

So

$$I_A = \frac{\alpha_2 I_G}{[1 - (\alpha_1 + \alpha_2)]} = \left[\frac{0.5 \times 60 \times 10^{-3}}{1 - (0.5 + 0.4)}\right] A$$

$$I_A = \frac{0.5 \times 60 \times 10^{-3}}{0.1} = 300 \times 10^{-3} A = 300 \text{ mA}.$$

Example 4.2 Calculate one cycle surge current rating and $I^2 t$ rating of an SCR having half cycle surge current rating of 3000 A for 50 Hz supply.

Solution Let I_{sb} and I be the subcycle and one cycle surge current ratings of the SCR. Equating the energies involved in these, we get

$$I^2 T = I_{sb}^2 \cdot t$$

or

$$I^2 \times \frac{1}{100} = (3000)^2 \times \frac{1}{200}$$

$$I = \frac{3000}{\sqrt{2}} = 2121.32 \text{ A}$$

$$I^2 t \text{ rating} = I^2 \times \frac{1}{2f} = (2121.32)^2 \times \frac{1}{100}$$

$$= 45000 \text{ Amp}^2 \text{ sec.}$$

Example 4.3 The reverse-biased junction capacitance of an SCR is 25 pF. The device can be turned on if the charging current flowing through junction capacitor is 5 mA. Calculate dv/dt capability of the device.

Solution The required dv/dt to produce a charging current of 5 mA is 2×10^8 V/sec. Since a current of 5 mA in layer P_2 will turn on the devices, the dv/dt capability is slightly less than 200 V/μsec.

Example 4.4 An SCR has a continuous current rating I_{av} of 25 and a dynamic resistance R of 1 Ω. If the casing temperature is decreased from 40° to 30°C by efficient cooling, calculate the percentage increase in rating. State the necessary approximations.

Solution It is assumed that the total internal power dissipation P_d is due to forward conduction loss and maximum permissible junction temperature is 125°C. Therefore, P_d of the device is equal to $I_{av}^2 R$. This is proportional to the rise in junction temperature. Let the new current rating be I. Power dissipation is proportional to I^2 since R is the same. Thus, we have

$$25^2 R = 125 - 40 = 85$$
$$I^2 R = 125 - 30 = 95$$
$$\frac{I^2 R}{25^2 R} = \frac{95}{85} \text{ or } I^2 = \frac{25^2 \times 95}{85} = 26.4 \text{ A}$$

Hence, the increase in rating of device is about 6%.

Example 4.5 A thyristor has an $\int i^2 dt$ rating of 15 amp$^2 \cdot$ sec and is being used to supply the circuit shown from a 120 V AC supply when a fault occurs, short circuiting the 10 Ω resistors to earth (Fig. E4.5). What is the shortest fault clearance time to be achieved if the damage to the thyristor is to be prevented?

Solution Worst case fault occurs when voltage is at maximum. Assume that the voltage is at its maximum value for the duration of the fault, then

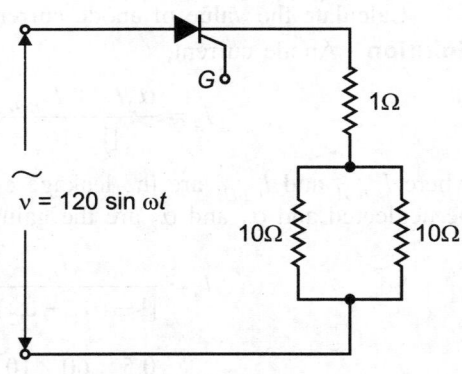

Fig. E4.5 Circuit diagram

$$\int_0^{t_c} i^2 dt = \int_0^{t_c} 120^2 dt = 15$$

Fault clearance time = $15/120^2 = 1.04$ μsec. The thyristor dv/dt rating is therefore chosen to prevent turn on in this way.

Example 4.6 Calculate the value of resistor R across series connected SCRs for equal sharing of voltage.

Solution The forward and reverse leakage current through the SCR varies due to production spread. The worst condition of voltage sharing takes place in two SCRs connected in series, when the forward or reverse leakage currents of the two devices are furthest apart. The above mentioned condition takes place, as seen in Fig. E4.6 when Th1 has $I_{b \text{ min}}$ (leakage current) and Th2 had $I_{b \text{ max}}$ (leakage current). The range of leakage current variation is $\Delta I_b = I_{b \text{ max}} - I_{b \text{ min}}$.

Let E_{max} is the maximum blocking voltage across Th1. From Fig. E4.6, $I_1 > I_2$. Hence, $E_{\text{max}} = I_1 R$.

Further $\qquad E_s = E_{\text{max}} + (n_s - 1) R I_2$

where E_s is the source voltage across the series string,
n_s: the number of thyristors in series string

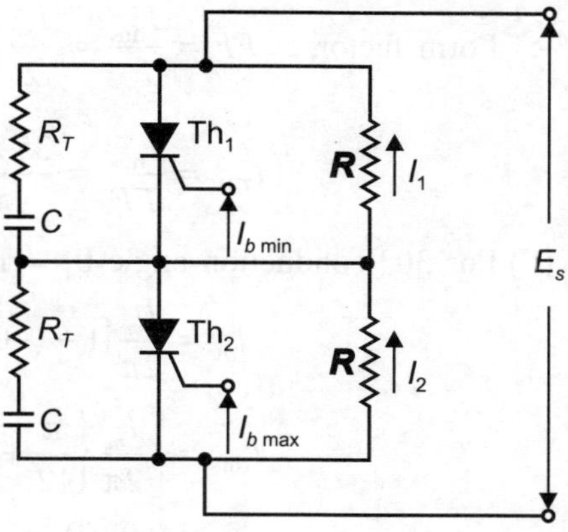

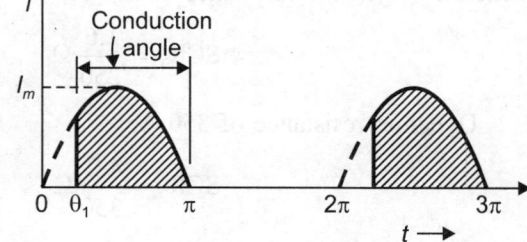

Fig. E4.6 Circuit diagram

$$I_2 = I_1 - \Delta I_b$$

Here $\qquad n = 2$

$\therefore \qquad E_s = E_{max} + I_2 R$

or $\qquad E_s = E_{max} + (I_1 - \Delta I_b)R$

$\qquad\qquad = E_{max} + I_1 R - \Delta I_b R$

$\qquad\qquad = 2E_{max} - \Delta I_b R$

Hence, $\qquad R = \dfrac{2E_{max} - E_s}{\Delta I_b}$ ...(i)

In manufacturer's data sheet only $I_{b\,max}$ is specified, for the worst case, if we assume $I_{b\,min} = 0$, then Equation (i) reduces to

$$R \le \frac{2E_{max} - E_s}{I_{b\,max}}$$

The power dissipated in the equalizing resistor 'R' is minimum when resistor is large. Hence, in calculating R, ΔI_b should be used instead of $I_{b\,max}$.

The transient sharing of voltages is achieved by connecting capacitor across each thyristor as shown. The resistor R_T in series with capacitor is to prevent large discharge current through thyristor during turn on.

Example 4.7 The specification sheet for an SCR gives maximum rms on-state current is 35 A. If the thyristor is used in a resistive circuit, calculate average on-state current rating for half sine wave current for conduction angles of: (i) 180°, and (ii) 30°.

Solution For half sine wave of current as shown in Fig. E4.7

Fig. E4.7 Waveform of a thyristor

$$I_{av} = \frac{1}{2\pi} \int_{\theta_1}^{\pi} I_m \sin\theta \, d\theta = \frac{I_m}{2\pi}(1 + \cos\theta_1)$$

$$I_{rms} = \left[\frac{1}{2\pi} \int_{\theta_1}^{\pi} I_m^2 \sin^2\theta \, d\theta \right]^{1/2} = \left[\frac{I_m^2}{2\pi} \left\{ \frac{\pi - \theta_1}{2} + \frac{1}{4}\sin 2\theta_1 \right\} \right]^{1/2}$$

(i) For 180° conduction angle,

$\qquad \theta_1 = 0°$

$$I_{av} = \frac{I_m}{2\pi}(1 + \cos 0°) = \frac{I_m}{\pi}$$

$$I_{rms} = \left[\frac{I_m^2}{2\pi} \left\{ \frac{\pi}{2} + \frac{1}{4}(0) \right\} \right]^{1/2} = \frac{I_m}{2}$$

Form factor, $FF = \dfrac{I_{\text{rms}}}{I_{av}} = \dfrac{I_m}{2} \cdot \dfrac{\pi}{I_m} = \dfrac{\pi}{2}$

$$I_{TAV} = \dfrac{I_{\text{rms}}}{FF} = \dfrac{35 \times 2}{\pi} = 22.28 \text{ A}$$

(ii) For 30° conduction angle $\theta_1 = 150°$

$$I_{av} = \dfrac{I_m}{2\pi}[1 + (-0.866)] = 0.02132 I_m$$

$$I_{\text{rms}} = \left[\dfrac{I_m}{2\pi}\left\{\dfrac{\pi}{12} + \dfrac{1}{4}(-0.866)\right\}\right]^{1/2}$$

$$= 0.0849 I_m$$

Form factor, $FF = \dfrac{I_{\text{rms}}}{I_{av}} = \dfrac{0.849 I_m}{0.02132 I_m} = 3.981$

$$I_{TAV} = \dfrac{35}{3.981} = 8.78 \text{ A}.$$

Example 4.8 It is required to operate 250 A, SCR in parallel with 350 A, SCR with their respective on-state voltage drops of 1.6 V and 1.2 V. Calculate the value of resistance to be inserted in series with each SCR so that they share the total load of 600 A in proportion to their current ratings.

Solution Dynamic resistance of 250 A,

$$SCR_1 = \dfrac{1.6}{250}\,\Omega$$

Dynamic resistance of 350 A,

$$SCR_2 = \dfrac{1.2}{350}\,\Omega$$

Let R' be the resistance inserted in series with each SCR.
With this, current shared by SCR_1

$$= 600\,\dfrac{\dfrac{1.6}{250} + R}{\text{Total resistance}} = 250 \qquad\qquad \text{...(i)}$$

Current shared by SCR_2 $= 600\,\dfrac{\dfrac{1.2}{350} + R}{\text{Total resistance}} = 350 \qquad\qquad \text{...(ii)}$

From Equations (i) and (ii)

$$\dfrac{\dfrac{1.6}{250} + R}{\dfrac{1.2}{350} + R} = \dfrac{250}{350} = \dfrac{5}{7} \;\Rightarrow\; R = 0.0063\,\Omega$$

The resistance to be inserted in series with each SCR is $0.00633\,\Omega$.

Example 4.9 A thyristor having equivalent capacitance of depletion layer of reversed biased junction as 30 pF, can be fixed with dv/dt of 150 V/μsec. Calculate the capacitive current flowing through the junction.

Solution

$$I_c = C \, dv/dt$$

Given

$$dv/dt = 150 \text{ V/μsec} = 150 \times 10^6 \text{ V/sec}$$

$$C = 30 \text{ pF} = 30 \times 10^{-12} \text{ Farad}$$

$$I_c = 30 \times 10^{-12} \times 150 \times 10^6 \text{ A} = 4.5 \text{ mA}.$$

Example 4.10 The peak inverse voltage of the thyristor is 1500 V for a six-phase, thyristor converter connected to 415 V. Calculate the voltage safety factor.

Solution Voltage safety factor,

$$V_f = V_{PIV}/\sqrt{2} \text{ V}$$

V is the rms value of operating voltage

Given

$$V_{PIV} = 1500 \text{ V}, \, V = 415 \text{ V}$$

$$V_f = \frac{1500}{\sqrt{2} \times 415} = 2.56.$$

Example 4.11 A silicon-controlled rectifier is rated for 650 V, PIV. Calculate the voltage for which the device can be operated if the voltage safety factor is 2.

Solution Voltage safety factor,

$$V_f = \frac{PIV}{\sqrt{2} \times \text{rms value of operating voltage}}$$

$$\text{rms value of operating voltage} = \frac{PIV}{V_f \times \sqrt{2}} = \frac{650}{2 \times 1.41} = 230.4 \text{ V}.$$

Example 4.12 An SCR with 650 V, PIV rating is connected in series with an electrical load consisting resistance of 160 Ω and inductance of 39 mH. Calculate the current flowing through the load when connected to 50 Hz supply with voltage safety factor 2.

Solution Voltage safety factor,

$$V_f = \frac{PIV}{\sqrt{2} \times \text{operating voltage}}$$

$$2 = \frac{650}{\sqrt{2} \times V}$$

or   operating voltage,

$$V = \frac{650}{\sqrt{2} \times 2} = 230 \text{ V}$$

Given

$$R = 16 \, \Omega$$

$$L = 39 \text{ mH,}$$

or

$$X_L = 2\pi f L = 2 \times \pi \times 50 \times 0.039 = 12.26 \, \Omega$$

Load impedance

$$= \sqrt{16^2 + 12.26^2} = 20.05 \, \Omega$$

Current through the load

$$= \frac{V}{Z_L} = \frac{230}{20.5} = 11.5 \text{ A}.$$

Example 4.13 Find the power rating of SCR if the rms value of current through SCR is 20 A when connected across ($415 \sin 314t$) volts supply.

Solution Supply voltage $= 415 \sin 314t$

Peak Inverse voltage $= \sqrt{3}V_m = \sqrt{3} \times 415 = 718.8$ V

rms value of current $= 20$ A

$\therefore$ Average value of current $= \dfrac{20}{1.11} = 18$ A

Power rating of SCR $= (PIV) \times (\text{Average Value of Current}) \cdot \text{Watt}$
$= 718.8 \times 18 = 12939$ Watt
$= 12.939$ kW.

EXERCISE

Multiple Choice Questions

1. Thyristors are basically:
   (A) SCRs
   (B) Triacs
   (C) both SCRs and Triacs
   (D) all pnpn devices

2. The number of p-n junctions in a thyristor is/are:
   (A) 1   (B) 2
   (C) 3   (D) 4

3. When anode is positive with respect to cathode in an SCR, the number of blocked p-n junctions is/are:
   (A) 1   (B) 2
   (C) 4   (D) none of these

4. When cathode is positive with respect to anode in an SCR, the number of blocked p-n junctions is/are:
   (A) 1   (B) 2
   (C) 3   (D) 4

5. In a thyristor, anode current is made up of:
   (A) electrons only
   (B) electrons or holes
   (C) electrons and holes
   (D) none of these

6. A thyristor, when triggered, will change from forward blocking state to conduction-state, if its anode to cathode voltage is equal to:
   (A) peak repetitive off-state forward voltage
   (B) peak working off-state forward voltage
   (C) peak working off-state reverse voltage
   (D) none of these

7. An SCR can be brought to forward conducting state with gate-circuit open when the applied voltage exceeds:
   (A) the forward breakover voltage
   (B) reverse breakdown voltage
   (C) 1.5 V
   (D) none of these

8. In a thyristor, holding current is:
   (A) more than latching current I_L
   (B) less than latching current
   (C) equal to latching current
   (D) very small

9. When a thyristor gets turned on, the gate drive:
   (A) should not be removed as it will turn off the SCR
   (B) may or may not be removed
   (C) should be removed
   (D) should be removed in order to avoid increased losses and higher junction temperature

10. An SCR is a:
    (A) two-layer two-junction device
    (B) three-layer two-junction device
    (C) four-layer three-junction device
    (D) four-layer four-junction device

11. An SCR has:
    (A) one terminal   (B) two terminals
    (C) three terminals   (D) four terminals

12. An SCR can be operated:
    (A) only on reverse-biased condition
    (B) only on forward-biased condition
    (C) both forward and reverse-biased conditions
    (D) none of these

13. While constructing a thyristor, silicon wafer together with the diffused discrete layers is mounted on a supporting base made of:
    (A) copper   (B) molybdenum
    (C) aluminium   (D) steel

14. Normally the value of turn on time of commonly used thyristors is approximately:
    (A) 10 msec   (B) 100 msec
    (C) 50 μsec   (D) 200 μsec

15. The capacitive current flowing through the junction of a thyristor is:
    (A) $I_c = C\ dv/dt$ (B) $I_C = \dfrac{1}{C} \cdot \dfrac{dv}{dt}$
    (C) $I_c = C \cdot di/dt$ (D) none of these

16. For normal SCRs, turn on time is:
    (A) less than turn off time, t_q
    (B) more than turn off time, t_q
    (C) equal to turn off time, t_q
    (D) none of these

17. The forward voltage drop during SCR on-state is 1.5 V. This voltage drop:
    (A) remains constant and is independent of load current
    (B) increases slightly with load current
    (C) decreases slightly with load current
    (D) varies linearly with load current

18. A thyristor can be termed as:
    (A) DC switch
    (B) AC switch
    (C) both (A) or (B) are correct
    (D) square-wave switch

19. The SCR ratings di/dt in A/μsec and dv/dt in V/μsec, may vary respectively between:
    (A) 20 to 500, 10 to 100
    (B) both 20 to 500
    (C) both 10 to 100
    (D) 50 to 300, 20 to 500

20. In a thyristor:
    (A) latching current I_L is associated with turn off process and holding current I_H with turn on process
    (B) both I_L and I_H are associated with turn on process
    (C) I_H is associated with turn off process and I_L with turn on process
    (D) both I_L and I_H are associated with turn on process

21. Once SCR starts conducting a forward current, its gate looses control over:
    (A) anode circuit voltage only
    (B) anode circuit current only
    (C) anode circuit voltage and current
    (D) none of these

22. On-state voltage drop across a thyristor used in a 250 V supply system is of the order of:
    (A) 100–110 V   (B) 240–250 V
    (C) 1–1.5 V   (D) none of these

23. In a thyristor, ratio of latching current to holding current is:
    (A) 0.4   (B) 1.0
    (C) 2.5   (D) none of these

24. During forward blocking state, a thyristor is associated with:
    (A) large current, low voltage
    (B) low current, large voltage
    (C) medium current, large voltage
    (D) none of these

25. Gate characteristic of a thyristor:
    (A) is a straight line passing through origin
    (B) is of the type $V_g = a + bI_g$
    (C) is a curve between V_g and I_g
    (D) has a spread between two curves of $V_g - I_g$

26. In an SCR, anode current flows over a narrow region near the gate during:
    (A) delay time, t_d
    (B) rise time, t_r and spread time, t_p
    (C) t_d and t_p
    (D) t_d and t_r

27. Turn on time for an SCR is 10 μsec. If an inductance is inserted in the anode circuit, then the turn on time will be:
    (A) 10 μsec

(B) less than 10 µsec

(C) more than 10 µsec

(D) about 10 µsec

28. Turn off time of an SCR is measured from the instant:

(A) anode current becomes zero

(B) anode voltage becomes zero

(C) anode voltage and anode current become zero at the same time

(D) gate current becomes zero

29. A forward voltage can be applied to an SCR after its:

(A) anode current reduces to zero

(B) gate recovery time

(C) reverse recovery time

(D) anode voltage reduces to zero

30. For an SCR with turn on time of 5 micro-second, and ideal trigger pulse should have:

(A) short rise time with pulse width = 3 µsec

(B) long rise time with pulse width = 6 µsec

(C) short rise time with pulse width = 6 µsec

(D) long rise time with pulse width = 3 µsec

31. Turn on time of an SCR in series with RL circuit can be reduced by:

(A) increasing circuit resistance R

(B) decreasing R

(C) increasing circuit inductance L

(D) decreasing L

32. Turn on time of an SCR can be reduced by using a:

(A) rectangular pulse of high amplitude and narrow width

(B) rectangular pulse of low amplitude and wide width

(C) triangular pulse

(D) trapezoidal pulse

33. Static $V–I$ characteristics of an SCR with different gate drives applied to the gate are indicated by:

(A) $I_{g2} > I_{g1} > I_{g0}$ (B) $V_{g2} > V_{g1} > V_{g0}$

(C) $P_{g2} > P_{g1} > P_{g0}$ (D) either (A) or (B)

34. In a thyristor, the magnitude of anode current will:

(A) increase, if gate current is increased

(B) decrease, if gate current is decreased

(C) increase, if gate current is decreased

(D) not change with any variation in gate current

35. The average on-state current for an SCR is 20 A for a conduction angle of 120°. Its average on-state current for 60° conduction angle would be:

(A) 20 A   (B) 10 A

(C) less than 20 A   (D) none of these

36. The average on-state current for an SCR is 20 A for a resistive load. If an inductance of 5 mH is included in the load, then average on-state current would be:

(A) more than 20 A

(B) less than 20 A

(C) 15 A

(B) none of these

37. Specification sheet for an SCR gives its maximum rms on-state current as 35 A. This rms rating for a conduction angle of 120° would be:

(A) more than 35 A

(B) less than 35 A

(C) 35 A

(D) none of these

38. Surge current rating of an SCR specifies the maximum:

(A) repetitive current with sine wave

(B) non-repetitive current with rectangular wave

(C) non-repetitive current with sine wave

(D) repetitive current with triangular wave

39. The di/dt rating of an SCR is specified for its:

(A) decaying anode current

(B) decaying gate current

(C) rising gate current

(D) rising anode current

40. For an SCR, dv/dt protection is achieved through the use of:

(A) RL in series with SCR

(B) RL across SCR

(C) L in series with SCR

(D) all of these

41. The number of gates in a silicon-controlled switch is:

(A) one   (B) two

(C) three   (D) four

42. A light-activated silicon-controlled rectifier (LASCR) is a:
    (A) two-layer two-junction device
    (B) four-layer three-junction device
    (C) three-layer two-junction devices
    (D) none of these

43. Diac is:
    (A) three-layer two-junction device
    (B) four-layer two-junction device
    (C) four-layer three-junction device
    (D) none of these

44. Diac has:
    (A) one terminal   (B) two terminals
    (C) three terminals   (D) four terminals

45. Diac can be fired by:
    (A) positive half cycle of the supply only
    (B) negative half cycle of the supply only
    (C) positive and negative half cycles of the supply
    (D) none of these

46. Diac can be used to:
    (A) trigger a triac
    (B) protect a triac
    (C) increase the efficiency of a triac
    (D) none of these

47. Breakover voltage for commonly used diacs is about:
    (A) 16 V   (B) 32 V
    (C) 48 V   (D) 60 V

48. Triac is a:
    (A) three-layer two-junction device
    (B) four-layer three-junction device
    (C) four-layer four-junction device
    (D) none of these

49. Triac can be considered as:
    (A) two SCRs connected in antiparallel with a common gate
    (B) two SCRs connected in parallel with common gate
    (C) two transistors connected in antiparallel
    (D) none of these

50. Triac has:
    (A) one terminal   (B) two terminals
    (C) three terminals   (D) four terminals

51. Triac can be fired by:
    (A) positive half cycle of the supply only
    (B) negative half cycle of the supply only

(C) positive and negative half cycles of the supply
(D) none of these

52. Diac–triac built-in the same chip is called:
    (A) ignition   (B) thermistor
    (C) thyristor   (D) quadrac

53. Triac can be used as a static switch for frequencies up to:
    (A) 1 kHz   (B) 10 kHz
    (C) 100 kHz   (D) 400 kHz

54. LASCR is a:
    (A) vacuum device   (B) unilateral device
    (C) bilateral device   (D) none of these

55. Which of the following devices have similar V–I characteristics in shape:
    (A) SCRs and diacs
    (B) SCSs and triacs
    (C) SCRs and UJTs
    (D) SCRs, SCSs, SUSs and LASCRs

56. In a GTO, anode current begins to fall when gate current:
    (A) is negative peak at time $t = 0$
    (B) is negative peak at t = storage time + fall time
    (C) is negative peak at t = storage time
    (D) none of these

57. Latching current of thyristor is associated with:
    (A) turn on process
    (B) turn off process
    (C) turn on and turn off proceses
    (D) none of these

58. Ratio of latching current to holding current of thyristor is generally of the order of:
    (A) two to three times
    (B) four to five times
    (C) five to seven times
    (D) none of these

59. If a voltage is applied to the thyristor, a displacement of charges occurs. Which of the following movements takes place?
    (A) The holes to the positive electrode
    (B) The electrons to the negative electrode
    (C) The holes to the negative electrode and the electrons to the positive electrode
    (D) None of these

60. The thyristor is turned off:
    (A) by switching off the voltage applied to the principal circuit and changing the thyristor into the reverse blocking state
    (B) by switching off the voltage applied to the gate circuit only
    (C) by changing the thyristor into forward blocking state
    (D) none of these

61. The magnitude of anode current can be controlled in the case of the thyristor only by:
    (A) changing the amplitude of trigger pulse
    (B) changing the width of the trigger pulse
    (C) varying the timing of the trigger pulse with respect to the positive half cycle of the applied AC voltage
    (D) none of these

62. For the transition from the forward blocking to the 'on' state the thyristor must be triggered.

This takes place by feeding a gate current into the gate electrode for:
    (A) the duration of the whole triggering process
    (B) the duration of the whole triggering and conducting process
    (C) a fraction of the period of the conducting process
    (D) none of these

63. From the following which one is an operational state of the thyristor?
    (A) Triggering   (B) Firing
    (C) Regulating   (D) Reverse blocking

64. Two operational states of a thyristor are: when anode is positive with respect to cathode:
    (A) reverse blocking and conducting
    (B) reverse blocking and forward blocking
    (C) conducting and forward blocking
    (D) none of these

ANSWER KEY

1. (D)	**2.** (C)	**3.** (A)	**4.** (B)	**5.** (C)	**6.** (B)	**7.** (A)	**8.** (B)
9. (D)	**10.** (C)	**11.** (C)	**12.** (B)	**13.** (D)	**14.** (C)	**15.** (A)	**16.** (A)
17. (B)	**18.** (A)	**19.** (B)	**20.** (C)	**21.** (C)	**22.** (C)	**23.** (C)	**24.** (B)
25. (D)	**26.** (D)	**27.** (C)	**28.** (A)	**29.** (B)	**30.** (C)	**31.** (D)	**32.** (A)
33. (A)	**34.** (D)	**35.** (C)	**36.** (A)	**37.** (C)	**38.** (C)	**39.** (D)	**40.** (B)
41. (B)	**42.** (B)	**43.** (B)	**44.** (B)	**45.** (C)	**46.** (A)	**47.** (B)	**48.** (B)
49. (A)	**50.** (C)	**51.** (C)	**52.** (D)	**53.** (D)	**54.** (B)	**55.** (D)	**56.** (C)
57. (A)	**58.** (A)	**59.** (C)	**60.** (A)	**61.** (C)	**62.** (C)	**63.** (D)	**64.** (C)

Review Questions

1. What are the advantages of semiconductor devices over conventional electrical devices?
2. What is the V–I characteristic of thyristor?
3. What is an on-state condition of thyristor?
4. What is latching current of thyristors?
5. Draw the symbols for the following thyristors: (a) SCR, (b) diac, (c) triac, (d) SCS, (e) SUS, and (f) LASCR.
6. Explain how the anode, cathode and gate connections are made in a thyristor?
7. Explain why an SCR is developed only by using silicon and germanium?
8. Explain why an SCR is operated only in the forward-biased condition?
9. Explain the two transistor analogy of an SCR.
10. Compare thyristor with transistor.
11. What are the types of thyristors?
12. What is the difference between an SCR and triacs?

13. What is turn off characteristics of thyristor?

14. What are the advantages and disadvantages of GTOs?

15. What are the advantages and disadvantages of SITHs?

16. What are the advantages and disadvantages of RCTs?

17. What are the advantages and disadvantages of LASCRs?

18. What is the common technique for voltage sharing of series connected thyristors?

19. What are the common techniques for current sharing of parallel-connected thyristors?

20. What is the effect of reverse recovery time on the transient voltage sharing of parallel connected thyristors?

21. What is a derating factor of series connected thyristors?

22. List the names of the methods adopted for triggering a thyristor.

5

Thyristor Triggering Devices

5.0 INTRODUCTION

Trigger circuits for SCRs and triacs employ a wide variety of trigger or gate switching devices. Since the SCR was introduced in the late 1950s, gate switching techniques and devices have been in a constant state of development. Multilayer thyristor research and advances in technology have resulted in a sophisticated family of two, three and four terminal devices. Variations in fabricating p-n junctions have made possible the production of devices with varied characteristics and operational capabilities.

The present trend in the AC and DC power control and switching field appears to be focusing on the use of optical or optoisolator triggering employing digital and/or automated control. The introduction of the microprocessor, and related digital control circuits, has opened a new and promising era. Many of the devices covered in this chapter will be used in automated control systems. Others will be used as replacement items in existing control systems.

5.1 TRIGGERING DEVICES

Switching a thyristor ON, is called its *triggering* or *firing*. There can be many methods by which a thyristor can be triggered such as gate triggering, radiation triggering and voltage triggering, etc. which are discussed in subsequent chapters. To achieve this, various triggering devices are employed like:

1. Unijunction transistor (UJT)
2. Programmable unijunction transistor (PUT)
3. Bidirectional diode (DIAC)
4. Silicon-controlled switch (SCS)
5. Silicon unilateral switch (SUS)
6. Silicon bilateral switch (SBS)
7. Shockley diode
8. Opto-isolator

5.2 UNIJUNCTION TRANSISTOR

The most applied method of producing pulses for gate triggering is by means of unijunction transistor (UJT) relaxation oscillator.

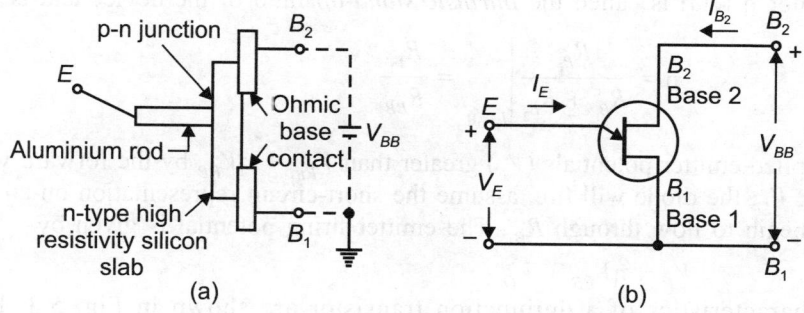

Fig. 5.1 UJT and symbol

The UJT is a three terminal device having the basic construction shown in Fig. 5.1 (a). A slab of lightly doped (increased resistance characteristic) n-type silicon material has two base contacts attached to both ends of one surface and an aluminium rod alloyed to the opposite surface. The p-n junction of the device is formed at the boundary of the aluminium rod and the n-type silicon slab. The single p-n junction accounts for the terminology unijunction. It is also called a duo (double) base diode due to the presence of two base contacts. The aluminium rod is alloyed to the silicon slab at a point closer to the base 2-contact than the base 1 contact and that the base 2 terminal is made positive with respect to the base 1 terminal by V_{BB} volts. The effect of each will become evident in the following text.

The symbol for the unijunction transistor is shown in Fig. 5.1 (b) with emitter leg drawn at an angle to the vertical line representing the slab of n-type material. The arrowhead is pointing in the direction of conventicnal current (hole) flow when the device is in the forward-biased, active or conducting state.

The circuit equivalent of the UJT is shown in Fig. 5.2 with two resistors (one fixed, one variable) and a single diode. The resistance R_{B_1} is shown as a variable resistor since its magnitude will vary with the current I_E. In fact, for a representative unijunction transistor, R_{B_1} may vary from 5 kΩ down to 50 Ω for a corresponding change of I_E from 0 to 50 μA. The interbase resistance R_{BB} is the resistance of the device between terminals B_1 and B_2 when $I_E = 0$. is given by the equation:

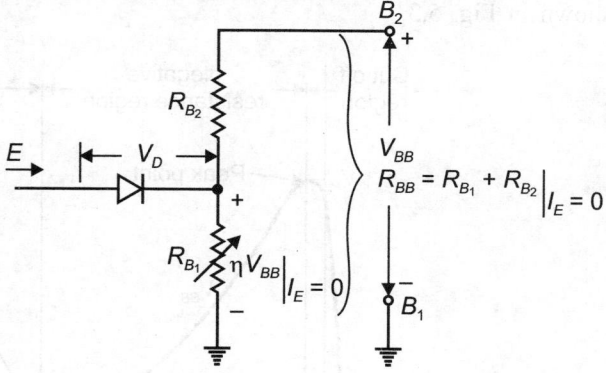

Fig. 5.2 The circuit equivalent of the UJT with two resistors

$$R_{BB} = (R_{B_1} + R_{B_2})|_{I_E = 0} \qquad ...(5.1)$$

(R_{BB} is typically within the range of 4 to 10 kΩ. Then position of the aluminium rod of Fig. 5.1 (a) will determine the relative values of R_{B_1} and R_{B_2} with $I_E = 0$. The magnitude of $V_{R_{B_1}}$ (with $I_E = 0$) is given by

$$V_{R_{B_1}} = \frac{R_{B_1} V_{BB}}{R_{B_1} + R_{B_2}} = \eta V_{BB} \Big|_{I_E=0} \qquad ...(5.2)$$

The letter η (eta) is called the *intrinsic stand-off* ratio of the device and is defined by

$$\eta = \frac{R_{B_1}}{R_{B_1} + R_{B_2}}\bigg|_{I_E=0} = \frac{R_{B_1}}{R_{BB}} \qquad \qquad ...(5.3)$$

For applied emitter potentials (V_E) greater than $V_{R_{B_1}} = \eta V_{BB}$ by the forward voltage drop of the diode, V_D, the diode will fire, assume the short-circuit representation on an ideal basis, and I_E will begin to flow through R_{B_1}. The emitter firing potential is given by

$$V_P = \eta V_{BB} + V_D \qquad \qquad ...(5.4)$$

The characteristics of a unijunction transistor are shown in Fig. 5.3. For emitter potentials to the left of the peak point the magnitude of I_E is never greater than I_{EO}. The current I_{EO} corresponds very closely with the reverse leakage current I_{CO} of the convention bipolar transistor. This region is called the cut off region. Once conduction is established at $V_E = V_P$, the emitter potential V_E will drop with increase in I_E. This corresponds exactly with the decreasing resistance R_{B_1} for increasing current I_E. This device, therefore, has a *negative resistance* region which is stable enough to be used with a great deal of reliability in the areas of application listed earlier. Eventually, the valley point will be reached, and a further increase in I_E will place the device in the saturation region.

The decrease in resistance in the active region is due to the holes injected into the n-type slab from the aluminium p-type rod when conduction is established. The increased hole content in the n-type material will result in an increase in the number of free electrons in the slab, producing an increase in conductivity (G) and a corresponding drop in resistance. Three other important parameters for the unijunction transistor are I_P, V_V and I_V, which are shown in Fig. 5.3.

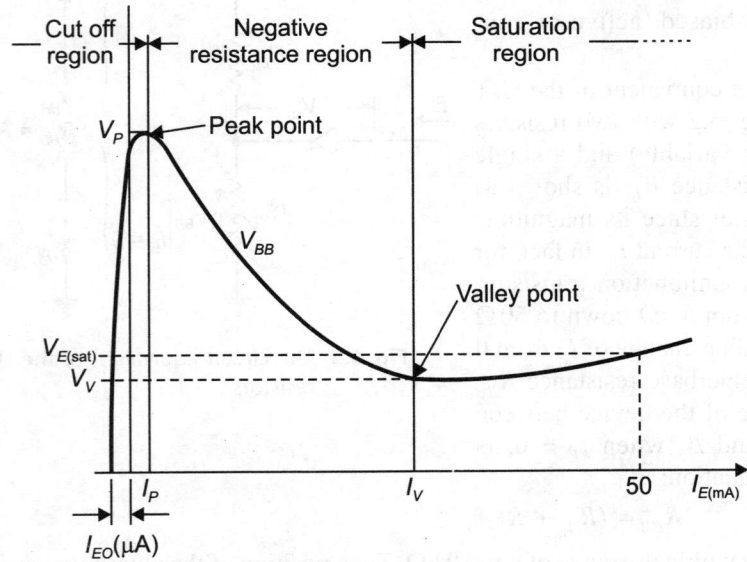

Fig. 5.3 Three other important parameters for the unijunction transistor I_P, V_V, and I_V

The emitter characteristics are provided in Fig. 5.4. The intersection of each curve with the vertical axis is the corresponding value of V_P. For fixed values of η and V_D, the magnitude of V_P will vary as V_{BB}, given by

$$V_P = \eta V_{BB} + V_D$$

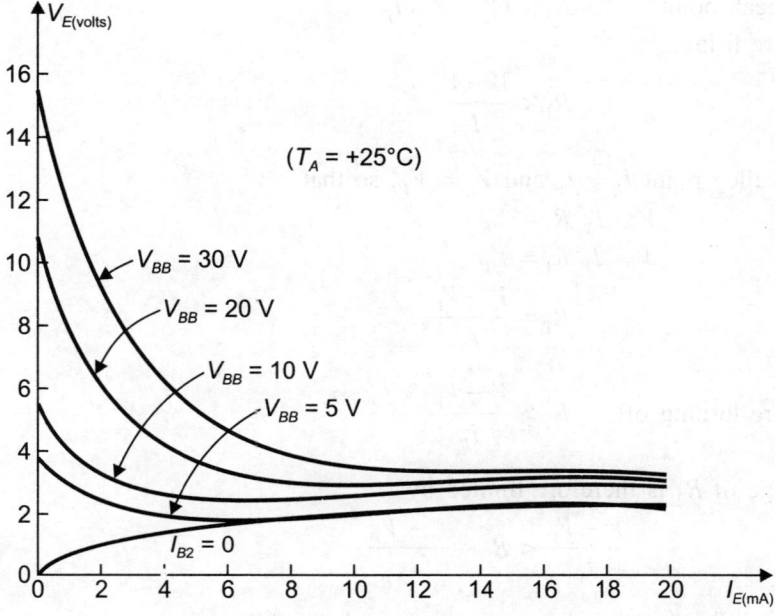

Fig. 5.4 Emitter characteristics

One common application of the UJT is in the triggering of other devices such as the SCR. The basic elements of such a triggering circuit are shown in Fig. 5.5 (a). The resistor R_1 must be so chosen to ensure that the load line determined by R_1 passes through the device characteristics in the negative resistance region, i.e. to the right of the peak point but to the left of the valley point as shown in Fig. 5.5 (b). If the load line fails to pass to the right of the peak point, the device cannot turn on. An equation for R_1 that will ensure a turn on condition can be established if we consider the peak point at which $I_{R_1} = I_P$ and $V_E = V_P$. The equality $I_{R_1} = I_P$ is valid since the charging current of the capacitor, at this instant, is zero. In other words, at this instant the capacitor changing from a charging to a discharging state. Then

$$V - I_{R_1} R_1 = V_E$$

or

$$R_1 = (V - V_E)/I_{R_1}$$

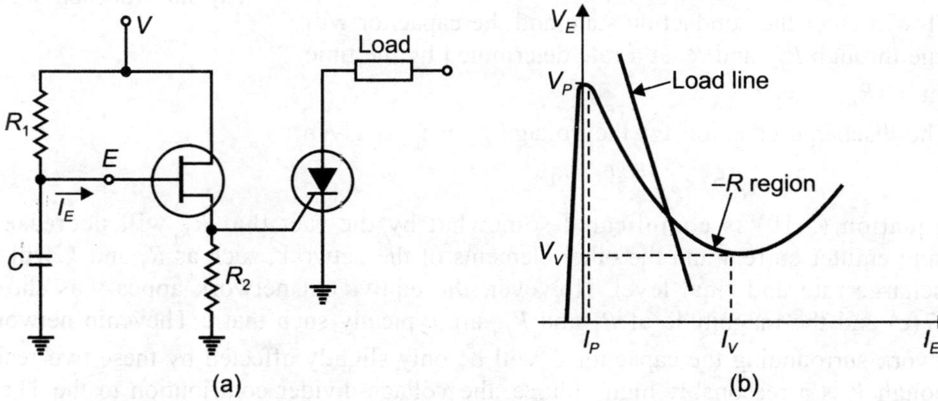

Fig. 5.5 Triggering circuit and characteristics

At the peak point, $\quad R_1 = (V - V_P)/I_P$

To ensure firing,

$$R_1 < \frac{V - V_P}{I_P} \qquad \qquad ...(5.5)$$

At the valley point $I_E = I_V$ and $V_E = V_V$, so that

$$V - I_{R_1} R_1 = V_E$$

becomes $\qquad V - I_V R_1 = V_V$

and $\qquad\qquad R_1 = \dfrac{V - V_V}{I_V}$

To ensure turning off, $\quad R_1 > \dfrac{V - V_V}{I_V} \qquad\qquad ...(5.6)$

The range of R_2 is therefore limited by

$$\frac{V - V_V}{I_V} < R_1 < \frac{V - V_P}{I_P} \qquad\qquad ...(5.7)$$

Resistance R_1 must be chosen small enough to ensure that the SCR is not turned on by the voltage V_{R_2} of Fig. 5.6, when $I_E \cong 0$ A.

Voltage, $\qquad V_{R_2} \cong \left. \dfrac{R_2 V}{R_2 + R_{BB}} \right|_{I_E \cong 0 \, A} \qquad ...(5.8)$

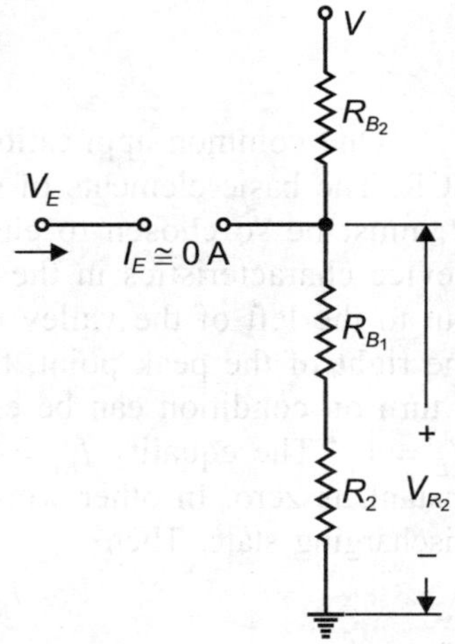

The capacitor C will determine the time interval between triggering pulses and the time span of each pulse.

At the instant the DC supply voltage V is applied, the voltage $v_E = v_C$ will charge toward V volts from V_V as shown in Fig. 5.7 with a time constant $\tau = R_1 C$.

The general equation for the charging period is

$$v_C = V_V + (V - V_V)(1 - e^{t/R_1 C}) \qquad ...(5.9)$$

As seen in Fig. 5.7, the voltage across R_2 is determined by Equation (5.8) during this charging period. When $v_C = v_E = V_P$, the UJT will enter the conduction state and the capacitor will discharge through R_{B_1} and R_2 at a rate determined by the time constant $= (R_{B_1} + R_2) C$.

Fig. 5.6 Function of UJT

The discharge equation for the voltage $v_C = v_E$ is given by

$$v_C \cong V_P e^{-t/(R_{B_1} + R_1) C} \qquad\qquad ...(5.10)$$

Equation (5.10) is complicated somewhat by the fact that R_{B_1} will decrease with increasing emitter current and the other elements of the network, such as R_1 and V will affect the discharge rate and final level. However, the equivalent network appears as shown in Fig. 5.7 (c) and the magnitude of R_1 and R_{B_2} are typically such that a Thevenin network for the network surrounding the capacitor C will be only slightly affected by these two resistors. Eventhough V is a reasonably high voltage, the voltage-divider contribution to the Thevenin voltage can be ignored on an approximate basis.

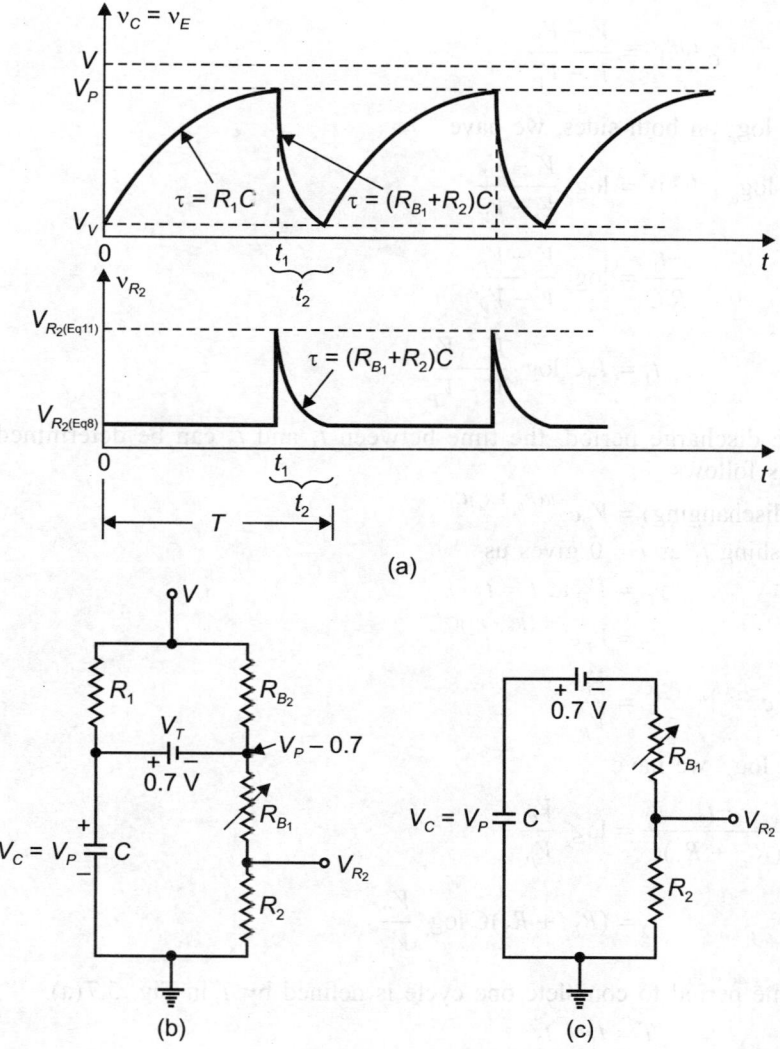

Fig. 5.7 Waveforms of UJT

Using the reduced equivalent of Fig. 5.7 (c), the discharge phase will result in the following approximation for the peak value of V_{R_2}:

$$V_{R_2} \cong \frac{R_2 (V_P - 0.7)}{R_2 + R_{B_1}} \qquad \qquad ...(5.11)$$

The period t_1 of Fig. 5.7 (a) can be determined in the following manner:

$$v_C \text{ (changing)} = V_V + (V - V_V) (1 - e^{-t/R_1C})$$
$$= V_V + V - V_V - (V - V_V) e^{-t/R_1C}$$
$$= V - (V - V_V) e^{-t/R_1C}$$

when $v_C = V_P$, $t = t_1$ and $V_P = V - (V - V_V) e^{-t_1/R_1C}$, or

$$\frac{V_P - V}{V - V_V} = -e^{-t_1/R_1C}$$

and
$$e^{-t_1/R_1C} = \frac{V - V_P}{V - V_V}$$

Taking $\log_e$ on both sides, we have

$$\log_e e^{-t_1/R_1C} = \log_e \frac{V - V_P}{V - V_V}$$

and
$$\frac{-t_1}{R_1C} = \log_e \frac{V - V_P}{V - V_V}$$

with
$$t_1 = R_1C \log_e \frac{V - V_V}{V - V_P} \qquad \qquad ...(5.12)$$

For the discharge period, the time between t_1 and t_2 can be determined from Equation (5.10) as follows:

$$v_C \text{ (dischanging)} = V_P e^{-t/(R_{B_1} + R_2)C}$$

Establishing t_1 as $t = 0$ gives us

$$v_C = V_V \text{ at } t = t_2$$

and
$$V_V = V_P e^{-t_2/(R_{B_1} + R_2)C}$$

or
$$e^{-t_2/(R_{B_1} + R_2)C} = \frac{V_V}{V_P}$$

Taking $\log_e$, we have

$$\frac{-t_2}{(R_{B_1} + R_2)C} = \log_e \frac{V_V}{V_P}$$

$$t_2 = (R_{B_1} + R_2)C \log_e \frac{V_P}{V_V} \qquad \qquad ...(5.13)$$

The time period to complete one cycle is defined by T in Fig. 5.7(a).

$$T = t_1 + t_2 \qquad \qquad ...(5.14)$$

If the SCRs were dropped from the configuration, the network would behave as a relaxation oscillator, generating the waveform of Fig. 5.7(a). The frequency of oscillation is determined by

$$f_{osc} = \frac{1}{T} \qquad \qquad ...(5.15)$$

In many systems $t_1 \gg t_2$ and

$$T \cong t_1 = R_1 C \log_e \frac{V - V_V}{V - V_P}$$

As $V \gg V_V$ in many instances,

$$T \cong t_1 = R_1 C \log_e \frac{V}{V - V_P}$$

$$= R_1 C \log_e \frac{1}{1 - V_P/V}$$

Since $\eta = V_P/V$. If we ignore the effects of V_D in Equation (5.4) and $\eta = V_P/V$, we have

$$T \cong R_1 C \log_e \frac{1}{1 - \eta}$$

$$f \cong \frac{1}{R_1 C \log_e[1/(1 - \eta)]} \qquad \qquad ...(5.16)$$

5.3 CHARACTERISTICS AND APPLICATIONS OF UJT

The important features of UJT are:

1. A stable triggering voltage (V_P) which is a fixed fraction of applied interbase voltage.
2. A very low value of firing current.
3. A negative resistance characteristics.
4. A high pulse current capability.
5. Low cost.

Uses: The above characteristics make the UJT advantageous in oscillators, timing circuits, pulse generators, a trigger circuit or as a saw tooth generator. One common application of the UJT is in the triggering of SCR.

5.3.1 Complementary Unijunction Transistor (CUJT)

CUJT characteristics are like those of a standard UJT except that the currents and voltages applied to it are of opposite polarity. The opposite polarity has been chosen so that standard n-p-n planar passivated transistor processing techniques can be used. CUJTs can be used in most applications that use standard UJTs. Their unique stability and uniform properties make them ideal for stable oscillators, timers, and frequency dividers. Figures 5.8 (a) and (b) shows the standard and complementary unijunction transistor.

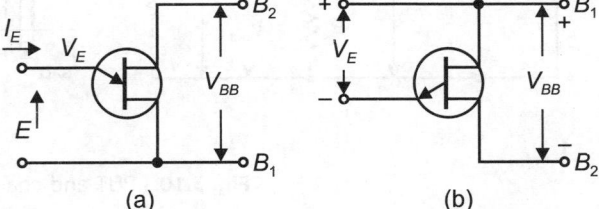

(a)   (b)

Fig. 5.8 CUJT circuit

5.4 PROGRAMMABLE UNIJUNCTION TRANSISTOR (PUT)

The programmable unijunction transistor (PUT) is similar in operational characteristics to the UJT. It is a four-layer three terminal device designated as anode, gate and cathode. This recent addition to the thyristor family looks and acts like an improved UJT. However, a microscopic examination of the internal structure of the PUT shows it to be another four-layer pn-pn silicon device [Fig. 5.9 (a)]. The symbol for the device is shown in Fig. 5.9 (b).

The major advantage of the PUT over UJT is the capability of programming key operating parameters with external resistors. A voltage divider network connected to the gate terminal will provide control of the peak value (V_P) of the output signal.

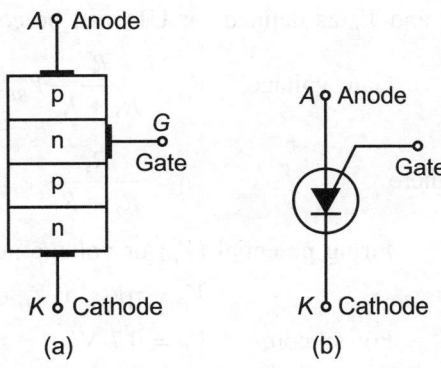

(a)   (b)

Fig. 5.9 PUT circuit and symbol

5.4.1 Theory of Operation

Figure 5.10(a) illustrates the operation of the PUT in a test circuit voltage divider network R_2–R_3 provides a fixed gate voltage V_G to the gate terminal. Potentiometer R_1 is connected as a variable voltage divider to control the anode voltage. When the anode voltage is less than the gate voltage, V_G the PUT is in a state of non conduction, and only a negligible reverse leakage current flows through PUT. This is due to anode-gate p-n junction being reverse biased. The V–I characteristic in Fig. 5.10(b) shows the leakage current as I_{GAO}. When the anode voltage is raised above the gate voltage by approx. 0.7 V, the anode gate p-n junction is forward-biased and permits turn on of I_A. This value of anode voltage, called the peak voltage, V_P. At this point, the PUT latches into full turn on and exhibits negative resistance characteristics over a portion of V–I characteristic. By adjusting the ratio of gate voltage divider network, peak voltage, V_P can be varied, which in turn controls the value of I_P.

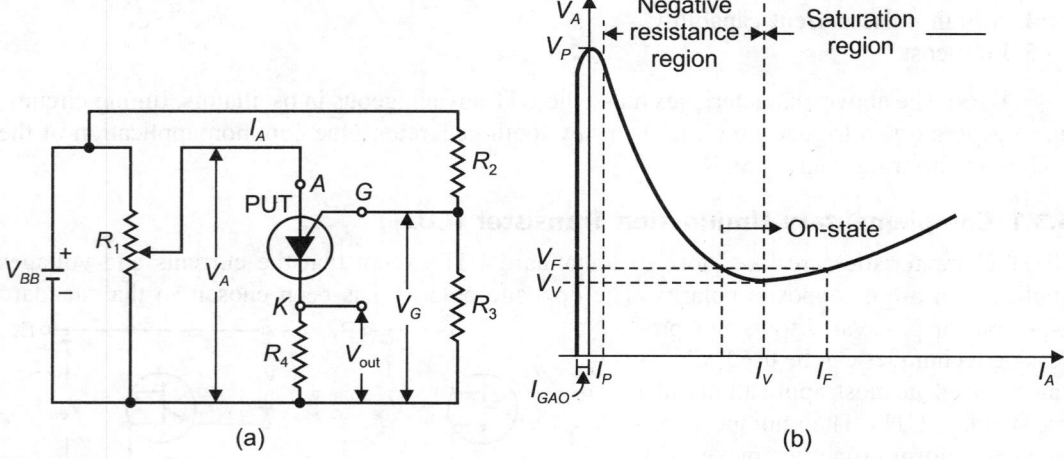

(a)   (b)

Fig. 5.10 PUT and characteristics

The output from a PUT circuit is normally taken at the cathode terminal. In the off-state condition, V_{out} is normally zero. When the PUT is triggered into conduction V_{out} suddenly rises to a positive value controlled by the circuit design when used as thyristor trigger, V_{out} is adjusted to 2–6 V. The turn on time is extremely fast. It can be as low as 80 nsec.

The term 'programmable' is applied because resistance of the device between the terminals, η and V_P as defined for UJT can be controlled through R_2, R_3 and supply voltage V_{BB}.

$$\text{Gate voltage,} \quad V_G = \frac{R_3}{R_2 + R_3} V_{BB} = \eta V_{BB} \qquad \qquad ...(5.17)$$

where

$$\eta = \frac{R_3}{R_2 + R_3}$$

Firing potential (V_P) or voltage necessary to fire the device is given by

$$V_P = \eta V_{BB} + V_D \qquad \qquad ...(5.18)$$

For silicon, $\quad V_D = 0.7$ V

$$V_P = \eta V_{BB} + 0.7 \qquad \qquad ...(5.19)$$

5.4.2 Relaxation Oscillator Using PUT

Main application of the PUT is in the relaxation oscillator of Fig. 5.11 (a). The instant supply is connected, the capacitor will begin to charge toward V_{BB} volts, since there is no anode current at this point. The charging curve appears in Fig. 5.11 (b).

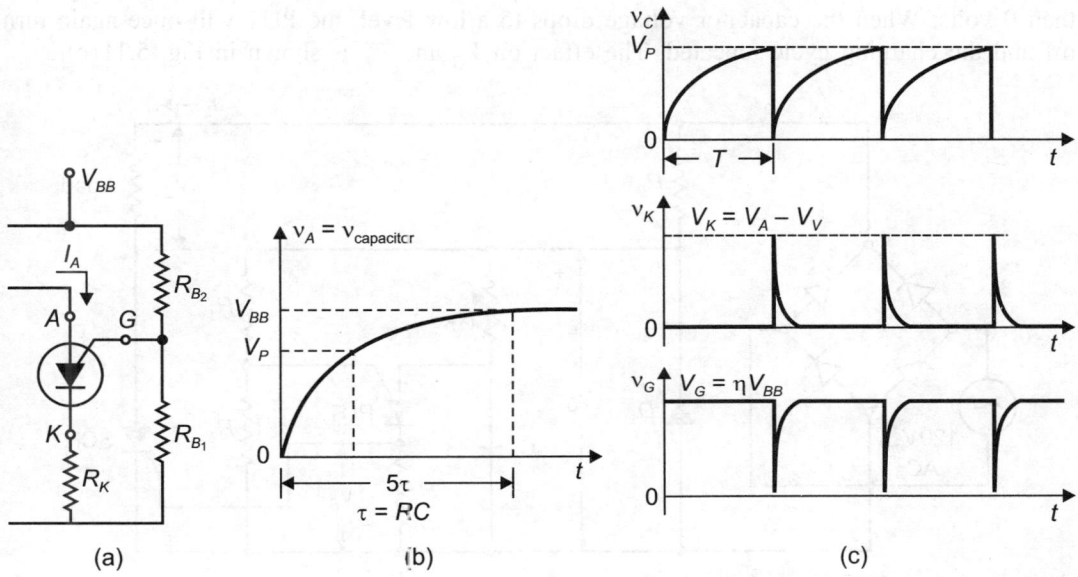

Fig. 5.11 Relaxation oscillator circuit using PUT and waveforms

The period T required to reach the firing potential V_P is given by

$$T \cong RC \log_e \frac{V_{BB}}{V_{BB} - V_P} \qquad \qquad ...(5.20)$$

when $V_P \cong \eta V_{BB}$, $\qquad T \cong RC \log_e \left(1 + \frac{R_{B_1}}{R_{B_2}} \right) \qquad ...(5.21)$

The instant the voltage across the capacitor equals V_P, the device will fire and a current $I_A = I_P$ established through the PUT. If R is too large, the current I_P cannot be established and the device will not fire. At the point of transition

$$I_P R = V_{BB} - V_P$$

and $\qquad \qquad R_{max} = \frac{V_{BB} - V_P}{I_P} \qquad \qquad ...(5.22)$

The subscript is included to indicate that any R greater than R_{max} will result in a current less than I_P. The level of R must also be such to ensure it is less than I_V if oscillations are to occur. In other words, it is required the device to enter the unstable region and then return to the off-state.

Similarly, $\qquad \qquad R_{min} = \frac{V_{BB} - V_V}{I_V} \qquad \qquad ...(5.23)$

For an oscillatory system,

$$R_{min} < R < R_{max} \qquad \qquad ...(5.24)$$

The waveforms of v_A, v_G, and v_K appear in Fig. 5.11 (c). Time period T determines the maximum voltage v_A. Once the device fires, the capacitor will rapidly discharge through the PUT and resistor R_K. The voltage v_G will rapidly drop down from V_G to a level just greater than 0 volts. When the capacitor voltage drops to a low level, the PUT will once again turn off and the charging cycle repeated. The effect on V_G and V_K is shown in Fig. 5.11 (c).

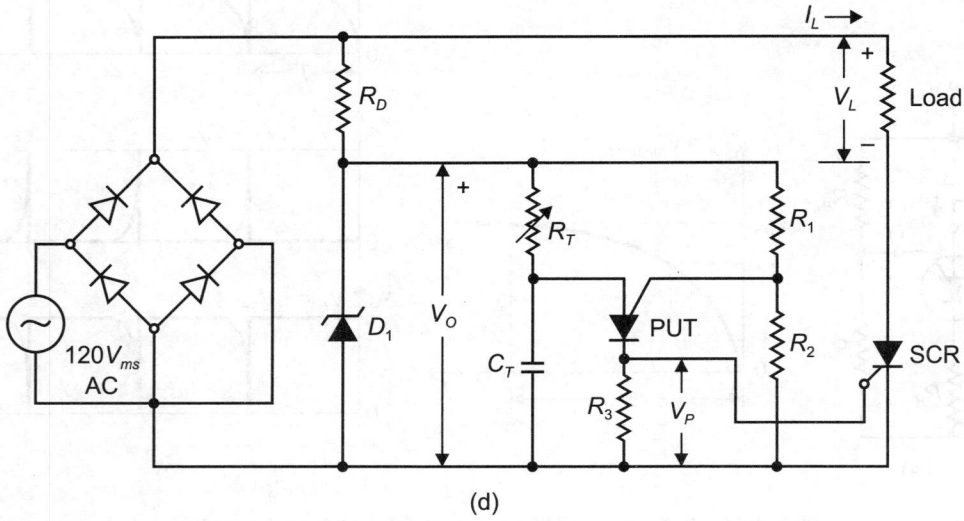

(d)

Fig. 5.11 Triggering circuit using PUT

5.4.3 Applications

Like the UJTs, the PUT is used in applications where a high speed, low power electronic switch is required. These applications include relaxation oscillators, timers and phase control circuits. The relaxation oscillator is useful for triggering high power SCRs, since it is capable of producing pulsed high current trigger signals.

A single-phase full-wave circuit using a PUT as a trigger circuit is shown in Fig. 5.11 (d). A similar circuit could be designed using UJT instead of PUT. The terminals of the PUT are equivalent to the UJT terminals as follows:

PUT	UJT
Anode	Emitter
Cathode	Base 1
GATE	Base 2

5.5 DIAC

The diac is basically a two terminal parallel-inverse combination of semiconductor layers that permits triggering in either direction. This possibility of an 'on condition' in either direction can be used to its fullest advantage in AC applications. The diac is used extensively as a trigger device for triac gate circuits.

The basic arrangement of the semiconductor layers of the diac is shown in Fig. 5.12 along with its graphic symbol. Note that neither terminal is referred to as the cathode.

MT_1 and MT_2 are the main terminals of the device. When MT_1 is positive with respect to MT_2 the semiconductor layers of particular interest are p_1, n_2, p_2, and n_3. For terminal MT_2 positive with respect to terminal MT_1 the applicable layers are p_2, n_2, p_1, and n_1. The doping level at the two ends of the device is the same which leads to identical V–I characteristics in both first and third quadrant. Figure 5.13 shows the V–I characteristic of a diac.

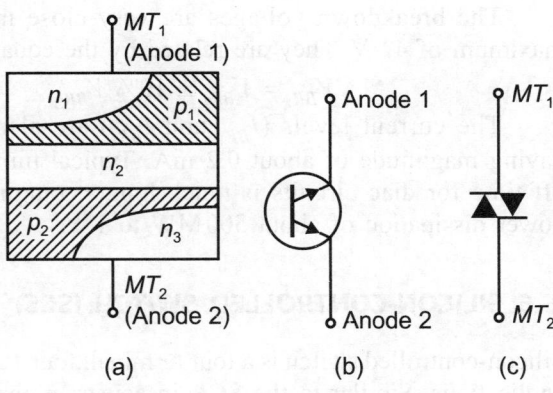

Fig. 5.12 Diac circuit and symbol

5.5.1 Theory of Operation

Diac resembles a bipolar transistor with no base connections. Figure 5.13 shows a V–I characteristics. The device can be triggered by either positive or negative half of an AC cycle. For the positive half cycle, at voltages less than the breakover voltage, a very small amount of current called the leakage current flows through the device. This current is produced due to the drift of electrons and holes at the depletion region and is not sufficient to cause conduction, hence device remains practically in nonconducting mode called the blocking state. If the voltage level reaches the break over voltage V_{BR_1}, the device starts conducting and the device exhibits negative resistance characteristic. The current through the device starts increasing and voltage across it starts decreasing. This region is known as conduction state. For the positive half cycle the characteristic is obtained in first quadrant and similar characteristic for the negative half cycle is obtained in the third quadrant.

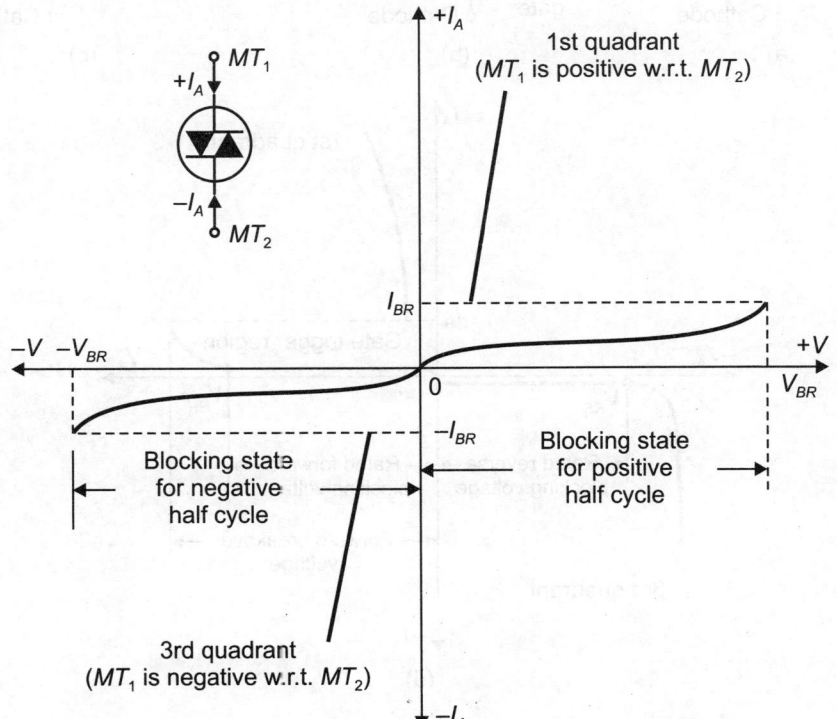

Fig. 5.13 V–I characteristics of diac

The breakdown voltages are very close in magnitude but may vary from 28 V to a maximum of 42 V. They are related by the equation provided in the specification sheet.

$$V_{BR_1} = V_{BR_2} \pm 10\% \; V_{BR_2} \qquad \qquad ...(5.25)$$

The current levels (I_{BR_1} and I_{BR_2}) are also very close in magnitude for each device having magnitude of about 0.2 mA. Typical turn on time range from 50 to 500 msec. Turn off time for diac circuits is much longer, ranging up to about 1000 ns. Most diacs have a power dissipation of about 300 MW to 1 W.

5.6 SILICON-CONTROLLED SWITCH (SCS)

Silicon-controlled switch is a four terminal, four-layer pn-pn device used in low power switching applications. Similar to the SCR in construction, the SCS has two gate terminals [i.e. anode gate and cathode gate shown in Fig. 5.14(a)]. Each gate can be used to turn on or turn off the main current through the device.

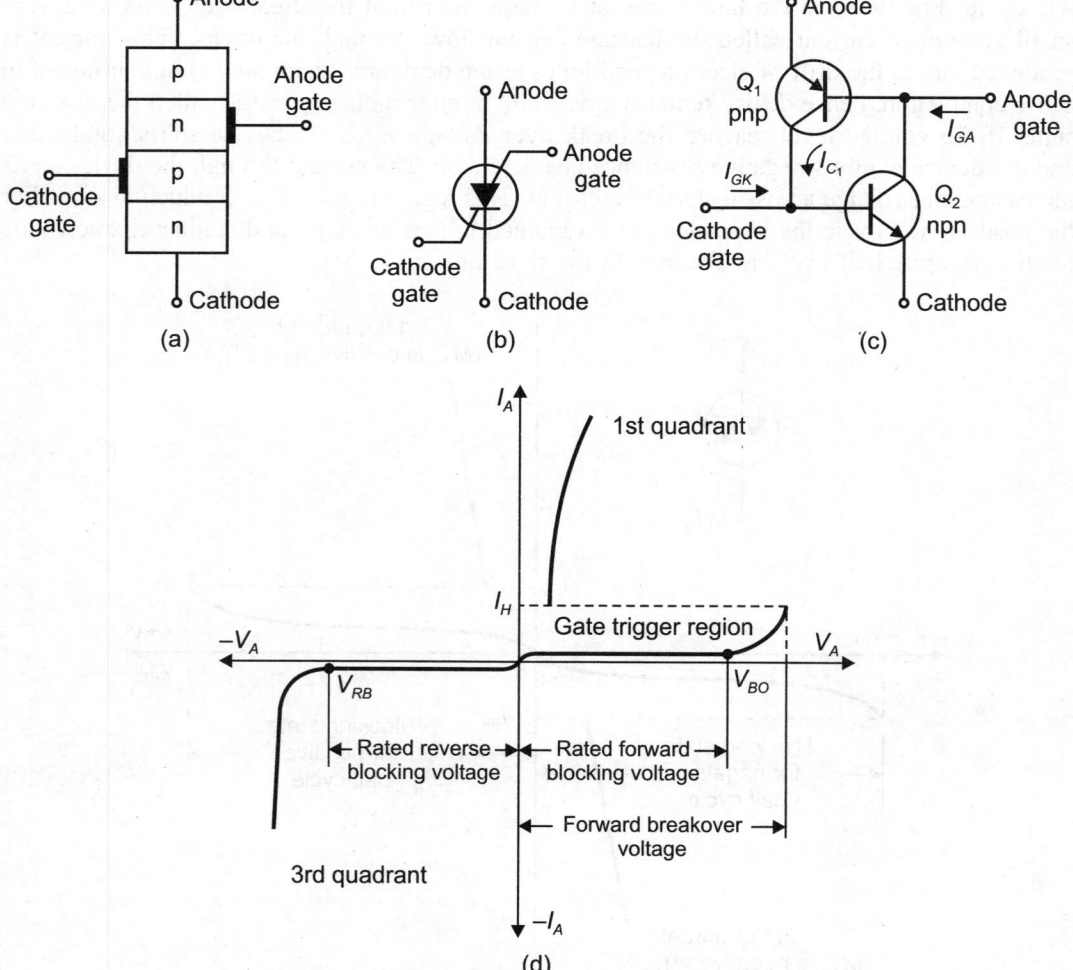

(a)

(b)

(c)

(d)

Fig. 5.14 Circuit diagram of four terminal SCS and characteristics

Theory of operation: An equivalent two-transistor switch configuration can be used to describe the operation of the SCS. *V–I* characteristic for an SCS is given in Fig. 5.14 (d).

The graphic symbol and transistor equivalent circuit are shown in Figs 5.14 (b) and (c). The characteristics of the device is essentially the same as that for the SCR. The higher, the anode gate current, the lower, the required anode-to-cathode voltage to turn the device on.

The device can either be turn on or turn off using the anode gate connection. To turn on the device, a negative pulse must be applied to the anode gate terminal, while a positive pulse is required to turn off the device. This can be demonstrated using the circuit of Fig. 5.14 (c). A negative pulse at the anode gate will forward-bias the base-to-emitter junction of Q_1, turning it on. The resulting heavy collector current I_{C_1} will turn on transistor Q_2, resulting in a regenerative action and the on-state for the SCR device. A positive pulse at the anode gate will reverse-bias the base-to-emitter junction of transistor Q_1, turning it off, resulting in the open circuit off-state of the device. In general, the triggering (turn on) anode gate current is larger in magnitude than the required cathode gate current. For SCS device, the triggering anode gate current is 1.5 mA while the required cathode gate current is 1 μA. The required turn on gate current at either terminal is affected by many factors such as: (1) the operating temperature, (2) anode-to-cathode voltage load placement, (3) type of cathode, and (4) gate-to-anode or anode gate-to-anode connection (short-circuit, open-circuit, bias, load, etc.).

Three fundamental types of turn off circuits for the SCS are shown in Fig. 5.15. When a pulse is applied to the circuit of Fig. 5.15 (a), the transistor conducts heavily, resulting in a low-impedance characteristic between collector and emitter. This low-impedance branch diverts anode current away from the SCS, dropping it below the holding value and consequently turning it off. Similarly, the positive pulse at the anode gate of Fig. 5.15 (b) will turn the SCS off. The circuit of Fig. 5.15 (c) can be turned either *off* or *on* by a pulse of the proper magnitude at the cathode gate. The turn off characteristic is possible only if the correct value of R_A is employed. It will control the amount of regenerative feedback, the magnitude of which is critical for this type of operation.

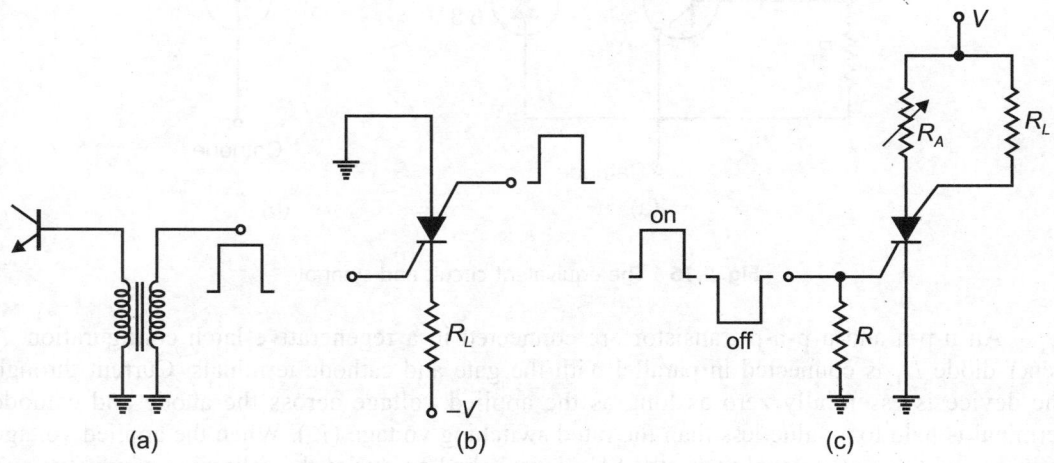

Fig. 5.15 Turn-off circuits of SCS

Advantages: The main advantage of the SCS over a corresponding SCR is the reduced turn off time, typically within the range 1 to 10 μsec for the SCS and 5 to 30 μsec for the SCR. Other advantages of the SCS over an SCR include increased control and triggering sensitivity and a more predictable firing situation. At present, however, the SCS is limited to low power, current and voltage ratings. Typical maximum anode currents range from 100 to 300 mA with dissipation (power) ratings of 100 to 500 mW.

Applications: SCSs are used in a wide variety of computer circuits like counters, registers, and timing circuits, pulse generators, voltage sensors, oscillators, lamp, and relay drives.

5.7 SILICON UNILATERAL SWITCH (SUS)

Silicon unilateral switch and silicon bilateral switch among the newest and most advanced members of the thyristor family. These three terminal devices are actually small integrated circuits (IC) containing transistors, zener diodes and resistors. This type of construction provides improved performance as well as reduced costs.

The SUS and SBS possess negative resistance switching characteristics similar to three and four-layer diodes and unijunction transistors. They are capable of producing fast rising, high current triggering signals for power thyristors such as SCRs and triacs. Maximum power dissipation ratings for the SUS and SBS are in the range of 300 mW to 350 mW.

5.7.1 Theory of Operation of SUS

The equivalent circuit and symbol for this device is shown in Fig. 5.16. The SUS is designed as a unilateral device in which current flows from anode to cathode. Reverse current, if permitted, can result in damage to the SUS.

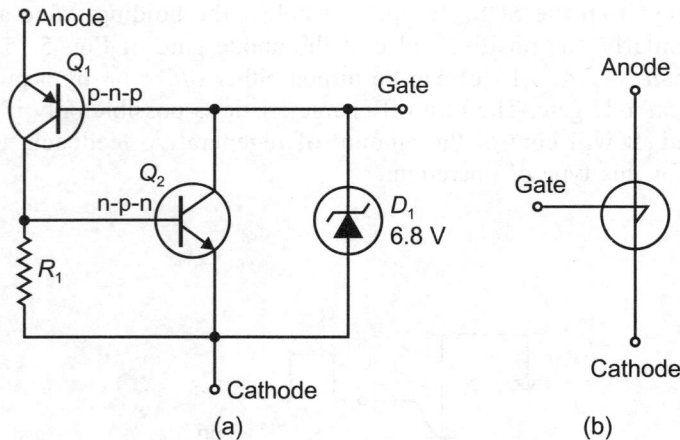

(a)   (b)

Fig. 5.16 The equivalent circuit and symbol

An n-p-n and a p-n-p transistor are connected in a regenerative latch configuration. A zener diode D_1 is connected in parallel with the gate and cathode terminals. Current through the device is essentially zero as long as the applied voltage across the anode and cathode terminal is held to a value less than the rated switching voltage (V_s). When the applied voltage is increased to a value equal to V_s, the SUS is switched on due to the following conditions and actions:

1. The value of V_s must be equal to the emitter-base forward-bias voltage of Q_1 plus the zero breakdown voltage of zener diode D_1. This is approximately 0.7 V plus 6.8 V for the circuit shown. Manufacturers may rate this total of 7.5 V as 8 V to ensure reliable turn on.

2. Current from the base of transistor Q_1 turns on the p-n-p transistor. High gain p-n-p transistors are usually employed in this circuit to ensure fast and reliable turn on characteristics. Turn on time for SUS is typically in the order of 1 μsec.

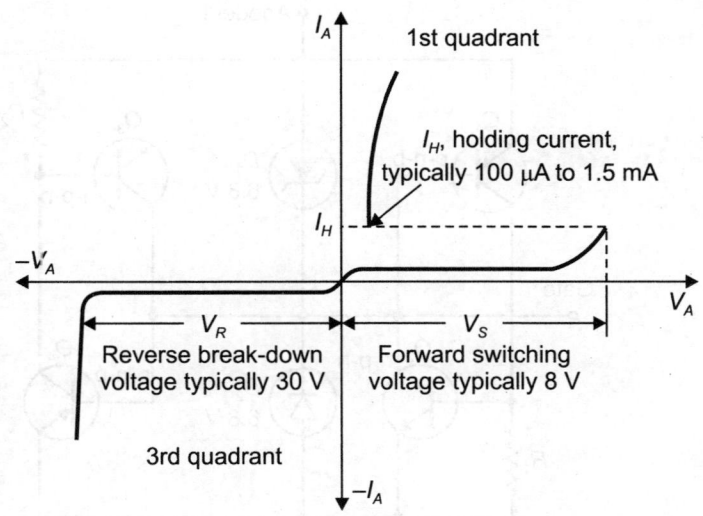

3. Collector current from Q_1 flows into the base terminal of Q_2 and collector resistor R_1 the current into Q_2 turns on the n-p-n transistor.

4. The collector current of each transistor provides base current for the other transistor and this regenerative action quickly drives both transistors into saturation.

Fig. 5.17   *V–I characteristics of SUS*

V–I characteristics are shown in Fig. 5.17.

Once turned on, SUS will remain in conduction until the anode current is reduced below the level of holding current. Turn off methods used in SCR circuits are also applicable to SUS. As with SCR, SUS turn off time is longer than turn on time. The gate terminal may be used to turn on the SUS when the applied forward voltage is below V_s.

5.8 SILICON BILATERAL SWITCH (SBS)

The SBS simply consists of two SUS circuits installed on the same IC chip and connected for bilateral current flow. Figure 5.18 shows the circuit configuration and commonly used symbol.

Theory of operation: Transistor Q_1 and Q_2 are connected as one of the two transistor latching circuits on the chip with anode-1 positive, diode D_1 is forward-biased and acts as conventional diode rectifier. As the applied voltage across the anode-1 and anode-2 terminals is increased to V_s, diode D_2 undergoes zener breakdown and turn on Q_1. This action, in turn causes Q_2 to conduct and both transistors are quickly switched into saturation, the primary conduction path for a positive voltage on anode-1 is through Q_1 and Q_2. Transistors Q_3 and Q_4 remain off during this interval.

For reverse current, operation within the SBS is a mirror image of operation for forward current. When the applied voltage to anode-2 is positive, for example during the negative half cycle of the AC voltage Q_4 and Q_3 form the primary conduction path. During this half cycle, D_2 acts as a forward-biased diode and D_1 undergoes zener breakdown when the applied voltage equals V_s.

Figure 5.19 illustrates typical SBS, *V–I* characteristics. For most SBS devices, V_s ranges from 6 V to 10 V. Due to uniform IC construction techniques SBS exhibits excellent symmetrical characteristics. Like SUS the SBS can be controlled or turned on by gate terminal. The

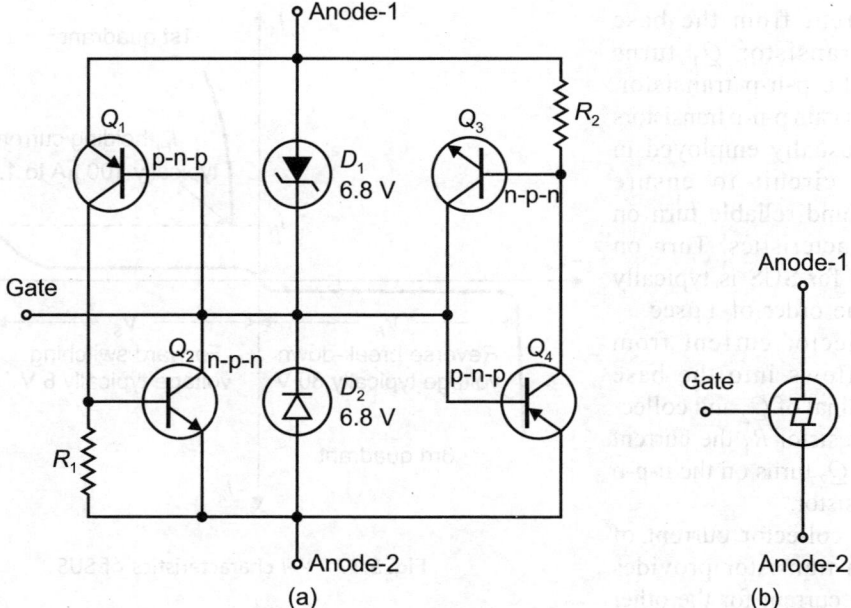

Fig. 5.18 Circuit diagram of SBS and symbol

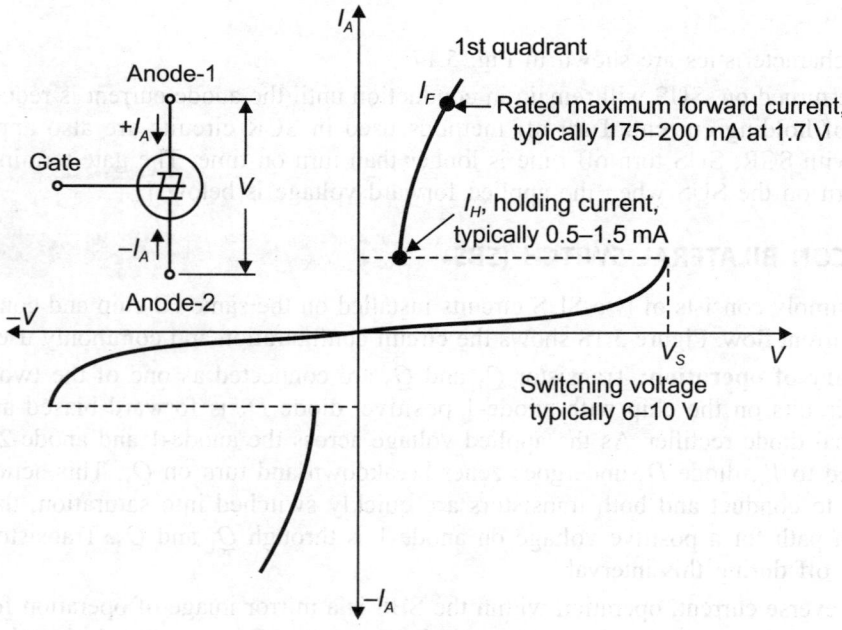

Fig. 5.19 V–I characteristics of SBS

maximum gate current for the off-state condition is about 5 mA. Maximum gate current for the on-state condition is limited only by the power dissipation rating of the device.

Applications: The SUS and SBS devices are designed for high speed signal switching applications where switching voltage stability and low cost are required. In addition to serving as triggers for power thyristors, the SUS and SBS are employed in digital circuits involving frequency dividers, ring counters, bistable memory circuits and pulse generators. Other related applications include voltage sensing in circuits, like electronic crowbar or over voltage protection for DC power supplies.

5.9 SHOCKLEY DIODE

The Shockley diode is a four-layer pn-pn diode with only two external terminals as shown in Fig. 5.20 (a) with its graphic symbol. The characteristics shown in Fig. 5.20 (b) of the device are exactly the same as those for the SCR with $I_G = 0$. As indicated by the characteristics, the device is in the off-state until the breakover voltage is reached, at which time avalanche conditions develop and the device turns on.

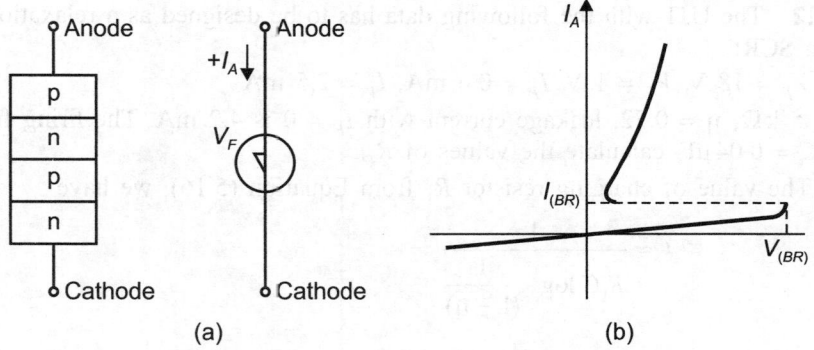

Fig. 5.20 Symbol and characteristic of Shockley diode

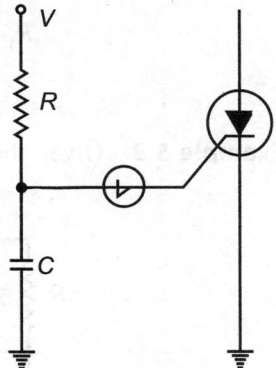

The Shockley diode shown in Fig. 5.21 is employed as a trigger switch for an SCR. When the circuit is energized, the voltage across the capacitor will begin to change toward the supply voltage. Eventually, the voltage across the capacitor will be sufficiently high to first turn on the Shockley diode and then the SCR.

5.10 OPTO-ISOLATORS

The opto-isolator is a device that incorporates many of the characteristics. It is simply a package that contains both an infrared LED and a photodetector such as silicon diode, transistor Darlington pair, or SCR. The wavelength response of each device is identical to permit the highest measure of coupling possible. The transparent insulating cap between each set of elements embedded in the structure permit the passage of light. They are designed with response times so small that they can be used to transmit data in the MHz range.

Fig. 5.21 The Shockley diode

SOLVED EXAMPLES

Example 5.1 A UJT has one base resistance of 4.8 kΩ and intrinsic stand off ratio of 0.6. If the interbase voltage of 10 V is applied across the two bases, calculate the value of V_c and I_B.

Solution Given

$$R_{B_1} = 4.8 \text{ k}\Omega, \; \eta = 0.6, \; V_{BB} = 10 \text{ V}$$
$$V_P = \eta V_{BB} = 0.6 \times 10 = 6 \text{ V}$$

We know

$$\eta = \frac{R_{B_1}}{R_{BB}} \text{ where } R_{BB} \text{ is the interbase resistance}$$

or

$$0.6 = \frac{4.8 \times 1000}{R_{BB}}$$

or

$$R_{BB} = 8000 \, \Omega$$

Current

$$I_B = \frac{V_{BB}}{R_{BB}} = \frac{10}{8000} = 1.25 \text{ mA}.$$

Example 5.2 The UJT with the following data has to be designed as a relaxation oscillator for firing an SCR:

Data: $V_P = 18$ V, $V_V = 1$ V, $I_P = 0.6$ mA, $I_V = 2.5$ mA

$R_{BB} = 5$ kΩ, $\eta = 0.72$, leakage current with $I_E = 0$ is 4.2 mA. The firing frequency is 2 kHz and $C = 0.04 \, \mu$F, calculate the values of R_1.

Solution The value of charging resistor R_1 from Equation (5.16), we have

$$f \approx \frac{1}{R_1 C \log_e \dfrac{1}{(1 - \eta)}}$$

or

$$R_1 \approx \frac{1}{f \cdot C \log_e[1/(1 - \eta)]} = \frac{1}{2 \times 10^3 \times 0.04 \times 10^{-6} \times \log_e \left(\dfrac{1}{1 - 0.72} \right)}$$

$$R_1 = \frac{10^6}{2000 \times 0.04 \times \log_e \left(\dfrac{1}{0.28} \right)} = 9.82 \text{ k}\Omega.$$

Example 5.3 Given the relaxation oscillator of Fig. E5.3 (a):

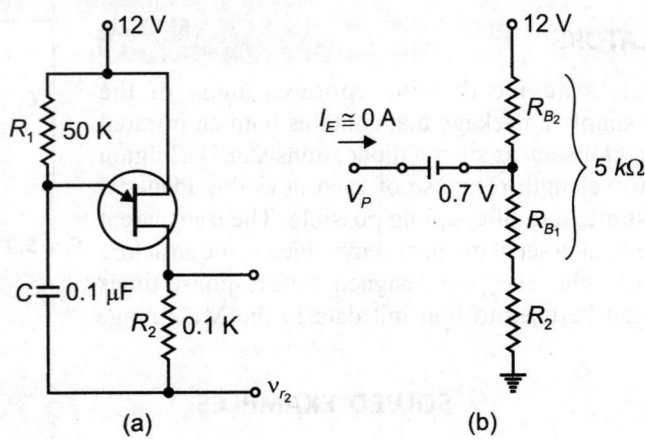

Fig. E5.3 Circuit diagram

(a) Determine R_{B_1} and R_{B_2} at $I_E = 0$ A.
(b) Calculate V_P, the voltage necessary to turn on the UJT.

(c) Determine whether R_1 is within the permissible range of values to insure firing of the UJT.

(d) Determine the frequency of oscillation if $R_{B_1} = 100\,\Omega$ during the discharge phase.

(e) Sketch the waveform of v_C for a full cycle.

(f) Sketch the waveform of v_{R_2} for a full cycle.

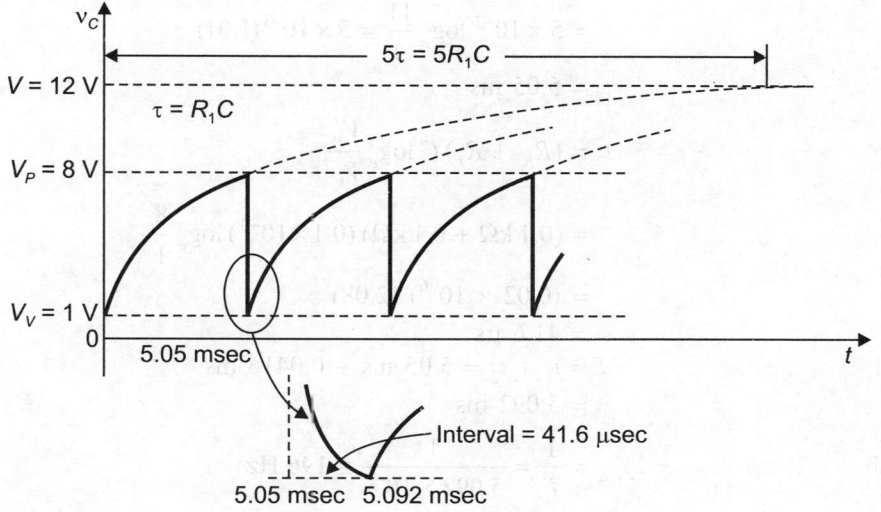

Fig. E5.3   Circuit diagram

Solution

(a)
$$\eta = \frac{R_{B_1}}{R_{B_1} + R_{B_2}}$$

$$0.6 = \frac{R_{B_1}}{R_{BB}}$$

$$R_{B_1} = 0.6 R_{BB} = 0.6 \ (5 \ \text{k}\Omega) = 3 \ \text{k}\Omega$$

$$R_{B_2} = R_{BB} - R_{B_1} = 5 \ \text{k}\Omega - 3 \ \text{k}\Omega = 2 \ \text{k}\Omega$$

(b) At the point where $v_C = V_P$, if $I_E = 0$ A, the network of Fig. E5.3(b) will result where

$$V_P = 0.7 + \frac{(R_{B_1} + R_2)12}{R_{B_1} + R_{B_2} + R_2}$$

where
$$R_{BB} = R_{B_1} + R_{B_2}$$

$$= 0.7 + \frac{(3 \ \text{k}\Omega + 0.1 \ \text{k}\Omega)12}{5 \ \text{k}\Omega + 0.1 \ \text{k}\Omega} = 0.7 + 7.294$$

$$\cong 8 \ \text{V}$$

(c) Using the equation $\dfrac{V - V_V}{I_V} < R_1 < \dfrac{V - V_P}{I_P}$ to ensure firing of UJT,

$$\frac{12 - 1}{10 \ \text{mA}} < R_1 < \frac{12 - 8}{10 \ \mu\text{A}}$$

$$1.1 \ \text{k}\Omega < R_1 < 400 \ \text{k}\Omega$$

The resistance $R_1 = 50 \ \text{k}\Omega$ falls within this range.

(d)
$$t_1 = R_1 C \log_e \frac{V - V_V}{V - V_P}$$

$$= (50 \times 10^3)(0.1 \times 10^{-6}) \log_e \frac{12 - 1}{12 - 8}$$

$$= 5 \times 10^{-3} \log_e \frac{11}{4} = 5 \times 10^{-3}(1.01)$$

$$= 5.05 \text{ ms}$$

$$t_2 = (R_{B_1} + R_2) C \log_e \frac{V_P}{V_V}$$

$$= (0.1 \text{ k}\Omega + 0.1 \text{ k}\Omega)(0.1 \times 10^{-6}) \log_e \frac{8}{1}$$

$$= (0.02 \times 10^{-6})(2.08)$$

$$= 41.6 \text{ μs}$$

and
$$T = t_1 + t_2 = 5.05 \text{ ms} + 0.0416 \text{ ms}$$

$$= 5.092 \text{ ms}$$

with
$$f_{osc} = \frac{1}{T} = \frac{1}{5.092 \times 10^{-3}} \cong 196 \text{ Hz}$$

Using the equation,
$$f = \frac{1}{R_1 C \log_e [1/(1 - \eta)]}$$

$$= \frac{1}{5 \times 10^{-3} \log_e 2.5}$$

$$= 218 \text{ Hz}$$

(e) See Fig. E5.3(c).

Example 5.4 For the circuit shown in Fig. E5.4, the parameter are: $V = 40$ V, $\eta = 0.6$, $V_V = 1$ V, $I_V = 8$ mA, and $I_P = 10 \, \mu$A, determine the range of R_1 for the triggering network.

Solution To ensure triggering

$$\frac{V - V_P}{I_P} > R_1$$

where
$$V_P = \eta V_{BB} + V_D$$
$$V_P = 0.6 \times 40 + 0.7$$
$$= 24.7 \text{ V}$$

or
$$\frac{40 - 24.7}{10 \times 10^{-6}} > R_1$$

or
$$1.53 \times 10^6 \Omega > R_1$$

To ensure turning off,

$$R_1 > \frac{V - V_V}{I_V}$$

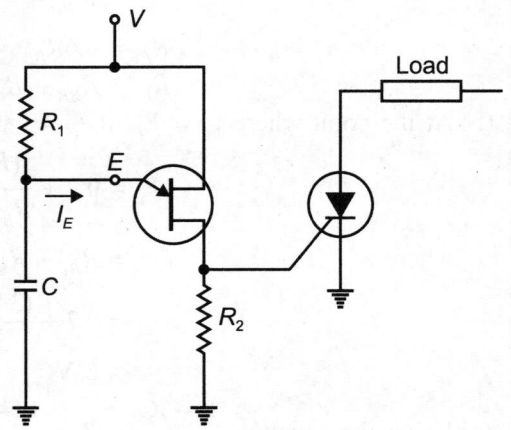

Fig. E5.4 Circuit diagram

or
$$R_1 > \frac{40 - 1}{8 \times 10^{-3}} = 4.875 \text{ k}\Omega < R_1$$

Thus, the range of R_1 for triggering is given by 1.53 M$\Omega > R_1 > 4.875$ kΩ.

Example 5.5 For a UJT with $V_{BB} = 20$ V, $\eta = 0.65$, $R_{B_1} = 2$ kΩ ($I_E = 0$) and $V_D = 0.7$ V, determine: (a) R_{B_2}, (b) R_{BB}, (c) V_{RB_1}, (d) V_P.

Solution

(a)
$$\eta = \frac{R_{B_1}}{R_{B_1} + R_{B_2}}\bigg|_{I_E=0} \Rightarrow 0.65 = \frac{2000}{2000 + R_{B_2}}$$

or
$$R_{B_2} = 1080 \,\Omega$$

(b)
$$R_{BB} = (R_{B_1} + R_{B_2})\big|_{I_E=0} = 2000 + 1080 = 3080 \,\Omega$$

(c)
$$V_{RB_1} = \eta V_{BB} = 0.65 \,(20) = 13 \text{ V}$$

(d)
$$V_P = \eta V_{BB} + V_D = 13 + 0.7 = 13.7 \text{ V}.$$

Example 5.6 Given the relaxation oscillator of Fig. E5.6:

(a) Determine R_{B_1} and R_{B_2} at $I_E = 0$ A.
(b) Find the value of V_P necessary to turn on UJT.
(c) Determine whether the range of R_1 is within the permissible range or not.
(d) Determine the frequency of oscillation if $R_{B_1} = 200 \,\Omega$ during the discharge phase.
(e) Determine the frequency and compare the value determined in part (d). Account for any major differences.

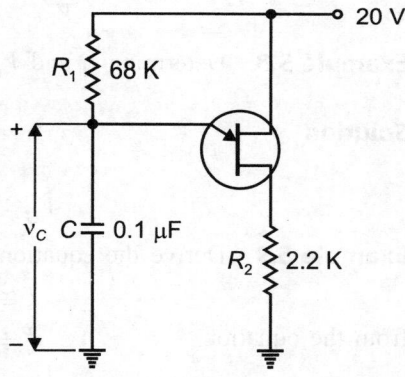

Fig. E5.6 Circuit diagram of oscillator

Solution

(a)
$$\eta = \frac{R_{B_1}}{R_{BB}}\bigg|_{I_E=0} \Rightarrow 0.55 = \frac{R_{B_1}}{10000}$$

or
$$R_{B_1} = 0.55 \times 10000 = 5500 \,\Omega = 5.5 \text{ k}\Omega$$
$$R_{BB} = R_{B_1} + R_{B_2} \Rightarrow 10 \text{ k}\Omega = 5.5 \text{ k}\Omega + R_{B_2}$$

or
$$R_{B_2} = 4.5 \text{ k}\Omega.$$

(b)
$$V_P = \eta V_{BB} + V_D = (0.55 \times 20) + 0.7 = 11.7 \text{ V}$$

(c) The range of R_1 is given by
$$R_1 < \frac{V - V_P}{I_P} = \frac{20 - 11.7}{50 \times 10^{-6}} = 166 \text{ k}\Omega$$

As
$$R_1 = 68 \text{ k}\Omega \text{ is less than } 166 \text{ k}\Omega$$
It is OK.

(d)
$$t_1 = R_1 C \log_e\left(\frac{V - V_V}{V - V_P}\right)$$

$$= (68 \times 10^3)(0.1 \times 10^{-6}) \log_e\left(\frac{18.8}{8.3}\right)$$

$$= 5.56 \text{ msec}$$

$$t_2 = (R_{B_1} + R_{B_2}) \, C \log_e \frac{V_P}{V_V}$$

$$= (0.2 \text{ k}\Omega + 2.2 \text{ k}\Omega)(0.1 \times 10^{-6}) \log_e (11.7/1.2) = 0.546 \text{ msec}$$

$$T = t_1 + t_2 = 5.56 + 0.546 = 6.106 \text{ msec}$$

Frequency $\qquad f = \dfrac{1}{T} = \dfrac{1}{6.106 \times 10^{-3}} = 163.7 \text{ Hz}$

(e) $\qquad f = \dfrac{1}{R_1 C \log_e (1/1 - \eta)} = \dfrac{1}{(6.8 \times 10^3)(0.1 \times 10^{-6}) \log_e 2.22}$

$$f = 184.16 \text{ Hz}$$

Difference partly due to the fact that $t_2 \approx 10\%$ of t_1.

Example 5.7 If V_{BR_2} is 6.4 V for diac, determine the range for V_{BR_1}.

Solution $\qquad V_{BR_1} = V_{BR_2} \pm 10\% \ V_{BR_2}$

$$V_{BR_1} = 6.4 \pm 0.10 \times 6.4 = 6.4 \pm 0.64$$

$\Rightarrow \qquad\qquad V_{BR_1} = 7.04 \text{ and } V_{BR_1} = 5.76 \text{ V}.$

Example 5.8 Determine η and V_G for a PUT with $V_{BB} = 20$ V and $R_{B_1} = 3R_{B_2}$.

Solution $\qquad \eta = \dfrac{R_{B_1}}{R_{B_1} + R_{B_2}} = \dfrac{3R_{B_2}}{3R_{B_2} + R_{B_2}} = \dfrac{3}{4} = 0.75$

$$V_G = \eta V_{BB} = 0.75 \ (20) = 15 \text{ V}.$$

Example 5.9 Derive the equation

$$T \cong RC \log_e [1 + (R_{B_1}/R_{B_2})]$$

from the equation $\qquad T \cong RC \log_e [V_{BB}/(V_{BB} - V_P)]$

Solution Given $\qquad T \cong RC \log_e \left(\dfrac{V_{BB}}{V_{BB} - V_P} \right) \qquad\qquad …(i)$

As $V_P = \eta V_{BB} + V_D$, neglecting the value of V_D, we have $V_P = \eta V_{BB}$. Putting the value of V_P in Equation (i)

$$T \cong RC \log_e \left(\dfrac{V_{BB}}{V_{BB} - \eta V_{BB}} \right) = RC \log_e \left(\dfrac{1}{1 - \eta} \right) \qquad\qquad …(ii)$$

Using the value of $\qquad \eta = \dfrac{R_{B_1}}{R_{B_1} + R_{B_2}}$ in Equation (ii)

$$T \cong RC \log_e \left(\dfrac{1}{1 - \dfrac{R_{B_1}}{R_{B_1} + R_{B_2}}} \right)$$

$$T \cong RC \log_e \left(\dfrac{R_{B_1} + R_{B_2}}{R_{B_2}} \right)$$

$$T = RC \log_e (1 + R_{B_1}/R_{B_2}).$$

Example 5.10 Determine R_3 and V_{BB} for a silicon PUT for $\eta = 0.8$, $V_P = 10.3$ V and $R_2 = 5$ kΩ.

Solution From the equation

$$\eta = \frac{R_3}{R_3 + R_2} = 0.8$$

$$R_3 = 0.8\,(R_3 + 5\text{ kΩ})$$

or

$$0.2R_3 = 5 \times 0.8 \text{ kΩ}$$

$$0.2R_3 = 4 \times 1000$$

$$R = \frac{4000}{0.2} = 20000\,Ω$$

From equation,

$$V_P = \eta V_{BB} + V_D$$

$$10.3 = (0.8)\,V_{BB} + 0.7$$

$$V_{BB} = 12 \text{ V}.$$

Example 5.11 The Data for PUT, relaxation oscillator are: $V_{BB} = 12$ V, $R = 20$ kΩ, $C = 1$ μF, $R_K = 100\,Ω$, $R_3 = 10$ kΩ, $R_2 = 5$ kΩ, $I_P = 100$ μA, $V_V = 1$ V, and $I_V = 5.5$ mA. Determine: (a) V_P, (b) R_{max} and R_{min}, and (c) T and frequency of oscillation.

Solution

(a) From the equation

$$V_P = \eta V_{BB} + V_D$$

$$V_P = \frac{R_3}{R_3 + R_1}\,V_{BB} + V_D$$

$$= \frac{10000 \times 12}{10000 + 5000} + 0.7 = 8.7 \text{ V}$$

(b) From the equation

$$R_{max} = \frac{V_{BB} - V_P}{I_P} = \frac{12 - 8.7}{100 \times 10^{-6}} = 33 \text{ kΩ}$$

$$R_{min} = \frac{V_{BB} - V_V}{I_V} = \frac{12 - 1}{5.5 \times 10^{-3}} = 2 \text{ kΩ}$$

∴

$$R : 2 \text{ kΩ} < 20 \text{ kΩ} < 33 \text{ kΩ}$$

(c)

$$T = RC \log_e \frac{V_{BB}}{V_{BB} - V_P}$$

$$= (20 \times 10^3)\,(1 \times 10^{-6}) \log_e \left(\frac{12}{12 - 8.7} \right)$$

$$= 20 \times 10^{-3} \log_e \left(\frac{12}{3.3} \right)$$

$$= 20 \times 10^{-3} \times (1.29)$$

$$= 25.8 \text{ msec}$$

$$f = \frac{1}{T} = \frac{1}{25.8 \times 10^{-3}} = 38.8 \text{ Hz}.$$

EXERCISE

Multiple Choice Questions

1. In a UJT with V_{BB} as the voltage across two base terminals, the emitter potential, at peak point is given by:
   (A) ηV_{BB}
   (B) ηV_D
   (C) $\eta V_{BB} + V_D$
   (D) $\eta V_D + V_{BB}$

2. A UJT exhibits negative resistance region:
   (A) before the peak point
   (B) between peak and valley points
   (C) after the valley point
   (D) both (A) and (B) are correct

3. In a UJT, maximum value of charging resistance is associated with:
   (A) peak point
   (B) valley point
   (C) any point between peak and valley points
   (D) after the valley point

4. When a UJT is used for triggering of an SCR, the wave shape of the voltage obtained from UJT circuit is a:
   (A) sine wave
   (B) saw-tooth wave
   (C) trapezoidal wave
   (D) square wave

5. For a UJT employed for the triggering of an SCR, stand-off ratio $\eta = 0.64$ and DC voltage source V_{BB} is 20 V. The UJT would trigger when the emitter voltage is:
   (A) 12.8 V
   (B) 13.1 V
   (C) 10 V
   (D) none of these

6. An SCR can withstand a maximum junction temperature of 120°C with an ambient temperature of 75°C. If this SCR has thermal resistance from junction to ambient as 1.5°C/W, the maximum internal power dissipation allowed is:
   (A) 30 W   (B) 60 W
   (C) 80 W   (D) 100 W

7. In synchronized UJT triggering of an SCR voltage v_C across capacitor reaches UJT threshold voltage thrice in each half cycle so that there are three firing pulses during each half cycle. The firing angle of the SCR can be controlled:
   (A) once in each half cycle
   (B) thrice in each half cycle
   (C) twice in each half cycle
   (D) four times in each half cycle

8. The function of connecting a zener diode in a UJT circuit, used for the triggering of SCRs, is to:
   (A) expedite the generation of triggering pulses
   (B) delay the generation of triggering pulses
   (C) provide a constant voltage to UJT to prevent erratic firing
   (D) provide a variable voltage to UJT as the source voltage changes

9. SCR can be turned on by:
   (A) applying anode voltage at a sufficiently fast rate
   (B) applying sufficiently large anode voltage
   (C) increasing the temperature of SCR to a sufficiently large value
   (D) all of these

10. Thyristor A has rated gate current of 2 A and thyristor B, a rated gate current of 100 mA:
    (A) thyristor A is a GTO and B is a conventional SCR
    (B) thyristor B is a GTO and A is a conventional SCR
    (C) thyristor B may operate as a transistor
    (D) none of these

11. Which of the following devices does *not* have negative resistance characteristics?
    (A) UJT
    (B) Tunnel diode
    (C) SCR
    (D) FET

12. Which of the following statements is true for firing a triac?
    (A) Either T_1 or T_2 is negative and there is current pulse into the gate
    (B) T_1 is negative and there is current pulse out of the gate
    (C) T_2 is negative and there is a current pulse out of the gate
    (D) Either T_1 or T_2 is positive and there is a current pulse out of the gate

13. As compared to UJT, SUS:
    (A) triggers only in one direction
    (B) does not have negative resistance characteristics
    (C) needs definite polarity of the applied voltage
    (D) triggers only at one particular voltage

14. In its application, an SUS behaves in the same way as:
    (A) UJT
    (B) SCR
    (C) tunnel diode
    (D) none of these

15. Which of the following pn-pn devices has two gates?
    (A) Triac   (B) SCS
    (C) SUS   (D) Diac

16. For thyristors, pulse triggering is preferred to DC triggering because:
    (A) gate dissipation is low
    (B) pulse system is simpler
    (C) triggering signal is required for a very short duration
    (D) all of these

17. Which of the following pn-pn devices has a terminal for synchronising purpose?
    (A) SCS   (B) Triac
    (C) Diac   (D) SUS

18. Which of the following device is a three-layer device?
    (A) SCS   (B) SUS
    (C) Triac   (D) Diac

19. Which of the following devices cannot be triggered with voltage of either polarity?
    (A) SCS   (B) SUS
    (C) Triac   (D) Diac

20. Which of the following methods will turn SCS off?
    (A) Applying negative pulse to the anode
    (B) Applying a positive pulse to the anode gate
    (C) Applying negative pulse to the cathode gate
    (D) All of these

21. In application, a programmable conjunction transistor resembles more to:
    (A) special purpose UJT
    (B) anode-fired SCR
    (C) triac
    (D) diac

22. Which of the following pn-pn devices does not have a gate terminal?
    (A) Triac
    (B) SCS
    (C) SUS
    (D) Complementary SCR

ANSWER KEY

1. (C)	2. (B)	3. (A)	4. (B)	5. (B)	6. (A)	7. (A)	8. (C)
9. (D)	10. (A)	11. (D)	12. (C)	13. (D)	14. (A)	15. (B)	16. (D)
17. (D)	18. (D)	19. (A)	20. (D)	21. (B)	22. (C)		

Review Questions

1. Explain the construction and working principle of a UJT with the help of a diagram.

2. Define intrinsic stand-off ratio of UJT.

3. Explain the terms: (i) peak point voltage V_P, (ii) peak point current I_P, (iii) valley point voltage V_V, and (iv) valley point current I_V.

4. Explain the V–I characteristics and applications of a UJT.

5. Explain the working of a UJT relaxation oscillator.

6. Draw the V–I characteristics of diac and explain its working principle.

7. Explain the construction and working principle of PUT with the help of diagram.

8. Explain with the circuit diagram and working of a PUT relaxation oscillator.

9. Draw the symbols for the following devices: (i) SCS, (ii) SUS, (iii) SBS, and (iv) Shockley diode.

10. Mention the main differences between SCS and SCR.

Thyristor Firing Circuits—
Turn-on Systems

6.0 INTRODUCTION

Thyristor controlled circuits can be roughly divided into: (1) power circuit, (2) trigger circuit, and (3) controller. Power circuits consist of thyristors and diodes mounted on heatsinks with protective circuitry and transformer. Controller consists of amplifiers, logic gates and transducers which has a direct relation to the controlled machine and process. A trigger circuit is the link between the controller and the power circuits. In the absence of a controller circuitry it is possible to control the converter circuit using only the trigger/firing circuit.

6.1 REQUIREMENTS FOR TRIGGERING CIRCUITS

To turn on the thyristor, these conditions are required to be satisfied:

1. Thyristor should be forward-biased.
2. The load impedance should not be too high so that if the thyristor turn on the current in the thyristor should reach more than the latching current.
3. The gate should be positive with respect to cathode.

Three different types of triggering systems are in use depending on the application, economy and simplicity.

1. **Pulse triggering:** In this process, pulse signals are applied to a forward-biased thyristor. For the application chosen, each pulse is sufficient to trigger the thyristor and the pulse duration is chosen to give a sufficient duration for SCR to latch. Once the thyristor has been latched, no triggering pulse is required. The main advantage of this method is that there is no need of applying continuous signals, thus reducing the gate power losses. Electrical isolation is also provided between the main device supply and its gating signals. The most common method for producing pulses for gate triggering is by a unijunction transistor (UJT) relaxation oscillator.
2. **DC triggering:** In this type of triggering a DC voltage of proper magnitude and polarity is applied between the gate and cathode of the device in such a way that the gate becomes positive w.r.t. the cathode. When the applied voltage is sufficient to produce the required gate current, the device starts conducting. This method is inferior to pulse triggering as there is no isolation between power and control circuits. Another drawback

is that a continuous DC signal has to be applied at the gate causing more gate power loss.

3. **AC gate triggering:** In this scheme a single pulse is used to trigger thyristor. The pulse width is sufficiently large to suit resistive and moderately inductive loads. This technique provides proper isolation between the power and control circuits.

Two types of circuits are generally applied for AC triggering:

1. Resistance triggering (R-triggering)
2. Resistance–capacitance triggering (R–C triggering)

6.2 FIRING CIRCUIT DESIGN CONSIDERATIONS

To drive a typical thyristor gate circuit from a zero input impedance source at such a voltage level as to trigger the least sensitive device, i.e. 2 V, could well prove catastrophic for many of the devices of lower input impedance. Figure 6.1 (a) shows a triggering circuit feeding power to gate cathode circuit.

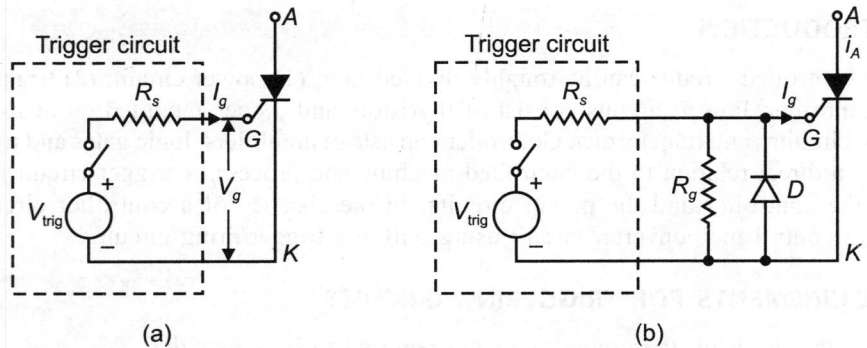

Fig. 6.1 Triggering circuit feeding power to gate cathode circuit

For this circuit,
$$V_{\text{trig}} = V_g + I_g R_s \qquad \qquad ...(6.1)$$
where
V_{trig} = gate source voltage, V_g = gate cathode voltage
I_g = gate current and R_s = gate source resistance

The resistance R_s of trigger source should be such that current (V_{trig}/R_s) is not harmful to the source as well as to the gate circuit when the SCR is turned on. In case, R_s is low, an external resistance in series with R_s must be connected.

A resistor R_g is connected across gate-cathode terminals so as to provide an easy path to the flow of leakage current between SCR terminals. Also the reverse voltage rating of the gate (usually 20 V) must not be exceeded either by the signal or by any interference. Figure 6.1 (b) shows a typical circuit which protects the device from such hazards. Diode 'D' prevents the gate voltage from being more than (about) 1 V below the cathode voltage. The necessary value of R_s to ensure the triggering of all the devices of a given type, including the least sensitive, is not affected by diode 'D' but is a function of R_g, as follows:

$$V_{\text{trig}} = \left(I_{g\,\text{min}} + \frac{V_{g\,\text{min}}}{R_g} \right) R_s + V_{g\,\text{min}} \qquad \qquad ...(6.2)$$

where $I_{g\,\text{min}}$ and $V_{g\,\text{min}}$ are the minimum gate current and gate voltage to turn on SCR. With the typical values, $I_{g\,\text{min}} = 100$ mA, $V_{g\,\text{min}} = 2$ V, $R_g = R_s = 30\,\Omega$, then

$$V_{\text{trig}} = \left(0.1 + \frac{2}{30}\right) 30 + 2 = 7 \text{ V}$$

A device of limited current sensitivity but lower input resistance will receive more current, but at lower voltage. A device of limited voltage sensitivity but higher input resistance will receive a greater voltage.

For selecting the operating point, usually a load line of the gate source voltage $V_{\text{trig}} = OA$ is drawn as AD in Fig. 6.2. Here OD = short circuit triggering current = V_{trig}/R_s. The relationship in the steady state between the gate voltage V_g and the gate current I_g is defined by the values of V_{trig} and R_s along the load line shown. When the firing signal is initiated, the gate current will grow along the line of the characteristic for that thyristor until in the steady state the load line at point P is reached.

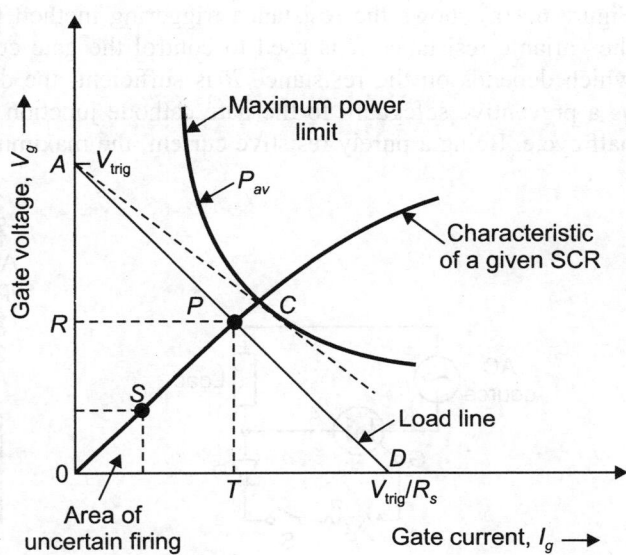

Fig. 6.2 Gate source voltage

Thus, for this SCR, gate voltage = PT and gate current = OT. The gradient of the load line AD (OA/OD) will give the required gate source resistance R_s. The minimum value of gate source series resistance is obtained by drawing a line AC tangent to power curve, P_{av}. However, before point P is reached, the thyristor will have turned on; most likely in the region of point S. The parameters of the firing network must be so chosen that the load line is above 'S' but within the maximum power limit. Typically, the value of V_{trig} will be 5 to 10 V having an associated maximum current of 0.5 to 1A.

6.3 REQUIREMENTS OF FIRING CIRCUITS

To turn on a thyristor positively in the shortest time, it is desirable to have a gate current with a fast rise time up to the maximum permitted value. This rise time is best achieved by pulse techniques, where the firing circuit generates a fast rise pulse of sufficient length to allow the anode current enough time to reach its latching value. The advantage of pulse is that much less power is dissipated in the gate compared to a continuous current, and the instant of firing can be accurately timed (Fig. 6.3).

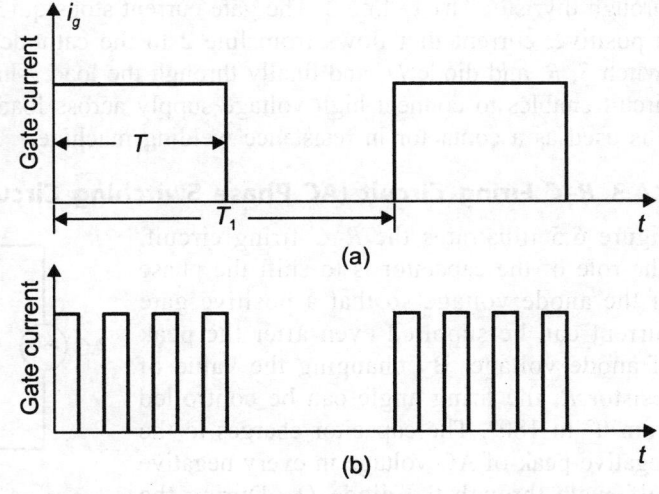

Fig. 6.3 Firing circuit requirement using pulse technology

6.4 THYRISTOR FIRING CIRCUITS

6.4.1 Resistance Triggering (R-Triggering)

Figure 6.4(a) shows the resistance triggering method for thyristor. As seen from the circuit, the variable resistance R is used to control the gate current. When the value of gate current which depends on the resistance R is sufficient, the device triggers. Blocking diode D acts as a preventive safeguard to the gate cathode junction from getting damaged in the negative half cycle. Being a purely resistive current, the maximum firing angle can be achieved of 90°.

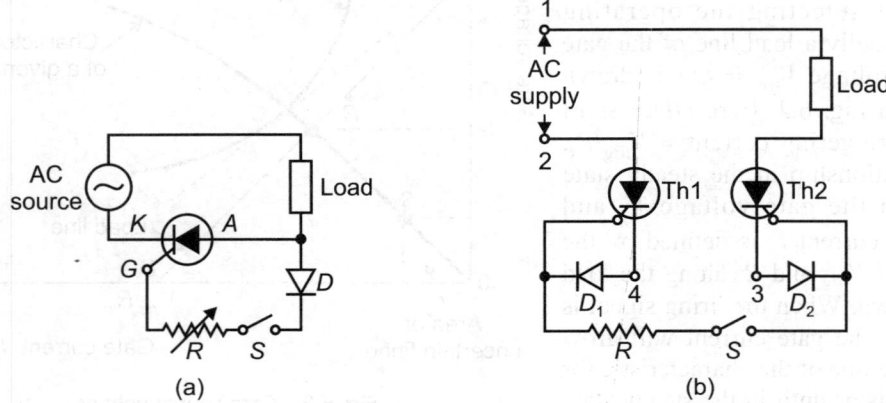

Fig. 6.4 The resistance triggering method for thyristor

6.4.2 Thyristor-Contactor Triggering

Simple circuit for switching the power to an AC loads of many amperes can be switched by only 10 mA flowing through switch S of Fig. 6.4(b). Resistor R serves to limit the peak gate current to less than 2 A. Since S connects the gate circuit to the anode, the gate voltage drops to about 1 V as soon as load current starts. Diodes D_1 and D_2 are added to prevent raverse gate current. When line 2 is positive, a small current flows from point 2 to point 3, the cathode of thyristor Th2, diode D_2, switch S and finally to point 2. This gate current quickly fires Th1 causing large number of electrons flow through the load, cathode to anode through thyristor Th1 to line 2. The gate current stops quickly. At half cycle later when line 1 is positive, current first flows from line 2 to the cathode of Th2 ant from its gate through switch S, R, and diode D_1 and finally through the load. This fires thyristor Th2. Basically this circuit enables to connect high voltage supply across load and only a small switch S. Thus, it is used as a contactor in resistance welding machine.

6.4.3 R–C Firing Circuit (AC Phase Switching Circuit)

Figure 6.5 illustrates the $R–C$ firing circuit. The role of the capacitor is to shift the phase of the anode voltage so that a positive gate current can be supplied even after the peak of anode voltage. By changing the value of resistor R, the firing angle can be controlled from 0° to 180°. The capacitor charges to the negative peak of AC voltage in every negative half cycle through the diode D_2. During the

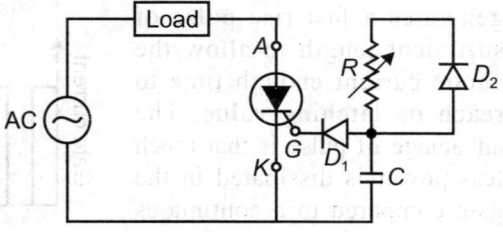

Fig. 6.5 R–C firing circuit

positive half cycle it begins to charge through the resistance R. When the voltage across the capacitor reaches the required positive value, the thyristor Th1 is triggered and the capacitor voltage remains almost constant. Diode D_1 prevents the breakdown of the gate to cathode junction during the negative half cycle.

In the range of power frequencies, for a typical thyristor, for zero output the value RC must be chosen as follows:

$$RC \geq \frac{1.3T}{2} = \frac{4}{\omega}, \ \omega = 2\pi f \qquad \qquad ...(6.3)$$

where $\qquad\qquad T = \dfrac{1}{f}$ the period of AC line frequency in secs.

The delayed phase angle is given by the equation

$$\theta = 90° + \tan^{-1} X/R \qquad \qquad ...(6.4)$$

where θ is the phase angle.

X is the capacitive reactance of C in Ohms and R is the resistance in Ohms.

When X and R are equal, the resulting phase angle is 45°.

This phase-delay gate trigger circuit suffers from two limitations. Variations in the AC supply voltage will impact on the thyristor firing angle, producing an undesired jitter. Furthermore the values of R and C may not be adequate for different AC power supply voltage levels. These problems can be minimised by employing a step down, centre tap transformer in the gate trigger circuit.

6.5 FULL-WAVE CONTROL OF AC WITH ONE THYRISTOR

The circuit in Fig. 6.6 employs a bridge rectifier D_1–D_4. Only one thyristor is required to control both the negative and positive half cycles of AC power to the load.

Switch S acts as on-off control. Also potentiometer R_2 may be eliminated and R_1 selected for proper gate current to the thyristor. This circuit is useful where

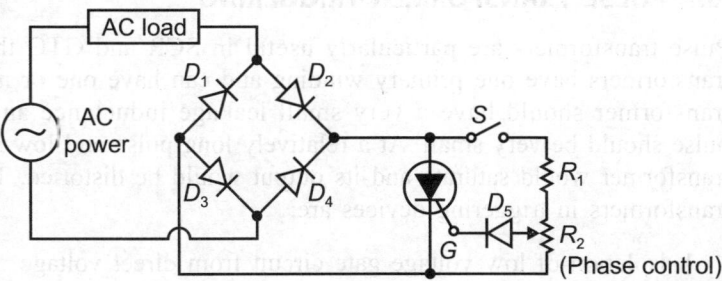

Fig. 6.6 The circuit employs a bridge rectifier D_1–D_4

low current switching for high power loads is required. A major advantage of the static switching circuits is that a small gate current can be used to control a large load current. Thyristor in these circuits acts as electronic relays without the disadvantage of arcing mechanical contacts, limited reliability and higher cost operation. Solid-state relays using thyristors are being employed in almost all phases of electrical control applications.

6.6 LIGHT ACTIVATED SCRs (LASCRs) CONTROL CIRCUIT

Commercially available LASCRs are limited to low power switching applications. This limitation is due to fabrication techniques required to produce light sensitive p-n junctions capable of switching the LASCR into conduction. Present day LASCRs can switch maximum rms currents of about 3 A. If higher power switching is required, the LASCR can be used as gate amplifier for conventional high power SCRs. Figure 6.7 shows two variations of this concept. Load power in Fig. 6.7(a) is switched on when light source is turned on. Note that the gate trigger current to the main thyristor is held to a safe operating level by resistor R_1. Resistor R_2 is simply a terminating resistor for LASCR.

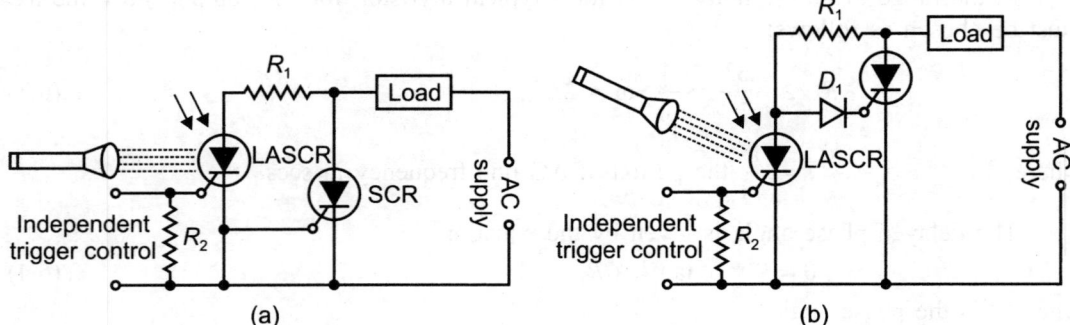

(a)　　　　　　　　　　　　　　　　　(b)

Fig. 6.7　LASCR control circuits

The main thyristor shown in Fig. 6.7(b) is held in the off-state as long as light source radiates the LASCR. During this time the LASCR turned on and shock out any gate signal to the main thyristor. When light source is removed, the LASCR ceases to conduct during the next positive half cycle. Hence, current through R_1 will be applied to the gate terminal of the main thyristor, turning it on and providing power to the load. LASCR can be controlled by the independent gate terminal connection. This is useful for emergency or back up operation.

6.7 PULSE TRANSFORMER TRIGGERING

Pulse transformers are particularly useful in SCR and GTO thyristor trigger circuits. Pulse transformers have one primary winding and can have one or more secondary windings. The transformer should have a very small leakage inductance and the rise time of the output pulse should be very small. At a relatively long pulse and low switching pulse frequency, the transformer would saturate and its output would be distorted. The advantages of using pulse transformers in triggering devices are:

1. isolation of low voltage gate circuit from direct voltage
2. two or more devices can be triggered from the same source very economically by using multi-filar secondary windings.

Either a full pulse may be faithfully transmitted or pulse transformer may be used to create a derivative processing of the input waveform so that essentially only the edges, which contain most of the high frequency content of the wave are transmitted. This condition is usually chosen as it has several advantages when pulse transformers are used in thyristor trigger circuits.

A suitable pulse transformer coupling circuit is shown in Fig. 6.8(a). Resistance R_L limits the current in the primary circuit of pulse transformer. Its equivalent circuit is drawn in Fig. 6.8(b). Here, L is the magnetising inductance of pulse transformer and R_G is the resistance of gate-cathode circuit of SCR.

In the simplified equivalent circuit of Fig. 6.8(c), the transfer of R_G to pulse transformer primary is given by $R' = n^2 R_G$ where n is the turn ratio. This circuit can be analysed using Thevenin's theorem at terminals a, b of Thevenin's equivalent shown in Fig. 6.8(d).

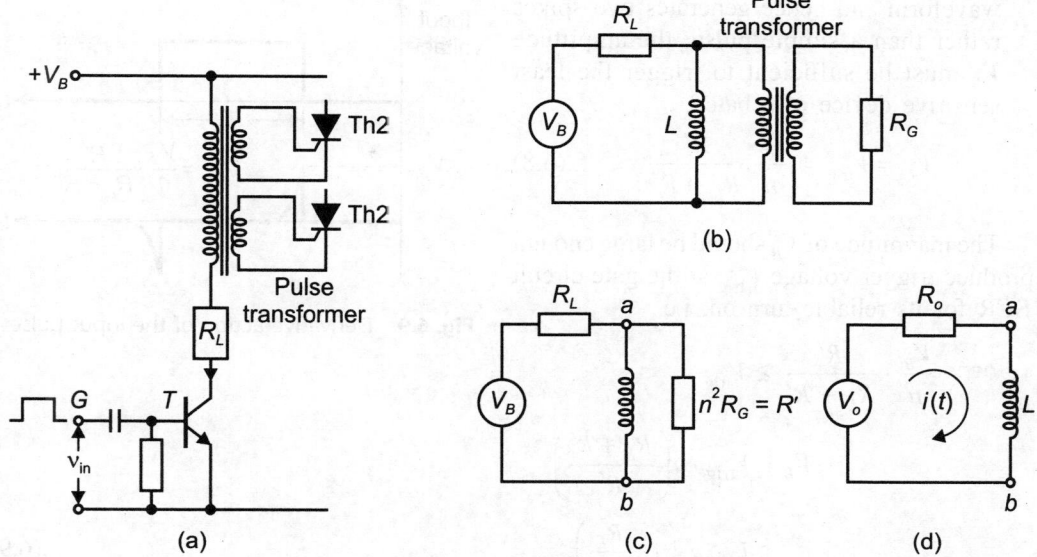

Fig. 6.8 Pulse transformer coupling circuit

$$V_0 = V_B \frac{R'}{R' + R_L} \text{ and } R_0 = \frac{R'R_L}{R' + R_L}$$

From Fig. 6.8(d), the voltage equation is

$$V_0 = R_0 i + L \, di/dt$$

or

$$\frac{V_B \cdot R'}{R' + R_L} = \frac{R'R_L}{R' + R_L} \cdot i + L \cdot \frac{di}{dt}$$

$$V_B = R_L i + L \left(\frac{R' + R_L}{R'} \right) \frac{di}{dt} \qquad \ldots (6.5)$$

which gives

$$i(t) = \frac{V_B}{R_L} \left[1 - e^{\frac{-R'R_L \cdot t}{L(R_L + R')}} \right] \qquad \ldots (6.6)$$

The transmitted voltage waveform is given by

$$e = L \cdot \frac{di}{dt}$$

$$= \frac{V_B R'}{(R' + R_L)} \cdot e^{-(R_T \cdot t/L)} \qquad \ldots (6.7)$$

where

$$R_T = R' R_L / (R' + R_L)$$

There are two operating conditions:

1. When R_T is small or L is large such that $(L/R_T) > 10T$, where T is the pulse width of the input to the base, then near faithful transmission of pulses is achieved.
2. When R_T is large or L is small such that say $(L/R_T) < T/10$, derivative action of the input pulse occurs as shown in Fig. 6.9, i.e. the transformer takes the derivative of the input

waveform and hence generates two spikes rather than a simple pulse, the amplitude V_B must be sufficient to trigger the least sensitive device of a batch.

$$V_G = V_{\text{trig}} = \frac{V_B}{n} \frac{R'}{R_L + R'} \qquad ...(6.8)$$

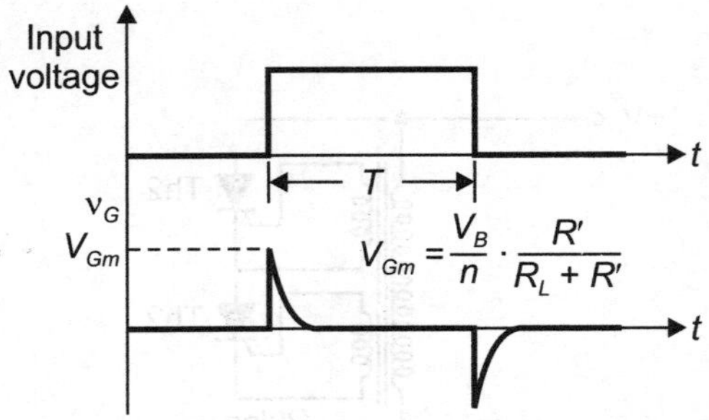

The magnitude of V_B should be large enough to produce trigger voltage V_{trig} at the gate circuit of SCR for its reliable turn on, i.e.

$$\frac{V_B}{n} \cdot \frac{R'}{R_L + R'} \geq V_{\text{trig}}$$

or

$$V_B \geq V_{\text{trig}} \cdot n \left(\frac{R_L + R'}{R'} \right)$$

$$\geq V_{\text{trig}} n \cdot \left(1 + \frac{R_L}{R'} \right). \qquad ...(6.9)$$

But

$$R' = n^2 \, (V_{\text{trig}}/I_{\text{trig}})$$

$$R' = n^2 R_{\text{trig}} \text{ of the least sensitive device}$$

$$V_B \geq \frac{V_{\text{trig}}}{n} \left[n^2 + (R_L/R_{\text{trig}}) \right] \qquad ...(6.10)$$

Fig. 6.9 Derivative action of the input pulse

Trigger pulses of Fig. 6.9 are preferred due to the reasons that the gate waveform is suitable to inject a large charge without significant heating of the gate.

6.8 CONTROL OF CONVERTER

The function of the controller is to control the firing angle of a converter symmetrically in response to a demand of DC voltage or current. The controller usually incorporates the following functions:

1. synchronising
2. firing angle control
3. advance limit control
4. retard limit control

The synchronising circuit helps to establish symmetrical firing angle control to all the thyristors with respect to the fixed angular position of the AC voltage wave. The firing control circuit alters the firing angle α in response to a variable input control voltage. The advance and retard limit controls restrict the α angle within safe limits. The advance limit angle can be established as early as $\alpha = 0$, but the retard limit angle should provide a sufficient margin so that a minimum turn off angle γ is maintained for successful commutation.

6.9 FIRING ANGLE CONTROL

It has been assumed that thyristors in the converters are fired at desired instant for proper operation. The firing instant is controlled by control signal so that the converter can provide a desired output performance. There are numerous variations of firing circuits and control logic circuits that can be used to control the firing of thyristors. In single-phase converters, the reference point is the zero crossing of the supply voltage. The firing angle is changed to vary the output DC voltage, which can control the speed of DC motor. There are two basic approaches to achieve this:

1. linear control of phase angle, α
2. cosine control of phase angle

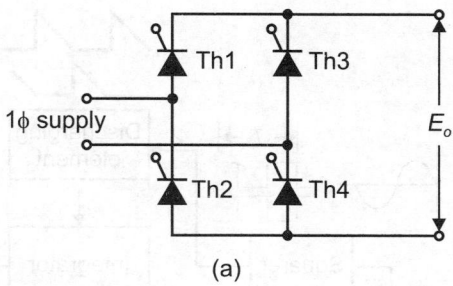

(a)

6.9.1 Linear Control of Phase Angle, α

In the scheme shown in Fig. 6.10 (b), a synchronising centre tapped transformer steps down the supply voltage to low voltages v_1 and v_2. The voltage v_1 is converted to a square voltage

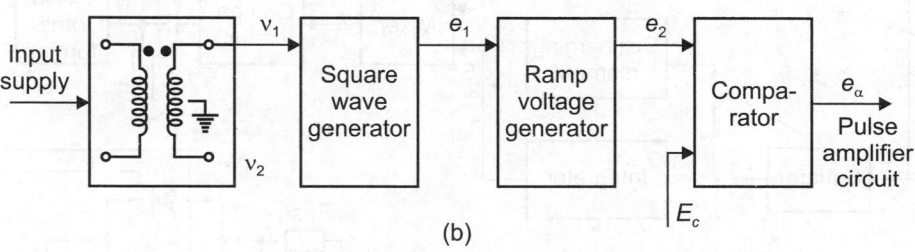

(b)

Fig. 6.10 Linear control of phase angle

e_1 and then to a ramp voltage e_2, which is then compared with a control voltage E_c. If e_2 is higher than E_c, a signal e_α is obtained at the output of the comparator. The time at which the rising edge of e_α occurs is proportional to E_c and defines the firing angle α. Thus, signal e_α is next fed to a pulse amplifier circuit and is used to fire thyristors Th1 and Th2 in the positive half cycle of the input supply. A similar circuit with input v_2 provides firing pulses

for thyristors Th3 and Th4 to be fired in the negative half cycle of the input supply of Figs 6.10(a), (c). The firing angle is given by

$$\alpha = K_1 E_c$$

The voltage of the converter,

$$E_o = E_m \cos \alpha = E_m \cos (k_1 E_c)$$

In this scheme a control voltage E_c changes linearly with the phase angle α.

Block diagram of firing circuit for phase angle control ($0 < \alpha < 180°$) is shown in Fig. 6.11. This shows the essential features of a precision analogue based firing controller for phase angle.

Phase control of the SCR: Phase control of the thyristor is the most commonly used form of control. A typical circuit using AC line voltage as the anode voltage for the thyristor for the input and output voltages as a function of time is shown in Fig. 6.12. A detailed analysis for single-phase load is given below.

In AC circuits, the SCR can be turned on by the gate at any angle α with respect to the

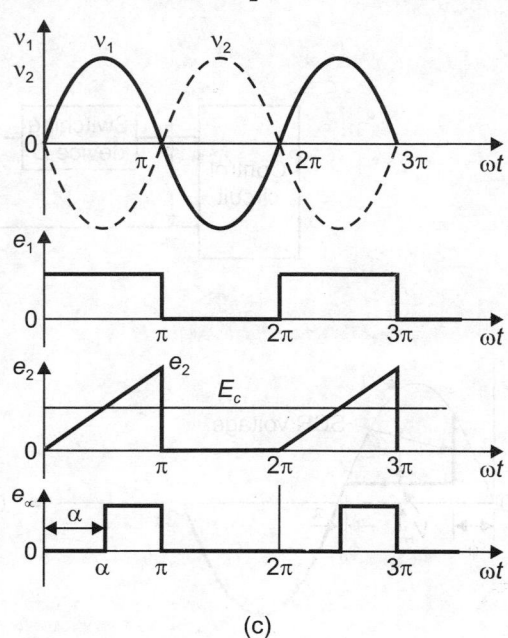

(c)

Fig. 6.10 Linear control of phase angle α

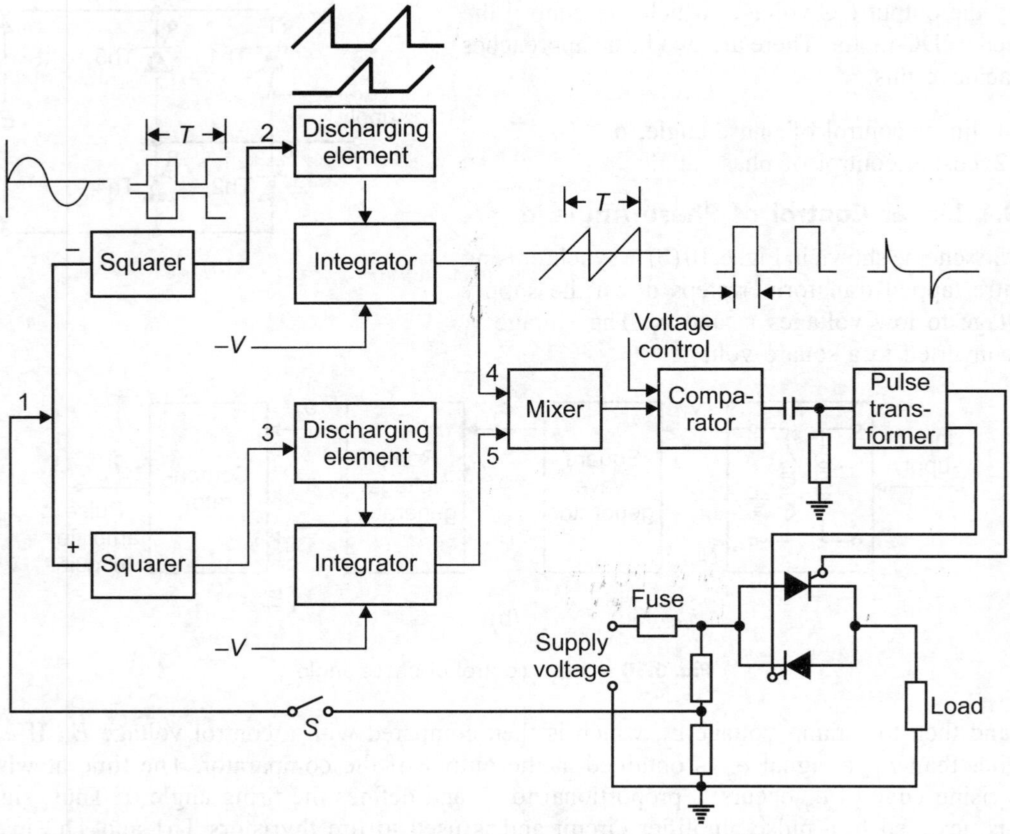

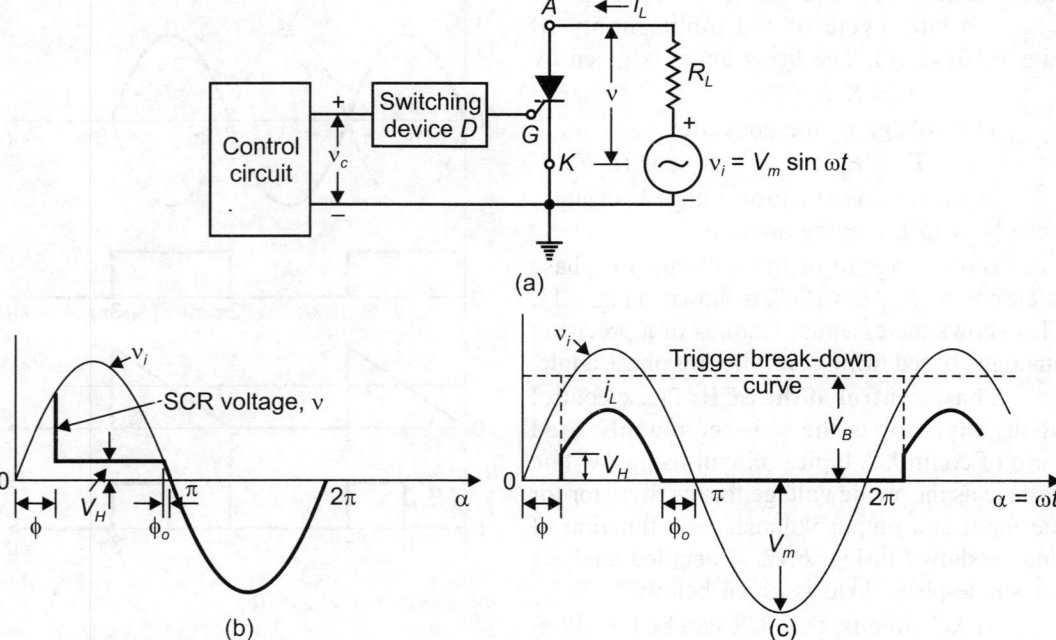

Fig. 6.11 Block diagram of firing circuit for phase angle control

Fig. 6.12 Circuit diagram of firing circuit for phase angle control and its characteristics

applied voltage. This angle α is the *firing angle*. Power control is obtained by varying the firing angle and this is known as *phase control*.

If a sinusoidal potential is applied to the SCR anode, the device will be turned OFF when the voltage falls below the holding voltage, provided it is triggered ON regularly. The average rectified current can be varied over wide limits by controlling the point in each half cycle at which the SCR is turned ON. In Fig. 6.12(a), AC line voltage is used as the anode voltage for the SCR. A switching device, such as pn-pn diode is connected in series with the control terminal. When the control voltage v_c exceeds the trigger breakdown voltage V_{BR} of the switching device D, then D goes to its low-resistance state and the current through the SCR gate triggers it ON. In Fig. 6.12(b), the line wave represents the input line voltage v_i as a function of time. The trigger breakdown curve is a straight line parallel to the time axis indicating that V_{BR} is independent of the anode potential.

Suppose that the control voltage v_c exceeds the trigger breakdown voltage by arranging the circuit at some angle ϕ, called the *delay angle*.

Conduction will start at this point in the cycle. The voltage drop across the SCR during conduction remains constant at a low value which is independent of current. This voltage drop V_H is of the order of 1 V. The current through a pure resistance load R_L during the time, SCR is conducting is given by

$$i_L = \frac{V_m \sin \omega t - V_H}{R_L} \qquad \text{...(6.11)}$$

where V_m is the maximum value of the applied potential. Figure 6.12(c) illustrates the resulting waveform of load current. The current rises abruptly at the point corresponding to the angle ϕ and then follows the sine variation given in Equation (6.11), until the supply voltage v_i falls below V_H at the phase $\pi - \phi_0$. The current will remain zero, until the phase ϕ is again reached in the next cycle.

The average current
$$I_{DC} = \frac{1}{2}\pi \int_{\phi}^{\pi-\phi_0} i_L d\alpha = \frac{V_m}{2\pi R_L} \int_{\phi}^{\pi-\phi_0} \left(\sin \alpha - \frac{V_H}{V_m} \right) d\alpha$$

$$I_{DC} = \frac{V_m}{2\pi R_L}\left[\cos \phi + \cos \phi_0 - \frac{V_H}{V_m}(\pi - \phi_0 - \phi) \right] \qquad \text{...(6.12)}$$

where $\alpha = \omega t$ and ϕ_0 is the smallest angle defined by the relation

$$V_H = V_m \sin \phi_0$$

If the ratio $\dfrac{V_H}{V_m}$ is very small, then ϕ_0 may be taken as zero and Equation (6.13) reduces to the form,

$$I_{DC} = \frac{V_m}{2\pi R_L}(1 + \cos \phi) \qquad \text{...(6.13)}$$

The maximum current is obtained when the SCR is triggered ON at the beginning of each cycle, and the minimum current is obtained when no conduction occurs. The voltage across SCR is shown in Fig. 6.12(c). The applied AC voltage v_i appears across the SCR until conduction begins. After breakdown, the SCR voltage drop is a constant equal to V_H. When the applied voltage falls below V_H, the SCR voltage is again equal to the applied voltage.

The reading of the DC voltmeter placed across the SCR will be

$$V_{DC} = \frac{1}{2\pi}\left[\int_{0}^{2\pi} v_d \alpha = \frac{1}{2\pi}\left[\int_{0}^{\phi} V_m \sin \alpha \, d\alpha + \int_{\phi}^{\pi-\phi_0} V_H d\alpha + \int_{\pi-\phi_0}^{2\pi} V_m \sin \alpha \, d\alpha \right] \qquad \text{...(6.14)}$$

or thus
$$V_{DC} = \frac{V_H}{2\pi}(\pi - \phi_0 - \phi) - \frac{V_m}{2\pi}(\cos\phi + \cos\phi_0)$$

If $V_m > V_H$ this reduces to

$$V_{DC} \approx -\frac{V_m}{2\pi}(1 + \cos\phi), \qquad\qquad ...(6.15)$$

the negative sign means that the cathode is more positive than the anode for most of the cycle.

6.9.2 Cosine Wave-crossing Method

In this scheme a control voltage E_c generates firing pulses at the crossing point of the control voltage and a cosine voltage derived from the supply voltage. The basic scheme is illustrated in Fig. 6.13. The phase angle α is given by

$$\alpha = \cos^{-1}\left[\frac{E_c}{e_{max}}\right]$$

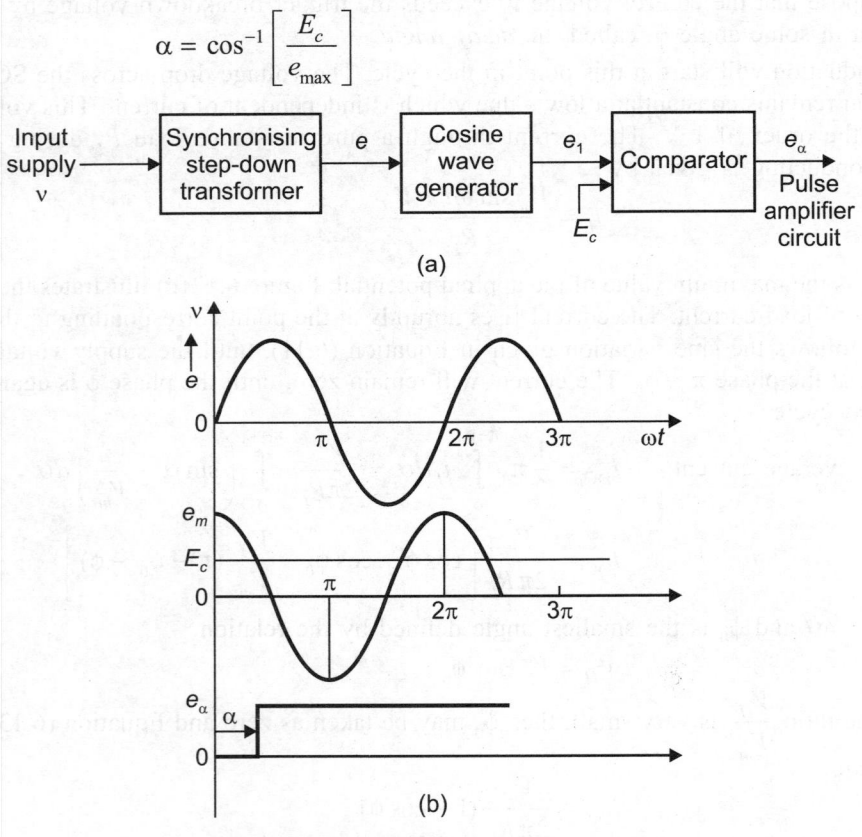

(a)

(b)

Fig. 6.13 Cosine control of phase angle

The output voltage of the converter is given by

$$E_o = E_{max}\cos\alpha$$

$$= E_{max}\cos\left[\cos^{-1}\frac{E_c}{e_{max}}\right]$$

$$= \frac{E_{max}}{e_{max}}E_c$$

$$= k_2 E_c$$

Therefore, the cosine firing scheme provides a linear transfer characteristic between the output voltage E_o and the control voltage E_c. This scheme is suitable for application in a closed-loop control drive. In fact, the cosine firing scheme is popular and widely used in industry.

Figure 6.14(a)(i) explains this method for a single-phase bridge converter. The supply voltage sine wave is phase shifted by 90° to generate the cosine wave and it is phase inverted every half cycle as shown in Fig. 6.14(a)(ii). The firing angle is generated by the crossover point of control voltage V_c and cosine wave as

$$\cos \alpha = \frac{V_c}{V_P} \qquad \ldots(6.16)$$

where V_P is the peak value of the cosine wave. Substituting Equation (6.16) in $V_d = V_{d0} \cos \alpha$ yields

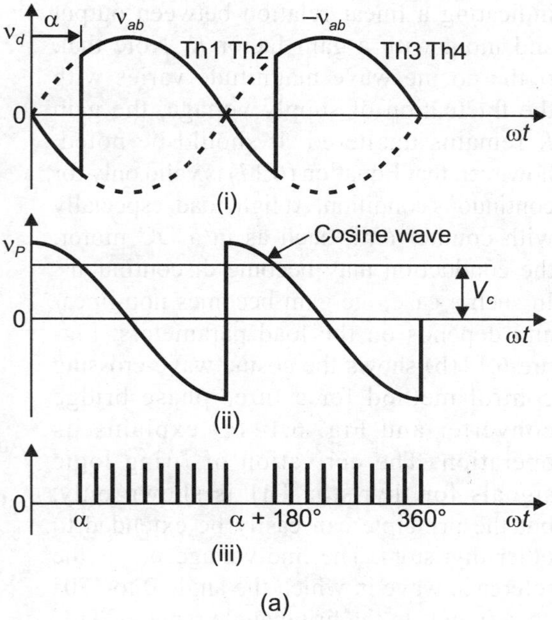

$$V_d = V_{d0}/V_P V_c = KV_c \qquad \ldots(6.17)$$

Fig. 6.14 Single-phase converter

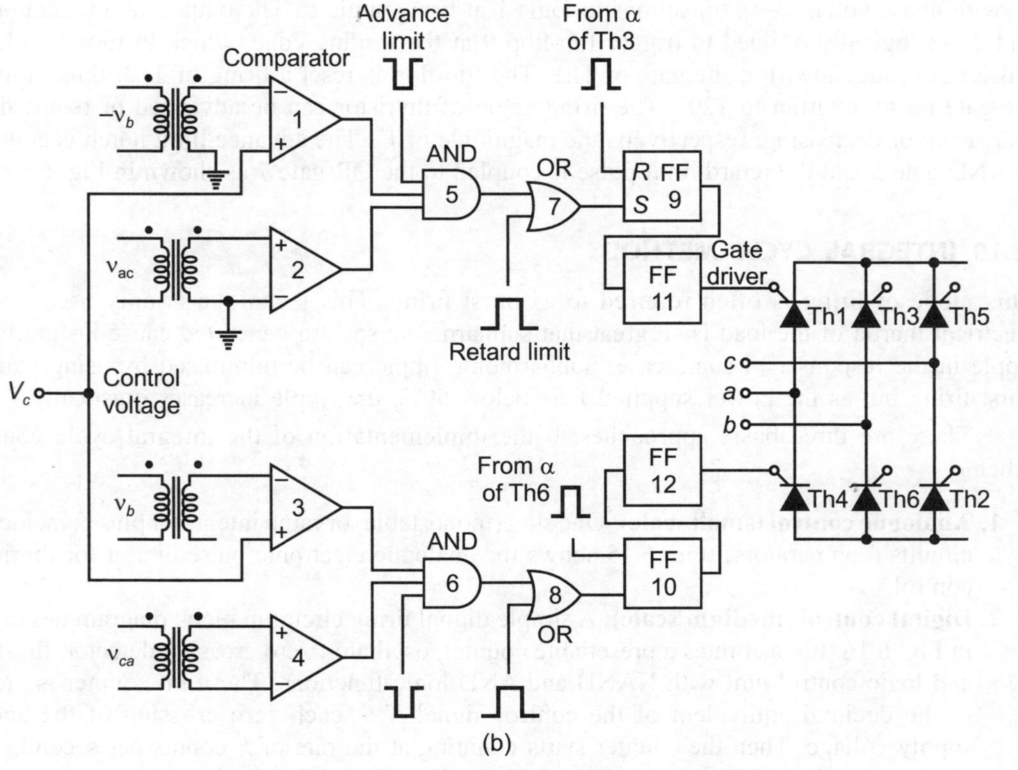

Fig. 6.14 Cosine wave-crossing method for three-phase bridge converter

indicating a linear relation between output and input with a gain factor K. Note that, if the cosine wave magnitude varies with the fluctuation of supply voltage, the gain K remains unaltered. It should be noted, however, that Equation (6.17) is valid only for continuous condition. At light load, especially with counter emf, such as in a DC motor, the conduction may become discontinuous. In such a case, the gain becomes non-linear and depends on the load parameters. Figure 6.14(b) shows the cosine wave-crossing control method for a three-phase bridge converter and Fig. 6.14(c) explains its operation. The derivation of firing logic signals for thyristor Th1 is shown only, but the principle can easily be extended to other thyristors. The line voltage v_{AC} is the reference wave in which the angle 0 to 180° corresponds to the firing angle range of Th1. The phase voltage $-v_b$ leads v_{AC} by 90° and constitutes the cosine reference wave

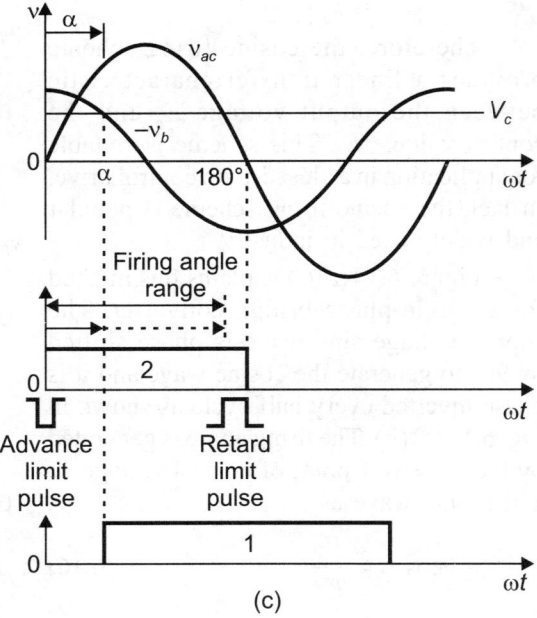

Fig. 6.14 Waveforms

for thyristor Th1. The phase and line voltages are stepped down through transformers and connected to the comparators as shown. The comparator 1, which compares the control voltage V_c with phase voltage $-v_b$, transitions to logic 1 at firing angle α. The output of comparators 1 and 2 are logically ANDed to trigger flip-flop 9 at the leading edge, which in turn couples a pulse train (not shown) to the gate of Th1. The flip-flop is reset a firing of Th3, thus limiting the gate pulse duration to 120°. The firing angle of thyristor can be advanced or retarded by increasing or decreasing, respectively, the magnitude of V_c. The advance limit notch is coupled to AND gate 5 and the retard limit pulse is coupled to the OR gate 7 as shown in Fig. 6.14(b).

6.10 INTEGRAL CYCLE METHOD

This mode of firing is often referred to as burst firing. This technique is only used where electrical inertia of the load is so great that subharmonic spectra generated cause insignificant ripple in the response. In some cases subharmonic ripple can be minimised by using pattern burst firing but as the power supplied falls below 50%, the ripple increases drastically.

There are three basic approaches to the implementation of the integral cycle control scheme:

1. **Analogue control (small scale):** One shot (monostable) or ramp integrator plus coincidence circuits (comparators), Fig. 6.15 shows the analogue electronic pulse circuit for thyristor control.
2. **Digital control (medium scale):** A simple digital firing circuit in block diagram described in Fig. 6.16. It constitutes a presettable counter, oscillator, zero crossing detector, flip-flop and logic control unit with NAND and AND logic functions. The n bit counter is preset to the decimal equivalent of the control signal N at each zero crossing of the anode supply voltage. Then the counter starts counting at the rate of f_c counts per second. The counter overflow signal max/min is processed to trigger the thyristor.

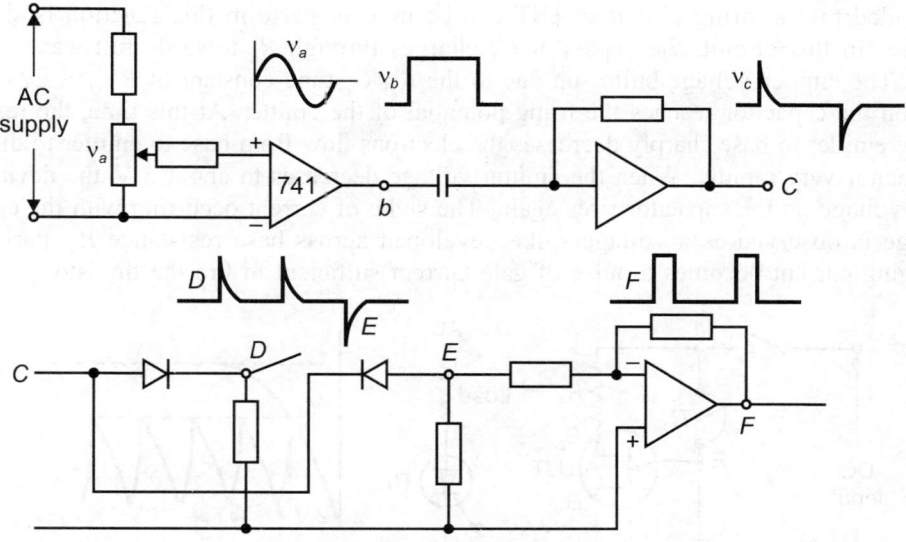

Fig. 6.15 The analogue electronic pulse circuit for thyristor control

With the help of flip-flop and a logic control unit and driver stage, the trigger pulses are produced which can be fed to thyristors thus by varying the preset input, it is possible to control the firing angle.

3. **Microprocessor or computer control (large scale):** The selected pulse pattern may depend on the values of several interrelated variables.

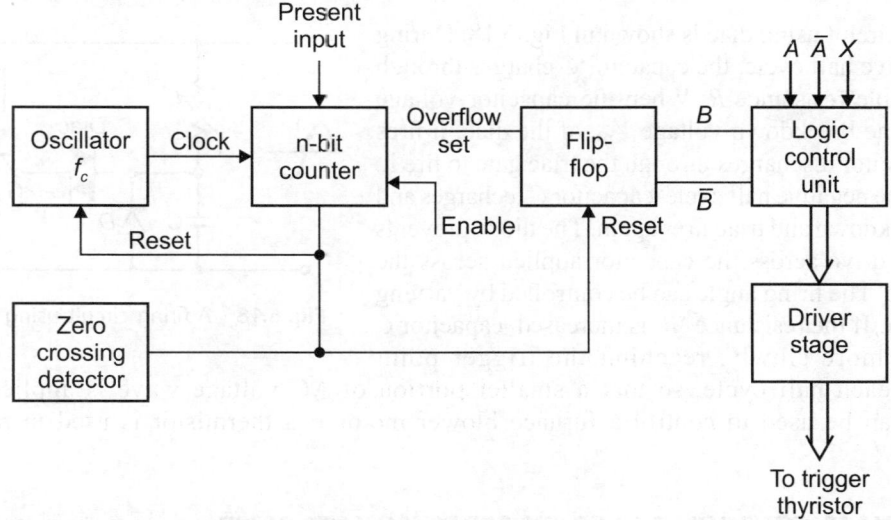

Fig. 6.16 A simple digital firing circuit

6.11 FIRING OF SCR BY UJT

To fire the thyristor Th short-time current pulse at its gate is to be applied. The delay of firing and controlling the instant at which the pulse occurs within each half cycle is to be provided by the firing circuit. A UJT can be used to perform this function is shown in Fig. 6.17. In this circuit, the capacitor C_1 charges through R_8 towards full-wave rectified voltage. The emitter voltage builds-up due to the R_3, C_1 time constant of R_3 and C_1 until the charge on the capacitor reaches the firing potential of the emitter. At this time, the resistance from the emitter to base sharply decreases the electrons flow from base to emitter to discharge the capacitor very rapidly. When the emitter voltage decreases to about 2 V, the device turns off and voltage on the capacitor rises again. The spike of current occurring with the capacitor discharge is observed as a voltage spike developed across base resistance R_5. Part of this discharging current becomes a pulse of gate current sufficient to fire the thyristor.

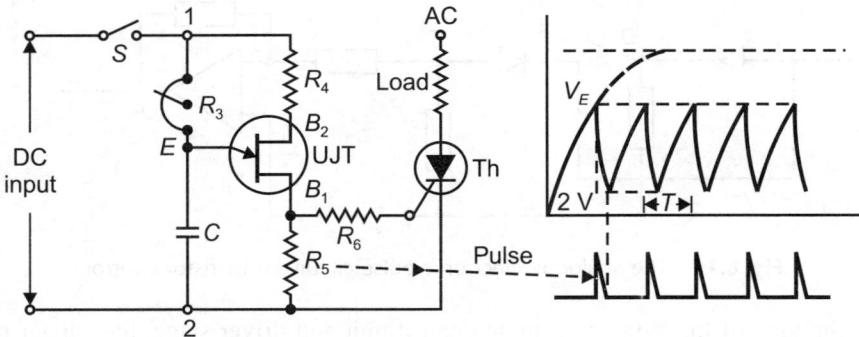

Fig. 6.17 Firing of SCR by UJT

6.12 TRIAC FIRING CIRCUIT

A firing circuit using diac is shown in Fig. 6.18. During the positive half cycle, the capacitor C charges through the variable resistance R. When the capacitor voltage reaches the breakdown voltage V_{BR} of the diac, it fires and capacitor discharges through the triac gate to fire it. During the negative half cycle, capacitor C recharges and diac breakdown and triac fires again. The diode prevents negative drive across the capacitor applied across the SCR gate. The firing angle can be controlled by varying resistor R. If the resistance 'R' is increased, capacitor C charges more slowly, reaching the trigger point

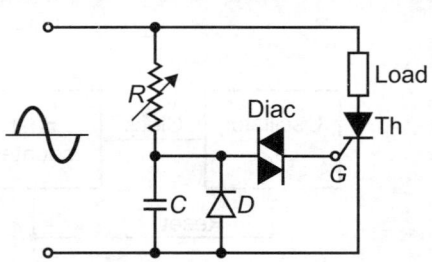

Fig. 6.18 A firing circuit using DIAC

later in each half cycle, so that a smaller portion of AC voltage wave is applied. This circuit can be used to control a furnace blower motor if a thermistor is used in place of rheostat R.

6.13 PHASE CONTROL OF SCR BY PEDESTAL AND RAMP

Figure 6.19 consists of voltage divider R_3, diode D_1 pot. P and the top of resistance R_5 is directly connected to the AC supply at line 1. During the each positive AC half cycle, zener diode Z limits the point 3 to a constant voltage of +20 V, and holds this constant voltage across the UJT. A part of the voltage appears across adjuster P, holding point 5 at say, 2 V. Although the top of capacitor C_1 is negative during the preceding half cycle; point 6 now rises rapidly as C_1 charges toward line 1 and electrons flow through diode D_1 and small resistors

R_5 and R_1. As line 1 rises in its half cycle and as point 6 reaches the +2 V-level of point 5 shown at A [Fig. 6.19(b)], diode D_1 becomes reverse biased and disconnects 6 from 5. Point 6 continues to rise, but much more slowly, as electrons flow through the large resistor R_5 toward the high potential of line 1. At point F the capacitor C_1 raises the voltage of the emitter E of the UJT above the V_E that triggers UJT, this occurs late in the half cycle, firing the thyristor, Th so that only the last portion of the AC, voltage wave is applied to the load. When the AC line voltage reverses, thyristor, Th is shut off, +20 V disappears, making the UJT current to stop; Capacitor C_1 charges through R_5 toward the potential of line 1.

In Fig. 6.19, if voltage divider P is turned clockwise, increasing its resistance and raising point 5 to 6 V shown at G. Capacitor C_1 charges rapidly to this adjustable DC value called the pedestal then follows the lesser slope called the ramp toward H, at which point the UJT is triggered thus firing the thyristor, Th about midway, i.e. 90° in the half cycle. When further increase of the P resistance raises the pedestal point 5 to +10 V shown at J, the slow rise of point 6 along its ramp triggers UJT at K, triggering thyristor, Th quite early in the AC half cycle. The manual rotating of pot P changes the amount of DC voltage between 2 and 5 and thereby varies the firing point of UJT so as to provide control of thyristor, Th. Similarly, a change in DC voltage caused by another device may produce phase control.

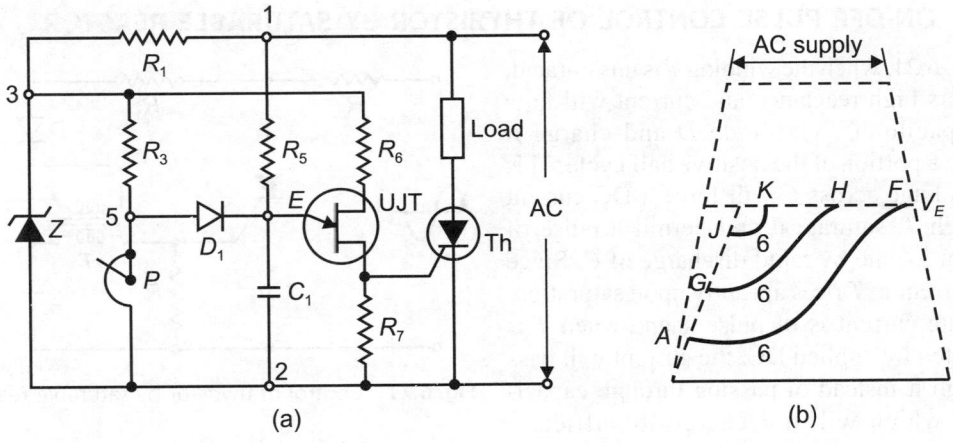

Fig. 6.19 Phase control of SCR using pedestal and ramp

6.14 SCR PHASE CONTROL BY TEMPERATURE OR LIGHT

In Fig. 6.20, a thermistor T senses a temperature and controls SCR so as to hold that temperature constant. A thermistor is a semiconductor made so that its internal resistance decreases greatly as its temperature rises, thus generating additional electron-hole pair. As the temperature of the R load increases, thermistor T senses this higher temperature and decreases the internal resistance of T. Increased current through T and R_2 lowers base 4 of n-p-n transistor Q, as emitter current through R_4 decreases, this lowers pedestal point 5 so as to trigger UJT later. Thus, thyristor, Th is fired later in its half cycle and decreases the average voltage applied to the R load.

A photo conductive cell may be substituted for thermistor T in Fig. 6.20. A great light reaches such a cell, increasing its electron flow and decreasing the cell resistance, SCR is fired later in its half cycle.

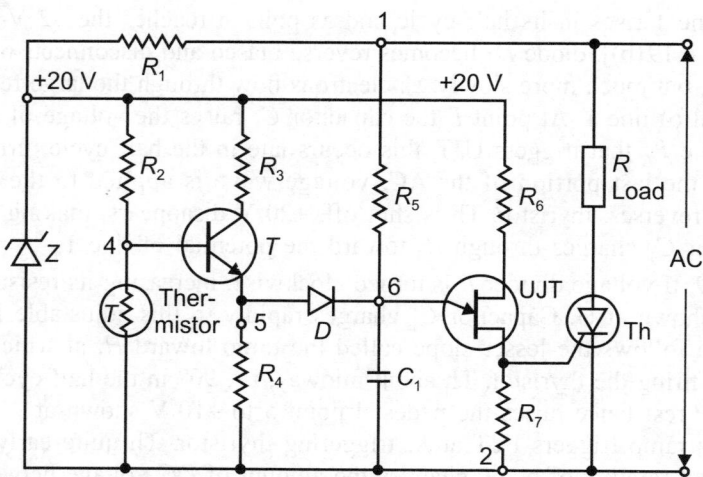

Fig. 6.20 A thermistor T senses the temperature and controls SCR

6.15 ON-OFF PULSE CONTROL OF THYRISTOR BY SATURABLE REACTOR

In Fig. 6.21, when the winding T is unsaturated, it offers high reactance and current will flow in capacitor C, via diode D and charge it during a portion of the positive half cycle. This DC voltage across C will force a DC current through T, saturate it and permit a pulse of current to gate by rapid discharge of C. Since the current in T rises abruptly upon saturation, the gate current is of pulse shape when T is saturated by applied DC, the current will pass through it instead of passing through capacitor C which will not charge to sufficient voltage to turn SCR-ON. Thus, thyristor acts as an AC switch controlled by DC.

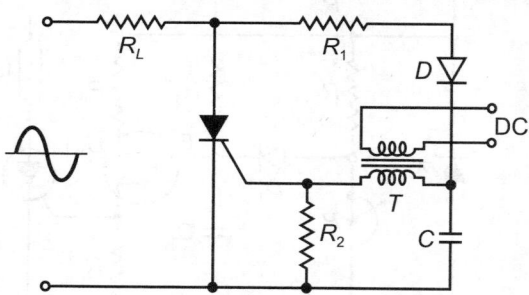

Fig. 6.21 Control of thyristor by saturable reactor

6.16 BLOCKING OSCILLATOR

Pulse transformer arrangement: In many circuits, particularly where thyristors are controlling the power to inductive loads, it is often considered desirable to trigger with a pulse train to ensure that the load current has risen above the sustaining value before the signal on the gate is removed. This can be achieved by a blocking oscillator circuit which is a combination of pulse transformer with feedback winding to create regeneration conveniently.

A blocking oscillator is a particularly convenient circuit for monostable or astable operation when the triggering of the thyristors is required because the timing element for the pulse is a transformer, which is thermally very stable.

Figure 6.22 depicts the astable circuit and shows its equivalent circuit when the transistor is saturated from the basic transformer theory,

$$v_1 = \frac{-n_1}{n_2} v_2$$

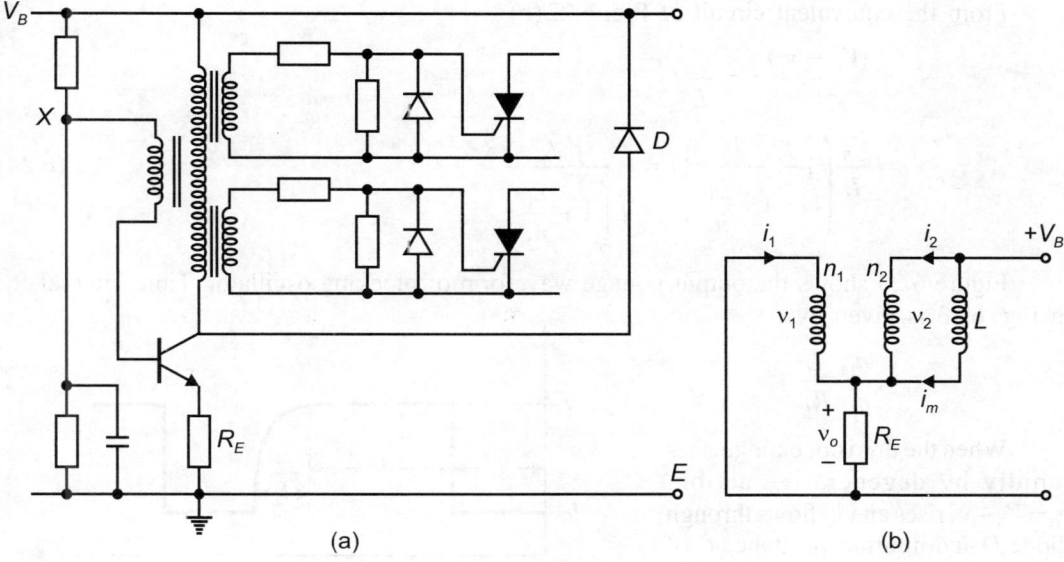

Fig. 6.22 Blocking oscillator circuit

and
$$i_1 = \frac{n_2}{n_1} i_2 \qquad \qquad \text{...(6.18)}$$

$\therefore$
$$V_B = v_2 - v_1 = v_2 - \left(\frac{-n_1}{n_2} \right) v_2$$

$$V_B = v_2 + \frac{n_1}{n_2} v_2 = v_2 \left(1 + \frac{n_1}{n_2} \right) \qquad \qquad \text{...(6.19)}$$

Also $\qquad (i_1 + i_2 + i_m) R_E = V_B - v_2 \qquad \qquad \text{...(6.20)}$

The transistor desaturates when

$$i_2 + i_m = (h_{FE} + 1) i_1 \approx h_{FE} i_1 \qquad \qquad \text{...(6.21)}$$

or $\qquad i_m = h_{FE} i_1 - i_2$

$$= h_{FE} \left(\frac{n_2}{n_1} \cdot i_2 \right) - i_2$$

$$i_m = i_2 \left(h_{FE} \frac{n_2}{n_1} - 1 \right)$$

$$i_m \approx \frac{i_m}{h_{FE}} \cdot \frac{n_2}{n_1} \qquad \qquad \text{...(6.22)}$$

Using Equations (6.18), (6.20), and (6.21), we have

$$\left[\left(1 + \frac{n_2}{n_1} \right) \frac{n_1}{n_2 h_{FE}} + 1 \right] i_m R_E = \frac{1}{\frac{n_2}{n_1} + 1} V_B \qquad \qquad \text{...(6.23)}$$

From the equivalent circuit of Fig. 6.22 (b)

$$i_m = \frac{(V_B - v_1)}{L} t = \frac{v_2}{L} t$$

$$= \frac{V_B}{L} \left(1 - \frac{1}{1 + \frac{n_1}{n_2}} \right) t = \frac{V_B t}{L \left(1 + \frac{n_1}{n_2} \right)} \qquad \qquad ...(6.24)$$

Figure 6.23 shows the output voltage waveform of blocking oscillator. Time interval 't_1' in Fig. 6.23 is given by

$$t_1 \approx \frac{n_1}{n_2} \frac{L}{R_E}$$

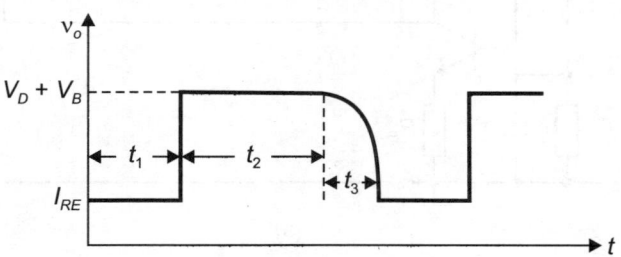

When the thyristor extinguishes rapidly by degenerative action, $v_0 = V_B - v_2$ rises and i_m flows through diode D and inturn capacitance C of the transformer. Voltage v_0 ceases at $V_B + V_D$ and current i_m starts to fall from I_{mo}

Fig. 6.23 The output voltage waveform of blocking oscillator

$$i_m = I_{mo} - \frac{V_D(t - t_1)}{L}$$

when i_m falls to zero, the time interval t_2 is given by

$$t_2 = L \frac{I_{mo}}{V_D} = \frac{\frac{n_1}{n_2} L \cdot V_B}{\left(\frac{n_1}{n_2} + 1 \right) R_E V_D}$$

When the diode ceases to conduct, the voltage falls to V_B by transformer resonating for a period of $T/4$ where

$$\frac{1}{T} = \frac{1}{2\pi\sqrt{LC}}$$

The time interval t_3 is given by

$$t_3 = 1.6\sqrt{LC}$$

The frequency of oscillation is

$$f_{osc} = \frac{1}{t_1 + t_2 + t_3} \text{ Hz}$$

It is worth noting that astable multivibrator can be conveniently constructed which can oscillate with very low duty cycle and powerful pulses can be generated.

SOLVED EXAMPLES

Example 6.1 The thyristor in Fig. E6.1 has a latching current level of 50 mA and is fired by a pulse of length 50 μsec show that without resistance R the thyristor will fail to remain on when firing pulse ends, and then find the maximum value of R to ensure firing. Neglect the thyristor volt drop.

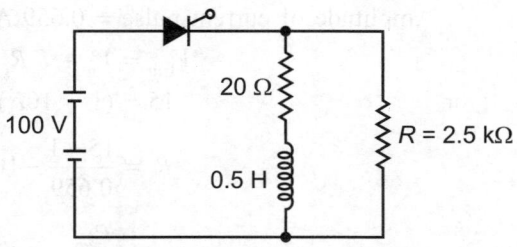

Fig. E6.1 Circuit diagram

Solution Without R, thyristor current i will grow exponentially, given after a time t as

$$i = I(1 - e^{-t/T})$$

where $I = 100/20 = 5$ A (time constant), $T = 0.5/20 = 0.025$ sec. Substituting values, after 50 μsec, $i = 10$ mA. Note that in this case as 50 μsec $\ll T$; the initial value of $di/dt = 100/0.5$ could have been assumed constant over 50 μsec, giving $i = 10$ mA. After 50 μsec, the thyristor has failed to reach its latching level, being $(50 - 10)$ mA below the required level of 50 mA. The addition of R will permit an immediate current of $100/R$ to be calculated.

$$\text{Required value of } R = \frac{100}{(50 - 10) \times 10^{-3}} = 2.5 \text{ k}\Omega.$$

Example 6.2 For the SCR, the gate cathode characteristic has a straight line slope of 130. For trigger source voltage of 15 V and allowable gate power dissipation of 0.5 W, compute the gate source resistance.

Solution Given $V_g I_g = 0.5$ W

Slope $\dfrac{V_g}{I_g} = 130 = R_g$, gate source resistance

∴ $I_g^2 \times 130 = 0.5 \text{ or } I_g = \left[\dfrac{0.5}{130}\right]^{1/2} = 62 \text{ mA}$

Gate voltage, $V_g = I_g R_g = 62 \times 10^{-3} \times 130 = 8.06$ V

For the gate circuit, $V_{trg} = I_g R_s + V_g = 0.062 R_s + 8.06$

or $15 = 0.062 R_s + 8.06$

or $R_s = \dfrac{15 - 8.06}{0.062} = 111.95 \ \Omega.$

Example 6.3 For an SCR, gate cathode characteristic is given by $V_g = 1 + 10 I_g$. Gate source voltage is a rectangular pulse of 15 V with 20 μsec duration. For an average gate power dissipation of 0.3 W and a peak gate drive power of 5 W, compute: (a) the resistance to be connected in series with the SCR gate, (b) triggering frequency, and (c) duty cycle of the triggering pulse.

Solution

(a) Given $V_g = 1 + 10 I_g$

As the gate pulse width is 20 μsec (less than 100 μsec) the DC data does not apply. If the gate pulse width has been more than 100 μsec, the relation $(1 + 10 I_g) I_g = 0.3$ W will hold good. But as the data does not apply, we have

$$(1 + 10 I_g) I_g = 5 \text{ W}$$

or $I_g + 10 I_g^2 - 5 = 0$

or $I_g = 0.659$ A

∴ Amplitude of current pulse = 0.659 A. During the pulse on period,

$$V_{trig} = V_g + I_g R_s$$

or

$$15 = (1 + 10 I_g) + I_g R_s$$

$$R_s = \frac{15 - 1}{0.659} - 10 = 11.245 \, \Omega$$

(b)

$$P_{gm} = \frac{P_{gav}}{fT} \quad \text{given } T = 20 \, \mu sec$$

Triggering frequency, $\quad f = \dfrac{0.3}{5 \times 20 \times 10^{-6}} = 3 \text{ kHz}$

(c) Duty cycle $\quad\quad \delta = fT = 3 \times 10^3 \times 20 \times 10^{-6} = 0.06.$

Example 6.4   The triggering circuit of a thyristor has a source voltage of 15 V and the load line has a slope of −120 V/amp. The minimum gate current to turn on the SCR is 25 mA, compute: (a) source resistance required in the gate circuit, and (b) trigger current and trigger voltage for an average power dissipation of 0.4 W.

Solution

(a) Slope of the load line gives the required gate source resistance from the load line, series resistance required in the gate circuit is $120 \, \Omega$.

(b) Given $\quad\quad\quad V_g I_g = 0.4 \text{ W or } V_g = \dfrac{0.4}{I_g}$

For the gate circuit, $\quad V_{trig} = V_g + I_g R_s$

$$15 = \frac{0.4}{I_g} + I_g \times 120$$

or $\quad\quad 120 I_g^2 - 15 I_g + 0.4 = 0$

or $\quad\quad\quad\quad I_g = 38.56 \text{ mA or } 86.44 \text{ mA}$

$$V_g = \frac{0.4}{38.56 \times 10^{-3}} = 10.37 \text{ V}$$

and $\quad\quad\quad V_g = \dfrac{0.4}{86.44 \times 10^{-3}} = 4.627 \text{ V}$

Take $\quad\quad\quad V_g = 4.627 \text{ V}$

and $\quad\quad\quad I_g = 86.44 \text{ mA for minimum gate current of 25 mA.}$

Example 6.5   An SCR has $V_g - I_g$ characteristic given as $V_g = 1.5 + 8 I_g$. In a certain application the gate voltage consists of rectangular pulses of 12 V and of duration 50 μsec with duty cycle of $\delta = 1/5$. (a) Calculate the value of R_g, series resistor in gate circuit to limit the peak power dissipation in the gate to 5 watts and (b) Calculate the power dissipation in the gate.

Solution   During the conduction,

$$V_{trig} = V_g + I_g R_g$$

$$12 = (1.5 + 8 I_g) + I_g R_g$$

$$10.5 = I_g (8 + R_g)$$

Peak power loss, $\quad V_g I_g = 5 \text{ W}$

or $\qquad (1.5 + 8I_g)\, I_g = 5$

or $\qquad (1.5I_g + 8I_g^2 - 5) = 0$

or $\qquad 8I_g + 1.5I_g - 5 = 0$

or $\qquad I_g = 0.7\ \text{A}$

$$R_g = \frac{10.5}{0.7} - 8 = 7\ \Omega$$

Average power loss $\qquad$ = Duty cycle × peak power loss

$$= \frac{1}{5} \times 5 = 1\ \text{W}.$$

Example 6.6 A thyristor has a forward characteristic which may be approximated over its normal working range to the straight line shown in Fig. E6.6. Estimate the mean power loss for: (a) a continuous on-state current of 23 A, (b) a half sine wave of mean value 18 A, (c) a level current of 39.6 A for one-half cycle, and (d) a level current of 48.4 A for one-third cycle.

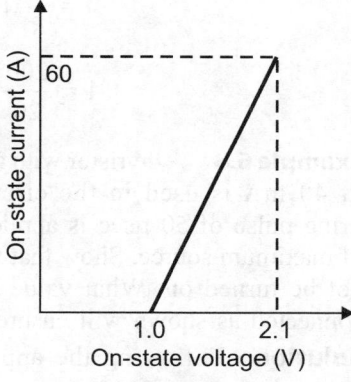

Fig. E6.6 Circuit diagram

Solution

(a) At 23 A, the on-state voltage for Fig. E6.6 is

$$1 + \frac{23 \times 1.1}{60} = 1.42\ \text{V}$$

Loss $\qquad$ = 23 × 1.42 = 32.7 W

Maximum value of sinewave

$$= 18\pi$$

From the figure at any current i,

$$v = 1.0 + \frac{1.1}{60}\, i$$

(b) Over one cycle, the total base length is 2π, from 0 to π, $i = 18\pi \sin x$ and from π to 2π, $i = 0$

Mean power $\qquad \displaystyle = \frac{1}{2\pi} \int_0^\pi vi\, dx = \frac{1}{2\pi} \int_0^\pi \left(1 + \frac{1.1}{60} \times 18\pi \sin x\right) 18\pi \sin x \cdot dx$

$$= 32.6\ \text{W}$$

(c) Mean power loss will be half the instantaneous power loss over the half cycle when the current is flowing

Mean power $\qquad \displaystyle = \frac{39.6\,[1.0 + (1.1/60)\,39.6]}{2} = 34.2\ \text{W}$

(d) Mean power $\qquad \displaystyle = \frac{48.5\,[1 + (1.1/60)\,48.5]}{3} = 30.5\ \text{W}.$

Example 6.7 Latching current for an SCR inserted inbetween a DC voltage source of 200 V and the load is 100 mA, compute the minimum width of gate pulse current required to turn on this SCR in case the load consists of: (a) $L = 0.2$ H, and (b) $L = 2.0$ H in series with $R = 20\,\Omega$.

Solution

(a) When load is pure inductance voltage equation is

$$V = L \cdot di/dt$$

or
$$di = \frac{V}{L} dt \quad \text{or} \quad i = \frac{V}{L} t$$

∴
$$0.100 = \frac{200}{0.2} t \quad \text{or} \quad t = \frac{0.1 \times 0.2}{200} = 100 \,\mu\text{sec}$$

(b) Voltage equation for R–L load is

$$V = Ri + L \, di/dt$$

$$i = \frac{V}{R} \left(1 - e^{\frac{-Rt}{L}} \right)$$

$$0.1 = \frac{200}{20} (1 - e^{-10t}) \quad \text{or} \quad t = 1005.03 \,\mu\text{sec}.$$

Example 6.8 A thyristor with a latching current of 40 mA is used in the circuit shown. If a firing pulse of 50 μsec is applied at the instant of maximum source. Show that the thyristor will not be turned on. What value of resistance R' connected as shown will ensure turn on?

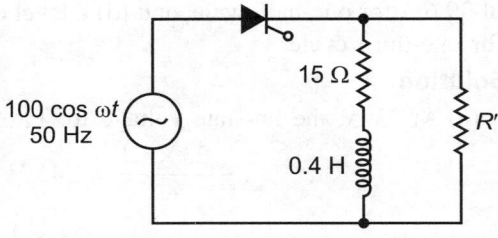

Fig. E6.8 Circuit diagram

Solution Following the application of gate pulse,

$$100 \cos \omega t = iR + L \, di/dt$$

Taking Laplace transform on both sides,

$$L \,[100 \cos \omega t] = L \,[iR + L \, di/dt]$$

$$100 \frac{s}{s^2 + \omega^2} = I(s)[R + s \, LI(s)] - i(0)$$

Using Laplace transforms, we get

$$i = \frac{100 [\cos (\omega t - \phi) - \cos \phi \exp (-Rt/L)]}{(R^2 + \omega^2 L^2)^{1/2}}$$

Hence,
$$\phi = \tan^{-1} \omega L/R = \tan^{-1} (2\pi \times 50 \times 0.4/15)$$
$$= 83.19° = 1.452 \text{ rad}$$

$$[R^2 + \omega^2 L^2]^{1/2} = \sqrt{15^2 + (2\pi \times 0.4)^2} = 126.6 \,\Omega$$

After 50 μsec, by substituting these values in equation for i, we get

$$i = 0.0124 \text{ A} = 12.4 \text{ mA}$$

After connecting a resistor R' as shown, let the current through R' is i',

Then current in thyristor, $= i_t = i + i'$

At turn on $\qquad i + i' = 40 \text{ mA} = 0.04 \text{ A}$

∴ $\qquad\qquad i' = 40 - 12.4 = 27.6 \text{ mA}$

Maximum value of $R' = 100 \cos(2\pi \times 50 \times 50 \times 10^{-6})/0.0276 = 3623 \,\Omega$.

As latching current of thyristor is 40 mA, hence thyristor fails to turn on.

Example 6.9 Design a triggering circuit using UJT with parameters: $V_{BB} = 30$ V, $\eta = 0.51$, $I_p = 10$ μA, $V_v = 3.5$ V, $I_v = 10$ mA. The frequency of oscillation is $f = 60$ Hz and width of triggering pulse is $t_g = 50$ μsec, $V_D = 0.5$ V.

Solution
$$T = \frac{1}{f} = \frac{1}{60} = 16.67 \text{ msec}$$

$$V_P = \eta V_{BB} + V_D = 0.51 \times 30 + 0.5 = 15.8 \text{ V}$$

To turn on limiting values of R as:

$$R < \frac{V_s - V_P}{I_P}$$

$$R < \frac{30 - 15.8}{10 \text{ μA}} = 1.42 \text{ m}\Omega$$

To turn off, limiting value of R is

$$R > \frac{V_s - V_v}{I_v} \Rightarrow R > \frac{30 - 3.5}{10 \text{ mA}} = 2.65 \text{ k}\Omega$$

We know
$$T = \frac{1}{f} \cong RC \log_e \left(\frac{1}{1 - \eta} \right)$$

Let $\quad\quad\quad\quad C = 0.5$ μF

$\Rightarrow \quad\quad\quad$ 16.67 msec $= R \times 0.5$ μF $\times \log_e [1/(1 - 0.51)]$

or $\quad\quad\quad\quad R = 46.7$ kΩ

which falls within the range.

Peak gate voltage $\quad V_{B1} = V_P = 15.8$ V

Width of triggering pulse, $t_g = R_{B1} C$

or
$$R_{B1} = \frac{t_g}{C} = \frac{50 \text{ μsec}}{0.5 \text{ μF}} = 100 \Omega$$

$$R_{B2} = \frac{10^4}{\eta V_{BB}} = \frac{10^4}{0.51 \times 30} = 654 \Omega$$

Example 6.10 The minimum and maximum values of gate current for a thyristor are 14 mA and 22 mA respectively. If the device is to be triggered by a 15 V supply, calculate the value of limiting resistance to be used in gate circuit.

Solution
$$I_{g \text{ min}} = 14 \text{ mA}, I_{g \text{ max}} = 22 \text{ mA}$$

Let average of $I_{g \text{ max}}$ and $I_{g \text{ min}}$ be the gate current I_g for safe triggering,

$$I_g = \frac{I_{g \text{ max}} + I_{g \text{ min}}}{2}$$

$$= \frac{14 + 22}{2} = 18 \text{ mA}$$

Supply voltage $\quad\quad = 15$ V

Limiting resistance $\quad\quad = \frac{15}{18 \times 10^{-3}} = 800 \Omega$

Example 6.11 Design the triggering circuit of Fig. E6.11. The parameters of the PUT are $V_s = 30$ V and $I_g = 1$ mA. The frequency of oscillation is $f = 60$ Hz. The pulse width t_g is 50 μsec and the peak triggering voltage is $V_{RK} = 10$ V.

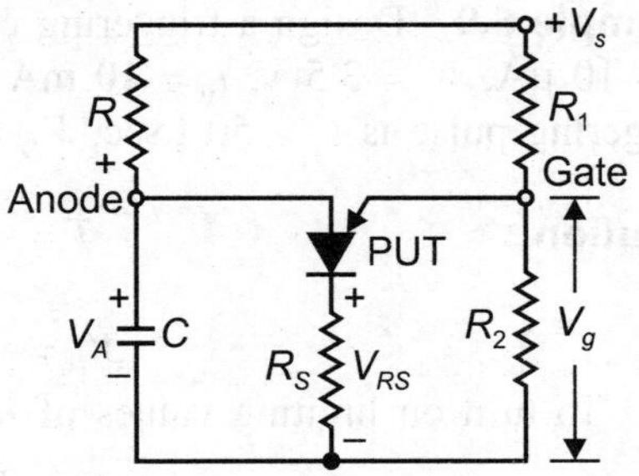

Fig. E6.11   Circuit diagram

Solution $T = \dfrac{1}{60} = 16.67\,\mu\text{sec}$

Peak triggering voltage,

$$V_{RK} = V_p = 10 \text{ V}$$

Let $C = 0.5\,\mu\text{F}$

$$R_k = \frac{t_g}{C} = \frac{50\,\mu\text{sec}}{0.5\,\mu\text{F}} = 100\,\Omega$$

Intrinsic stand off ratio, $\eta = \dfrac{V_p}{V_s} = \dfrac{10}{30} = \dfrac{1}{3}$

Period of oscillation, $T = \dfrac{1}{f} = RC \ln \dfrac{V_s}{V_s - V_p} = RC \ln \left(1 + \dfrac{R_2}{R_1}\right)$

or $16.67 \text{ msec} = R \times 0.5\,\mu\text{F} \times \ln \left(\dfrac{30}{30 - 10}\right)$

or $R = \dfrac{16.67 \text{ msec}}{0.5\,\mu\text{F} \times \ln(30/20)} = 82.2 \text{ k}\Omega$

Gate current, I_g at valley point,

$$I_g = (1 - \eta)\frac{V_s}{R_g}$$

Given $I_g = 1 \text{ mA}$

$$1 \text{ mA} = (1 - 1/3) \times 30/R_g$$

or $R_g = \left(\dfrac{2}{3}\right) \times \dfrac{30}{1 \text{ mA}} = 20 \text{ k}\Omega$

$$R_1 = \frac{R_g}{\eta} = \frac{20 \times 1000 \times 3}{1} = 60 \text{ k}\Omega$$

and $R_2 = \dfrac{R_g}{1 - \eta} = \dfrac{20}{1 - 1/3} = 30 \text{ k}\Omega.$

Example 6.12   Design the triggering circuit using UJT having parameters as; $\eta = 0.66$, $V_v = 2.5$ V, $I_p = 10\,\mu\text{A}$, $V_s = 20$ V, $I_v = 10$ mA, the frequency of oscillation is $f = 1$ kHz and the width of gate pulse is $t_g = 40\,\mu\text{sec}$, $V_D = 0.5$.

Solution $T = \dfrac{1}{f} = \dfrac{1}{10^3} = 1 \text{ msec}$

$$V_p = \eta V_s + V_D = 0.66 \times 20 + 0.5 = 14.2 \text{ V}$$

Let $$C = 0.5 \ \mu F$$

Limiting values of R are:

$$R < \frac{V_s - V_p}{I_p} < \frac{20 - 14.2}{10 \ \mu A} = 580 \ k\Omega$$

$$R > \frac{V_s - V_v}{I_v} > \frac{20 - 2.5}{10 \ mA} = 1.75 \ k\Omega$$

Period of oscillation, $$T = \frac{1}{f} \approx RC \ln \frac{1}{(1 - \eta)}$$

or

$$1 \ msec = R \times 0.5 \ \mu F \times \ln \left[\frac{1}{(1 - 0.66)} \right]$$

$\Rightarrow$

$$R = 1.86 \ k\Omega$$

This value falls within the limiting values

$$R_{B1} = \frac{t_g}{C} = \frac{40 \ \mu sec}{0.5 \ \mu F} = 80 \ \Omega$$

$$R_{B2} = \frac{10^4}{\eta V_s} = \frac{10^4}{(0.66 \times 20)} = 758 \Omega.$$

Example 6.13 A thyristor is connected in series with a load resistance. The main supply voltage of 240 V is applied across it. If the forward breakover voltage is 150 V, calculate: (a) firing angle, and (b) the conduction angle.

Solution

(a) Firing angle α, can be calculated from the equation.

$$v = V_m \sin \alpha \ or \ \alpha = \sin^{-1} v/V_m = \sin^{-1} (150/\sqrt{2} \times 240) = 26.2°$$

(b) Conduction angle, $\gamma = 180° - 26.2° = 153.8°$.

Example 6.14 The circuit of Fig. E6.14 has a peak input voltage of 30 V, $R_2 = 5.6 \ k\Omega$, $R_L = 12 \ \Omega$. The gate trigger current is $20 \ \mu A$, gate cathode voltage = 0.7 V, $V_D = 0.6$ V. Calculate value of R_1 for the thyristor to trigger at (a) 15°, and (b) 45°.

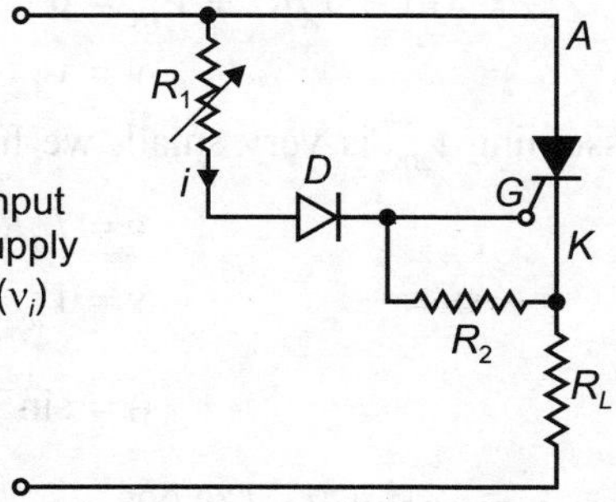

Solution Current,

$$I_1 = \frac{0.7}{5600} = 125 \ \mu A$$

Current through R_1 is

$$i = 125 + 20 = 145 \ \mu A$$

Fig. E6.14 Circuit diagram

(a) At 15°, the instantaneous value of input voltage

$$v_i = 30 \sin 15° = 7.77 \ V$$

Applying KVL, we get

$$i \cdot R_1 = [7.77 - 0.6 - 0.7 - 145 \times 10^{-6} \times 12]$$

$$R_1 = \frac{(7.77 - 1.3 - 145 \times 10^{-6} \times 12)}{145 \times 10^{-6}} = 44.61 \text{ k}\Omega$$

(b) At 45°, input voltage, $v_i = 30 \sin 45° = 21.21$ V

Applying KVL, we get

$$iR_1 = (21.21 - 0.6 - 0.7 - 145 \times 10^{-6} \times 12)$$

or
$$R_1 = \frac{(21.21 - 1.3 - 145 \times 10^{-6} \times 12)}{(145 \times 10^{-6})} = 137.3 \text{ k}\Omega.$$

Example 6.15 Determine the value of resistance R_1 to trigger a SCR to provide a minimum conduction angle of 150°. Data: I_G (max) = 25 mA, $V_{GK} = 1.5$ V.

Solution The maximum conduct angle for any rectifier is 180°. If the SCR is to conduct 150°, the firing angle will be = 180° − 150° = 30°. Thus,

$$v = V_m \sin \alpha = 170 \sin 30°$$
$$= 170 \times 0.5 = 85 \text{ V}$$

Applying KVL to gate circuit
$$-v + I_G R_1 + v_D + v_{GK} = 0$$

$$R_1 = \frac{v - v_D - v_{GK}}{I_G} = \frac{85 - 0.7 - 1.5}{25 \text{ mA}}$$

$$= 3.31 \text{ k}\Omega.$$

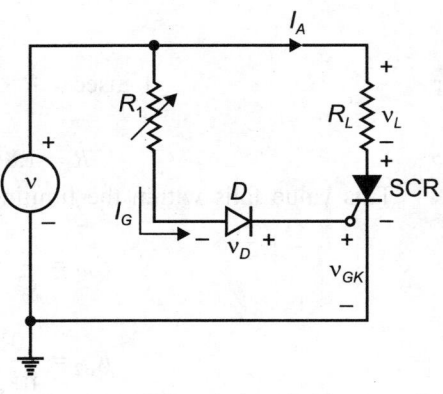

Fig. E6.15 Circuit diagram

Example 6.16 Determine the value of forward voltage that will turn off the SCR in above Fig. E6.15, what is the corresponding angle of v? According to manufacture's data, the gate trigger current i_{GT} has a typical value of 18 mA and a maximum value of 25 mA. The gate to cathode trigger voltage (v_{GT}) ranges from 0.9 to 1.5 V, $I_H = 30$ mA.

Solution Applying KVL around SCR's anode cathode circuit,

$$-v + I_A R_L + v_{AK} = 0$$

or
$$v = v_{AK} + I_A R_L$$

Assuming v_{AK} is very small, we have

$$v = I_A R_L = I_H R_L = (30 \text{ mA}) \times 100 = 3 \text{ V}$$
$$v = V_m \sin \theta, \ \sin \theta = v/V_m,$$

or
$$\theta = \sin^{-1}\left(\frac{v}{V_m}\right) = \sin^{-1}\left(\frac{3}{170}\right) = 1.01°$$

This corresponds to 178.99°.

Example 6.17 Design the triggering circuit using PUT with the following parameters: $V_s = 20$ V, $I_g = 1.5$ mA, frequency of oscillation $f = 1$ kHz, pulse width is $t_g = 40$ μsec and peak triggering pulse is $V_{RK} = 10$ V.

Solution
$$T = \frac{1}{f} = \frac{1}{1 \text{ kHz}} = 1 \text{ msec}$$

Peak triggering voltage, $V_{RK} = 10$ V = V_p

Let $\qquad C = 0.5\,\mu F$

$$R_K = \frac{t_g}{C} = \frac{40\,\mu sec}{0.5\,\mu F} = 80\,\Omega$$

$$\eta = \frac{V_p}{V_s} = \frac{10}{20} = 0.5$$

From the equation $\qquad T = \frac{1}{f} = RC\ln\frac{V_s}{V_s - V_p} = R\times 0.5\,\mu F\ln\left(\frac{20}{20-10}\right)$

$$R = 2.9\,k\Omega$$

Gate current, $\qquad I_g = (1-\eta)\frac{V_s}{R_g} = \left(1-\frac{1}{2}\right)\times\frac{20}{R_g}$

or $\qquad R_g = \left(1-\frac{1}{2}\right)\frac{20}{1.5\,mA} = 6.67\,k\Omega$

Value of resistance R_1 and R_2 are given by

$$R_1 = \frac{R_g}{\eta} = \frac{6.67\,k\Omega}{0.5} = 13.3\,k\Omega$$

$$R_2 = \frac{R_g}{(1-\eta)} = \frac{6.67}{(1-0.5)} = 13.33\,k\Omega.$$

EXERCISE

Multiple Choice Questions

1. The commonly used method for triggering the thyristor is:
   (A) gate triggering
   (B) thermal triggering
   (C) voltage triggering
   (D) all of these

2. A saw tooth wave generator can be obtained from UJT relaxation oscillator circuit by using a:
   (A) field effect transistor
   (B) junction field effect transistor
   (C) power transistor
   (D) none of these

3. Pulse gate triggering is achieved by:
   (A) UJT relaxation oscillator
   (B) rheostatic method
   (C) R–C method
   (D) none of these

4. R–C triggering is preferred over resistance triggering method because it provides:
   (A) larger value of firing angle
   (B) quick firing
   (C) accurate firing
   (D) all of these

5. In the forward-biased condition of a thyristor:
   (A) one of three junctions is reverse-biased
   (B) two out of three junctions are reverse-biased
   (C) all the three junctions are forward-biased
   (D) none of these

6. The moment a thyristor starts conducting:
   (A) it acts like a short circuit device
   (B) it acts like an open circuit device
   (C) heavy voltage drop takes place across the device
   (D) none of these

7. Intrinsic stand-off ratio of a UJT is given by:
   (A) $R_{B_1} + R_{B_2}$
   (B) $R_{B_1} \| R_{B_2}$
   (C) $R_{B_1}/(R_{B_1} + R_{B_2})$
   (D) none of these

8. The value of forward voltage drop of a UJT at 25°C is:
   (A) 0.5 V
   (B) 0.25 V
   (C) 1.0 V
   (D) 1.5 V

9. Series and parallel operations of thyristors require voltage and current sharing networks to protect them under:
   (A) steady state conditions only
   (B) transient state conditions only
   (C) steady state and transient state conditions
   (D) none of these

10. A pulse train for gating thyristor is used as it:
    (A) reduces thyristor loss
    (B) increases thyristor loss
    (C) reduces harmonics
    (D) none of these

11. Generally, for generating triggering pulses:
    (A) UJT are used
    (B) PUT are required
    (C) UJT and PUT are required
    (D) none of these

ANSWER KEY

1. (A)	2. (B)	3. (A)	4. (A)	5. (A)	6. (A)	7. (C)	8. (A)
9. (C)	10. (A)	11. (C)					

Thyristor Commutation Circuits— Turn-off Systems

7.0 INTRODUCTION

A thyristor is normally switched on by applying a gate pulse. When the thyristor is in the conducting mode, its voltage drop is small, ranging from 0.25 to 2 V, which can be neglected. Once the thyristor is turned-on, after its requirement, it is usually to turn it off. The turning-off means that the forward conduction of the thyristor has decreased and reapplication of a positive voltage to the anode will not cause current flow without applying the gate signal.

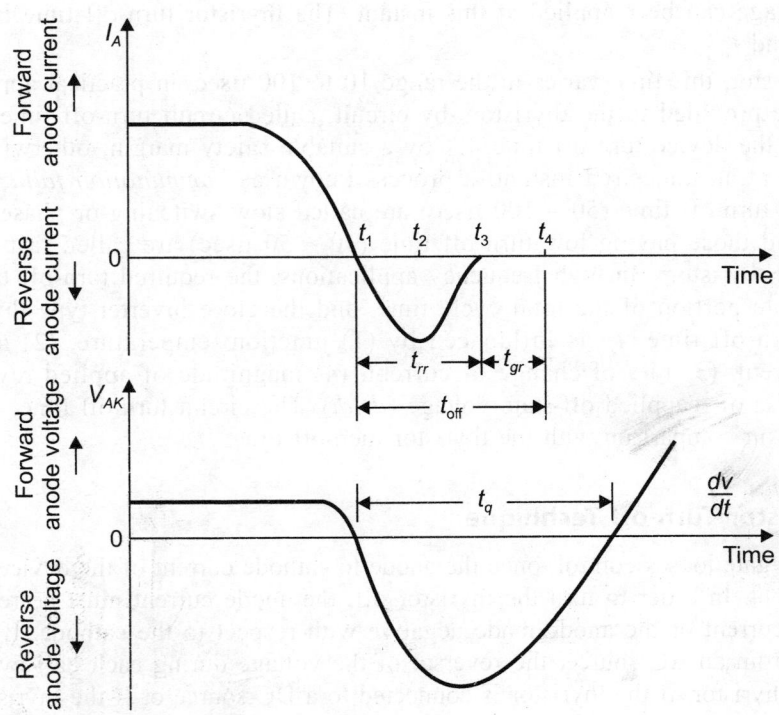

Fig. 7.1 Division of total turn-off time into t_{rr} and t_{gr}

Commutation is the process of turning-off a thyristor and it normally causes transfer of current flow to other parts of the circuit. A commutation circuit normally uses additional components to accomplish the turning-off. With the development of thyristors, many commutation circuits have been developed and the objectives of all the circuits is to reduce the turn-off process of the thyristors.

7.1 TURN-OFF MECHANISM

Once the thyristor starts conducting an appreciable forward current, the gate has no control of it. Thyristor will turn-off if the anode current becomes zero, called the natural commutation, or is forced to become zero, called force commutation. However, if a forward voltage is applied immediately after the anode current is reduced to zero, it will not block the forward voltage and will start conducting again although it is not triggered by a gate pulse. It is therefore necessary to keep the device reverse biased for a finite period before a forward anode voltage can be reapplied. Turn-off time t_{off} of the thyristor is defined as the minimum time interval between the instant the anode current becomes zero and the instant the device is capable of blocking the forward voltage. The total turn-off time t_{off} is divided into two time intervals, the reverse recovery time t_{rr} and gate recovery time t_{gr} as shown in Fig. 7.1. At instant, t_1 the anode forward current becomes zero. During the reverse recovery time, t_1 to t_3, the anode current flows in the reverse direction. At instant t_2 a reverse anode voltage is developed and the reverse recovery current continues to decrease. At instant 't_3' junction J_1 and J_3 are able to block a reverse voltage. However, the thyristor is not yet able to block a forward voltage because carries called trapped charges are still present at junction J_2. During the internal t_3 to t_4 these carriers recombine. At instant t_4 the recombination is complete, and therefore a forward voltage can be reapplied at this instant. The thyristor turn-off time is the interval between t_4 and t_1.

In thyristor, this time varies in the range 10 to 100 μsec, in practical applications, the turn-off time provided to the thyristors by circuit, called circuit turn-off time 't_q' must be greater than the device turn-off time, t_{off} by a suitable safety margin; otherwise the device will turn on at an undesired instant, a process known as *commutation failure*. Thyristors having large turn-off time (50 – 100 μsec) are called slow switching or phase control type thyristors and those having low turn-off time (10 – 50 μsec) are called fast switching or inverter type thyristors. In high frequency applications, the required turn-off time becomes an appreciable portion of the total cycle time, and therefore inverter-type thyristors must be used. Turn-off time 't_q' is influenced by (1) junction temperature, (2) magnitude of on-state current, (3) rate of change of current, (4) magnitude of applied reverse voltage (5) rate of rise of reapplied off-state voltage (dv/dt). The circuit turn-off time, 't_q' has to be large enough in comparison with the thyristor tuen-off time, 't_{off}'.

7.1.1 Thyristor Turn-off Technique

The thyristor gate looses control, once the anode to cathode current of the device is above the holding current. In order to turn the thyristor off, the anode current must be reduced below the holding current or the anode made negative with respect to the cathode. If the thyristor is supplied from an AC source, the reversal of the voltage during each half cycle serves to turn-off the thyristor. If the thyristor is connected to a DC source or if the thyristor has to be turned-off during a positive half cycle of an AC source, some sort of commutation (turn-off process) must be employed.

The current interruption through the thyristor for commutation could be achieved by opening the switch placed in two different positions as shown in Figs 7.2(a) and (b).

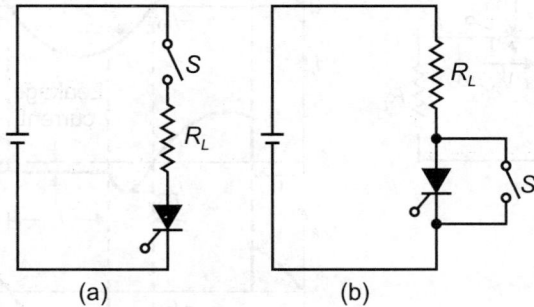

(a)   (b)

Fig. 7.2 Turn-off circuit of thyristor

The thyristor shown in Fig. 7.2(a) is turned-off by opening of the switch S, and in Fig. 7.2(b) by closing the switch S. After the thyristor has been commutated the switch is opened. These methods of commutation are seldom used except at low power levels. Applying a reverse voltage to the thyristor until it has turned-off requires circuitry more complicated than the switch used in the above turn-off circuits. The commutating circuit may be an integral part of the thyristor application circuit whose function is only to turn the thyristor off. There is a large variety of commutating circuits including:

1. underdamped $L–C$ circuits which are said to be self commutating because no other switch device is necessary,
2. capacitor commutated circuits where a charged capacitor is switched across the conducting thyristor,
3. combination of the above two schemes, and
4. external pulsed circuits.

The latter chapters will describe different commutation circuits.

7.1.2 Thyristor Turn-off Circuits

A thyristor, in the on-state, can be turned-off by reducing the level of forward current below the holding current, I_H. There are various techniques for turning-off a thyristor. In all the commutation techniques, the anode current is maintained below the holding current I_H for a sufficiently long time, so that all the excess carriers in the four layers are swept out or recombined. It is necessary to keep the device reverse-biased for a finite period before a forward anode voltage can be applied. This period is known as turn-off period of the thyristor.

When the gate pulse is applied, anode current builds and the voltage across the device falls. When the device is fully turned-on, the voltage across it is quite small (1 to 2.5 V). The device switches on very quickly, the turn-on time t_{on} being 1 to 3 µsec. The width of the gate pulse is in the range 10 to 50 µsec and its amplitude in the range 20 to 200 mA. In a line-commutated circuit where the input voltage is alternating as shown in Fig. 7.3 a reverse voltage appears across the thyristor immediately after the forward current goes through the zero value. This reverse voltage will accelerate the turn-off process, by sweeping out the excess carriers from p-n junctions J_1 and J_3.

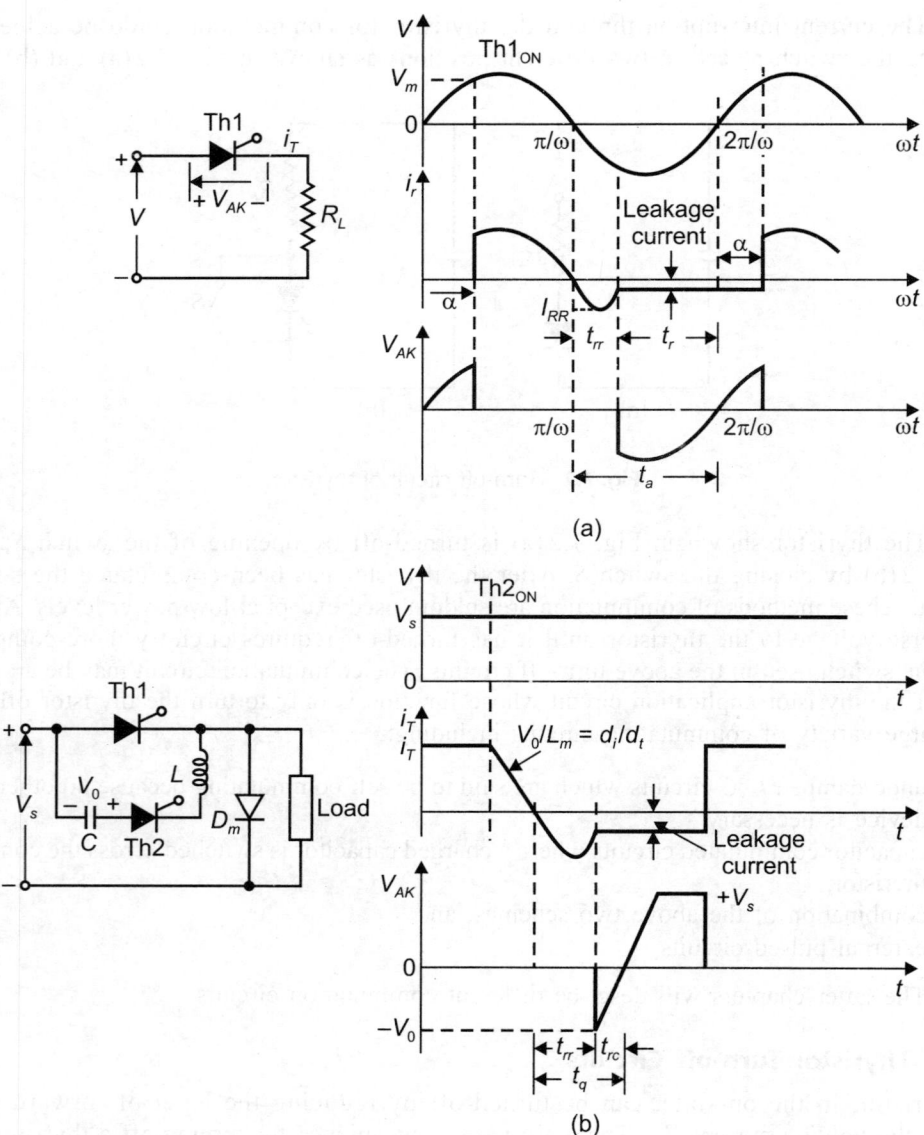

Fig. 7.3 Turn-off circuit and characteristics of thyristor

The p-n junction, J_2, will require a time known as recombination time, t_{rc} to recombine the excess carriers. A negative reverse voltage would reduce this recombination time which is dependent on the magnitude of the reverse voltage. The turn-off characteristics are shown in Figs 7.3 (a) and (b) for a line commutated circuit and forced-commutated circuit, respectively.

If the device is reverse-biased, its current falls and becomes zero then reverses and then becomes zero again. After the expiry of the turn-off time, the device is in a state capable of blocking the forward voltage. This time interval is known as turn-off time. Thus, the turn-off time, t_q, is the sum of reverse recovery time, t_{rr} and recombination time, t_{rc}. At the end of turn-off, a depletion layer develops across junction J_2 and it is at this time that the thyristor recovers its ability to withstand forward voltage. In all the commutation techniques, a reverse voltage is applied across the thyristor during turn-off process. Thyristors having large turn-off time (50–100 μsec) are called slow switching thyristors and those having a small turn-off time (10–50 μsec) are called fast switching thyristors.

7.2 COMMUTATION OF A THYRISTOR

The method of switching off a thyristor is known as *commutation of thyristor*. The thyristor can be turned-off:

- ❑ By reducing the forward current below its holding current or
- ❑ By applying a large reverse voltage across it. There are many techniques to commutate a thyristor. However, these can be broadly classified into *two* types:

    1. natural commutation
    2. forced commutation.

7.2.1 Natural Commutation

If the input voltage is AC, the thyristor current passes through a natural zero, and a reverse voltage appears across the thyristor which in turn automatically turned-off the device due to the natural behaviour of AC voltage source. This is known as natural commutation or line commutation. This type of commutation is applied in AC voltage controllers, phase controlled rectifiers and cycloconverters. In case of DC circuits this technique will not work as the DC current is unidirectional and does not change its direction, thus the reverse polarity voltage does not appear across the thyristor.

7.2.2 Forced Commutation

In source thyristor circuits, if the input voltage is DC, the forward current of the thyristor is forced to zero by an additional circuitry called commutation circuit to turn-off the thyristor. This technique is called forced commutations. Normally this method for turning-off the thyristor is applied in choppers (AC–DC converters) and inverters (DC–AC converters).

7.3 CONDITIONS FOR COMMUTATION

1. Thyristor forward current must be reduced to zero.
2. Reverse voltage must be applied across the thyristor for a duration more than the turn-off time of the device.
3. Critical rate of rise of voltage of the device should never be exceeded to avoid retriggering of the thyristor.
4. In case of inductive loads, care should be taken against retriggering of the device due to dissipation of stored energy. Generally, a freewheeling diode is used across the inductive load.

7.4 CLASSIFICATION OF FORCED COMMUTATION METHODS

There are six basic methods of commutation by which thyristors may be turned-off as class *A*, *B*, *C*, *D*, *E*, and *F* explained on the next page.

7.4.1 Class A—Commutation by Resonating Load

In class A turn-off, reverse voltage is applied to the load carrying thyristor from the over swinging of an underdamped L–C resonant circuit in series with the thyristor. Class A commutation circuit using L–C components in series with the load is shown in Figs 7.4(a) and (b).

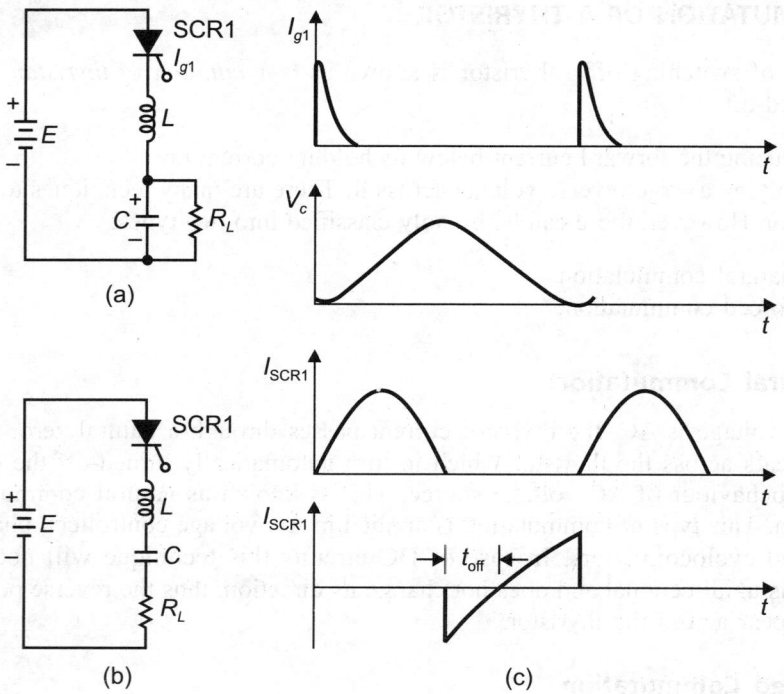

Fig. 7.4 Class A: Commutation circuit using L–C components in series with the load

Load R_L is connected in parallel with the capacitor and in series with L–C circuit shown in Figs 7.4(a) and (b). The underdamped L–C resonating circuit in series with the load applies a reverse voltage to SCR1 to turn it off as shown in Fig. 7.4(c). The capacitor supplies the commutation energy. The period of L–C resonating circuit and the switching rate of SCR1 determines the turn-off time. This type of commutation is used in series inverter circuits. The commutation process for Fig. 7.4(a) is shown in Fig. 7.4(c). When SCR1 is triggered by injecting gate current I_{g1}, the capacitor charges up to a voltage higher than E depending on losses in the circuit. The current $I_{SC\,R_1}$ falls to zero and SCR1 is reverse-biased, and therefore the SCR is turned-off. This circuit is suitable for frequencies above 400 Hz. At very low frequency the size of L–C circuit is large. The waveforms of the capacitor voltage, thyristor voltage and current are shown in Fig. 7.4(c).

7.4.2 Class B—Self-commutation by an L–C Circuit

In class B, turn-off reverse voltage is applied to the thyristor by over-swinging of an underdamped L–C circuit connected across the thyristor. The commutation circuit is shown in Fig. 7.5(a), and the associated waveforms are shown in Fig. 7.5(b). The turn-off time is decided by the period of L–C circuit and load impedance. Thyristor conducts for the period of the L–C circuit, the load impedance being positive. The circuit impedance between the load and the DC supply is low.

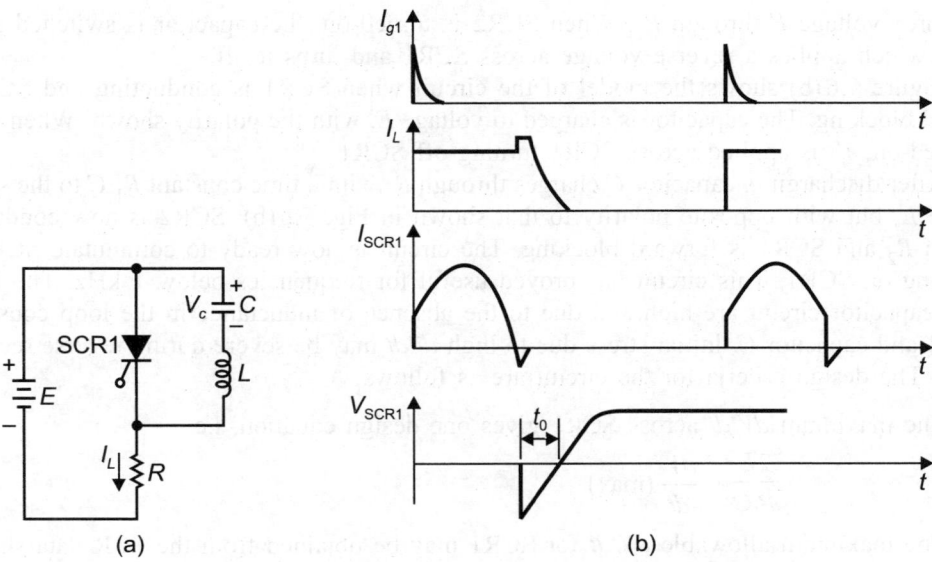

Fig. 7.5 Class B: Self-commutated circuit and waveforms

The capacitor C is initially charged to a voltage V_c with the polarity shown. When SCR1 is triggered, the capacitor discharges through them L–C resonant circuit and a reverse voltage is applied across the SCR, turning it off. Once the SCR is turned-on, it conducts for a definite period, and then it automatically turns off. SCR will fail to commutate if $I_L > E\sqrt{C/L}$. Large dv/dt can turn the SCR-ON without gate signal.

In class B commutation method, the commutating component does not carry the load current. This method of commutation is used in DC chopper circuit.

Both class A and class B turn-off circuits are self-commutating types, i.e. in both of these circuits, the SCR turns off automatically after it has been turned-on.

7.4.3 Class C—Turn-off a Charged Capacitor Switched by a Load-carrying SCR

In class C turn-off, reverse voltage for turn-off is obtained by turning on another SCR. Both SCRs are conducting the load current. However, in some cases the SCR used for turn-off may carry very small amount of current required for charging. A circuit using class C turn-off is shown in Fig. 7.6(a). In this circuit, the conducting SCR is turned-off when it is reverse-biased by switching the capacitor voltage across it through the second SCR.

In this figure assume that SCR1 is conducting, so current is flowing through R_1. At this instant SCR2 is blocking. So, capacitor C is charged to

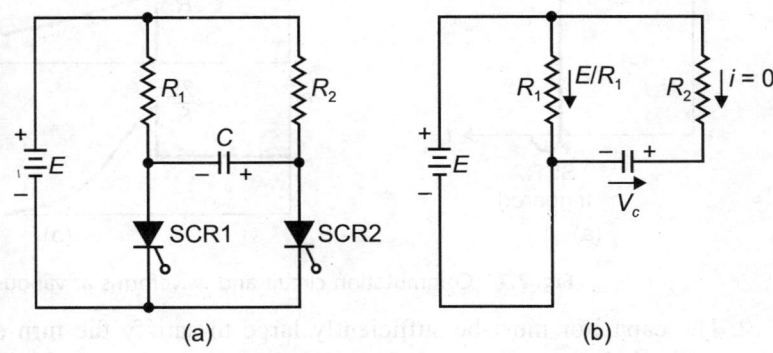

Fig. 7.6 Class C: Turn-off circuit of thyristor

the source voltage E through R_2. When SCR2 is turned-on, the capacitor is switched across SCR1, which applies a reverse voltage across SCR1 and turns it off.

Figure 7.6(b) shows the model of the circuit when SCR1 is conducting and SCR2 is forward blocking. The capacitor is charged to voltage E, with the polarity shown. When SCR2 is turned-on, V_c is applied across SCR1 turning-off SCR1.

After discharging, capacitor C charges through R_1 with a time constant $R_1 C$ to the supply voltage E, but with opposite polarity to that shown in Fig. 7.6(b). SCR2 is now conducting through R_2 and SCR1 is forward blocking. The circuit is now ready to commutate SCR2 by switching on SCR1. This circuit has proved useful for frequencies below 1 kHz. The losses in the capacitor circuit are high and due to the absence of inductance in the loop consisting of SCR and capacitor C, initial stress due to high di/dt may be severe during reverse recovery period. The design criteria for the circuit are as follows:

1. The maximum dV/dt across SCR1 gives one design equation, i.e.

$$\frac{2E}{R_1 C} < \frac{dV}{dt}(\text{max}) \qquad \qquad ...(7.1)$$

The maximum allowable dV/dt for SCR1 may be obtained from the SCR data sheet.

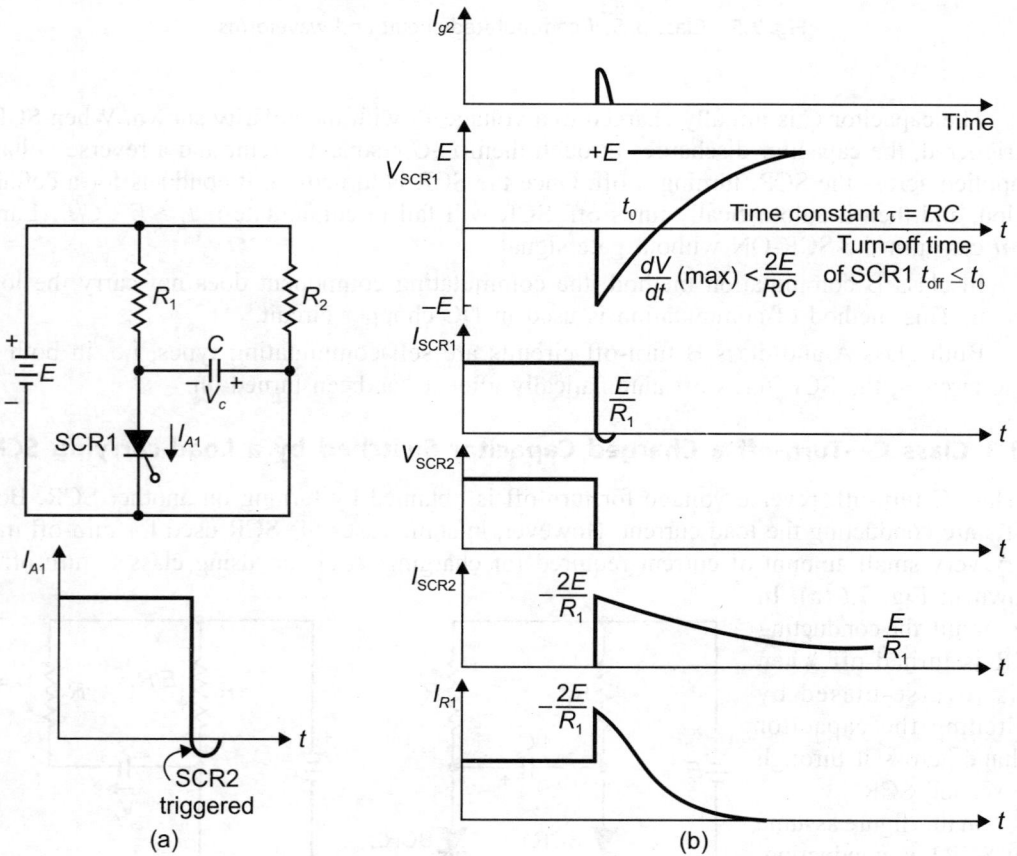

Fig. 7.7 Commutation circuit and waveforms at various points

2. The capacitor must be sufficiently large to satisfy the turn-off time, t_{off} of SCR1 as shown below:

$$V_c = E(1 - 2e^{-t/R_1 C}) \qquad \qquad ...(7.2)$$

when $V_c = 0$, let $t = t_{off}$ = turn-off time of the SCR

or $0 = 1 - 2e^{-t/R_1 C}$

$$t_{off} = 0.6931 R_1 C \text{ or } C = 1.43 \cdot \frac{t_{off}}{R_1} \qquad \qquad ...(7.3)$$

So, from Equation (7.3), R_1 and C must be such that t_{off}, the turn-off time of the SCR, is satisfied. The waveforms at various points on the commutation circuit are shown in Fig. 7.7(b).

7.4.4 Class D Turn-off—A Charged Capacitor Turned on by an Auxiliary Switching SCR

In class D turn-off, the reverse voltage for turn-off is obtained by switching another thyristor which cannot be used to conduct the load current but is used to apply the voltage from the charged capacitor to the conducting SCR. The conducting time of the load carrying SCR is from the start of current flow till the auxiliary SCR is fired. A typical class D commutation circuit is shown in Fig. 7.8(a), and the corresponding waveforms are shown in Fig. 7.8(b). In this circuit, SCR1 is the main load-carrying component and SCR2 is the auxiliary device.

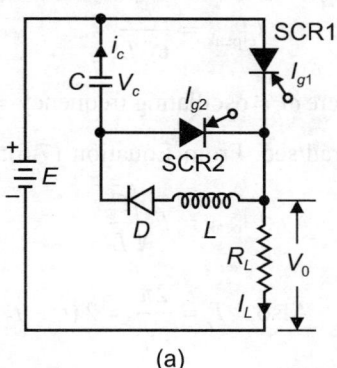

In Fig. 7.8(a), SCR2 is triggered first so that capacitor C is charged to a voltage v_c with the polarity shown. SCR2 turns off when the capacitor is fully charged and the current through SCR2 falls below the holding current. When SCR1 is triggered, the load current I_L, flows through SCR1 and R_L and the commutating current flows through SCR1, L, D, and C, and charges up the capacitor to the opposite polarity and is ready for commutating SCR1. At the instant when SCR2 is triggered, the capacitor voltage V_c reverse-biases SCR1 and turns it off. Jone's circuit is another example of this class.

(a)

Fig. 7.8 Class D: Commutation circuit

Design of the commutation circuit: The basic design of the commutation circuit involves the proper choice of capacitor C and inductor L.

Commutating capacitor: The value of the commutating capacitor is dependent on the following circuit parameters:

1. maximum load current to be commutated, I_L
2. turn-off time, t_0, of SCR1
3. the battery voltage, E

From the manufacturer's data sheet, the turn-off time, t_0, of SCR1 is obtained. The capacitor voltage changes from $-E$ to 0 during turn-off time, t_0. Assuming load current, I_L, remains constant during turn-off time, t_0,

$$CE = I_L t_0 \qquad \qquad ...(7.4)$$

$$C = \frac{I_L t_0}{E} \qquad \qquad ...(7.5)$$

Commutating inductor L: The design of the inductor L is dependent on two contradictory criteria as follows:

1. The acceptable maximum capacitor current I_c when SCR1 is triggered.
2. The time interval $(t_2 - t_1)$, during which capacitor voltage must reset to correct polarity for commutating SCR1.

The capacitor current I_c is the oscillatory current through SCR1, L, D, and C when SCR1 is triggered. The peak value of current I_c is given by the expression

$$I_c|_{peak} = \frac{E}{\omega_r L} \qquad ...(7.6)$$

where ω_r = oscillating frequency = $\dfrac{1}{\sqrt{LC}}$

in rad/sec. From Equation (7.6):

$$I_c|_{peak} = E\sqrt{\frac{C}{L}} \qquad ...(7.7)$$

Also, $\quad T_r = \dfrac{2\pi}{\omega_r} = 2\,(t_1 - t_2) \quad ...(7.8)$

where $\quad T_r$ = periodic time during oscillation.

Let the maximum current through SCR1 be limited to I_{Lm}. From Equation (7.7):

$$E = \sqrt{\frac{C}{L}} \le I_{Lm}$$

or $\qquad L \ge C\left(\dfrac{E}{I_{Lm}}\right)^2 \qquad ...(7.9)$

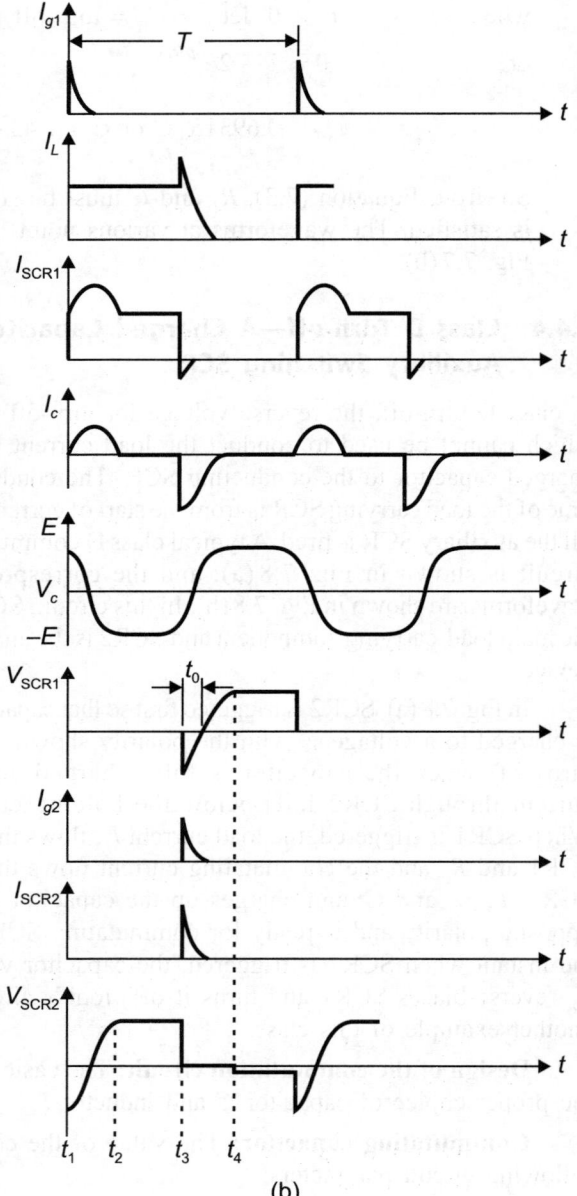

(b)

Fig. 7.8 Waveforms

The resetting time could be reduced by decreasing the value of L, but the peak capacitor current would increase as seen from Equation (7.7). A large resetting time would limit the minimum voltage available at the load, which means the range of voltage available at the load is reduced.

The minimum load voltage available is given by

$$V_0|_{min} = \frac{t_1 - t_2}{T}\,E \qquad\qquad ...(7.10)$$

where T is the chopper time period.

or $\qquad V_0|_{min} = \dfrac{\pi\sqrt{LC}}{T} \cdot E$ ...(7.11)

Therefore, $\qquad L \leq \left(\dfrac{V_0|_{min}}{E}\right)^2 \dfrac{T^2}{\pi^2 C} \cdot$ ...(7.12)

7.6.5 Class E Turn-off—with an External Source of Pulse for Commutation

In class E turn-off, the reverse voltage is applied to the load carrying SCR from an external source across or in series with conducting SCR. The width of the pulse has to be chosen such that the SCR is reverse-biased for a period greater than the turn-off time of the SCR. This ensures, a wide range of operation for this method. A typical class E commutation circuit is shown in Fig. 7.9 (a), and the associated waveforms are shown in Fig. 7.9 (b). In this particular case, the commutating pulse is applied in series with the SCR. The transformer must be capable of carrying the load current.

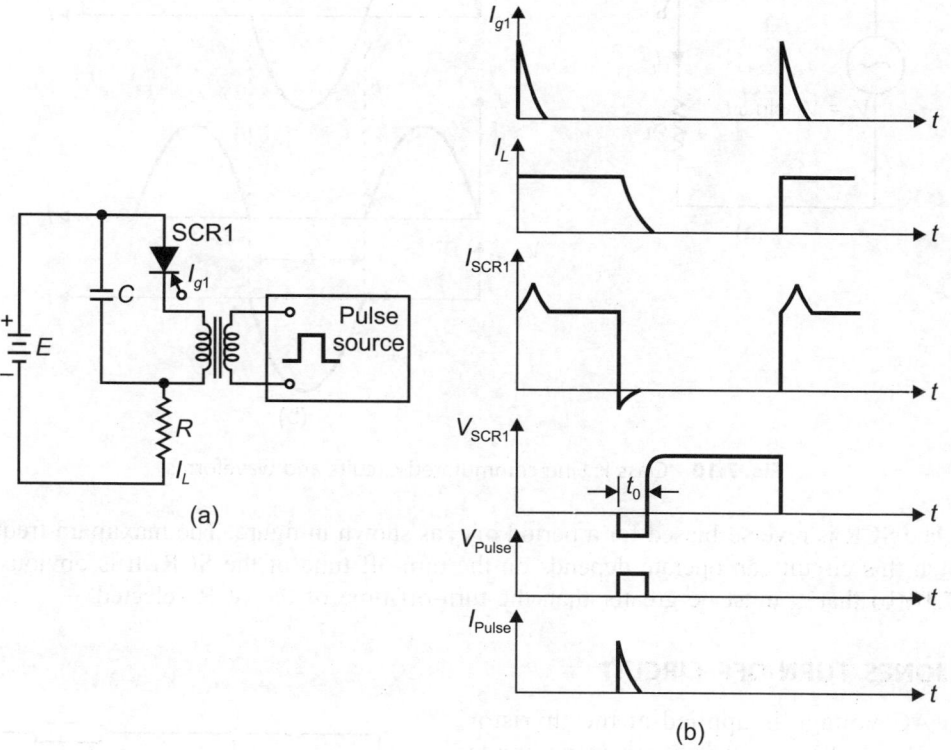

Fig. 7.9 Class E: Commutation circuit and waveforms

When SCR1 is triggered into conduction by injecting current I_{g_1} to its gate, the load current I_L, flows through SCR1, secondary of the pulse transformer, and the load R_L. To turn-off SCR1 a positive pulse from the pulse transformer is applied to the cathode of SCR1, which turns it off. This class of turn-off method is capable of high efficiency. Failure of commutation can be prevented using stronger pulses. Time and pulse width control are easily incorporated.

7.4.6 Class F—AC Line Commutation

If the supply is alternating, the reverse voltage is applied to the conducting SCRs during the negative half cycle. If the duration of negative cycle is longer the turn-off time, the SCR is turned-off. A typical line commutated circuit and its associated waveforms are shown in Figs 7.10(a) and (b). Class F turn-off also called line commutation is used in all rectifier and mains connected inverter circuits and cycloconverters. Various circuits can be modified and will be dealt with in more detail in the chapter on 'inverters'.

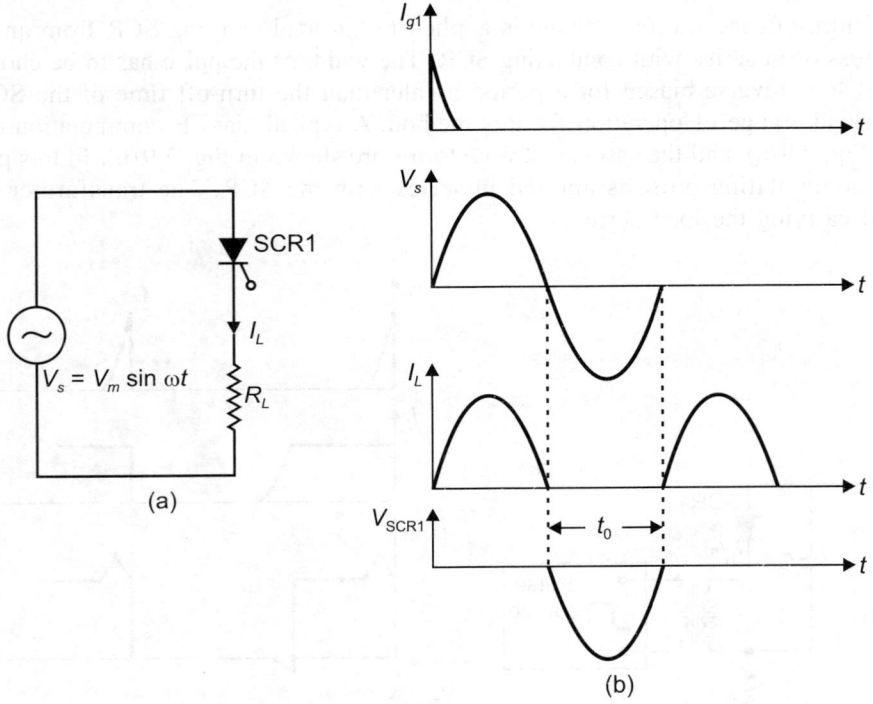

Fig. 7.10 Class F: Line commutated circuits and waveforms

The SCR is reverse-biased for a period of t_0 as shown in figure. The maximum frequency at which this circuit can operate depends on the turn-off time of the SCR. It is obvious from Fig. 7.10(b) that t_0 must be greater than the turn-off time of the SCR selected.

7.5 JONES TURN-OFF CIRCUIT

When AC voltage is applied at the thyristor anode, this voltage reverses and stops anode current during each negative half cycle. If higher frequency voltage is used, this voltage wave may return positive before the stored charges are fully removed. Thus, the thyristor again conducts, eventhough there is no gate current.

When the thyristor/SCR operates with DC anode voltage, a special circuit is added to turn-off the SCR by reducing its anode current

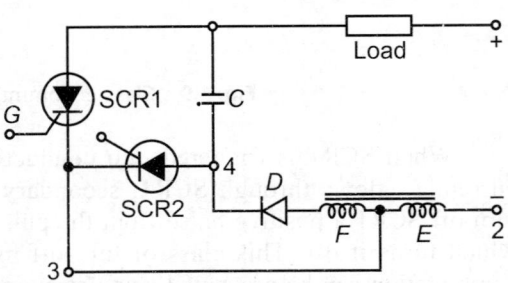

Fig. 7.11 Jones turn-off circuit

below the holding value. For this, the Jone's circuit is shown in Fig. 7.11. SCR1 conducts current to the load when its gate G receives a firing pulse from another firing circuit. SCR1 is turned-off by firing the other SCR2 as follows. Each time that SCR1 is fired, current increases rapidly through SCR1, winding E and the load. Since E and F are two windings on the same core acting as an autotransformer, this sudden current increase through winding E induces a voltage in winding F which is positive at its end connected to diode D. As diode D conducts, capacitor C will be charged, positive at terminal 4. This charge serves to hold anode of SCR2 more positive than SCR2 cathode at terminal 3. To turn-off SCR1, SCR2 is fired; the discharging of capacitor C forces electron flow from line 1 down through SCR1, and from cathode to anode SCR2. In SCR1, this flow opposes the load current, thereby momentarily reducing the total flow below the holding current value in SCR1, so that SCR1 stops conducting. The voltage across capacitor C reverses, thus stopping current in SCR2, the capacitor voltage is restored the next time SCR1 fires.

7.6 PERFORMANCE OF SCR WITH DIFFERENT LOADS

7.6.1 SCR with Resistive Load

In Fig. 7.12, the SCR is connected in series with load R_L across the AC supply voltage v_i. So long as the SCR receives no turn-on pulse at its gate, the entire supply voltage appears between anode and cathode. When gate current fires the SCR at an angle ϕ, the anode potential drops at once to a few volts at K. The anode current in SCR and resistance R_L rises abruptly at H, then follows a sine wave which reduces to zero at J, when supply voltage is zero. During the negative half cycle anode becomes negative, no current flows thus SCR turns off. The various waveforms are shown in Fig. 7.12.

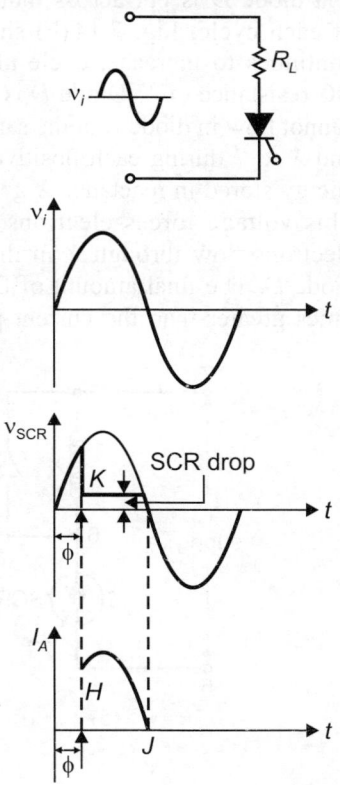

The DC current produced is given by

$$I_{DC} = \frac{1}{2\pi} \int_{\phi}^{\pi} \frac{V_m \sin d\omega t}{R_L} \, d\omega t$$

$$= \frac{V_m}{2\pi R_L} (1 + \cos \phi) \qquad ...(7.13)$$

Voltage across load R_L,

$$v_L = I_{DC} \cdot R_L$$

$$= \frac{V_m}{2\pi} (1 + \cos \phi). \qquad ...(7.14)$$

Fig. 7.12 Circuit of SCR with resistive load

7.6.2 SCR with Inductive Load

There are differences when the load is assumed to be inductive, such as transformer winding. In Fig. 7.13, the SCR is connected in series with inductive load X across the AC voltage. When the SCR is fired, the anode current through SCR and load X, rises more slowly and reaches a lower value, moreover, owing to energy stored because of this current flowing in the inductive load X, the current does not decrease to zero but continues to flow for some time after the supply has reversed. So long as this current flows, the voltage drop across SCR remains at a few volts, the anode potential remains positive far into the following half cycle, as shown at *Y* and at *Z*.

Starting at *B*, in the second cycle, the SCR current increases to a value before the supply voltage reverses. As long as current flows in the SCR, it connects load *X* to the AC voltage, even though that voltage has become negative, so that it opposes the current and finally decreases it to zero.

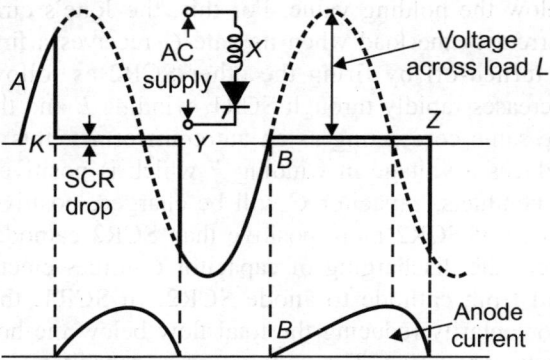

Fig. 7.13 Circuit of SCR with inductive load

7.6.3 SCRs with Diodes in AC Inductive Circuits

If a diode *D* is put across inductance *X* shown in Fig. 7.14 (a), and SCR is fired at the start of each cycle, Fig. 7.14 (b) shows that the current through SCR and inductive reactance *X* continues to increase, cycle after cycle, until the amount of current is limited only by the DC resistance of *X*. Diode *D* is connected opposite to SCR, so that any current flowing in SCR cannot flow in diode *D* at the same instant. Electrons from the AC line, flow from 5 through SCR and *X* to 7 during each positive half cycle. At point *P*, when the supply voltage reverses, the energy stored in reactance *X* generates a voltage which is positive at terminal 6 of reactance *X*. This voltage forces electrons to flow from terminal 7 through diode *D* to 6. Thus, these electrons flow through *X* in the same direction, whether they pass through SCR or through diode *D*. The final amount of DC current is limited only by load resistance and may be many times greater than the current pulses in Fig. 7.13.

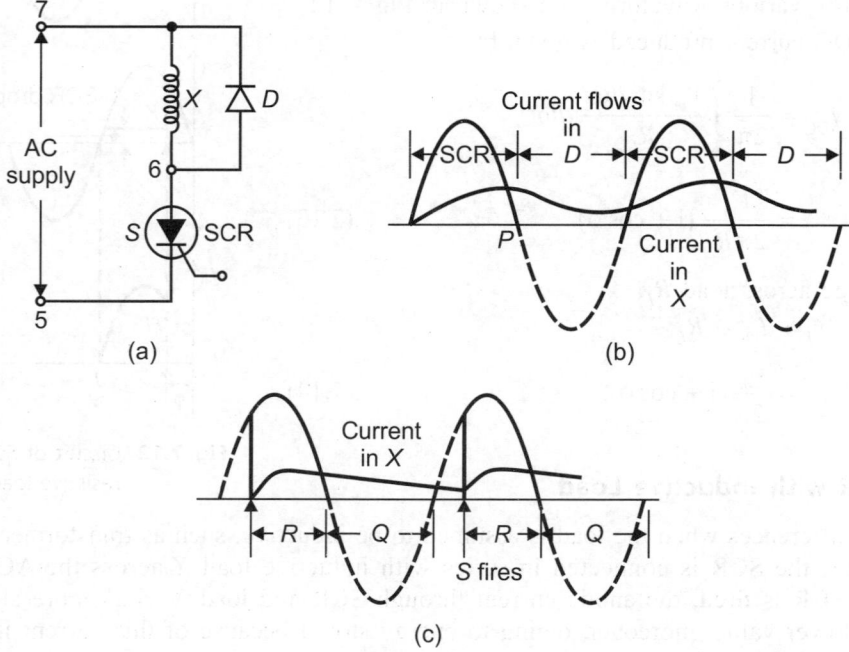

Fig. 7.14 Circuit of SCR with diodes in AC inductive load

If now SCR is fired later say at N in Fig. 7.14(c), the current increases in reactance X only during the small intervals R. During the longer intervals Q, diode D conducts, the diode current decreases slowly, owing to the resistance and a small heat loss in this circuit. The final magnitude of direct current in Fig. 7.14(c) is less than in Fig. 7.14(b) because of the firing point of SCR is delayed.

7.6.4 Voltage Change, *dv/dt*

When two SCRs are connected back to back as shown in Fig. 7.15(a). The anode to cathode voltage of SCR1 is also the cathode to anode voltage of SCR2. The phase control is added so as to delay by 90° the firing of each SCR, Fig. 7.15(b) shows the wave shapes of current and voltage across these SCRs while they energise an inductive load. At point A, SCR1 fires, its anode potential drops to about 1 V while anode current flows. This current does not stop when the AC voltage reverses at B but flows upto point C.

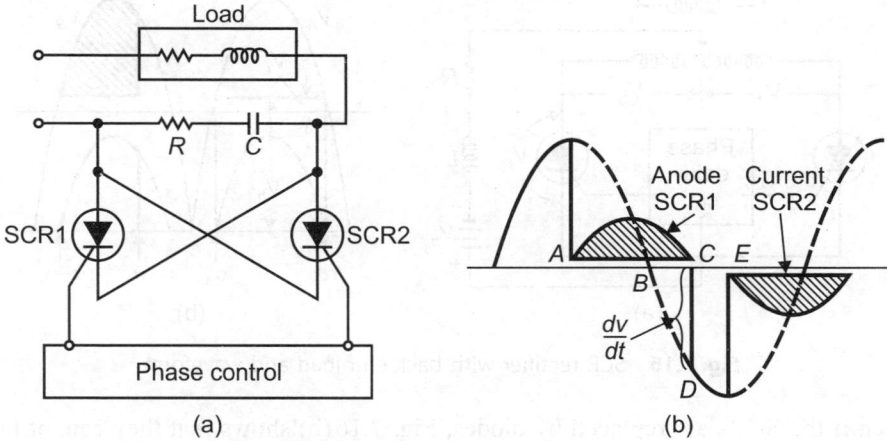

Fig. 7.15 Two SCRs connected back to back and its waveforms

At the instant, when current stops in SCR1, the voltage across both SCRs suddenly changes from C to D. As the anode of SCR1 becomes negative, the SCR2 anode becomes more positive. This large rate of change of voltage dv/dt, can fire SCR2 at point C instead of at the desired point 90°. When a forward voltage is applied to a thyristor, a depletion layer is established at the central junction. In order to establish this depletion layer, the junction capacitance C must be charged and a displacement current flows which is given by

$$i_D = C \frac{dv}{dt}$$

that acts as a gate current that can fire the SCR. As this current consists of holes which flow to the N-emitter and electrons which flow to the P-emitter, and these carriers can give rise to carrier emission in the normal way. If i_D is greater than the holding current of the thyristor, sufficient emission takes place for the thyristor to turn-on by regenerative action. To prevent unwanted firing dv/dt must be kept less than the upper limit. To limit such changes of voltages across SCR in AC inductive circuits, resistor R and capacitor C in series must be connected between anode and cathode of SCR.

7.6.5 Rectifier with Back EMF Load

When SCRs with phase control supply to the load only certain portions of the sine wave of AC voltage resulting average load current I_{DC} and its waveshape depend on the amount of load inductance. The biphase rectifier supplies power to a battery or to the armature of a DC motor, the back/counter emf V_b of the armature effects the magnitude of I_{DC} and its waveshape. The back emf is generated by the rotating motor armature as its winding yet are being driven through the field flux ϕ, using the energy stored in the flywheel. The polarity of back emf V_b always opposes the applied voltage V. In Fig. 7.16(a), each half wave of voltage V_a from anode to centre tap is applied across one SCR in series with armature, consisting of load R and L including back emf V_b. Meanwhile, the motor field is separately excited, since field flux ϕ is constant, the motor torque is proportional to I_{DC}. As back emf V_b increases with armature speed, it raises the cathode potential, so that the only potential that can force current through the load is the difference between the applied voltage $v = \sqrt{2}V \sin \omega t$ and $V_b + V_t$, where V_t is the small voltage across SCR.

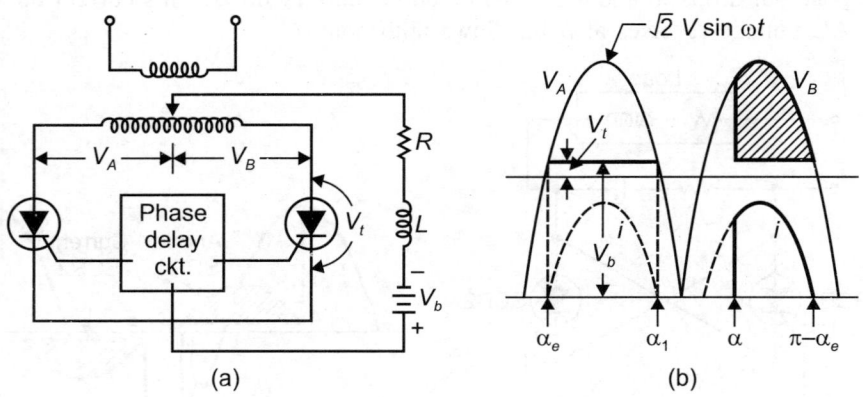

Fig. 7.16 SCR rectifier with back emf load and waveform

Even if the SCRs are replaced by diodes, Fig. 7.16(b) shows that they cannot fire earlier than the angle $\alpha_e = \sin^{-1}\left(\dfrac{V_b + V_t}{\sqrt{2}V}\right)$.

If the firing of each SCR is delayed (with $L = 0$) until α, the voltage that causes current flow in each half cycle is shown shaded in Fig. 7.16(b) and the average value of armature current is

$$I_{DC} = \frac{1}{\pi} \int_{\alpha}^{\pi-\alpha_e} \frac{\sqrt{2}V \sin \omega t - (V_b + V_t)}{R} \cdot d\omega t \qquad \qquad ...(7.15)$$

or $$I_{DC} = \frac{1}{\pi R}[\sqrt{2}V(\cos \alpha_e + \cos \alpha) - (V_b + V_t)(\pi - \alpha - \alpha_e)]$$

where α is the angle in radians.

SOLVED EXAMPLES

Example 7.1 In class C commutation circuit shown in Fig. 7.6(a), the service voltage $E = 120$ V, current through loads R_1 and R_2 is 20 A. Turn-off time of both the SCRs is 60 μsec. Find the value of C for successful commutation.

Solution Resistance $$R_1 = R_2 = \frac{120}{20} = 6\Omega$$

Turn-off time of SCR, $t_{\text{off}} = 0.6931 R_1 C$

$\therefore \qquad 60 \times 10^{-6} = 0.6931 \times 6 \times C$

or $\qquad C = \dfrac{60 \times 10^6}{0.6931 \times 6} = 14.4\,\mu F.$

Example 7.2 In class D commutation circuit shown in Fig. 7.8, the battery voltage $E = 50$ V, maximum load current, $I_L = 50$ A, t_{off} turn-off time of SCR1 = 30 μsec, chopper frequency is 500 Hz. Calculate the values of commutating capacitor C, the commutating inductor L. The load voltage variation required is 10% to 100%.

Solution For reliable commutation, assume 50% tolerance on turn-off time.

Hence, $t_o = 45$ μsec. From the equation

$$C = \frac{I_L \cdot t_o}{E} = \frac{50 \times 45 \times 10^{-6}}{50} = 45\,\mu F$$

Period, $\qquad T = \dfrac{1}{500\ \text{Hz}} = 2$ msec

From the equation, $\qquad L \le \left(\dfrac{V_0|_{\min}}{E}\right)^2 \cdot \dfrac{\pi^2}{\pi^2 C}$

$$\le (0.1)^2 \frac{(2 \times 10^{-3})^2}{\pi^2 \times 45 \times 10^{-6}}$$

$$\le 90\,\mu H$$

The value of inductor L is given by

$$L \ge C \left(\frac{E}{I_{LM}}\right)^2 = 45 \times 10^{-6} \left(\frac{50}{50}\right)^2$$

$$L \ge 45\,\mu H$$

Range of commutating inductor is $45\,\mu H < L < 90\,\mu H$. Choice of lower value of L would allow larger voltage variation at the load.

EXERCISE

Multiple Choice Questions

1. While the SCR anode is positive and gate current is zero, the voltage appears across just one of its three junction is:

    (A) nearly all the applied voltage

    (B) nearly half the applied voltage

    (C) zero voltage

    (D) none of these

2. For the control purpose of a thyristor:

    (A) a cathode gate requires a positive supply whereas the anode gate requires a negative supply

    (B) a cathode and anode gate requires a negative supply

    (C) a cathode gate as well as anode gate require a positive supply

    (D) none of these

3. In the reverse-biased condition of an SCR:

    (A) two out of three junctions are reverse-biased

    (B) one out of three junctions is reverse-biased

    (C) all the three junctions are reverse-biased

    (D) none of these

4. Method of switching off a thyristor is known as:
   (A) commutation (B) triggering
   (C) snubbing (D) none of these

5. Natural commutation method can be adopted for:
   (A) DC applications only
   (B) AC applications only
   (C) both AC and DC applications
   (D) none of these

6. Forced commutation method is meant for:
   (A) DC applications only
   (B) AC applications only
   (C) both DC and AC applications
   (D) none of these

7. Natural commutation method is applied in:
   (A) AC voltage controllers
   (B) choppers
   (C) inverters
   (D) none of these

8. Forced commutation method is generally used in:
   (A) controlled rectifiers
   (B) cycloconverter
   (C) AC voltage controller
   (D) choppers

9. Auxiliary commutation is a type of:
   (A) natural commutation method
   (B) forced commutation method
   (C) both (A) and (B) are correct
   (D) none of these

10. In case of inductive circuits, retriggering of an SCR is avoided by using a:
    (A) capacitor (B) inductor
    (C) freewheeling diode (D) none of these

11. In class A commutation:
    (A) the forward current flowing through the device is reduced to less than the level of holding current of the device
    (B) the current through the device is made maximum
    (C) the same current is flowing through the device as well as load
    (D) none of these

12. Class C commutation circuit should be designed such that the circuit recovery time must be:
    (A) more than the turn-off time of the main SCR
    (B) equal to the turn-off time of main SCR
    (C) less than the turn-off time of the main SCR
    (D) none of these

13. The main application of class B self commutation process is found in:
    (A) cycloconverter
    (B) DC chopper circuits
    (C) power rectifier circuits
    (D) none of these

14. In class E external pulse commutation method, the external pulse is obtained by:
    (A) transistor
    (B) zener diode
    (C) battery
    (D) none of these

15. The condition for underdamped oscillations in a series capacitor method is:
    (A) $R(4L/C)$
    (B) $R = (4L/C)$
    (C) $R^2 < (4L/C)$
    (D) $R^2 (4L/C)$

ANSWER KEY

1. (A) 2. (A) 3. (A) 4. (A) 5. (B) 6. (C) 7. (A) 8. (D)
9. (B) 10. (C) 11. (A) 12. (A) 13. (B) 14. (A) 15. (C)

AC to DC Converter
Controlled Rectifier

8.0 INTRODUCTION

In chapter 3, two line frequency diode rectifiers that are increasingly being used at the end of switch mode power electronic system to convert 50 Hz AC to an uncontrolled DC output voltage. In some applications such as a class of DC and AC motor drives, it is necessary that the DC output voltage be controlable the AC to controlled DC conversion is achieved in the line-frequency phase controlled converters by means of thyristors. By using thyristors, power transistors, power MOS, etc. additional control over the magnitude of the direct voltage can be achieved by varying the point-on-wave at which the device is placed into the conducting state. In this chapter, operation of rectifiers ranging from half-wave rectifier to complex multiphase, full-wave bridge circuits employing several thyristors will be considered.

8.1 CIRCUIT NOMENCLATURE OF RECTIFIERS

Rectifier circuits can be separated broadly into *three* classes:

  1. uncontrolled 2. fully-controlled, and
  3. half-controlled

1. *The uncontrolled* rectifier uses only diodes and the DC output voltage is fixed in amplitude by the amplitude of the AC supply.
2. *The fully-controlled* rectifier uses thyristors as the rectifying elements and the DC output voltage is a function of the amplitude of the AC supply voltage and the point on the wave at which the thyristors are fired.
3. *The half-controlled* rectifier contains a mixture of diodes and thyristors, allowing a more limited control of DC output voltage level than the fully-controlled rectifier. Half-controlled rectifier is cheaper than a fully-controlled rectifier of the same rating but has operational limitations.

Rectifiers are often described by their pulse number. This is the number of discrete switching operations involving load transfer (commutation) between individual diodes, thyristors, etc. during one cycle of AC supply waveform. Pulse number is therefore directly related to the repetition period of DC voltage waveform and is sometimes expressed in terms of ripple frequency of this waveform.

Uncontrolled and half-controlled rectifiers will perform power to flow only from the AC system to DC load and therefore referred to as unidirectional converters. However, with a fully-controlled rectifier it is possible, by control of the point on the wave at which switching takes place, to allow power to be transferred from the DC side of the rectifier back into the AC system. When this occurs, operation is said to be in the inverting mode. The fully-controlled converter may therefore be referred to as a bi-directional converter.

8.2 CLASSIFICATION OF CONVERTERS

The phase control converter can be classified into *two* types depending on the input supply:

1. single-phase converters   2. three-phase converters

Single-phase converter can be subdivided into:

(i) semiconverter   (ii) full converter
(iii) dual converter

Semiconverter is a one-quadrant converter and has one polarity of output voltage and current.

Full converter is a two-quadrant converter and the polarity of its output voltage can be either positive or negative. However, the output current of full converter has one polarity only.

A dual converter can operate in four-quadrants; and both the output voltage and current can be either positive or negative. In some applications converters are connected in series to operate at higher voltages and to improve the input power factor.

Method of Fourier series can be applied to analyse the performances of phase controlled converters with RL loads. To simplify the analysis, the load inductance can be assumed sufficiently high so that the load current is continuous and has negligible ripple.

8.3 INDUSTRIAL APPLICATIONS OF CONVERTERS

The class of converters discussed in this chapter uses the phase control principle which include the following applications:

1. Electrochemical processes, such as electroplating, anodising, metal refining and hydrogen production
2. HVDC conversion
3. DC motor speed control
4. DC supply for inverters
5. DC–AC conversion from fuel cells, solar cells and so on
6. Magnet power supply, such as machine excitation, fusion reactor supply, etc.
7. Portable hand tool drives
8. DC traction.

8.4 EFFECT OF R–L LOAD

If the load consists of resistance and inductance then the waveforms for load voltage v_L and load current i, are shown in Fig. 8.1, which is different from that obtained with a pure resistive load. In this case, as the AC voltage passes through zero after 180° the thyristor current may not reach zero and may continue to conduct even when the voltage wave is passing through the negative half cycle. This is due to the fact that current through the inductance cannot be reduced to zero. During negative voltage half cycle, current continues to flow till the energy

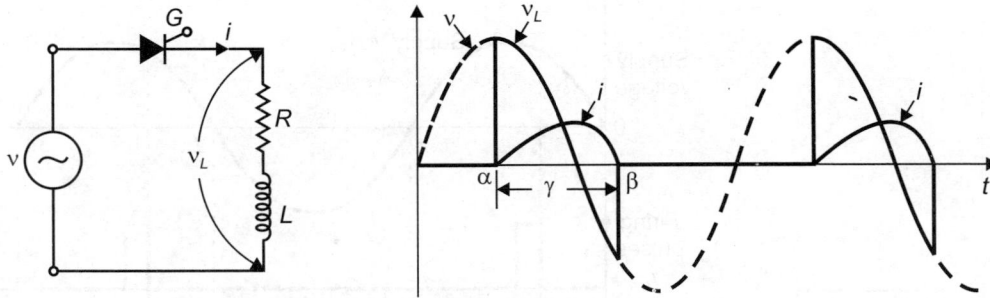

Fig. 8.1 The waveforms for load voltage v_L and load current i

stored in the inductance is dissipated in load resistor and a part of the energy is fed back to the source. As the current i reaches zero, the thyristor is turned off automatically and regains its blocking capability.

8.5 EFFECT OF COMMUTATING DIODE OR FREE-WHEELING DIODE

It is seen that if a thyristor or diode is used to supply an inductive load, then the load voltage will reverse during the conduction interval. This condition is shown in Fig. 8.1. The commutating diode or freewheeling diode D_{FW} of Fig. 8.5 is used particularly with uncontrolled or half-controlled converters to prevent this voltage reversal. Also by transferring current from the main converter, the commutating diode allows the main thyristors and diodes to resume their blocking state. The current through the commutating diode is maintained by the energy stored in the magnetic field of the load inductance.

Thus, commutating diode serves two functions:

1. It prevents reversal of load voltage except for small diode voltage drop.
2. It transfers the load current away from the main rectifier, thereby allowing all of its thyristors to regain their blocking state.

8.6 CONTROLLED RECTIFIERS

In this section, various controlled rectifiers are discussed.

8.6.1 Single-phase Controlled Rectifier Circuits

In single-phase rectifier circuits the input is a single-phase AC supply. Circuits using thyristors provide variable output voltage. The power circuits for the controlled rectifiers are the same whether natural or forced commutations are required.

8.6.2 Half-wave Controlled Rectifier with Resistive Load (One-quadrant)

In its simplest form, phase control can be described by considering the half-wave thyristor circuit with resistive load R_L shown in Fig. 8.2 (a). The circuit is energised by a line voltage or transformer secondary voltage $v_i = \sqrt{2}V \sin \omega t$. It is assumed that the peak supply voltage never exceeds the forward and reverse blocking ratings of thyristor. The thyristor is forward-biased during the interval $0 < \omega t < \pi$, $2\pi < \omega t < 3\pi$ and so on. A gate pulse is triggered at an angle α as shown in Fig. 8.2 (b). This angle is known as firing angle of thyristor. The thyristor current becomes zero at $\omega t = \pi$, 3π and so on. Thyristor conducts from α to $\pi (2\pi + \alpha)$ to 3π and so on. During the interval when the thyristor conducts, known as conduction interval, the load voltage is the same as the supply voltage.

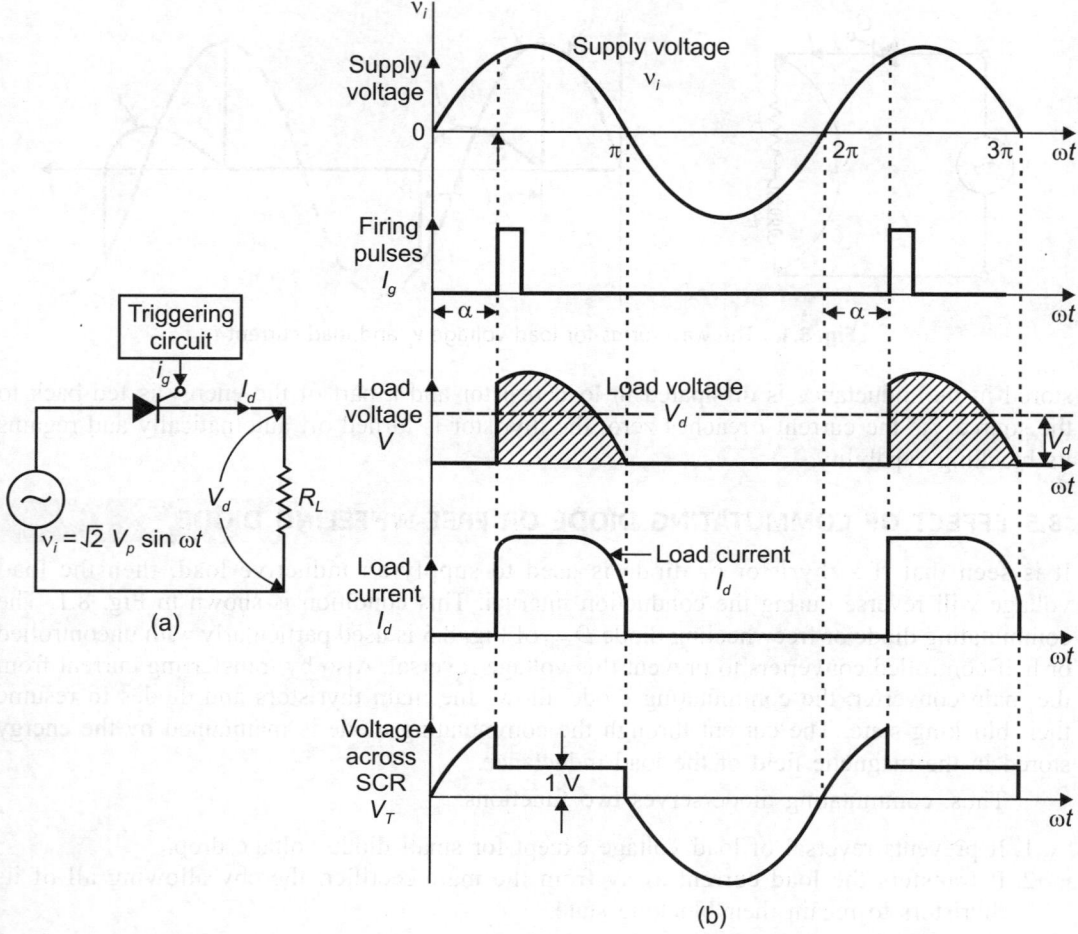

Fig. 8.2 Half-wave controlled rectifier with resistive load and waveforms

1. Average value of the load voltage is given by:

$$V_d = \frac{1}{2\pi} \int_\alpha^\pi \sqrt{2}V \sin \omega t \, d(\omega t)$$

$$= \frac{1}{2\pi} \sqrt{2}V \left[-\cos \omega t \right]_\alpha^\pi$$

$$= \frac{\sqrt{2}V}{2\pi} [1 + \cos \alpha] = \frac{V}{\sqrt{2}\pi} (1 + \cos \alpha) \qquad \qquad ...(8.1)$$

At $\alpha = 0$, $\qquad V_d = \frac{\sqrt{2}V}{\pi}$

which is the same as voltage obtained from the diode rectifier. If thyristor is fired at $\alpha = 0$, the thyristor circuit behaves like diode circuit.

2. Average value of load current with resistive load R_L is:

$$I_d = \frac{\sqrt{2}V}{2\pi R_L} (1 + \cos \alpha) \qquad \qquad ...(8.2)$$

3. DC load power

$$P_{DC} = I_d^2 R_L = \frac{V_m^2}{4\pi^2 R_L}(1 + \cos\alpha)^2 \qquad \qquad ...(8.3)$$

where $V_m = \sqrt{2}V$, V is the rms value of voltage.

4. rms value of load voltage is:

$$V_{rms} = \left[\frac{V^2}{2\pi}\int_\alpha^\pi (1 - \cos 2\omega t)\, d(\omega t)\right]^{1/2} = \left[\frac{V^2}{2\pi}\left\{\omega t - \frac{\sin 2\omega t}{2}\right\}_\alpha^\pi\right]^{1/2}$$

$$= \left[\frac{V^2}{2\pi}\left\{\pi - \alpha + \frac{1}{2}\sin 2\alpha\right\}\right]^{1/2} = \frac{V}{\sqrt{2}\sqrt{\pi}}\left\{\pi - \alpha + \frac{1}{2}\sin 2\alpha\right\}^{1/2}$$

$$= \frac{\sqrt{2}V}{\sqrt{2}\sqrt{2}\sqrt{\pi}}\left[\pi - \alpha + \frac{1}{2}\sin 2\alpha\right]^{1/2} = \frac{\sqrt{2}V}{2\sqrt{\pi}}\left(\pi - \alpha + \frac{1}{2}\sin 2\alpha\right)^{1/2}$$

$$= \frac{\sqrt{2}V}{2}\left[1 - \frac{1}{\pi}\left(\alpha - \frac{1}{2}\sin 2\alpha\right)\right]^{1/2} \qquad \qquad ...(8.4)$$

Variation of average and rms load voltage with delay angle α as expressed in Equations (8.1) and (8.3) is shown in Fig. 8.3 for half-wave thyristor phase controlled circuits with resistive load.

Half-wave phase controlled rectification is relatively simple, cheap and reliable.

8.6.3 Half-wave Controlled Rectifier with R–L Load

The operation of the half-wave controlled rectifier circuit is shown in Fig. 8.4 when inductive load changes slightly. Now when the thyristor is fired, at t_1 say, the load current will increase in a finite time through the inductive load. At t_2 the supply voltage reverses, but the thyristor is kept conducting while the load energy stored during time t_1 to t_2 is fed back to the supply. The load voltage goes negative, following the reverse half cycle of the supply voltage. At t_3 the load current falls below the holding current of the thyristor and is turned off.

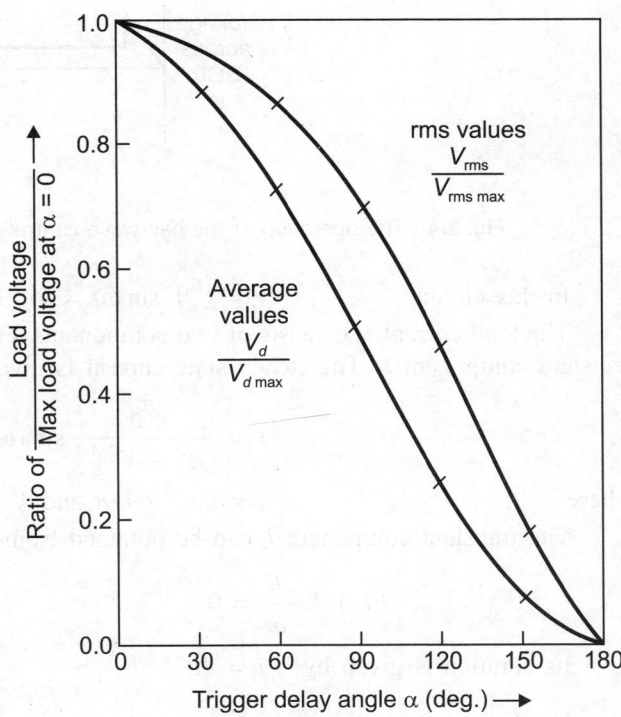

Fig. 8.3 Variation of average and rms load voltage with delay angle α

The half-wave circuit is not normally used because it produces a large output voltage ripple and is incapable of providing continuous load current.

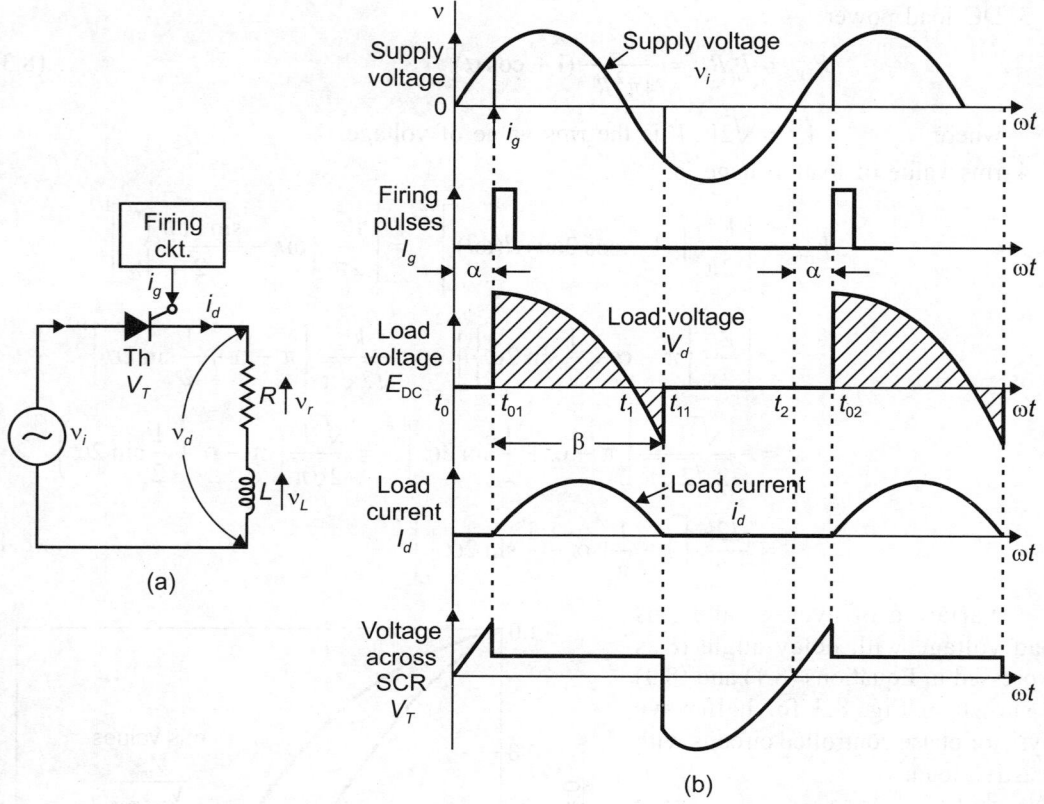

Fig. 8.4 The operation of the half-wave controlled rectifier circuit with RL load

In this circuit, $v_i = \sqrt{2}V \sin \omega t = Ri_d + Ldi_d/dt$

The load current i_d consists of two components. One steady state component i_s and other transient component i_t. The steady state current is given by:

$$i_s = \frac{\sqrt{2}V}{[R^2 + X^2]^{1/2}} \sin (\omega t - \phi)$$

where $\phi = \tan^{-1} \omega L/R$ and $X = \omega L$

The transient component i_t can be obtained from

$$Ri_t + L \frac{di_t}{dt} = 0.$$

Its solution is given by $i_t = Ae^{-(R/L)t}$

Load current, $i_d = i_s + i_t = \dfrac{\sqrt{2}V}{[R^2 + X^2]^{1/2}} \sin (\omega t - \phi) + Ae^{-(R/L)t}$

As the value of load current $i_d = 0$ at $\omega t = \alpha$ find the value of A and thus the value of load current,

$$i_d = \frac{\sqrt{2}V}{[R^2 + X]^{1/2}} \left\{ \sin (\omega t - \phi) - \sin (\alpha - \phi) \cdot e^{-\frac{R(\omega t - \alpha)}{\omega L}} \right\} \qquad ...(8.5)$$

It is seen from the waveform of i_d of Fig. 8.4(b) that when $\omega t = \beta$, load current $i_d = 0$. Equation (8.5) gives

$$\sin(\beta - \phi) = \sin(\alpha - \phi) \cdot e^{-(\beta-\alpha)R/\omega L}$$

β can be obtained by the solution of this transcendental equation. So, $\gamma = \beta - \alpha$

When β is known, average load voltage V_d is given by

$$V_d = \frac{1}{2\pi} \int_{\alpha}^{\beta} V_m \sin \omega t \, d(\omega t) = \frac{V_m}{2\pi}(\cos\alpha - \cos\beta) \qquad ...(8.6)$$

Average load current,

$$I_d = \frac{V_m}{2\pi R}(\cos\alpha - \cos\beta) \qquad ...(8.7)$$

rms load voltage,

$$V_{d\,rms} = \left[\frac{1}{2\pi}\int_{\alpha}^{\beta} V_m^2 \sin^2 \omega t \, d(\omega t)\right]^{1/2} = \frac{V_m}{2\sqrt{\pi}}\left[(\beta-\alpha) - \frac{1}{2}\{\sin 2\beta - \sin 2\alpha\}\right]^{1/2}$$

$$...(8.8)$$

8.6.4 Half-wave Rectifier with Inductive Load and Flywheeling Diode

With inductive loads, thyristor tends to conduct a continuous current on which trigger circuit has no control. This is overcome if thyristor is stopped from conducting at $\omega t = 180°$. This is achieved by connecting a diode of proper rating across the load. At some delay angle α, forward-biased SCR is triggered and source voltage v_i appears across load. At $\alpha = 180°$ input supply voltage is zero and just after this instant, as v_i tends to reverse, freewheeling diode is forward-biased through the conducting SCR. As a result, load current is immediately transferred from SCR to freewheeling diode as v_i tends to reverse. At the same time, SCR is subjected to reverse voltage and zero anode current, it is therefore turned off at $\omega t = 0$. It is assumed that during freewheeling period, load current does not decay to zero, until the SCR is triggered again at $(2\pi + \alpha)$.

Voltage drop across freewheeling diode is taken as almost zero, the load voltage is zero during freewheeling period. The advantages of using freewheeling diode are:

1. input power factor is improved
2. load current waveform is improved and thus the load performance is better.

The presence of commutating diode which prevents the load voltage reversing beyond the diode voltage drop value resulting in waveforms is shown in Fig. 8.5. During the thyristor on-period, when the voltage reverses, v_L is effectively zero and load current follows an exponential decay. If the current level decays below the diode holding level, then the load current becomes discontinuous. Figure 8.5 shows a continuous load current condition, where the decaying load current is still following when the thyristor is fixed in the next cycle. Analysis of load voltage waveform gives a mean voltage of

$$V_d = \frac{1}{2\pi}\int_{\alpha}^{\pi} V_m \sin \omega t \, d\omega t = \frac{V_m}{2\pi}(1 + \cos\alpha)$$

Inspection of the waveform shows clearly that the greater the firing delay angle α, lower is the mean load voltage, it falls to zero when $\alpha = 180°$. Waveforms with commutating diode

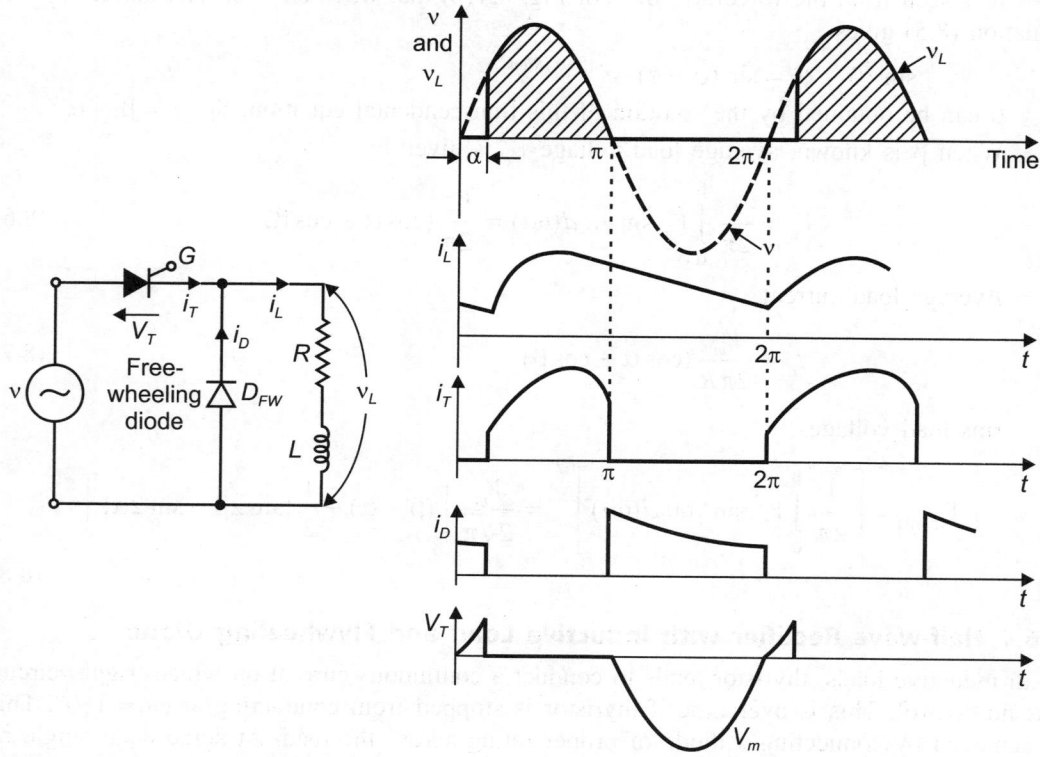

Fig. 8.5 Waveforms of half wave rectifier with inductive and flywheeling diode

shows the two roles of commutating diode, one to prevent load voltage and the other to allow the thyristor to regain its blocking state at the voltage zero by transferring the load current away from the thyristor.

8.7 BI-PHASE HALF-WAVE (SINGLE-WAY) OR SINGLE-PHASE FULL-WAVE CONTROLLED RECTIFIER (TWO QUADRANT)

In the controlled circuit a given thyristor can be fired during any time that its anode voltage is positive with respect to cathode.

8.7.1 Single-phase Full-wave Controlled Rectifier with Resistive Load

A single-phase full-wave phase-controlled rectifier is widely used in the control of DC motors. The single-phase circuit with centre tap transformer is shown in Fig. 8.6. The triggering of thyristors Th1 and Th2 can be controlled by varying the delay angle, α, over the full range of 0 to π. Thyristor Th1 conducts when the upper half of the transformer secondary is positive, and Th2 conducts when the lower half is positive. Each half of the input wave is applied across the load. Hence, the ripple frequency across the load is twice that of input supply frequency.

The output DC voltage, V_d, across the resistive load is given by

$$V_d = \frac{1}{\pi} \int_\alpha^\pi V_m \sin \omega t \, d(\omega t) = \frac{V_m}{\pi} \left[-\cos \omega t \right]_\alpha^\pi = \frac{V_m}{\pi} (1 + \cos \alpha) \quad \text{...(8.9)}$$

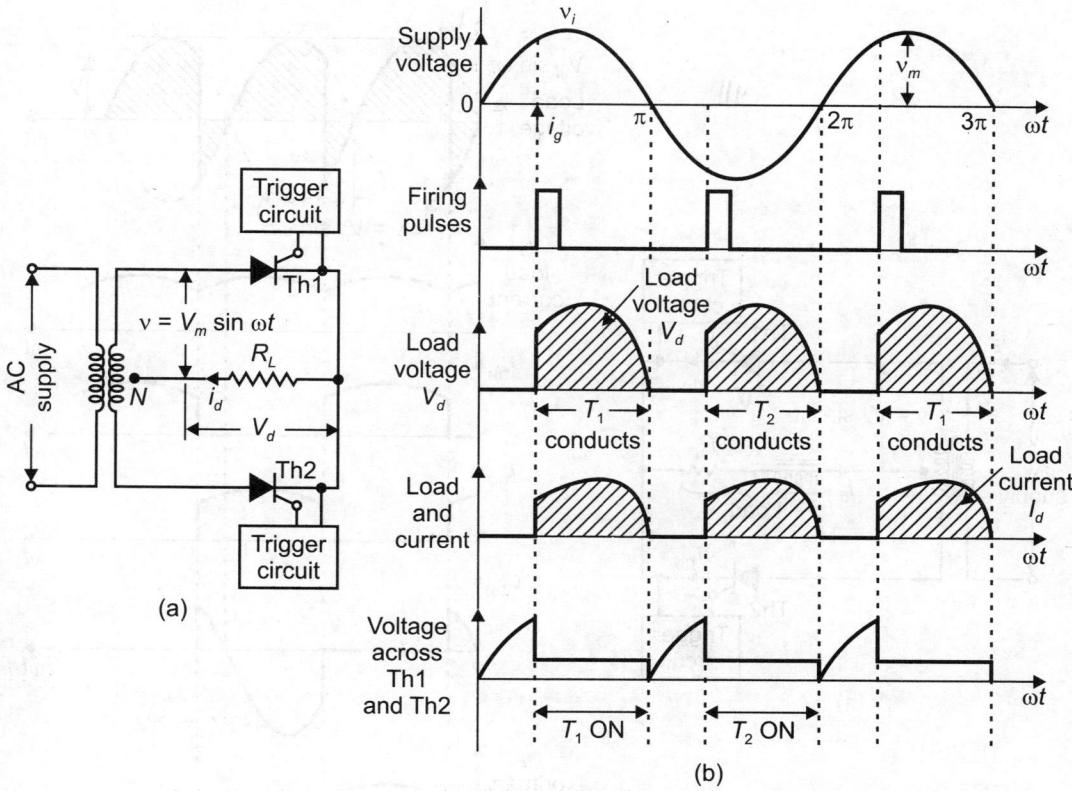

Fig. 8.6 Single-phase full wave controlled rectifier with R load

$$I_d = \frac{V_m}{\pi R_L}(1 + \cos \alpha) \qquad \qquad ...(8.10)$$

Also,
$$V_{\text{rms}} = \sqrt{\frac{1}{\pi} \int_{\alpha}^{\pi} V_m^2 \sin^2 \omega t \, d(\omega t)}$$

$$= \frac{V_m}{\sqrt{2\pi}}\left(\pi - \alpha + \frac{\sin 2\alpha}{2} \right)^{1/2}. \qquad ...(8.11)$$

8.7.2 Single-phase Full-wave Controlled Rectifier with *R–L* Load

A single-phase full-wave circuit with inductive load is shown in Fig. 8.7. It is assumed that the inductance is sufficiently large, so that each thyristor conducts for a period of 180° (conduction of current is continuous). Both thyristors are triggered with the same delay angle, hence they share the load current equally. The load voltage and current waveforms are shown in Fig. 8.7(b). Due to large inductance in the circuit and continuous current conduction, the thyristors continue to conduct even when their anode voltages are negative with respect to the cathode. The load current is shown to be constant DC.

The output DC voltage, V_d may be obtained as a function of delay angle α, as shown below:

$$V_d = \frac{1}{\pi} \int_{\alpha}^{\pi+\alpha} V_m \sin \omega t \, d(\omega t) = \frac{2V_m}{\pi} \cos \alpha \qquad ...(8.12)$$

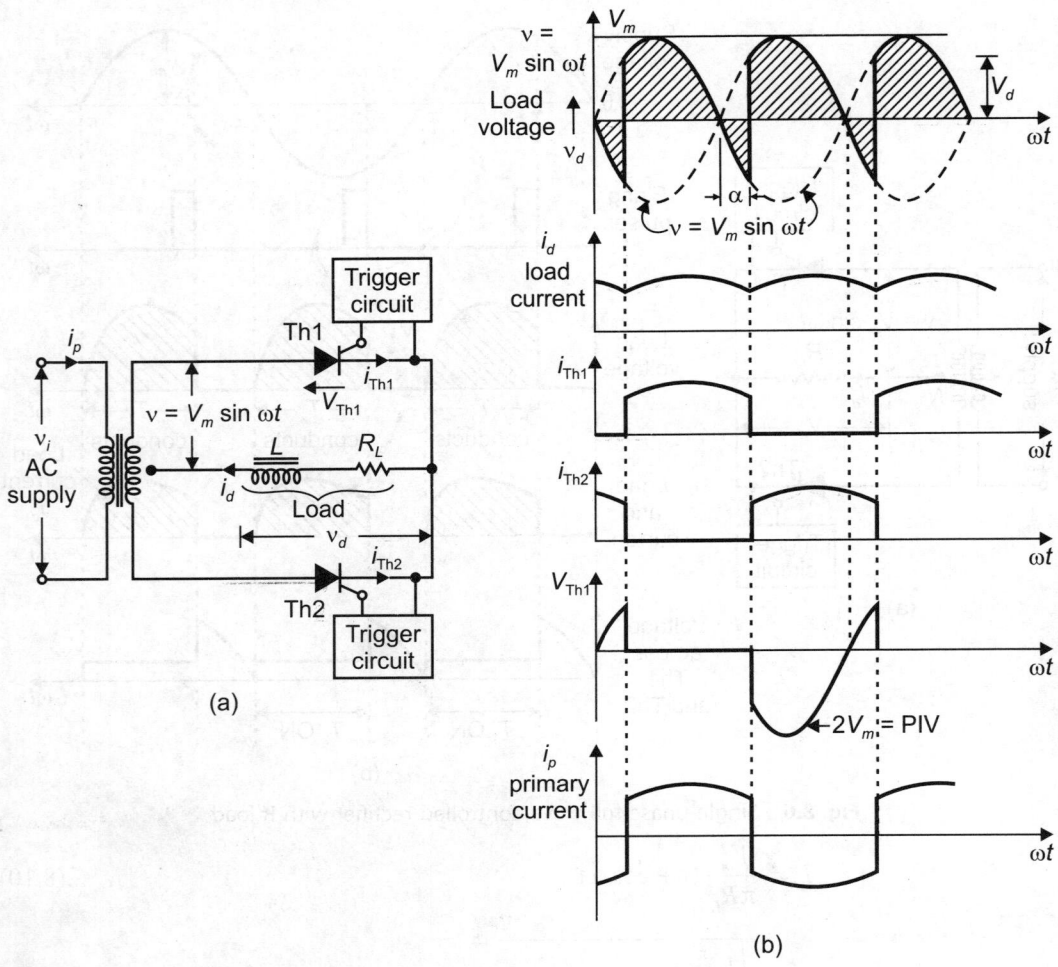

Fig. 8.7 Single-phase full wave controlled rectifier with R-L load

$$I_d = \frac{2V_m}{\pi R_L} \cos \alpha$$

and

$$P_d = I_d^2 R_L = \frac{4V_m^2}{\pi^2} \cdot R_L \cdot \cos^2 \alpha$$

The output voltage is maximum when $\alpha = 0°$, it is zero when $\alpha = 90°$ and negative maximum when $\alpha = 180°$.

8.7.3 A Single-phase Full-wave Phase-controlled Converter with Freewheeling Diode

Full-wave converter with a freewheeling diode with inductive load is shown in Fig. 8.8. The load voltage and current waveforms are also shown. The freewheeling diode D, is connected across the inductive load. The thyristors are triggered with a delay angle α. The α is varied to obtain a variable DC voltage at the load. It can be seen from Fig. 8.8 that as the input voltage goes through zero at 180°, the load voltage cannot be negative since the freewheeling

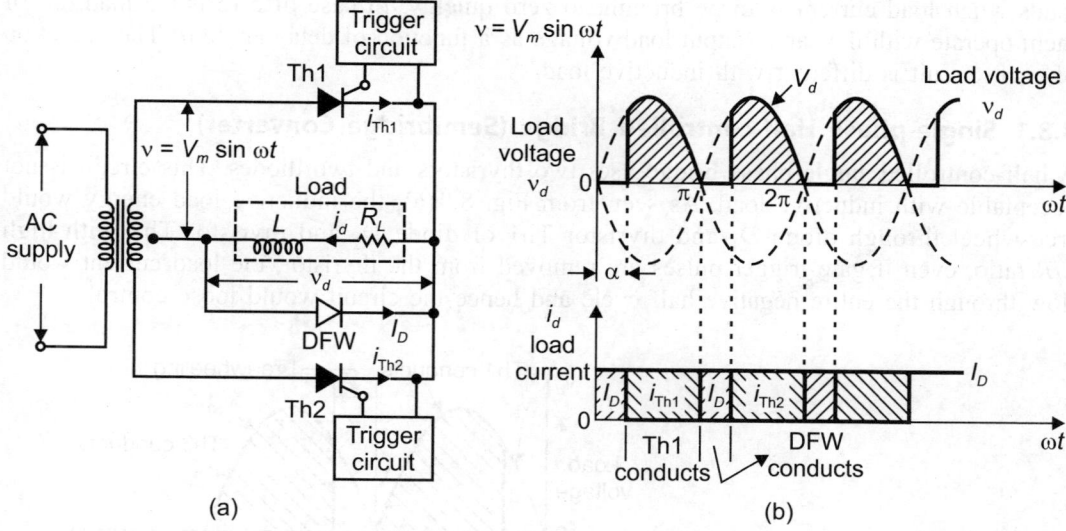

Fig. 8.8 Waveforms of single-phase full wave controlled rectifier with freewheeling diode

diode, D, starts conducting and clamps the load voltage to zero volts. A constant load current is maintained by freewheeling current through the diode. The inductive energy of the load circulates current through the feedback diode which decays depending on the time constant of the load.

The steady state duty of the freewheeling diode is dependent on the delay angle of firing of the thyristors in the circuit.

The DC load current I_d is given by:

$$I_d = \frac{1}{\pi} \int_{\alpha}^{\pi} V_m \sin \omega t \, d(\omega t)$$

$$I_d = \frac{V_m}{\pi R_L} (1 + \cos \alpha) \qquad \qquad ...(8.13)$$

The freewheeling diode, D_{FW} carries the load current during the delay period α when the thyristors are not conducting. Hence, the current through diode D_{FW} is I_{df}

$$I_{df} = \frac{I_d \alpha}{\pi} = \frac{V_m}{\pi R_L} (1 + \cos \alpha) \frac{\alpha}{\pi} = \frac{V_m}{\pi^2 R_L} (\alpha + \alpha \cos \alpha) \qquad ...(8.14)$$

It may be noted that each of the thyristor Th1 and Th2 has to withstand a maximum full winding voltage, i.e. $2V_m$.

8.8 SINGLE-PHASE FULL-WAVE PHASE CONTROLLED CONVERTER USING BRIDGE PRINCIPLE (DOUBLE WAY)

A single-phase bridge can be designed by: (1) using two thyristors and two diodes, (2) by using four thyristors, and (3) using one thyristor as discussed below. A rectifier bridge with two diodes and two thyristors is called half-controlled bridge and the other with four thyristors is called fully-controlled bridge. Fully controlled circuit has a capability to pump back stored magnetic energy into the AC supply and hence it is used for active motor and highly inductive loads when load current is to be brought to zero quickly. In case of a resistive load, all of them operate with the same output load voltage as a function of delay angle α. The operation of each circuit is different with inductive load.

8.8.1 Single-phase Half-controlled Bridge (Semibridge Converter)

A half-controlled single-phase bridge uses two thyristors and two diodes. This circuit is not acceptable with inductive load. As seen from Fig. 8.9(a), the inductive load energy would free-wheel through diode D_1 and thyristor Th1 or diode D_2 and thyristor Th2 with high L/R ratio, even if gate trigger pulses are removed from the thyristor, the load current would flow through the entire negative half cycle and hence the circuit would loose control.

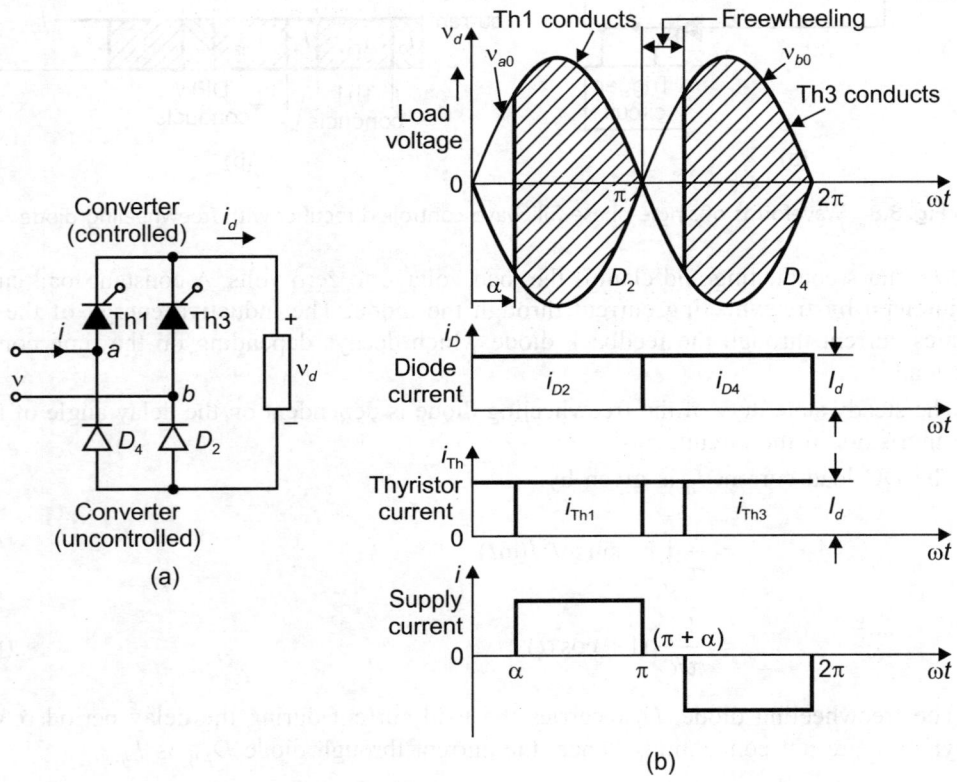

Fig. 8.9 Waveforms of single-phase half-controlled bridge

If a freewheeling diode is connected across the load of a bridge converter, it will not be able to sustain any negative voltage and therefore its inversion capability will be lost. Figure 8.9(b) shows waveforms for the circuit of Fig. 8.9(a), assuming a perfectly filtered load. The semibridge circuit can be considered as series of a positive controlled converter and a negative uncontrolled converter, and therefore the waveforms of Fig. 8.9(b) can be drawn by the superposition principle. The firing angle of positive converter can be controller in the range $0 \leq \alpha \leq 180°$, but a diode of the negative converter will conduct whenever its cathode is more negative than the other. The resulting load voltage wave is shown by the hatched areas. The current i_{Th1} becomes phase shifted with respect to current i_{D2} and there will be freewheeling between the series elements every half cycle, whenever the load voltage tends to be negative. During the freewheeling interval, the line does not contribute any current as shown. The DC load voltage V_d is:

$$V_d = \frac{1}{\pi} \int_{\alpha}^{\pi} V_m \sin \omega t \, d\omega t$$

or

$$V_d = \frac{V_m}{\pi} (1 + \cos \alpha) \qquad \qquad ...(8.15)$$

The DC current is $\quad I_d = \dfrac{V_d}{R_d} \cdot$ ...(8.16)

8.8.2 Bridge with Two Thyristors, Two Diodes and a Freewheeling Diode

In Fig. 8.10 case diode D_1, thyristor Th2 and diode D_2, thyristor Th1 conduct in pairs. When the load is inductive and thyristor Th1 is triggered, it will conduct with D_2 to pass current through the load. When the supply voltage is negative, load emf will drive current through diode D_1 and thyristor Th1. This is an exponentially decreasing current. When the new negative half cycle begins thyristor Th1 is in conduction and it would keep on conducting with D_2

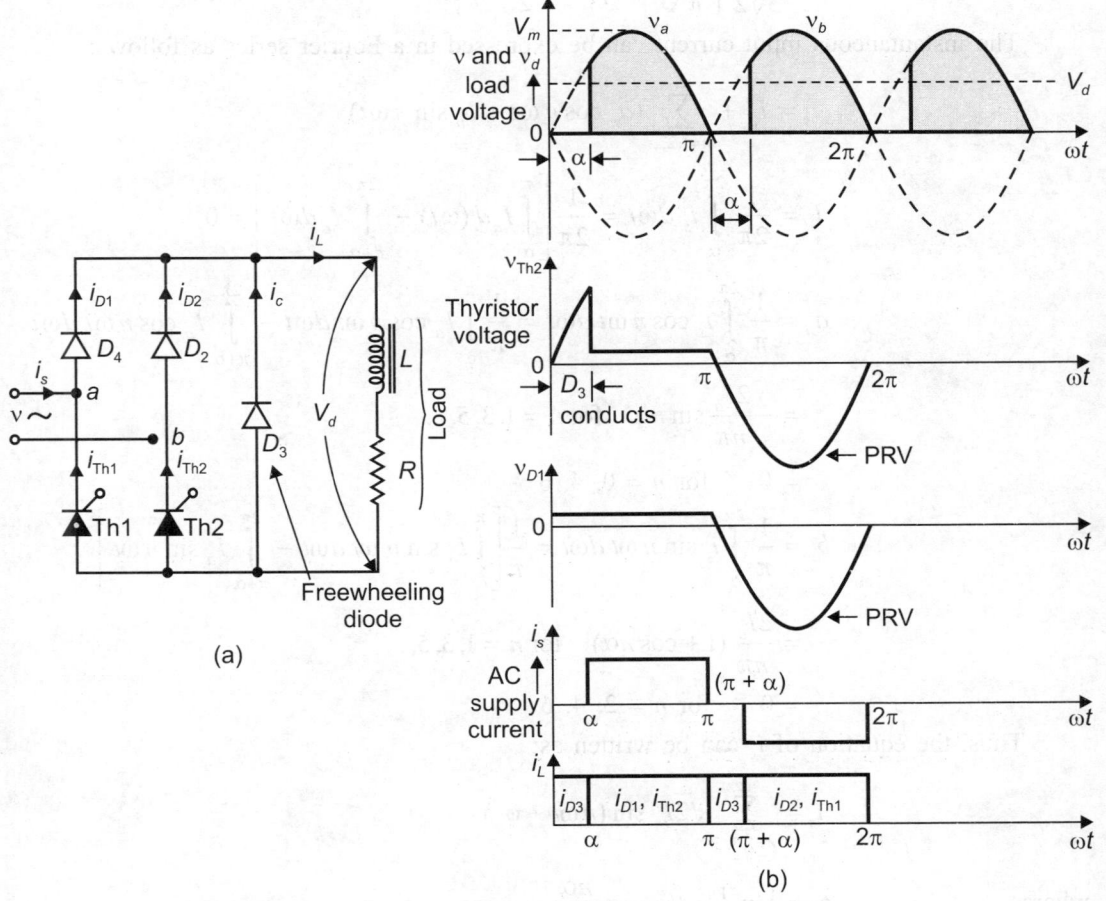

Fig. 8.10 Waveforms of the bridge circuit

as if triggered. In this case load may not receive the DC power. To ensure proper operation, at the beginning of positive half cycle, thyristor Th1 has to be turned off and similarly Th2 should be turned off when negative half cycle begins. This is achieved by the freewheeling diode D_3, which applies the negative voltage to Th1 and Th2 by bypassing the load.

The average output voltage V_d is given by

$$V_d = \frac{1}{\pi} \int_\alpha^\pi V_m \sin \omega t \, d\omega t = \frac{V_m}{\pi} (1 + \cos \alpha) \qquad \qquad ...(8.17)$$

V_d can be varied from $\dfrac{2V_m}{\pi}$ to 0 by varying α from 0 to π. The maximum average output voltage is $V_{dm} = 2V_m/\pi$ and normalised average voltage is $V_m = \dfrac{V_d}{V_{dm}} = 0.5\,(1 + \cos \alpha)$.

rms output voltage is given by

$$V_{rms} = \left[\frac{1}{\pi} \int_\alpha^\pi V_m^2 \sin^2 \omega t \, d\omega t \right]^{1/2}$$

$$= \frac{V_m}{\sqrt{2}} \left[\frac{1}{\pi} \left(\pi - \alpha + \frac{\sin 2\alpha}{2} \right) \right]^{1/2} \qquad \qquad ...(8.18)$$

The instantaneous input current can be expressed in a Fourier series as follows:

$$i_s = I_d + \sum_{n=1,2,...}^\alpha (a_n \cos n\,\omega t + b_n \sin n\,\omega t)$$

$$I_d = \frac{1}{2\pi} \int_\alpha^{2\pi} i_s \, d\omega t = \frac{1}{2\pi} \left[\int_\alpha^\pi I_a d\,(\omega t) - \int_{\pi+\alpha}^{2\pi} I_d d\omega t \right] = 0$$

$$a_n = \frac{1}{\pi} \int_\alpha^{2\pi} i_s \cos n\,\omega t \, d\omega t = \frac{1}{\pi} \int_\alpha^\pi I_a \cos n\,\omega t \, d\omega t - \int_{\pi+\alpha}^{2\pi} I_a \cos n\,\omega t \, d\omega t$$

$$= -\frac{2I_a}{n\pi} \sin n\alpha \quad \text{for } n = 1, 3, 5, ...$$

$$= 0 \qquad \text{for } n = 0, 4, 6$$

$$b_n = \frac{1}{\pi} \int_\alpha^{2\pi} i_s \sin n\,\omega t \, d\omega t = \frac{1}{\pi} \left[\int_\alpha^\pi I_a \sin n\,\omega t \, d\omega t - \int_{\pi+\alpha}^{2\pi} I_a \sin n\,\omega t \right]$$

$$= \frac{2I_a}{n\pi} (1 + \cos n\alpha) \quad \text{for } n = 1, 3, 5, ...$$

$$= 0 \qquad \text{for } n = 2, 4, 6$$

Thus, the equation of i_s can be written as:

$$i_s = \sum_{n=1,3,5}^\alpha \sqrt{2} I_n \sin\,(n\omega t + \phi_n)$$

where
$$\phi_n = \tan^{-1} a_n / b_n = -\frac{n\alpha}{2}$$

rms value of nth harmonic component of input current is derived as:

$$I_{sn} = \frac{1}{\sqrt{2}} (a_n^2 + b_n^2)^{1/2} = \frac{2\sqrt{2}I_a}{n\pi} \cos{(n\alpha/2)}$$

$$I_{s1} = \frac{2\sqrt{2}I_a}{\pi} \cos{\alpha/2}$$

rms value of

$$I_s = \left[\frac{2}{2\pi} \int_\alpha^\pi I_a^2 d\omega t \right]^{1/2} = I_a (1 - \alpha/\pi)^{1/2}$$

Harmonic factor,

$$HF = \left[(I_s/I_{s1})^2 - 1 \right]^{1/2}$$

or

$$HF = \left[\frac{\pi (\pi - \alpha)}{4 (1 + \cos{\alpha})} - 1 \right]^{1/2} \qquad ...(8.19)$$

Displacement factor, $DF = \cos{\phi_1} = \cos{(-\alpha/2)}$

Input power factor, $PF = \dfrac{V_s I_{s1}}{V_s I_s} \cos{\phi} = \dfrac{I_{s1}}{I_s} \cos{(\alpha/2)} = \dfrac{\sqrt{2}\,(1 + \cos{\alpha})}{[\pi (\pi - \alpha)]^{1/2}}$.

8.8.3 Single-phase Half-controlled Bridge with Two Thyristors and Two Diodes in the Same Arm

Figure 8.11 shows a single-phase full-wave half-controlled bridge. Two thyristors and two diodes are used in the same arm. For a given AC input, full-wave rectified DC is obtained

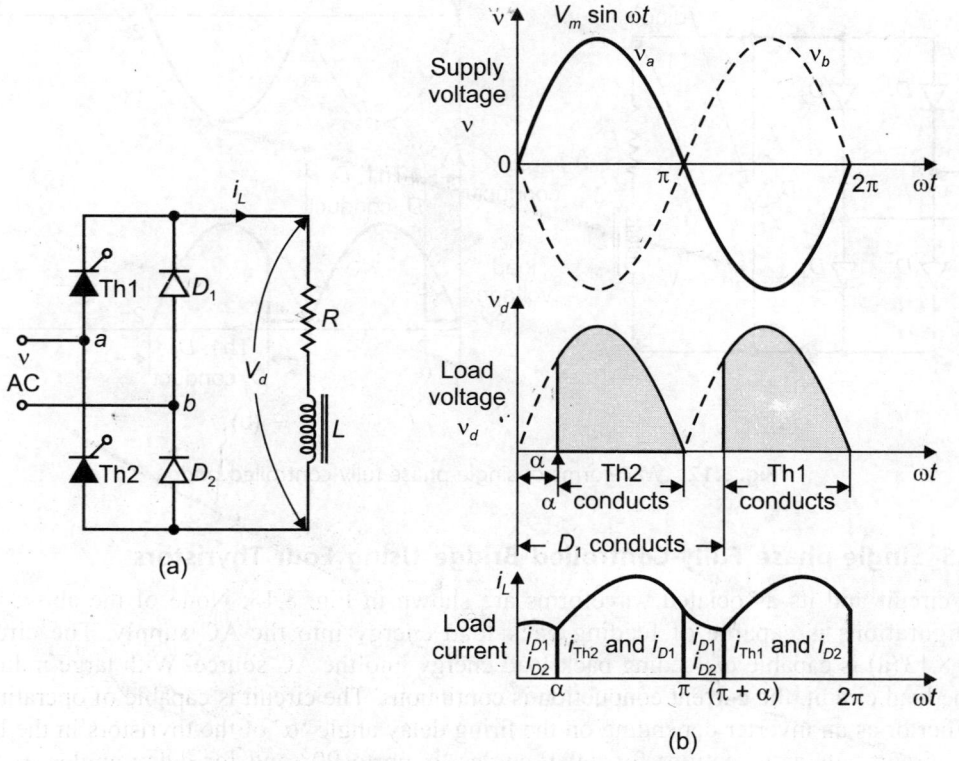

Fig. 8.11 Waveforms of single-phase half-controlled bridge

at its output. For the positive half of the input, thyristor Th1 and diode D_2 are conducting whereas, in the negative half cycle, thyristor Th2 and diode D_1 are conducting and thus full-wave rectification is achieved. In this configuration, freewheeling action in case of inductive loads is done by the pair of diodes D_1 and D_2. No separate diode is used for this purpose. The current duty on the diodes is large in this case. The voltage and current in the circuit are shown in Fig. 8.11(b).

8.8.4 Single-phase Full-wave Fully-Controlled Bridge

Figure 8.12(a) shows a bridge connection of single-phase full-wave fully-controlled bridge using five diodes and one thyristor. In this circuit each half of the input AC voltage is rectified by diode bridge. The rectified full-wave voltage is then controlled by thyristor Th1. As shown in Fig. 8.12(b), the output DC voltage is continuously controlled by phase delayed triggering of Th1. The freewheeling diode D_5 allows the circulation of stored load energy during the time that Th1 is not conducting. The freewheeling diode maintains a continuous load current even with large delay angle in triggering of ·Th1, required for a low output voltage. Continuous flow of load current is essential for proper control of DC motors with minimum torque and speed ripple. The freewheeling diode carries a large current when thyristor Th1 is operating with large delay angle. In many phase-controlled rectifier DC motor drives, this configuration of the bridge is used as shown below.

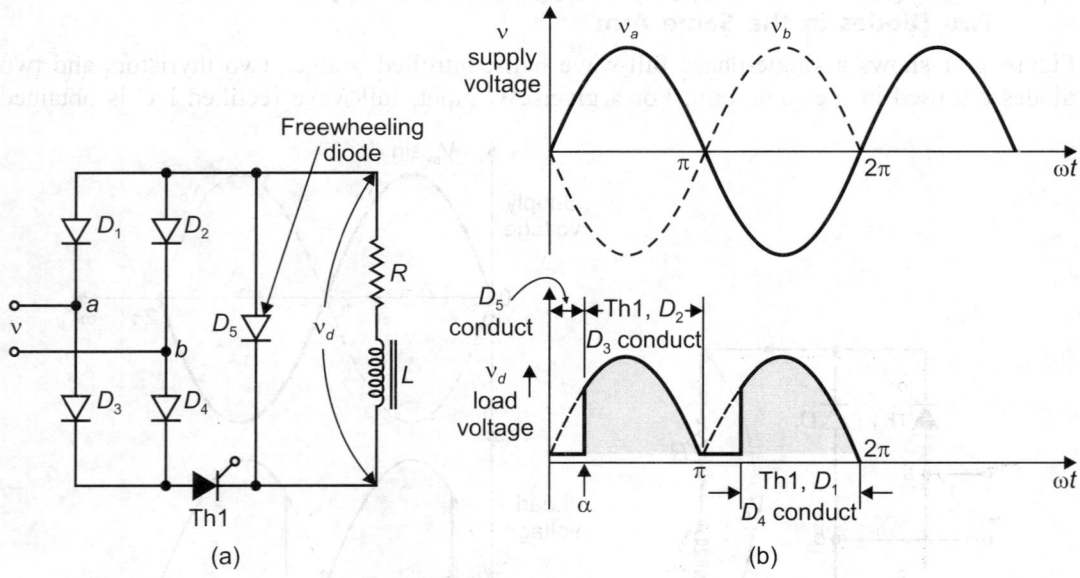

Fig. 8.12 Waveforms of single-phase fully-controlled bridge

8.8.5 Single-phase Fully-Controlled Bridge Using Four Thyristors

The circuit and its associated waveforms are shown in Fig. 8.13. None of the above bridge configurations is capable of feeding back load energy into the AC supply. The circuit of Fig. 8.13(a) is capable of feeding back load energy into the AC source. With large inductance in the load circuit, the current conduction is continuous. The circuit is capable of operating as a rectifier or as an inverter depending on the firing delay angle 'α' of the thyristors in the bridge. This circuit acts as a rectifier for delay angles, α up to 90° and for delay angles, α > 90°, the circuit operates as an inverter.

A single-phase AC supply is connected to the centre of the legs through transformer. The transformer is optional but can be used for electrical isolation and voltage level change. The DC output can be connected to passive resistance-inductance load in a bridge converter, diagonally opposite pairs of thyristors Th1, Th4 and Th3, Th4 conduct in sequence for half cycle intervals.

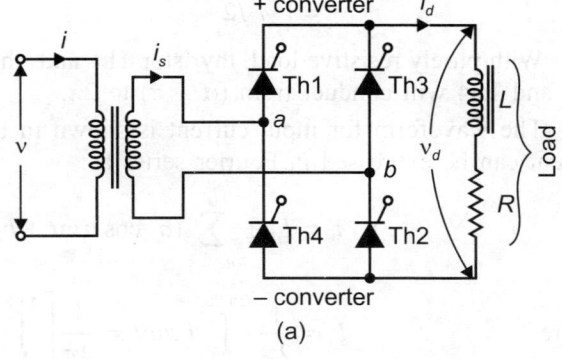

(a)

Figure 8.13 (b) shows waveforms for the load with large inductance to maintain continuous conduction. The thyristors Th1 or Th3 can be fired at any instant during the half cycle when its anode voltage is positive, and once fired the load voltage follows the profile of AC voltage as shown. The conduction of the thyristor will continue beyond 80° angle, i.e. when the anode voltage is negative, until the next thyristor is fired. At angle $(\alpha + \pi)$, when thyristor Th3 is fired a reverse voltage is impressed across the outgoing device turning it off, and then the conduction will be undertaken by the incoming device. The two devices will conduct in a symmetrical manner and average DC voltage is given by:

$$V_d = \frac{1}{\pi} \int_{\alpha}^{(\pi+\alpha)} V_m \sin \omega t \, d\omega t$$

$$= \frac{2V_m}{\pi} \cos \alpha \qquad ...(8.20)$$

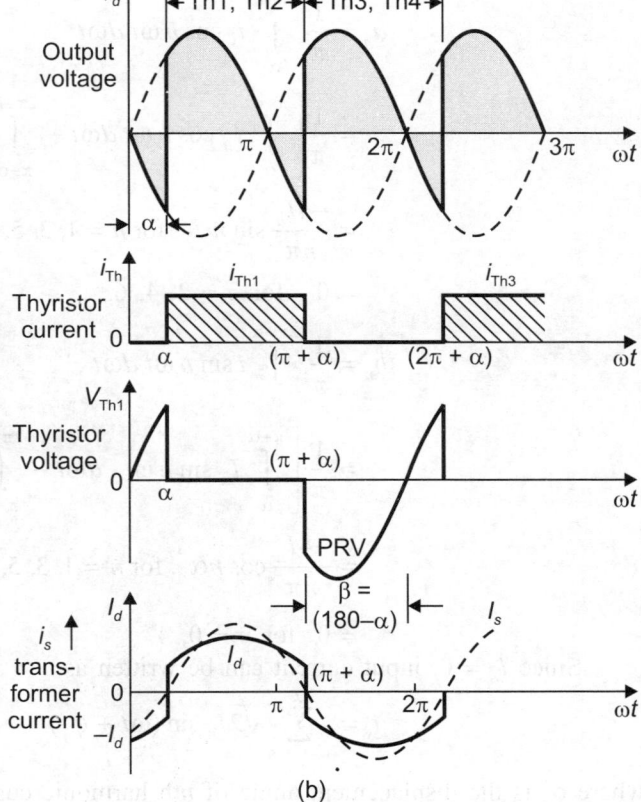

(b)

Fig. 8.13 Waveforms of single-phase fully-controlled bridge using four thyristors

The voltage V_d can be controlled by controlling the firing angle α. At $\alpha = 0$, $V_{dm} = \frac{2V_m}{\pi}$ and the normalised average output voltage $V_n = \frac{V_d}{V_{dm}} = \cos \alpha$.

The rms value of output voltage is:

$$V_{rms} = \left[\frac{2}{2\pi} \int_{\alpha}^{\pi+\alpha} V_m^2 \sin^2 \omega t \, d\omega t \right]^{1/2} = \left[\frac{V_m^2}{2\pi} \int_{\alpha}^{\pi+\alpha} (1 - \cos 2\omega t) \, d\omega t \right]^{1/2}$$

$$= V_m / \sqrt{2} \qquad \qquad ...(8.21)$$

With purely resistive load, thyristor Th1 and Th2 will conduct from α to π and thyristors Th3 and Th4 will conduct from $(\alpha + \pi)$ to 2π.

The waveform for input current is shown in Fig. 8.13 (b) and the instantaneous input current can be expressed in Fourier series as:

$$i_s = I_d + \sum_{n=1, 2}^{\alpha} (a_n \cos n\,\omega t + b_n \sin \omega t)$$

where

$$I_d = \frac{1}{2\pi} \int_{\alpha}^{2\pi+\alpha} i_s d\omega t = \frac{1}{2\pi} \left[\int_{\alpha}^{\pi+\alpha} I_a d\omega t - \int_{\pi+\alpha}^{2\pi+\alpha} I_a d\omega t \right] = 0$$

$$a_n = \frac{1}{\pi} \int_{\alpha}^{2\pi+\alpha} i_s \cos n\,\omega t\, d\omega t$$

$$= \frac{1}{\pi} \left[\int_{\alpha}^{\pi+\alpha} I_a \cos n\,\omega t\, d\omega t - \int_{\pi+\alpha}^{2\pi+\alpha} I_a \cos n\,\omega t\, d\omega t \right]$$

$$= \frac{-4I_a}{n\pi} \sin n\alpha, \quad \text{for } n = 1, 3, 5, ...$$

$$= 0 \quad \text{for } n = 2, 4, 6, ...$$

$$b_n = \frac{1}{\pi} \int_{\alpha}^{2\pi+\alpha} i \sin n\,\omega t\, d\omega t$$

$$= \frac{1}{\pi} \left[\int_{\alpha}^{\pi+\alpha} I_a \sin n\,\omega t \cdot d\omega t - \int_{\pi+\alpha}^{2\pi+\alpha} I_a \sin n\,\omega t\, d\omega t \right]$$

$$= \frac{-4I_a}{n\pi} \cos n\alpha, \quad \text{for } n = 1, 3, 5, ...$$

$$= 0 \quad \text{for } n = 0, 4$$

Since $I_d = 0$, input current can be written as

$$i_s = \sum_{n=1, 3, 5}^{\alpha} \sqrt{2} I_n \sin (\omega t + \phi_n)$$

where ϕ_n is the displacement angle of nth harmonic current and

$$\phi_n = \tan^{-1} \frac{a_n}{b_n} = -n\alpha$$

rms value of nth harmonic input current,

$$I_{sn} = \frac{1}{\sqrt{2}} (a_n^2 + b_n^2)^{1/2} = \frac{4I_a}{\sqrt{2} n\pi} = \frac{2\sqrt{2} I_a}{n\pi}$$

rms value of fundamental current,

$$I_{s1} = \frac{2\sqrt{2} I_a}{\pi}$$

rms value of input current,

$$I_s = \left[\frac{2}{2\pi} \int\limits_{\alpha}^{\pi+\alpha} I_a^2 d\omega t \right]^{1/2} = I_a$$

Harmonic factor, $\text{HF} = \left[\left(\frac{I_s}{I_{s1}} \right)^2 - 1 \right]^{1/2} = 0.483$

Displacement factor, $\text{DF} = \cos \phi_1 = \cos(-\alpha)$

Power factor, $\text{PF} = \dfrac{I_{s1}}{I_s} \cos(-\alpha) = \dfrac{2\sqrt{2}}{\pi} \cos \alpha$

The α angle can be retarded to the maximum value of 90° for the rectifier mode of operation when $V_d = 0$. This condition requires practically infinite load inductance to maintain continuous conduction. A converter that can operate both as a rectifier and as an inverter is defined as two quadrant converter. This is in contrast to the one-quadrant converter, which operates in the rectification mode only.

Figure 8.14 shows the characteristics of the per unit output voltage against angle of delay for single-phase circuits one with four thyristors (fully-controlled) and other with two thyristors and two diodes (half-controlled), both circuits operating as rectifiers into a resistive load through a smoothing reactor. For the former circuit, the output voltage theoretically comes to zero at $\alpha = 90°$, but in practice the output voltage closely follows the theoretical curve until the current becomes discontinuous when it departs from the theoretical curve and the voltage decreases more gradually as shown by the dotted line, coming to zero at $\alpha = 180°$. The point of discontinuity changes with the load conditions and depends upon the ratio L/R.

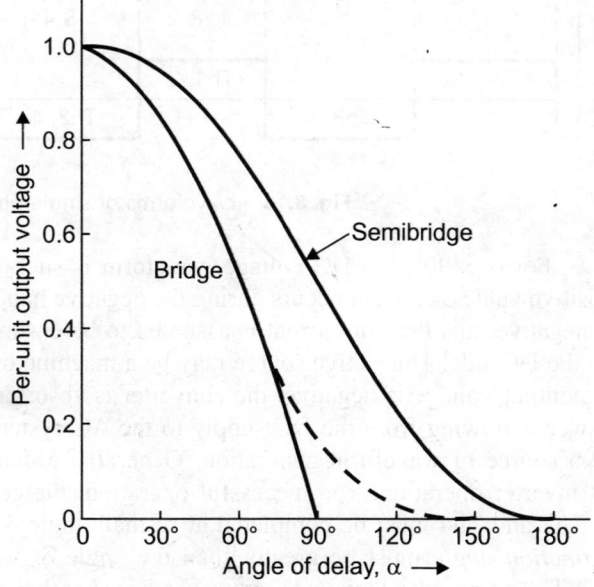

Fig. 8.14   The characteristics of the per unit output voltage

In the circuit with two thyristors and two diodes, there is no discontinuity of current, since during periods of freewheeling action, when the current falls, it does so exponentially. Thus, the output voltage follows the same characteristic for all load conditions, falling to zero at $\alpha = 180°$ as shown in Fig. 8.14. The harmonic content of the output voltage is substantially less in the mixed circuit, having a maximum value for h of 0.65 compared to the harmonic content of the output voltage in the four thyristor circuit with maximum $h = 1.11$. The power factor of AC current is also improved.

Clearly, if only rectifier operation is necessary, this full-wave mixed circuit is superior to the four thyristor circuit. Furthermore, on the basis of superior transformer utilisation, the full-wave mixed circuit is undoubtedly preferable to the biphase circuit with freewheeling diode.

8.9 SINGLE-PHASE BRIDGE INVERTER

Inverter operation: Inverters convert DC power to AC power at a desired output voltage and frequency and in most of the inverters both are required to be controlled. When $\alpha = 90°$, the output voltage consists of equal positive and negative portions. As α delayed angle is increased above $90°$, cos α becomes negative and the net output voltage becomes negative. Figure 8.15(a) shows such a waveform for $\alpha = 150°$, i.e. $\beta = 30°$. For $\alpha = 180°$, i.e. $\beta = 0$, the output voltage is $-V_d$.

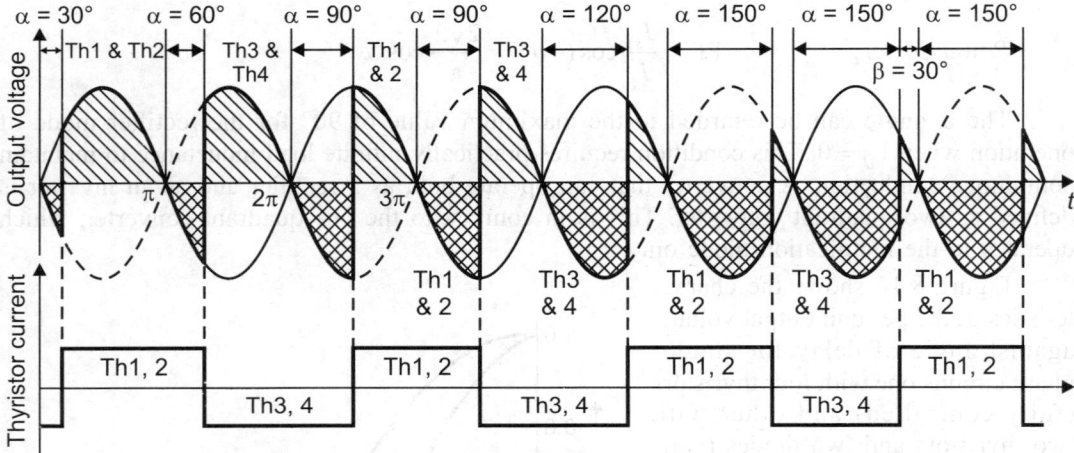

Fig. 8.15 Waveforms of single-phase bridge inverter

For $\alpha > 90°$, the DC voltage waveform is such that it has a negative average value. The positive value of current occurs during the negative half cycle of voltage. When the output voltage is negative, this flow of current is assumed to be ensured by some active source of DC supplied on the DC side. This active source may be a machine or a converter working as a rectifier. When the output voltage is negative, the converter is absorbing power, i.e. working as an inverter and power is flowing from the DC supply to the AC system. The AC system is assumed to have its own source of waveform generation. Generally, a converter cannot operate with $\alpha = 180°$, in its inverter operation. For successful operation, the commutation from thyristors Th1 and Th2 to Th3 and Th4 must be completed at a small angle δ before point 2π. This angle δ, called the *extinction angle* must be greater than the angle δ_0 which is just sufficient for thyristors Th1 and Th2 to recover their gate control; otherwise they will take the current back from Th3 and Th4 and result in a fault called *commutation failure*. δ_0 is called the *deionisation angle* and is of the order of $1°$. For safety purposes δ must be about $5°$ or more.

The commutation process takes a finite time which depends upon current and alternating voltage magnitudes and may be $10°$ to $30°$ for normal full-load operation. This means that the angle of advance, β, must be sufficiently large for commutation to be completed in good time. Unduly large values of β will, however, result in poorer utilisation of inverter capacity and excessive consumption of reactive power.

8.10 POWER FLOW

Power is not allowed to flow from AC to DC because of change in the polarity on the DC side. The current flowing must maintain its direction because of unidirectional property of the thyristors. The power delivered changes its direction of flow, i.e. it flows from DC to AC if there is an active source of proper polarity on the DC side to provide current in the required direction, power flows from DC to AC. The converter operates in the mode of inversion. It is now be cleared that the same gate-controlled converter can operate as either a rectifier or an inverter, just as an electrical machine can operate as either a generator or a motor.

This point can be understood by considering the two-converter system shown in Fig. 8.16(a). In the equivalent circuit of Fig. 8.16(b), each converter is represented by a variable–voltage battery, the voltage being controlled from positive to negative values (by gate-control). The unidirectional conduction feature is represented by series diodes. It will be seen that, with converter 1 having a positive output voltage V_{d1}, converter 2 having a negative output voltage $-V_{d2}$ and current I_d flowing in the only possible direction, power flows from converter 1 to converter 2 (battery 1 charging battery 2), i.e. converter 1 is operating as a rectifier and converter 2 as an inverter. The current flow is given by

$$I_d = \frac{V_{d1} - V_{d2}}{R_L}$$ ...(8.21)

where R_L is the resistance of the DC circuit.

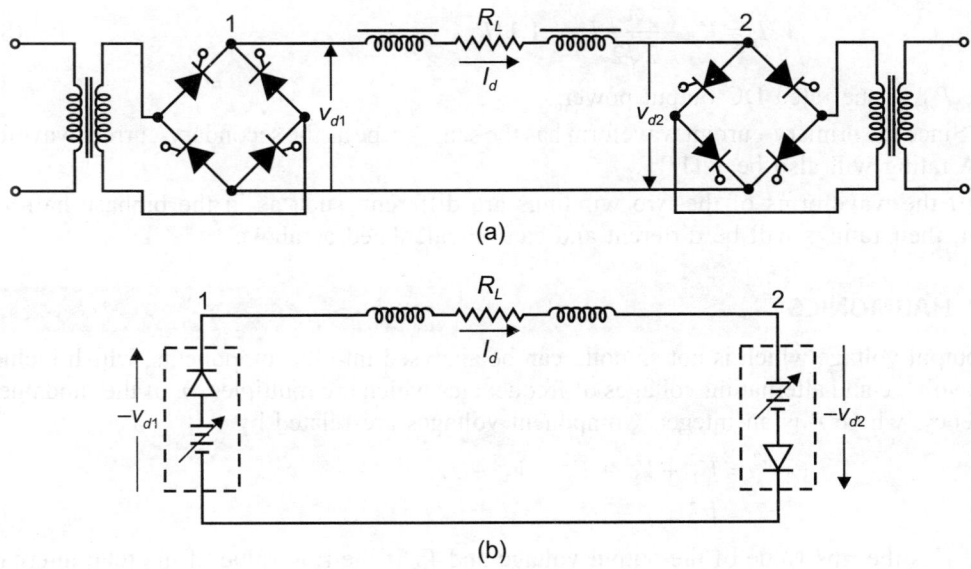

(a)

(b)

Fig. 8.16 Power flow in gate controlled converter as inverter or rectifier

If the voltages are reversed, the power flow is reversed. Also the two converter circuits may be of different types: single-phase or three-phase. The frequencies of the two AC systems may also be different.

The fundamental component of transforms current I_s is obtained by equating AC power P_a, and DC power P_d:

$$P_a = V_s I_s \cos \phi$$

$$P_d = V_d I_d = \frac{2\sqrt{2}V_s}{\pi} \cos \alpha \cdot I_d = \frac{2}{\pi} V_m \cos \alpha \cdot I_d$$

whence $$I_s = \frac{2\sqrt{2}}{\pi} I_d$$ ...(8.22)

and $\qquad \cos \phi = \cos \alpha$ (lagging)

The power $V_s I_s \cos \phi$, flows from the AC system to DC system. As α is increased above 90°, the current lags behind the voltage by an angle greater than 90°, and this means that power flows into the AC system.

8.11 TRANSFORMER RATING

The transformer has to be rated for the rms value of the current flowing through its windings. This rms current can be calculated from its waveform, refer Fig. 8.13 (b).

$$I_s = \left[\frac{1}{\pi} \int_0^\pi I_d^2 \, d\omega t \right]^{1/2} = I_d \qquad \qquad ...(8.23)$$

The transformer-secondary volt-ampere rating $= V_s \, I_s$

As $\qquad V_{do} = \dfrac{2\sqrt{2}V_s}{\pi}$ or $V_s = \dfrac{V_{do} \cdot \pi}{2\sqrt{2}}$

$$V_s I_s = V_{do} \cdot \frac{\pi}{2\sqrt{2}} I_d = 1.11 P_{do} \qquad \qquad ...(8.24)$$

where P_{do} is the rated DC output power.

Since the primary-current waveform has the same shape as the secondary-current waveform, its VA rating will also be $1 \cdot 11 P_{do}$.

If the waveforms of the two windings are different, such as in the biphase half-wave circuit, their ratings will be different and can be calculated as above.

8.12 HARMONICS

The output voltage which is not smooth, can be analysed into its components, which include a direct voltage and alternating voltages of frequencies which are multiples, n, of the fundamental frequency, where n is an integer. Component voltages are related by

$$V^2 = V_d^2 + V_1^2 + V_2^2 + V_3^2 + ...$$
$$= V_d^2 + V_h^2 \qquad \qquad ...(8.25)$$

where V is the rms value of the output voltage and V_h is the rms value of the total alternating content of the output voltage.

$$V_h^2 = V_1^2 + V_2^2 + V_3^2 ... V_n^2 \qquad \qquad ...(8.26)$$

Thus, a knowledge of V_h is a convenient measure of harmonic content and hence of the smoothness of the output voltage. The quotient of V_h by V_{do} is called the *harmonic factor*, h and is given by

$$h = \frac{V_h}{V_{do}} = \left[\frac{V^2 - V_d^2}{V_{do}^2} \right]^{1/2} \qquad \qquad ...(8.27)$$

The rms output voltage

$$V_{rms} = \left[\frac{1}{\pi} \int_\alpha^{\pi+\alpha} (\sqrt{2}V_s \sin \omega t)^2 \, d\omega t \right]^{1/2} = V_s \qquad \qquad ...(8.28)$$

$$h = \left[\frac{\pi^2}{8} - \cos^2 \alpha \right]^{1/2} \qquad \qquad ...(8.29)$$

h has a minimum value of 0.483 for $\alpha = 0$ and a maximum value of 1.11 for $\alpha = 90°$. The harmonic content of the alternating current can be calculated similarly:

$$\frac{I_h}{I_{(1)}} = \left[\frac{I_s^2 - I_{(1)}^2}{I_{(1)}} \right]^{1/2} = \left[\left(\frac{I_s}{I_{(1)}} \right)^2 - 1 \right]^{1/2} \qquad \qquad ...(8.30)$$

For the single-phase full-wave circuit, Equations (8.22), (8.23), and (8.30) give

$$\frac{I_h}{I_{(1)}} = \left[\frac{\pi^2}{8} - 1 \right]^{1/2} = 0.483.$$

8.13 DISCONTINUOUS CONDUCTION

The converter circuits discussed so far were assumed to operate in continuous conduction. The term continuous means that load current never ceases but continues to flow through thyristor/ diode or their combination. In practice, a converter may also operate in the discontinuous conduction mode, which is more likely with a counter emf load or at high values of firing angle or at low values of load current. The term discontinuous is applied to the conduction when load current reaches zero during each half cycle before the next thyristor in sequence is fired. A two quadrant converter may have discontinuous conduction both in rectification and in inversion. It is preferable to operate DC load in continuous current mode. This is promoted by having freewheeling action and using an external inductor in series with load.

8.14 SINGLE-PHASE FULL-WAVE CONVERTER WITH DISCONTINUOUS CURRENT

Figure 8.17 shows the circuit and waveforms for a bridge converter under discontinuous conduction with R_d, L_d and counter emf load V_c. The thyristor pair Th1, Th2 can be fired

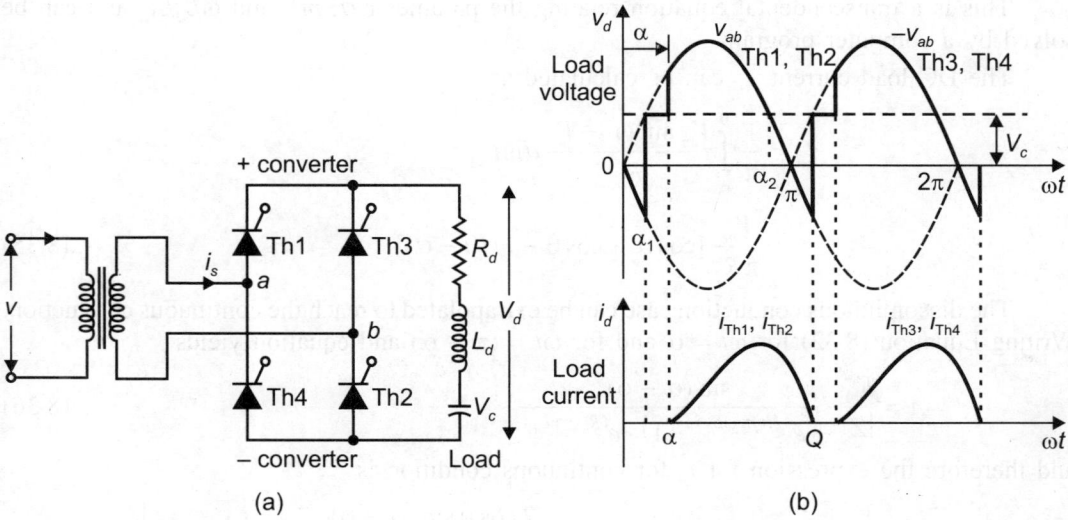

Fig. 8.17 Waveforms for bridge converter with discontinuous current

to conduct when the supply voltage v_{ab} is positive and exceeds the counter emf. The load current i_d will grow up to an angle α_2 but will continue to angle θ due to the load inductance effect. Then v_d will equal V_c until thyristor Th3, Th4 pair is fired symmetrically at angle 'α' in the next half cycle. Obviously, the range of α can be given as $\alpha_1 \leq \alpha \leq \alpha_2$, where $\alpha_1 = \sin^{-1}(V_c/V_m)$ and $\alpha_2 = \pi - \alpha_1$. The conduction may change from discontinuous to continuous if α or V_c is low or the load inductance L_d is high. The equation during conduction can be given as:

$$L_d \frac{di_d}{dt} + R_d i_d = V_m \sin \omega t - V_c \qquad \qquad ...(8.31)$$

which can be solved for i_d as

$$i_d = Ae^{-(R_d/L_d)t} + \frac{V_m}{|Z|} \sin(\omega t - \phi) - \frac{V_c}{R_d} \quad \text{for } i_d \geq 0 \qquad ...(8.32)$$

where $\qquad \qquad |Z| = \sqrt{R_d^2 + \omega^2 L_d^2}$
and $\qquad \qquad \phi = \tan^{-1}(\omega L_d/R_d)$

Noting that $i_d = 0$ at $\omega t = \alpha$, Equation (8.32) can be written as

$$i_d = \frac{V_m}{R_d} \left\{ \frac{R_d}{|Z|} \sin(\omega t - \phi) - m + \left[m - \frac{R_d}{|Z|} \sin(\alpha - \phi) \right] e^{-(R_d/\omega L_d)(\omega t - \alpha)} \right\}$$

$$...(8.33)$$

where $\qquad \qquad m = \frac{V_c}{V_m} = \text{Counter emf coefficient}$

Equation (8.33) is valid in the range $\alpha \leq \omega t \leq \theta$ for discontinuous conduction. From the waveform, $i_d = 0$ again at $\omega t = \theta$. Substituting this condition in Equation (8.33) yields.

$$e^{(R_d/\omega L_d)\theta} \cdot \frac{\cos \phi \sin(\theta - \phi) - m}{\cos \phi \sin(\alpha - \phi) - m}$$

$$= e^{(R_d/\omega L_d)\alpha} \qquad \qquad ...(8.34)$$

This is a transcendental equation relating the parameter α, $m\theta$ and $\omega L_d/R_d$ and can be solved by a computer program.

The DC load current, I_d can be calculated as

$$I_d = \frac{1}{\pi} \int_{\alpha}^{\theta} \frac{V_m \sin \omega t - V_c}{R_d} d\omega t$$

$$= \frac{V_m}{\pi R_d} [\cos \alpha - \cos \theta - m(\theta - \alpha)] \qquad \qquad ...(8.35)$$

The discontinuous conduction case can be extrapolated to reach the continuous conduction. Writing Equation (8.32) for $\omega t = \alpha$ and for $\omega t = (\pi + \alpha)$ and equation yields

$$A = \frac{V_m}{|Z|} \cdot \frac{\sin(\alpha - \phi)}{\left(e^{-(R_d \pi/\omega L_d)} - 1 \right) \cdot e^{(R_d \alpha/\omega L_d)}} \qquad \qquad ...(8.36)$$

and therefore the expression for i_d for continuous condition is

$$i_d = \frac{V_m}{R_d} \left[\cos \phi \sin(\omega t - \phi) - m - \frac{2 \cos \phi \sin(\alpha - \phi)}{1 - e^{-(R_d \pi/\omega L_d)}} e^{-(R_d/\omega L_d)(\omega t - \alpha)} \right] \qquad ...(8.37)$$

DC load current, I_d can be determined from Equation (8.37) as

$$I_d = \frac{1}{\pi} \int\limits_{\alpha}^{\alpha+\pi} i_d \, d\omega t = \frac{V_m}{\pi R_d} (2 \cos \alpha - m\pi). \qquad \qquad ...(8.38)$$

8.15 LOAD VOLTAGE AND HARMONICS

When the positive and negative converters are superimposed to constitute the full bridge converter, all the waveforms remain the same except that the load voltage magnitude becomes double. The instantaneous load circuit equation of bridge can be written as

$$v_d = R_d i_d + L_d \frac{di_d}{dt} + V_c \qquad \qquad ...(8.39)$$

where R_d, L_d and V_c are the resistance, inductance, and counter emf of the load, respectively. Equation (8.39) can be averaged to derive the DC load circuit equation as

$$V_d = \frac{1}{\pi} \int\limits_{\alpha}^{\alpha+\pi} \left(R_d i_d + L_d \frac{di_d}{dt} + V_c \right) d\omega t$$

$$V_d = I_d R_d + V_c$$

or

$$I_d = \frac{V_d - V_c}{R_d} \qquad \qquad ...(8.40)$$

The average power delivered to the load is:

$$P_0 = \frac{1}{\pi} \int\limits_{\alpha}^{\alpha+\pi} v_d i_d \, d\omega t = I_0^2 R_d + V_c I_d \qquad \qquad ...(8.41)$$

where I_0 is the rms load current. The load voltage v_d contains harmonics and can be given by general Fourier expression

$$v_d = V_d + \sum_{n=1}^{\alpha} (a_n \cos n \omega t + b_n \sin n \omega t) \qquad \qquad ...(8.42)$$

where

$$a_n = \frac{2}{\pi} \int\limits_{\alpha}^{\alpha+\pi} v_d \cos n \omega t \, d\omega t$$

$$b_n = \frac{2}{\pi} \int\limits_{\alpha}^{\alpha+\pi} v_d \sin n \omega t \, d\omega t$$

The unsymmetrical nature of the v_d waveform indicates that it contains even harmonics and its fundamental frequency is twice that of the AC source. The harmonics can be attenuated by an inductance, capacitance or inductance-capacitance filter. The harmonics flowing in the utility system are undesirable, because of unnecessary loading of the power equipment and interference to neighbouring telephone lines. The line voltage may also be distorted to some extent, depending on source impedance. It can be shown that the line current will approach a square wave if the load is assumed to be perfectly filtered. Figure 8.18 shows the line voltage and current waves under this condition.

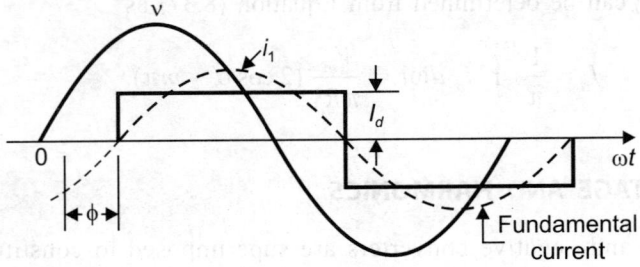

Fig. 8.18 Line voltage and current waves under this condition

8.15.1 Distortion Factor

The degree of line current distortion can be determined by the distortion factor. It is defined as:

$$DF = \frac{\text{rms value of fundamental current}}{\text{rms value of total current}} \qquad ...(8.43)$$

$$= \frac{I_1}{\sqrt{I_1^2 + \sum\limits_{n=2}^{\alpha} I_n^2}}$$

For square wave current, as shown in Fig. 8.18,

$$DF = \frac{(4/\pi)(1/\sqrt{2}) I_d}{I_d} = \frac{2\sqrt{2}}{\pi}.$$

8.15.2 Displacement Factor

The fundamental line current of a phase-controlled converter always lags the line voltage by a displacement angle ϕ, i.e. the line must supply lagging reactive power demanded by the converter. Displacement factor can be defined as:

Displacement factor

$$= \frac{\text{Average input power}}{\text{rms fundamental voltage} \times \text{rms fundamental current}} \qquad ...(8.44)$$

$$= \frac{P_1}{VI_1} = \frac{VI_1 \cos \phi}{VI_1} = \cos \phi$$

In Fig. 8.18, i_1 is the fundamental line current and therefore ϕ is the true displacement angle. It may be noted here that the displacement angle always equals the firing angle α for a perfectly filtered load. Again, assuming the converter as lossless, the input power must balance the output power.

8.15.3 Power Factor

Input power factor can be defined as:

$$PF = \frac{\text{Average input power}}{\text{rms supply voltage} \times \text{rms supply current}} \qquad ...(8.45)$$

$$= \frac{P_1}{V\sqrt{I_1^2 + \sum_{n=2}^{\infty} I_n^2}}$$

Substituting Equations (8.43) and (8.44) in Equation (8.45) yields

$$\text{PF} = \frac{VI_1 \cos \phi}{V\sqrt{I_1^2 + \sum_{n=2}^{\infty} I_n^2}} = \cos \phi \frac{I_1}{\sqrt{I_1^2 + \sum_{n=2}^{\infty} I_n^2}}$$

$$\text{PF} = \text{Dis. } F \times \text{DF} \qquad \qquad ...(8.46)$$

For a sine current wave DF = 1 and therefore power factor is the same as the displacement factor. For a distorted wave, DF is less than unity and therefore PF < Dis. F. Note that for a diode bridge converter, Dis. $F = 1$ and PF = DF, which is always less than unity.

8.16 THREE-PHASE THYRISTOR CONVERTER

For a load power requirement of several kilowatts or more, it is desirable to use a three-phase converter. A polyphase converter, in general, not only imposes balanced loading on the utility system, but provides considerable improvement in load voltage and line current, harmonics. Load current is mostly continuous in three-phase converters. As a result, the harmonic filtering requirement becomes a nominal problem. Various types of three-phase phase controlled converters are:

1. three-phase half-wave converter,
2. three-phase semiconverter,
3. three-phase full converter,
4. three-phase bridge converter, and
5. three-phase dual converter.

Three-phase half-wave converter is rarely used in industry because it introduces DC components in the supply current. Semiconductors and full converters are quite common in industrial applications. A dual converter is used when reversible DC drives with power ratings of several MW are required.

8.16.1 Three-phase Half-wave Converters

Figure 8.19 (a) shows the circuit of a three-phase half-wave converter, whose mean load voltage can be adjustable by control of firing delay angle α. The firing circuit is not shown, but it may be assumed that each thyristor has a firing circuit connected to its gate and cathode, producing a firing pulse relative in position to its own phase voltage. A master control will ensure that the three gate pulses are displaced by 120° relative to each other, giving the same firing delay angle to each thyristor, thyristors Th1, Th2, and Th3 conduct symmetrically, each for 120° through the load and provide a common return to the transformer neutral point N. A thyristor can be fired to conduct when its anode voltage is positive with respect to cathode and conduction will continue until the subsequent thyristor is fired after a 120° angle. The firing delay angle α is defined such that it is zero when the output mean voltage is a maximum, that is the diode case. Hence, the firing delay angle α shown in Fig. 8.19 (b) is defined relative to the instant when the supply phase voltage cross and diodes would commutate naturally, not the supply voltage zero. For a resistive load and $\alpha > \pi/6$, the load current would be discontinuous and each thyristor is self commutated when the polarity of its phase voltage is reversed. The frequency of output ripple voltage is $3f_s$. This converter is not used in practical systems, because the supply currents contain DC components.

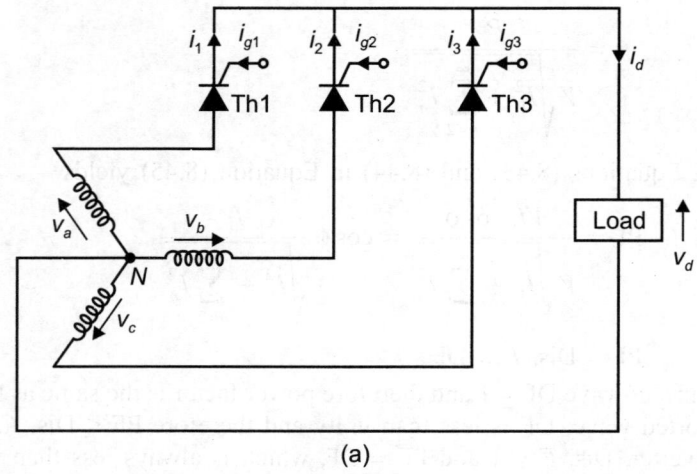

(a)

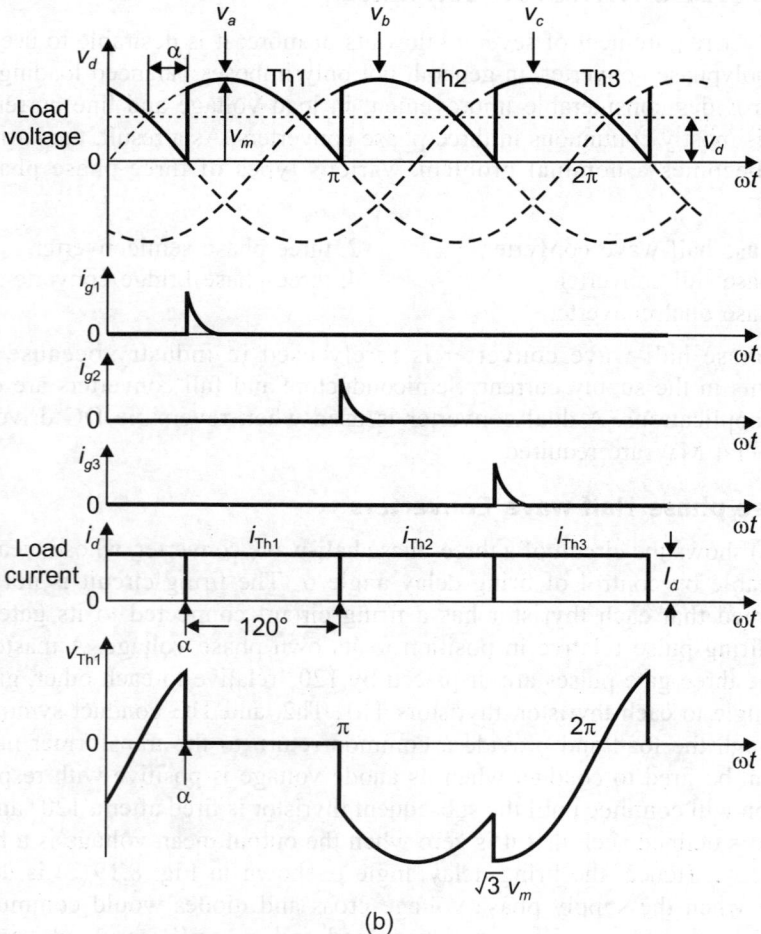

(b)

Fig. 8.19 Waveforms of three-phase half-wave converters

If the phase voltage is $v = V_m \sin \omega t$, the average output voltage for a continuous current is

$$V_d = \frac{3}{2\pi} \int_{\pi/6+\alpha}^{5\pi+\alpha} V_m \sin \omega t \, d\omega t = \frac{3\sqrt{3}}{2\pi} V_m \cos \alpha = \frac{3}{\pi} \times 0.866 V_m \cos \alpha \qquad ...(8.47)$$

less by the single thyristor volt drop.

The maximum average output voltage that occurs at a delay angle $\alpha = 0$ is

$$V_{dm} = \frac{3\sqrt{3} V_m}{2\pi} \qquad ...(8.48)$$

and the normalised average output voltage is

$$V_n = \frac{V_d}{V_{dm}} = \cos \alpha$$

rms output voltage, $\quad V_{rms} = \left[\frac{3}{2\pi} \left\{ \int_{\pi/6+\alpha}^{5\pi/6+\alpha} V_m^2 \sin^2 \omega t \, d\omega t \right\} \right]^{1/2}$

$$V_{rms} = \sqrt{3} V_m \left(\frac{1}{6} + \frac{\sqrt{3}}{8\pi} \cos 2\alpha \right)^{1/2} \qquad ...(8.49)$$

For a resistive load and $\alpha = \pi/6$:

$$V_d = \frac{3}{2\pi} \int_{\pi/6+\alpha}^{\pi} V_m \sin \omega t \, d\omega t = \frac{3V_m}{2\pi} [1 + \cos (\pi/6 + \alpha)]$$

$$= \frac{3}{\pi} V_m [0.5 + 0.5 \cos (\pi/6 + \alpha)] \qquad ...(8.50)$$

$$V_n = \frac{V_d}{V_{dm}} = \frac{1}{\sqrt{3}} [1 + \cos (\pi/6 + \alpha)] \qquad ...(8.51)$$

$$V_{rms} = \left[\frac{3}{2\pi} \int_{\pi/6+\alpha}^{\pi} V_m^2 \sin^2 \omega t \, d\omega t \right]^{1/2}$$

$$= \sqrt{3} V_m \left[\frac{5}{24} - \frac{\alpha}{4\pi} + \frac{1}{8\pi} \sin (\pi/3 + 2\alpha) \right]^{1/2} \qquad ...(8.52)$$

Three-phase half-wave thyristor converter with a freewheeling diode: The circuit voltage and current diagrams are shown in Fig. 8.20. For an angle of delay 'α' less than 30°, the output voltage is same as for the circuit without a freewheeling diode, because the output voltage remains positive at every instant.

For $\quad 0 \le \alpha \le \pi/6$,

$$V_d = \frac{3}{2\pi} \int_{\pi/6+\alpha}^{5\pi/6+\alpha} (\sqrt{2} V_p) \sin \omega t \, d\omega t$$

$$= \frac{3\sqrt{6}}{2\pi} V_p \cos \alpha = \frac{3\sqrt{3}}{2\pi} V_m \cos \alpha = V_{d0} \cos \alpha \qquad ...(8.53)$$

where $\quad V_{d0} = \frac{3\sqrt{3}}{2\pi} V_m$ and V_p is the rms phase voltage.

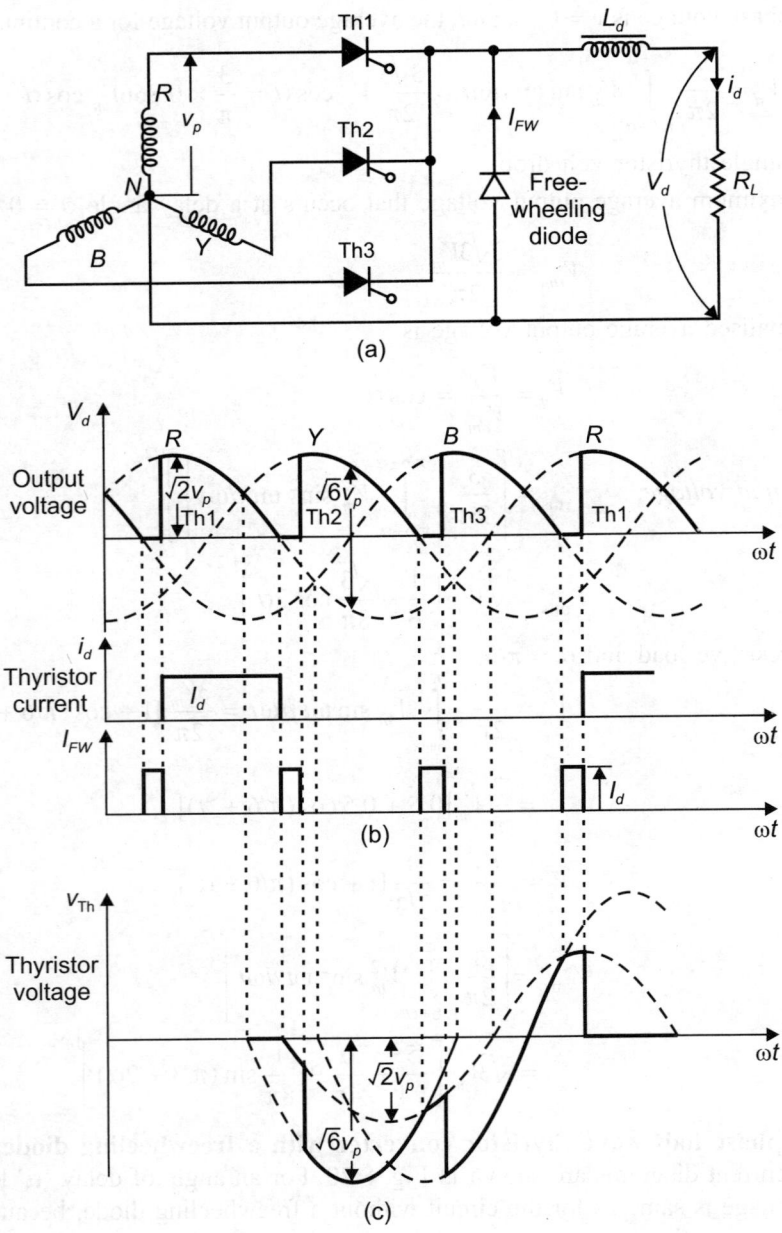

Fig. 8.20 Waveforms of three-phase half-wave thyristor converter with freewheeling diode

However, for an angle of delay α greater than 30°, the freewheeling diode eliminates the negative voltage by bypassing the current during these periods.

Thus, $\pi/6 \leq \alpha \leq 5\pi/6$,

$$V_d = \frac{3}{2\pi} \int_{\pi/6+\alpha}^{\pi} \sqrt{2}V_p \sin \omega t \, d\omega t$$

$$= \frac{3\sqrt{2}V_p}{2\pi}[1 + \cos{(\pi/6 + \alpha)}] = \frac{V_{d0}}{\sqrt{3}}[1 + \cos{(\pi/6 + \alpha)}]$$

$$= \frac{3}{2\pi}V_m[1 + \cos{(\pi/6 + \alpha)}] \qquad\qquad ...(8.54)$$

For $\alpha > 30°$, the output voltage is smoother compared to the circuit without a freewheeling diode, and continuous current is ensured.

It may be noted from Fig. 8.20 (c) that the reverse voltage has a peak value of $\sqrt{6}V_p = \left(\frac{2\pi}{3}\right)V_{d0}$. The same as for this ϕ circuit without a freewheeling diode but the peak forward voltage is only $\sqrt{2}V_p = \left(\frac{2\pi}{3\sqrt{3}}\right)V_{d0}$. Thus, a diode with a voltage rating of $(\sqrt{6} - \sqrt{2})V_p$ may be put in series with each thyristor to permit the provision of a thyristor with a voltage rating of $\sqrt{2}V_p$ only.

8.16.2 Three-phase Semiconverters

Three-phase semiconverters are used in industrial applications up to 120 kW load where one-quadrant operation is required. In this converter as the delay angle increases, the power factor decreases. These converters are better than half-wave converters. Figure 8.21 (a) shows a three-phase semiconverter with highly inductive load and the load current has a negligible ripple content. Figure 8.21 (b) shows the waveforms for input voltages, output voltages, current through diodes. The frequency of output voltages is $3f_s$, the delay angle α can be varied from 0 to π. During the period $\pi/6 \leq \omega t \leq 7\pi/6$, thyristor Th1 is forward-biased. If thyristor Th1 is fired at $(\pi/6 + \alpha)$, Th1 and D_1 conduct and the line voltage appears across the load. At $\omega t = 7\pi/6$, line voltage starts to be negative and the freewheeling diode D_{FW} conducts. The load current continues to flow through D_{FW}; thyristor Th1 and diode D_1 are turned off.

If there were no freewheeling diode, thyristor Th1 would continue to conduct until thyristor Th2 fired at $\omega t = 5\pi/6 + \alpha$ and the freewheeling action would be accomplished through thyristor Th1 and diode D_2. If $\alpha \leq \pi/3$, each thyristor conducts for $2\pi/3$ and the freewheeling diode, D_{FW} does not conduct.

Let the three line to neutral voltage are as follows:

$$v_{an} = V_m \sin \omega t$$
$$v_{bn} = V_m \sin (\omega t - 2\pi/3)$$
and
$$v_{cn} = V_m \sin (\omega t + 2\pi/3)$$

The corresponding line to line voltages are:

$$v_{ac} = v_{an} - v_{cn} = V_m [\sin \omega t - \sin (\omega t + 2\pi/3)]$$

$$= V_m \left[2 \cos \frac{1}{2}\{\omega t + (\omega t + 2\pi/3)\} \cdot \sin \frac{1}{2}\{\omega t - (\omega t + 2\pi/3)\} \right]$$

$$= V_m \left[2 \cos \frac{1}{2}\{2\omega t + 2\pi/3\} \right] \cdot \sin (-\pi/3)$$

$$= V_m [2 \cos (\omega t + \pi/3) \cdot \sin \pi/3]$$

$$= V_m \sin \pi/3 [2 \cos (\omega t + \pi/3 + \pi/6 - \pi/6)]$$

$$= V_m \cdot \frac{\sqrt{3}}{2}[2 \cos (\pi/2 + \omega t - \pi/6)] = \sqrt{3}V_m \sin (\omega t - \pi/6)$$

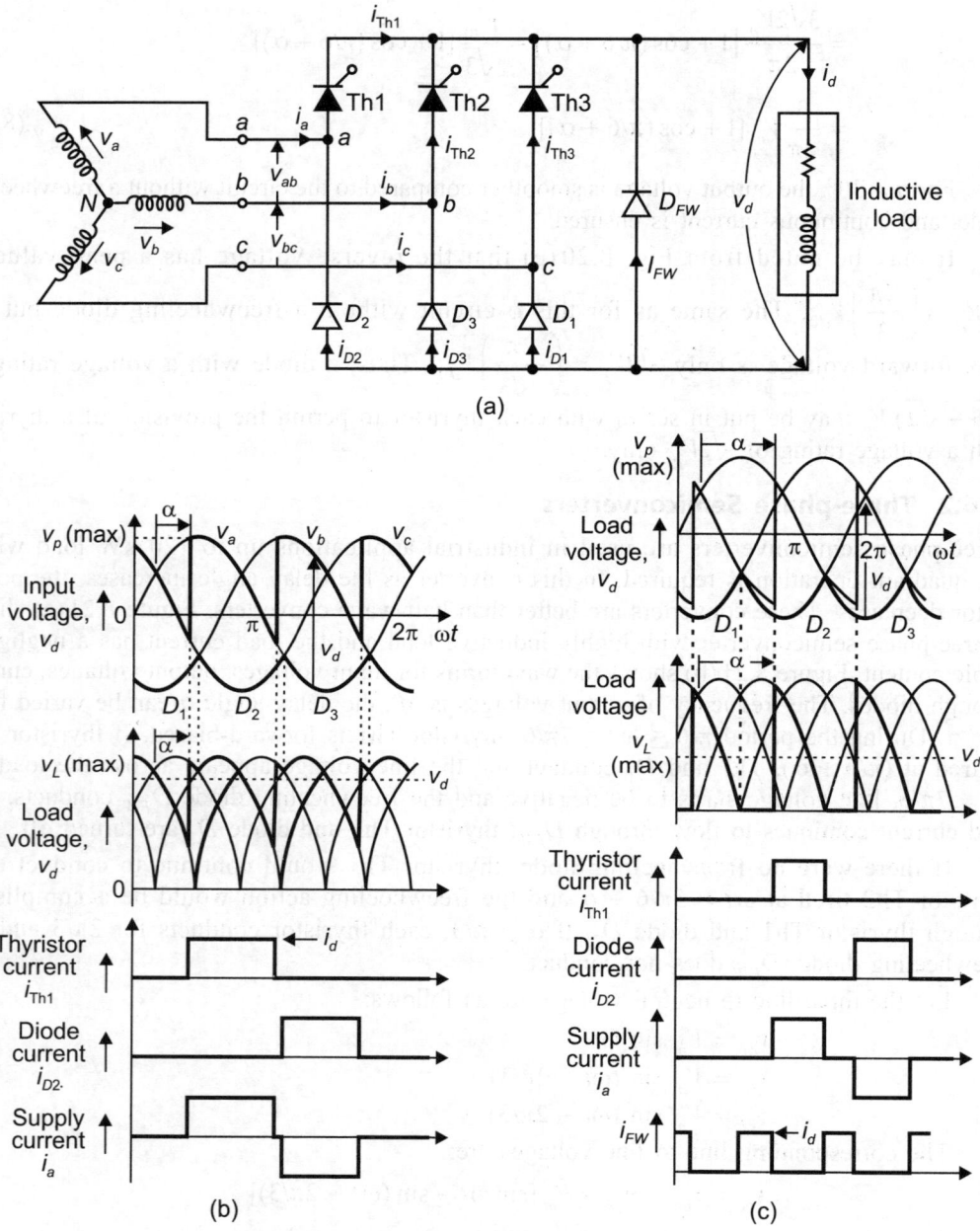

Fig. 8.21 Waveforms of three-phase semiconverters

Similarly,
$$v_{ba} = v_{bn} - v_{an} = \sqrt{3}V_m \sin(\omega t - 5\pi/6)$$

$$v_{cb} = v_{cn} - v_{bn} = \sqrt{3}V_m \sin(\omega t + \pi/2)$$

$$v_{ab} = v_{an} - v_{bn} = \sqrt{3}V_m \sin(\omega t + \pi/6)$$

where V_m is the peak phase voltage, $V_m = \sqrt{2}V_p$ and line voltage $V_{L\,(max)} = \sqrt{3}V_m$.

For $\alpha \geq \pi/3$, and discontinuous output voltage, average output voltage V_d, is given by

$$V_d = \frac{3}{2\pi} \int_{\pi/6+\alpha}^{\pi/6} v_{ac} d\omega t = \frac{3}{2\pi} \int_{\pi/6+\alpha}^{\pi/6} \sqrt{3} V_m \sin(\omega t - \pi/6) \, d\omega t$$

$$V_d = \frac{3\sqrt{3} V_m}{2\pi} (1 + \cos \alpha) = \frac{3}{2\pi} V_{L\,(max)} (1 + \cos \alpha) \qquad \qquad ...(8.55)$$

At $\alpha = 0$, average output voltage is maximum, i.e.

$$V_{dm} = \frac{3\sqrt{3} V_m}{\pi} \text{ and normalised average output voltage is}$$

$$V_n = \frac{V_d}{V_{dm}} = 0.5 \, (1 + \cos \alpha)$$

rms output voltage,

$$V_{rms} = \left[\frac{3}{2\pi} \int_{\pi/6+\alpha}^{\pi/6} 3V_m^2 \sin^2(\omega t - \pi/6) \, d\omega t \right]^{1/2}$$

$$= \sqrt{3} V_m \left[\frac{3}{4\pi} \left(\pi - \alpha + \frac{1}{2} \sin 2\alpha \right) \right]^{1/2}$$

For $\alpha \leq \pi/3$, and continuous output voltage:

$$V_{dc} = \frac{3}{2\pi} \left[\int_{\pi/6+\alpha}^{\pi/2} v_{ab} d\omega t + \int_{\pi/2}^{5\pi/6+\alpha} v_{ac} d\omega t \right] = \frac{3\sqrt{3} V_m}{2\pi} (1 + \cos \alpha)$$

$$= \frac{3}{2\pi} V_{L\,(max)} (1 + \cos \alpha)$$

Normalised average output voltage,

$$V_n = \frac{V_d}{V_{dm}} = 0.5 \, (1 + \cos \alpha)$$

$$V_{rms} = \left[\frac{3}{2\pi} \int_{\pi/6+\alpha}^{\pi/2} v_{an}^2 \, d\omega t + \int_{\pi/2}^{5\pi/6+\alpha} v_{ac}^2 \, d\omega t \right]^{1/2}$$

$$= \sqrt{3} V_m \left[\frac{3}{4\pi} \left(\frac{2\pi}{3} + \sqrt{3} \cos^2 \alpha \right) \right]^{1/2} \qquad \qquad ...(8.56)$$

or

$$V_{rms} = \sqrt{3} \cdot \sqrt{2} V_p \left[\frac{3}{4\pi} \left(\frac{2\pi}{3} + \sqrt{3} \cos^2 \alpha \right) \right]^{1/2}$$

$$= \sqrt{2} \, (\sqrt{3} V_p) \sqrt{\frac{3}{2\pi}} \sqrt{\frac{\pi}{3} + \frac{\sqrt{3}}{4} (1 + \cos 2\alpha)}$$

Form factor, $\quad FF = \dfrac{V_{rms}}{V_{av}} = \dfrac{\sqrt{2} V_L \sqrt{\dfrac{3}{2\pi}} \sqrt{\dfrac{\pi}{3} + \dfrac{\sqrt{3}}{4} (1 + \cos 2\alpha)}}{\dfrac{3\sqrt{2} V_L}{2\pi} (1 + \cos \alpha)}$

$$= \frac{\sqrt{\frac{2\pi}{3} \left[\frac{\pi}{3} + \frac{\sqrt{3}}{4} (1 + \cos 2\alpha) \right]^{1/2}}}{(1 + \cos \alpha)}$$

Form factor, $\alpha \dfrac{\left[\dfrac{\pi}{3} + \dfrac{\sqrt{3}}{4} (1 + \cos 2\alpha) \right]^{1/2}}{(1 + \cos \alpha)}$

Figure 8.21 (b) shows the waveforms of three-phase semibridge for small firing delay. Figure 8.21 (c) shows the waveforms of three-phase semibridge converter with larger firing delay angle.

8.16.3 Three-phase Full Converters

Three-phase converters are extensively used in industrial applications up to the 120 kW, where two-quadrant operation is required. Figure 8.22 (a) shows a full converter circuit with a highly inductive load this circuit is known as a three-phase bridge. The thyristors are fired at an interval of $\pi/3$. The frequency of output ripple voltage is $6f_s$. At $\omega t = (\pi/6 + \alpha)$, thyristor Th6 is already conducting and thyristor Th1 is turned on. During interval $(\pi/6 + \alpha) \le \omega t \le (\pi/2 + \alpha)$, thyristors Th1 and Th6 conduct and the line to line voltage $v_{ab} = (v_{an} - v_{bn})$ appears across the load. At $\omega t = (\pi/2 + \alpha)$, thyristor Th2 is fired and thyristor Th6 is reverse-biased immediately. Thyristor Th6 is turned off due to natural commutation. During interval $(\pi/2 + \alpha) \le \omega t \le (5\pi/6 + \alpha)$, thyristors Th1 and Th2 conduct and the line to line voltage v_{AC} appears across the load. If the thyristors are numbered as shown, the firing sequence is 12, 23, 34, 45, 56 and 61. Figure 8.22 (b) shows the waveforms for input voltage, output voltage, input current and currents through thyristors with a small firing delay angle.

When the circuit is connected to the AC supply, firing gate pulse will be delivered to the thyristors in the correct sequence but, if only a single firing gate pulse is used, no current will flow, as the other thyristor in the current path will be in the off-state. Hence, in order to start the circuit functioning two thyristors must be fired at the same time in order to commence current flow. With reference to Fig. 8.22 (b) supply is connected when v_a is at its peak value, the next firing pulse will be to the thyristor Th2. For starting purposes, the firing circuit must produce a firing pulse 60° after its first pulse. Once the circuit is running normally, the second pulse will have no effect, as the thyristor will already be in the on-state.

When the firing delay is large, with the load voltage having negative periods, it is difficult to visualise the load voltage waveform from the two three-pulse pictures. Figure 8.22 (c) shows the six line voltages $v_a - v_b$, $v_a - v_c$, $v_b - v_c$, $v_b - v_a$, $v_c - v_a$, $v_c - v_b$ give a direct picture of the load-voltage waveform and clearly show that zero mean voltage is reached when the firing delay angle is 90°.

Let the line to neutral voltages are:

$$v_{an} = V_m \sin \omega t$$
$$v_{bn} = V_m \sin (\omega t - 2\pi/3)$$
$$v_{cn} = V_m (\omega t + 2\pi/3)$$

The corresponding line to line voltages are:

$$v_{ab} = v_{an} - v_{bn} = \sqrt{3} V_m \sin (\omega t + \pi/6)$$
$$v_{bc} = v_{bn} - v_{cn} = \sqrt{3} V_m \sin (\omega t - \pi/2)$$
$$v_{ca} = v_{cn} - v_{an} = \sqrt{3} V_m (\omega t + \pi/2)$$

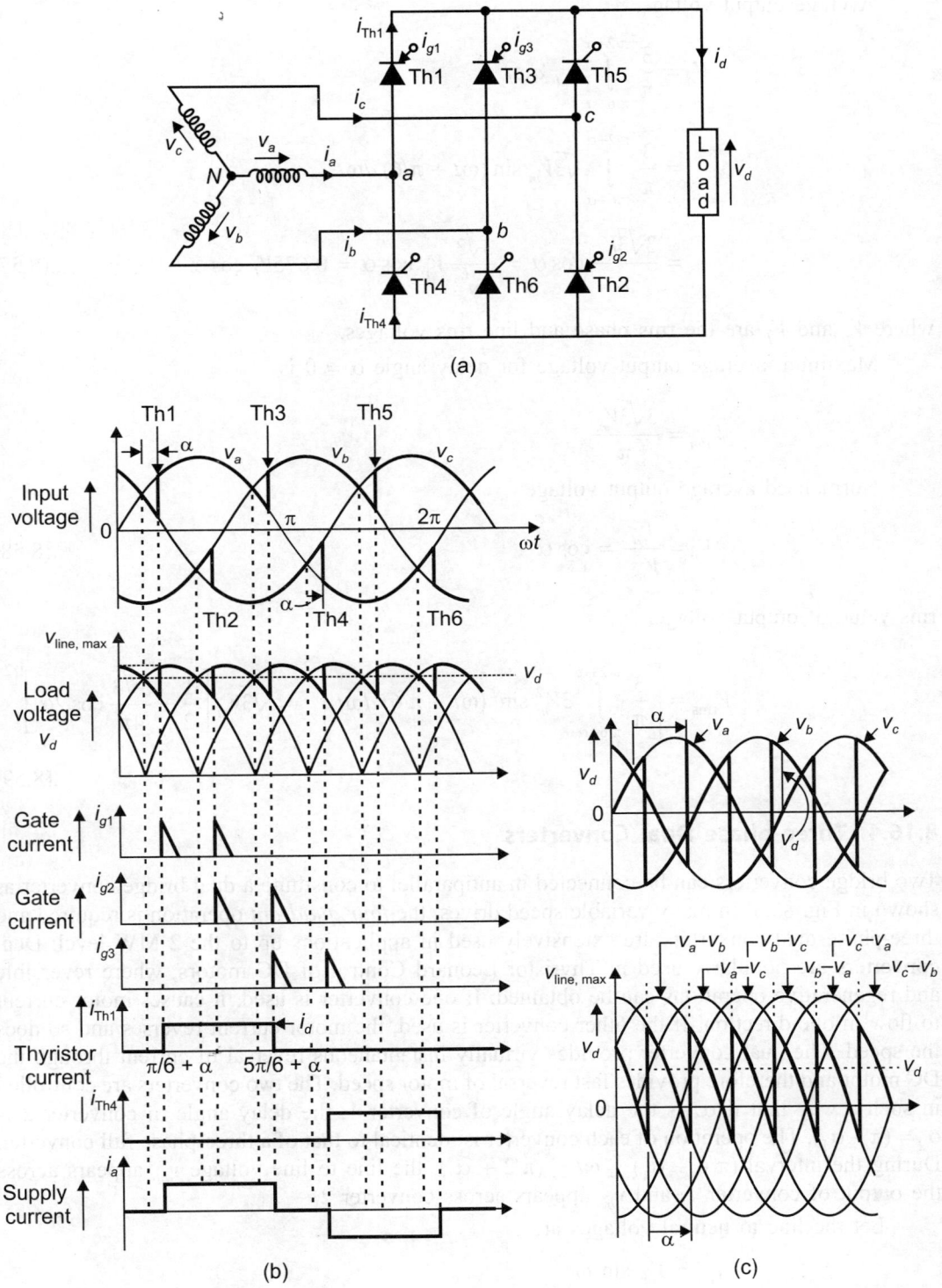

Fig. 8.22 Waveforms of three-phase full converters

Average output voltage,

$$V_d = \frac{3}{\pi} \int_{\pi/6+\alpha}^{\pi/2+\alpha} v_{ab} \, d\omega t$$

$$= \frac{3}{\pi} \int_{\pi/6+\alpha}^{\pi/2+\alpha} \sqrt{3} V_m \sin(\omega t + \pi/6) \, d\omega t$$

$$= \frac{3\sqrt{3} V_m}{\pi} \cos \alpha = \frac{3\sqrt{3}}{\pi\sqrt{2}} V_p \cos \alpha = 0.675 V_L \cos \alpha \qquad \dots(8.57)$$

where V_p and V_L are the rms phase and line rms voltages.

Maximum average output voltage for delay angle $\alpha = 0$ is

$$V_{dm} = \frac{3\sqrt{3} V_m}{\pi}$$

Normalised average output voltage

$$V_n = \frac{V_d}{V_{dm}} = \cos \alpha \qquad \dots(8.58)$$

rms value of output voltage,

$$V_{rms} = \left[\frac{3}{\pi} \int_{\pi/6+\alpha}^{\pi/2+\alpha} 3V_m^2 \sin^2(\omega t + \pi/6) \, d\omega t \right]^{1/2} = \sqrt{3} V_m \left[\frac{1}{2} + \frac{3\sqrt{3}}{4\pi} \cos 2\alpha \right]^{1/2}$$

$$\dots(8.59)$$

8.16.4 Three-phase Dual Converters

Two bridge converters can be connected in antiparallel to constitute a dual bridge converter as shown in Fig. 8.23. In many variable speed drives, the *four-quadrant* operation is required and three-phase dual converters are extensively used in applications up to the 2 MW level. Dual converters are popularly used in Thyristor Leonard Control of DC motors, where reversible and regenerative operations can be obtained. If one converter is used, it causes motor current to flow in one direction. If the other converter is used, the motor current reverses and so does the speed. The dual converter provides virtually instantaneous reversal of current through the DC motor and therefore provides fast reversal of motor speed. The two converters are controlled in such a way that if α_1 is the delay angle of converter 1, the delay angle of converter 2 is $\alpha_2 = (\pi - \alpha_1)$. The operation of each converter is identical to that of a three-phase full converter. During the interval $(\pi/6 + \alpha_1) \le \omega t \le (\pi/2 + \alpha_1)$, the line to line voltage v_{ab} appears across the output of converter 1, and v_{bc} appears across converter 2.

Let the line to neutral voltages are:

$$v_{an} = V_m \sin \omega t$$

$$v_{bn} = V_m \sin(\omega t - 2\pi/3)$$

$$v_{cn} = V_m \sin(\omega t + 2\pi/3)$$

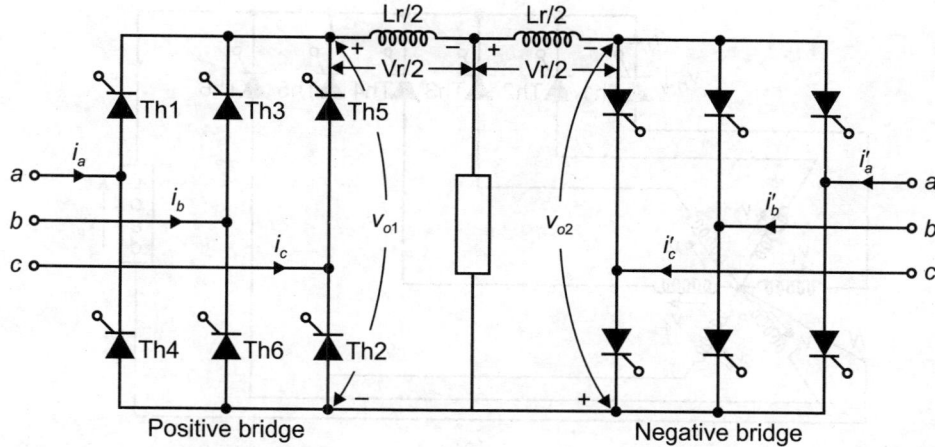

Fig. 8.23 Three-phase dual converter

Line to line voltages are:

$$v_{ab} = v_{an} - v_{bn} = \sqrt{3}V_m \sin(\omega t + \pi/6)$$
$$v_{bc} = v_{bn} - v_{cn} = \sqrt{3}V_m \sin(\omega t - \pi/2)$$
$$v_{ca} = v_{cn} - v_{an} = \sqrt{3}V_m \sin(\omega t + 5\pi/6)$$

If v_{o1} and v_{o2} are the output voltages of converters 1 and 2, respectively, the instantaneous voltage across the inductor during interval $(\pi/6 + \alpha_1) \leq \omega t \leq (\pi/2 + \alpha_1)$ is

$$v_r = v_{o1} + v_{o2} = v_{ab} - v_{bc}$$
$$= \sqrt{3}V_m [\sin(\omega t + \pi/6) - \sin(\omega t - \pi/2)]$$
$$= 3V_m \cos(\omega t - \pi/6) \qquad \qquad ...(8.60)$$

Due to the instantaneous voltage differences between the output voltages of converters, a circulating current flows through the converters. The circulating current is normally limited by circulating reactor, L_r.

The circulating current,

$$i_r(t) = \frac{1}{\omega L_r} \int_{\pi/6+\alpha_1}^{\omega t} v_r \, d\omega t$$

$$= \frac{1}{\omega L_r} \int_{\pi/6+\alpha_1}^{\omega t} 3V_m \cos(\omega t - \pi/6) \, d\omega t = \frac{3V_m}{\omega L_r} [\sin(\omega t - \pi/6) - \sin \alpha_1]$$

$$...(8.61)$$

The circulating current depends on delay angle α_1 and on inductance L_r. This current becomes maximum when $\omega t = 2\pi/3$ and $\alpha_1 = 0$.

In this case, the circuit can be controlled so as to have circulating current between the positive bridge and the negative bridge, i.e. one operates as rectifier and the other as an inverter with equal output voltage.

8.17 SIX-PHASE HALF-WAVE (OR SINGLE-WAY)

The connections with waveforms of the six-phase half-wave circuit using a simple star supply are shown in Fig. 8.24. The theory of the connection is an extension of the three-phase half-wave circuit, each thyristor conducting for one-sixth cycle.

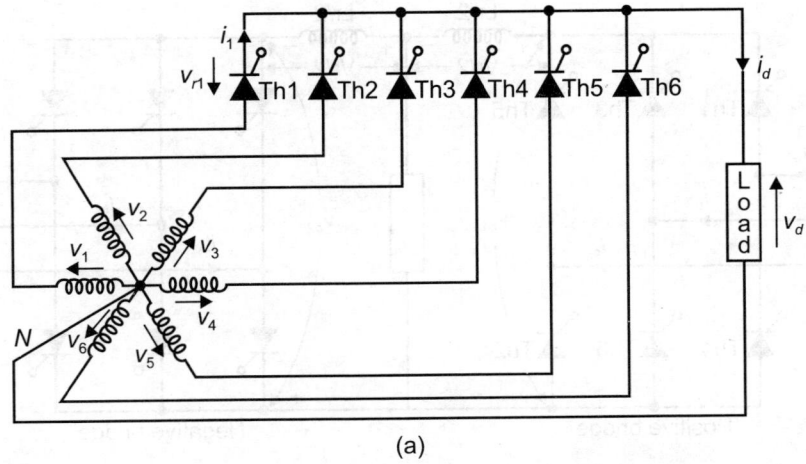

(a)

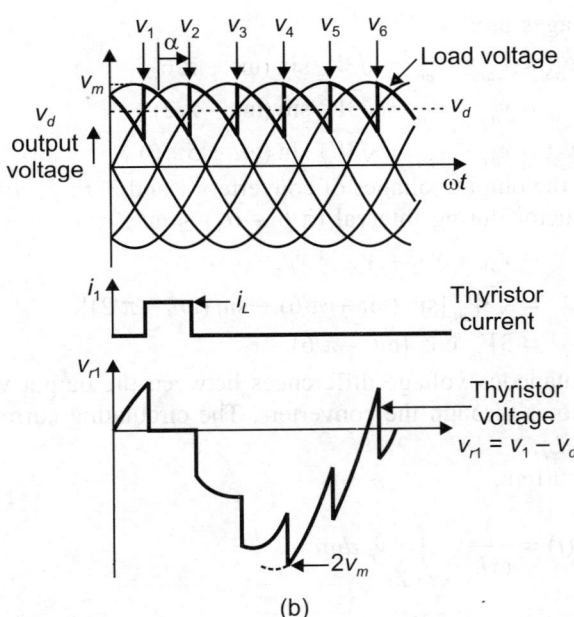

(b)

Fig. 8.24 Waveforms of six-phase half-way circuit

The top of the six-phase voltages delayed by the firing delay angle α is shown in Fig. 8.24 (b). The load-voltage waveform is of six-pulse characteristic, having a small ripple at a frequency of six times the supply frequency. Including firing delay, the average value of the load voltage is

$$V_{av} = \frac{1}{2\pi/6} \int_{(\pi/3+\alpha)}^{(2\pi/3+\alpha)} V_m \sin \omega t \, d\omega t = \frac{3}{\pi} V_m \cos \alpha \qquad ...(8.62)$$

The thyristor voltage waveform of Fig. 8.24 (b) shows the peak reverse voltage equal to twice the maximum value of the phase voltage. The diode is inefficiently used as it conducts for only one-sixth of the cycle, giving an rms value of

$$I_{rms} = I_d/\sqrt{6} \qquad\qquad ...(8.63)$$

for a level load current I_d.

The simple star connection of Fig. 8.24 (a) is not used in practice as the currents reflected into the primary winding have a large third-harmonic component. To eliminate the third-harmonic component, the fork connection can be used, but more frequently the double-star connection is used. The double-star connection is essentially two independent three-phase half-wave circuits operating in parallel to give a six-pulse output. The two star points are linked via an interphase transformer, which is best considered as a reactor rather than a transformer. The load current is returned to the centre of the reactor.

8.18 EFFECT OF TRANSFORMER LEAKAGE INDUCTANCE

So far we have neglected the source leakage inductance and assumed that transfer of current from the outgoing to the incoming thyristor, occurs instantaneously. Such a conduction is practically impossible, and a finite amount of source leakage inductance will permit current transfer to occur only gradually, as shown in Fig. 8.25. During commutation, overlap angle μ as shown, the line-to-line voltage is shorted and the supply volt-seconds area is absorbed by the two leakage L_c in series until the current transfer is completed. During this period, the load voltage dwells at intermediate level between the two-phase voltages. To explain the phenomenon associated with transfer of current, the three-phase half-wave rectifier connection will be used. Figure 8.25 (a) shows the three-phase supply to the three voltages, each in series with an inductance I_c. Reference to the waveform in Fig. 8.25 (b) shows that at commutation there is an angular period γ during which both the outgoing thyristor and incoming thyristor are conducting. This period is known as overlap period and γ is defined as commutation angle. During this period the load voltage is two mean of two conducting phases, the effect of overlap being to reduce the mean value it can be seen that with a firing delay angle α, a finite voltage is present from the start of commutation. We can write the following equations:

$$v_a = L_c \frac{di_{Th1}}{dt} + v'_d \qquad\qquad ...(8.64)$$

$$v_b = L_c \frac{di_{Th3}}{dt} + v'_d \qquad\qquad ...(8.65)$$

During commutation, the load current I_d can be assumed as constant. Therefore,

$$i_{Th1} + i_{Th3} = I_d \qquad\qquad ...(8.66)$$

That is
$$\frac{di_{Th1}}{dt} + \frac{di_{Th3}}{dt} = 0 \qquad\qquad ...(8.67)$$

Combining Equations (8.64), (8.65) and (8.67) yields

$$v'_d = \frac{v_a + v_b}{2} \qquad\qquad ...(8.68)$$

The load voltage is the mean of two-phase voltages. From Fig. 8.25 (b), it is evident that every $120°$ interval some volt-second area is lost during commutation, and as a result, the DC voltage V'_d will be reduced. Combining Equations (8.64) and (8.68) yields

$$\frac{di_{Th1}}{dt} = -\frac{1}{2L_c}(v_b - v_a) \qquad\qquad ...(8.69)$$

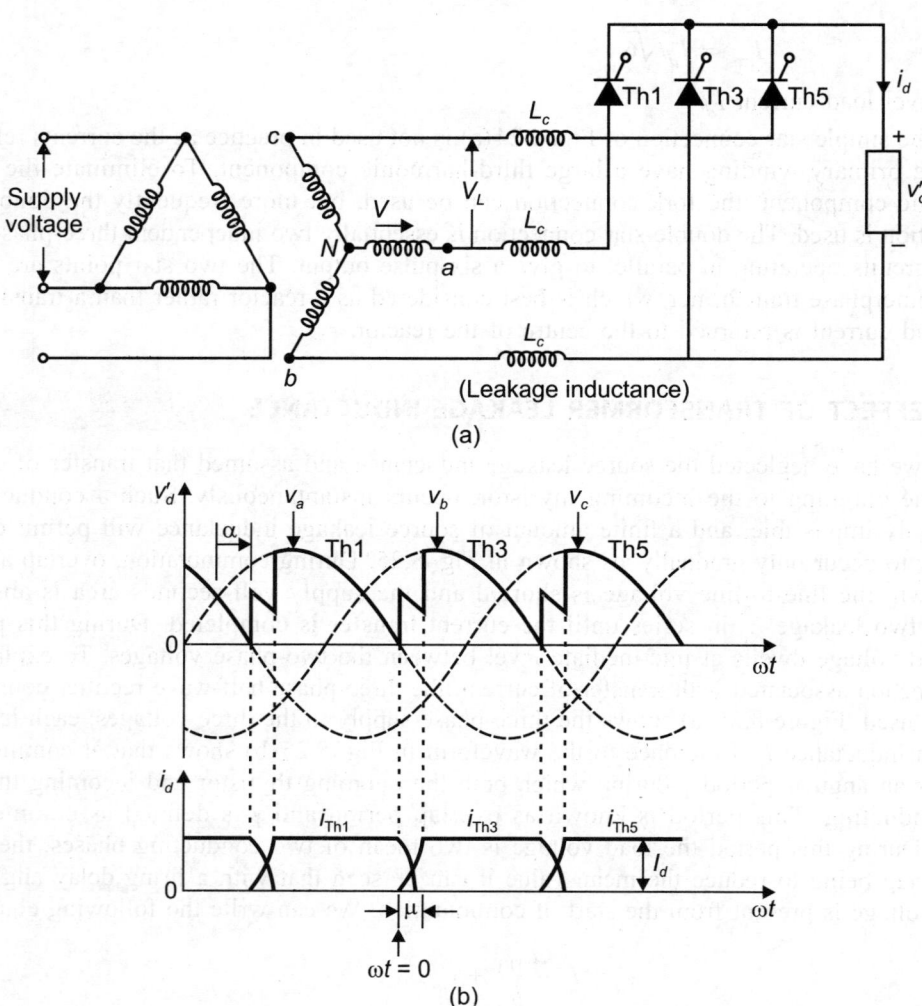

Fig. 8.25 Effect of transformer leakage inductance

or
$$i_{\text{Th1}} = -\frac{1}{2L_c} \int (v_b - v_a)\, d\omega t \qquad \qquad ...(8.70)$$

Putting the line-to-line voltage $v_{ba} = v_b - v_a = \sqrt{3}\sqrt{2}V \sin(\omega t + \alpha)$ in Equation (8.70) and solving, we get

$$i_{\text{Th1}} = \frac{\sqrt{16}V}{2\omega L_c} \cos(\omega t + \alpha) + A, \qquad \qquad ...(8.71)$$

where A is constant.

Assuming $\omega t = 0$ at the beginning of commutation, where $i_{\text{Th1}} = I_d$, the constant A can be given as

$$A = I_d - \frac{\sqrt{6}V}{2\omega L_c} \cos \alpha \qquad \qquad ...(8.72)$$

Substituting this in Equation (8.71) yields

$$i_{Th1} = I_d - \frac{\sqrt{6}V}{2\omega L_c}[\cos \alpha - \cos (\omega t + \alpha)] \qquad ...(8.73)$$

Using Equation (8.73) in Equation (8.66), we have

$$i_{Th3} = \frac{\sqrt{6}V}{2\omega L_c}[\cos \alpha - \cos (\omega t + \alpha)] \qquad ...(8.74)$$

Using $i_{Th3} = I_d$ at $\omega t = \mu$ in Equation (8.74) gives

$$\cos \alpha - \cos (\mu + \alpha) = \frac{2\omega L_c I_d}{\sqrt{6}V} \qquad ...(8.75)$$

The commutation angle μ can be expressed as

$$\mu = \cos^{-1}\left(\cos \alpha - \frac{2\omega L_c I_d}{\sqrt{6}V}\right) - \alpha \qquad ...(8.76)$$

Equation (8.76) shows that the overlap angle will increase if L_c or I_d increases or as α deviates from the middle of half cycle. The mean DC voltage loss due to commutation match can be given as

$$V_x = \frac{3}{2\pi} \int_0^\mu \frac{1}{2}(v_b - v_a)\, d\omega t$$

$$= \frac{3}{4\pi} \int_0^\mu \sqrt{3}\sqrt{2}V \sin (\omega t + \alpha)\, d\omega t \qquad ...(8.77)$$

$$= -\frac{3\sqrt{3}\sqrt{2}V}{4\pi}[\cos (\mu - \alpha) - \cos \alpha]$$

Substituting Equation (8.75) in Equation (8.77) gives

$$V_x = L_c I_d \frac{3\omega}{2\pi} = 3L_c I_d f \qquad ...(8.78)$$

where f is the supply frequency in hertz. Therefore, with loading the DC voltage V_{d1} can be given as

$$V_{d1} = V_d' - V_x = \frac{3\sqrt{3}V}{\sqrt{2}\pi}\cos \alpha - 3L_c I_d f \qquad ...(8.79)$$

Equation (8.79) indicates that the load voltage is reduced linearly with DC current and the Thevenin resistance of the converter is given as $R_{Th} = 3L_c f$.

Figure 8.26 shows inverter operation of a three-phase half-wave positive converter, where the commutation notch makes the DC voltage more negative. The overlap angle has more significance in inverter operation since it determines how far α angle can be increased. From the figure,

$$\beta = \mu + \gamma \qquad ...(8.80)$$

where γ is the turn off angle as shown. Substituting $\alpha = 180° - (\mu + \gamma)$ in Equation (8.75) yields

$$\cos \gamma - \cos (\mu + \gamma) = \frac{2\omega L_c I_d}{\sqrt{6}V} \qquad ...(8.81)$$

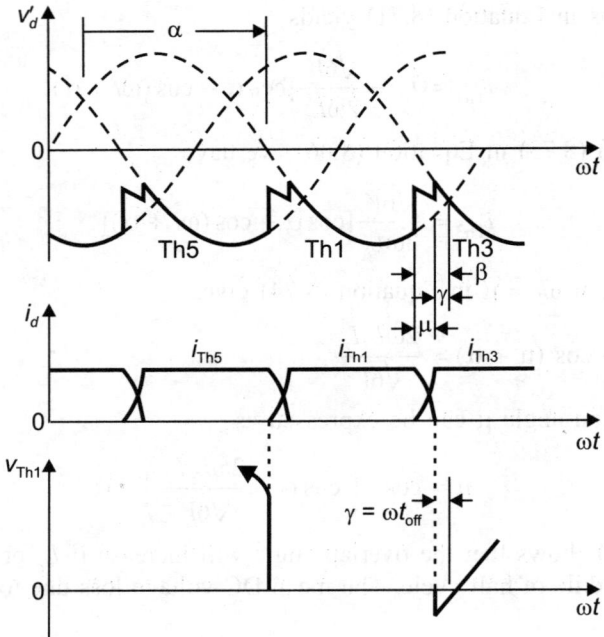

Fig. 8.26 Inverter operation of a three-phase half-wave positive converter

The thyristors require minimum turn off time t_{off} for successful commutation, which correspondingly determines the turn off angle $\gamma = \omega t_{off}$.

8.19 REGULATION

The term regulation is used to describe the characteristic of equipment as it is loaded.

There are three main sources for the loss of output voltage:

1. The voltage drop across the thyristors.
2. The resistance of the AC supply source and conductors.
3. The AC supply source inductance.

These three voltage drops can be represented respectively by the three resistors R_1, R_2 and R_3 in the equivalent circuit of Fig. 8.26. The open-circuit voltage is given by V_0 and the actual load voltage by V_d. If the load current I_d is taken to be a pure direct current, then any voltage drop can only be represented by resistors.

The voltage drop across the thyristors and diodes can be taken as a constant value. In circuits containing a mixture of thyristors and diodes, the volt-drop and equivalent resistance attributed to this cause may depend on the degree of firing delay.

The resistance of the leads and AC source resistance can be considered constant in most cases. If throughout a cycle the current is always flowing in two of the supply phases, then the resistance per phase can be doubled and added to the DC lead resistance to give the value for the equivalent circuit.

The voltage drop due to the AC supply source inductance is the overlap effect, and can be calculated for the three-pulse fully-controlled case is given by:

$$V_{d1} = \frac{3\sqrt{3}V_p \cos\alpha}{\sqrt{2}\pi} - 3L_c I_d f = \frac{3\sqrt{3}V_m}{2\pi}\cos\alpha - \frac{3\omega}{2\pi}L_c I_d, \text{ where } V_m = \sqrt{2}V_p.$$

The above equation shows that load voltage is reduced by $(3\omega L_c/2\pi)\, I_d$ for the three-pulse output: Hence, this voltage can be represented in the equivalent circuit of Fig. 8.27 as a resistance of value $\left(\dfrac{3\omega L_c}{2\pi}\right)$. Unlike the other equivalent circuit resistances, this value does not represent any power loss, but merely represents the voltage drop due to overlap.

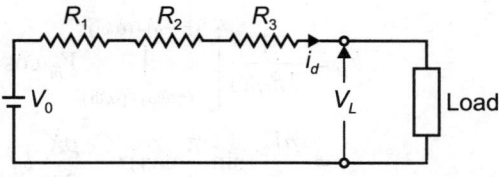

Fig. 8.27 Euivalent circuit representing voltage drops

The voltage drop due to overlap is changed to a higher value if overlap continues into the period of the next commutation. Simple overlap is known as a mode 1 condition, while overlap involving three elements is known as a mode 2 condition.

8.20 EQUATIONS FOR *p*-PULSE CONVERTER

A general expression for the mean load voltage of *p*-pulse fully-controlled rectifier, including the effects of the overlap angle γ, can be determined by reference to Fig. 8.28, where the average voltage is given by:

$$V_d = \frac{1}{2\pi/p}\left[\int_{-(\pi/p)+\alpha+\gamma}^{(\pi/p)+\alpha} V_m \cos \omega t\, d\omega t + \int_{\alpha}^{\alpha+\gamma} V_m \cos \frac{\pi}{p}\cos \omega t\, d\omega t\right]$$

$$= \frac{pV_m}{2\pi}\left[\sin\left(\frac{\pi}{p}+\alpha\right) - \sin\left\{-\frac{\pi}{p}+(\alpha+\gamma)\right\} + \cos\frac{\pi}{p}\sin(\alpha+\gamma) - \cos\frac{\pi}{p}\sin\alpha\right]$$

$$\therefore \quad V_d = \frac{pV_m}{2\pi}\sin\frac{\pi}{p}[\cos\alpha + \cos(\alpha+\gamma)] - \text{device voltage drop} \qquad \text{...(8.82)}$$

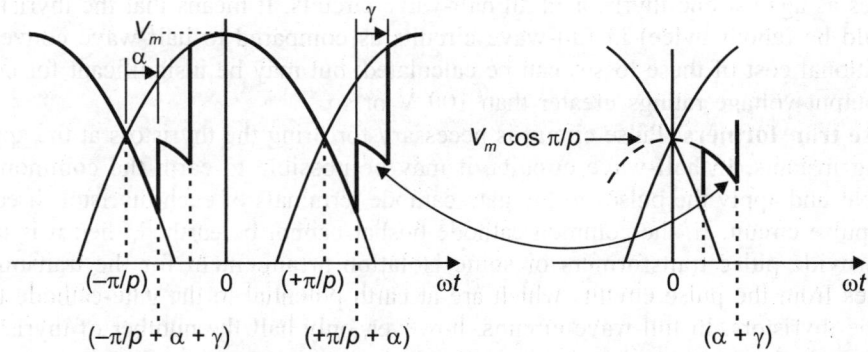

Fig. 8.28 Waveform of *p*-pulse converter

By substituting $\alpha = \pi - \beta$ and calling the average voltage V_d positive then, for the inverting mode.

$$V_d = \frac{pV_m}{2\pi}\sin\frac{\pi}{p}[\cos\beta + \cos(\beta-\gamma)] \qquad \text{...(8.83)}$$

The voltage drop due to the supply commutating reactance $X\Omega$/phase when a *p*-pulse rectifier supplies a load of current I_d can be determined from Fig. 8.28. Taking the base as time, but allowing for the loss of area due to overlap as $L_c I_d$.

$$V_d = \frac{1}{2\pi/p\omega} \left[\int_{-(\pi/p\omega)+(\alpha/\omega)}^{(\pi/p\omega)+(\alpha/\omega)} V_m \cos \omega t \, dt - L_c I_d \right]$$

$$= \frac{pV_m}{\pi} \sin \frac{\pi}{p} \cos \alpha - \frac{pX}{2\pi} I_d \qquad\qquad ...(8.84)$$

where $\qquad X = \omega L_c$

The relationship between the overlap angle γ, load current I_d, supply voltage V_m and commutating reactance X for a p-pulse rectifier operating at any firing delay angle α can be determined by equating Equations (8.82) and (8.84), yielding

$$XI_d = V_m \sin \frac{\pi}{p} [\cos \alpha - \cos (\alpha + \gamma)]. \qquad\qquad ...(8.85)$$

8.21 FACTORS AFFECTING THE CHOICE OF ANY CONVERTER CIRCUITS

There are two decisions involved, which will reduce the choice to a particular circuit: (1) half-wave or full-wave circuit, and (2) single-phase or three-phase circuit.

Single-phase full converters are primarily of two types, namely fully-controlled and half-controlled converters. For large power DC loads, three-phase AC to DC converters are commonly used. The various types of three-phase controled converters are three-phase half-wave and three-phase full converters. Three-phase half-wave converter is rarely used in industry because it introduces DC component in the supply current.

It will become clear from the following discussion that, full-wave converters are a much better choice than half-wave converters, and therefore, the choice is narrowed to single-phase bridge or three-phase bridge converters.

1. **Thyristor losses:** In full-wave converters, there are always two thyristors conducting in series as against one thyristor in all half-wave circuits. It means that the thyristor losses would be (about twice) in full-wave circuits as compared to half-wave converter. The additional cost of these losses can be calculated, but may be insignificant for converters of output voltage ratings greater than 100 V or so.

2. **Pulse transformers:** Pulse circuit is necessary for firing the thyristors at the appropriate firing instants. In half-wave circuits, it may be possible to earth the common cathode busbar and apply the pulses to the gate-cathode terminals of each thyristor directly from the pulse circuit. If, the common-cathode busbar cannot be earthed, then it is necessary to provide pulse transformers or some isolation arrangement for the transmission of pulses from the pulse circuits which are at earth potential to the gate-cathode terminals of the thyristors. In full-wave circuits, however, only half the number of thyristors have their cathodes connected to a common terminal and pulse transformers are necessary for isolating those thyristors whose cathodes are not earthed. The cost of these pulse transformers has to be taken into account, although considering the insignificant power requirement for firing a thyristor, the extra cost is small compared to the cost of pulse circuits.

3. **Number of thyristors and their ratings:** Generally the number of thyristors required for a converter and their ratings are a deciding factor in choosing a circuit for low and medium-power converters. For example, three-phase bridge circuits requires six thyristors as against three for the three-phase half-wave circuit; the latter, however, requires three thyristors of twice the voltage rating of the former. Usually, two thyristors of half the

voltage rating are cheaper than one thyristor of the full voltage rating. This is true for thyristors at voltage ratings close to the maximum available voltage range, in which case the choice is clearly in favour of the bridge circuit. The same argument is applicable to single-phase circuit.

4. **Transformer rating:** Half-wave converters have a poorer transformer utilisation and therefore higher transformer cost compared to full-wave circuits. For example, the biphase half-wave circuit requires a transformer with its secondary rating of $1 \cdot 57 P_d$ as compared to $1 \cdot 11 P_d$ for the single-phase bridge converter. Among three-phase circuits, the three-phase bridge has the lowest transformer rating of $1 \cdot 047 P_d$ except for low-power converters of rating less than 5 kW. The transformer cost should be of primary importance.

   The transformers in half-wave converter have to withstand the ill-effects of unidirectional current and hence the associated flux linkages and mechanical forces in the transformers. This difficulty is generally overcome by providing fork-type secondary windings, which increase the cost of transformer.

5. **Harmonics:** In three-phase converters, the ripple frequency of the converter output voltage is higher than in single-phase converter. Consequently, the filtering requirement for smoothing out the load current are less. It is the consideration of permissible harmonics that often decides the choice of the circuit, particularly between three-phase and single-phase. Clearly, the output voltage of the single-phase circuit has a much greater harmonic content and would require a much larger smoothing reactor compared to a three-phase converter. If the single-phase current harmonics injected in the AC system are acceptable, then a single-phase circuit particularly the bridge is preferable. If the converter output harmonics have to be significantly reduced, the additional cost of reducing harmonics to acceptable values in a single-phase bridge circuit has to be balanced against the saving in the cost of thyristors and the need to provide a single-phase transformer as against a three-phase transformer of slightly reduced rating for a three-phase bridge converter. The three-phase circuit would be generally preferable in such cases.

   The load current is mostly continuous in three-phase converters. The load performance, when three-phase converters are used, is therefore, superior as compared to when single-phase converters are used.

6. **Difficulties of three-phase supply:** In some cases, the non-availability of three-phase supply, or the cost of bringing three-phase supply to the load terminals, may dictate a clear choice in favour of single-phase converters. Railway electrification, in which the power is supplied by single-phase alternating current and rectified on the train, is a clear example of this.

SOLVED EXAMPLES

Example 8.1 The half-wave circuit is supplied at 120 V line to neutral. Determine the average load voltage for firing angles of 0°, 30°, 60° and 90°, assuming the load current to be continuous and level with a constant 2.5 V drop on each thyristor.

Solution The average value of voltage:

$$V_d = \frac{1}{p} \int_{\alpha}^{\pi+\alpha} \sqrt{2} V_p \sin \omega t \, d\omega t = \frac{2\sqrt{2} V_p}{\pi} \cos \alpha$$

$$V_d = \frac{2 \times \sqrt{2} \times 120 \cos \alpha}{\pi} - \text{thyristor voltage drop}$$

As thyristor drop = 2.5 V

For $\alpha = 0$,　　$V_d = \dfrac{2 \times \sqrt{2} \times 120}{\pi} \cos 0° - 2.5 \text{ V} = 105.5 \text{ V}$

For $\alpha = 30°$,　　$V_d = \dfrac{2 \times \sqrt{2} \times 120}{\pi} \cos 30° - 2.5 \text{ V} = 91.1 \text{ V}$

For $\alpha = 60°$,　　$V_d = \dfrac{2 \times \sqrt{2} \times 120 \cos 60°}{\pi} - 2.5 \text{ V} = 51.5 \text{ V}$

For $\alpha = 90°$,　　$V_d = \dfrac{2 \times \sqrt{2} \times 120 \cos 90°}{\pi} - 2.5 \text{ V} = 0 \text{ V}$.

Example 8.2　A single-phase fully-controlled bridge using two diodes is supplied at 120 V. Determine the average load voltage, assuming each diode to have a volt-drop of 0.78 V.

Solution　Using the equation with $\alpha = 0°$,

$$V_d = \frac{2}{\pi} \sqrt{2} V_p \cos \alpha - \text{total diode drop}$$

$$= \frac{2}{\pi} \times \sqrt{2} \times 120 - (2 \times 0.8)$$

$$V_d = 106.4 \text{ V}.$$

Example 8.3　A single-phase fully-controlled bridge is supplied at 120 V. Determine the average load voltage for firing delay angles of 0°, 45° and 90°, assuming continuous load current. Allow a thyristor volt drop of 1.5 V. Determine the required peak voltage of each thyristor.

Solution　Given　　　　　　$V_p = 120 \text{ V}$

Average load voltage　　$V_d = \dfrac{2}{\pi} \sqrt{2} V_p \cos \alpha - \text{thyristors voltage drop}$

for $\alpha = 0°$,　　　　　　$V_d = \dfrac{2}{\pi} \sqrt{2} \times 120 \cos 0° - (2 \times 1.5) \text{ V} = 105 \text{ V}$

for $\alpha = 45°$,　　　　　　$V_d = \dfrac{2}{\pi} \sqrt{2} \times 120 \cos 45° - (2 \times 1.5) \text{ V} = 73.4 \text{ V}$

for $\alpha = 90°$,　　　　　　$V_d = \dfrac{2}{\pi} \sqrt{2} \times 120 \cos 90° - (2 \times 1.5) \text{ V} = 0 \text{ V}$

Peak voltage across each thyristor

$$= \sqrt{2} V_p = \sqrt{2} \times 120 = 170 \text{ V}.$$

Example 8.4　The half-controlled single-phase bridge circuit is supplied at 120 V. Neglecting volt drops, determine the mean load voltage at firing delay angles of 0°, 60°, 90°, 135° and 180°. If the load is highly inductive taking 20 A, determine the required device ratings.

Solution　　　　　$V_d = \dfrac{\sqrt{2} V_p}{\pi} (1 + \cos \alpha) = \dfrac{\sqrt{2} \times 120}{\pi} (1 + \cos \alpha)$

for $\alpha = 0°$,　　　　　$V_d = \dfrac{\sqrt{2} \times 120}{\pi} (1 + \cos 0°) = 108 \text{ V}$

for $\alpha = 60°$, $\qquad V_d = \dfrac{\sqrt{2} \times 120}{\pi}(1 + \cos 60°) = 81$ V

for $\alpha = 90°$, $\qquad V_d = \dfrac{\sqrt{2} \times 120}{\pi}(1 + \cos 90°) = 54$ V

for $\alpha = 135°$, $\qquad V_d = \dfrac{\sqrt{2} \times 120}{\pi}(1 + \cos 135°) = 16$ V

for $\alpha = 180°$, $\qquad V_d = \dfrac{\sqrt{2} \times 120}{\pi}(1 + \cos 180°) = 0$ V

Each thyristor and diode must withstand

$$V_m = \sqrt{2}V_p$$
$$V_m = \sqrt{2} \times 120 = 170 \text{ V}$$

The bridge components conduct for a maximum of one-half cycle, hence for level current,

$$I_{\text{rms}} = \frac{20}{\sqrt{2}} = 14.16 \text{ A}$$

The freewheeling diode will conduct for almost the complete cycle when $\alpha = 180°$, therefore it must be rated to 20 A.

Example 8.5 The single-phase half-wave rectifier with a freewheeling diode is used to supply a heavily inductive load up to 15 A from a 240 V AC supply. Determine the mean load voltage for firing angles of $0°$, $45°$, $90°$, $135°$ and $180°$, neglecting the thyristor and volt-drops. Specify the required rating of the thyristor and diode.

Solution The average load voltage,

$$V_d = \frac{\sqrt{2}V_p}{2\pi}(1 + \cos \alpha)$$

for $\alpha = 0°$, $\qquad V_d = \dfrac{\sqrt{2} \times 240}{2\pi}(1 + \cos 0°) = 108$ V

for $\alpha = 45°$, $\qquad V_d = \dfrac{\sqrt{2} \times 240}{2\pi}(1 + \cos 45°) = 92$ V

for α $90°$, $\qquad V_d = \dfrac{\sqrt{2} \times 240}{2\pi}(1 + \cos 90°) = 54$ V

for $\alpha = 135°$, $\qquad V_d = \dfrac{\sqrt{2} \times 240}{2\pi}(1 + \cos 135°) = 16$ V

for $\alpha = 180°$, $\qquad V_d = \dfrac{\sqrt{2} \times 240}{2\pi}(1 + \cos 180°) = 0$ V

Thyristor rating: Peak reverse voltage

$$V_m = \sqrt{2}V_p = \sqrt{2} \times 240 = 340 \text{ V}.$$

The thyristor will conduct for a maximum duration when $\alpha = 0°$ of one-half cycle. Using two equal time intervals the rms current rating can be calculated as

$$I_{rms} = \left(\frac{15^2 + 0^2}{2} \right)^{1/2} = 10.6 \text{ A}$$

Diode rating: Peak reverse voltage (PRV), $V_m = \sqrt{2}V_p = 340$ V where V_p is the rms value of voltage. As the firing delay approaches 180°, the diode will conduct for almost the whole cycle, hence the required current rating would be 15 A.

Example 8.6 A single-phase half-wave rectifier with freewheeling diode, feed a low voltage load supplied by a 20 V AC supply. Assuming continuous load current, calculate average load voltage when firing delay angle is 60°. Taking forward volt drops of 1.5 V and 0.9 V across the thyristor and diode, respectively.

Solution Neglecting volt drops, the average load voltage

$$V_d = \frac{\sqrt{2}V_p}{2\pi} (1 + \cos \alpha)$$

$$= \frac{\sqrt{2} \times 20}{2\pi} (1 + \cos 60°) = 6.752 \text{ V}$$

The thyristor will conduct for duration (180° − 60°) giving an average volt-drop over the cycle of $\frac{120°}{360°} \times 1.5 = 0.5$ V. The diode when conducting imposes a 0.7 V across the load, in this case it averages over the cycle to $\frac{0.9 \times (180° + 60°)}{360°} = 0.6$ V.

Average load voltage $= 6.752 − 0.5 − 0.6 = 5.652$ V.

Example 8.7 The single-phase half-wave circuit with commutating diode is used to supply a heavily inductive load up to 15 A from a 240 V AC supply. Determine the average load voltage for delay angles of 45° and 90°. Neglecting the thyristor and diode volt-drops, give the ratings of thyristor and diode.

Solution Average load voltage, V_d at $\alpha = 45°$:

$$V_d = \frac{V_m}{2\pi} (1 + \cos \alpha)$$

$$= \frac{\sqrt{2} \times 240}{2\pi} (1 + \cos 45°) = 92 \text{ V}$$

At $\alpha = 90°$

$$V_d = \frac{\sqrt{2} \times 240}{2\pi} (1 + \cos 90°) = 54 \text{ V}$$

Thyristor rating: Peak forward voltage $= V_m$

$$\text{PFV} = \text{PRV} = \sqrt{2} \times 240 = 340 \text{ V}$$

Thyristor will conduct for a maximum duration at $\alpha = 0°$ of one-half cycle and if one assumes level current, then using two equal time intervals, the rms current rating is given by

$$I_{rms} = \left(\frac{15^2 + 0^2}{2} \right) = 10.6 \text{ A}$$

Diode rating: PRV = V_m = 340 V. As the firing approaches 180°, diode will conduct for whole cycle. Hence, the required rating would be 15 A.

Example 8.8   Using the single-phase half-wave controlled circuit with commutating diode, a low-voltage load is supplied by a 20 V AC supply. Assuming continuous load current, calculate the average load voltage when the firing delay angle is 60°. Assuming forward volt-drops of 1.5 V and 0.7 V across the thyristor and diode, respectively.

Solution   Average load voltage,

$$V_d = \frac{V_m}{2\pi}(1 + \cos \alpha)$$

$$V_d = \frac{\sqrt{2} \times 20}{2\pi}(1 + \cos 60°) = 6.752 \text{ V}$$

Thyristor will conduct for (180° − 60°). Thus, the average voltage drop over the cycle of $\frac{120°}{360°} \times 1.5 = 0.5$ V. The diode when conducting imposes a 0.7 V drop across the load, it averages over the cycle to

$$0.7 \times \frac{(180° + 60°)}{360°} = 0.467 \text{ V}$$

Thus, the average load voltage

$$= 6.752 - 0.5 - 0.467 = 5.78 \text{ V}.$$

Example 8.9   The half-controlled single-phase bridge circuit is supplied at 120 V. Determine the average load voltage at firing delay angle of 90°. If the load is highly inductive taking 25 A, determine the required device ratings. Assuming thyristor and diode drops of 1.5 and 0.7 V, respectively.

Solution   At α = 90°, the components conduct for half the time, hence over one cycle, their volt-drops will reduce the mean voltage by (1.5 + 0.7)/2 = 1.1 V. The commutating diode imposes a 0.7 V negative voltage on the load for other half of the time, averaging over to the cycle to 0.7/2 = 0.35 V

$$V_d = \frac{\sqrt{2} \times 120}{\pi}(1 + \cos 90°) - 1.1 - 0.35$$

$$= 52.6 \text{ V}.$$

Example 8.10   A single-phase half-wave controlled rectifier has a purely resistive load, R_L and delay angle is α = π/2. Find: (i) the rectification efficiency, (ii) form factor, (iii) ripple factor, (iv) transformer utilisation factor, TUF, and (v) peak inverse voltage of thyristor.

Solution   At α = π/2, $$V_d = \frac{V_m}{2\pi}(1 + \cos \alpha) = \frac{V_m}{2\pi}$$

$$I_d = \frac{V_m}{2\pi R_L}$$

rms value of voltage, $$V_{\text{rms}} = \frac{V_m}{2}\left[\frac{1}{\pi}\left(\pi - \alpha + \frac{\sin 2\alpha}{2}\right)\right]^{1/2}$$

$$= \frac{V_m}{2}\left[1 - \frac{\pi/2}{\pi} + \frac{\sin 2\pi/2}{2}\right]^{1/2} = \frac{V_m}{2}\left[1 - \frac{1}{2} + 0\right]^{1/2}$$

$$= \frac{V_m}{2} \frac{1}{\sqrt{2}}$$

$$I_{rms} = \frac{V_m}{2\sqrt{2} \cdot R_L}$$

DC power, $\qquad P_d = V_d I_d = \dfrac{V_m}{2\pi} \cdot \dfrac{V_m}{2\pi R_L} = \dfrac{V_m^2}{4\pi^2 R_L}$

AC power, $\qquad P_{AC} = V_{rms} \cdot I_{rms} = \dfrac{V_m}{2\sqrt{2}} \cdot \dfrac{V_m}{2\sqrt{2}R_L}$

(i) Rectification efficiency, $\eta = \dfrac{V_m^2}{4\pi^2 R_L} \times \dfrac{2\sqrt{2} \times 2\sqrt{2}R_L}{V_m \cdot V_m} = 20.27\%$

(ii) Form factor, $FF = \dfrac{V_m}{2\sqrt{2}} \dfrac{2\pi}{V_m} = 2.22$

(iii) Ripple factor, $RF = \sqrt{FF^2 - 1} = \sqrt{2.22^2 - 1} = 1.983$

(iv) rms voltage of transformer secondary,

$$V_s = \frac{V_m}{\sqrt{2}}$$

rms value of transformer secondary current is the same as that of load,

$$I_s = \frac{V_m}{2\sqrt{2}R_L}$$

Volt ampere rating of transformer,

$$VA = V_s I_s$$

$$= \frac{V_m}{\sqrt{2}} \cdot \frac{V_m}{2\sqrt{2}R_L} = \frac{V_m^2}{4R_L}$$

$$TUF = \frac{P_{DC}}{V_s I_s} = \frac{V_m^2 \times 4R_L}{4\pi^2 R_L \times V_m^2} = \frac{1}{\pi^2} = 0.101$$

(v) Peak inverse voltage, $PIV = V_m$.

Example 8.11 A single-phase half-wave converter is operated from a 120 V, 60 Hz supply. If the load is resistive of value 10Ω and delay angle is $\alpha = \pi/3$. Determine: (i) The efficiency, (ii) form factor, (iii) ripple factor, (iv) transformer utilisation factor, and (v) peak inverse voltage of thyristor.

Solution

$$V_d = \frac{\sqrt{2}V_p}{2\pi}(1 + \cos\alpha) = \frac{\sqrt{2} \times 120}{2\pi}\left(1 + \cos\frac{\pi}{3}\right) = 40.36 \text{ V}$$

$$I_d = \frac{V_d}{R} = \frac{40.36}{10} = 4.036 \text{ A}$$

$$V_{rms} = \frac{V_m}{2}\left[\frac{1}{\pi}\left(\pi - \alpha + \frac{\sin 2\alpha}{2}\right)\right]^{1/2} = 76.11 \text{ V}$$

$$I_{rms} = \frac{V_{rms}}{R} = \frac{7 - 6.11\,A}{10} = 7.611\,A$$

$$P_d = V_d I_d = 40.36 \times 4.036 = 162.89\,W$$
$$P_{AC} = V_{rms} I_{rms} = 76.11 \times 7.611 = 579.2\,W$$

(i) Rectification efficiency, $\eta = \dfrac{162.89}{579.2} \times 100 = 28.32\%$

(ii) Form factor, $FF = \dfrac{\text{rms value}}{\text{Average value}} = \dfrac{76.11}{40.36} = 1.88$

(iii) Ripple factor, $RF = \sqrt{(FF)^2 - 1} = \sqrt{(1.88)^2 - 1} = 1.59$

(iv) rms value of transformer secondary

$$V_s = \frac{V_m}{\sqrt{2}} = 0.707 V_m$$

rms value of transformer secondary is the same as load current,

$$I_s = 7.611\,A$$

Rating VA of transformer $= V_s I_s = 0.707 V_m \times 7.611$

$$TUF = \frac{P_{DC}}{V_s I_s} = \frac{162.89}{7.611 \times 0.707 \times 169.7} = 0.179$$

(v) Peak inverse voltage, $PIV = V_m = 169.7\,V$.

Example 8.12 A single-phase half-controlled rectifier is supplied from a 120 V, 60 Hz supply and a freewheeling diode is connected across the load. The load consists of series connected resistance $R = 10\,\Omega$, inductance $L = 5$ mH and a battery voltage $E = 20$ V. (a) Express the instantaneous output voltage in a Fourier series, and (b) determine the rms value of the lowest order output harmonic content.

Solution Given

$$V_p = 120\,V, f = 60\,Hz, R = 100\,\Omega, E = 20\,V \text{ and } L = 5\,mH,$$
$$\omega = 2\pi f = 2\pi \times 60 = 377\,rad/sec$$

$$v_o = V_d + \sum_{n=1,2}^{\alpha} (a_n \cos n\,\omega t + b_n \sin n\,\omega t) - E$$

$$V_d = \frac{1}{2\pi} \int_{\alpha}^{2\pi} V_m \sin \omega t\, d\omega t = \frac{V_m}{2\pi}(1 + \cos \alpha)$$

$$a_n = \frac{1}{\pi} \int_{\alpha}^{\pi} V_m \sin \omega t \cos n\,\omega t\, d\omega t$$

$$= \frac{V_m}{2\pi}\left[-\frac{\cos(1-n)\pi}{1-n} - \frac{\cos(1+n)\pi}{1+n} + \frac{\cos(1-n)\alpha}{1-n} + \frac{\cos(1+n)}{1+n} \right], n \geq 2$$

$$b_n = \frac{1}{\pi} \int_{\alpha}^{\pi} V_m \sin n\,\omega t \sin \omega t\, d\omega t$$

$$= \frac{V_m}{2\pi}\left[\frac{\sin(1+n)\alpha}{1+n} - \frac{\sin(1-n)\alpha}{1-n}\right], n \geq 2$$

$$a_1 = \frac{1}{\pi}\int_\alpha^\pi V_m \sin\omega t \cos\omega t \, d\omega t = -\frac{V_m}{2\pi}\sin^2\alpha$$

$$b_1 = \frac{1}{\pi}\int_\alpha^\pi V_m \sin\omega t \sin\omega t \, d\omega t = \frac{V_m}{2\pi}\left[\pi - \alpha + \frac{\sin 2\alpha}{2}\right]$$

rms value of fundamental component is

$$I_1 = \frac{[a_1^2 + b_1^2]^{1/2}}{\sqrt{2}\,[R^2 + (\omega L)^2]^{1/2}}$$

for $\alpha = \pi/6$, $a_1 = 6.752$, $b_1 = 82.4$

$$Z_1 = \sqrt{10^2 + (377 \times 0.5 \times 10^{-3})} = 10.176\,\Omega$$

$$I_1 = \frac{[6.752^2 + 82.4^2]^{1/2}}{\sqrt{2}\,[10.176]} = 5.74 \text{ A.}$$

Example 8.13 A single-phase semiconverter is operated from a 120 V, 60 Hz supply. The load current with an average value of I_a is continuous with negligible ripple content. The turns ratio of the transformer is unity. If the delay angle is $\alpha = \pi/3$, calculate: (i) the harmonic factor of input current, (ii) the displacement factor, and (iii) input power factor.

Solution Given $\quad V_p = 120 \text{ V}, f = 60 \text{ Hz}, \alpha = \pi/3, V_m = \sqrt{2}V_s = \sqrt{2} \times 120 = 169.7 \text{ V}$

From the equation $\quad I_1 = (2\sqrt{2}I_a/\pi)\cos(\pi/6) = 0.7797I_a$

$$I_s = I_a[1 - (\alpha/\pi)]^{1/2} = 0.8165I_a$$

(i) Harmonic factor, HF $= [(I_s/I_1)^2 - 1]^{1/2} = 0.3108$

(ii) Displacement factor, DF $= \cos(-\pi/6) = 0.866$

(iii) PF $= \left(\dfrac{I_1}{I_s}\right)$ DF $= (0.7797/0.8165) \times 0.866 = 0.827$ lag.

Example 8.14 A single-phase semiconverter is operated from a 120 V, 60 Hz supply and load is resistive of $R = 10\,\Omega$. If the average output voltage is 25% of the maximum possible average output voltage, calculate: (i) Delay angle, (ii) rms and average output currents, and (iii) average and rms thyristor currents.

Solution Given

$$V_p = 120 \text{ V}, f = 60 \text{ Hz}, V_m = \sqrt{2}V_p = \sqrt{2} \times 120 = 169.7 \text{ V}, V_n = 0.25 \text{ pu}$$

(i) From the equation $\quad V_n = 0.5(1 + \cos\alpha)$

$\Rightarrow \qquad\qquad\quad 0.25 = 0.5(1 + \cos\alpha)$

or $\qquad\qquad\qquad \alpha = 120°$

(ii) Using the equation $\quad V_d = \dfrac{V_m}{\pi}(1 + \cos\alpha)$

$$V_d = \frac{169.7}{\pi}(1 + \cos 120°) = 27 \text{ V}$$

or
$$I_d = \frac{V_d}{R} = \frac{27}{10} = 2.7 \text{ A}$$

$$V_{rms} = \frac{V_m}{\sqrt{2}} \left[\frac{1}{\pi} \left(\pi - \alpha + \frac{\sin 2\alpha}{2} \right) \right]^{1/2}$$

$$= \frac{V_m}{\sqrt{2}} \left[\frac{1}{\pi} \pi \left(\pi - \frac{2\pi}{3} + \frac{\sin 4\pi/3}{2} \right)^{1/2} \right] = 53.1 \text{ V}$$

$$I_{rms} = \frac{V_{rms}}{R} = \frac{53.1}{10} = 5.31 \text{ A}$$

(iii) $I_{av} = I_d/2 = 2.7/2 = 1.35$ A, $I_R = \dfrac{I_{rms}}{\sqrt{2}} = 3.75$ A.

Example 8.15 A single-phase full-wave converter is operated from a 120 V, 60 Hz for a resistive load of 10 Ω. If the average output voltage is 25% of maximum possible average output voltage, find: (i) The delay angle, (ii) average and rms output currents, and (iii) average and rms thyristor currents.

Solution Given
$$V_p = 120 \text{ V}, \ R_L = 10\Omega, \ V_n = 0.25 \text{ pu},$$
$$V_m = \sqrt{2} V_p = \sqrt{2} \times 120 = 169.7 \text{ V}$$

For a single-phase full converter

(i)
$$V_n = \cos \alpha$$

or
$$0.25 = \cos \alpha$$

or
$$\alpha = 75.5°$$

(ii) Average output voltage, $V_d = \dfrac{2V_m}{\pi} \cos \alpha$

$$= \frac{2 \times 169.7}{\pi} \cos 75.5° = 27 \text{ V}$$

$$I_{dc} = \frac{V_d}{R_L} = \frac{27}{10} = 2.7 \text{ A}$$

$$V_{rms} = \frac{V_m}{\sqrt{2}} = 120, \ I_{rms} = \frac{V_{rms}}{R_L} = \frac{120}{10} = 12 \text{ A}$$

(iii)
$$I_{av} = \frac{I_d}{2} = \frac{2.7}{2} = 1.35 \text{ A}$$

and
$$I_R = \frac{I_{rms}}{\sqrt{2}} = \frac{12}{\sqrt{2}} = 8.49 \text{ A}.$$

Example 8.16 The dual converter is operated from a 120 V, 60 Hz supply and delivers ripple free average current of $I_d = 20$ A. The circulating inductance is $L_r = 5$ mH and delay angles $\alpha_1 = 30°$ and $\alpha_2 = 150°$. Calculate the peak circulating current and peak current of converter 1.

Solution Given
$$V_p = 120 \text{ V}, \ L_c = 5 \text{ mH},$$
$$I_d = 20 \text{ A}, \ f = 60 \text{ Hz},$$

$$\omega = 2\pi f = 377 \text{ rad/sec,}$$
$$\alpha_1 = 30°, \ \alpha_2 = 150°$$
$$V_m = \sqrt{2}V_p = \sqrt{2} \times 120 = 169.7 \text{ V}$$

For
$$\omega t = 2\pi \text{ and } \alpha_1 = \pi/6$$

Peak circulating current is

$$\widetilde{I_{c \text{ (max)}}} = \frac{2V_m}{\omega L_c}(\cos \omega t - \cos \alpha_1)$$

$$= \frac{2 \times 169.7}{(377 \times 0.005)}(1 - \cos \pi/6) = 24.12 \text{ A}$$

Peak load current, $\qquad I_p = 20 \text{ A}$

Peak current of converter 1

$$= (20 + 24.12) = 44.12 \text{ A.}$$

Example 8.17 A three-phase half-wave controlled rectifier has a supply of 150 V/phase. Determine the load voltage for firing delay angle of 0°, 30°, 60°, assuming a thyristor volt drop of 1.5 V and continuous load current.

Solution (a) Output voltage,

$$V_d = \frac{3}{2\pi} \int_{\pi/6+\alpha}^{5\pi/6+\alpha} V_m \sin \omega t \, d\omega t$$

Here $\qquad V_m = \sqrt{2}V_p,$

$$V_d = \frac{3 \times \sqrt{3}}{2\pi} \cdot V_m \cos \alpha = \frac{3 \times \sqrt{3}}{2\pi} \times \sqrt{2} \times 120 \times \cos 60°$$

$$V_d = 70.2 \text{ V.}$$

Example 8.18 Three-phase half-wave controlled rectifier consists of a resistance and a very large inductance. The inductance is so large that the output current I_d can be assumed to be continuous and ripple free. For $\alpha = 60°$, (a) determine the average value of output voltage if phase voltage, $V_p = 120$ V, and (b) draw the waveforms of output voltage v_d and i_d.

Solution

(a) Output voltage, $\qquad V_d = \frac{3}{2\pi} \int_{\pi/6+\alpha}^{5\pi/6+\alpha} V_m \sin \omega t \, d\omega t$

Here $\qquad V_m = \sqrt{2}V_p,$

$$V_d = \frac{3 \times \sqrt{3}}{2\pi} \cdot V_m \cos \alpha = \frac{3 \times \sqrt{3}}{2\pi} \times \sqrt{2} \times 120 \times \cos 60°$$

$$V_d = 70.2 \text{ V}$$

(b) The supply voltages and the firing instants of the thyristors are shown in Fig. E8.18(a), the output current i_d is constant and is shown in Fig. E8.18(b). During the interval $(\pi/6 + \alpha) \leq \omega t \leq (\pi/6 + \alpha + 5\pi/6)$ thyristor Th1 conducts the load current and during this interval $v_d = v_{an}$ as shown in Fig. E8.18(c) the output voltage waveform v_d is also shown in the figure.

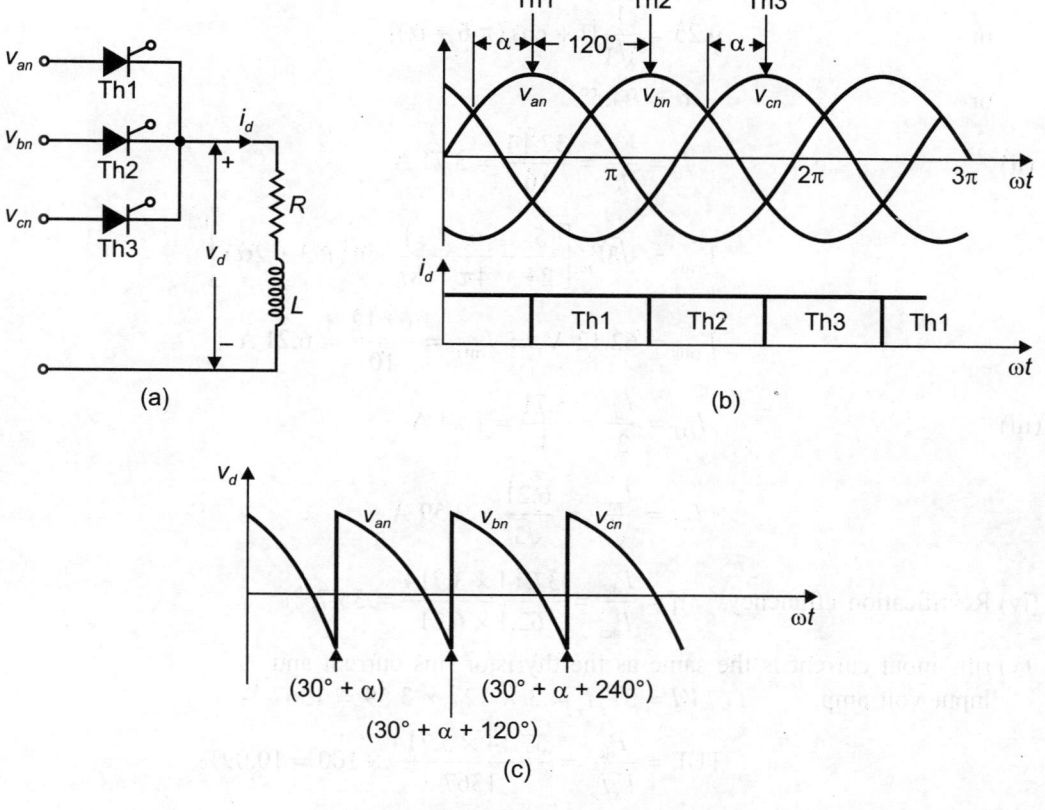

Fig. E8.18 Circuit diagram

Example 8.19 The three-phase half-wave converter is operated from a three-phase Wye-connected 220 V, 60 Hz supply and load resistance R_L is 10Ω. If the average output voltage is 25% of the maximum possible average voltage, calculate: (i) Delay angle, (ii) rms and average output currents, (iii) average and rms thyristor currents, (iv) rectification efficiency, (v) transformer utilisation factor, and (vi) input power factor.

Solution

(i) Given

$$V_L = 220 \text{ V}, f = 60 \text{ Hz}, R_L = 10\Omega$$

$$V_p = 220/\sqrt{3} = 127 \text{ V}$$

$$V_m = \sqrt{2}V_p = \sqrt{2} \times 127 = 179.6 \text{ V}$$

$$V_n = 0.25, V_d = \frac{3\sqrt{3}V_m}{2\pi} \cos\alpha$$

$$V_{dm} = \frac{3\sqrt{3}V_m}{2\pi} = \frac{3\sqrt{3} \times 179.6}{2\pi} = 148.5 \text{ V}$$

$$V_d = V_n V_{dm} = 0.25 \times 148.5 = 37.14 \text{ V}$$

From equation

$$V_n = \frac{1}{\sqrt{3}}[1 + \cos(\pi/6 + \alpha)]$$

or
$$0.25 = \frac{1}{\sqrt{3}}[1 + \cos(\pi/6 + \alpha)]$$

or
$$\alpha = 94.5°$$

(ii)
$$I_d = \frac{V_d}{R} = \frac{37.14}{10} = 3.71 \text{ A}$$

$$V_{\text{rms}} = \sqrt{3}V_m\left[\frac{5}{24} - \frac{\alpha}{4\pi} + \frac{1}{8\pi}\sin(\pi/3 + 2\alpha)\right]^{1/2}$$

$$V_{\text{rms}} = 62.12 \text{ V} \Rightarrow I_{\text{rms}} = \frac{62.12}{10} = 6.21 \text{ A}$$

(iii)
$$I_{DT} = \frac{I_d}{3} = \frac{3.71}{3} = 1.24 \text{ A}$$

$$I_{RT} = \frac{I_{\text{rms}}}{\sqrt{3}} = \frac{6.21}{\sqrt{3}} = 3.59 \text{ A}$$

(iv) Rectification efficiency, $\quad \eta = \dfrac{P_{\text{dc}}}{P_{\text{ac}}} = \dfrac{37.14 \times 3.714}{62.1 \times 6.21} = 35.75\%$

(v) rms input current is the same as the thyristor rms current and input volt amp, $\quad V.I = 3V_s I_s = 3 \times 127 \times 3.59 = 1367 \text{ W}$

$$\text{TUF} = \frac{P_{\text{dc}}}{V_s I_s} = \frac{37.14 \times 3.714}{1367} \times 100 = 10.09\%$$

(vi)
$$P_0 = I_{\text{rms}}^2 R = (6.21)^2 \times 10 = 385 \text{ W}$$

Input form factor $\quad = \dfrac{P_0}{\text{Input } VA} = \dfrac{385}{1367} = 0.282 \text{ lag.}$

Example 8.20 A three-phase, half-wave converter is operated from a three-phase γ connected 208 V, 60 Hz supply and load resistance $R_L = 10\,\Omega$. It is required to obtain an average output voltage of 50% of maximum possible output voltage. Calculate: (i) Delay angle α, (ii) rms and average output currents, (iii) average and rms thyristor currents, (iv) rectification efficiency, (v) transformer utilisation factor, and (vi) input power factor.

Solution Phase voltage,

$$V_p = \frac{208}{\sqrt{3}} = 120 \text{ V}, V_m = \sqrt{2}V_p = \sqrt{2} \times 120 = 169.8 \text{ V}$$

Given
$$V_n = \frac{V_d}{V_{dm}} = 0.5, R_L = 10\Omega$$

So, using
$$V_{dm} = \frac{3\sqrt{3}V_m}{2\pi} = \frac{3\sqrt{3} \times 169.8}{2\pi}$$

$$V_{dm} = 140.4 \text{ V}$$

∴
$$V_d = V_n \times V_{dm} = 0.5 \times 140.4$$
$$= 7.2 \text{ V}$$

(i) For a resistive load, load current is continuous if $\alpha \le \pi/6$

$$V_n = \frac{1}{\sqrt{3}}[1 + \cos(\pi/6 + \alpha)] \text{ gives}$$

$$V_n \ge 86.6\%$$

with resistive load and 50% output, the load current is discontinuous.

From the equation $\quad 0.5 = \left(\frac{1}{\sqrt{3}}\right)[1 + \cos(\pi/6 + \alpha)]$

Delay angle, $\quad\quad\quad \alpha = 67.7°$

(ii) $\quad\quad\quad\quad\quad I_d = \frac{V_d}{R_c} = \frac{70.2}{10} = 7.02 \text{ A}$

$$V_{rms} = \sqrt{3}V_m\left[\frac{5}{24} - \frac{\alpha}{4\pi} + \frac{1}{8\pi}\sin(\pi/3 + 2\alpha)\right]^{1/2}$$

$$= \sqrt{3} \times 169.7\left[\frac{5}{24} - \frac{67.7°}{4\pi} + \frac{1}{8\pi}\sin(\pi/3 + 2 \times 67.7)\right]^{1/2}$$

$$= 94.7 \text{ V}$$

$$I_{rms} = \frac{94.7}{10} = 9.47 \text{ A}$$

(iii) Average current of thyristor,

$$I_{AT} = \frac{I_d}{3} = \frac{7.02}{3} = 2.34 \text{ A}$$

$$I_{RT} = \frac{I_{rms}}{\sqrt{3}} = \frac{9.47}{\sqrt{3}} = 5.47 \text{ A}$$

(iv) Rectification efficiency, $\eta = \dfrac{70.23 \times 7.02 \times 100}{(94.74 \times 9.47)} = 54.95\%$

(v) Input *VA* rating, $\quad VI = 3V_pI_{RT} = 3 \times 120 \times 5.47 = 1970.8 \text{ W}$

$$\text{TUF} = \frac{70.23 \times 7.02 \times 100}{1970.8} = 25\%$$

(vi) Input power factor $= \dfrac{I_{rms}^2 \cdot R}{3V_pI_{RT}} = \dfrac{9.47^2 \times 10}{3 \times 120 \times 5.47} = 0.455 \text{ lag.}$

Example 8.21 A three-phase half-wave controlled rectifier has a supply of 150 V/phase. Determine the average load voltage for firing angle of 60°. Assuming a thyristor volt-drop of 1.5 V and continuous load current.

Solution Average load voltage, V_d

$$= \frac{3\sqrt{3}}{2\pi}V_m\cos\alpha - \text{voltage drop}$$

$$= \frac{3\sqrt{3}}{2\pi} \times \sqrt{2} \times 150\cos 60° - 1.5 = 86.2 \text{ V}.$$

Example 8.22 A commutating diode is placed across the load of three-phase half-wave controlled rectifier. Plot a curve of average load voltage against firing delay angle. Take the supply voltage of 100 V/phase and neglecting device voltage drops.

Solution The effect of commutating diode is to prevent reversal of the load voltage, leading to the waveform is shown in Fig. E8.22(a).

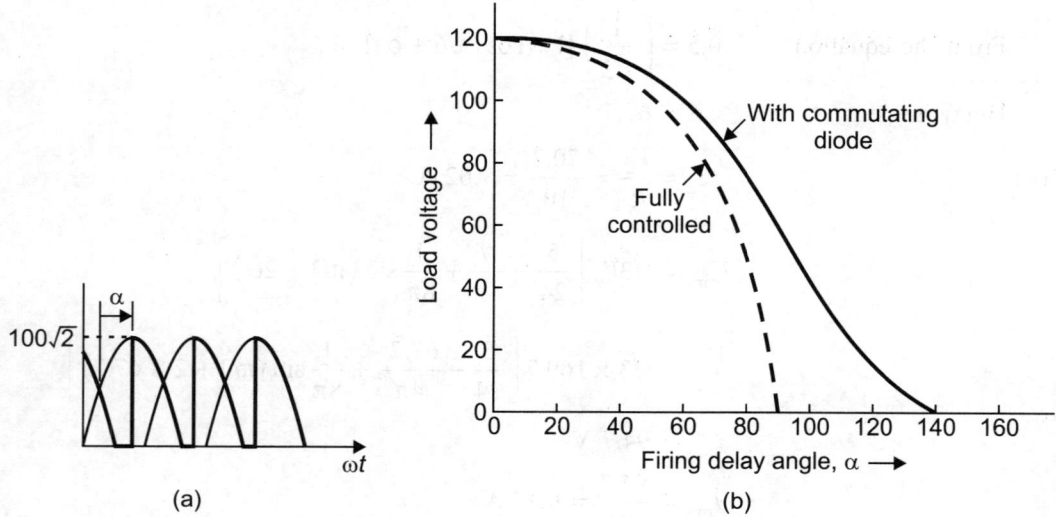

(a)   (b)

Fig. E8.22 Circuit diagram

From inspection of the waveform

$$V_d = \frac{3}{2\pi} \int_{\pi/6+\alpha}^{\pi} \sqrt{2} \times 100 \sin \omega t \; d\omega t \quad \text{when } \alpha \geq \pi/6$$

But for firing delay angles below $\pi/6$, this equation applies

$$V_d = \frac{3}{2\pi} \int_{\pi/6+\alpha}^{5\pi/6+\alpha} V_m \sin \omega t \; d\omega t = \frac{3\sqrt{3}}{2\pi} V_m \cos \alpha$$

The values of average load voltage against firing delay angle are shown in the plot [Fig. E8.22(b)] together with the fully-controlled values with continuous current and no commutating diode.

Example 8.23 The three-phase semiconverter is operated from a three-phase Wye-connected 220 V, 60 Hz supply and the load resistance is $R_L = 10\,\Omega$. If the average output voltage is 25% of the maximum possible average output voltage, calculate: (i) Delay angle, (ii) rms and average output currents, (iii) average and rms thyristor currents, (iv) the rectification efficiency, and (v) transformer utilisation factor.

Solution Given $\qquad V_{\text{line}} = 220 \text{ V}, f = 60 \text{ Hz},$
$$R_L = 10\,\Omega$$

Phase voltage $\qquad V_p = \dfrac{220}{\sqrt{3}} = 127 \text{ V}$

Peak voltage $\qquad V_m = \sqrt{2}V_p = \sqrt{2} \times 127 = 179.6 \text{ V}$

$$V_{dm} = \frac{3\sqrt{3}V_m}{\pi} = \frac{3\sqrt{3} \times 179.6}{\pi} = 297 \text{ V}$$

$$V_n = \frac{V_d}{V_{dm}} = 0.25$$

So $\qquad V_d = V_n \times V_{dm} = 0.25 \times 297 = 74.26 \text{ V}$

(i) Using the equation, $\qquad V_n = 0.5 \, (1 + \cos \alpha)$

or $\qquad\qquad\qquad \alpha = 120°$

(ii) $\qquad\qquad I_{DC} = \frac{V_{dc}}{R} = \frac{74.26}{10} = 7.426 \text{ A}$

From the equation, $\qquad V_{rms} = \sqrt{3}V_m \left[\frac{3}{4\pi} \left(\frac{2\pi}{3} + \sqrt{3} \cos^2 \alpha \right) \right]^{1/2}$

$$V_{rms} = \sqrt{3} \times 179.6 \left[\left(\frac{3}{4\pi} \right) \left(\pi - \frac{2\pi}{3} + 0.5 \sin 2 \times 120° \right) \right]^{1/2}$$

$$= 119.1 \text{ V}$$

$$I_{rms} = \frac{V_{rms}}{R_L} = \frac{119.1}{10} = 11.91 \text{ A}$$

(iii) Average thyristor current,

$$I_{AT} = I_d/3 = \frac{7.426}{3} = 2.46 \text{ A}$$

rms thyristor current, $\qquad I_{RT} = \frac{I_{rms}}{\sqrt{3}} = \frac{11.91}{\sqrt{3}} = 6.88 \text{ A}$

(iv) Rectification efficiency, $\eta = \dfrac{V_d \cdot I_d}{V_{rms} \cdot I_{rms}} = \dfrac{74.26 \times 7.426}{119.1 \times 11.91} = 100 = 38.8\%$

(v) Transformer utilisation factor,

$$\text{TUF} = \frac{P_d}{V_s I_s}$$

$$I_s = I_{rms} \sqrt{2/3} = 9.73 \text{ A}$$

$$VI = 3V_s I_s = 3 \times 127 \times 9.73 = 3705 \text{ VA}$$

$$\text{TUF} = \frac{74.26 \times 7.426}{3705} = 0.1488.$$

Example 8.24 Three-phase dual converter is operated from a three-phase Wye-connected 220 V, 60 Hz and load resistance $R_L = 10\,\Omega$, the circulating inductance $L_r = 5$ mH and delay angles are $\alpha_1 = 60°$ and $\alpha_2 = 120°$. Calculate the peak circuiting current and peak current of converters.

Solution Given $\qquad V_L = 220 \text{ V}, f = 60 \text{ Hz}, R_L = 10\,\Omega, L_r = 5 \text{ mH}$

$$\alpha_1 = 60°, \alpha_2 = 120°,$$

$$\omega = 2\pi f = 375 \text{ rad/sec}$$

$$V_p = \frac{210}{\sqrt{3}} = 127 \text{ V}, V_m = \sqrt{2} \times 127 = 179.6 \text{ V}$$

From the equation; $\quad i_r(t) = \frac{3V_m}{\omega L_r}[\sin(\omega t - \pi/6) - \sin \alpha_1]$

$$= \frac{3 \times 179.6}{377 \times 0.05}[\sin(\omega t - \pi/6) - \sin 60°]$$

$$i_r(t) = 285.8 [\sin(\omega t - \pi/6) - \sin 60°]$$

Current will be maximum when $(\omega t - \pi/6) = \pi/2$ and peak circulating current is

$$I_{r(max)} = 285.8 [1 - 0.866] = 38.3 \text{ A}$$

At $\qquad (\omega t - \pi/6) = 3\pi/2$

$$I_{r(min)} = 285.5 [-1 - 0.866] = -539.1 \text{ A}$$

Peak load current $\qquad = \dfrac{\sqrt{2} \times 220}{10} = 31.11 \text{ A}$

Current of converter 1 $\qquad = 31.11 + 38.3 = 69.41 \text{ A}$

Example 8.25 The holding current of thyristors in the three-phase full converter is 200 mA and the delay time is 2.5 μsec. The converter is supplied from a three Wye-connected 208 V, 60 Hz supply and has a load of $L = 8$ mH and $R_L = 2\,\Omega$, it is operated with a delay angle of $\alpha = 60° = (\pi/3)$. Determine the minimum width of gate pulse width, t_g. If $L = 0$ determine t_g.

Solution Given $I_H = 200$ mA $= 0.2$ A, $t_d = 2.5$ μsec, $\alpha = 60° = \pi/3$, $L = 8$ mH, $R = 2\,\Omega$, $V_{line} = 208$ V.

(i) Phase voltage, $\qquad V_p = \dfrac{208}{\sqrt{3}} = 120 \text{ V}$

Peak voltage, $\qquad V_m = \sqrt{2}V_p = \sqrt{2} \times 120 = 169.7 \text{ V}$

Instantaneous value, $\qquad v = \sqrt{3}V_m \sin(\omega t + \pi/6)$.

At $\qquad \omega t = \alpha = \pi/3, V_1 = \sqrt{3}V_m \sin(\pi/3 + \pi/6) = 294.2 \text{ V}$

The rate of rise of anode current, di/dt at the instant of triggering is

$$di/dt = \frac{V_1}{L} = \frac{294.2}{8 \times 10^{-3}} = 36.77 \text{ kA/sec}$$

If di/dt is assumed constant for a short time after the gate triggering, the time required for the anode current to rise to the level of holding current, t_1 is calculated from

$$I_H = t_1 \times (di/dt) = t_1 \times 36770 = 0.2$$

or $\qquad t_1 = \dfrac{0.2}{36770} = 5.439 \text{ μsec}$

Minimum width of gate pulse is

$$t_g = t_1 + t_d = 5.439 + 2.5$$

$$= 7.933 \text{ μsec}$$

(ii) Rate of rise of anode current, di/dt at the instant of triggering is

$$di/dt = \frac{V_1}{L} = \frac{294.2}{0} = \infty \text{ A/sec}$$

If di/dt is assumed constant for a short time after the gate triggering, the time required for the anode current to rise to the level holding current, t_1 is calculated from

$$t_1 \times (di/dt) = I_H$$

or
$$t_1 = \frac{0.2}{\infty} = 0 \,\mu sec$$

∴ Minimum width of gate pulse is

$$t_g = t_1 + t_d = 0 + 2.5 = 2.5 \,\mu sec$$

Example 8.26 Show that for a generated AC to DC converter operating from a polyphase supply, with a highly inductive load, the average output voltage, $V_{av} \propto \cos \alpha$ where α is the firing angle delay.

Solution Figure E8.26 shows the general polyphase converter operating from m-phase balanced supply. The load contains a very large inductance so that the ripple in the output current is insignificant and I_d can be assumed constant. It is assumed that all the supply voltages have zero output impedances and commutation is instantaneous. Each thyristor conducts for a period $2\pi/m$ radian when rhyristor is delayed by an angle α.

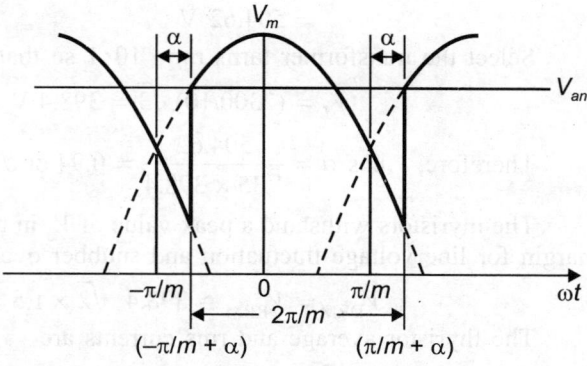

Fig. E8.26 Waveform of AC-DC converter

Average output voltage is given by,

$$V_{av} = \frac{1}{2\pi/m} \int_{(\pi/p)+\alpha}^{(\pi+p)+\alpha} V_m \cos \omega t \, d\omega t$$

$$= \frac{mV_m}{2\pi}\left[\sin\left(\frac{\pi}{m}+\alpha\right) - \sin\left(-\frac{\pi}{m}+\alpha\right)\right]$$

$$= \frac{mV_m}{2\pi}[\sin \pi/m \cos \alpha + \cos \pi/m \sin \alpha - \sin(-\pi/m)\cos \alpha + \cos(-\pi/m)\sin \alpha]$$

$$= \frac{mV_m}{\pi}\sin \pi/m \cos \alpha = \left(\frac{mV_m}{\pi}\sin \pi/m\right)\cos \alpha = V_{do}\cos \alpha.$$

Example 8.27 Determine the rms value of level current I that flows for $1/m$ of each cycle.

Solution Divide the cycle into m intervals than the current in one interval is I and zero in the other intervals.

The sum of the squares for each interval $= I^2$

Average value of the sum of the squares $= I^2/m$

rms value of the current $= \left(\dfrac{I^2}{m}\right)^{1/2} = \dfrac{I}{\sqrt{m}}.$

Example 8.28 A 2300 V, 60 Hz three-phase power supply is connected to a bridge converter through a delta-star transformer. The converter supplies a DC load current of 90 A at a voltage varying from +500 to –500 V. Design the converter, for the turns ratio of 10:1.

Solution Assume that the commutating inductance per phase, $L_c = 50$ μH. Then

$$V_m = 3fL_cI_d = 3 \times 60 \times 50 \times 10^{-6} \times 90 = 0.81 \text{ V}$$

Assume a thyristor conduction drop of 1.5 V, if V_L is line rms voltage. Then

$$V_d = \frac{3\sqrt{3}}{\sqrt{2}\pi} V_p \cos \alpha = 1.35 V_L \cos \alpha = 500 + 2 \times 0.81 + 1.5 \times 2$$

$$= 504.62 \text{ V}$$

Select the transformer turns ratio 10:1 so that

$$V_L = (2300/10)\sqrt{3} = 398.4 \text{ V}$$

Therefore, $\cos \alpha = \dfrac{504.62}{1.35 \times 398.4} = 0.94$ or $\alpha = 19.9°$

The thyristors withstand a peak value of V_c in the reverse direction. Allow a 50% voltage margin for line voltage fluctuation and snubber overshoot. Therefore, the voltage rating is

$$V_{DRM} = V_{RRM} \cong 398.4 \sqrt{2} \times 1.5 = 845 \text{ V}$$

The thyristor average and rms currents are

$$I_{av} = \frac{90}{3} = 30 \text{ A}, I_{rms} = \frac{90}{\sqrt{3}} = 51.96 \text{ A}$$

Select thyristor from the GE series which has rating of 900 V and 63 A (rms).

In the inverting mode, $V_d = -500 + 1.62 + 3 = 495.38$ V, which gives $\cos \alpha = -0.921$. Therefore,

$$\cos (\alpha + \mu) = \cos \alpha - \frac{2\omega L_c I_d}{\sqrt{6}V}$$

$$= -0.921 - \frac{2 \times 377 \times 50 \times 10^{-6} \times 90 \times \sqrt{3}}{\sqrt{6} \times 398.4}$$

$$= -0.926$$

(i.e. $\alpha + \mu = 157.8°$). Hence, $\gamma = 180 - (\alpha + \mu) = 22.21°$

or $$t_{off} = \frac{22.21 \times 10^3}{377 \times 57.3} = 1.028 \text{ ms}$$

which is adequate. Assuming with a 33% duty cycle for $I_{av} = 30$ A:

$$T_{c \text{ max}} = 104°C \text{ and } P_{av} = 43 \text{ W}$$

The thermal resistance $\theta_{CA} = \theta_{CS} + \theta_{SA} = (T_{c \text{ max}} - T_A)/P_{av}$. Assuming that $T_A = 25°C$ and $\theta_{CS} = 0.075°C/W$

$$\theta_{SA} = \frac{104 - 25}{43} - 0.075 = 1.76°C/W$$

The transformer currents are

$$I_d\sqrt{\frac{2}{3}} = 90\sqrt{\frac{2}{3}} = 73.48 \text{ A rms in secondary}$$

and
$$\frac{I_d}{n}\sqrt{\frac{2}{3}} = \frac{90}{10}\sqrt{\frac{2}{3}} = 7.35 \text{ A rms in primary}$$

The transformer VA rating

$$= 3VI_{ph} = \frac{3 \times 398.4}{\sqrt{3}} \times 73.48 = 50.7 \text{ kVA}$$

The line rms fundamental current is

$$I_{Lf} = \frac{3}{\pi}\frac{2I_d}{n}\frac{1}{\sqrt{2}} = \frac{3 \times \sqrt{2} \times 90}{\pi \times 10} = 12.16 \text{ A}$$

and the rms current is $I_L = \dfrac{\sqrt{2}I_d}{n} = \dfrac{\sqrt{2} \times 90}{10} = 12.73 \text{ A}$

In the rectification mode

Dis. $F = \cos \alpha = 0.94$

$$DF = \frac{12.16}{12.73} = 0.96$$

and
$$PF = 0.94 \times 0.96 = 0.90.$$

The fundamental input power $P_{in} = \sqrt{3} \times 2300 \times 12.16 \times 0.94 = 45.53$ kW and output power $P_d = V_dI_d = 500 \times 90 = 45$ kW. Therefore, the efficiency, $\eta = 45/45.53 = 98.8\%$ where the losses due to conduction drop only has been considered.

EXERCISES

Multiple Choice Questions

1. Single-phase half-wave controlled bridges use:
   (A) one thyristor
   (B) two thyristors
   (C) three thyristors
   (D) none of these

2. A freewheeling diode is used in a controlled rectifier circuit in case of:
   (A) resistive load   (B) inductive loads
   (C) capacitive loads   (D) none of these

3. A single-phase full-wave fully-controlled bridge uses:
   (A) two thyristors   (B) three thyristors
   (C) four thyristors   (D) none of these

4. A single-phase half-wave controlled rectifier has 400 sin 314t as the input voltage and R as the load. For a firing angle of 60° for the thyristor, the average output voltage is:
   (A) 400/π (B) 300/π
   (C) 240/π (D) none of these

5. A single-phase full-wave half-controlled bridge uses:
   (A) one SCR
   (B) two SCRs
   (C) four SCRs
   (D) six SCRs

6. A three-phase semiconverter can work as:
   (A) converter for $\alpha = 0°$ to 180°
   (B) converter for $\alpha = 0°$ to 90°
   (C) inverter for $\alpha = 90°$ to 180°
   (D) inverter for $\alpha = 0°$ to 90°

7. A four quadrant operation requires:
   (A) two full converters in series
   (B) two full converters connected back to back
   (C) two full converters connected in parallel
   (D) two converters connected back to back

8. A single-phase one-pulse controlled circuit has resistance and counter emf load and 400 sin 314t as the source voltage. For a

load counter emf of 200 V, the range of firing angle control is:

(A) 30° to 15°   (B) 60° to 180°

(C) 60° to 120°   (D) none of these

9. In a single-phase half-wave circuit with R–L load, and a freewheeling diode across the load, extinction angle β is more than π. For a firing angle α, the thyristor and freewheeling diode would conduct respectively, for:

(A) $(\pi - \alpha)$, β   (B) $(\beta - \alpha)(\pi - \alpha)$

(C) $(\pi - \alpha)$, $(\beta - \pi)$ (D) $(\pi - \alpha)$, $(\pi - \beta)$

10. In a single-phase one-pulse circuit with R–L load and a freewheeling diode, extinction angle β is less than π. For a firing angle α, the SCR and freewheeling diode would, respectively, conduct for:

(A) $(\beta - \alpha)$, 0°   (B) $(\pi - \alpha)$, $(\pi - \beta)$

(C) α $(\beta - \alpha)$ (D) $(\beta - \alpha)$, α

11. A single-phase full-wave midpoint thyristor converter uses a 230/200 V transformer with centre tap on the secondary side. The PIV for each thyristor is:

(A) 100 V   (B) 141.4 V

(C) 200 V   (D) 282.8 V

12. A single-phase two-pulse bridge converter has an average output voltage and power output of 500 V and 10 kW, respectively. The thyristor used in the two-pulse bridge converter is now re-employed to form a single-phase two-pulse midpoint converter. This new controlled converter would give, respectively, an average output voltage and power output of:

(A) 500 V, 10 kW   (B) 250 V, 5 kW

(C) 250 V, 10 kW   (D) 500 V, 5 kW

13. In a single-phase full converter bridge, the average output voltage is given by:

(A) $\dfrac{1}{\pi} \displaystyle\int_{\alpha}^{\pi+\alpha} V_m \cos \omega t \, d\omega t$

(B) $\dfrac{1}{\pi} \displaystyle\int_{\alpha}^{\alpha+\pi} V_m \cos \omega t \, d\omega t$

(C) $\dfrac{1}{\pi} \displaystyle\int_{\alpha-(\pi/2)}^{\alpha+(\pi/2)} V_m \cos \omega t \cdot d\omega t$

(D) $\dfrac{1}{\pi} \displaystyle\int_{\alpha}^{\pi+\alpha} V_m \cos \omega t \, d\omega t$

14. In a single-phase semiconverter, the average output voltage is given by:

(A) $\dfrac{1}{\pi} \displaystyle\int_{\alpha}^{\pi} V_m \cos \omega t \cdot d\omega t$

(B) $\dfrac{1}{\pi} \displaystyle\int_{\alpha}^{\pi+\alpha} V_m \cos \omega t \, d\omega t$

(C) $\dfrac{1}{\pi} \displaystyle\int_{\alpha-(\pi/2)}^{\alpha+(\pi/2)} V_m \cos \omega t \cdot d\omega t$

(D) $\dfrac{1}{\pi} \displaystyle\int_{\alpha-(\pi/2)}^{\pi} V_m \cos \omega t \cdot d\omega t$

15. For continuous conduction, in a single-phase full converter each pair of SCRs conduct for:

(A) $\pi - \alpha$ (B) π

(C) α   (D) $\pi + \alpha$

16. For discontinuous load current and extinction angle β > π, in a single-phase full converter each thyristor conducts for:

(A) α   (B) $\beta - \alpha$

(C) β   (D) $\alpha + \beta$

17. In a single-phase semiconverter with resistive load and for a firing angle α, each thyristor freewheeling diode conduct, respectively, for:

(A) α, 0°

(B) $\pi - \alpha$, α

(C) $\pi + \alpha$, α

(D) $\pi - \alpha$, 0°

18. In a single-phase full converter with resistive load and for a firing angle α, the load current is zero and non-zero, respectively, for:

(A) α, $\pi - \alpha$ (B) $\pi - \alpha$, α

(C) α, $\pi + \alpha$ (D) α, π

19. In a single-phase full converter, if α and β are the firing and extinction angles, the load current is:

(A) discontinuous if $(\beta - \alpha) < \pi$

(B) discontinuous if $(\beta - \alpha) > \pi$

(C) discontinuous if $(\beta - \alpha) = \pi$

(D) continuous if $(\beta - \alpha) < \pi$

20. With discontinuous conduction and extinction angle, β < π, in a single-phase semiconverter, freewheeling diode conducts for:

(A) α   (B) $\pi - \beta$

(C) $\beta - \pi$ (D) zero degree

21. With discontinuous conduction and extinction angle $\beta > \pi$, freewheeling diode conducts for:
    (A) α (B) $\beta - \pi$
    (C) $\pi + \alpha$ (D) β

22. For a continuous conduction, freewheeling diode, in a single-phase semiconverter, conducts for:
    (A) α (B) $\pi - \alpha$
    (C) π (D) $\pi + \alpha$

23. For discontinuous conduction and extinction angle $\beta < \pi$, in a single-phase semiconverter, each thyristor conducts for:
    (A) $\pi - \alpha$ (B) $\beta - \alpha$
    (C) α (D) β

24. For discontinuous conduction and extinction angle $\beta > \pi$ in a single-phase semiconverter, each thyristor conducts for:
    (A) $\pi - \alpha$ (B) $\beta - \pi$
    (C) α (D) β

25. For continuous conduction, in a single-phase semiconverter, each thyristor conducts for:
    (A) α (B) π
    (C) $\alpha + \pi$ (D) $\pi - \alpha$

26. In controlled rectifiers, the nature of load current:
    (A) does not depend on type of load and firing angle delay
    (B) depends both on the type of load and firing angle delay
    (C) depends only on the type of load
    (D) depends only on the firing angle delay

27. In a single-phase full converter, if output voltage has peak and average values of 325 V and 133 V respectively, then the firing angle is:
    (A) 40° (B) 140°
    (C) 50° (D) 130°

28. In a single-phase semiconverter, if output voltage has peak and average values of 325 V and 133 V respectively, the firing angle is:
    (A) 40° (B) 140°
    (C) 73.49° (D) 80°

29. For a single-phase, phase-controlled rectifier, with a freewheeling diode across the load:
    (A) the instantaneous output voltage v_0 is always positive
    (B) v_0 may be positive or zero
    (C) v_0 may be positive, zero or negative
    (D) none of these

30. In a single-phase full converter, if load current is I, and ripple free, then average thyristor current is:
    (A) $\frac{1}{2}I$ (B) $\frac{1}{3}I$
    (C) $\frac{1}{4}I$ (D) I

31. In a single-phase full converter, the number of SCRs conducting during overlap is:
    (A) 1 (B) 2
    (C) 3 (D) 4

32. In a single-phase full converter, the output voltage during overlap is equal to:
    (A) zero
    (B) source voltage
    (C) source voltage minus the inductance drop
    (D) none of these

33. The frequency of the ripple in the output voltage of a three-phase semiconverter depends upon:
    (A) firing angle and load resistance
    (B) firing angle and load inductance
    (C) the load circuit parameters
    (D) firing angle and the supply frequency

34. In a three-phase full converter, if load current is I and ripple free, then average thyristor current is:
    (A) $\frac{1}{2}I$ (B) $\frac{1}{3}I$
    (C) $\frac{1}{4}I$ (D) I

35. The effect of source inductance on the performance of single-phase and three-phase full converters is to:
    (A) reduce the ripples in the load current
    (B) make discontinuous current as continuous
    (C) reduce the output voltage
    (D) increase the load voltage

36. In a three-phase full converter, the output voltage during overlap is equal to:
    (A) zero
    (B) source voltage
    (C) source voltage minus the inductance drop
    (D) average value of the conducting-phase voltages

37. The total number of thyristors conducting simultaneously in a three-phase full converter with overlap considered has the sequence of:
    (A) 3, 3, 2, 2   (B) 3, 3, 3, 2
    (C) 3, 2, 3, 2   (D) 2, 2, 2, 3

38. A three-phase full converter has an average output of 200 V for zero degree firing angle and for resistive load. For a firing angle of 90°, the output voltage would be:
    (A) zero   (B) 50 V
    (C) 100 V   (D) 26.8 V

39. A converter which can operate in both three-pulse and six-pulse modes is a:
    (A) one-phase full converter
    (B) three-phase half-wave converter
    (C) three-phase semiconverter
    (D) three-phase full converter

40. In a three-phase semiconverter, for firing angle less than or equal to 60°, each thyristor and diode conduct, respectively, for:
    (A) 60°, 60°
    (B) 90°, 30°
    (C) 120°, 120°
    (D) none of these

41. In a three-phase semiconverter, for firing angle less than or equal to 60°, freewheeling diode conducts for:
    (A) 30°   (B) 60°
    (C) 90°   (D) zero degree

42. In a three-phase semiconverter, for firing angle equal to 90°, and for continuous conduction, each thyristors and diode conduct, respectively, for:
    (A) 30°, 60°   (B) 60°, 30°
    (C) 60°, 60°   (D) 30°, 30°

43. In a three-phase semiconverter, for a firing angle equal to 90° and for continuous conduction, freewheeling diode conducts for:
    (A) 30°   (B) 60°
    (C) 90°   (D) zero degree

44. In a three-phase semiconverter, or firing angle equal to 120° and extinction angle equal to 110°, each thyristor and diode conducts, respectively, for:
    (A) 30°, 60°   (B) 60°, 60°
    (C) 90°, 30°   (D) 110°, 30°

45. In a three-phase semiconverter, for firing angle equal to 120° and extinction angle equal to 110°, freewheeling diode conducts for:
    (A) 10°   (B) 30°
    (C) 50°   (D) 110°

46. In a three-phase semiconverter, for firing angle equal to 120° and extinction angle equal to 100°, none of the bridge elements conduct for:
    (A) 10°   (B) 20°
    (C) 30°   (D) 60°

47. In a three-phase semiconverter, the three thyristors are triggered at an interval of:
    (A) 60°   (B) 90°
    (C) 120°   (D) 180°

48. In a three-phase full converter, the six thyristors are fired at an interval of:
    (A) 30°   (B) 60°
    (C) 90°   (D) 120°

49. In a three-phase full converter, the output voltage pulsates at a frequency equal to:
    (A) supply frequency, f (B) $2f$
    (C) $3f$ (D) $6f$

50. The three-phase AC to DC converter which requires neutrals point connection is:
    (A) three-phase semiconverter
    (B) three-phase full converter
    (C) three-phase half-wave converter
    (D) three-phase full converter with diodes

51. In a three-phase semiconverter, frequency of the ripple in the output voltage may be:
    (A) 3 times the supply frequency f for firing angle α less than 60°
    (B) $3f$ for α greater than 60°
    (C) $6f$ for α less than 60°
    (D) $6f$ for α greater than 60°

52. In circulating-current type of dual converter, the nature of voltage across reactor is:
    (A) alternating   (B) pulsating
    (C) direct   (D) triangular

53. The peak inverse voltage in AC to DC converter systems is highest in:
    (A) single-phase full-wave midpoint converter
    (B) single-phase full converter
    (C) three-phase bridge converter
    (D) three-phase half-wave converter

54. In a dual converter, converters 1 and 2 work as under:

   (A) 1 as rectifier, 2 as inverter

   (B) both as rectifiers

   (C) both as inverters

   (D) none of these

55. For the same AC voltage and load impedance, which of the following statements about rectifiers are correct?

   (A) The average load current in a full-wave rectifier is π times that in a half-wave rectifier

   (B) Half-wave rectifier will have bigger sized transformer compared to full-wave rectifier

   (C) Half-wave rectifier will have a smaller sized transformer compared to a full-wave rectifier

   (D) None of these

56. Output voltage expression for single-phase semiconverter is:

   (A) $\dfrac{3V_m}{2\pi}(1 + \cos\alpha)$ (B) $\dfrac{2V_m}{\pi}\cos\alpha$

   (C) $\dfrac{V_m}{\pi}(1 + \cos\alpha)$ (D) $\dfrac{3\sqrt{3}}{2\pi}V_m\cos\alpha$

ANSWER KEY

1. (A)	**2.** (B)	**3.** (B)	**4.** (B)	**5.** (B)	**6.** (A)	**7.** (B)	**8.** (A)
9. (C)	**10.** (A)	**11.** (D)	**12.** (B)	**13.** (C)	**14.** (C)	**15.** (B)	**16.** (B)
17. (D)	**18.** (A)	**19.** (A)	**20.** (D)	**21.** (B)	**22.** (A)	**23.** (B)	**24.** (A)
25. (D)	**26.** (B)	**27.** (C)	**28.** (C)	**29.** (B)	**30.** (A)	**31.** (D)	**32.** (A)
33. (D)	**34.** (B)	**35.** (C)	**36.** (D)	**37.** (C)	**38.** (C)	**39.** (C)	**40.** (C)
41. (D)	**42.** (C)	**43.** (C)	**44.** (B)	**45.** (C)	**46.** (B)	**47.** (C)	**48.** (B)
49. (D)	**50.** (C)	**51.** (C)	**52.** (A)	**53.** (A)	**54.** (A)	**55.** (C)	**56.** (C)

Review Questions

1. A single-phase half-wave thyristor circuit feeds power to a resistive load. Draw waveforms for source voltage, load voltage and load current for a given firing angle α. Derive expressions for average and rms load voltages.

2. Describe the operation of a single-phase full-wave thyristor converter feeding power to resistive load.

3. Explain with the help of circuit and waveforms, the function of a freewheeling diode.

4. Explain with the help of a wave diagram, the effect of an inductive load on the performance of a half-wave rectifier using one thyristor.

5. Explain the difference between a half-controlled bridge and a fully-controlled bridge.

6. Explain the working principle of a single-phase full-wave half-controlled bridge rectifier using two thyristors and two diodes.

7. Explain why a separate freewheeling diode is not needed in case of a single-phase full-wave half-controlled bridge rectifier?

8. Draw and explain the connection of a three-phase full-wave half-controlled rectifier bridge.

9. Describe the operation of a single-phase two-pulse midpoint converter with waveforms. Discuss how each thyristor is subjected to reverse voltage equal to double the supply cottage in case turns ratio form primary to secondary is unity.

10. Show that the average output voltage for a phase full-converter connected to a resistive load is given by $V_{av} = \dfrac{3V_m}{\pi}\cos\alpha$ for $0 < \alpha < \pi/3$ and $\dfrac{3V_m}{\pi}[1 + \cos(\alpha + \pi/3)]$ for $\pi/3 < \alpha \le 2\pi/3$.

11. Sketch output voltage waveform for a three-phase semiconverter for a firing angle 75°. Obtain an expression for the average output voltage by using both sine and cosine functions for the supply voltages.

12. Discuss the effect of source inductance on the performance of a single-phase full converter indicating clearly the conduction of various thyristors during one cycle.

13. Discuss the effect of source inductance on the performance of a three-phase full converter with the help of a phase voltage waveforms. Derive an expression for its output voltage in turns of supply voltage, source inductance and load current.

14. For a three-phase dual converter derive an expression for the circulating current in terms of supply voltage, rector inductance, firing angle delay, etc.

15. A three-phase full converter thyristor bridge feeds a resistive load, sketch the waveform of output current for a firing angle of 30°.

DC-to-DC Converters— Choppers

9.0 INTRODUCTION

In many industrial applications, it is required to convert a fixed DC voltage into a variable DC voltage source to control the speed of industrial drives. Classically a variable DC voltage is obtained from a fixed DC voltage by the following methods: (1) Motor generator set, (2) Resistance controller, (3) Inverter rectifier, and (4) DC chopper. A DC chopper converts a fixed voltage DC supply to a variable voltage DC supply. A chopper can be considered as DC equivalent to an AC transformer with a continuously variable turns ratios. Like a transformer, it can be used to step-up or step-down a DC voltage source.

Choppers are widely used for traction motor control in electric traction, trolley cars, marine hoists, mine haulers, electric braking and for power factor improvement of AC/DC thyristor converters. Choppers provide smooth acceleration control, high efficiency and fast dynamic response and regeneration.

9.1 DC CHOPPER

DC chopper converts directly from DC to DC and is relatively a new technology. In programming DC to DC conversion, its behaviour is similar to that of a continuously variable turn ratio transformer.

9.1.1 Step-down Chopper (Buck Converter)

Principle of step-down operation: A simplified diagram of a step-down chopper is shown in Fig. 9.1 (a) that supplies a chain of pulses of DC voltage to the load. The fixed DC voltage can be converted to a variable, average voltage on a load by placing a high speed switch between DC source and the load. The high speed static switch is called the *chopper*.

When the switch S is turned-on, say at $t = 0$ [Fig. 9.1 (b)], the supply is connected to the load and $v_0 = V$. The load current i_0 builds-up when the switch S is turned-off at $t = T_{ON}$, the load current free-wheels through D_{FW} and $v_0 = 0$. At $t = T$, switch S is turned-on again and the cycle repeats. The waveforms of load voltage v_0 and load current i_0 are shown in Fig. 9.1 (b). It is assumed that i_0 is continuous. Note that the output voltage v_0 is a chopped voltage derived from supply voltage V. Hence, the name chopper.

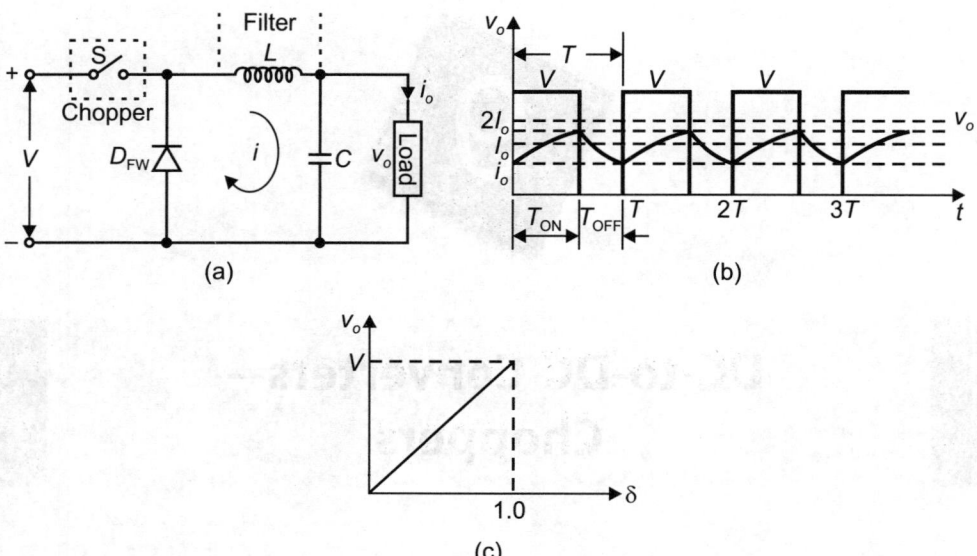

Fig. 9.1 Step down chopper (buck converter), waveforms, and characteristics

Average output voltage, V_o is

$$V_o = \frac{T_{ON}}{T_{ON} + T_{OFF}} \cdot V = \frac{T_{ON}}{T} \cdot V = T_{ON} \cdot f \cdot V \qquad ...(9.1)$$

$$V_o = \delta \cdot V \qquad ...(9.2)$$

where chopping period, $\quad T =$ on time + off-time

or chopping frequency, $\quad f = \dfrac{1}{T}$

$$\delta = \frac{T_{ON}}{T} = \text{duty cycle and}$$

$$\frac{T_{ON}}{T_{OFF}} = \text{mark space ratio}$$

rms value of output voltage V_o

$$= \left[\frac{1}{T} \int_0^{\delta T} v_0^2 \, dt \right]^{1/2} = \sqrt{\delta} \, V \qquad ...(9.3)$$

For a lossless chopper, input to the chopper is the same as output power and is expressed by

$$P_i = \frac{1}{T} \int_0^{\delta T} v_0 \, i \, dt = \frac{1}{T} \int_0^{\delta T} \frac{v_0^2}{R} \, dt = \frac{\delta V^2}{R} \qquad ...(9.4)$$

The duty cycle δ can be varied from 0 to 1 by varying T_{ON}, T or f. Thus, the output voltage can be varied from 0 to V by controlling δ and power can be controlled as shown in Fig. 9.1 (c).

The average output voltage can be varied in two different ways:

1. Variable frequency control
2. Constant frequency control

1. Variable frequency control: In this, control the chopping period T is varied. Either on time T_{ON} or off-time T_{OFF} is kept constant. This is called *frequency modulation.* The frequency has to be varied over a wide range to provide the full output voltage range which results in discontinuity and sluggish at low frequencies. This type of control would generate harmonics at unpredictable frequencies which would produce interference with signalling and telephone lines.

2. Constant frequency control: In this, control chopping frequency f is kept constant and the on time T_{ON} is varied. The width of the pulse is varied and this type of control is known as pulse–width modulation (PWM). The constant frequency control gives low ripple and requires smaller size of filter and has fast response. Thus, this scheme is preferred for chopper drives.

Basic chopper feeding resistive load through choke filter: The filter (L–C) plays an important role in the operation of chopper. The inductor stores energy during the on-period and delivers it during off-period of each cycle. The flywheel diode provides a path for the current through filter L during the off-period shown in Fig. 9.1 (a). The value of capacitor is sufficiently large so that the reactance X_C is much smaller than the load resistance at the operating frequency. The energy stored in the capacitor is much larger than that in the inductor L. If the time constant $R–C$ is very large as compared to the time period of switching cycle, the output voltage may be assumed constant.

The flywheel diode maintains current through load during the off-period thus maintaining the capacitor voltage constant. If the resistance in the loop of L, C and D_{FW} are neglected, applying KVL equation in this loop, we have

$$\frac{L di}{dt} + v_c + v_{DFW} = 0 \qquad \qquad ...(9.5)$$

where v_c is the capacitor voltage and v_{DFW} is the voltage across the diode, assuming v_c and v_{DFW} to be constants.

For maintaining the current flow during the off-period, the value of critical inductance should be such that conduction is maintained just up to the expiring of the off-period [Fig. 9.1 (b)]. The initial current at the beginning of the off-period must be equal to $2I_0$ so that the average load current $I_0 \left[= \dfrac{2I_0 + 0}{2} \right]$.

If the load maintains a constant voltage during off-period, the following equation holds good:

$$v_0 = L \, di_o/dt$$

For this critical operation, energy received by the inductor during on-period must be equal to that lost during off-period.

During ON-period:

$$\frac{L_c di}{dt} + v_0 = v$$

Hence,

$$L_c \cdot \frac{2I_0}{T_{ON}} = (V - V_0)$$

or

$$L_c = \frac{(V - V_0) T_{ON}}{2I_0}$$

From Equation (9.1), $V_0 = T_{\text{ON}} fV$ or $T_{\text{ON}} = \dfrac{V_0}{fV}$

and output power, $P_o = V_0 I_0$ or $I_0 = \dfrac{P_o}{V_o}$

or $L_c = \dfrac{(V - V_0) V_0}{2 I_0 fV}$

$$= \dfrac{(V - V_0) V_0}{2 (P_o/V_o) fV} = \dfrac{(V - V_0) V_0^2}{2 P_o fV}$$

$$L_c = \dfrac{V_0^2 (V - V_0)}{2 fV P_o} \qquad \qquad ...(9.6)$$

The actual design calls for higher inductance than the critical value to ensure proper damping otherwise undesirable oscillation will take place during switching.

If the switch S is ideal and the losses of the inductor and capacitor are zero, the theoretical efficiency becomes 100%. But in actual fact, the switch S is either a thyristor or transistor and the inductor and capacitor are not lossfree and hence the overall efficiency is reduced to 75% to 90%. If the losses are neglected, input power becomes equal to output power.

$\therefore$ $VI_1 = V_0 I_0$, where I_1 and I_0 are the average values of input neglecting the ripples and V_0 is taken as constant

Hence, $I_1 = \dfrac{V_0}{V} I_0.$ $...(9.7)$

9.1.2 Step-up Chopper (Boost Converter)

The chopper can be used to step-up a DC voltage and an arrangement for step-up operation is shown in Fig. 9.2 (a).

When the switch S is closed or open, the inductor is connected to supply V, and the energy from the supply is stored in it. When the chopper or switch is off, the inductor current is forced to flow through the diode and the load. The induced voltage v_L across the inductor is negative. The inductor voltage adds to the source voltage to force the inductor current into load. Thus, the energy stored in the inductor is released to the load. The waveforms of v_0, i_L and v_L are shown in Fig. 9.2 (b).

If the ripple in the source current is neglected, then during the time, the chopper is on (t_{ON}), the energy input to the inductor from the source is

$$W_i = V \cdot I \cdot t_{\text{ON}}$$

During the time t_{OFF}, when the chopper is off, the energy released by the inductor to the load is

$$W_0 = (V_0 - V) \, I \cdot t_{\text{OFF}}$$

For a lossless system in the steady state, these two energies will be same. Thus,

$$VI \cdot t_{\text{ON}} = (V_0 - V) \, I \cdot t_{\text{OFF}}$$

or $$Vt_{\text{ON}} = V_0 t_{\text{OFF}} - Vt_{\text{OFF}}$$

or $$V_0 = \dfrac{V (t_{\text{ON}} + t_{\text{OFF}})}{t_{\text{OFF}}} \qquad \qquad ...(9.8)$$

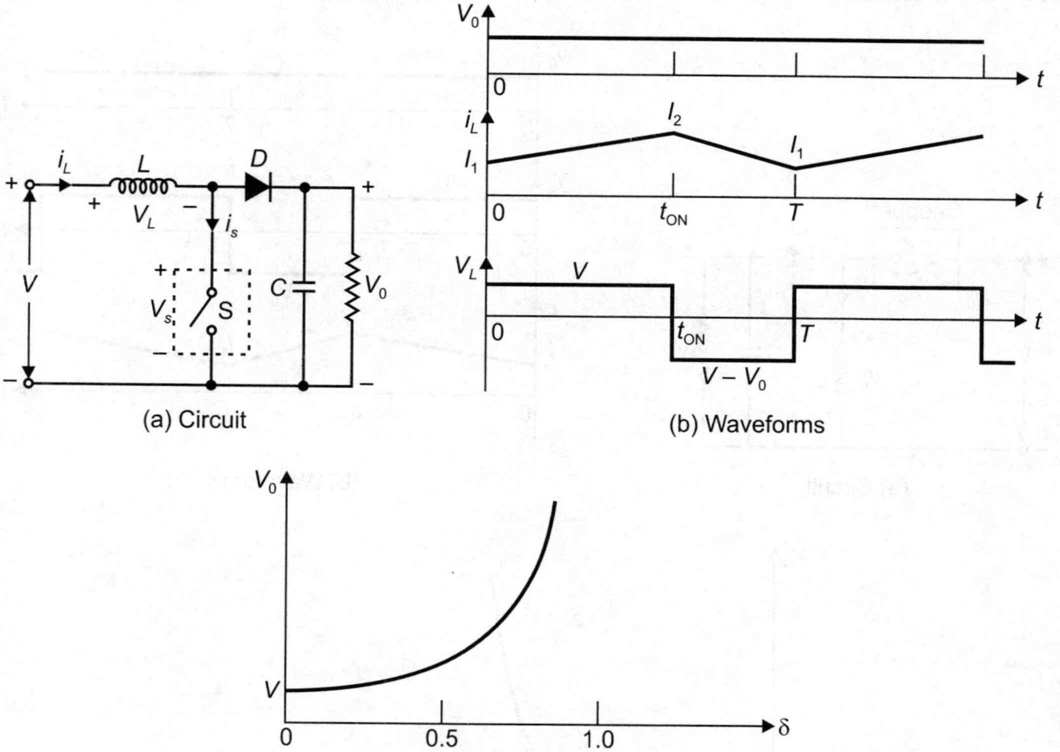

(a) Circuit (b) Waveforms

(c) V_0 versus δ

Fig. 9.2 Boost converter

$$= \frac{VT}{T - t_{ON}} = \frac{V}{1 - t_{ON}/T} = \frac{V}{1 - \delta} \qquad \qquad ...(9.9)$$

This principle is utilised in the regeneration braking of a DC motor. Thus, for a variation of δ in the range $0 < \delta < 1$, the voltage V_0 varies in the range $V < V_0 < \delta$, thus the variation of V_0 with δ is shown in Fig. 9.2 (c). If V represents the armature voltage of the machine, and V_0 the DC supply, power can be fed back from the decreasing motor voltage V to the fixed supply voltage V_0 by proper adjustment of the duty cycle δ.

If $t_{ON} = t_{OFF} = T$,

then from Equation (9.8), we have

$$V_0 = \frac{V(t_{ON} + t_{OFF})}{t_{OFF}} = \frac{V(T + T)}{T}$$

$$= \frac{2VT}{T} = 2 \text{ V} \qquad \qquad ...(9.10)$$

Thus, the output voltage obtained is twice the input voltage. This circuit is very useful for obtaining higher output voltages. The large inductor L also helps in minimising ripples at the output.

9.1.3 Step-down and Step-up Chopper (Buck–Boost Converter)

A chopper configuration which will provide load voltages lower and higher than the supply voltage is shown in Fig. 9.3 (a).

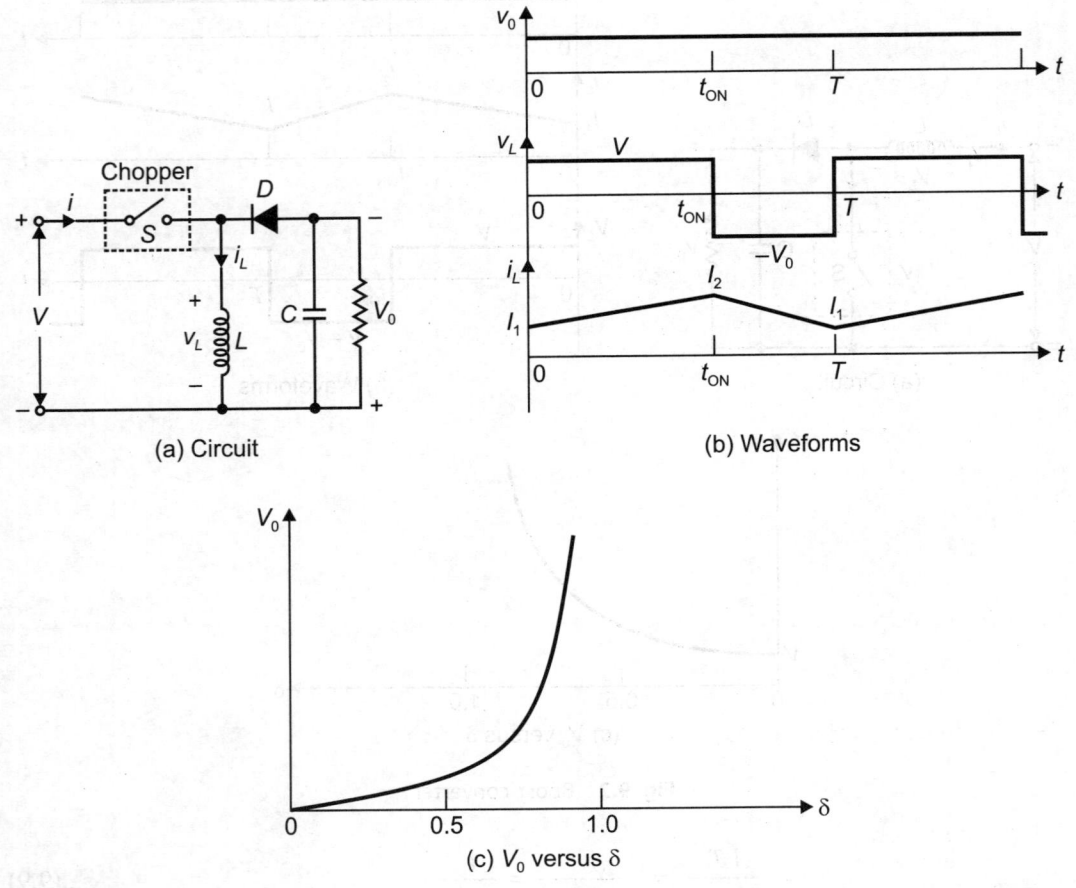

(a) Circuit

(b) Waveforms

(c) V_0 versus δ

Fig. 9.3 Buck–boost converter, waveforms and characteristics

When the chopper is on, the inductor is connected to the supply voltage V and $v_L = V$, the inductor increases linearly. When the chopper is off, the inductor current flows through the load and the diode. The inductor voltage is $v_L = -V_0$ and the inductor current decreases linearly. The waveforms of v_0, i_L and v_L are shown in Fig. 9.3(b). Note that the polarity of v_0 is reversed with respect to supply voltage.

During t_{ON}:

$$v_L = V = L\frac{(I_2 - I_1)}{t_{ON}} = \frac{L\Delta I}{t_{ON}} \qquad \text{...(9.11)}$$

where

$$\Delta I = \frac{V}{L} \cdot t_{OFF} \qquad \text{...(9.12)}$$

During

$$t_{OFF} = T - t_{ON}$$

$$v_L = -V_0 = \frac{L(I_1 - I_2)}{t_{OFF}} = -\frac{L\Delta I}{t_{OFF}}$$

where

$$\Delta I = \frac{V_0}{L} \cdot t_{OFF} \qquad \text{...(9.13)}$$

Duty cycle, $\qquad \delta = \dfrac{t_{ON}}{T}$

From Equations (9.12) and (9.13)

$$V_0 = \frac{t_{ON}}{t_{OFF}} \cdot V = \frac{t_{ON}}{T - t_{ON}}$$

$$= \frac{\delta}{1 - \delta} \cdot V$$

Thus, for a variation of δ in the range $0 < \delta < 1$, the output voltage V_0 will vary in the range $0 < V_0 < \delta$. This variation of V_0 with δ is shown in Fig. 9.3 (c).

9.2 CHOPPER CLASSIFICATION

Depending on the directions of current and voltage, choppers can be classified into following classes:

1. Class A chopper   2. Class B chopper
3. Class C chopper   4. Class D chopper
5. Class E chopper

1. **Class A chopper:** The load current flows into load. Both the load voltage and current are positive as shown in Fig. 9.4 (a). This is a single quadrant chopper and is said to be operated as a rectifier.

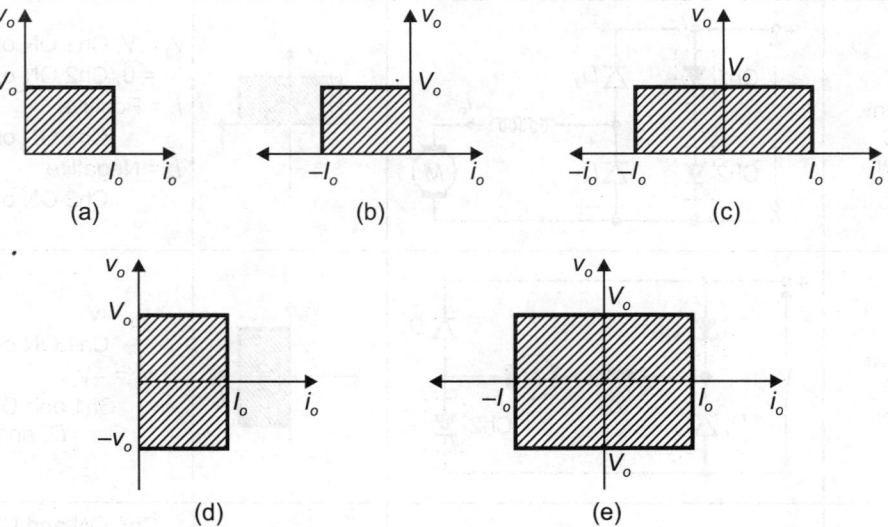

Fig. 9.4 Classification of choppers

2. **Class B chopper:** The output load current flows out of the load. The load voltage is positive, but the load current is negative as shown in Fig. 9.4 (b). This is a single quadrant chopper but operated in the second quadrant and is said to be operated as inverter.
3. **Class C chopper:** The load current is either positive or negative as shown in Fig. 9.4 (c). This is known as two quadrant chopper. Class A and class B choppers can be combined to form class C chopper. Class C chopper can operate either as a rectifier or as an inverter.

4. **Class D chopper:** The load current is always positive. The load voltage is either positive or negative as shown in Fig. 9.4(d). It can also operate either a rectifier or an inverter.

5. **Class E chopper:** The load current is either positive or negative. This is known as *four quadrant chopper*. Two Class C choppers can be combined to form a Class E chopper as shown in Fig. 9.4(e).

9.3 CHOPPER CONFIGURATIONS

Various types of chopper configurations and their operation in different quadrants are shown in Fig. 9.5. In the first quadrant chopper $v_0 = V$ when the chopper is on, and $v_0 = 0$ when the

Type	Chopper Configuration	Characteristics	Function
First quadrant chopper class A	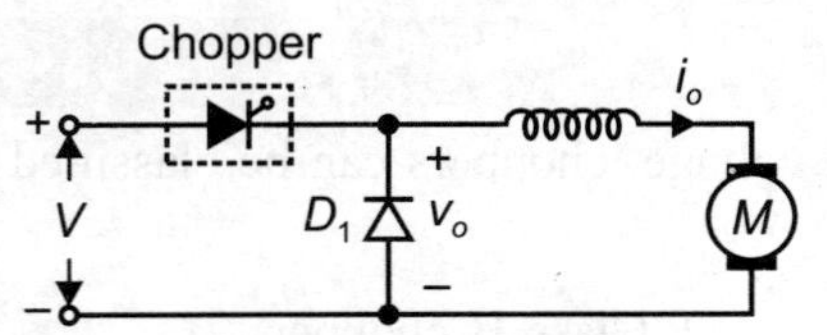	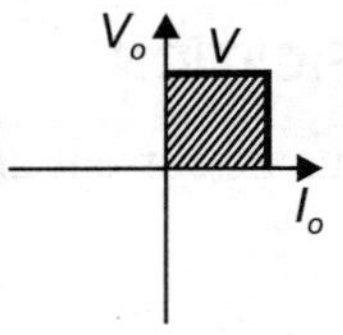	$v_0 = V$, Ch1 ON $= 0$, Ch1 OFF D_1 ON
Second quadrant or regenerative chopper class B	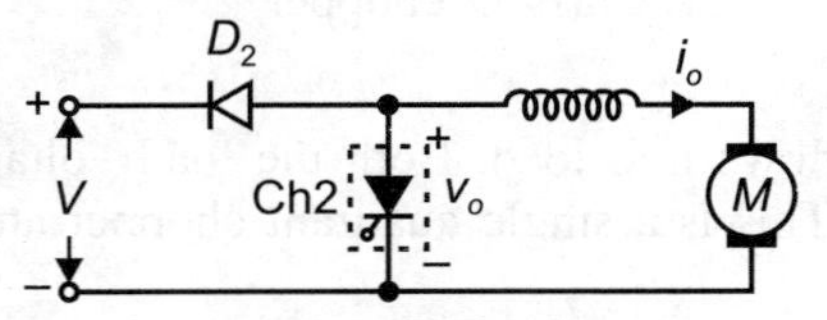	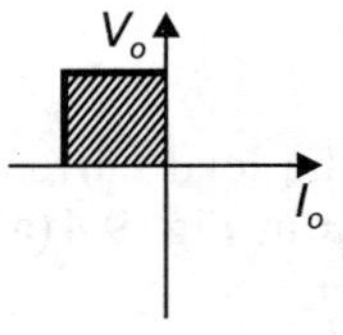	$v_0 = 0$, Ch2 ON $= V$, Ch2 OFF D_2 ON
Two quadrant class C chopper	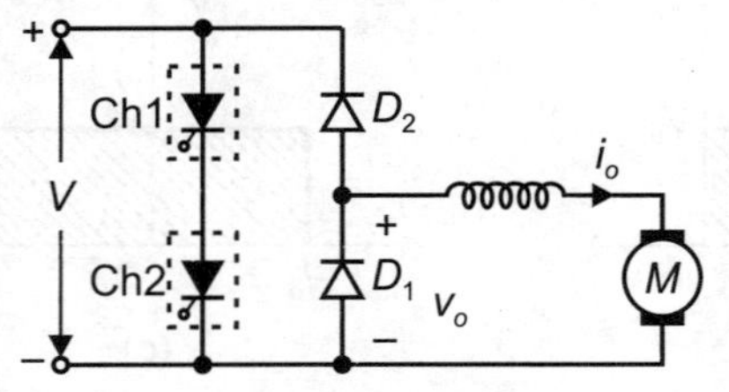	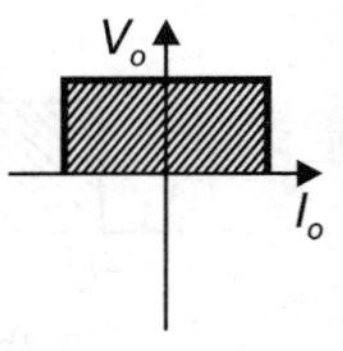	$v_0 = V$, Ch1 ON or D_2 ON $= 0$, Ch2 ON or D_1 ON $i_0 = $ Positive Ch1 ON or D_1 ON $i_0 = $ Negative Ch2 ON or D_2 ON
Two quadrant class D chopper	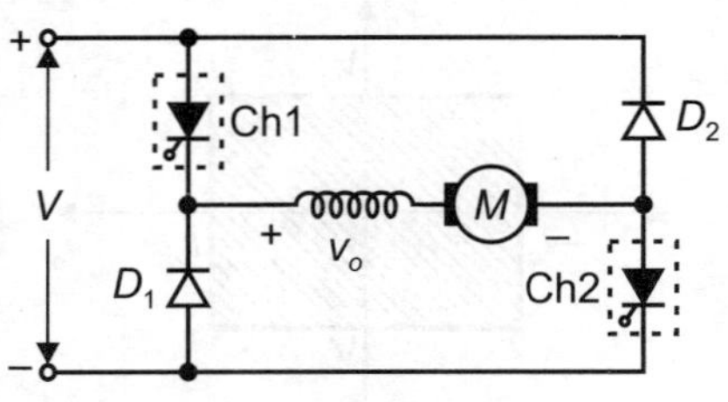	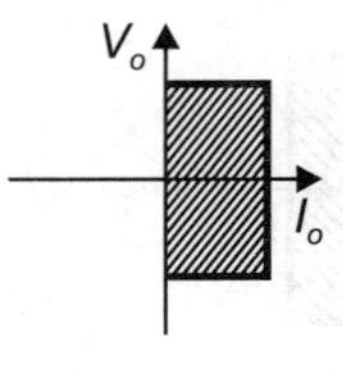	$v_0 = V$ Ch1 ON or D_2 ON $v_0 = -V$ Ch1 and Ch2 OFF D_1 and D_2 ON
Four quadrant chopper class E	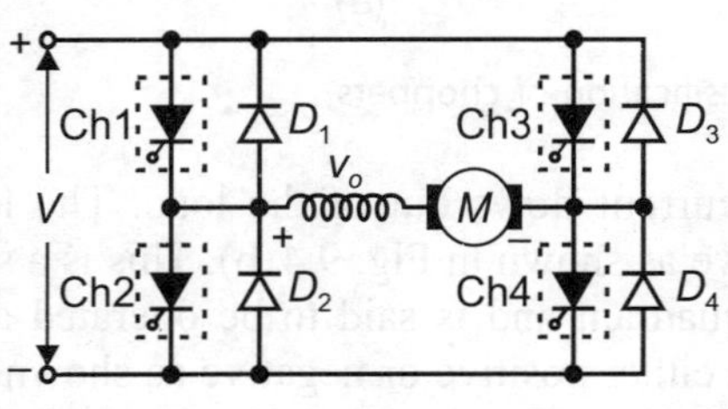	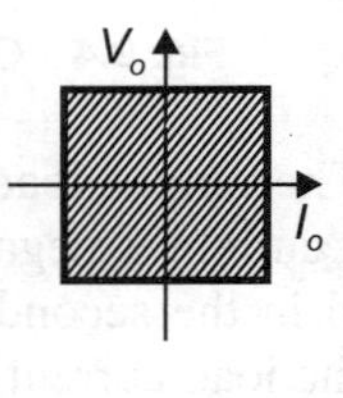	Ch4 ON and Ch3 OFF Ch1 and Ch2 operated $v_o = $ Positive $I_o = $ Reversible Ch2 ON and Ch1 OFF Ch3 and Ch4 operated $V_0 = $ Negative $I_0 = $ Reversible

Fig. 9.5 Various types of chopper configurations and their operation in different quadrants

chopper is off. Therefore, both the average load voltage V_0 and current I_0 are positive and it is used for motoring operation. In the second quadrant or regenerative chopper V_0 is positive, I_0 is negative. The power flow is from load to source. This configuration is used for regenerative braking of DC motor.

The above two circuits are combined in the two quadrant chopper, type A. Here $v_0 = 0$, if chopper Ch2 or diode D_1 conducts and $v_0 = V$, if chopper Ch1 or diode D_2 conducts, V_0 is positive. However, i_0 can reverse the direction. It is positive if Ch1 is on or D_1 conducts and negative if Ch2 is on or D_2 conducts. However, when a two quadrant chopper is used, a simple reversal of field or armature terminal of DC motor will make the drive reversible and regenerative. This chopper configuration may be used for both motoring and regenerative braking of a DC motor.

In the two quadrant type B, chopper I_0 is positive and V_0 is reversible, power flow is reversible. This configuration may be used for both motoring and regenerative braking of DC motor. In regenerative mode of operation, the motor voltage has to be reversed. In four quadrant chopper both V_0 and I_0 are reversible. This configuration can be used for a reversible regenerative braking DC drive.

9.4 TYPES OF CHOPPER

According to the mode of communication, the chopper is classified as series turn-off or parallel turn-off circuits. A thyristor is turned-off when its current drops below its holding current value for a specified duration. This time duration can be reduced if the anode voltage of the thyristor is reversed during turn-off. During the turn-off process, the time gap between the zero-crossing of the thyristor current and the zero-crossing of the reapplied anode voltage must be greater than the recovery time of the device if it has to remain in the turn-off state. The recovery time depends on various factors, such as temperature, current prior to switch-off and the negative anode voltage. The recovery time varies due to spread of the thyristors and high current devices have larger recovery times.

9.4.1 Series Turn-Off Chopper

There are two thyristors Th1 and Th2 shown in Fig. 9.6. One main thyristor and another auxiliary thyristor employed to commutate the main thyristor. One commonly used series-turn-off chopper circuit is shown in Fig. 9.6. The series turn-off circuit is based on the idea of incorporating a reverse voltage between the thyristor and the load. The magnitude of the reverse voltage should be sufficiently high to reduce the thyristor current to zero. This reverse voltage is produced in an inductor L_1 as shown in Fig. 9.6. During the off-period, the load current is maintained through the flywheel diode D_3. When both the thyristors Th1 and Th2 are off, and the supply is on, the capacitor C starts charging through L_2 and D_2 with the polarity as shown. As the charging circuit resistance is negligibly small, the circuit may be assumed to be an undamped resonant circuit. The capacitor voltage charge up to approximately twice the supply voltage when the charging current reduces to zero. The charging current is interrupted after this instant by the diode D_2, because it does not allow

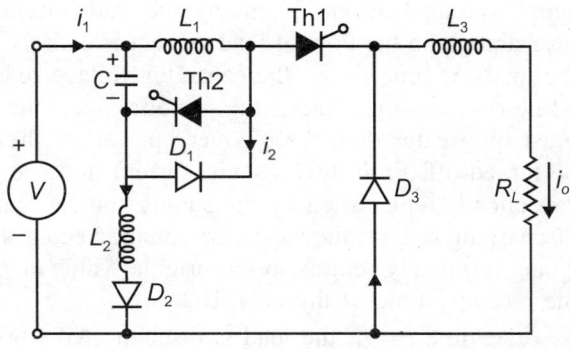

Fig. 9.6 Thyristors Th1 and Th2

a reverse current to flow and the capacitor C remains charged to approximately twice the supply voltage. When thyristor Th1 is fired, load current rises from zero almost linearly to the maximum value due to the presence of inductors L_1 and L_2 connected in the load circuit. The load current is supplied by the source but the voltage of the capacitor C remains unaltered.

The moment, current through capacitor becomes equal to load current, thyristor Th1 is turned-off. During the turning-on of thyristor Th2, the turn-off process of thyristor Th1 starts. The relevant voltage and current waveforms during the turn-off process are shown in Fig. 9.7. The turn-off process starts at the instant t_1. When the negative voltage of C is impressed at the anode of thyristor Th1, the capacitor C discharges through L_1 and thyristor Th2. Since the capacitor voltage is greater than that of the supply voltage, the current through thyristor Th1 reduces to zero from the full-load current. However, the load current is maintained by the stored energy in L_3 through the diode D_3.

The discharged circuit of the capacitor consists of L_2, thyristor Th2, and diode D_1. This forms an undamped resonant circuit which starts to oscillate. The polarity of the capacitor voltage changes when current flows through thyristor Th2 and returns to the original polarity when diode D_1 conducts. During the conduction-period of diode D_1, thyristor Th2 is turned-off due to the reverse voltage of the saturated diode across it. The capacitor cannot be discharged again through thyristor Th2 or diode D_2 due to their off condition and reverse polarity, respectively.

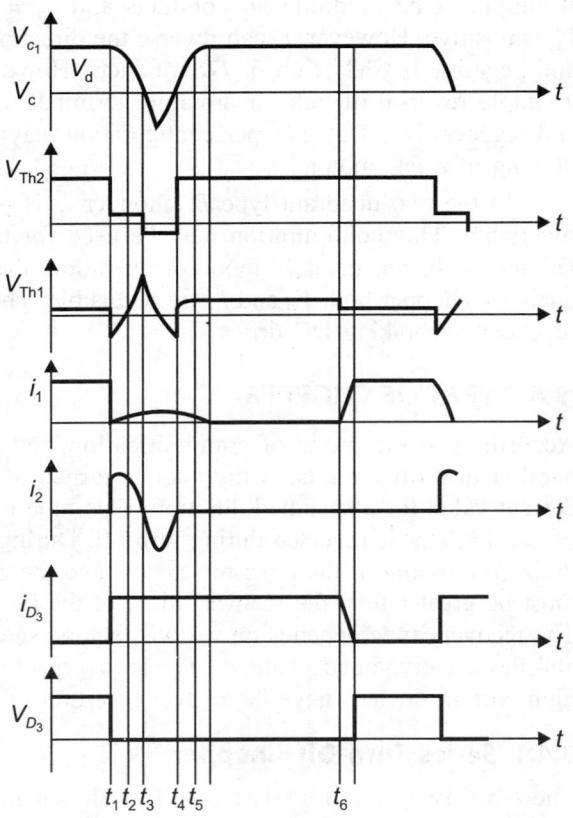

Fig. 9.7 The relevant voltage and current waveforms during the turn-off process

After a finite duration of time, thyristor Th1 is again fired which then takes over to supply the load current. Normally, the load current is reduced for a while the capacitor is fully discharged and the flow of load current is entirely due to inductor L_3, connected in series with the load. At time $t = t_2$, the capacitor voltage reduces to a value equal to the supply voltage while discharging. During the period $t_2 - t_1$, the thyristor Th1 is reverse-biased and $t_2 - t_1$ must be greater than the recovery period of thyristor Th1, otherwise thyristor Th1 cannot be turned-off. From this instant, current flows to the diode D_2 and the loss of charge in the capacitor is replenished by the supply source. During the oscillatory period of the capacitor-discharging circuit, the capacitor voltage becomes zero at t_3, goes negative and again zero at t_4 and ultimately returns to the original value at t_5. The period $t_4 - t_3$ should be greater than the recovery time of thyristor Th2.

At time $t = t_1$, the load is disconnected from the supply and thyristor Th2 is turned-on and remains in that condition till thyristor Th1 is retriggered at t_6. During the commutation-period of thyristor Th1, diode D_3 conducts and the voltage across it is the saturated voltage of the diode.

During the conduction of thyristor Th2, the current in L_1 is equal to the load current and the initial capacitor voltage is V_{C_1} which is greater than V. If the capacitor discharge path is assumed to have zero resistance, the current i_2 is oscillatory in nature because of the undamped resonant circuit. The equation of current i_2 can be written as

$$i_2 = I_m \sin{(\omega_0 t + \psi)} \qquad \qquad ...(9.14)$$

where,
$$\omega_0 = 1/\sqrt{L_1 C} \qquad \qquad ...(9.15)$$

The voltage developed across L_1 is

$$V_{L_1} = \omega_0 L_1 I_m \cos{(\omega_0 t + \psi)} \qquad \qquad ...(9.16)$$

The voltage across the capacitor is approximately equal to V_{L_1}, the capacitor voltage being slightly higher than the inductor voltage by an amount equal to the saturated voltage of thyristor Th2.

At time $t = 0$,

$$I_2 = I_m \sin{\psi} = I_L \qquad \qquad ...(9.17)$$

and
$$V_{L_1} = \omega_0 L_1 I_m \cos{\psi} = V_{C_1} - V_{\text{sat}} \qquad \qquad ...(9.18)$$

where, V_{sat} is the saturated voltage of thyristor Th2 or diode D_1.

From Equations (9.17) and (9.18)

$$I_m = \left[I_L^2 + \frac{(V_{C_1} - V_{\text{sat}})^2}{\omega_0^2 (L_1)^2} \right]^{1/2} \qquad \qquad ...(9.19)$$

and
$$\tan{\psi} = \frac{\omega_0 L_1 I_L}{V_{C_1} - V_{\text{sat}}} \qquad \qquad ...(9.20)$$

In Fig. 9.7, it is seen that the capacitor voltage starts to decrease from the instant t_1 and at t_2 it becomes equal to V. During the period $t_2 - t_1$, thyristor Th1 is reverse-biased and the voltage across it is

$$V_{\text{Th1}} = V - v_{L_1} + V_{\text{sat}} = v_C \qquad \qquad ...(9.21)$$

In Equation (9.21), V_{sat} is the saturated voltage of the flywheeling diode D_3 which starts to conduct as soon as thyristor Th1 is turned-off.

One of the main conditions for the turn-off of thyristor Th1 is that $(t_2 - t_1) > t_q$ where, t_q is the turn-off time of thyristor Th1.

Now, from Equation (9.16) and subsequent discussion, the capacitor voltage is

$$v_C = \omega_0 L_1 I_m \cos{(\omega_0 t + \psi)} + V_{\text{sat}}$$
$$= \frac{V_{C_1} - V_{\text{sat}}}{\cos{\psi}} \cos{(\omega_0 t + \psi)} + V_{\text{sat}} \qquad \qquad ...(9.22)$$

At the time $t_2 - t_1$, v_C becomes equal to V and $V_{\text{Th1}} = 0$ and in the limiting case, if it is assumed that $t_2 - t_1 = t_q$, then

$$V_{C_1 t = t_2 - t_1} = V = \frac{V_{C_1} - V_{\text{sat}}}{\cos{\psi}} \cos{(\omega_0 t_q + \psi)} + V_{\text{sat}} \qquad \qquad ...(9.23)$$

If t_q is known, the other components of the commutation circuit can be calculated.

The initial voltage across the capacitor V_{C_1} depends upon the oscillatory circuit. During the discharge, when the capacitor becomes equal to the supply voltage, the diode D_2 begins to conduct and the capacitor current becomes equal to the difference between i_2 and the current through the diode D_2. The energy stored in the capacitor at the beginning of the discharge cycle is

$$W_C = \frac{1}{2} C V_c^2 \qquad \qquad ...(9.24)$$

The energy is the same at the beginning of each discharge cycle, i.e. the capacitor is charged to the same voltage every time.

The energy stored in the inductance L_1 prior to switch-off is

$$W_{L_1} = \frac{1}{2} L_1 I_L^2 \qquad \qquad ...(9.25)$$

The energy is either converted into an electrostatic form during oscillation in the capacitor or dissipated. If the total energy is not dissipated in one off-period, the voltage across the capacitor increases until the loss generated by the increased current causes an equilibrium. On the other hand, if the circuit dissipation is too large, the loss is made up by the source through D_2. The average dissipation of the circuit depends upon the operating frequency, the value of L_1, and the load current. However, the minimum value of the inductance should be determined by the recovery time of the thyristor.

Advantages: The advantage of the series turn-off circuit is that the turn-off circuit is independent of the load current. As a result, variable load current does not affect the turn-off process. The turn-off circuit can handle the maximum load current as long as the turn-off-period is longer than the time required to recharge the capacitor. Therefore, the turn-off time must be greater than the time required to recharge the capacitor but the on time may be as short as necessary and is practically independent of load current. As a result, the duty ratio can be reduced to almost zero.

The circuit is recommended for low supply voltages because the reverse voltage developed by the inductance may exceed the supply voltage by several factors and the required PIV of the thyristor should be much higher.

9.4.2 Parallel Capacitor Turn-Off Chopper

A commonly used basic chopper circuit is shown in Fig. 9.8. It has a main thyristor Th1 which carries the load current for the period it is on and voltage is applied across the load. For high output power, the parallel capacitor turn-off chopper is generally used. This above circuit finds wide application in control systems where load fluctuation is not very large. Thyristor Th1 is the main power switch and diode D_2, the flywheel diode connected across the inductive load. The capacitor C along with Th2, diode D_1 and L_1 form the turn-off circuit of thyristor Th1. Initially, thyristor Th2, takes over conduction when it is fired and applies the capacitor voltage as a negative voltage across the main thyristor to turn it off.

Capacitor C is charged through thyristor Th2 and the load to the voltage

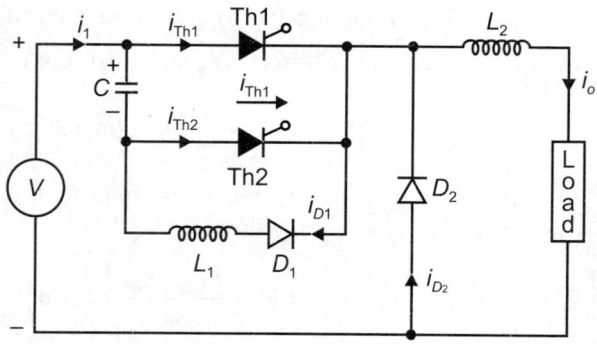

Fig. 9.8 A basic chopper circuit

almost equal to the supply with the polarity as shown in Fig. 9.8. After the charge is completed, thyristor Th2 is turned-off. The load current continues to flow through diode D_2. When thyristor Th1 is fired, the load current commutate from D_2 to the main switch thyris-tor Th1. The rate of rise of current through thyristor Th1 depends on the load inductance. The capacitor discharges simultaneously through thyristor Th1, diode D_1 and L_1, to zero voltage in a definite amount of time called the turn-off time. The capacitor continues to charge in the opposite polarity with the lower plate positive, to a voltage approximately equal to the supply voltage. After lapse of t_{off}, the main thyristor is switched on. The load current get transferred to the main thyristor from the freewheeling circuit.

The relevant voltage and current waveforms are shown in Fig. 9.9. At the instant t_1, thyristor Th2 is turned-on and the capacitor voltage of about $-V$ is impressed across Th1 and turns it off. The load current continues to be supplied by the supply source through the capacitor C and thyristor Th2. Due to the presence of inductance in the load, the load current remains almost constant and the capacitor voltage rises linearly from $-V$ to $+V$. The capacitor voltage is governed by the equation

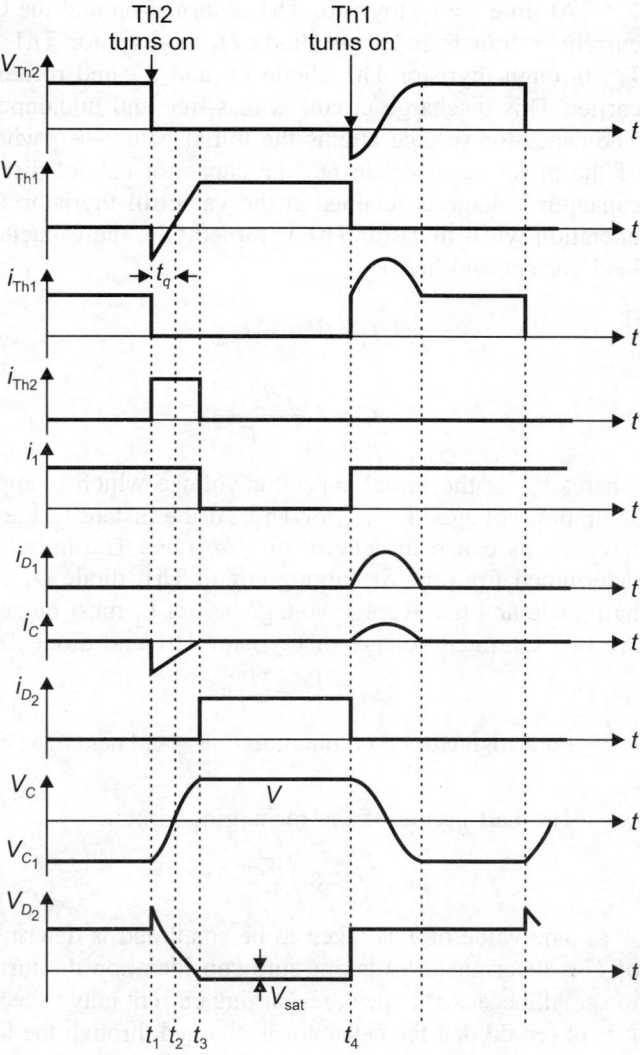

Fig. 9.9 Voltage and current waveforms

$$V_C = \frac{1}{C} I_L t + V_{C_1} \qquad \qquad ...(9.26)$$

where, V_{C_1} is the initial capacitor voltage and is equal to about $-V$. The voltage across diode D_2 is the sum of the supply voltage and the capacitor voltage and shoots up to about 2 V at the instant t_1 and decreases linearly as the capacitor voltage decreases. From t_1 to t_3, the capacitor voltage rises from $-V$ to $+V$ and at the instant t_2 the capacitor voltage is zero. The voltage across thyristor Th1 is also zero at t_2. During $t_2 - t_1$, thyristor Th1 is reverse-biased and in order to turn it off $(t_2 - t_1)$ must be greater than the recovery time t_q. At the instant t_3, the anode voltage of thyristor Th1 attains a value $V + V_{sat}$, where V_{sat} is the saturated voltage of diode D_2 and the capacitor voltage becomes equal to that of the supply. The input current I_1 reduces to zero and load current flows through the flywheel diode D_2. The load voltage becomes equal to $-V_{sat}$, where V_{sat} is the saturated voltage of diode D_2. The circuit is in the OFF state.

At time $t = t_4$, thyristor Th1 is turned-on and the ON state of the circuit starts. The load current is transferred from diode D_2 to thyristor Th1, the capacitor discharges from $+V$ to V_{sat} through thyristor Th1, diode D_1 and L_1, and recharges in the opposite polarity as stated earlier. This discharge circuit is loss-free and undamped, the current is therefore sinusoidal. The capacitor voltage attains the initial value $-V_{C_1}$ when current through it is zero. Because of the presence of diode D_1, the capacitor cannot discharge through the same path and the capacitor voltage is retained at the value till thyristor Th2 is fired again. During the turn-off operation when thyristor Th1 is turned ON, the capacitor is charged by the constant value of load current and hence

$$\frac{1}{C} I_L t_q = V_{C_1} - V_{sat} \qquad \qquad ...(9.27)$$

or
$$C = \frac{I_L t_q}{V_{C_1} - V_{sat}} \qquad \qquad ...(9.28)$$

where, V_{C_1} is the initial capacitor voltage which is approximately equal to V and V_{sat} is the saturation voltage of thyristor Th2. At the instant t_2, the capacitor voltage is zero. The interval $t_1 \cdot t_2 = t_c$ is called the *circuit turn-off time*. The initial value of the capacitor voltage can be determined from the recharging circuit Th1, diode D_1 and L_1. The capacitor is charged in the half-cycle and the average voltage across L_1 must be zero. During recharging operation, there are two saturated voltage of thyristor Th1 and diode D_1, and hence

$$V_{C_1} = V - 3V_{sat} \qquad \qquad ...(9.29)$$

For satisfactory commutation $t_c > t_q$ and hence the capacitance chosen must be $C > \dfrac{I_L \cdot t_q}{V_{C_1}}$

The half-period of the recharging circuit,

$$\frac{T}{2} = \pi \sqrt{L_1 C} \qquad \qquad ...(9.30)$$

The value of T is taken to be small and is determined by inductance L_1, since the value of C is determined by taking into consideration the turn-off time. But L_1 should not be taken too small because the peak recharging current may exceed the value allowable by thyristor Th1. It is observed that the capacitor is charged through the load after turning off thyristor Th1. The period $(t_3 - t_1)$ is inversely proportional to the load current and consequently the circuit cannot operate at no-load. At light loads, this period becomes large and determines the maximum value of the operating frequency.

Another disadvantage of the circuit is that the required value of the capacitor is rather large for low supply voltage because its initial voltage is less than the supply. This disadvantage can be partly overcome by introducing a reactor between the supply source and the circuit and by replacing diode D_1 by thyristor Th3 as shown in Fig. 9.10.

9.4.2.1 Effects of Source Inductance on the Performance of Chopper

A chopper circuit with a load and source inductance of L_1 is shown in Fig. 9.10. One of the effects of this inductance is to make the voltage of the capacitor available to turn-off the main thyristor less than the supply voltage. The energy stored in the inductance also causes over voltages of capacitor in the negative direction. This type of charging of capacitor leads to uncertain turn-off times as higher ratings of thyristors. Due to the presence of inductance L_1, the capacitor C is charged to a higher voltage than the supply when thyristor Th2 is turned-on. This is because the capacitor C also stores the energy of the inductance L_1 at the end of the turn-off process. The capacitor is charged from a negative voltage to a positive voltage higher than the supply. When the capacitor voltage becomes equal to the supply, the flywheel diode begins to conduct. But the load current is maintained to flow through the capacitor by the inductance L_1 until the total load current begins to flow through the diode D. During this period, the capacitor is charged to a higher voltage than the supply due to this current. The magnitude of the capacitor voltage rise is a function of load current.

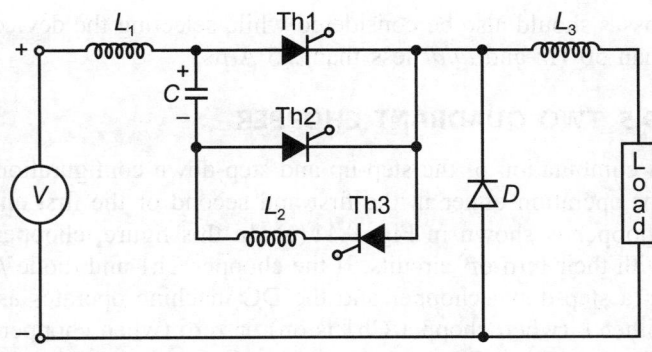

Fig. 9.10 A chopper circuit with a load and source inductance of L_1

If the load current is high, the voltage rise is more. In this condition, if there were a diode instead of thyristor Th3, it would conduct and the capacitor would discharge partially towards the supply source and finally the initial voltage of the capacitor would have been less than the supply. But since thyristor Th3 is OFF, the capacitor cannot discharge and capacitor voltage remains at the higher value. Thyristors Th1 and Th3 are turned-on simultaneously during the turn-on process and the capacitor is charged through them to a voltage almost of the same magnitude but in the opposite polarity. The minimum value of the ON time should be sufficiently large for complete recharge of the capacitor. In this circuit, the inverse voltage impressed across the thyristor may exceed the supply voltage, particularly if the supply inductance is large. The inverse voltage across the flywheel diode is about twice the supply voltage and may be greater if the commutating capacitor is charged more than the supply voltage. The dynamic voltage rating of the main thyristor depends upon the commutating circuit. In the parallel capacitor commutated circuit, the dv/dt of the main thyristor is approximately equal to V/t_q. Therefore, for successful operation, the thyristor with higher value of dv/dt should be selected.

The circuit rating of the thyristor and the flywheel diode depend upon the duty ratio. If the duty ratio is DR then

$$\frac{I_{\text{Th}(av)}}{I_L} = DR \text{ and } \frac{I_{\text{Th}(rms)}}{I_L} = \sqrt{DR}$$

Similarly, for the flywheel diode,

$$\frac{I_{D(av)}}{I_L} = 1 - DR \text{ and } \frac{I_{D(rms)}}{I_L} = \sqrt{1 - DR}$$

In the calculations of thyristor and flywheel diode, the current waveforms are taken as square waves. It is needless to mention that for constant load current, the thyristor current becomes maximum and the flywheel diode current becomes zero for unity duty ratio and maximum for zero duty ratio.

The commutating thyristor carries the load current for a brief period of time which is approximately equal to t_q. Therefore, in selecting an appropriate thyristor for this purpose, the allowable peak repetitive current rating of the device should be carefully considered. In addition to the above factors the switching losses may be neglected then the turn-on and turn-off losses should also be considered while selecting the devices. If the operating frequency is less than 50 Hz and di/dt less than 2.5 A/μs.

9.5 TWO QUADRANT CHOPPER

A combination of the step-up and step-down configurations can form a two quadrant chopper for operation either in the first and second or the first and fourth quadrants. A two quadrant chopper is shown in Fig. 9.11 (a). In this figure, choppers Ch1 and Ch2 represent thyristor with their turn-off circuits. If the chopper Ch1 and diode D_1 are operated, the system operates as a step-down chopper and the DC machine operates as a motor. The output voltage V_0 is either V (when chopper Ch1 is on) or zero (when chopper Ch1 is off and diode D_1 conducts). The average value of the output voltage is positive and the output current i_0 flows in the positive direction. The chopper operates in the first quadrant as shown in Fig. 9.11 (b). If the chopper Ch2 and diode D_2 are operated, the system operates as a step-up chopper with E_a as source and the DC machine operates in the regenerative braking mode. The output voltage V_0 is either zero (when chopper Ch2 is on) or V (when chopper Ch2 is off and diode D_2 conducts). The average value of the output voltage is positive, but the output current now flows in the negative direction. The chopper then operates in the fourth quadrant as shown in Fig. 9.11 (b).

The chopper shown in Fig. 9.11 (a) can thus be operated in either the first or fourth quadrant and hence known as a *two quadrant chopper*.

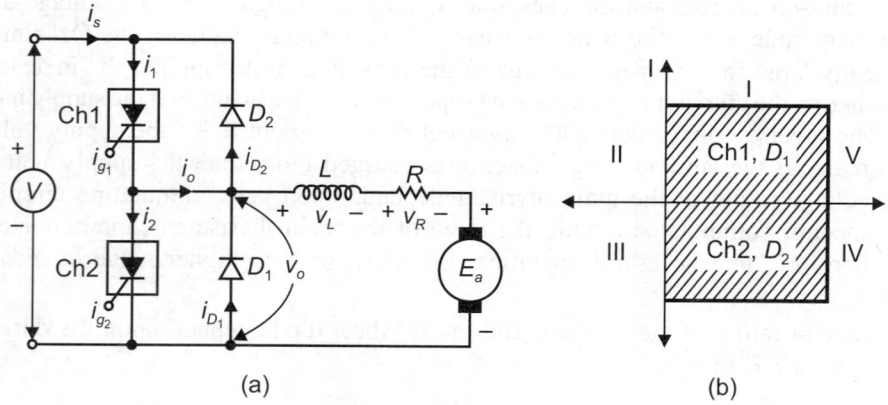

(a)   (b)

Fig. 9.11 Two quadrant chopper and quadrant operation

Analysis of two quadrant chopper circuit: Gating signal for two thyristors are shown in Fig. 9.12 and it is assumed that with these gating signals the values of V and E_a are such that the average output current I_0 is positive and that the converter is therefore operating in first quadrant of the diagram V_0 versus I_0.

After the converter has been switched on for a short time, it may be found to be operating with time variations of the circuit variables such as shown in Fig. 9.12. The waveform i_0 may be broken up into four segments, and in Fig. 9.12, the semiconductor device current identical with i_0 for any interval is indicated on the i_0 waveform and is given by

$$i_0 = \frac{V - E_a}{R}(1 - e^{t/\tau}) + I_{min}e^{-t/\tau} \quad 0 \le t \le t_{ON} \qquad ...(9.31)$$

where, $\tau = L/R$

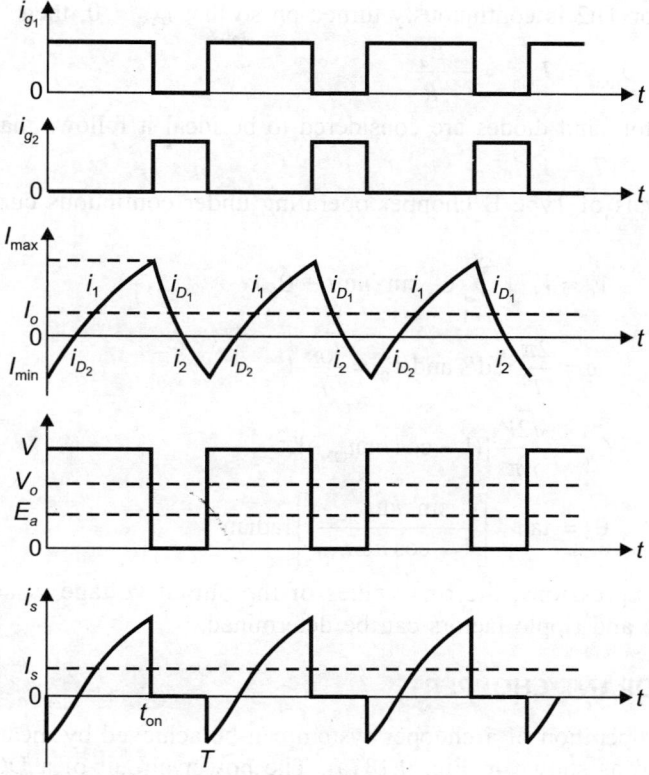

Fig. 9.12 Waveforms

$$i_0 = -\frac{E_a}{R}(1 - e^{-t'/\tau}) + I_{max} e^{-t'/\tau} \qquad t_{ON} \leq t \leq T \qquad \ldots(9.32)$$

where, $\qquad t' = t - t_{ON}$

Solving for currents I_{max} and I_{min}, we have

$$I_{max} = \frac{V[1 - e^{-t_{ON}/\tau}]}{R[1 - e^{-T/\tau}]} - \frac{E_a}{R} \qquad \ldots(9.33)$$

$$I_{min} = \frac{V[e^{t_{ON}/\tau} - 1]}{R[e^{T/\tau} - 1]} - \frac{E_a}{R} \qquad \ldots(9.34)$$

If the independent variables are such that $I_{max} > 0$ and $I_{min} < 0$, then the result may be first quadrant operation as illustrated in Fig. 9.11, where $I_0 > 0$. If the independent variables are such that I_{max} has a smaller positive value and I_{min} a larger negative value so that $I_0 < 0$, then the result will be second quadrant operation.

If $V_0 > E_a$ then the average power flows to the load circuit and the converter operates in the first quadrant. If however, $V_0 < E_a$ then the average power flows from the load circuit to source V and the converter operates in the second quadrant. These conditions are shown on the voltage waveform of Fig. 9.12.

When thyristor Th1 is continuously turned-on so that $t_{ON} = T$, then

$$I_{max} = I_{min} = \frac{V - E_a}{R} \qquad \ldots(9.35)$$

When thyristor Th2 is continuously turned-on so that $t_{ON} = 0$, then

$$I_{max} = I_{min} = -\frac{E_a}{R} \qquad \qquad ...(9.36)$$

As the thyristors and diodes are considered to be ideal it follows that

$$VI_s = V_0 I_0$$

Fourier analysis of Type B chopper operating under continuous current condition are described by

$$v_0 = V_0 + \sum_{n=1}^{\alpha} C_n \sin(n\omega t + \theta_n) \qquad \qquad ...(9.37)$$

where

$$\omega = \frac{2\pi}{T} \text{ rad/s and } V_0 = \frac{t_{ON}}{T} V$$

$$C_n = \frac{\sqrt{2}V}{n\pi}[(1 - \cos n\omega t_{ON})]^{1/2}$$

and

$$\theta_n = \tan^{-1}\left[\frac{\sin n\omega t_{ON}}{1 - \cos n\omega t_{ON}}\right] \text{radian} \qquad \qquad ...(9.38)$$

From these expressions, the rms values of the output voltage v_0 and current i_0, rms values of harmonic and ripple factors can be determined.

9.6 FOUR QUADRANT CHOPPER

The four quadrant operation of a chopper system can be achieved by means of two quadrant choppers connected as shown in Fig. 9.13 (a). The power circuit of a DC-to-DC converter

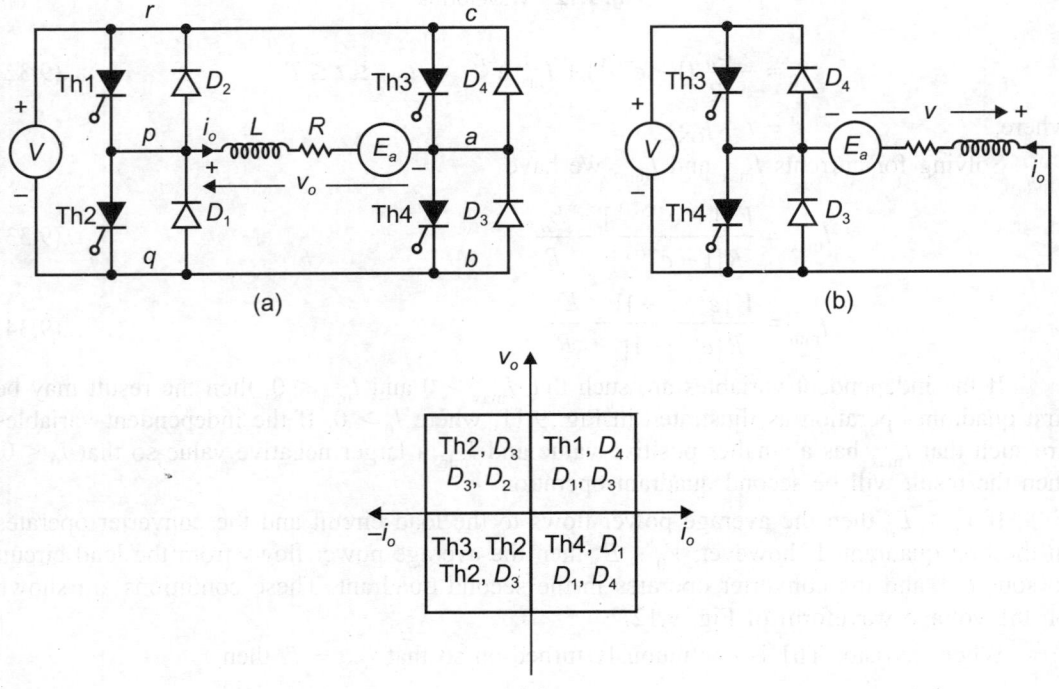

(a)

(b)

(c)

Fig. 9.13 Four quadrant chopper and operation

capable of operating in any of the four quadrants of V_0 versus I_0. If the thyristor Th4 is turned-on continuously, then the antiparallel connected pair of devices Th4 and D_3 constitute a short circuit of nodes 'a' and 'b' as in Fig. 9.13 (a), clearly thyristor Th3 may not be turned-on, since this would short circuit source V. The antiparallel connected pair of devices Th3 and D_4 this constitutes an open circuit between nodes 'a' and 'c'. The remainder of the circuit now forms a converter operating in the first and second quadrant with positive value of V_0.

If the thyristor Th2 in the circuit of Fig. 9.13 (a) is turned-on continuously then Th2 and D_1 short circuit nodes p and q. Thyristor Th1 may not be turned-on, thus thyristor Th1 and diode D_2 constitute an open circuit between nodes 'p' and 'r'. The remainder of the circuit may be rearranged as shown in Fig. 9.13 (b) to constitute chopper operating in third and fourth quadrant with negative values of V_0.

9.7 MORGAN CHOPPER

The Morgan chopper is shown in Fig. 9.14 which consists of a single thyristor chopper with a saturable reactor SR in the commutation circuit. The exciting current of the saturable reactor is very small and can be neglected. When thyristor Th1 is off, the capacitor C is charged to the supply voltage V with the polarity as shown in the Fig. 9.14 and the saturable reactor is placed in the positive saturation condition. When thyristor Th1 is turned-on, the capacitor voltage appears across the saturable reactor and the core flux is driven from positive saturation towards

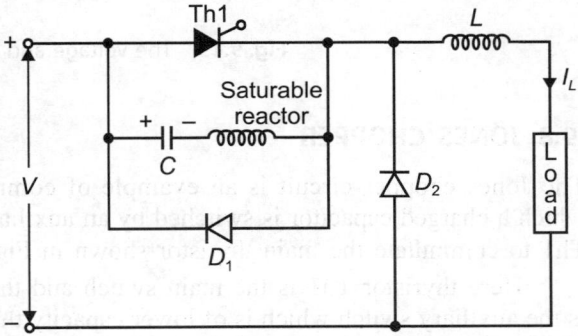

Fig. 9.14 Morgan chopper

negative saturation. Till the negative saturation point is reached, the capacitor voltage remains essentially constant having the same polarity as before. This is due to the negligible exciting current of the SR. When the core flux reaches the negative saturation, the capacitor discharges through thyristor Th1, and the post-saturation inductance of SR. This resonant circuit has a discharging time $T = \pi\sqrt{L_S C}$ seconds, where L_S is the post-saturation inductance of the reactor. The time of discharge of the capacitor is thus comparatively small and the reversal of the polarity of the capacitor takes place very quickly. After this, the capacitor voltage which is $-V$ now is impressed on the saturable reactor in the reverse direction and the core is driven from negative saturation towards positive saturation. After a fixed interval of time, the core flux reaches the positive saturation after which the capacitor discharges very quickly through thyristor Th1 in the reverse direction and the post-saturation inductance as before. The discharged current first passes through thyristor Th1 turning it off and then through diode D_1. During the whole operation, the load current is delivered through the thyristor Th1 by the supply voltage. When thyristor Th1 is turned-off, the load current begins to flow through the flywheel diode D_2. This state continues till thyristor Th1 is again turned-on. Since the volt-time integral to saturate the core is constant, the ON period of thyristor Th1 is fixed. The ON period is a function of $L_S C$ and the average output voltage can be altered only by varying the operating frequency. However, the ON period can be controlled by varying the volt-time product of the saturable reactor by means of DC controlled current through it. The total ON time of thyristor Th1 is determined by the time required for the saturable reactor to move from positive saturation to negative saturation and back to positive saturation again.

The voltage and current waveforms are shown in Fig. 9.15. In this figure, thyristor Th1 is turned-on at t_1 and t_3 and is automatically turned-off at t_2 and t_4.

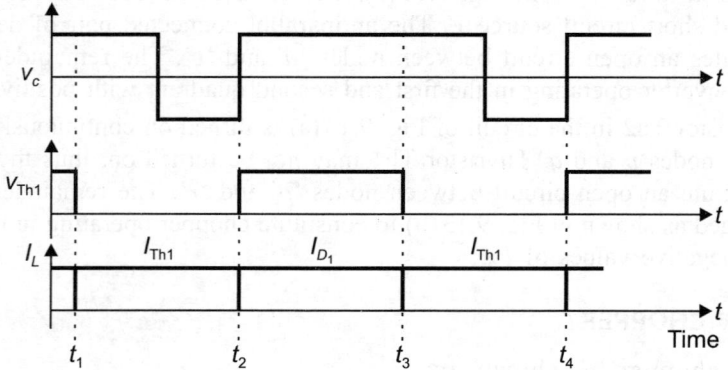

Fig. 9.15 The voltage and current waveforms

9.8 JONES CHOPPER

The Jones chopper circuit is an example of commutation in which a charged capacitor is switched by an auxiliary thyristor Th1 to commutate the main thyristor shown in Fig. 9.16.

Here thyristor Th1 is the main switch and thyristor Th2 is the auxiliary switch which is of lower capacity than thyristor Th1 and is used to commutate thyristor Th1 by a reverse voltage developed across capacitor C. The special feature of the circuit is the tapped autotransformer T through a portion of which the load current flows. Since L_1 and L_2 are closely coupled, the capacitor always gets sufficient energy to turn thyristor Th1 OFF.

When thyristor Th1 is turned-on, the capacitor C discharges through thyristor Th1, L_1 and D_1. This discharge current does not flow through L_2 and back to the battery because of the transformer action of T. The load current is picked up by thyristor Th1, and the flywheel diode D_1 is reverse-biased and its current reduced to zero. As the capacitor voltage swings negative, the reverse bias on diode D_2 decreases. This continues up to a time $\pi\sqrt{L_1 C}$. The capacitor assumes a polarity as shown in Fig. 9.16. When Th2 is turned-on, the negative voltage on

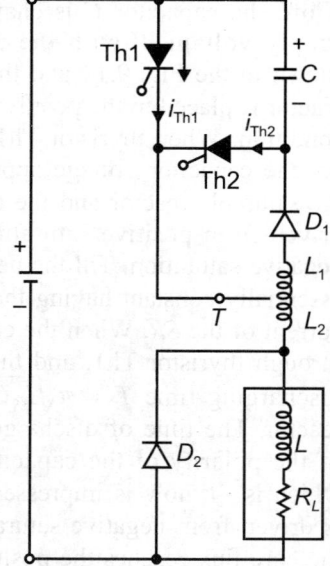

Fig. 9.16 Jones chopper circuit

capacitor C is applied across thyristor Th1 which is turned-off after its recovery current becomes zero. The load current which is nominally constant starts to flow in thyristor Th2 and capacitor C. The di/dt of thyristor Th2 is limited by the circuit stray inductance. Capacitor C is charged positively up to a voltage equal to the supply voltage V. The flywheel diode becomes forward-biased and begins to pick up the load current and capacitor current starts to reduce. The energy stored in the inductance $L_2 = \dfrac{1}{2}L_2 I^2$. This energy is forced into the capacitor C charging it positively. The capacitor current continues to decrease increasing the current through diode D_2 till the capacitor current reduces below the holding current of thyristor Th2 when it is turned-off. The cycle repeats when thyristor Th1 is again turned-on.

The voltage and current waveforms in Jones chopper are shown in Fig. 9.17.

Advantages: The main advantage of the Jones chopper over other circuits is that it allows the use of higher voltage and lower microfarad commutating capacitors because the trapped energy in the inductor L_2 can be forced into the commutating capacitor 'C' rather than simply charging the capacitor by the supply voltage. In Jones circuit, there is no starting problem and any one of the thyristors can be turned-on initially.

There is greater flexibility in control, because both the on time and off-time can be varied individually.

The maximum voltage across the thyristors and the capacitor is somewhat greater than the supply voltage,

$$V_m \leq V + I_L \sqrt{\frac{L_2}{C}} \qquad ...(9.39)$$

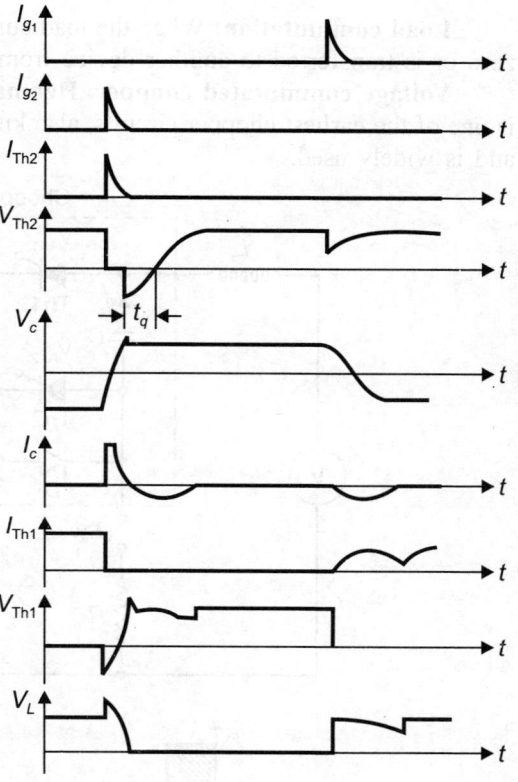

Fig. 9.17 Voltage and current waveforms in Jones chopper

The commutating capacitor however does not retain all its energy and is negatively charged at the instant of commutation to ηv_m where, η lies between 0.5 to 0.9.

The commutating capacitor C must divert the load current long enough so that before C is charged to zero voltage, thyristor Th1 is OFF. Therefore, the minimum value of

$$C = \frac{I_L t_q}{\eta V_m} . \qquad ...(9.40)$$

9.9 THYRISTOR COMMUTATION IN CIRCUITS

The ideal on/off switch represented by chopper (dotted rectangle) is, in fact, a main power thyristor switch together with the commutation circuit to turn it off. Various methods have been developed for commutation which may be divided into two groups.

9.9.1 Forced Commutation

When the current through thyristor is forced to become zero to turn the thyristor off, there are two ways in which this can be achieved:

1. **Voltage commutation:** When a charged capacitor momentarily reverse biases the conducting thyristor and turn it off.
2. **Current commutation:** When a current pulse is made to flow in the reverse direction through the conduction thyristor and the net thyristor current is zero, it is turned-off.

Load commutation: When the load current flowing through the thyristor either becomes zero or is transferred to another device from the conducting thyristor.

Voltage commutated chopper, Henman's chopper: The circuit shown in Fig. 9.18 (a) is one of the earliest chopper circuits, also known as *Henman's chopper* or *oscillation chopper* and is widely used.

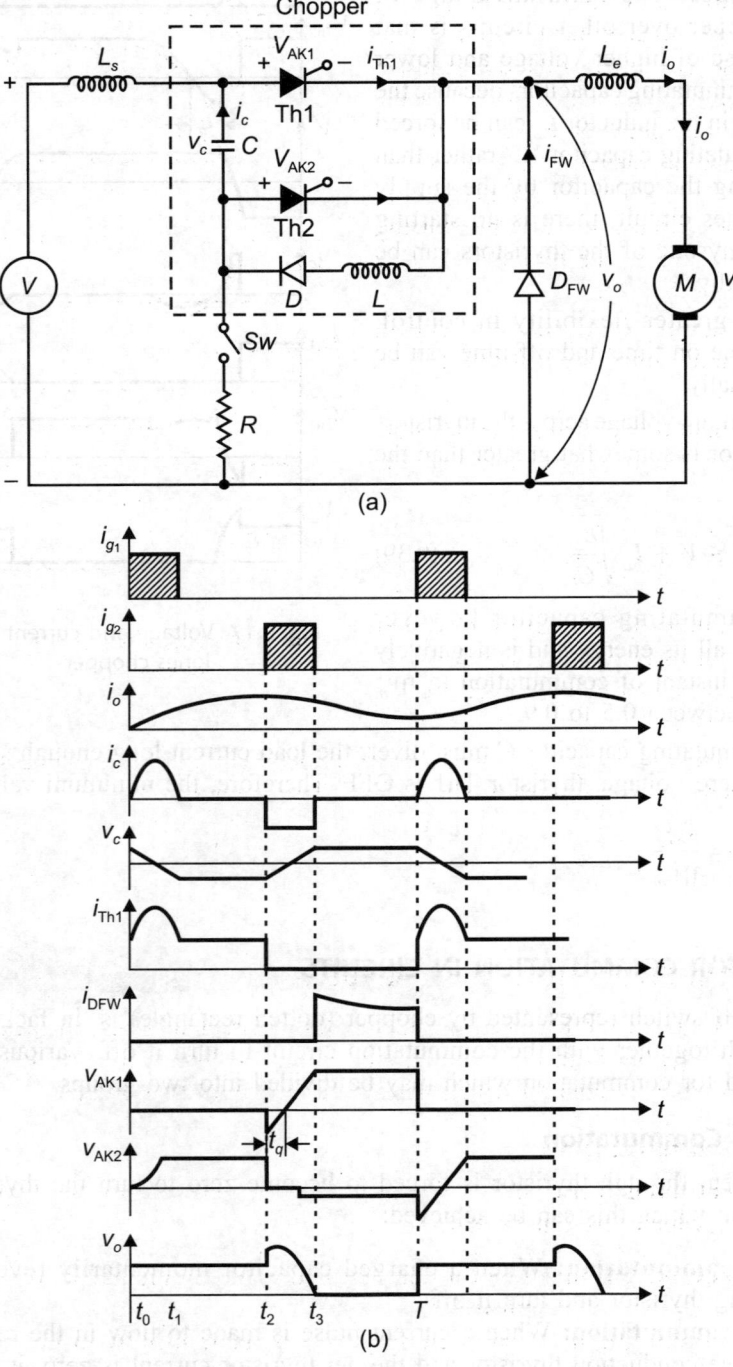

Fig. 9.18 Voltage commutated chopper

In this circuit, thyristor Th1 is the main power switch. Commutation circuit for this chopper is made up of an auxiliary thyristor Th2, capacitor C, diode D and inductor L. Freewheeling diode D_{FW} is connected across the load which is a DC machine. In the circuit, capacitor C is charged with the polarity shown in one of two ways: (1) By closing and then opening the starting switch S, and (2) by triggering the auxiliary thyristor Th2 first, the charging will be through source V, capacitor C, thyristor Th2 and load. Thyristor Th2 will turn-off as the charging current decays to zero.

With capacitor C charged, the chopper circuit is ready for operations. Simplifying, assumptions for this chopper are: (1) load current is constant, and (2) thyristors and diodes are ideal elements. The current and voltage waveforms are shown in Fig. 9.18(b). The main thyristor Th1 is triggered at t_0. Current flows in two paths, load current i_0 flows through thyristor Th1 and the commutation current i_c flows through C, thyristor Th1, inductor L and diode D. The charge on C is reversed at t_1 and held by the retaining diode D. To turn-off the main thyristor Th1, the auxiliary thyristor Th2 is triggered at a desired instant t_2. The capacitor C is placed across thyristor Th1, reverse biases it and turns it off. After thyristor Th1 is turned-off, C provides load current through source V, capacitor C, thyristor Th2 and load. Thyristor Th2 turns off naturally at t_3 when the capacitor current decays to zero. The load current now freewheels through D_{FW}. At $t = T$, thyristor Th1 is fired again and the cycle repeats. This is a simple chopper which is widely used. It suffers from the following disadvantages: (1) Thyristor Th2 is load commutated and requires a starting circuit or a logic circuit such that Th2 is triggered first. (2) The load voltage jumps to twice the supply voltage when the commutation is initiated. It cannot work at no load, as capacitor would not get charged from $-V$ to V when auxiliary thyristor Th2 is fired for commutating the main thyristor Th1.

Design considerations: The values of commutating components C and L can be obtained as follows.

Commutating capacitor C: Its value depend upon the turn-off time t_q available for thyristor Th1. During this time t_q, the capacitor voltage changes linearly from $-V$ to zero. Assuming a constant load current I_{om} during commutation.

$$CV = I_{om} t_q$$

or
$$C = \frac{I_{om} t_q}{V} \qquad \qquad ...(9.41)$$

where, I_{om} is the maximum load current.

For successful commutation, circuit turn-off time t_q must be greater than the thyristor turn-off time t_{off}.

Let
$$t_q = t_{off} + \Delta t \text{ or } C = \frac{I_{om}(t_{off} + \Delta t)}{V}$$

So,
$$C \geq \frac{t_{off} \cdot I_{om}}{V} \qquad \qquad ...(9.42)$$

Commutating inductor L: It can be designed from a consideration of the oscillatory current established when main thyristor is triggered. The value of commutating inductor L can be designed from a compromise between two requirements: (1) peak value of the capacitor current i_c which flows through thyristor Th1 when it is fired, and (2) time $t_1 - t_0$ during which the capacitor current pulse lasts. This time is required to reset the capacitor voltage polarity so that the capacitor is ready to commutate thyristor Th1 when thyristor Th2 is fired. The capacitor current i_c when thyristor Th1 is fired a current in an LC oscillating circuit and is given by

$$i_c = V\sqrt{\frac{C}{L}} \sin \omega_r t \qquad \qquad ...(9.43)$$

Oscillating frequency, $\omega_r = (1/LC)^{1/2} = \dfrac{2\pi}{T_r}$ where, $T_r = 2(t_1 - t_0)$

The peak capacitor current

$$i_c\big|_{\text{peak}} = V\sqrt{\frac{C}{L}} = \frac{V}{\omega_r L} \qquad \qquad ...(9.44)$$

This current flows through thyristor Th1 when it is turned-on. Thyristor Th1 has to carry the load current as well as the peak capacitor current. Let this peak capacitor current be equal to or less than the load current so that peak current through thyristor Th1 is not very large, i.e. $i_{c/\text{peak}} \le I_{om}$

$$V\sqrt{\frac{C}{L}} \le I_{om} \ \text{ or } L \ge C\left(\frac{V}{I_{om}}\right)^2 \qquad \qquad ...(9.45)$$

The resetting time $t_1 - t_0$ can be decreased by selecting a small value of L. However, a small value of L will increase the peak value of i_c. On the other hand, if $t_1 - t_0$ is large, the range of load voltage is reduced. The minimum load voltage is given by

$$V_0\big|_{\text{min}} = \frac{t_1 - t_0}{T} V, \qquad \qquad ...(9.46)$$

where T is the time period of chopper.

If the load voltage is varied over a wide range, the minimum value V_0 should be equal to or less than 10% of the supply voltage V.

So,
$$\frac{t_1 - t_0}{T} \le 0.1$$

or
$$\frac{\pi\sqrt{LC}}{T} \le 0.1$$

or
$$L \le \frac{0.01T^2}{\pi^2 C} \qquad \qquad ...(9.47)$$

9.9.2 Current Commutated Chopper

The power circuit for current commutated chopper is shown in Fig. 9.19(a). In this circuit, thyristor Th1 is the main thyristor. The other components, namely auxiliary thyristor Th2, inductor L, capacitor C, diodes D_1 and D_2 constitute a commutation circuit. D_{FW} is the freewheeling diode and R is the charging resistor. Thyristor Th1 of the chopper is commutated by a current pulse generated in the commutation circuitry. Capacitor C is charged through source V, the capacitor C and the charging resistor R. Thyristor Th1 is fired at $t = t_0$ when the load current flows through thyristor Th1. At $t = t_1$, commutation of thyristor Th1 is initiated by firing thyristor Th2.

Like voltage commutated chopper, energy for current commutation comes from the energy stored in a capacitor. Therefore, capacitor C is charged to a voltage V so that energy for commutation process is available. An oscillatory current flows in the circuit consisting of capacitor C, thyristor Th2 and inductor L when i_c reverses at t_2, thyristor Th2 turns off (natural commutation) and the oscillatory current flows through diode D_2 and thyristor Th1. In thyristor Th1, it tries to flow in the opposite direction and so decreases the current i_{Th1}. At t_3, $i_c = i_{\text{Th1}}$ and thyristor Th1 is turned-off as the net current through it is zero. As i_{Th1} reduces to zero diode D_1 begins to conduct the current $i_c - i_0$ and keeps thyristor Th1 reverse

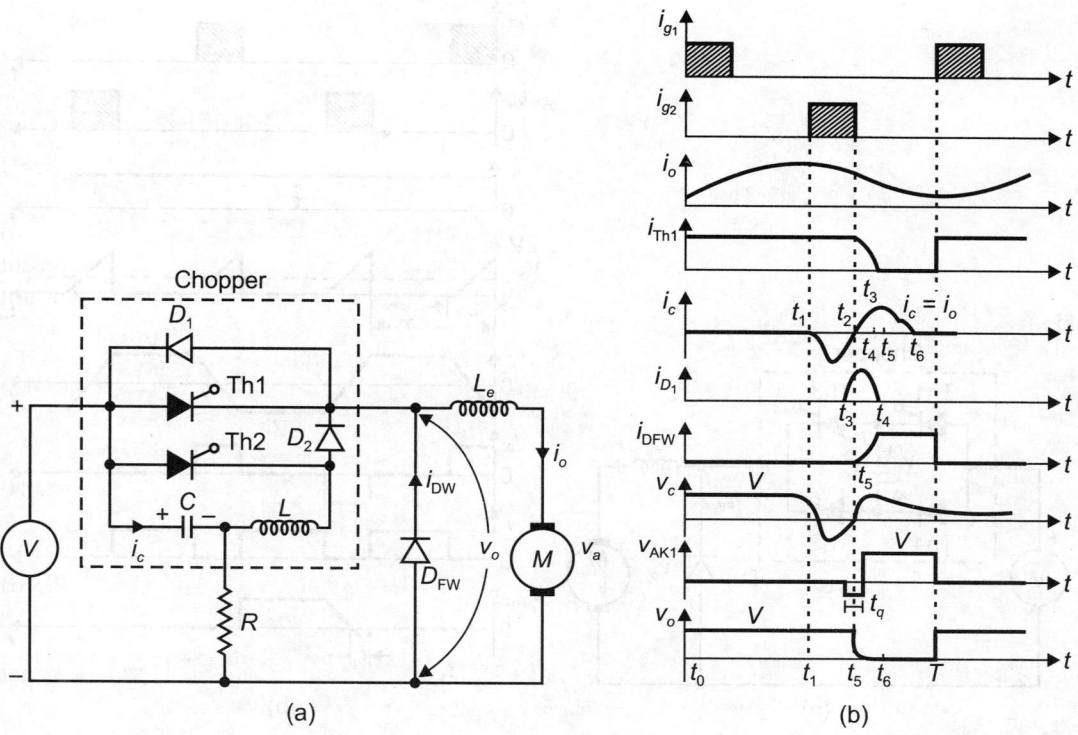

Fig. 9.19 The power circuit for current commutated chopper

biased. At t_4, $i_c = i_0$ and $i_{D_1} = 0$. At t_5, D_{FW} starts to conduct i_0. Capacitor current i_c decays to zero at t_6. This chopper has been used in traction cars. Commutation is reliable here and the capacitor always remains charged with the right polarity.

Advantages:

1. In this type of chopper, commutation is reliable so long as the load current is less than the peak commutating current.
2. Capacitor is always charged with correct polarity.
3. Auxiliary thyristor Th2 is naturally commutated as its commutating current passes through zero value in the oscillatory circuit formed by L and C.

9.9.3 Load Commutated Chopper

Figure 9.20 (a) shows a load commutated chopper. The chopper has four thyristors that function both as main and commutation thyristors. Thyristor pairs alternately conduct the load current. The sequence of operation is as follows, and some important waveforms are shown in Fig. 9.20 (b).

When thyristor Th1, Th2 are conducting, these act as main thyristors and thyristors Th3, Th4 and C as commutating components. Likewise with the conduction of Th3, Th4, these become the main thyristors and Th1, Th2 and C as the commutating components. Initially the capacitor is charged to a voltage V with upper plate negative and lower plate positive as shown in Fig. 9.20 (a). In this chopper, the output voltage is varied by varying the chopper frequency. At $t = t_0$, thyristors Th1 and Th2 are triggered. Prior to this instant, the capacitor

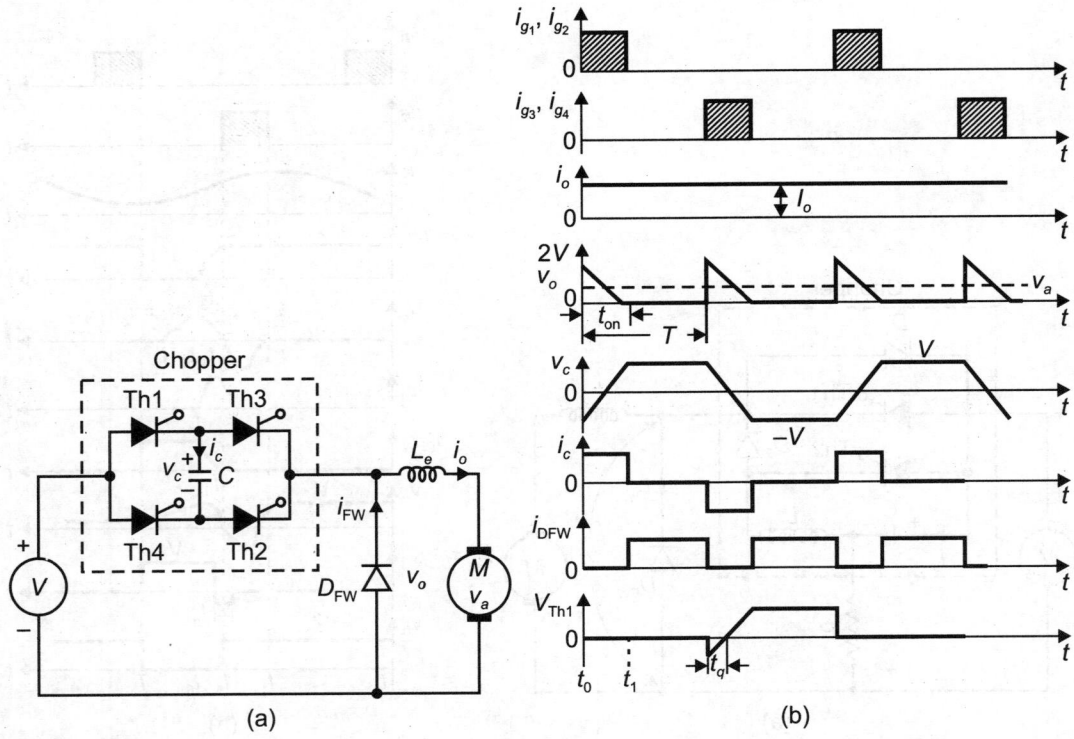

Fig. 9.20 Load commutated chopper and waveforms

voltage was negative ($v_C = -V$) due to conduction by thyristors Th3 and Th4. The load voltage jumps to 2 V. The load current now flows through V, Th1, C, Th2 and the load. The load current flowing through C charge it, and at t_1 the capacitor voltage becomes $+V$. The freewheeling diode D_{FW} becomes forward-biased, and the load current transferred from thyristors Th1 and Th2 to freewheeling diode D_{FW}. From t_1 onward, the freewheeling diode D_{FW} conducts the load current. At $t = T$, the other pair of thyristors Th3 and Th4 are triggered. This places the capacitor across thyristors Th1 and Th2, reverse-biases them, and turns them off. The cycle repeats.

The output voltage in this chopper is varied by changing the chopper frequency. It is, therefore, a frequency-modulated chopper.

Design of commutating capacitance: For a constant load current I_0, the on time of the chopper

$$t_{ON} = \frac{2VC}{I_0} \qquad \qquad ...(9.48)$$

The output voltage $\quad V_o = V\dfrac{t_{ON}}{T} = Vt_{ON}f$

$$= \frac{2V^2Cf}{I_0} \qquad \qquad ...(9.49)$$

where, $\qquad\qquad\qquad f$ = frequency of the chopper.

At the maximum, chopper frequency

$$f = f_{max} \qquad \qquad ...(9.50)$$

$$V_0\big|_{f_{max}} = V$$

From Equations (9.49) and (9.50),

$$V = \frac{2V^2C}{I_0} f_{max} \qquad \qquad ...(9.51)$$

or

$$f_{max} = \frac{I_0}{2VC} \qquad \qquad ...(9.52)$$

The value of the capacitor is chosen for maximum load current I_{om}. From Equation (9.21), this value is

$$C = \frac{I_{om}}{2Vf_{max}} \qquad \qquad ...(9.53)$$

This chopper has very good features. It is capable of commutating any amount of current. It does not require a commutating conductor that is normally costly and noisy. It can operate at very high frequencies in the order of kilocycles.

Disadvantages: This chopper has several minor disadvantages:

1. Peak load voltage is twice the supply voltage but load filtering may reduce this peak voltage.
2. For high power applications, efficiency may become low because of higher switching losses at high frequencies and losses in the two conducting thyristors in series with the load.
3. The freewheeling diode must be a fast recovery type because twice the supply voltage appears across it in a short time.

Advantages: This chopper has some very good features:

1. It is capable of commutating any amount of current.
2. It does not require a commutating inductor which is normally costly and noisy.
3. It can operate at high frequencies of the order of kilocycles and therefore filtering requirements to smooth out motor current are minimal.

9.10 MULTIPHASE CHOPPER

A multiphase chopper is one that consists of two or more choppers, connected in parallel. The two chopper configuration shown in Fig. 9.21 (a) is called a two-phase chopper. If two or more choppers are operated in parallel and phase shifted from each other, the ripple amplitude decreases and ripple frequency increases. As a result, the supply harmonic current is reduced. A two-chopper arrangement is shown in Fig. 9.21 (a), with waveform for α in phase and phase shift operation in Figs 9.21 (b) and (c). Multiphase chopper operation may be advantageous for larger drives, particularly, if the load current is large. Use of single chopper in such a case will involve the task of matching the static and dynamic conduction characteristics of the devices connected in parallel. In a multiphase chopper system, separate inductors provide an effective means for paralleling. The disadvantages of multiphase chopper system are: (1) the need of additional motor connections, (2) additional external inductors, (3) additional commutating components, and (4) additional complexity in the control logic. A multiphase chopper is used where large load current is required. Its input current has reduced ripple amplitude and increased ripple frequency.

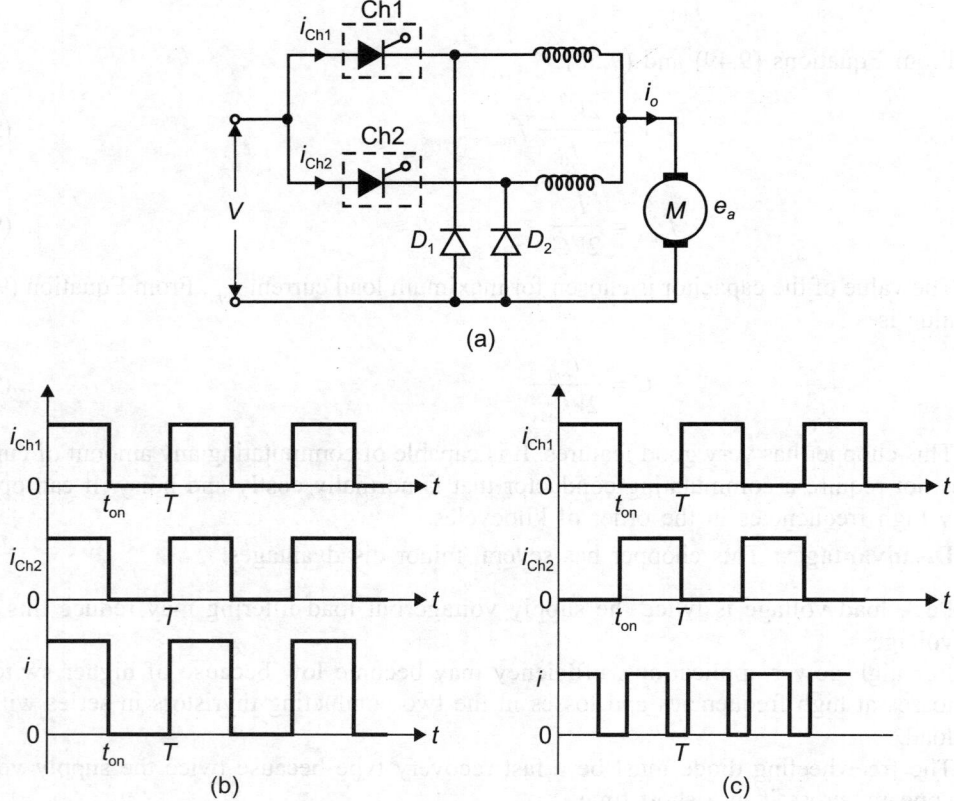

Fig. 9.21 Multiphase chopper and waveforms

9.11 EFFECTS OF SOURCE AND LOAD INDUCTANCE

The effect of source inductance plays an important role on the operation of chopper and this inductance should be as small as possible to limit the transient voltage within an acceptable level. It is clear that the commutation capacitor is overcharged due to the source inductance L_s (Fig. 9.18) and the semiconductor devices will be subjected to capacitor voltages. If the minimum value of source inductance cannot be generated an input filter is required. In practice the stray inductance always exists and its value depends on the type of wiring and layout of components. Because of source inductance L_s, the capacitor gets always overcharged.

Due to the presence of source inductance L_s and diode D as in Fig. 9.18, the capacitor C also gets undercharged and this may cause commutation problem of chopper. Load ripple current is an inverse function of load inductance and chopping frequency. Hence, the peak load current is dependent on the load inductance therefore the performance of the chopper is influenced by the load inductance. A smoothing choice is normally connected in series with the load to limit the ripple current.

9.12 SWITCHING MODE REGULATORS

DC choppers can be used as switching mode regulators to convert the unregulated DC input into a controlled DC output at a desired voltage level. The regulation is normally achieved by pulse width modulation at a fixed frequency and the switching device is normally a MOSFET or power BJT. The elements of switching mode regulators are shown in Fig. 9.22. Control voltage v_c is obtained by comparing the output voltage with the desired value. v_c can be compared with a saw tooth voltage v_r to generate the PWM control signal for the DC chopper. This is shown in Fig. 9.22(b).

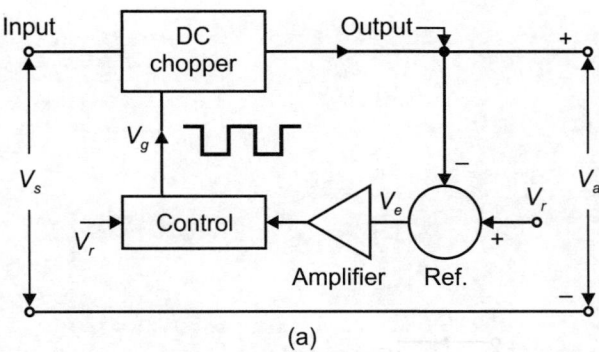

(a)

Advantages of switched mode regulators: The advantages of switched mode regulators over linear regulators are: (1) very less losses, (2) smaller size filter components, (3) light weight, and (4) noiseless operation.

There are four basic switching regulators:

1. Step-down or buck regulator
2. Step-up or boost regulator
3. Buck–boost regulator
4. Cuk regulator.

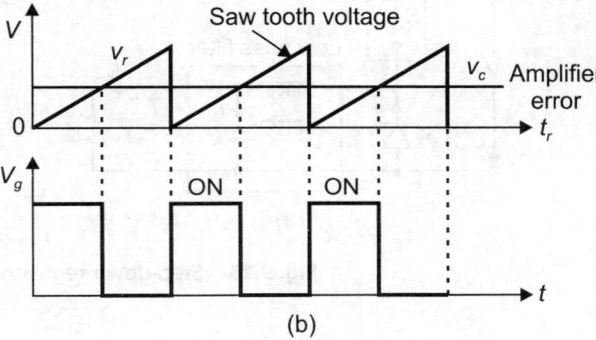

(b)

Fig. 9.22 Switching mode regulator

9.12.1 Step-Down or Buck Regulator

As the name applies, a step-down converter produces a lower average output voltage than the DC input voltage V_d. Its main application is the regulated DC power supply and speed control of DC motor.

The basic circuit of Fig. 9.22(a) constitutes a step-down regulator for a purely resistive load. Assuming the instantaneous output voltage depends on the switch position. From Fig. 9.23(b), the average output voltage can be calculated in terms of switch duty ratio.

$$V_0 = \frac{1}{T_s} \int_0^{T_s} v_i(t) \, dt = \frac{1}{T} \left(\int_0^{t_{ON}} V \, dt + \int_{t_{ON}}^{T} 0 \, dt \right)$$

$$= \frac{t_{ON}}{T} V = \delta V \qquad \qquad ...(9.54)$$

By varying the duty ratio (t_{ON}/T) of the switch V_0 can be controlled.

In actual application the circuit has *two drawbacks*: (1) In practice the load would be inductive. Even with a resistive load, there would always be certain associated stray inductance. This means that the switch would have to absorb the inductive energy and therefore it may be destroyed. (2) The output voltage fluctuates between 0 and V which is not acceptable in most applications. The problem of stored inductive energy is overcome by using a diode as shown in Fig. 9.23(a). The output voltage fluctuations are very much diminished by using a low pass filter consisting of an inductor and a capacitor. Figure 9.23(b) shows the waveforms of input voltage v_i, input current i_s, load current i_0, i_D, i_C, etc.

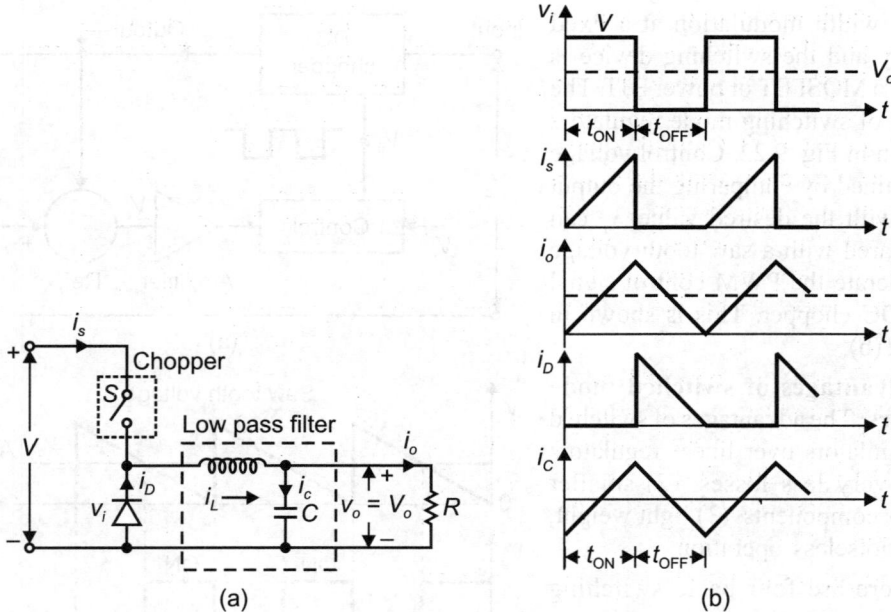

Fig. 9.23 Step-down regulator and waveforms

9.12.2 Step-Up or Boost Regulator

In a boost regulator, the output voltage is greater than the input voltage. Figure 9.24 (a) shows a step-up converter. When the switch is on, the diode is reverse-biased, thus isolating the output stage. The input supplies energy to the inductor. When the switch is off, the output stage receives energy from the inductor as well as from the input. In the steady state analysis, the output filter capacitor is assumed to be large to ensure constant output voltage $v_0 \approx V_0$.

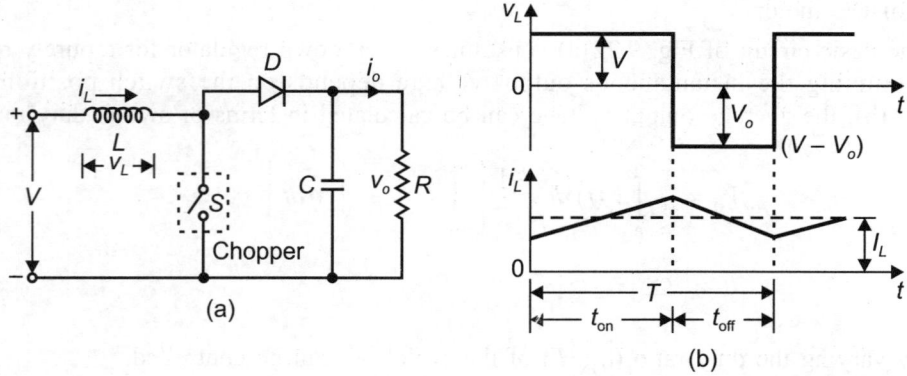

Fig. 9.24 Step-up regulator

Figure 9.24 (b) shows the steady state waveforms for this mode of conduction when the inductor current flows continuously $i_L > 0$. Since in steady state, the time integral of the inductor voltage over one time period must be zero.

$$Vt_{ON} + (V - V_0)\, t_{OFF} = 0 \qquad \qquad ...(9.55)$$

Dividing both sides by T and rearranging the terms, we have

or
$$\frac{V_0}{V} = \frac{T}{t_{OFF}} = \frac{1}{1-\delta} \qquad ...(9.56)$$

Assuming a lossless circuit,

$$P = P_0$$
$$VI = V_0 I_0$$

and
$$\frac{I_0}{I} = (1-\delta). \qquad ...(9.57)$$

9.12.3 Buck–boost Regulator

The buck–boost regulator provides output voltage which may be less than or greater than the input voltage, hence the name buck–boost. The output voltage polarity is opposite to that of input voltage. This regulator is known as *inverting regulator*.

A buck–boost converter can be obtained, by cascade connection of the two basic converters: (1) the step-down, and (2) step-up. In steady state, the output to input voltage conversion ratio is the product of conversion ratios of the two converters in cascade assuming that switches in both converters have the same duty ratio.

$$\frac{V_0}{V} = \delta \cdot \frac{1}{1-\delta} \qquad ...(9.58)$$

This allows the output voltage to be higher or lower than the input voltage based on duty ratio δ. The cascade connection of step-down and step-up converters can be combined into a single buck–boost converter shown in Fig. 9.25 (a). When the switch is closed, the input provides energy to the inductor and the diode is reverse-biased. When the switch is open, the energy stored in the inductor is transferred to the output. No energy is supplied by the input during this interval. In the steady state analysis output capacitor is assumed to be very large, which results in a constant output voltage $v_0(t) = V_0$.

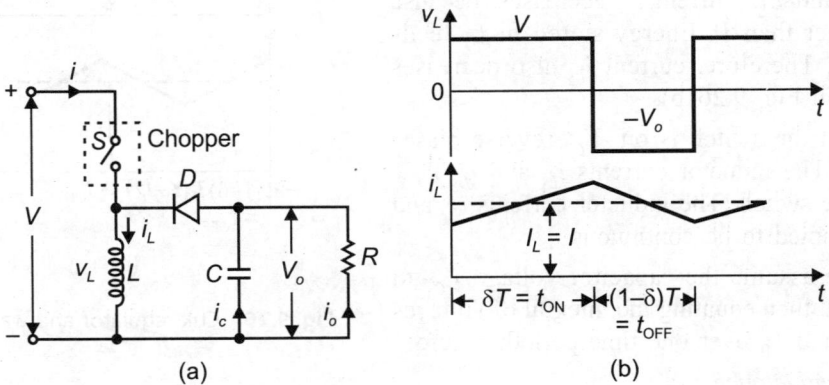

(a)   (b)

Fig. 9.25 Buck-boost regulator and waveforms

In continuous conduction mode, the inductor current flows continuously as shown in Fig. 9.25 (b).

Equating the integral of the inductor voltage over one time period to zero.

$$V\delta T + (-V_0)(1-\delta)T = 0$$

and
$$\frac{I_0}{I} = \frac{(1-\delta)}{\delta} \text{ assuming } (P = P_0)$$

Equation $\frac{V_0}{V} = \frac{\delta}{(1-\delta)}$ implies that depending on the duty ratio, the output voltage can be higher or lower than the input.

9.12.4 Cuk Regulator

Cuk regulator provides an output voltage which is less than or greater than the input voltage but output voltage polarity is opposite to that of input voltage.

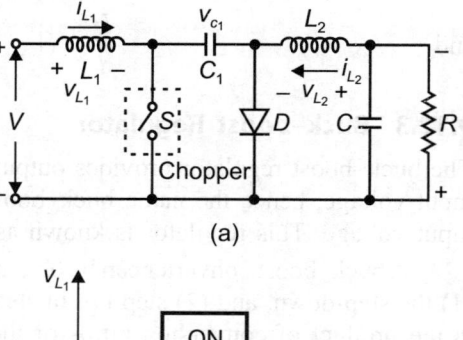

Here the capacitor C_1 acts as the primary means of storing and transferring energy from the input to the output.

In steady state, the average inductor voltages V_{L_1} and V_{L_2} are zero. Therefore, inspection of Fig. 9.26(a) results

$$V_{C_1} = V + V_0$$

Therefore, V_{C_1} is greater than both V and V_0. Assuming C_1 to be sufficiently large, in steady state the variation in v_{C_1} from its average value V_{C_1} can be assumed to negligibly small, i.e. $v_{C_1} = V_{C_1}$, even though it stores and transfers energy from the input to the output.

When the switch is off, the inductor currents i_{L_1} and i_{L_2} flow through the diode. Capacitor C_1 is charged through diode by energy from both the input and L_1. Current i_{L_1} decreases, because V_{C_1} is larger than V. Energy stored in L_2 feeds the output. Therefore, current i_{L_2} also decreases as shown in Fig. 9.26(b).

When the switch is on, V_{C_1} reverse biases the diode. The inductor currents i_{L_1} and i_{L_2} flow through the switch. The inductor currents i_{L_1} and i_{L_2} are assumed to be continuous.

If we assume the capacitor voltage V_{C_1} to be constant, then equating the integral of voltages across L_1 and L_2 over one time period to zero.

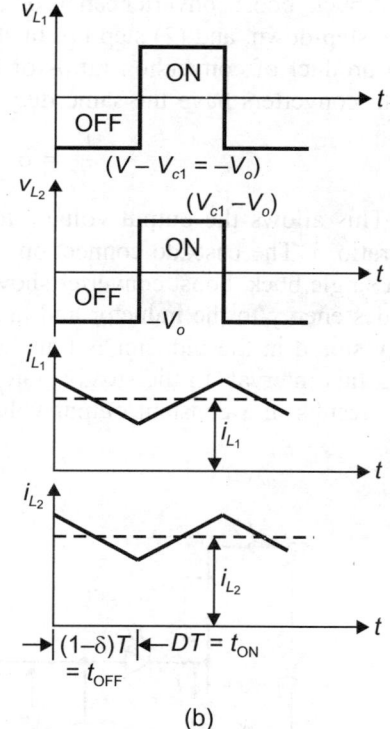

Fig. 9.26 Cuk regulator and waveforms

For inductance L_1:

$$V\delta T + (V - V_{C_1})(1 - \delta) T = 0$$
$$V\delta T + VT - V_{C_1}T - V\delta T + V_{C_1}\delta T = 0$$
$$V\delta T + VT - V\delta T = V_{C_1}(1 - \delta) T$$

$$V_{C_1} = \frac{V}{1-\delta} \qquad \qquad ...(9.59)$$

For inductance L_2:

$$(V_{C_1} - V_0) \, \delta T + (-V_0) \, (1 - \delta) \, T = 0$$

$$\therefore \qquad V_{C_1} = \frac{1}{\delta} V_0 \qquad \qquad ...(9.60)$$

From Equations (9.59) and (9.60)

$$\frac{V_0}{V} = \frac{\delta}{1 - \delta} \qquad \qquad ...(9.61)$$

Assuming

$$\frac{I_0}{I} = \frac{1 - \delta}{\delta}$$

where, $\qquad \qquad I_{L_1} = I$ and $I_{L_2} = I_0$

An advantage of this circuit is that both the input current and the current feeding the output stage are reasonably ripple free.

SOLVED EXAMPLES

Example 9.1 A dc chopper is turned-on for 30 μsec and off for 10 μsec. Find its: (i) duty cycle, and (ii) chopping frequency.

Solution

(i) $\qquad \qquad t_{ON} = 30$ μsec, $t_{OFF} = 10$ μsec

Total time period $\qquad T = t_{ON} + t_{OFF} = (30 + 10) = 40$ μsec

Duty cycle (δ) $\qquad \delta = \dfrac{t_{ON}}{t_{OFF} + t_{ON}} = \dfrac{30 \times 10^{-6} \text{ sec}}{(30 \times 10^{-6} + 10 \times 10^{-6} \text{ sec})}$

$$\delta = \frac{3}{4} = 0.75$$

(ii) Chopper frequency $f = \dfrac{1}{T} = \dfrac{1}{40 \times 10^{-6} \text{ sec}} = 25$ kHz.

Example 9.2 A dc chopper of input voltage 200 V remains on for 25 msec and off for 10 msec. Determine the average voltage which appears across the load.

Solution

$$t_{ON} = 25 \text{ msec} = 25 \times 10^{-3} \text{ sec}$$

$$t_{OFF} = 10 \text{ msec} = 10 \times 10^{-3} \text{ sec}$$

Total time period $\qquad T = t_{ON} + t_{OFF} = (25 + 10) \times 10^{-3}$ sec $= 35 \times 10^{-3}$

Duty cycle $\qquad \delta = \dfrac{t_{ON}}{T} = \dfrac{25 \times 10^{-3}}{35 \times 10^{-3}} = 0.71$

Voltage across load $\qquad V_L = \delta V = 0.7 \times 200 = 140$ V.

Example 9.3 A step-up chopper has input voltage of 220 V and output voltage of 660 V. If the nonconducting time of thyristor is 100 μsec. Compute the pulse width of output voltage. If the pulse width is halved for a constant frequency operation, find the new output voltage.

Solution $\qquad \qquad V_0 = V \cdot \dfrac{1}{1 - \delta}$

or $\qquad \qquad 660 = \dfrac{220 \times 1}{1 - \delta}$ or $\delta = \dfrac{2}{3} = \dfrac{t_{ON}}{T}$

or
$$t_{ON} = \frac{2T}{3}$$

and
$$t_{OFF} = T - t_{ON} = T - \frac{2}{3}T = \frac{1}{3}T$$

or
$$100 \ \mu sec = \frac{1}{3}T$$

or
$$T = 300 \ \mu sec$$

and
$$t_{ON} = \frac{2}{3}T = \frac{2}{3} \times 300 = 200 \ \mu sec$$

When pulse width is halved, $t_{ON} = \frac{1}{2} \times 200 = 100 \ \mu sec$ for constant frequency operation, $T = 300 \ \mu sec$

$$t_{OFF} = T - t_{ON} = 300 - 100 = 200 \ \mu sec$$

$$\delta = \frac{t_{ON}}{T} = \frac{100}{300} = \frac{1}{3}$$

New output voltage, $V_0 = \dfrac{V \times 1}{1 - \delta} = \dfrac{220 \times 1}{1 - 1/3} = 330 \ V.$

Example 9.4 A step-up chopper is required to deliver a load voltage of 660 V from a 220 V DC source. If the nonconduction time of the thyristor is 100 μsec, find the required pulse width.

Solution As
$$V_0 = V \frac{t_{ON} + t_{OFF}}{t_{OFF}}$$

or
$$\frac{V_0}{V} = \frac{t_{ON} + t_{OFF}}{t_{OFF}} \Rightarrow \frac{660}{220} = \frac{t_{ON} + 100}{100}$$

or
$$t_{ON} + 100 = 300$$
$$t_{ON} = 200 \ \mu sec.$$

Example 9.5 A battery operated electric car is controlled by a chopper as shown in Fig. E9.5, the battery voltage is 100 V, starting current is 100 A, thyristor turn-off time is 20 μsec, chopper frequency is 400 Hz. Calculate the values of commutating capacitor C and commutating inductor L.

Solution For reliable commutation, let

$$\Delta t = t_{OFF} = 20 \ \mu sec,$$
Circuit turn-off time,
$$t_q = t_{OFF} + \Delta t$$
$$= 20 + 20$$
$$= 40 \ \mu sec$$

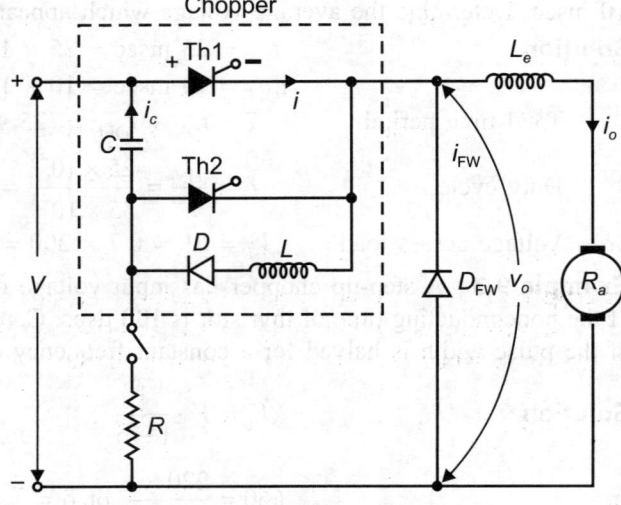

Fig. E9.5 Circuit diagram

From Equation (9.41), $C = \dfrac{I_{om}t_q}{V} = \dfrac{100 \times 40 \times 10^{-6}}{100} = 40\,\mu F$

Commutating inductor $L \geq C\left(\dfrac{V}{I_{om}}\right)^2$

$$L \geq 40 \times 10^{-6}\left(\dfrac{100}{100}\right)^2 > 40\mu H$$

Now the chopper period $T = 2.5$ msec

From the relation $L \leq \dfrac{0.01T^2}{\pi^2 C}$ Henry

$$L \leq 0.01\dfrac{(2.5 \times 10^{-3})^2}{\pi^2(40 \times 10^{-6})}\ \text{Henry}$$

$$L \leq 158.5\,\mu H.$$

The value of commutating inductor is in the range $40 \leq L \leq 158.5\,\mu H$. A value close to $40\,\mu H$ will be a good choice because it will allow the minimum load voltage about 50% of supply voltage, thereby giving a wider range of variation in the load voltage.

Example 9.6 In Example 9.5, if the chopper used is as load commutated chopper as shown in Fig. E9.6 and maximum chopper frequency is 5 kHz. Calculate the value of the commutating capacitor C.

Solution From the relation

$$C = \dfrac{I_{om}}{2Vf_{max}}$$

$$C = \dfrac{100}{2 \times 100 \times 5000} = 100\,\mu F.$$

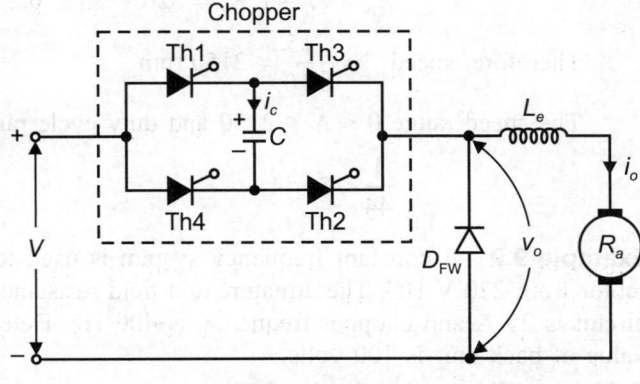

Fig. E9.6 Circuit diagram

Example 9.7 The speed of a separately excited DC motor is controlled by a chopper. The supply voltage is 120 V, armature circuit resistance is $R_a = 0.5\,\Omega$, armature circuit inductance is $L_e = 20$ mH and motor constant is $K_a\Phi = 0.05$ V/rpm. The motor drives a constant torque load requiring an average armature current of 20 A. Assume motor current is continuous. Determine: (i) the range of speed control, (ii) range of the duty cycle δ.

Solution

(i) Minimum speed is zero at which back emf $E_b = 0$.

∴ $\qquad V_0 = I_a R_a = 20 \times 0.5 = 10$ V

From the relation $\qquad V_0 = \delta V$ or $10 = 120\delta$

or $\qquad\qquad\qquad \delta = \dfrac{1}{12}$

Maximum speed corresponds to $\delta = 1$ at which

$$V_0 = V = 120\ \text{V}$$

Therefore, $\qquad E_b = V_0 - I_a R_a = 120 - (20 \times 0.5) = 110$ V

(ii) Speed $N = \dfrac{E_b}{K_a \Phi} = \dfrac{110}{0.05} = 2200$ rpm. The range of speed is $0 < N < 2200$ rpm and range

of duty cycle is $\dfrac{1}{12} < \delta < 1$.

Example 9.8 A DC chopper is used to control the speed of a separately excited DC motor. The DC supply voltage is 220 V, armature resistance $R_a = 0.2\,\Omega$ and motor constant $K_a \Phi = 0.10$ V/rpm. The motor drives a constant torque load requiring an average armature current of 25 A. Determine: (i) the range of duty cycle δ, and (ii) the range of speed control. Assume the motor current to be continuous.

Solution When back emf is zero, speed is zero

$$V_0 = E_b + I_a (R_a + R_{se}) = 0 + 25 \times 0.2 = 5 \text{ V}$$

$$V_0 = \frac{V \cdot t_{ON}}{T} = V \cdot \delta$$

$$\delta = \frac{V_0}{V} = \frac{5}{220} = \frac{1}{44}$$

The speed will be maximum when total voltage V is applied, i.e. $\delta = 1$

$$E_b = V_0 - I_a R_a = 220 - 25 \times 0.2 = 215 \text{ V}$$

Therefore, speed $\quad = \dfrac{215}{0.10} = 2150$ rpm

The speed range $0 < N < 2150$ and duty cycle range

$$\frac{1}{44} < \delta < 1.$$

Example 9.9 A constant frequency system is used to control the speed of a DC traction motor from 220 V DC. The armature and field resistance is $0.2\,\Omega$. The average current in the circuit is 25 A and chopper frequency is 400 Hz. Determine the pulse width if the average value of back emf is 100 volts.

Solution The average load voltage

$$\begin{aligned} V_0 &= E_b + I_a (R_a + R_{se}) \\ &= 100 + 25 \times 0.2 = 105 \text{ V} \\ &= V \cdot \frac{t_{ON}}{T} = V \cdot t_{ON} \cdot f \end{aligned}$$

where f is the chopping frequency $\left(f = \dfrac{1}{T} \right)$

$$105 = 220 \cdot t_{ON} \times 400$$

$$t_{ON} = \frac{105 \times 1000}{220 \times 400} = 1.19 \text{ msec.}$$

Example 9.10 An electric rapid transit system uses a DC series motor as shown in Fig. E9.10. The solid-state speed controller uses a Jones chopper circuit. The motor is rated at 200 HP, 1000 V, 1000 rpm, armature resistance $R_a = 0.05\,\Omega$, inductance in series with the armature is L_e (armature inductance, series field inductance and external inductance), and efficiency of the motor is 90%. The chopper frequency is 500 Hz. Find the inductance L_e required to limit the current swing in the armature under the worst condition to 5 amperes.

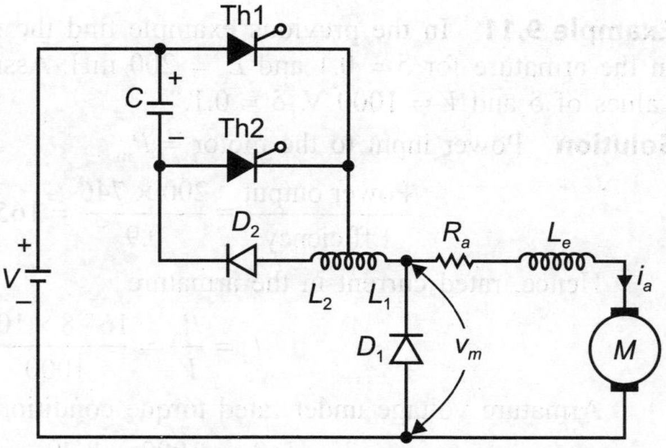

Solution The voltage

$$V_0 = \frac{t_{ON}}{t_{ON} + t_{OFF}} \cdot V = V \cdot \delta$$

The voltage across the inductor L_e is given by $(V - \delta V)$.

Fig. E9.10 Circuit diagram

Then,

$$L_e \frac{di_a}{dt} = (V - \delta V)$$

or

$$di_a = \frac{(V - \delta V)}{L_e} dt$$

From Fig. E9.10, $dt = t_{ON}$

$\therefore$

$$di_a = \frac{V - \delta V}{L_e} t_{ON} \qquad \ldots(i)$$

The worst condition of current swing would depend on a particular value of δ. Equation (i) could be written as

$$di_a = \frac{V - \delta V}{L_e} \cdot \frac{t_{ON}}{T} \cdot T$$

or

$$di_a = \frac{V - \delta V}{L_e} \delta \cdot T \qquad \ldots(ii)$$

Differentiating the above equation with respect to δ

$$\frac{di_a}{d\delta} = \frac{V}{L_e} \cdot T - \frac{2\delta V}{L_e} \cdot T$$

The worst condition $\dfrac{di_a}{d\delta} = 0$

or

$$(1 - 2\delta)\frac{V_B}{L_e} \cdot T = 0$$

$\therefore$

$$2\delta = 1$$

or

$$\delta = 0.5 \text{ is the worst condition.}$$

Using $\delta = 0.5$ in Equation (ii)

$$di_a = 5 = \frac{1000 - 0.5 \times 1000}{L_e} \times 0.5 \times 2 \times 10^{-3}$$

$\therefore$

$$L_e = \frac{500 \times 0.5 \times 2 \times 10^{-3}}{2} = 100 \text{ mH}.$$

Example 9.11 In the previous example find the steady state speed and the current swing in the armature for $\delta = 0.1$ and $L_e = 200$ mH. Assume rated current in the armature for all values of δ and $V = 1000$ V, $\delta = 0.1$.

Solution Power input to the motor = P_{in}

$$\frac{\text{Power output}}{\text{Efficiency}} = \frac{200 \times 746}{0.9} = 165.8 \text{ kW}$$

Hence, rated current in the armature

$$I_a = \frac{P_{in}}{V} = \frac{165.8 \times 10^3}{1000} = 165.8 \text{ A}$$

Armature voltage under rated torque condition is

$$V - I_a R_a = 1000 - 8.29 = 991.71 \text{ V}$$

At the above armature voltage, the speed is 1000 rpm.

For $\delta = 0.1$, voltage at the armature is given by

$$\delta \times 1000 - I_a R_a = 100 - 8.29 = 91.71 \text{ V}$$

∴ Speed of the motor, $N = \dfrac{91.71 \times 1000}{991.71} = 92.5$ rpm

Current swing $\Delta i_a = \dfrac{V - \delta V}{L_e} \cdot \delta T$

$$= \frac{1000 - 0.1 \times 1000 \times 0.1 \times 2 \times 10^{-3}}{200 \text{ mH}}$$

$$\Delta i_a = 0.9 \text{ A}.$$

Example 9.12 In the two quadrant chopper circuit, $V = 110$ V, $L = 0.2$ mH, $R = 0.25\,\Omega$, $E = 40$ V, $T = 2500$ μsec, $t_{ON} = 1250$ μsec. Calculate the average output current I_0 and average output voltage V_0. Calculate the maximum and minimum value of instantaneous output current I_{max} and I_{min}.

Solution Using equation

$$V_0 = \frac{t_{ON}}{T} V = \frac{1250}{2500} \times 110 = 55 \text{ V}$$

$$I_0 = \frac{V_0 - E_a}{R} = \frac{55 - 40}{0.25} = 60 \text{ A}$$

We know $\tau = \dfrac{L}{R} = \dfrac{0.2 \times 10^{-3}}{0.25} = 800 \, \mu\text{sec}$

Also $\dfrac{t_{ON}}{\tau} = \dfrac{1250}{800} = 1.562$

$$\frac{T}{\tau} = \frac{2500}{800} = 3.125$$

Using the equations $I_{max} = \dfrac{V}{R} \dfrac{[1 - e^{-t_{ON}/\tau}]}{[1 - e^{-T/\tau}]} - \dfrac{E_a}{R} = \dfrac{110}{0.25} \dfrac{(1 - e^{-1.562})}{(1 - e^{-3.125})} - \dfrac{40}{0.25} = 204 \text{ A}$

$$I_{min} = \frac{V}{R} \frac{[e^{-t_{ON}/\tau} - 1]}{[e^{T/\tau} - 1]} - \frac{E_a}{R} = \frac{110}{0.25} \frac{(e^{1.562} - 1)}{(e^{3.125} - 1)} - \frac{40}{0.25} = -83.6 \text{ A}.$$

Example 9.13 In a buck–boost regulator operating at 20 kHz, L is 0.05 mH. The output capacitor is sufficiently large and $V = 15$ V. The output is to be regulated at 10 V and the converter is supplying a load of 10 W. Calculate the initial estimate of duty ratio δ.

Solution
$$I_0 = \frac{P}{V} = \frac{10}{10} = 1 \text{ A}$$

Assuming current at the edge of continuous conduction
$$\frac{V_0}{V} = \frac{\delta}{1-\delta} = \frac{10}{15}$$
or
$$\delta = 0.4.$$

Example 9.14 The two quadrant chopper shown in Fig. E9.14 is used to control the speed of DC motor and also for regenerative braking of motor. The motor constant is $k\Phi = 0.1$ V/rpm. The chopping frequency is $f_C = 250$ Hz and motor armature resistance is $R_a = 0.2\,\Omega$. The inductance L_a is sufficiently large and the motor current i_0 can be assumed to be ripple free. The supply voltage is 120 V. (a) Chopper S_1 and diode D_1 are connected to control the speed of motor. At $n = 400$ rpm and $i_0 = 100$ A (ripple free). (i) determine the turn-on time t_{ON} of the chopper. (ii) find the power developed by the motor absorbed by R_a.

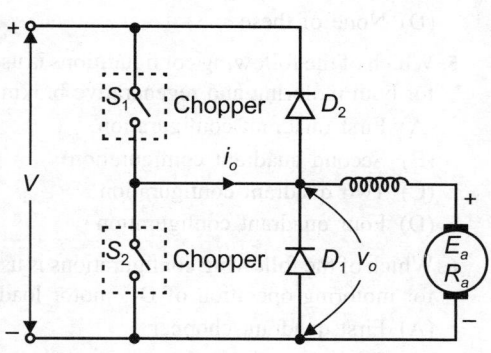

Fig. E9.14 Circuit diagram

Solution

(i) We know
$$E_a = K\Phi n = 0.1 \times 400 = 40 \text{ V}$$
$$V_0 = E_a + I_a R_a = 40 + 100 \times 0.2 = 60 \text{ V}$$
$$V_0 = \frac{t_{ON}}{T} V = \frac{t_{ON}}{T} \times 120 = 60$$
or
$$t_{ON} = \frac{T}{2}$$

(ii) Power absorbed by motor
$$P_m = E_a I_0 = 40$$
$$= 40 \times 100 = 4000 \text{ W}$$
Power absorbed $\quad P_a = i_0^2 R_a = 100^2 \times 0.2 = 2000 \text{ W}.$

EXERCISES

Multiple Choice Questions

1. DC chopper converts:
   - (A) DC to DC
   - (B) AC to DC
   - (C) DC to AC
   - (D) none of these

2. The features of the chopper drives are:
   - (A) smooth control but slow response
   - (B) smooth control but fast response
   - (C) fast response with smooth control but less efficient
   - (D) none of these

3. Choppers can be used in future electric automobiles:
   (A) for speed control only
   (B) for braking only
   (C) for speed control and braking
   (D) none of these

4. Which of the following system is preferred for chopper drives?
   (A) Constant frequency system
   (B) Variable frequency system
   (C) Constant voltage system
   (D) None of these

5. Which of the following configurations is used for both motoring and regenerative braking?
   (A) First quadrant configuration
   (B) Second quadrant configuration
   (C) Two quadrant configuration
   (D) Four quadrant configuration

6. Which of the following configurations is used for motoring operation of DC motor load?
   (A) First quadrant chopper
   (B) Second quadrant chopper
   (C) Fourth quadrant chopper
   (D) None of these

7. If two or more choppers are operated in parallel and shifted from each other, the ripple frequency will:
   (A) remain same   (B) increase
   (C) decrease   (D) none of these

8. Which of the following systems has the greater possibility of interference with signalling and telephone lines?
   (A) Constant frequency system
   (B) Variable frequency system
   (C) Both are correct
   (D) None of these

9. Constant frequency system gives:
   (A) low ripple and fast response
   (B) high ripple and slow response
   (C) low ripple and slow response
   (D) none of these

10. Chopper controlled DC motor used in underground traction with regenerative braking, the power consumption will be reduced to:
    (A) 35–40%   (B) 50–60%
    (C) 60–70%   (D) none of these

11. In DC choppers, if T_{ON} is the on-period and f is the chopping frequency, then output voltage in terms of input voltage V is given by:
    (A) $V \cdot T_{ON}/f$ (B) $V \cdot f/T_{ON}$
    (C) $V/f \cdot T_{ON}$ (D) $V \cdot f \cdot T_{ON}$

12. In DC choppers, the waveforms for input and output voltages are respectively:
    (A) discontinuous, continuous
    (B) both continuous
    (C) both discontinuous
    (D) continuous, discontinuous

13. For type A chopper, V is the source voltage, R is the load resistance and δ is the duty cycle. The average output voltage for this chopper is:
    (A) δV (B) $(1 - \alpha) V$
    (C) V/α (D) $V/(1 - \alpha)$

14. A chopper has V as the source voltage, R as the load resistance and δ as the duty cycle. For this chopper, rms value of output voltage is:
    (A) δV (B) $\sqrt{\delta} \cdot V$
    (C) $V/\sqrt{\delta}$ (D) $\sqrt{1 - \delta} \cdot V$

15. For a chopper, V is the source voltage, R is the load resistance and δ is the duty cycle. Rms and average values of thyristor currents for this chopper are:
    (A) $\delta \cdot \dfrac{V}{R}, \sqrt{\delta} \cdot \dfrac{V}{R}$ (B) $\sqrt{\delta} \cdot \dfrac{V}{R}, \sqrt{\delta} \cdot \dfrac{V}{R}$
    (C) $\sqrt{\delta} \dfrac{V}{R}, \delta \dfrac{V}{R}$ (D) none of these

16. In DC chopper, per unit ripple is maximum when duty cycle 'δ' is:
    (A) 0.2   (B) 0.5
    (C) 0.7   (D) none of these

17. A thyristor chopper is preferred over other choppers because it:
    (A) provides static switching
    (B) is economical
    (C) can be easily manufactured
    (D) none of these

18. Duty cycle of chopper circuit is expressed by:
    (A) $t_{ON}/(t_{ON} + t_{OFF})$ (B) t_{ON}/t_{OFF}
    (C) $\dfrac{t_{OFF}}{t_{ON}}$ (D) T/t_{ON}

19. Chopper frequency is:

(A) $\dfrac{1}{t_{ON}}$

(B) $\dfrac{1}{t_{ON} + t_{OFF}}$

(C) $\dfrac{t_{OFF}}{t_{ON}}$

(D) none of these

20. Load voltage of DC chopper for duty cycle δ, and input voltage V:

(A) $\delta \cdot V$

(B) V/δ

(C) δ/V

(D) none of these

21. In single thyristor chopper circuit, the capacitor discharge current is:

(A) oscillating but decaying in nature

(B) pulsating in nature

(C) alternating in nature

(D) none of these

22. In DC chopper circuit, the load voltage is governed by:

(A) number of thyristors used in the circuit

(B) duty cycle of the circuit

(C) DC voltage applied to circuit

(D) none of these

23. A voltage commutated chopper has the following parameters:

$V = 200$ V, load circuit parameter: 1Ω, 2 mH, 50 V

Commutation circuit parameters, $L = 25\,\mu H$, $C = 50\,\mu F$.

For a constant load current of 100 A, the effective on-period and peak current through the main thyristor are respectively:

(A) 1000 μs, 200 A

(B) 700 μs, 382.8 A

(C) 700 μs, 282.8 A

(D) none of these

24. A load commutated chopper, fed from 200 V DC source, has a constant load current of 50 A. For a duty cycle of 0.4 and a chopping frequency of 2 kHz, the value of commutating capacitor and the turn-off time for one thyristor pair are respectively:

(A) 25 μF, 50 μs

(B) 50 μF, 50 μs

(C) 25 μF, 25 μs

(D) 50 μF, 25 μs

25. A DC battery is charged from a constant DC source of 200 V through a chopper. The DC battery is to be charged from its internal emf of 90 to 120 V. The battery has internal resistance of 1Ω. For constant charging current of 10 A, the range of duty cycle is:

(A) 15 to 65

(B) 65 to 0.8

(C) 0.8 to 0.95

(D) none of these

26. For type A chopper V, R, I_0 and δ are respectively the DC source voltage, load resistance, constant load current and duty cycle. For this chopper, average and rms values of freewheeling diode currents are:

(A) $\delta I_0, \sqrt{\delta} \cdot I_0$

(B) $(1 - \delta) I_0, \sqrt{1 - \delta} \cdot I_0$

(C) $\delta \cdot V/R, \sqrt{\delta} \cdot V/R$

(D) none of these

27. A step-up chopper has V as the source voltage and δ as the duty cycle. The output voltage for this chopper is given by:

(A) $V (1 + \delta)$

(B) $V/(1 - \delta)$

(C) $V (1 - \delta)$

(D) none of these

28. A DC chopper is fed from 100 V DC. Its load voltage consists of rectangular pulses of duration 1 msec in an overall cycle time of 3 msec. The average output voltage and ripple factor for this chopper are respectively:

(A) 25 V, 1

(B) 50 V, 1

(C) 33.33, 1.5

(D) none of these

29. When a series LC circuit is connected to a DC supply of V volts through a thyristor, then the peak current through thyristor is:

(A) $V \cdot \sqrt{CL}$

(B) $V/\sqrt{CL}$

(C) $V \cdot \sqrt{C/L}$

(D) $V \cdot \sqrt{L/C}$

30. In DC choppers if T is the chopping period, then output voltage can be controlled by PWM by varying:

(A) T keeping T_{ON} constant

(B) T_{ON} keeping T constant

(C) T keeping T_{OFF} constant

(D) none of these

31. In DC choppers, for chopping period T, the output voltage can be controlled by FM by varying:

(A) T keeping T_{ON} constant

(B) T_{ON} keeping T constant

(C) T_{OFF} keeping T constant

(D) none of these

ANSWER KEY

1. (A)	**2.** (B)	**3.** (C)	**4.** (A)	**5.** (C)	**6.** (A)	**7.** (B)	**8.** (B)
9. (A)	**10.** (A)	**11.** (D)	**12.** (D)	**13.** (C)	**14.** (B)	**15.** (C)	**16.** (B)
17. (A)	**18.** (A)	**19.** (B)	**20.** (A)	**21.** (A)	**22.** (B)	**23.** (B)	**24.** (A)
25. (A)	**26.** (B)	**27.** (B)	**28.** (C)	**29.** (C)	**30.** (B)	**31.** (A)	

Review Questions

1. What is a DC chopper? Give various types of chopper configurations with appropriate figures.

2. Explain with suitable circuit, the operation of a switched mode power supply. What makes it superior to a power supply with a regulating transistor?

3. Draw the circuit diagram of Jones chopper. Explain its operation with particular emphasis on the commutation process.

4. Show that for a basic DC to DC converter, the critical inductance 'L' of the filter choke is given by $L = \dfrac{V_0^2 (V_1 - V_0)}{2fV_1P_0}$, where f is the switching frequency, V_0 and V_1 are the output and input voltages, P_0 is the output power.

5. Draw the circuit diagram with waveform and explain the principle of a parallel capacitor turn-off chopper.

6. Discuss the working of a load commutated chopper with waveform of voltage and current.

7. What is a multiphase chopper? Discuss the difference between in-phase operation and phase shifted operation of a multiphase chopper.

8. Discuss the advantages and disadvantages of multiphase chopper.

9. Explain the term: (i) duty cycle, and (ii) chopper frequency.

10. Explain the working principle of single thyristor chopper.

11. Explain with the help of a diagram the working of Morgan chopper circuit.

12. Derive an expression for the output voltage of a step-up chopper circuit.

13. State the advantages of Morgan chopper circuit.

AC Voltage Controllers

10.0 INTRODUCTION

AC voltage controllers convert a fixed AC supply into a variable AC supply. They are equivalent to autotransformers. AC controllers can be used to control the speed of induction motors. There are some applications of variable AC voltage at line frequency. The applications include control of heating in load circuit, controlled lighting, induction heating, control of single-phase and three-phase motors. The conversion equipments used for this purpose are called *AC voltage controllers*. The load voltage is varied in a stepless manner and flow of current in the load can be controlled. The source has fixed voltage and frequency, and load voltage and current are variable having the same frequency as the source. AC voltage controller is installed between the source and the load. Voltage controller provides a variable voltage across the load.

Industrial applications: There are some applications of variable AC voltage at line frequency. The applications include:

1. Lighting controls and for soft start of motors
2. Industrial heating
3. Induction heating of metals
4. Speed control of single-phase and three-phase AC drives
5. Transformer tap changing
6. Starting of three-phase induction motors
7. Primary transformer control for electro-chemical processes

Thyristorised AC voltage controllers have high efficiency, flexibility in control and require less maintenance. The main drawback of AC voltage controllers is the introduction of objectionable harmonics in the supply current and load voltage waveforms particularly at reduced output voltage levels. In this chapter, principle of working and gating signal requirements of AC voltage controller are discussed.

10.1 TYPES OF AC VOLTAGE CONTROLLERS

There are two methods of control:

1. ON-OFF control   2. Phase control

In ON-OFF control, the thyristors are employed as switches to connect the load circuit to the source for a few cycles of the source voltage and then disconnect it for a comparable period. The thyristor thus acts as high speed contactor. In phase control, the thyristors are employed as switches to connect the load circuit to the voltage source for a chosen portions of each cycle of the load voltage. Here more attention is concentrated on phase control methods.

10.2 AC PHASE VOLTAGE CONTROLLERS

AC voltage controllers are classified into two categories:

    1. Single-phase AC voltage controllers   2. Three-phase AC voltage controllers.

    The power circuits and output voltage or current waveforms of a half-wave and full-wave voltage controller supplying a resistive load are shown in Figs 10.1 (a) and (b). It is seen in Fig. 10.1 (a), that positive half-cycle is not identical with negative half-cycle. As a result the DC component is introduced in the supply and load circuit which is undesirable. The waveform of the full-wave controller has alternating symmetry and therefore has no direct component. For this reason the full-wave controller are more suited to practical circuits than single-phase half-wave circuits.

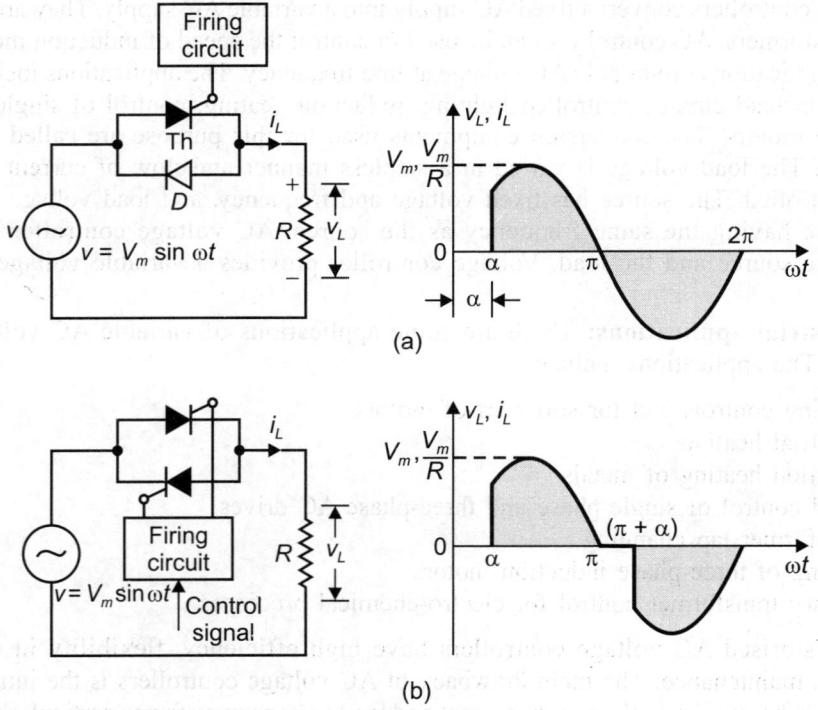

Fig. 10.1 Single-phase AC voltage controllers and waveforms

10.2.1 Single-phase Half-wave Controller with Resistive Load

The principle of phase control can be explained with reference to Fig. 10.1 (a). The power flow to the resistive load is controlled by firing the thyristor Th and the waveform for the load voltage is shown in Fig. 10.1 (a) which is asymmetrical and contains a DC component. Since the power flow is controlled during the positive half-cycle of input voltage, this is also known as *unidirectional controller*.

Mathematical analysis: Set the input voltage,

$$v = V_m \sin \omega t = \sqrt{2}V \sin \omega t$$

rms output load voltage V_L

$$= \left[\frac{1}{2\pi} \left\{ \int_{\alpha}^{\pi} (\sqrt{2}V \sin \omega t)^2 \, d\omega t + \int_{\pi}^{2\pi} (\sqrt{2}V \sin \omega t)^2 \, d\omega t \right\} \right]^{1/2}$$

$$= \left[\frac{2V^2}{4\pi} \left\{ \int_{\alpha}^{\pi} (1 - \cos 2\omega t) d\omega t + \int_{\alpha\pi}^{2\pi} (1 - \cos 2\omega t) d\omega t \right\} \right]^{1/2}$$

$$= V \left[\frac{1}{2\pi} \left(2\pi - \alpha + \frac{\sin 2\alpha}{2} \right) \right]^{1/2}$$

Average value of output voltage V_{av}

$$= \frac{1}{2\pi} \left[\int_{o}^{\alpha} \sqrt{2}V \sin \omega t \cdot d\omega t + \int_{\pi}^{2\pi} \sqrt{2}V \sin \omega t \, d\omega t \right]$$

or $$V_{av} = \frac{\sqrt{2}V}{2\pi} (\cos \alpha - 1)$$

If α is varied from o to π, V_L varies from V to $\dfrac{V}{\sqrt{2}}$ and V_{av} varies from o to $\dfrac{-\sqrt{2}V}{\pi}$.

Circuits of three-phase half-wave and full-wave controller for star connected load are shown in Figs 10.2 (a) and (b). The half-wave controller economizes in cost of devices and does not give rise to DC components in any part of the system. However, it introduces harmonics into the line current than does the full-wave controller and this is a serious disadvantage.

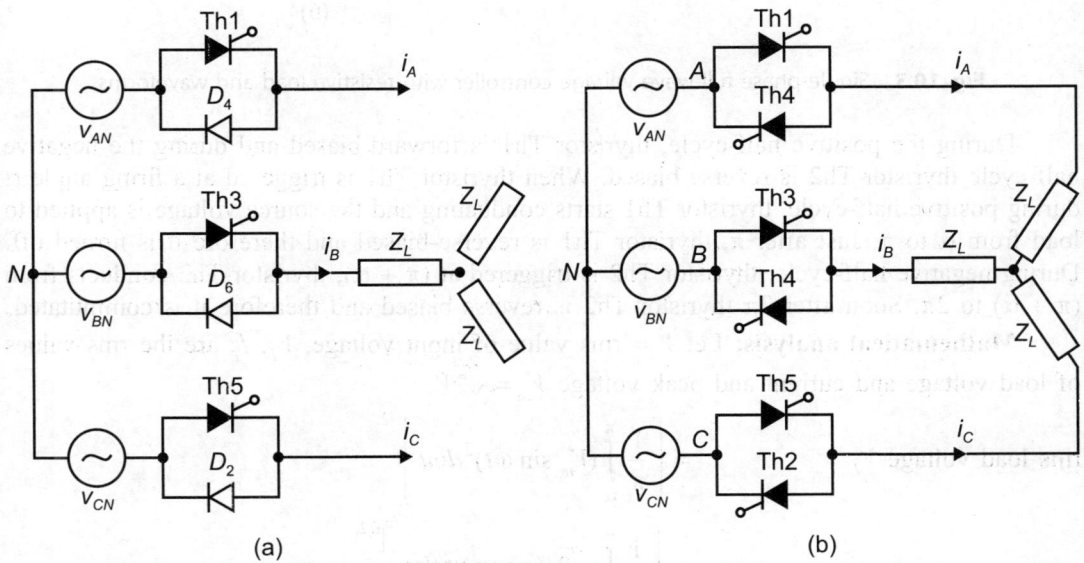

(a)   (b)

Fig. 10.2 Three-phase ac voltage controller circuit

10.2.2 Single-phase Full-wave Voltage Controller with Resistive Load

Figure 10.3 (a) shows a single-phase voltage controller with resistive load. The various waveforms at various points in the circuit are shown in Fig. 10.3 (b).

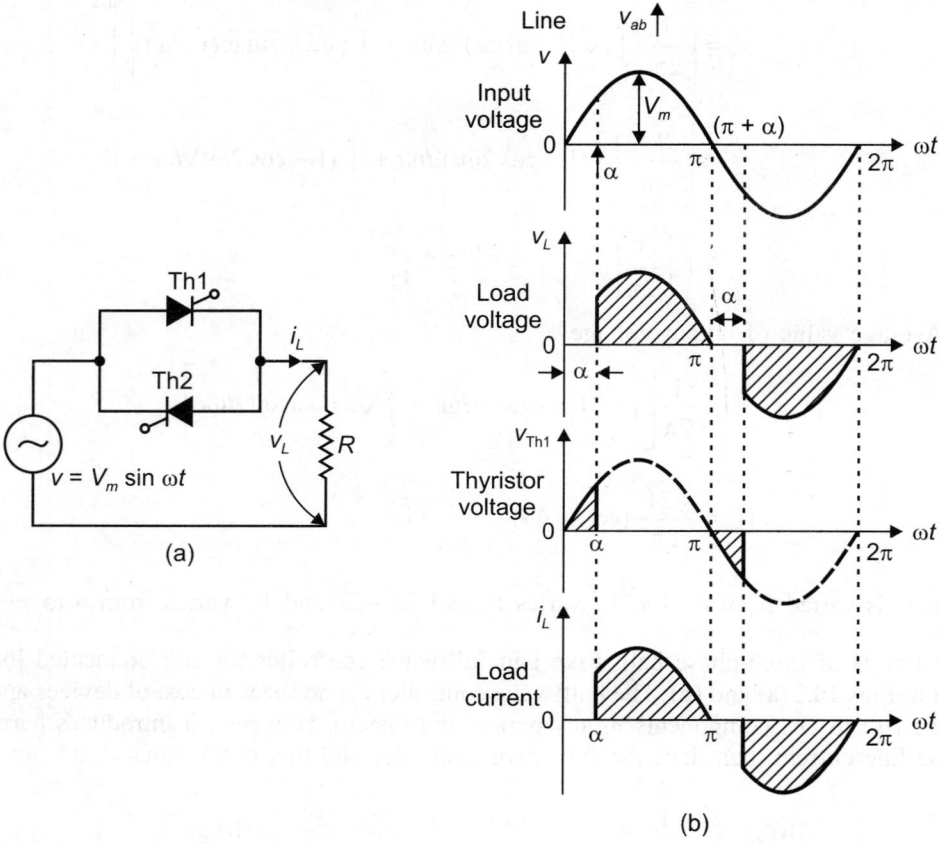

Fig. 10.3 Single-phase full-wave voltage controller with resistive load and waveforms

During the positive half-cycle, thyristor Th1 is forward-biased and during the negative half-cycle thyristor Th2 is reverse-biased. When thyristor Th1 is triggered at a firing angle α during positive half-cycle, thyristor Th1 starts conducting and the source voltage is applied to load from α to π. Just after π, thyristor Th1 is reverse-biased and therefore it is turned off. During negative half-cycle, thyristor Th2 is triggered at $(\pi + \alpha)$, thyristor Th2 conducts from $(\pi + \alpha)$ to 2π. Soon after 2π thyristor Th2 is reverse-biased and therefore it is commutated.

Mathematical analysis: Let V = rms value of input voltage, V_L, I_L are the rms values of load voltage and current and peak voltage $V_m = \sqrt{2}V$.

rms load voltage V_L

$$= \left[\frac{1}{\pi} \int_{\alpha}^{\pi} (V_m \sin \omega t)^2 d\omega t \right]^{1/2}$$

$$= \left[\frac{1}{\pi} \int_{\alpha}^{\pi} (\sqrt{2}V \sin \omega t)^2 d\omega t \right]^{1/2}$$

$$= V \left[1 - \frac{\alpha}{\pi} + \frac{\sin 2\alpha}{2\pi} \right]^{1/2} = V \sqrt{\frac{1}{2\pi} [2(\pi - \alpha) + \sin 2\alpha]} \qquad ...(10.1)$$

Load current $\quad I_L = \dfrac{V}{R} \left[1 - \dfrac{\alpha}{\pi} + \dfrac{\sin 2\alpha}{2\pi} \right]^{1/2}$ $\qquad\qquad$ ...(10.2)

$$= \frac{V_m}{R} \sqrt{\frac{1}{2\pi} \left[(\pi - \alpha) + \frac{\sin 2\alpha}{2} \right]}$$

rms value of thyristor current is given by

$$I_{\mathrm{Th}} = \frac{1}{R} \left[\frac{1}{2\pi} \int_{\alpha}^{\pi} (\sqrt{2} V \sin \omega t)^2 \, d\omega t \right]^{1/2}$$

$$= \frac{V}{\sqrt{2} R} \left[1 - \frac{\alpha}{\pi} + \frac{\sin 2\alpha}{2\pi} \right]^{1/2} \qquad\qquad ...(10.3)$$

Average thyristor current:

$$I_{A\,\mathrm{Th}} = \frac{1}{2\pi R} \int_{\alpha}^{\pi} V_m \sin \omega t \, d\omega t = \frac{V_m}{2\pi R} (1 + \cos \alpha)$$

Average value of load voltage: The mean average value of load voltage is zero over any complete number of cycles. The half-cycle average of the waveform v_L as in Fig. 10.3 (b) is given by

$$V_{av} = \frac{1}{\pi} \int_{\alpha}^{\pi} v_L d\omega t = \frac{1}{\pi} \int_{\alpha}^{\pi} V_m \sin \omega t \, d\omega t$$

$$= \frac{V_m}{\pi} (1 + \cos \alpha) \qquad\qquad ...(10.4)$$

Average power: $P = \dfrac{1}{\pi} \displaystyle\int_{\alpha}^{\pi} \dfrac{V_m^2}{R} \sin^2 \omega t \, d\omega t$

$$= \frac{V_m^2}{4\pi R} [2(\pi - \alpha) + \sin 2\alpha]$$

$$= \frac{V^2}{2\pi R} [2(\pi - \alpha) + \sin 2\alpha] \qquad\qquad ...(10.5)$$

Form factor: As form factor

$$= \frac{I_{\mathrm{rms}}}{I_{av}}$$

From Equation (10.2), the value of rms current is given by

$$I_{\mathrm{rms}} = I_L = \frac{V_m}{R} \sqrt{\frac{1}{2\pi} \left[(\pi - \alpha) + \frac{\sin 2\alpha}{2} \right]}$$

Value of average current,

$$I_{av} = \frac{V_{av}}{R} = \frac{V_m}{\pi R} (1 + \cos \alpha)$$

$$\text{Form factor} \quad = \frac{V_m/R \sqrt{\dfrac{1}{2\pi}\left[(\pi - \alpha) + \dfrac{\sin 2\alpha}{2}\right]}}{V_m/\pi R\,(1 + \cos \alpha)}$$

$$= \frac{\sqrt{\dfrac{1}{2\pi} \times \pi^2 \left[(\pi - \alpha) + \dfrac{\sin 2\alpha}{2}\right]}}{(1 + \cos \alpha)}$$

$$= \frac{\sqrt{\dfrac{\pi}{2}\left[(\pi - \alpha) + \dfrac{1}{2}\sin 2\alpha\right]}}{(1 + \cos \alpha)}.$$

10.3 HARMONICS OF OUTPUT VOLTAGE AND CURRENT FOR RESISTIVE LOAD

From Fig. 10.3 (b), it is clear that output quantities like output voltage and current are nonsinusoidal. These quantities can be represented by Fourier series. As the positive and negative half-cycles are identical, DC components and even harmonics are absent. The output voltage can be represented by Fourier series as:

$$v_L = \sum_{n=1,3,5}^{\alpha} a_n \sin n\omega t + \sum_{n=1,3,5}^{\alpha} b_n \cos n\omega t \qquad \qquad ...(10.6)$$

where,

$$a_n = \frac{2}{\pi} \int_0^\pi v_L \sin n\omega t \cdot d\omega t \qquad \qquad ...(10.7)$$

and

$$b_n = \frac{2}{\pi} \int_0^\pi v_L \cos n\omega t \cdot d\omega t \qquad \qquad ...(10.8)$$

Load voltage during the first half-cycle is

$$v_L = V_m \sin \omega t$$

$$\alpha < \omega t < \pi$$

$$a_n = \frac{2V_m}{\pi} \int_\alpha^\pi \sin \omega t \cdot \sin n\omega t \cdot d\omega t$$

$$= \frac{V_m}{\pi} \int_\alpha^\pi [\cos (n - 1)\,\omega t - \cos (n + 1)\,\omega t]\, d\omega t$$

$$= \frac{V_m}{2\pi} \left[\frac{2}{n+1} \sin (n + 1)\,\alpha - \frac{2}{n-1} \sin (n - 1)\,\alpha\right] \qquad ...(10.9)$$

$$b_n = \frac{2V_m}{\pi} \int_\alpha^\pi \sin \omega t \cdot \cos n\omega t \cdot d\omega t$$

$$= \frac{V_m}{2\pi} \left[\frac{2}{n+1} \{\cos (n + 1)\,\omega t - 1\} - \frac{2}{n-1} \{\cos (n - 1)\,\omega t - 1\}\right] \qquad ...(10.10)$$

where, $\qquad V_m = \sqrt{2} V$

The peak amplitude of the nth harmonic output voltage V_{nm} and phase angle ϕ are given by

$$V_{nm} = \sqrt{a_n^2 + b_n^2} \text{ and } \phi = \tan^{-1} \frac{b_n}{a_n} \qquad \text{...(10.11)}$$

nth harmonic current, $\qquad I_{nm} = \dfrac{V_{nm}}{R}$

Equations (10.9), (10.10) and (10.11) can be used to find the magnitude of harmonics for which n = 3, 5, ... It cannot be used to calculate the fundamental component by putting n = 1. For n = 0 and n = even (i.e. DC component) coefficients a_n, b_n are zero. The Fourier spectrum of load voltage therefore contains only odd harmonics. At ωt = 0, the load voltage is sinusoidal and therefore contains no higher harmonic components. For fundamental component, i.e. n = 1. Equations (10.9) and (10.10) leads to undefined expressions, thus V_{1m} and ϕ_1 cannot be obtained. In present case, this difficulty is overcome by putting n = 1 in the following equations of a_n and b_n.

$$a_n = \frac{2V_m}{\pi} \int_0^\pi \sin \omega t \sin n\omega t \cdot d\omega t$$

$$a_1 = \frac{2V_m}{\pi} \int_\alpha^\pi \sin^2 \omega t \cdot d\omega t$$

$$= \frac{V_m}{2\pi} [\sin 2\alpha + 2 (\pi - \alpha)] \qquad \text{...(10.12)}$$

$$b_n = \frac{2V_m}{\pi} \int_\alpha^\pi \sin \omega t \cos n\omega t \cdot d\omega t$$

$$b_1 = \frac{2V_m}{\pi} \int_\alpha^\pi \sin \omega t \cos \omega t \cdot d\omega t = \frac{V_m}{2\pi} [\cos 2\alpha - 1] \qquad \text{...(10.13)}$$

Coefficients a_1 and b_1 are combined to give the peak amplitude c_1 and phase angle ϕ_1 of the fundamental component of load voltage as follows:

$$c_1 = \sqrt{a_1^2 + b_1^2} \text{ and } \phi_1 = \tan^{-1} \frac{b_1}{a_1}$$

or $\qquad c_1 = \dfrac{V_m}{2\pi} \sqrt{(\cos 2\alpha - 1)^2 + [\sin 2\alpha + 2 (\pi - \alpha)]^2} \qquad \text{...(10.14)}$

and $\qquad \phi_1 = \tan^{-1} \left[\dfrac{\cos 2\alpha - 1}{\sin 2\alpha + 2 (\pi - \alpha)} \right] \qquad \text{...(10.15)}$

When α = 90°, it is found that

$$a_1 = \frac{V_m}{2} \text{ and } b_1 = -\frac{V_m}{\pi} \text{ and } c_1 = 0.59 V_m \text{ and } \phi_1 = -32.5°$$

The fundamental component of the discontinuous load voltage is defined, by relationship

$$v_L = 0.59 V_m \sin (\omega t - 32.5°)$$

Power factor: Irrespective of waveform, the power factor of a circuit is the factor by which the apparent volt ampere must be multiplied in order to give the average power.

$$\text{Power factor PF} = \frac{\text{Real power}}{\text{Apparent power}} = \frac{V_L^2/R}{VI_{\text{rms}}} = \frac{V_L^2/R}{V \cdot V_L/R} = \frac{V_L}{V}$$

$$= \frac{V\left[1 - \frac{\alpha}{\pi} + \frac{\sin 2\alpha}{2\pi}\right]^{1/2}}{V} = \left[1 - \frac{\alpha}{\pi} + \frac{\sin 2\alpha}{2\pi}\right]^{1/2} \qquad ...(10.16)$$

Displacement factor: The phase angle between the voltage and current of load at any harmonic frequency is a function only of load itself, not of the excitation voltage waveform. Phase angle ψ, which represents the phase displacement between the fundamental components of voltage and current at the circuit terminals, the factor $\cos \psi$ is given the special name *displacement factor*. For circuits with sinusoidal supply voltage and nonsinusoidal current, displacement factor is significant.

$$\text{Distortion factor} = \frac{\text{Power factor}}{\text{Displacement factor}}. \qquad ...(10.17)$$

10.4 SINGLE-PHASE VOLTAGE CONTROLLER WITH R-L LOAD

Figure 10.4 (a) shows the circuit of a controller supplying a load possessing resistance and inductance, i.e. (R-L load). If the load is inductive, the following effects are observed: (1) Each thyristor conducts for less than 180°, if the firing angle is greater than the load phase angle $\left(\tan^{-1}\dfrac{\omega L}{R}\right)$, (2) If the firing angle is equal to the load phase angle, the conduction angle is 180°, and (3) Thyristor conduction angle will exceed 180° for triggering angle less than load phase angle. In this case only one thyristor will conduct resulting into rectification if larger triggering pulses is not used.

With an inductive load, current flow is maintained through the thyristor even after the input voltage has reversed polarity and goes negative. The duration of this interval is determined by load power factor. Typical waveforms are shown in Fig. 10.4 (b). The expression for thyristor current Th1 can be obtained as follows:

Let the input voltage be $v = \sqrt{2}V \sin \omega t = V_m \sin \omega t$

Applying KVL equation in the circuit of Fig. 10.4 (a)

$$v = V_m \sin \omega t = \frac{L di_L}{dt} + Ri_L$$

and load current, $i_L = i_{ss} + i_t$ where, i_{ss} and i_t are the steady state and transient currents.

Solution of this equation is of the form as:

or

$$i_L = \frac{\sqrt{2}V}{Z} \sin (\omega t - \phi) + Ae^{-t/T}$$

where,

$$Z = \sqrt{R^2 + \omega^2 L^2} \text{ Ohm}$$

and power factor angle $\phi = \tan^{-1}(\omega L/R)$ and $\tau = \dfrac{L}{R}$

At $\omega t = \alpha$, $i_L = 0$, the above relation becomes:

$$0 = \frac{\sqrt{2}V}{Z} \sin (\omega t - \phi) + Ae^{-(R/\omega L)\alpha}$$

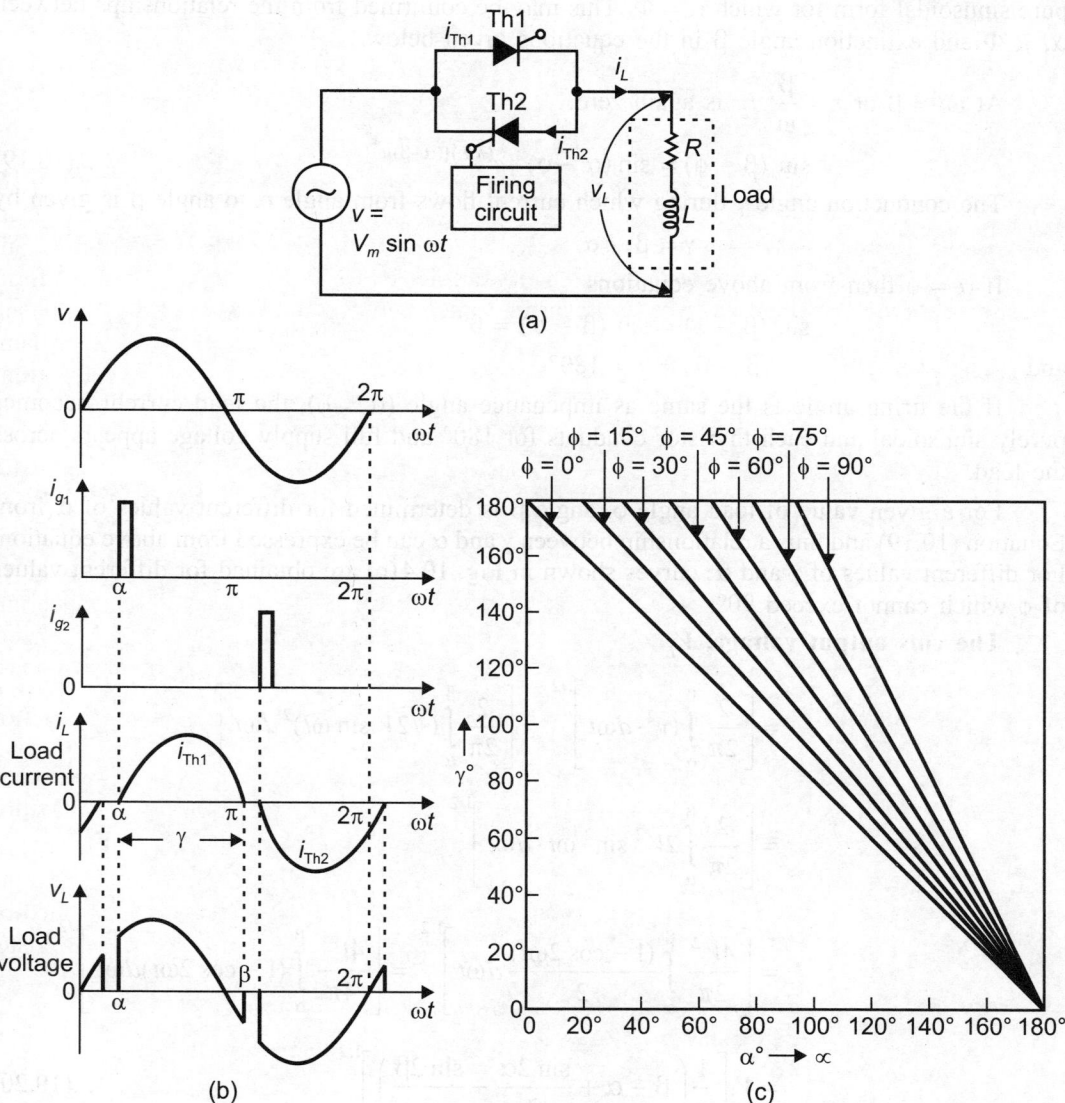

Fig. 10.4 Single-phase voltage controller with R-L load and waveforms circuit

or
$$A = -\frac{\sqrt{2}V}{Z} \sin(\omega t - \phi) \, e^{(R/\omega L)\alpha}$$

Putting $V_m = \sqrt{2}V$, thus the solution of the load current equation is given by

$$i_L = \frac{V_m}{Z} \left[\sin(\omega t - \phi) - \sin(\alpha - \phi) \, e^{(R/L)\left(\frac{\alpha}{\omega} - t\right)} \right] \qquad ...(10.18)$$

Each thyristor conducts during one-half-cycle with the constraint that the conduction angle γ cannot exceed 180°, if the triggering circuit is properly designed. From the waveform it may be clearly seen that as α is reduced until $\gamma = 180°$, the waveforms of i_L and v_L approach the pure sinusoidal form for which $\alpha = \Phi$. This may be confirmed from the relationships between α, γ, Φ and extinction angle β in the equations given below.

At $\omega t = \beta$ or $t = \dfrac{\beta}{\omega}$, i_L is again zero.

$$\sin (\beta - \phi) = \sin (\alpha - \phi) \cdot e^{(R/L)[(\alpha-\beta)/\omega]} \qquad \ldots(10.19)$$

The conduction angle γ during which current flows from angle α to angle β is given by

$$\gamma = \beta - \alpha$$

If $\alpha = \phi$ then from above equations

$$\sin (\beta - \phi) = \sin (\beta - \alpha) = 0°$$

and

$$(\beta - \alpha) = \gamma = 180°$$

If the firing angle is the same as impedance angle ($\alpha = \phi$), the load current becomes purely sinusoidal and each thyristor conducts for 180° and full supply voltage appears across the load.

For a given value of load angle ϕ, angle β is determined for different values of α from Equation (10.19) and thus a relationship between γ and α can be expressed from above equation. For different values of γ and α, curves shown in Fig. 10.4(c) are obtained for different values of ϕ which cannot exceed 90°.

The rms output voltage, V_L

$$= \left[\frac{2}{2\pi} \int_{\alpha}^{\beta} (v^2 \cdot d\omega t) \right]^{1/2} = \left[\frac{2}{2\pi} \int_{\alpha}^{\beta} (\sqrt{2} V \sin \omega t)^2 \, d\omega t \right]^{1/2}$$

$$= \left[\frac{2}{2\pi} \int_{\alpha}^{\beta} 2V^2 \sin^2 \omega t \cdot d\omega t \right]^{1/2}$$

$$= \left[\frac{4V^2}{2\pi} \int_{\alpha}^{\beta} \frac{(1 - \cos 2\omega t)}{2} \, d\omega t \right]^{1/2} = \left[\frac{4V^2}{4\pi} \int_{\alpha}^{\beta} (1 - \cos 2\omega t) d\omega t \right]^{1/2}$$

$$= V \left[\frac{1}{\pi} \left(\beta - \alpha + \frac{\sin 2\alpha}{2} - \frac{\sin 2\beta}{2} \right) \right]^{1/2} \qquad \ldots(10.20)$$

Here V is the rms value of input voltage.

rms value of thyristor current, I_{Th1} is given by

$$I_{\text{Th1}} = \left[\frac{1}{2\pi} \int_{\alpha}^{\beta} i_L^2 \cdot d\omega t \right]^{1/2}$$

Putting the value of i_L from Equation (10.18) in the above equation, we get

$$I_{\text{Th1}} = \left[\frac{1}{2\pi} \int_{\alpha}^{\beta} \frac{V_m^2}{Z^2} \left\{ \sin (\omega t - \phi) - \sin (\alpha - \phi) \, e^{(R/L)(\alpha/\omega - t)} \right\}^2 \cdot d\omega t \right]^{1/2}$$

Here $\qquad V_m = \sqrt{2} V$

$$I_{Th1} = \frac{V}{Z}\left[\frac{1}{\pi}\int_{\alpha}^{\beta}\left\{\sin(\omega t - \phi) - \sin(\alpha - \phi)e^{R/L(\alpha/\omega - t)}\right\}^2 d\omega t\right]^{1/2} \quad ...(10.21)$$

rms value of load current: I_L is the combination of rms value of two thyristors currents I_{Th1} and I_{Th2} as

$$I_L = \sqrt{I_{Th1}^2 + I_{Th2}^2} = \sqrt{2}\,I_R \quad [I_{Th1} = I_{Th2} = I_R]$$

Average value of thyristor current, I_A

$$= \frac{1}{2\pi}\int_{\alpha}^{\beta} i_{Th1}\cdot d\omega t$$

$$= \frac{1}{2\pi}\int_{\alpha}^{\beta}\frac{V_m}{Z}[\sin(\omega t - \phi) - \sin(\alpha - \phi)e^{(R/L)(\alpha/\omega - t)}]\,d\omega t \quad ...(10.22)$$

Equation (10.18) indicates that the load current i_L will be sinusoidal if the delay angle α, is less than the load angle ϕ, if α is greater than ϕ, the load current would be discontinuous and nonsinusoidal.

10.5 HARMONIC ANALYSIS OF SINGLE-PHASE FULL-WAVE CONTROLLER WITH R-L LOAD

The load or line current waveform may be described by Fourier series

$$i_L = \sum_{n=1}^{\alpha} a_n \sin n\,\omega t + \sum_{n=1}^{\alpha} b_n \cos n\,\omega t \quad ...(10.23)$$

where,

$$a_n = \frac{1}{\pi}\int_0^{2\pi} i_L \sin n\,\omega t\cdot d\omega t = \frac{2}{\pi}\int_0^{\pi} i_L \sin n\,\omega t\cdot d\omega t$$

and

$$b_n = \frac{1}{\pi}\int_0^{2\pi} i_L \cos n\,\omega t\cdot d\omega t = \frac{2}{\pi}\int_0^{\pi} i_L \cos n\,\omega t\cdot d\omega t$$

In Fig. 10.4 (a), $i_L = i_{Th1}$ amp
rms value of nth harmonic is

$$I_{nR} = \frac{1}{\sqrt{2}}[a_n^2 + b_n^2]^{1/2} \text{ amp} \quad ...(10.24)$$

The rms value of nth harmonic of output voltage is given by

$$V_{nR} = I_{nR}[R^2 + (n\omega L)^2]^{1/2} \text{ volts} \quad ...(10.25)$$

From Equations (10.23) to (10.25), the harmonic content of line current for any value of α and ϕ may be determined by obtaining γ from Fig. 10.4 (c). If, for simplicity, a purely resistive load circuit is considered, so that $\phi = 0$, then from Equation (10.18),

$$i_L = \frac{V_m}{R}\sin \omega t \quad (\alpha < \omega t < \pi)$$

In general, current harmonic amplitudes are reduced when load circuit inductance is increased.

10.6 GATING SIGNALS

Pulse gating is suitable for resistive load as discussed but it is not suitable for R-L load circuits shown in Fig. 10.4(a) because of the reason as shown in Fig. 10.5.

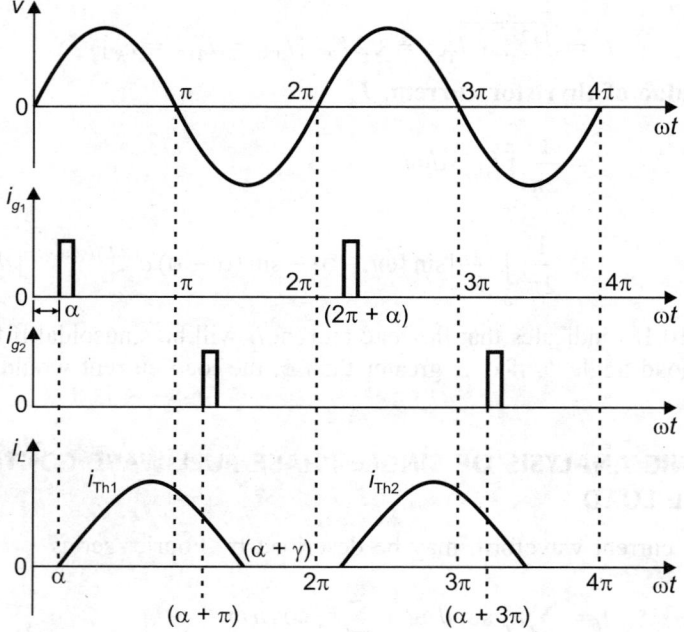

Fig. 10.5 Pulse-gating signals

When thyristor Th2 is fired, at $\omega t = (\pi + \alpha)$ thyristor Th1 is still conducting due to the inductance of the load. Voltage drop across thyristor Th1 reverse-biases Th2 at $(\pi + \alpha)$ which does not allow thyristor Th2 to turn on. At $(\alpha + \gamma)$, thyristor current i_{Th1} decays to zero shown in Fig. 10.4(b) and thyristor Th1 stops conducting as a result thyristor Th2 gets forward-biased but gate pulse I_{g2} applied to thyristor Th2 at $(\pi + \alpha)$ is already zero and therefore thyristor Th2 does not get turned on, thus the controller operates with an asymmetrical waveform due to conduction of thyristor Th1 only, and this produces an undesirable DC component of load and source current.

This difficulty could be removed by using continuous gating, i.e. by making gating pulse last for a period of $(\pi - \alpha)/\omega$ second so that as soon as thyristor current i_{Th1} fell to zero, thyristor Th2 would then turn on. However, due to the need to isolate the gating signals of the two thyristors, it is desirable that these signals should be supplied to the two thyristors via isolating transformers. Such transformers are small when only a short pulse must be transmitted but become large when a long pulse is required.

The technique which ensures turn on of thyristor Th2 and at the same time requires only a small isolating transformer, with high frequency carrier gating in which a series of short pulses lasting throughout the intervals $\alpha < \omega t < \pi$ for thyristor Th1 and $(\alpha + \pi) < \omega t < 2\pi$ for thyristor Th2 are applied. These pulses normally have a frequency of the order of 30 kHz. Three types of gating signals are illustrated namely: (1) pulse gating, (2) continuous gating, and (3) high frequency carrier gatting shown in Fig. 10.6.

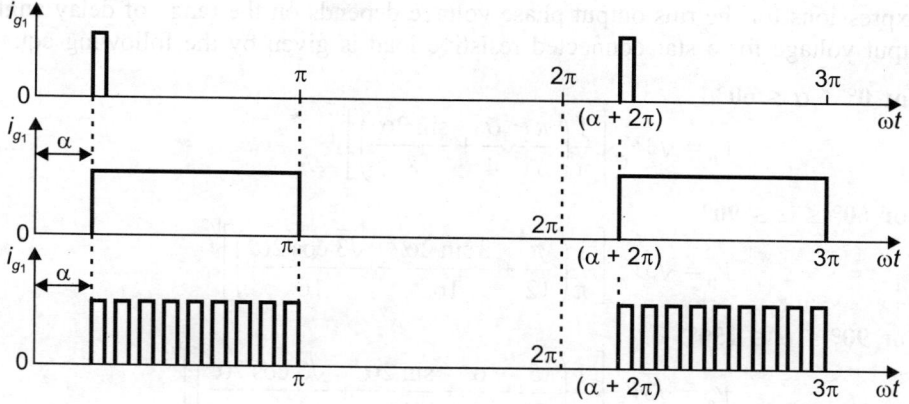

Fig. 10.6 High frequency carrier gating

10.7 THREE-PHASE FULL-WAVE CONTROLLER

The circuit diagram of a three-phase full-wave controller is shown in Fig. 10.7(a) with a star connected load. The firing sequence of thyristors are Th1, Th2, Th3, Th4, Th5 and Th6. In the other circuit [Fig. 10.7(b)], thyristors are connected in series with the loads to form a delta connection.

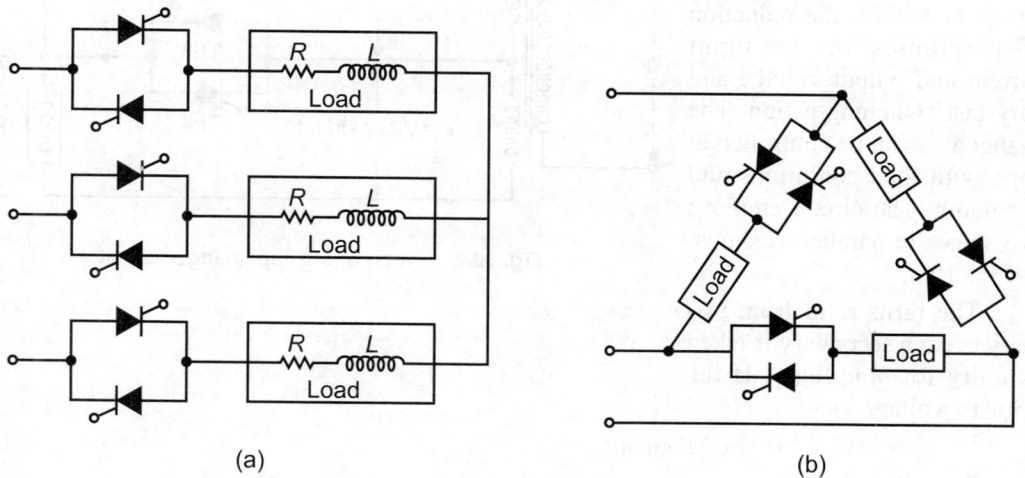

(a)   (b)

Fig. 10.7 Three-phase full-wave controllers

Analysis of the star connected three-phase controller is complex as operation of one-phase is dependent on the operation of the other phases. However, the operation of delta connected controller can be studied on a perphase basis because each phase is connected across a known supply voltage.

For high power loads such as three-phase induction motors, pumps and driving fans, three-phase full wave controllers are generally used. Due to the presence of DC input current and higher harmonic content in the voltage waveform, three-phase half-wave controllers are not used in AC motor drives.

Expressions for the rms output phase voltage depends on the range of delay angles. The rms output voltage for a star connected resistive load is given by the following equations:

1. For $0° \leq \alpha \leq 60°$

$$V_0 = \sqrt{3}V_m \left[\frac{1}{\pi} \left(\frac{\pi}{6} - \frac{\alpha}{4} + \frac{\sin 2\alpha}{8} \right) \right]^{1/2} \qquad \qquad ...(10.26)$$

2. For $60° \leq \alpha \leq 90°$

$$V_0 = \sqrt{3}V_m \left[\frac{1}{\pi} \left(\frac{\pi}{12} + \frac{3\sin 2\alpha}{16} + \frac{\sqrt{3}\cos 2\alpha}{16} \right) \right]^{1/2} \qquad ...(10.27)$$

3. For $90° \leq \alpha \leq 150°$

$$V_0 = \sqrt{3}V_m \left[\frac{1}{\pi} \left(\frac{5}{24} - \frac{\alpha}{4} + \frac{\sin 2\alpha}{16} + \frac{\sqrt{3}\cos 2\alpha}{16} \right) \right]^{1/2}. \qquad ...(10.28)$$

10.8 SYNCHRONOUS TAP CHANGER (SINGLE-PHASE TRANSFORMER TAP CHANGER)

Thyristors can be used as static switches for on-load tap changing of transformers. The static tap changers are employed for the improvement of system power factor, for the reduction of harmonics in the input current and output voltage and very fast switching action. The changeover can be controlled to cope with load conditions and is smooth. The circuit employs two stages in parallel as shown in Fig. 10.8.

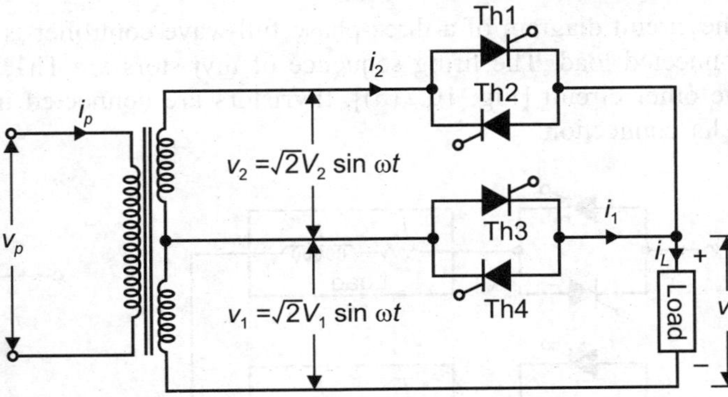

Fig. 10.8 Synchronous tap changer circuit

The turns ratio from primary to each secondary is taken as unity for simplicity. If the primary voltage is

$$v_p = \sqrt{2}V \sin \omega t$$

Secondary instantaneous voltages are:

$$v_1 = \sqrt{2}V_1 \sin \omega t \text{ and } v_2 = \sqrt{2}V_2 \sin \omega t$$

If $V_1 = V_2 = V$, then sum of two secondary voltages is $2\sqrt{2}V \sin \omega t = 2V_m \sin \omega t$.

If the full output voltage is required, only thyristors Th1 and Th2 are alternately fired with delay angle of $\alpha = 0$ and the full voltage is

$$v_L = V_1 + V_2$$

Thyristors are fired to control the load voltage, output voltage can be varied within three possible ranges (Fig. 10.9):

1. $0 < V_L < V_1$
2. $0 < V_L < (V_1 + V_2)$
3. $V_1 < V_L < (V_1 + V_2)$

Case I: $0 < V_L < V_1$: To vary the load voltage within this range, thyristors Th1 and Th2 are turned off and thyristors Th3 and Th4 are made to operate as single-phase voltage controller. The instantaneous load voltage and current are shown in Fig. 10.9(c) for a resistive load. The rms value of load voltage as derived in the previous articles is given by

$$V_L = V_1 \left[\frac{1}{\pi} \left(\pi - \alpha + \frac{\sin 2\alpha}{2} \right) \right]^{1/2} \quad \text{... for } 0 \le \alpha \le \pi \qquad \text{...(10.29)}$$

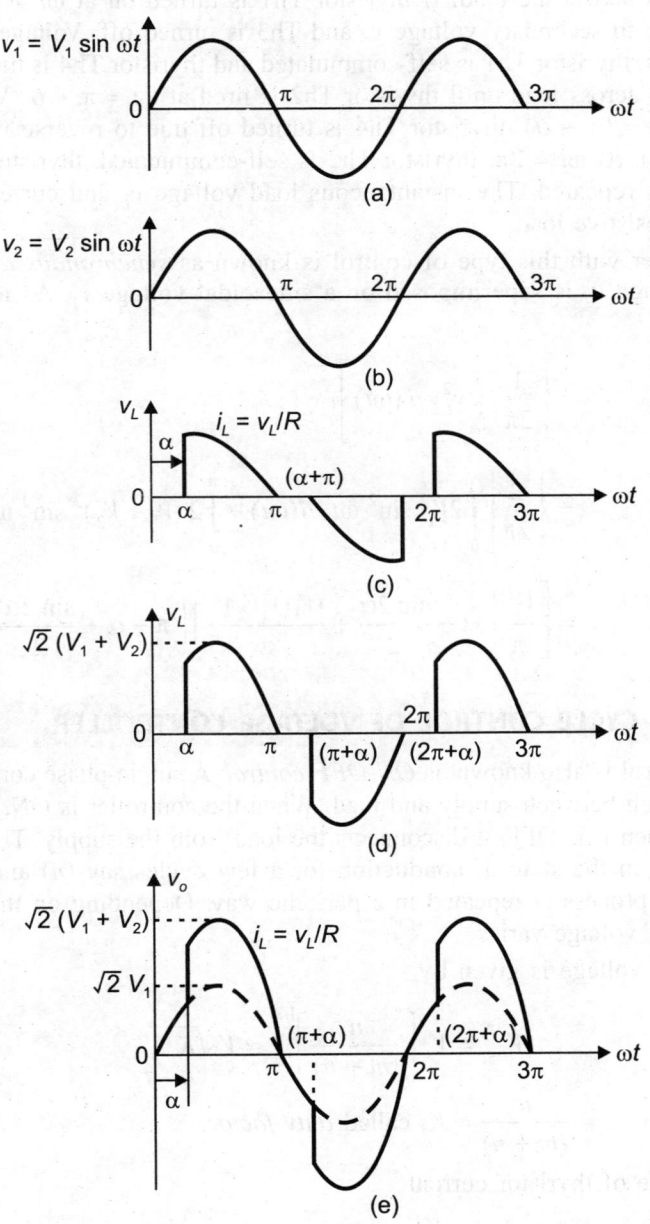

Fig. 10.9 Waveforms of synchronous tap-changer

Case II: $0 < V_L < (V_1 + V_2)$: Thyristors Th1 and Th2 operate as single-phase voltage controller and thyristors Th3 and Th4 are turned off. Load voltage and current for resistance load are shown in Fig. 10.9(d).

rms value of load voltage for range of delay angle $0 \leq \alpha \leq \pi$ is as follows:

$$V_L = (V_1 + V_2)\left[\frac{1}{\pi}\left(\pi - \alpha + \frac{\sin 2\alpha}{2}\right)\right]^{1/2} \qquad \ldots(10.30)$$

Case III: $V_1 < V_L < (V_1 + V_2)$: Thyristor Th3 is turned on at $\omega t = 0$ and the secondary voltage v_1 appears across the load. If thyristor Th1 is turned on at $\omega t = \alpha$, thyristor Th3 is reverse-biased due to secondary voltage v_2 and Th3 is turned off. Voltage across the load is $(v_1 + v_2)$. At $\omega t = \pi$, thyristor Th1 is self-commutated and thyristor Th4 is turned on. Secondary voltage v_1 appears across load until thyristor Th2 is fired at $\omega t = \pi + \alpha$. When thyristor Th2 is turned on at $\omega t = (\pi + \alpha)$, thyristor Th4 is turned off due to reverse voltage v_2 and load voltage is $(v_1 + v_2)$. At $\omega t = 2\pi$, thyristor Th2 is self-commutated, thyristor Th3 is turned on again and cycle is repeated. The instantaneous load voltage v_L and current i_L are shown in Fig. 10.9(e) for resistive load.

A tap changer with this type of control is known as *synchronous tap changer*. A part of secondary voltage v_2 is superimposed on a sinusoidal voltage v_1. As result the harmonic contents are less.

$$V_L = \left[\frac{1}{2\pi}\int_0^{2\pi} v_L^2 \cdot d(\omega t)\right]^{1/2}$$

$$= \left\{\frac{2}{2\pi}\left[\int_0^\alpha 2V_1^2 \sin^2 \omega t \cdot d(\omega t) + \int_\alpha^\pi 2(V_1 + V_2)^2 \sin^2 \omega t \cdot d(\omega t)\right]\right\}^{1/2}$$

$$= \left[\frac{V_1^2}{\pi}\left(\alpha - \frac{\sin 2\alpha}{2} + \frac{(V_1 + V_2)^2}{\pi}\right)\left(\pi - \alpha + \frac{\sin 2\alpha}{2}\right)\right]^{1/2}. \qquad \ldots(10.31)$$

10.9 INTEGRAL CYCLE CONTROL OF VOLTAGE CONTROLLER

Integral cycle control is also known as *ON-OFF control*. A single-phase controller supplying a load acts as a switch between supply and load. When the controller is ON, it connects supply to the load and when it is OFF, it disconnects the load from the supply. The controller is ON at $\alpha = 0$ and kept in the state of conduction for a few cycles say (n) and it switched OFF for m cycles. The process is repeated in a periodic way. Depending on the values of m and n the rms value of voltage varies.

rms value of voltage is given by

$$V_{rms} = V_s\left[\frac{n}{(m + n)}\right]^{1/2} = V\sqrt{K}$$

Ratio $$\frac{n}{(m + n)} = K, \text{ called } duty\ factor.$$

Average value of thyristor current

$$= \frac{KI_m}{\pi}$$

rms value of thyristor current

$$= I_m \sqrt{K}$$

Power factor $\qquad = \sqrt{K}$

This can be applied to heating loads. When applied to lighting, there will be flicker. When applied to motor loads, there will be coasting and speed oscillations.

10.10 FEATURES OF PHASE CONTROL OF AC VOLTAGE CONTROLLERS

Single-phase and three-phase controllers employ phase control. The features of this control may be summarized as follows:

1. Control is accomplished by varying the firing instant during the half-cycle when thyristor has positive voltage.
2. The output voltage and current are distorted and rich in harmonic content.
3. For R-L load, control is not possible for firing angles less than impedance angle.
4. Thyristor voltage increases with increase in firing angle. For firing, at $\alpha = 0$, all the voltages are available across the load, the forward thyristor voltage is zero.
5. The dv/dt of the incoming thyristor is high when current goes naturally to zero for inductive loads. The power factor control angle affects the rise of voltage.
6. In pure resistive load, high di/dt may also occur for R-L load, it is decided by the load.

10.10.1 Gate Pulse Requirements to Trigger Thyristor of AC Voltage Controllers

To bring a thyristor to the conducting state, it is sufficient to provide a narrow pulse at the gate of thyristor whenever its voltage is positive.

If the load is purely resistive, thyristor of AC voltage controllers can be triggered with short narrow pulses. If the controller supplies an inductive load, short or narrow pulses may not be able to trigger thyristors. When the load is inductive and if the firing angle is less than the impedance angle, no control on the output side is possible.

In a single-phase or three-phase AC voltage controllers, a control pulse of long duration equal to the conduction period of thyristor is used. This is to make sure that the firing pulse is available at the gate whenever a positive current is anticipated in the thyristor. If because of some circuit condition the current goes to zero, the thyristor turns off. Only a lengthy pulse can bring it into conduction if there is a possibility of forward current in the thyristor.

For antiparallel thyristor, the two firing pulse must be isolated from each other to avoid the connection of two cathodes of thyristors. In three-phase controller conduction can be established only if two controllers in different lines receive pulses simultaneously.

SOLVED EXAMPLES

Example 10.1 Show that in a single-phase controlled AC circuit, the firing angle α, extinction angle β, and load power factor angle ϕ are related by the expression

$$\sin (\beta - \phi) = \sin (\alpha - \phi) \cdot \exp [(\alpha - \beta)/\tan \phi]$$

Solution A single-phase AC controller is shown in Fig. E10.1.

When thyristor Th1 is triggered with a delay angle of α. Writing KVL equation as in Fig. E10.1.

$$Ri_L + L\frac{di_L}{dt} = v_i = \sqrt{2}V \sin \omega t$$

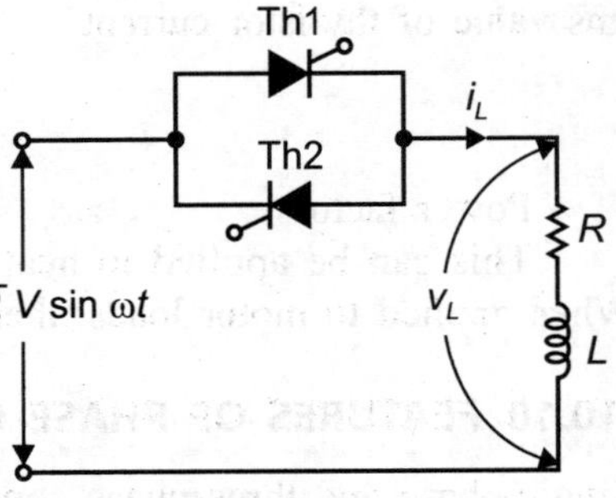

The load current $i_L = i_s + i_t$, solution of this equation is given by:

$$i_L = \frac{\sqrt{2}V}{Z} \sin(\omega t - \phi) + Ae^{-t/\tau} \qquad ...(i)$$

where, $\qquad Z = \sqrt{R^2 + (\omega L)^2}$

$$\phi = \tan^{-1} \omega L/R \text{ and } \tau = L/R$$

At $\omega t = \alpha,\qquad i_L = 0$

From Equation (i)

$$0 = \frac{\sqrt{2}V}{Z} \sin(\omega t - \phi) + Ae^{-(R/\omega L)\alpha}$$

$$A = -\frac{\sqrt{2}V}{Z} \sin(\alpha - \phi) e^{(R/\omega L)\alpha}$$

$$= -\frac{\sqrt{2}V}{Z} \sin(\alpha - \phi) e^{\alpha/\tan\phi}$$

$$i_L = \frac{\sqrt{2}V}{Z} \left[\sin(\omega t - \phi) - \frac{\sin(\alpha - \phi) e^{-(R/\omega L)\omega t}}{e^{-\alpha/\tan\phi}} \right]$$

$$i_L = \frac{\sqrt{2}V}{Z} \left[\sin(\omega t - \phi) - \frac{\sin(\alpha - \phi)}{e^{-\alpha/\tan\phi}} e^{-\omega t/\tan\phi} \right] \qquad ...(ii)$$

When the conduction ceases at $\omega t = \beta$, the above equation gives

$$0 = \frac{\sqrt{2}V}{Z} \left[\sin(\beta - \phi) - \frac{\sin(\alpha - \phi)}{e^{-\alpha/\tan\phi}} e^{-\beta/\tan\phi} \right]$$

or $\qquad \sin(\beta - \phi) = \sin(\alpha - \phi) e^{(\alpha - \beta)/\tan\phi}$

$\qquad\qquad\qquad = \sin(\alpha - \phi) \cdot \exp[(\alpha - \beta)|\tan\phi].$

Fig. E10.1 Circuit diagram

$v_i = \sqrt{2} V \sin \omega t$

$Th1$

$Th2$

i_L

v_L

R

L

Example 10.2 A single-phase full-wave AC controller has a resistive load of $10\,\Omega$ and an input voltage of 120 V at 50 Hz. The firing angles of thyristors Th1 and Th2 is $\pi/2$. Calculate: (i) rms value of load voltage, (ii) input power factor, (iii) average value of current of thyristor, and (iv) rms current of thyristor I_R.

Solution

(i) Here $\qquad V = 120$ V (rms),

$$V_m = \sqrt{2} \times 120 = 169.7 \text{ V}$$

$$R = 10\,\Omega, \alpha = \pi/2$$

rms value of load voltage V_L

$$= V\left[\frac{1}{\pi}\left(\pi - \alpha + \frac{\sin 2\alpha}{2}\right)\right]^{1/2} = 120\left[\frac{1}{\pi}\left(\pi - \frac{\pi}{2} + \frac{\sin \pi}{2}\right)\right]^{1/2}$$

$$= 120\left[\frac{1}{\pi} \cdot \frac{\pi}{2} + 0\right]^{1/2} = \frac{120}{\sqrt{2}} = 84.85 \text{ V}$$

(ii) rms value of load current I_L

$$= \frac{V_L}{R} = \frac{84.85}{10} = 8.485 \text{ A}$$

Load power, $\qquad P_L = I_L^2 R = 8.485^2 \times 10 = 719.9 \text{ W}$

Since the input current is the same as load current, input volt-ampere rating

$$V_A = VI = VI_L = 120 \times 8.485$$

Input power factor, $\qquad \text{PF} = \dfrac{P_L}{VA} = \dfrac{I_L^2 R_L}{VI_L} = \dfrac{I_L R_L}{V} = \dfrac{V_L}{V}$

$$= \frac{V\left[\dfrac{1}{\pi}(\pi - \alpha) + \dfrac{\sin 2\alpha}{2}\right]}{V}$$

$$= \left[\frac{1}{\pi}\left(\pi - \alpha + \frac{\sin 2\alpha}{2}\right)\right] = \frac{V_L}{V} = \frac{84.85}{120}$$

$$= 0.707 \text{ (lagging)}$$

(iii) average thyristor current, $I_{A\,\text{Th}}$

$$(I_{\text{Th}})_{AV} = \frac{V_m}{2\pi R}(1 + \cos\alpha)$$

$$= \frac{\sqrt{2} \times 120}{2\pi \times 10} = 2.7 \text{ A}$$

(iv) rms current of thyristor, I_R

$$I_{\text{Th(rms)}} = \frac{V}{\sqrt{2}R}\left[1 - \frac{\alpha}{\pi} + \frac{\sin 2\alpha}{2\pi}\right]^{1/2}$$

$$= \frac{120}{\sqrt{2} \times 10}\left[1 - \frac{\pi/2}{\pi} + \frac{\sin 180°}{2\pi}\right]^{1/2}$$

$$= \frac{120}{10 \times \sqrt{2}}\left[\frac{1}{\sqrt{2}}\right] = \frac{120}{10 \times 2} = 6 \text{ A.}$$

Example 10.3 Two thyristors connected back, to back have a load resistance of $400\,\Omega$ and a supply of 110 V AC. If the firing angle is 60°, find the rms output voltage and average power.

Solution Given $\qquad V = 110 \text{ V}, \ \alpha = 60° = \pi/3$

$$V_m = \sqrt{2} \times 110 \text{ V}$$

(i) rms output voltage,

$$V_L = \sqrt{\frac{1}{2\pi}\int_0^{2\pi} v_L^2 \cdot d\omega t}$$

$$= \sqrt{\frac{1}{2\pi}\left[\int_{\pi/3}^{\pi}(110 \times \sqrt{2}\sin\omega t)^2\,d\omega t + \int_{\pi+\pi/3}^{2\pi}(\sqrt{2} \times 110\sin\omega t)^2\,d\omega t\right]}$$

$$= \frac{\sqrt{2} \times 110}{\sqrt{2\pi}} \sqrt{2 \int_{\pi/3}^{\pi} \sin^2 \omega t \cdot d\omega t}$$

$$= \frac{\sqrt{2} \times 110}{\sqrt{2} \times \sqrt{\pi}} \sqrt{2 \int_{\pi/3}^{\pi} \left(\frac{1 - \cos 2\omega t}{2} \right) d\omega t}$$

$$= \frac{110}{\sqrt{\pi}} \sqrt{\left[\omega t - \frac{\sin 2\omega t}{2} \right]_{\pi/3}^{\pi}} = \frac{110}{\sqrt{\pi}} \sqrt{\left(\pi - \frac{\pi}{3} \right) - \frac{1}{2} (\sin 2\pi - \sin 2\pi/3)}$$

$$V_L = \frac{110}{\sqrt{\pi}} \sqrt{\frac{2\pi}{3} + \frac{\sqrt{3}}{4}} = \frac{110}{\sqrt{\pi}} \times 1.59 = 98.7 \text{ volts.}$$

(ii) Average power
$$P = \frac{V^2}{2\pi R} [2(\pi - \alpha) + \sin 2\alpha]$$

$$= \frac{110^2}{2\pi \times 400} \left[2 \left(\pi - \frac{\pi}{3} \right) + \sin 120° \right]$$

$$= \frac{110 \times 110}{2\pi \times 400} \left[\frac{4\pi}{3} + \frac{\sqrt{3}}{2} \right] = \frac{121}{8\pi} [5.05] = 24.334 \text{ W}$$

$$= 24.33 \text{ watts.}$$

Example 10.4 Derive an expression for the rms load current in a single-phase resistive circuit in which the load voltage is controlled by symmetrical phase angle triggering of a pair of antiparallel thyristors. If the voltage is given by $e = 100 \sin \omega t$ and $R = 50\Omega$, what is the rms load current at $\alpha = 0°, 30°, 60°$?

Solution The instantaneous current for half-cycle, is given by:

$$i(\omega t) = \frac{V_m}{R} \sin \omega t, \quad 0 < \omega t < \pi$$

rms value of every half-cycle,

$$I_{rms} = \left[\frac{1}{\pi} \int_{\alpha}^{\pi} \left(\frac{V_m \sin \omega t}{R} \right)^2 d\omega t \right]^{1/2}$$

$$I_{rms}^2 = \frac{V_m^2}{\pi R^2} \int_{\alpha}^{\pi} \frac{(1 - \cos 2\omega t)}{2} d\omega t$$

$$= \frac{V_m^2}{\pi R^2} \left[\frac{\omega t}{2} - \frac{\sin 2\omega t}{4} \right]_{\alpha}^{\pi} = \frac{V_m^2}{2\pi R^2} \left[(\pi - \alpha) + \frac{\sin 2\alpha}{2} \right]$$

$$I_{rms} = \frac{V_m}{R} \sqrt{\frac{1}{2\pi} \left[\pi - \alpha) + \frac{\sin 2\alpha}{2} \right]}$$

$$= \frac{V}{R} \sqrt{\frac{1}{2\pi} [2 (\pi - \alpha) + \sin 2\alpha]}$$

For $\alpha = 0°$, $\quad I_{rms} = \dfrac{100}{50} \sqrt{\dfrac{1}{2\pi}[2(\pi - 0) + \sin 0]} = \dfrac{100/\sqrt{2}}{50} = 1.4142$ A

For $\alpha = 30°$, $\quad I_{rms} = \dfrac{V}{R}\sqrt{\dfrac{1}{2\pi}[2(\pi - \alpha) + \sin 2\alpha]}$

$$= \dfrac{100/\sqrt{2}}{50}\sqrt{\dfrac{1}{2\pi}[2(\pi - 30°) + \sin 60°]} = 1.39 \text{ A}$$

For $\alpha = 60°$, $\quad I_{rms} = \dfrac{100/\sqrt{2}}{50}\sqrt{\dfrac{1}{2\pi}[2(\pi - 60°) + \sin 120°]} = 1.27$ A.

Example 10.5 The single-phase full wave AC voltage controller has a resistive load $R = 5\,\Omega$ and an input voltage 120 V at 50 Hz. The angle of firing for thyristors Th1 and Th2 is $2\pi/3$. Determine: (i) rms output voltage, (ii) input power factor, (iii) average current of thyristor, and (iv) rms current of thyristor.

Solution   Given
$$V = 120 \text{ V,}$$
$$V_m = \sqrt{2} \times 120 = 169.7 \text{ V}$$
$$R = 5\,\Omega, \ \alpha = \frac{2\pi}{3}$$

rms output voltage,
$$V_L = V_m\left[\frac{1}{\pi}\left(\pi - \alpha + \frac{\sin 2\alpha}{2}\right)\right]^{1/2}$$

$$= 169.7\left[\frac{1}{\pi}\left(\pi - \frac{2\pi}{3} + \frac{\sin 4\pi/3}{2}\right)\right]^{1/2}$$

$$= 169.7\left[\frac{1}{3} + \frac{\sin 4\pi/3}{2\pi}\right]^{1/3}$$

$$= 53.06 \text{ V}$$

Load current, $\quad I_L = V_L/R = 53.06/5 = 10.61$ A (rms)

Output power $\quad P_o = I_L^2 R = 10.61^2 \times 5 = 563$ W

$$VA = VI = VI_L = 120 \times 10.61 = 1273.2$$

$$\text{PF} = \frac{P_o}{VI_L} = \frac{563}{1273.2} = 0.442$$

Average current of thyristor,

$$I_{Th} \text{ (average)} = \frac{\sqrt{2}V}{2\pi R}(1 + \cos \alpha)$$

$$= \frac{\sqrt{2} \times 120}{2\pi \times 5}(1 + \cos 2\pi/3) = 2.7 \text{ A}$$

Thyristor current (rms), $\quad I_R = \dfrac{V}{\sqrt{2}R}\left[\dfrac{1}{\pi}\left(\pi - \alpha + \dfrac{\sin 2\alpha}{2}\right)\right]^{1/2} = 7.5$ A.

Example 10.6 The single-phase full-wave voltage controller has a resistive load of $1.50\,\Omega$ with input voltage 120 V at 50 Hz. If the desired output power is 7.5 kW. Determine: (i) firing angles of thyristors, (ii) rms output voltage, (iii) input power factor, (iv) average current of thyristors, and (v) rms value of current of thyristor.

Solution

(i) Given
$$V = 120 \text{ V}, V_m = \sqrt{2} \times 120 = 169.7 \text{ V}$$

Output power
$$P_o = 7500 \text{ W}, R = 1.5\,\Omega$$

$$P_o = 7500 = \frac{V_L^2}{R} = \frac{V^2}{R}\left[\frac{1}{\pi}\left(\pi - \alpha + \frac{\sin 2\alpha}{2}\right)\right]$$

$$7500 = \frac{120^2}{1.5}\left[\frac{1}{\pi}\left(\pi - \alpha + \frac{\sin 2\alpha}{2}\right)\right]$$

Iteration yields
$$\alpha = 62.7°.$$

(ii) Output voltage
$$V_L = V\left[\frac{1}{\pi}\left(\pi - \alpha + \frac{\sin 2\alpha}{2}\right)\right]^{1/2}$$

$$= 120\left[\frac{1}{\pi}\left(\pi - 67.7° + \frac{\sin 2 \times 67.7°}{2}\right)\right]^{1/2}$$

$$V_L = 106.1 \text{ V}$$

(iii) PF $= \dfrac{P_o}{VI_L} = \dfrac{V_L^2/R}{VI_L} = \dfrac{106.1^2/1.5}{120 \times 106.1/1.5} = \dfrac{7.498}{8520} = 0.88$

(iv) Average current of thyristor,

$$I_A = \frac{\sqrt{2}V(1 + \cos\alpha)}{2\pi R}$$

$$= \frac{\sqrt{2} \times 120(1 + \cos 62.7°)}{2\pi \times 1.5} = 26.25 \text{ A}$$

(v) rms value of thyristor current

$$I_R = \frac{I_L}{\sqrt{2}} = \frac{V_L/R}{\sqrt{2}} = \frac{106.1/1.5}{\sqrt{2}} = 49.99 \text{ A.}$$

Example 10.7 The three-phase full-wave rectifier supplies a star-connected resistive load of $R = 10\,\Omega$ per phase and line to line voltage is 208 V, the firing angle is $\alpha = \pi/3$. Determine: (i) the rms output phase voltage V_L, and (ii) input power factor.

Solution

(i) Load line voltage
$$= 208 \text{ V}$$

Voltage/phase
$$= 208\sqrt{3} = 120 \text{ V}$$
$$\alpha = \pi/3, R = 10\,\Omega$$

Output load voltage
$$V_L = \sqrt{3}V_m\left[\frac{1}{\pi}\left(\frac{\pi}{6} - \frac{\alpha}{4} + \frac{\sin 2\alpha}{8}\right)\right]^{1/2}$$

$$= \sqrt{3} \times \sqrt{2} \times 120\left[\frac{1}{\pi}\left(\frac{\pi}{6} - \frac{\pi/3}{4} + \frac{\sin 2\pi/3}{8}\right)\right]^{1/2}$$

$$= \sqrt{3} \times \sqrt{2} \times 120 \left[\frac{1}{\pi} \left(\frac{\pi}{12} + \frac{\sin 120°}{8} \right) \right]^{1/2} = 100.9 \text{ V}$$

(ii) Phase current $\quad I_a = 100.9/10 = 10.09$ A

Output power $\quad P_o = 3I_a^2 R = 3 \times 10.09^2 \times 10 = 3054.24$ W

Line current $\quad$ = phase current = 10.09 A

$$VA = 3VI_L = 3 \times 120 \times 10.09 = 3632.4 \text{ V-A}$$

$$\text{PF} = \frac{P_0}{VA} = \frac{3054.2}{3632.4} = 0.84 \text{ (lag)}.$$

Example 10.8 The circuit is controlled as a synchronous tap changer. Primary voltage is 208 V at 60 Hz. The secondary voltages are $V_1 = 120$ V and $V_2 = 88$ V. If load resistance is $5\,\Omega$ and load voltage is 180 V, determine: (i) the delay angles of thyristors, (ii) rms current of thyristors Th_1 and Th_2, and (iii) rms current of thyristors Th3 and Th4.

Solution

(i) Given $V_1 = 120$ V, $V_2 = 88$ V, $V_L = 180$ V, primary voltage, $V_p = 208$ V and $R = 5\,\Omega$

$$V_L = \left[\frac{V_1^2}{\pi} \left(\alpha - \frac{\sin 2\alpha}{2} \right) + \frac{(V_1 + V_2)^2}{\pi} \left(\pi - \alpha + \frac{\sin 2\alpha}{2} \right) \right]^{1/2}$$

$$180 = \left[\frac{120^2}{\pi} \left(\alpha - \frac{\sin 2\alpha}{2} \right) + \frac{(120 + 88)^2}{\pi} \left(\pi - \alpha + \frac{\sin 2\alpha}{2} \right) \right]^{1/2}$$

Using iterative technique, the α is given by

$$\alpha = 79°$$

(ii) rms current of thyristors Th1 and Th2 is

$$I_R = \frac{V_1 + V_2}{\sqrt{2}R} \left[\frac{1}{\pi} \left(\pi - \alpha - \frac{\sin 2\alpha}{2} \right) \right]^{1/2}$$

$$= \frac{120 + 88}{\sqrt{2} \times 5} \left[\frac{1}{\pi} \left(\pi - 79° + \frac{\sin 2 \times 79°}{2} \right) \right]^{1/2} = 23.18 \text{ A}$$

(iii) rms current of thyristors Th3 and Th4

$$I_{R3} = \frac{V_1}{\sqrt{2}R} \left[\frac{1}{\pi} \left(\alpha - \frac{\sin 2\alpha}{2} \right) \right]^{1/2} = 6.5 \text{ A}.$$

EXERCISES

Multiple Choice Questions

1. A single-phase voltage controller is connected to resistance of $10\,\Omega$ and a supply of $200 \sin 314t$ Volts. For a firing angle of $90°$, the average thyristor current is:

   (A) 10 A $\qquad$ (B) $5\sqrt{2}/\pi$ A

   (C) $10/\pi$ A $\qquad$ (D) $5\sqrt{2}$ A

2. A single-phase voltage controller, using two back to back thyristors, is found to be operating as a controlled rectifier, because:

   (A) load is R and pulse gating is used

   (B) load is R-L and pulse gating is used

(C) load is R and high frequency carrier gating is used

(D) load is R-L and continuous gating is used

3. A single-phase AC voltage controller fed from 50 Hz system supplies a load having resistance and inductance of $2.0\,\Omega$ and 6.36 mH respectively. The control range of firing angle for this regulator is:

(A) $0° < \alpha < 180°$ (B) $45° < \alpha < 180°$

(C) $90° < \alpha < 180°$ (D) $0° < \alpha < 45°$

4. In a single-phase voltage controller with R-L load, with firing angle α the load phase angle ϕ and β is the extinction angle. For this voltage controller, output power can be controlled if $\alpha > \phi$ and

(A) $(\beta - \alpha) = \pi$ (B) $(\beta - \alpha) < \pi$

(C) $\beta < \pi$ (D) none of these

5. A single-phase voltage controller feeds power to a resistance of $10\,\Omega$. The source voltage is 200 V. For a firing angle of 90°, the rms value of thyristor current is:

(A) 3 A   (B) 4 A

(C) 10 A   (D) 15 A

6. In a single-phase voltage controller with R-L load, AC output power can be controlled if:

(A) firing angle $\alpha > \phi$ (load phase angle) and conduction angle $\gamma = \pi$

(B) $\alpha > \phi$ and $\gamma < \pi$

(C) $\alpha > \phi$ and $\gamma = \pi$

(D) none of these

7. A load, consisting of $R = 10\,\Omega$ and $L = 10\,\Omega$, is being fed from 230 V, 50 Hz source through a single-phase voltage controller.

For a firing angle of 30°, the rms value of load current would be:

(A) 23 A

(B) $\dfrac{23}{\sqrt{2}}$ A

(C) more than $\dfrac{23}{\sqrt{2}}$ A

(D) less than $\dfrac{23}{\sqrt{2}}$ A

8. A single-phase voltage controller is employed for controlling the power flow from 260 V, 50 Hz source into a load consisting of $R = 5\,\Omega$ and $X = 12\,\Omega$. The value of maximum rms load current and the firing angle are respectively:

(A) 20 A, 0°   (B) 26 A, 0°

(C) 20 A, 90°   (D) 25 A, 60°

9. A resistance of $10\,\Omega$ is fed through a single-phase voltage controller from a voltage source of $200 \sin 314t$. For a firing angle of 90°, the power delivered to load is:

(A) 0.5 kW   (B) 0.75 kW

(C) 1 kW   (D) none of these

10. A single-phase voltage controller feeds an induction motor and a heater:

(A) in both the loads, fundamental and harmonics are useful

(B) in an induction motor only fundamental and in heater only harmonics are useful

(C) in an induction motor only fundamental and in heater harmonics as well as fundamental are useful

(D) in an induction motor only harmonics and in heater only fundamental are useful

ANSWER KEY

1. (C)   **2.** (B)   **3.** (B)   **4.** (B)   **5.** (C)   **6.** (B)   **7.** (B)   **8.** (A)

9. (C)   **10.** (C)

Review Questions

1. What is an AC voltage controller?

2. Give the industrial applications of AC voltage controllers.

3. Draw the possible configurations of single-phase voltage controller and compare them.

4. Analyse the output voltage waveform into various harmonics with fourier series and derive the expressions for amplitude of nth harmonic.

5. Show that power factor for a single-phase AC controller for resistive load is as follows:

$$PF = \left[\frac{1}{\pi} \left\{ (\pi - \alpha) + \frac{1}{2} \sin 2\alpha \right\} \right]^{1/2}.$$

6. Show that the form factor of thyristor current for single-phase inverse parallel thyristor connection with resistive load and α triggering angle is given by

$$FF = \frac{1}{(1 + \cos \alpha)} \left[\pi \left(\pi - \alpha + \frac{1}{2} \sin 2\alpha \right) \right]^{1/2}.$$

7. What is the control range of delay angle for single-phase bidirectional controllers?

8. What are the advantages of transformer tap changers?

AC-to-AC Converters— Cycloconverters

11.0 INTRODUCTION

A cycloconverter directly converts AC power at one input frequency to output power at a different (normally lower) frequency. It is essentially a dual power converter which operates in such a way to produce an alternating output voltage.

A cycloconverter is a frequency changer that converts AC power at one output frequency to output power at a different frequency without any intermediate DC link.

These converters are basically meant for producing low frequency AC voltage. The main application of cycloconverter is in the electric traction where low frequency is preferred. The ruggedness and low weight of solid state cycloconverters also makes it attractive for aircraft electric systems which require the production of a constant output frequency from a variable speed alternator.

Industrial applications: In addition to general speed control of induction and synchronous motors, the industrial applications of cycloconverters include:

1. Used to link the supply systems of different frequencies
2. Gearless cement and ball mill drives
3. Static VAR generation
4. Variable speed, constant frequency power generation for shipboard or aircraft power supply
5. Speed control of high power AC drives
6. HVDC transmission
7. Electric traction
8. Wind power systems
9. Induction heating

Major problems in cycloconverter is that their output wave shapes are not very good. For higher frequencies, the wave shapes are distorted. Another disadvantage of using a cycloconverter for motor speed control is that the power factor of the system is reduced.

11.1 TYPES OF CYCLOCONVERTERS

Cycloconverter can be categorised as follows:

1. Single-phase cycloconverter   2. Three-phase cycloconverter

Principle of cycloconverter:
A cycloconverter can be considered
to be composed of two converters
connected back to back as shown in
Fig. 11.1 (a) with a pure resistive load.
A positive (P) and negative (N) centre
tap converter are connected in parallel
so that the voltage and current of either
polarity can be supplied to the load.
The load waveforms of Fig. 11.1 (b)
show that the instantaneous power
flows in the load fall into one of
the four periods. The two periods,
when the product of load voltage
and current is positive, require power
flow into the load dictating a situation
where the converter groups rectify,
the positive and negative groups
conducting respectively during the
appropriate positive and negative load
current periods.

The other two periods represent
times when the product of load
voltage and current is negative, hence
the power flow is out of the load,
demanding that the converter operates
in the inverting mode. A positive and
a negative centre-tap converter groups
are connected in parallel so that voltage
and current of either polarity can be
supplied to the load. The waveforms
are drawn assuming a resistive load. In
Fig. 11.1 (c) an integral half-cycle output
wave is fabricated with fundamental

frequency $f_o = \left(\dfrac{1}{n}\right) f_i$, where n is

the number of input half-cycles per
half-cycle of the output. The firing
angle can be modulated to control
the voltages as well as its harmonic
content as shown in Fig. 11.1 (d). Step
up conversion is one in which $f_o > f_i$
shown in Fig. 11.1 (e).

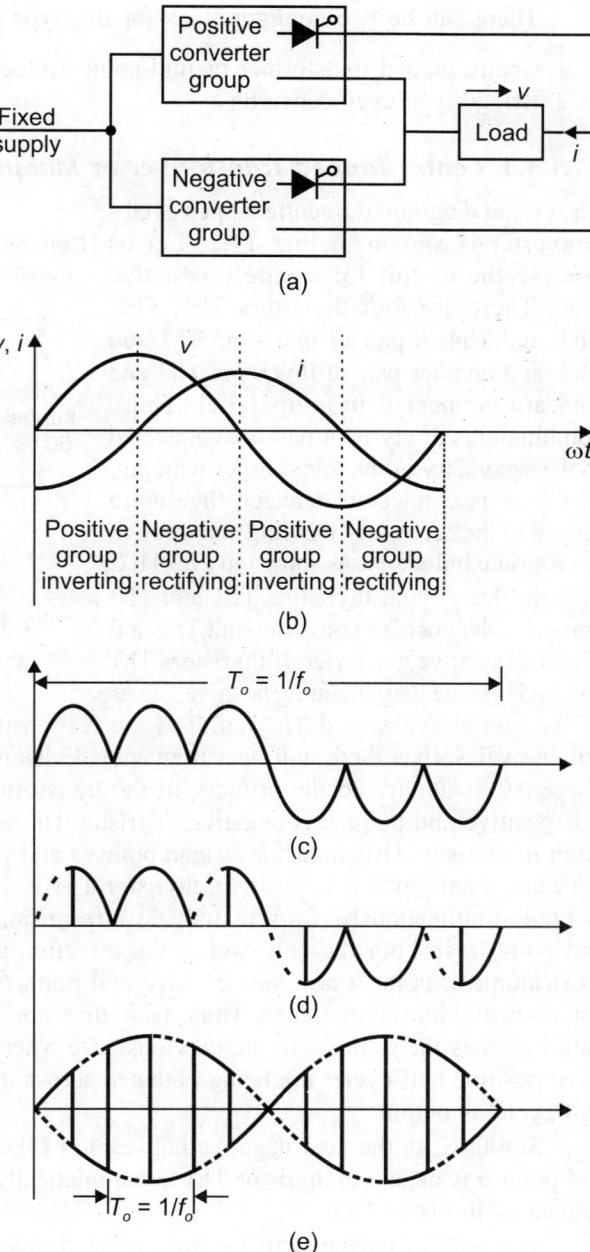

Fig. 11.1 Waveforms of two converters connected back
to back

11.1.1 Single-phase Cycloconverter

It is a single-phase converter whose input and output are single-phase AC. The input
AC voltage of supply frequency is converted into lower frequency AC output.

There can be two configurations for this type of cycloconverter, namely:

1. Centre tapped transformer or mid-point cycloconverter
2. Bridge type cycloconverter

11.1.1.1 *Centre Tapped Transformer or Mid-point Cycloconverter*

The circuit diagram of a centre tapped cyclo-converter is shown in Fig. 11.2. Let us analyse the circuit for a purely resistive load. There are four thyristors Th1, Th2, Th3, and Th4. A pair of thyristors Th1 and Th2, and another pair of thyristors Th3 and Th4 are connected in antiparallel. These combinations of thyristor pair are connected to the secondary of the transformer winding. The load is connected between the centre tapped of the secondary winding and through appropriate inductor L as shown in Fig. 11.2.

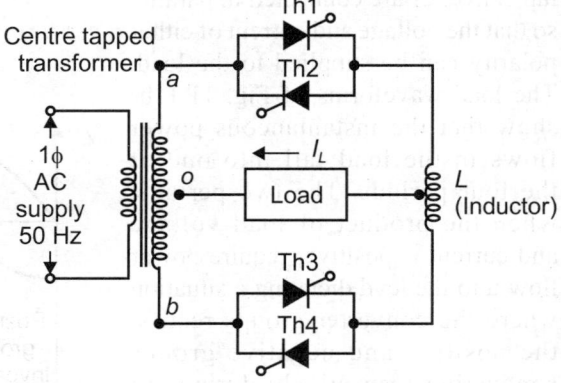

Fig. 11.2 Circuit diagram of centre tapped cyclo-converter

In this circuit, thyristors Th1 and Th3 comprise the positive converter and Th2 and Th4 the negative converter. If thyristors Th1 and Th3 operate for generating positive halves of the output cycles and Th2 and Th4 are responsible for producing negative halves of the output cycles, then there will be a frequency division by three at the output. Assuming single-phase AC is applied to the primary of the transformer. During the positive half-cycle, point a is positive and point b is negative, thyristor Th1 is turned on, then the current passes from point a, thyristor Th1, inductor L, load point O and point b. In the negative half-cycle, point a is negative and point b is positive, thyristor Th1 is automatically turned off and thyristor Th3 is fired simultaneously. Current will flow from point b, thyristor Th3, inductor L, point O and point a. In both the half-cycles, the direction of current in the load remains the same. Next moment, point a becomes positive and point b becomes negative, thus thyristor Th3 is automatically line commutated. Thus, again thyristor Th1 turned on simultaneously. The current path becomes the same as in the previous case when thyristor Th1 was conducting. Thus, the three positive half-cycles are being obtained across the load to produce one combined positive half-cycle as output.

Similarly, in the next negative half-cycle of the AC input when point b is again positive and point a is negative, thyristor Th1 is automatically switched off. Now thyristor Th2 is fired instead of thyristor Th3.

The path of current will be from point b, load, inductor L, thyristor Th2 and point a. Thus, the direction of flow of current through the load is reversed. In the next positive half-cycle point a is positive and point b is negative. Thyristor Th2 is automatically turned off, thyristor Th4 which is in the conducting mode is simultaneously turned on. Now current will follow the path from point a, load, inductor L, thyristor Th4 and back to point b. Thus, the direction of current through the load remains the same. In the next negative half-cycle, point b is positive and point a is negative, thyristor Th4 is automatically switched off and thyristor Th2 is fired. The direction of current through the load remains again the same. Hence, one negative half-cycle at the output is produced by combining three negative half-cycles of the input cycle. Three positive half-cycles of the input are combined to produce one positive half-cycle at the output by thyristors Th1 and Th3. Similarly three negative half-cycles of the input AC are combined to produce one negative cycle at the output by thyristors Th2 and Th4. This process is repeated again for continuous production of voltage at the output. The input and output waves are shown in Fig. 11.3. The output voltage may be changed by varying the firing angle of the thyristors. For resistive load each thyristor will conduct for $(\pi - \alpha)$ and turn off naturally by the supply voltage reversal at the end of each half-cycle. If the output frequency is a submultiple of input frequency, the last conducting thyristor of the positive or negative converter is turned off when the voltage goes through zero, and those of the other group are turned on. If the output frequency is not submultiple of the input frequency, the thyristors in the negative converter are fired even before the thyristor in the positive converter are turned off, and this results in a short circuit of the input source and the output voltage is zero. The short circuit condition continues for less than a half-cycle and occurs at every half-cycle of the output frequency.

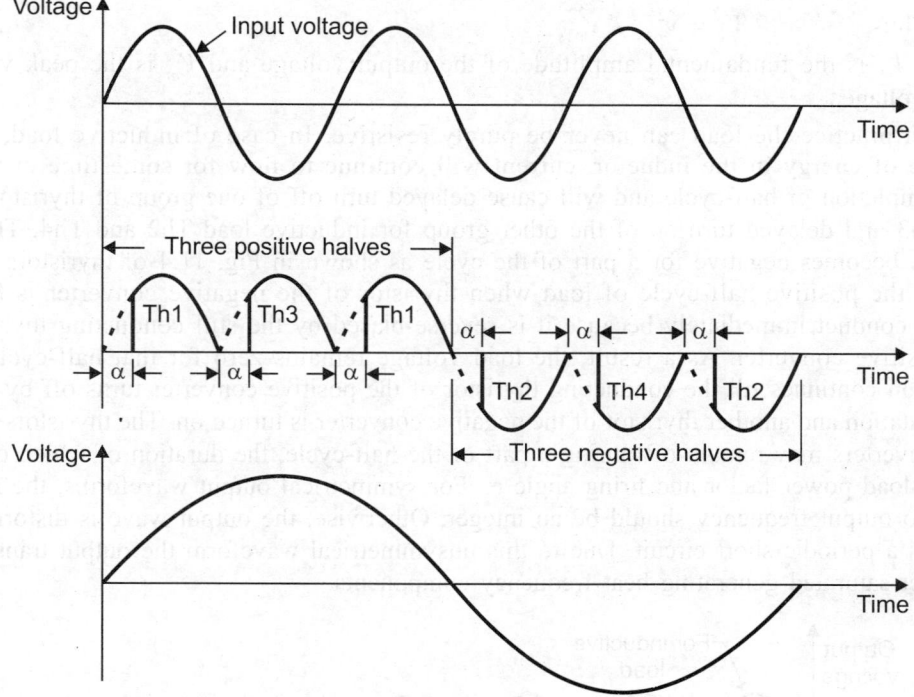

Fig. 11.3 The input and output waves

Mathematical analysis of fundamental amplitude of output: Assume the circuit shown in Fig. 11.2, which has a resistive load and has output frequency of $(1/m)$ times the input frequency. Thus, there will be m half-cycles input in each half period of the output. Let the firing angle be α. The rms value of output voltage will be

$$(V_0)_{\text{rms}} = \sqrt{\frac{1}{2} \frac{V_m^2}{\pi} \left(\pi - \alpha + \frac{1}{2}\sin 2\alpha\right)} \qquad \qquad ...(11.1)$$

Fundamental amplitude $\quad v_{o1s} = \dfrac{2V_m}{m\pi}\left[\displaystyle\int_\alpha^\pi \sin\omega t \sin\left(\frac{\omega t}{m}\right) d\omega t + \int_\alpha^\pi \sin\omega t \sin\left(\frac{\omega t + \pi}{m}\right) d\omega t\right.$

$$\left. + ... + \int_\alpha^\pi \sin\omega t \sin\left(\frac{\omega t + (m-1)\pi}{m}\right) d\omega t\right]$$

$$= \frac{V_m}{m\pi} \sum_{n=1}^{m} \left[\frac{m}{m-1} \left\{ \sin\frac{n\pi}{m} - \sin\left(\alpha - \frac{\alpha + (n-1)\pi}{m} \right) \right\} \right.$$

$$\left. + \frac{m}{m+1} \left\{ \sin\frac{n\pi}{m} + \sin\left(\alpha - \frac{\alpha + (n-1)\pi}{m} \right) \right\} \right] \quad ...(11.2)$$

Similarly, $\quad v_{olc} = \dfrac{V_m}{m\pi} \displaystyle\sum_{n=1}^{m} \left[\dfrac{m}{m+1} \left\{ \dfrac{\cos n\pi}{m} + \cos\left(\alpha + \dfrac{\alpha + (n-1)\pi}{m} \right) \right\} \right.$

$$\left. + \frac{m}{m-1} \left\{ \frac{\cos n\pi}{m} + \cos\left(\frac{\alpha + \alpha + (n-1)\pi}{m} \right) \right\} \right] \quad ...(11.3)$$

Thus, $\qquad V_1^2 = v_{ols}^2 + v_{ols}^2 \qquad\qquad\qquad ...(11.4)$

where, V_1 is the fundamental amplitude of the output voltage and V_m is the peak value of input voltage.

In practice the load can never be purely resistive. In case of inductive load, due to storage of energy in the inductor, current will continue to flow for some time even after the completion of half-cycle and will cause delayed turn off of one group of thyristors, Th1 and Th3 and delayed turn on of the other group for inductive load Th2 and Th4. The load voltage becomes negative for a part of the cycle as shown in Fig. 11.4 of thyristors. At the end of the positive half-cycle of load when thyristor of the negative converter is fired, it cannot conduct immediately because it is reverse-biased by the still conducting thyristor of the positive converter. As a result, the load voltage remains zero for that half-cycle. This condition continues till the conducting thyristor of the positive converter turns off by natural commutation and another thyristor of the negative converter is turned on. The thyristors of both the converters may remain off during a part of the half-cycle, the duration of which depends on the load power factor and firing angle α. For symmetrical output waveforms, the ratio of input to output frequency should be an integer. Otherwise, the output wave is distorted and there is a periodic short circuit. Due to this unsymmetrical waveform the output transformer may get saturated generating beat frequency components.

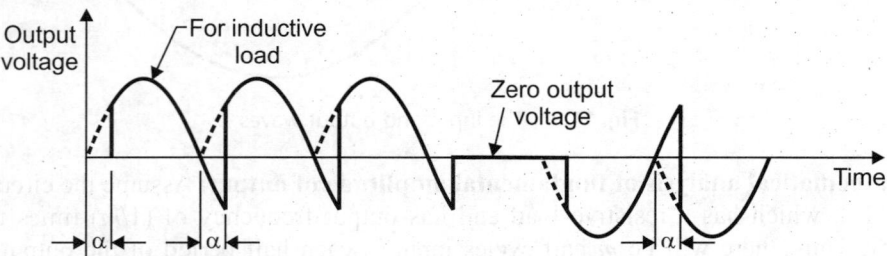

Fig. 11.4 The load voltage becomes negative for a part of the cycle

If the frequency is a small fraction of the total time period then the distortion is less. Therefore in a cycloconverter, the ratio of output to input frequency should be at least less than three.

11.1.1.2 *Bridge Type Cycloconverter*

Bridge type cycloconverter is a fully controlled bridge circuit both for positive and negative group converters as shown in Fig. 11.5.

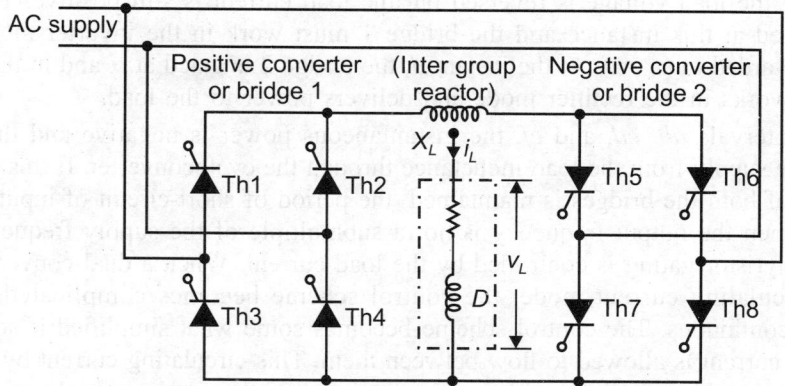

Fig. 11.5 Bridge type cycloconverter is a fully controlled bridge circuit both for positive and negative group converters

Bridge 1 is formed by four thyristors, Th1, Th2, Th3 and Th4, similarly, bridge 2 is formed by four thyristors, Th5, Th6, Th7 and Th8. In this circuit, the positive converter supplies load current in the positive half of the output cycle and the negative converter supplies current in the negative half-cycle. The two bridges should not conduct simultaneously because that would produce short circuit at the input. Since only one converter supplies the load current at a given time, it is not necessary to provide any gating pulse to the other converter. The other converter is off and there is no circulating current. This can also be achieved by shifting the firing angles of the 'idle' converter to such a value that there is no possibility of the circulating current to flow.

It is clear from the above that the circuit functions like a circulating current mode dual converter. Depending upon the firing sequence of the thyristors, the output frequency may be varied. The sequence of firing is repeated again and again to produce continuous output waves. The firing angles of thyristors of both the converters should be the same to produce a symmetrical output. For resistive load, the load current is discontinuous but for inductive load, the load current may be continuous or discontinuous, depending upon the firing angle and load power factor.

The output voltage and current are in phase for resistive load and both the positive and negative converters work in the rectifier mode. But for inductive load, the load current lags behind the load voltage shown in Fig. 11.6. The load current i_L lags behind the load voltage V_L by an angle ϕ. During the period ab, the load voltage is positive but the load current is negative, and so load current is supplied by bridge 2 and bridge 1 cannot be fired till the load current becomes zero. During this period, bridge 1 must work in the inverter mode with the thyristor firing angle starting at $\pi-\alpha$.

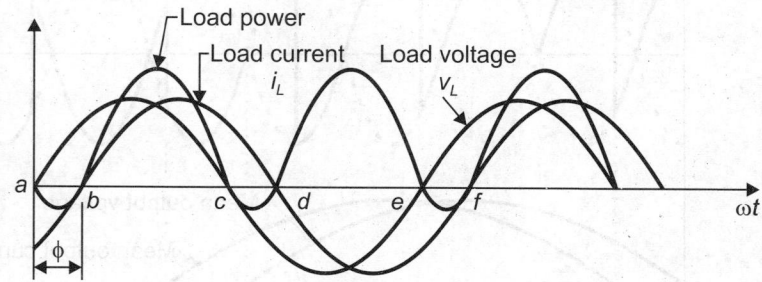

Fig. 11.6 The load current lags behind the load voltage

During the period bc the bridge 1 is triggered at angle α, both the load voltage and current are positive, and it works in the rectifier mode delivering power to the load. During the period *cd*, the load voltage is reversed but the load current is still positive. The bridge 2 cannot be gated at this instance and the bridge 1 must work in the inverter mode with the firing angle retarded to $(\pi - \alpha)$. At the instant *d*, the bridge 2 is gated at α and in the period *de*, the bridge 2 works in the rectifier mode and delivers power to the load.

In the intervals *ab*, *cd*, and *ef*, the instantaneous power is negative and the energy is returned to the supply from the load inductance through the cycloconverter. If this sequence of firing angles of both the bridges is maintained, the period of short-circuit of input supply can be avoided when the output frequency is not a submultiple of the supply frequency. This is because the thyristor gating is controlled by the load current. When a dual converter operates in the noncirculating current mode, the control scheme becomes complicated if the load current is discontinuous. The control scheme becomes some what simplified if some amount of circulating current is allowed to flow between them. This circulating current by itself keeps both the converters in virtually continuous conduction over the whole control range. This type of operation is called the *circulating-current mode of operation*.

The firing angles of the back to back converters are such that $\alpha_P + \alpha_N = 180°$. These values are controlled depending on the voltage required. For higher average voltage, the firing angle of the converter in the rectifying mode is advanced and that of the converter in the invertering mode is retarded. In that case, the positive converter operates in the rectifier mode while the negative converter operates in the inverter mode. The firing angles of the individual converter are controlled so that each produces the same mean DC terminal voltage with the same circuit polarity. Thus, the mean output voltage of the rectifier group is maintained equal to the mean back-emf of the inverter group. Thus, the circulation of large low frequency currents between groups is avoided.

But the instantaneous voltages of the two groups are not identical and large harmonic currents circulate unless limited by some external impedance. The flow of harmonic current may be reduced by the introduction of an intergroup reactor. The lowfrequency output current of the cycloconverter is reduced by the reactance $4k\,X_L$ where, X_L is the reactance of half of the reactor and k is the order of harmonic. In the cycloconverter application, the delay angles of α_P and α_N are continuously modulated maintaining the definite relation of $\alpha_P + \alpha_N = 180°$. The output voltage and current of a single-phase bridge-type cycloconverter are shown in Fig. 11.7.

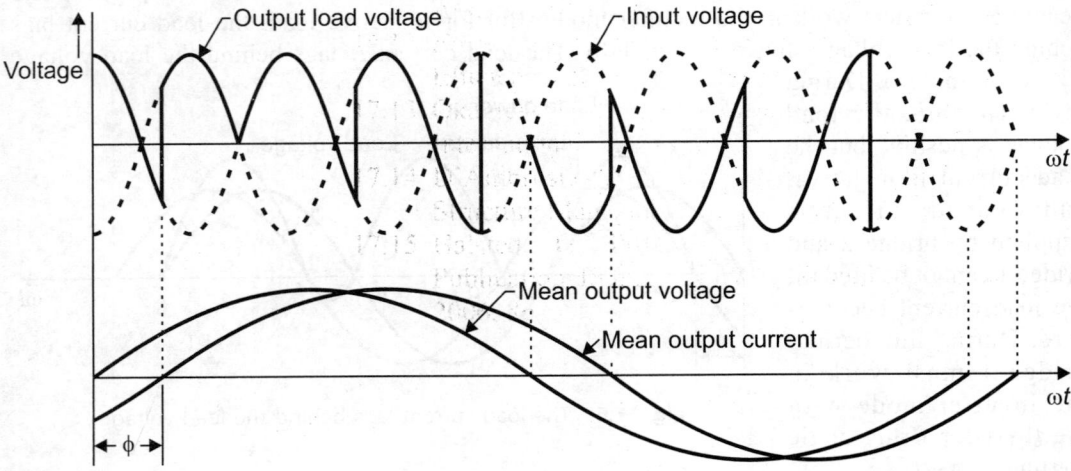

Fig. 11.7 The output voltage and current of a single-phase bridge-type cycloconverter

Circulating current mode: The internal impedance of converters is very small and hence these circulating currents may build-up to dangerously high values. Current limiting reactors are introduced in the circuit. The problem of circulating current can be completely avoided by operating one converter at a time. The circulating current mode has two important effects upon the system operation. In the power circuit, the load current is free to flow in either direction at any time. As a result, the reversal of load current becomes smooth and is not affected by any control function. Secondly, a cosine relationship can be maintained between the output voltage and the firing angle, irrespective of the type of load current, i.e. the relationship between the output voltage and the firing angle is not hampered whether the load current is continuous or discontinuous.

11.1.2 Three-phase Cycloconverters

A three-phase cycloconverter works on the same principle as a single-phase cycloconverter. The type of a three-phase cycloconverter depends on the number of pulses used. A single-phase mid-point cycloconverter has two pulses. The amount of ripple content can be reduced by increasing the number of pulses as used in three-phase cycloconverters. In this section we will study how single-phase low frequency output voltage is fabricated from the segments of three-phase input voltage waveform.

11.1.2.1 *Three-phase to Single-phase Cycloconverter*

The operation of three-phase to single-phase cycloconverter with resistive and inductive loads alongwith associated waveforms are discussed here. Figure 11.8(a) shows the connection of a three-phase cycloconverter feeding single-phase load having positive and negative group. If at any instant thyristors are conducting in both positive and negative groups then a short circuit exists on the supply via thyristors. To avoid this eventually, a reactor can be inserted as shown in Fig. 11.8(b) between the groups to limit the circulating current or the firing control circuitry can be arranged so that neither group is fired, while current is flowing in the other group.

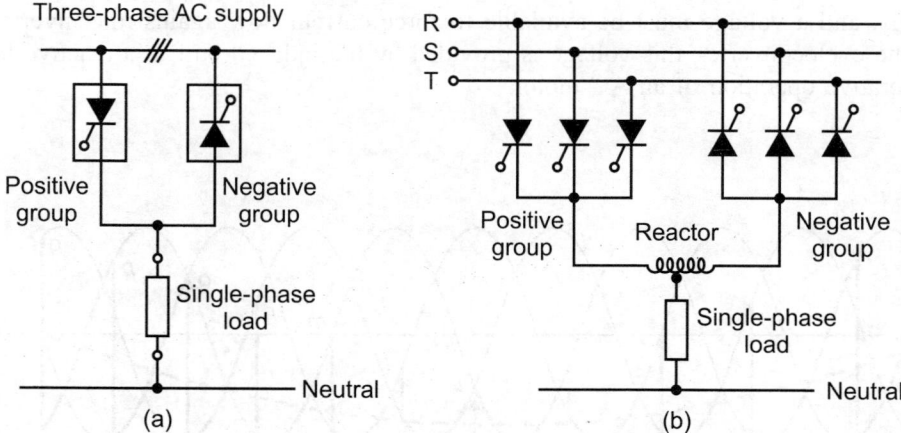

Fig. 11.8 Circuit diagram of three-phase to single-phase cycloconverter with load

For converting three-phase supply to single-phase supply at a lower frequency, the basic principle is to vary progressively the firing angle of three thyristors of a three-phase half-circuit. Average direct voltage output is given by

$$V_d = V_{d_0} \cos \alpha \qquad \qquad ...(11.5)$$

where, α is the firing angle or delay angle, V_{d_0} is the mean output voltage with zero firing delay. Assume the rectifier firing angle is slowly varied as shown in Fig. 11.9. At point a, there is zero delay and the mean output voltage has its maximum value V_{d_0}. At point b, the output voltage is slightly reduced by the introduction of a small firing delay. Further reduction is obtained at c, d and e, while at f, the firing angle is $\pi/2$ and V_d is zero. Thus, if the gating circuitry is suitably designed, a low frequency sinusoidal variation may be superimposed on the output voltage, V_d. In Fig. 11.9, the rectifier is shown conducting during intervals of negative output voltage. This means that the circuit is temporarily inverting, and returning reactive energy from the load to the AC supply. However, for firing angles between zero and $\pi/2$, the net power flow is from the AC supply into the load.

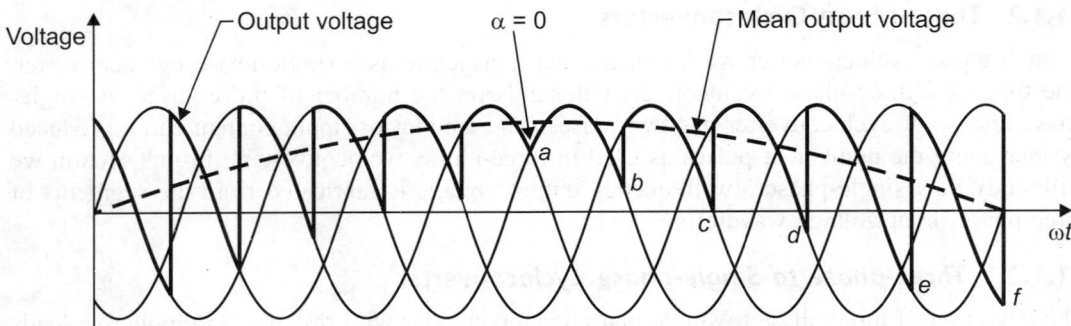

Fig. 11.9 The rectifier is shown conducting during intervals of negative output voltage

The average back emf of an inverter may also be controlled in a sinusoidal manner by suitable variation of the firing angle between $\pi/2$ and π. In Fig. 11.10, the back emf has its maximum value, $-V_{d_0}$, at point k where, $\alpha = \pi$. By reducing α, the back emf is also reduced as shown at points l, m, n, o, p and q. For inverter operation, the net power flow V_s into the AC supply, and a voltage must be available to force current flow against the inverter back emf. In the cycloconverter, this voltage is provided by the induced emf in a reactive load, or by regenerative operation of an AC motor.

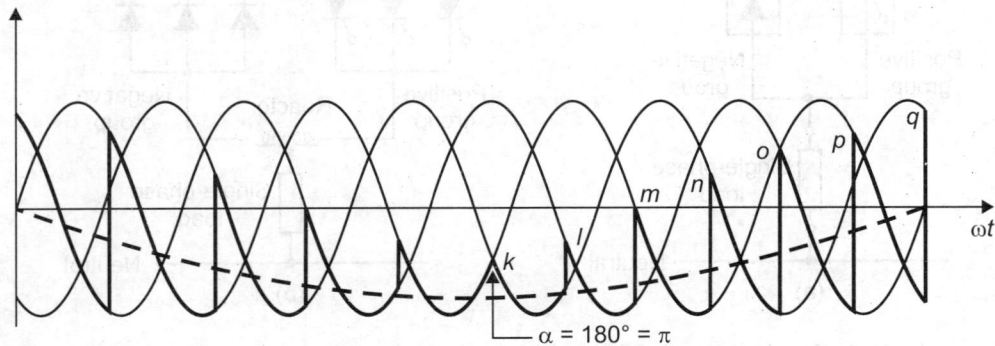

Fig. 11.10 Waveforms as firing angle

If the firing angle is varied from zero to π and back again to zero, one complete cycle of the low frequency variation is superimposed on the average output voltage. The superimposed frequency is determined solely by the rate of variation of α and is independent of the supply

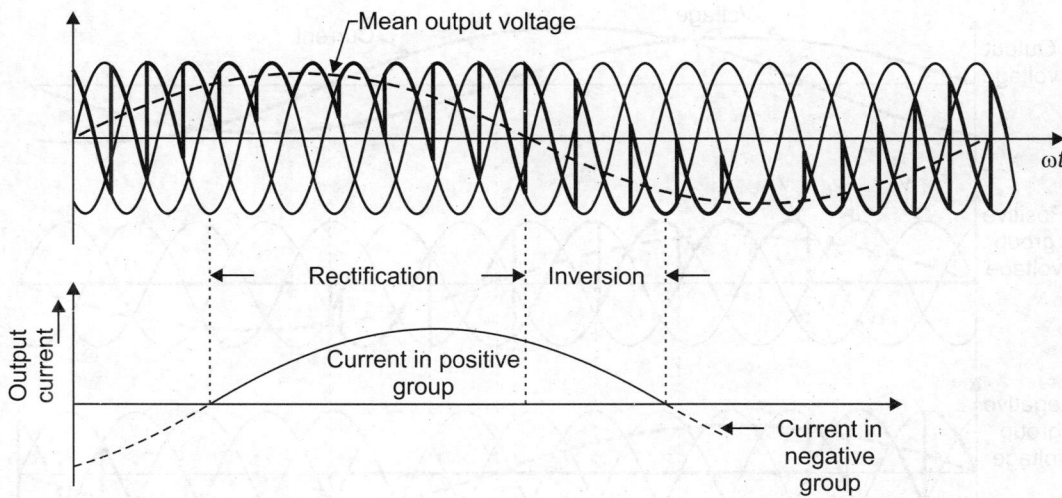

Fig. 11.11 The production of a complete cycle of the low frequency waveform

frequency. Figure 11.11 shows the production of a complete cycle of the low frequency waveform. This diagram emphasizes the fact that the cycloconverter is basically a switching arrangement. Each thyristor switch opens and closes at suitable instants so that a low frequency output waveshape is fabricated from segments of the input waveform. The harmonic content of the output voltage decreases as the ratio of output to input frequency is reduced and as the number of supply phases is increased.

The output voltage of a cycloconverter can be varied sinusoidal, it is not free from distortion. The distortions are inherent in the phase-controlled converter, because the output voltage is obtained from segments of input voltage waves. However, the distortion is less in a three-phase system as compared to that in a single-phase system and can be minimised by adequate filters at the output. The distortion increases if the ratio of the output and input frequency increases. The cycloconverter can convert power from a single-phase supply, producing either single-phase power and also from a three-phase supply, giving out either single-phase or three-phase power.

Since the positive and negative groups are connected in inverse parallel, their average output voltages must always be equal in magnitude and opposite in sign. This is done in order to avoid large circulating currents at output frequency. This is achieved when the delay angles of the positive and negative groups, α_P and α_N, are related by the formula $\alpha_P = \pi - \alpha_N$. However, the instantaneous output voltages of the two groups are quite different, and large harmonic currents will circulate around the low impedance circuit unless they are limited by using a centre-tapped reactor to combine the outputs. Alternatively, the circulating currents may be suppressed by removing the gating signals from each group for the half-cycle during which the group is not delivering load current. When the load is inductive the waveforms are shown in Fig. 11.12. The load current will lag the voltage, the group on periods are delayed relative to the desired output voltage. The group thyristors are fired at such angles to achieve an output as close as a possible sine wave, but now the lagging load current takes each group into the inverting mode. The group will cease conducting when the load current reverses.

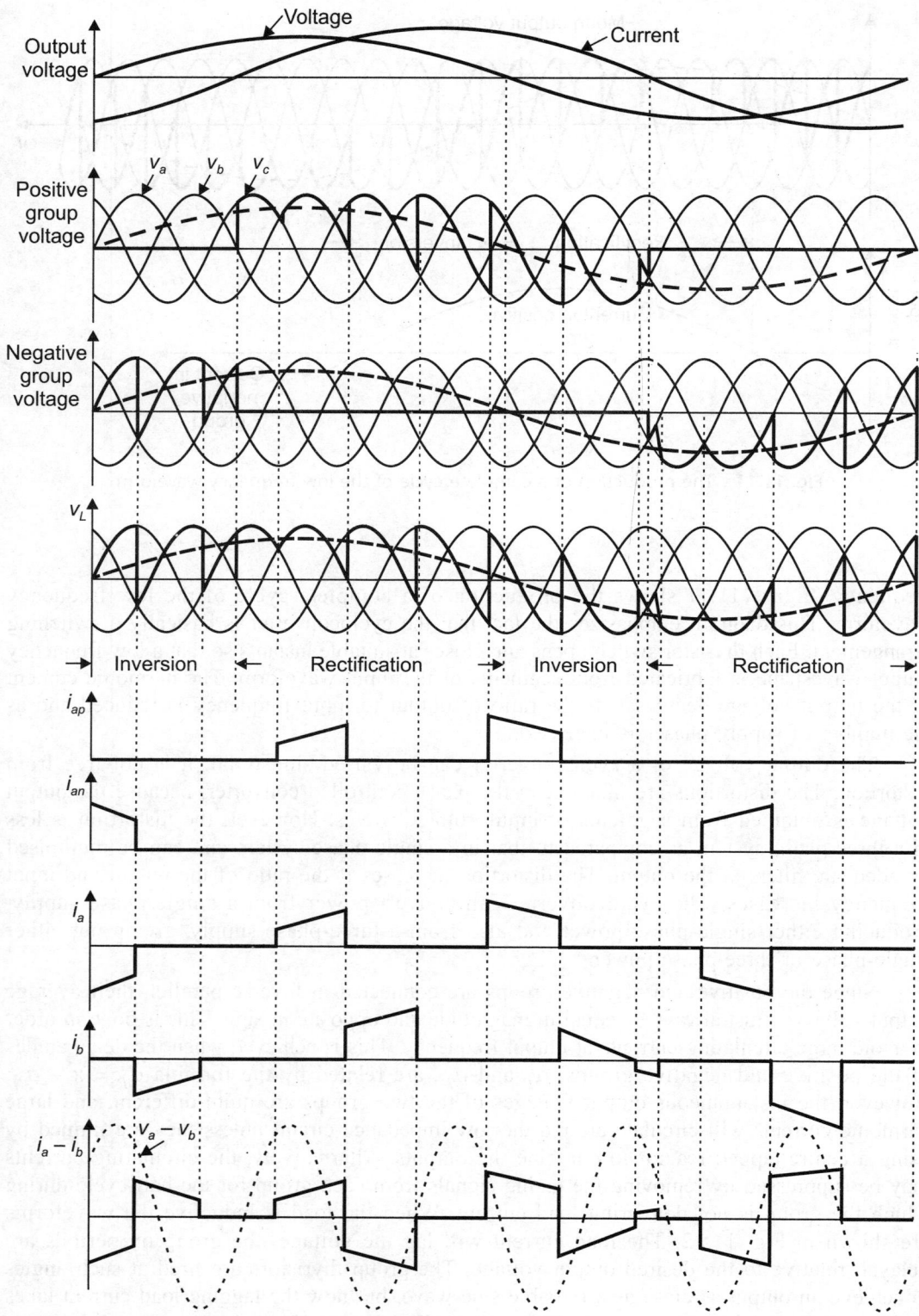

Fig. 11.12 Waveforms when the load is inductive

11.1.2.2 *Three-phase to Three-phase Cycloconverter*

When a three-phase output is required, three single-phase cycloconverters with a phase displacement of 120° between their outputs are connected, as shown in Fig. 11.13(a). With a balanced load, the neutral connection is no longer necessary and may be omitted. The simplest arrangement using three-phase half-wave circuits is shown in Fig. 11.13(b). This circuit requires eighteen thyristors, but if each group consists of a three-phase bridge or six-phase circuit or six pulse bridges then thirty-six thyristors are required.

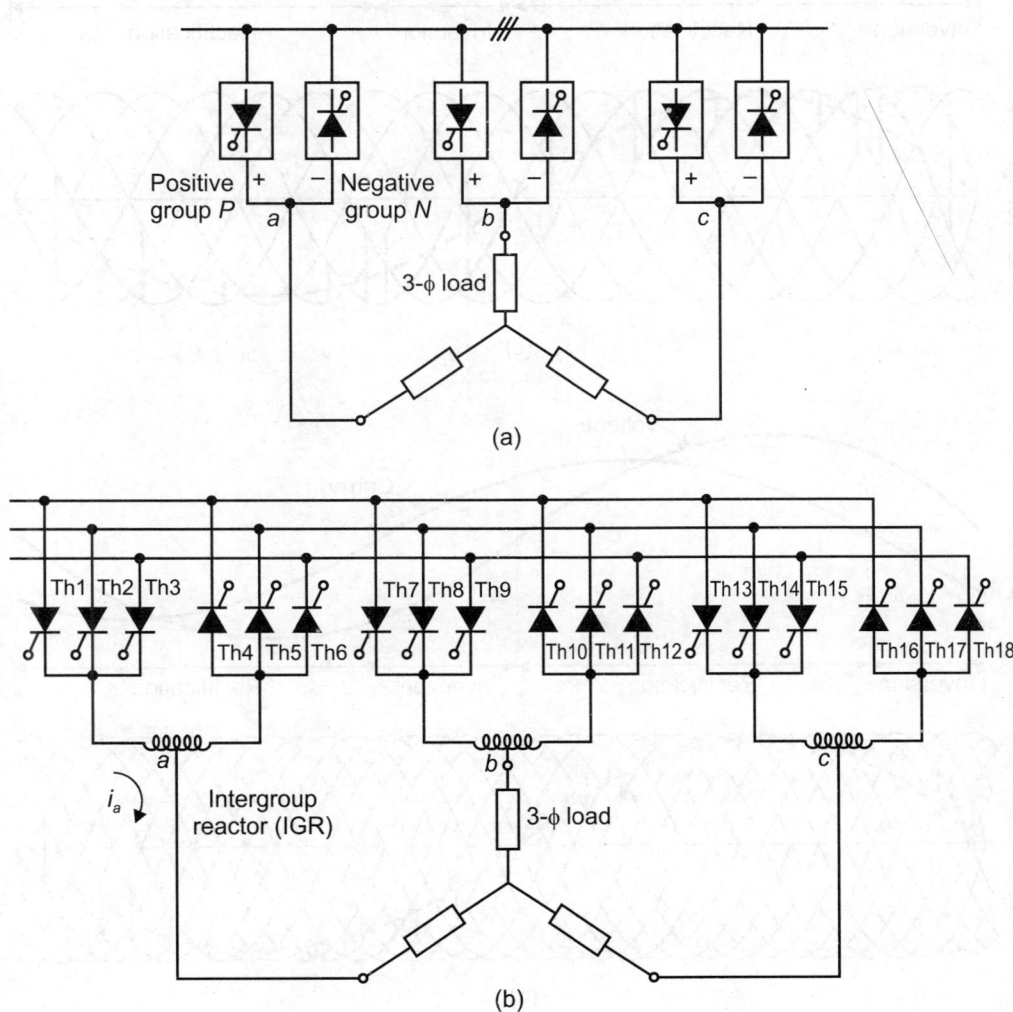

Fig. 11.13 Three-phase to three-phase cycloconverter

It is considered that the output frequency of the cycloconverter is less than the supply frequency. However, the cycloconverter is also capable of operation as a step-up frequency changer, but the output power is limited and circuit losses are high. In practice, with the basic cycloconverter circuit of Fig. 11.13(b), reasonable power output and efficiency are only obtained in the step-down region with output frequencies from zero to about one-third of the input frequency. As the cycloconverter output frequency approaches the supply frequency, the harmonic distortion in the output voltage increases, since the output voltage waveform is

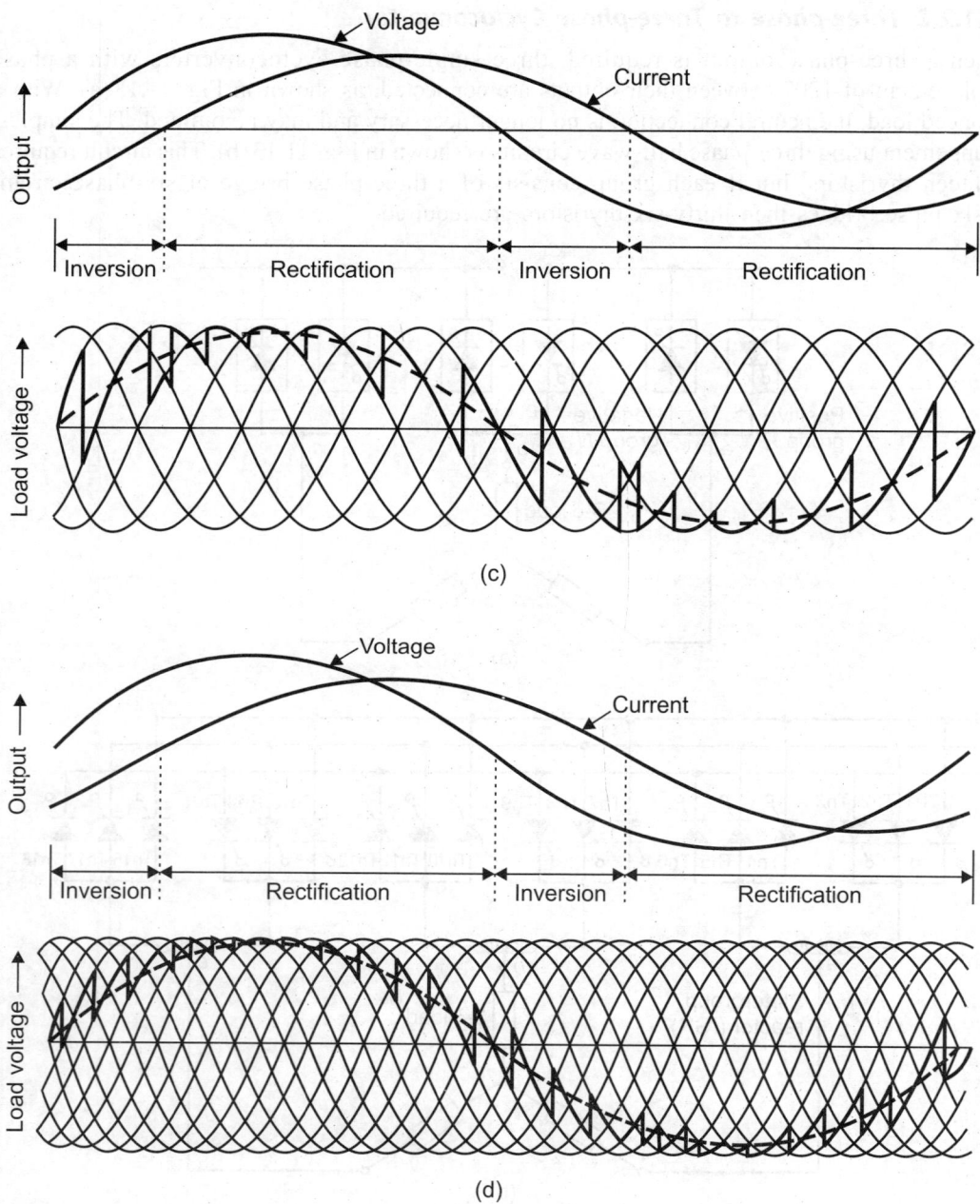

(c)

(d)

Fig. 11.13 Cycloconverter output waveforms for higher pulse connection

composed of fewer segments of the supply voltage. As a result, the losses in the cycloconverter and in the AC motor become excessive, and there is a drop in overall efficiency. By using more complex rectifier circuits, the output voltage waveform is improved and the maximum useful ratio of output to input frequency is increased to about one-half. The AC motor normally presents a high impedance at the ripple frequency, and, hence, the output current is nearly sinusoidal and no additional filtering is necessary.

Operation with inductive loads: In a cycloconverter, the positive and negative rectifier groups must each supply one-half-cycle of the low frequency output current. When feeding a resistive load, there is no necessity for inverter operation, since the positive group supplies load current during the positive half-cycle of output voltage, and the negative group conducts for the negative half-cycle. When inductive loads are being supplied, the low frequency output current lags the output voltage. This means that each rectifier group must continue to conduct after its output voltage changes sign. During this period, the group functions as an inverter, and power is returned to the AC supply. Inverter operation continues until such time as the load current reduces to zero and reverses, when the other group starts to conduct. In this manner, energy can be transferred in either direction through the cycloconverter.

The flow of reactive power associated with an inductive load occurs directly in an AC circuit, but when a cycloconverter separates the load from the supply, the reactive power transfer must take place through the cyclocon-verter. At certain parts of the low frequency output cycle this demands inverter operation, which occurs automatically whenever the voltage and current of a rec-tifier group have opposite signs. Thus, the cycloconverter is able to deliver low frequency AC to any type of reactive load, and the phase displacement between the half-cycles of current conduction and output voltage determines the periods of rectifier and inverter operation. Figures 11.13 (c) and (d) shows the cycloconverter output waveforms for higher pulse con-nections with an output frequency of one-third of the input frequency. It is clear from the waveform that the higher the pulse number, the closer is the output waveform to the desired sinusoidal waveform. Figure 11.13 (e) shows the relation between firing angles of positive and negative converters for the same output voltage under con-tinuous conduction.

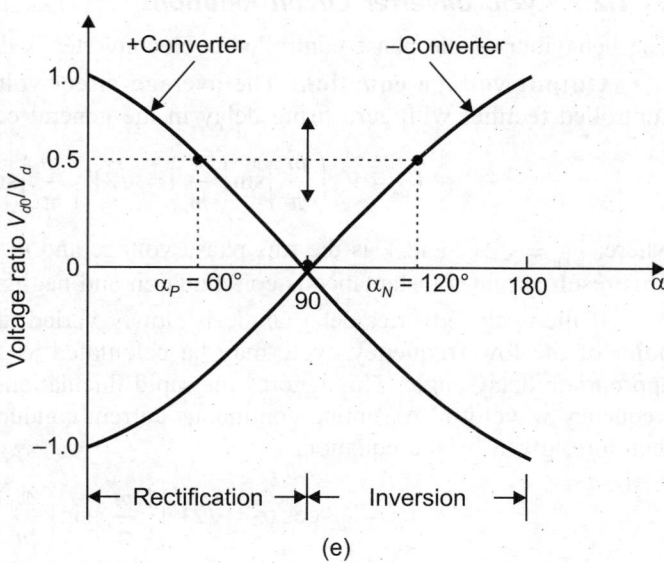

Fig. 11.13 Curve showing the relation of firing angle of converter

An interchange of energy between load and source occurs in any simple AC circuit with a reactive load, refer to Fig. 11.14. The instantaneous output power is

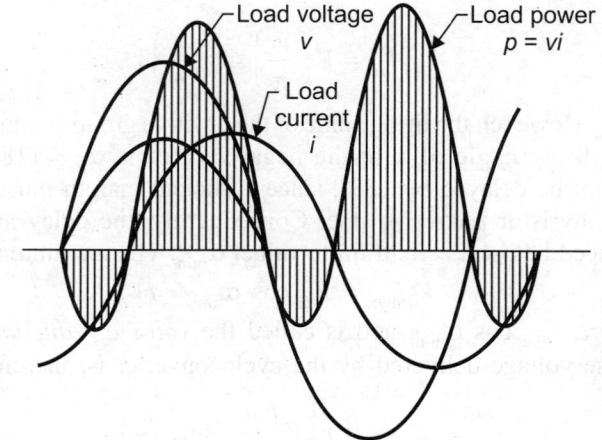

Fig. 11.14 An interchange of energy between load and source occurs in any simple AC circuit with a reactive load

the product of the instantaneous values of load current and voltage. During part of the AC cycle, the instantaneous power is positive and the load absorbs energy from the supply. Later in the cycle, the instantaneous power is negative, and energy is returned to the supply from the magnetic field of the load inductance (or from the electric field of the load capacitance). The difference between the two energies is the net energy delivered to the load. Real power and energy are associated with the component of current which is in phase with the applied voltage, and reactive power is associated with the current component which is 90° out of phase with the applied voltage. Thus, reactive power represents real power, which oscillates between source and load and has an average value of zero.

11.1.2.3 *Cycloconverter Circuit Relations*

The behaviour of the phase-controlled cycloconverter is discussed in more detail here.

Output voltage equation: The average direct voltage output of a half-wave phase-controlled rectifier with zero firing delay in the general equation is given by

$$V_{d_0} = V_m \left(\frac{m}{\pi} \right) \sin \left(\frac{\pi}{m} \right) = \sqrt{2} V \left(\frac{m}{\pi} \right) \sin \left(\frac{\pi}{m} \right) \qquad ...(11.6)$$

where, $V_m = \sqrt{2} V$ and V is the rms phase voltage and m is the number of secondary phases. This result assumes instantaneous commutation and negligible thyristor forward voltage drop.

If the cycloconverter delay angle is slowly varied, the output voltage per phase at any point of the low frequency cycle may be calculated as the average output voltage for the appropriate delay angle. This ignores the rapid fluctuations superimposed on the average low frequency waveform. Assuming continuous current conduction, the average output voltage is, therefore, given by the equation:

$$V_d = V_{d_0} \cos \alpha = \sqrt{2} V \left[\left(\frac{m}{\pi} \right) \sin \left(\frac{\pi}{m} \right) \right] \cos \alpha \qquad ...(11.7)$$

If V_{or} is the fundamental rms output voltage per phase of the cycloconverter, then the peak output voltage corresponding to zero delay is

$$\sqrt{2} V_{or} = V_{d_0} = \sqrt{2} V \left(\frac{m}{\pi} \right) \sin \left(\frac{\pi}{m} \right)$$

$$V_{or} = V \left(\frac{m}{\pi} \right) \sin \left(\frac{\pi}{m} \right) \qquad ...(11.8)$$

However, the firing angle of the positive group cannot be reduced to zero as this corresponds to a firing angle of π in the negative groups $\alpha_P = (180° - \alpha_N)$. In practice, inverter firing cannot be delayed by 180°, since sufficient margin must be allowed for commutation overlap and thyristor recovery time. Consequently, the delay angle of the positive group cannot be reduced below a certain finite value, α_{min}. The maximum output voltage per phase is therefore

$$V_{d\,max} = V_{d_0} \cos \alpha_{min} = r V_{d_0} \qquad ...(11.9)$$

where, $r = \cos \alpha_{min}$, and is called the *voltage reduction factor*. The expression for the rms phase voltage delivered by the cycloconverter is, therefore, modified to

$$V_{or} = r \left[V \left(\frac{m}{\pi} \right) \sin \left(\frac{\pi}{m} \right) \right] \qquad ...(11.10)$$

Since α_{min} is necessarily greater than zero, the voltage reduction factor r is always less than unity. By deliberately increasing α_{min} and thereby reducing range of variation of about 90° value, the output voltage V_{or} can be reduced and a static method of voltage control is obtained. In practice, the output voltage is less than the theoretical value given by Equation (11.10) due to commutation overlap and the circulating currents between positive and negative groups.

11.2 INPUT DISPLACEMENT FACTOR

A cycloconverter demands a lagging reactive current at the input because of the phase-control mechanism for output voltage fabrication. The cycloconverter consists essentially of a number of phase-controlled rectifiers operating with a variable firing angle. As already explained, the AC supply currents are nonsinusoidal in a rectifier installation, and one must distinguish between the power factor, the displacement factor and the distortion factor. The power factor λ in such cases, is defined as the ratio of the real input power in watts to the total apparent power in volt-amp. The displacement factor, $\cos \phi$ is the fundamental power factor, since ϕ is the phase displacement between the fundamental phase current and the sinusoidal phase voltage. The distortion factor, μ is the ratio of the fundamental rms current to the total rms current. As $\lambda = \mu \cos \phi$. If the commutation overlap between phases is negligible, the displacement factor, $\cos \phi$ equals $\cos \alpha$, where α is the firing angle. Thus, when α is 90°, the displacement factor is zero. In practice, the presence of overlap tends to increase ϕ and so reduces the displacement factor and power factor of the system. The large lagging reactive current requirement at the input together with the harmonic distortion of cycloconverter makes the input power factor poor, which is indeed a great disadvantage of cycloconverter application.

Voltage control in cycloconverter is obtained by reducing the variation of the delay angle about the 90° value. At low output voltages, the average phase displacement between input current and voltage is large, and the cycloconverter has a low displacement factor. The input current always lags the supply voltage, since phase delay is always present, irrespective of the nature of the load. The cycloconverter cannot transmit leading reactive power, and the lagging reactive power drawn from the supply is always greater than that delivered to the load. A capacitive load consumes leading reactive power but this appears as a demand for lagging reactive power on the input side of the cycloconverter. Thus, the displacement factor has its greatest value when the load is purely resistive, and a capacitive load with a given leading power factor will reduce the displacement factor by exactly the same amount as an inductive load with the same lagging power factor.

Mathematical expression for input displacement factor of a cycloconverter: Consider that the positive converter of phase '*a*' of an 18-thyristor cycloconverter is conducting only. Figure 11.15 (a) shows the input phase current at retard angle α_P with the corresponding phase voltage. Assume that the output frequency ratio is low so that the current pulse width always remains 120° wide. The Fourier series of the current wave can be given by

$$i = \frac{i_o}{3} + \frac{\sqrt{3}}{\pi} i_o \left[\cos(\omega_i t - \alpha_P) - \frac{1}{2} \cos 2(\omega_i t - \alpha_P) \right.$$

$$\left. - \frac{1}{4} \cos 4(\omega_i t - \alpha_P) - \frac{1}{5} \sin 5(\omega_i t - \alpha_P) - \dots \right] \quad \dots(11.11)$$

where, i is the instantaneous value of input current, i_o the instantaneous value of output current, and ω_i the supply frequency. The fundamental of input current lags the phase voltage by the angle of phase retard. Since the harmonics do not contribute to the real and reactive power

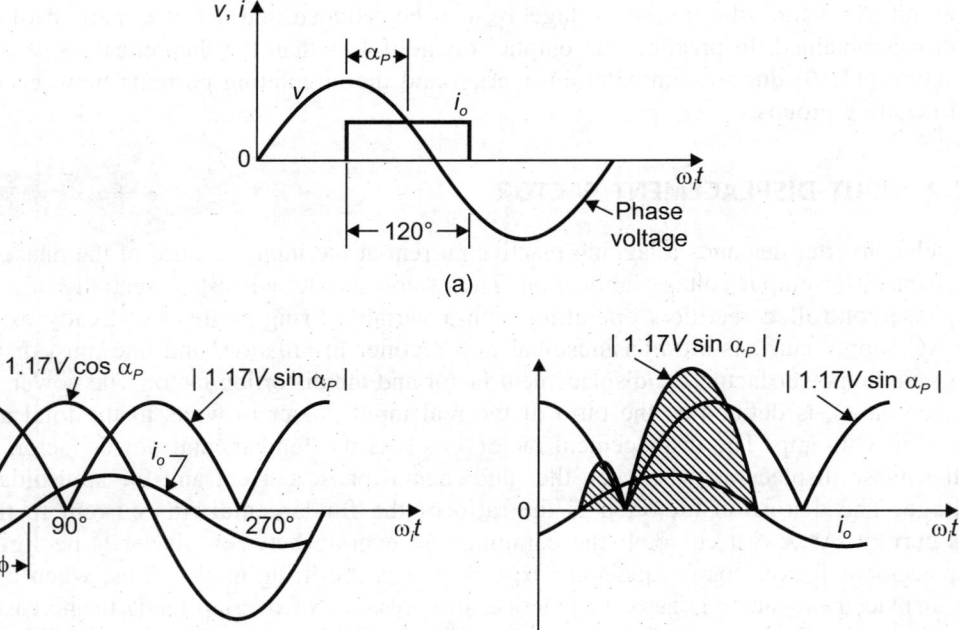

Fig. 11.15 Output voltage and current and reactive power waveforms

when the supply voltage is a sine wave, the instantaneous real power and reactive power of the supply are given by

$$p_i = 3V \left(\frac{\sqrt{3} i_o}{\sqrt{2} \pi} \right) \sin \alpha_P = (1.17V \cos \alpha_P) i_o \qquad \qquad ...(11.12)$$

$$q_i = 3V \left(\frac{\sqrt{3} i_o}{\sqrt{2} \pi} \right) \sin \alpha_P = (1.17V \sin \alpha_P) i_o \qquad \qquad ...(11.13)$$

where, V is the rms supply phase voltage. Since the angle α_P is continuously modulated, p_i and q_i in the above equations are being modulated accordingly. The mean output voltage given by $1.17V \cos \alpha_P$ is plotted in relation to output current i_o in Fig. 11.15 (b), where, ϕ is the load power factor angle. The voltage wave $1.17V \sin \alpha_P$ can be determined from the $1.17V \cos \alpha_P$ wave by a phase shift of 90° and noting that the α_P range is $0 < \alpha_P < 180°$. The instantaneous input reactive power = $(1.17V \sin \alpha_P) i_o$ is shown in Fig. 11.15 (c). The average input reactive power is given as

$$Q_i = \frac{1}{\pi} \int_0^\pi q_i d\omega_i t = \frac{2P_o}{\pi} \cos^2 \phi + \frac{2Q_o}{\pi} \left(\phi + \frac{1}{2} \sin 2\phi \right) \qquad ...(11.14)$$

where, P_o and Q_o are the cycloconverter active and reactive power, respectively, at the output. Since the active power at the output is the same as that of the input. Thus,

$$P_i + jQ_i = P_i + j \frac{2}{\pi} \left[P_i \cos^2 \phi + Q_o \left(\phi + \frac{1}{2} \sin 2\phi \right) \right] \qquad ...(11.15)$$

$$\text{Input displacement factor} = \frac{P_i}{P_i + jQ_i}$$

$$= \cfrac{1}{1 + j\,(2/\pi)\left[\cos^2\phi + (Q_o/P_i)\left(\phi + \frac{1}{2}\sin 2\phi\right)\right]} \qquad ...(11.16)$$

$$= \frac{1}{1 + j\,(2/\pi)\,(1 + \phi\tan\phi)}.$$

11.2.1 Features of Cycloconverters

Voltage waveform: The quality of output waveform of cycloconverter in its harmonic content, the capacity of cycloconverter to handle all types of loads and power factor, capacity to handle power flow in both the direction have enabled its application for low speed reversible AC drives. The disadvantage of these cyloconverters is very poor power factor at low voltages. To improve the performance of cycloconverter with respect to output waveform, power factor, etc., it is necessary to study various aspects of the control of cycloconverters which affects the performance. These aspects are:

1. Continuous and discontinuous conduction
2. Circulating and noncirculating current operations
3. The effect of source impedance and pulse number
4. The ratio of output/input frequencies

Generally, the voltage waveform of cycloconverter contains harmonics which is impressed across the machine, cause additional heating and torque pulsation problems. The harmonics are generally influenced by

1. Pulse number
2. Circulating or noncirculating mode of operation
3. Output-to-input frequency ratio
4. Load power factor
5. Continuous or discontinuous conduction
6. Commutation overlap effect
7. Feedback control method and its bandwidth

The three-phase, phase-controlled converter contains triplen harmonics at the output. Since in a cycloconverter the firing angles are sinusoidally modulated, it is expected that the output will contain sidebands or beat-frequency components in the form $Mf_i \pm Nf_o$.

Some of the lower sideband frequencies, although having reduced voltage magnitude, may fall below the fundamental frequency and cause reasonably large subharmonic currents. The problem may be especially severe if the frequency ratio is high.

11.2.2 Load and Source Harmonics

The output voltage of cycloconverter fabricated from the segments of input voltages. If larger number of segments are available at the output, the ripple decreases. Therefore, output voltage waveform approaches a sine wave if there are larger number of pulses. Larger the number of pulses, smaller is the harmonic content of cycloconverter.

Circulating current mode of operation of cycloconverter is preferred if there is a possibility of discontinuous conduction, this improves the voltage waveform to a great extent even eleminating the distorsion at cross-over currents.

Output to input frequency ratio: The performance limits of cycloconverter are decided by the harmonic distortion of the output voltage wave. The output voltage will have more harmonic distorsion if the ratio of output frequency to input frequency is large. Therefore, a cycloconverter is preferred for output frequencies which are fraction of input frequency. The load power factor and the method of controlling the firing instants also affect the output voltage distortion. The firing angle modulation techniques are only designed towards improving the voltage waveform.

Discontinuous conduction of converter causes more distortion of voltages. During discontinuous conduction, both load voltage and load current are zero and the voltage waveform has zero values. These increase the distortion due to cross-over of currents and effects on the voltages due to discontinuous conduction that can be reduced by closed loop control of converter. The distortion introduced at the instants of current cross-over cannot be eliminated.

Any nonlinear voltage relationships that may occur due to abnormal performance of converter may cause deviations of voltage from the desired sine wave. High inductance loads cause continuous conduction where the voltage relationships are linear and the distortion is less.

11.3 CIRCULATING CURRENT MODE OF OPERATION OF CYCLOCONVERTER

Back to back connected phase controlled converters forming a cycloconverter are so controlled that the load voltage is of the desired polarity at a given instant of time. Circulating currents are also present due to trapped mmf of these inductors. These are called *self-induced circulating currents*. The current limiting inductance is also effective in smothening the load current. The time integral of the voltage divided by the inductance is the circulating current. The circuit may be operating in a circulating or noncirculating current mode. The circulating mode is one in which, there is flow of harmonic current between positive and negative group of rectifiers. The noncirculating mode is one in which flow of current is blocked between the two groups. The positive group has a delay angle α_P and delivers positive current to the load. The negative group has a delay angle α_N and permits current flow in the opposite or negative direction. If the forward voltage drop of a thyristor is negligible, the delay angles must be controlled so that $\alpha_P = (180° - \alpha_N)$. In this manner, the mean output voltage of the rectifier group is maintained equal to the mean back emf of the inverter group, and the circulation of large low frequency currents between groups is avoided. However, the instantaneous voltages of the two groups are not identical, and large harmonic currents will circulate unless they are limited or suppressed. These intergroup currents are undesirable, since they increase the circuit losses and impose a heavier loading on the thyristors. They also reduce the displacement factor of the system. The flow of harmonic current may be reduced by two techniques: (1) by the introduction of an intergroup reactor, and (2) suppression by intergroup blanking or by removing the gating pulses from the non conducting group.

1. By introduction of intergroup reactor: The reactor is connected between the two groups in order to limit the flow of harmonic current, and the load is connected to a centre-tap as in Fig. 11.8(b). The low frequency output current of the cycloconverter is opposed by the reactance X due to one-half of the intergroup reactor. If the two halves of the reactor are tightly coupled, the flow of harmonic current between groups is opposed by a reactance of $4kX$, where k is the order of the harmonic. A suitable choice of inductance will restrict the flow of harmonic current without seriously affecting the fundamental output current.

Figure 11.16 shows the difference of the instantaneous voltage appears as a ripple voltage in the closed circuit of back to back connected converters. The low impedance of converters causes a circulating current. These circulating currents flow in the converters alone but not in the load. The load current flowing through the converter is superimposed by the circulating current. Thyristors have to be over-rated. These currents are effective in reducing the power factor. A centre-tapped inductor with tight coupling is used to limit the currents. In Fig. 11.16, the voltage and current waveforms are shown for the three-phase half-wave antiparallel connection of Fig. 11.8(b), assuming a highly inductive load. The positive group operates as a rectifier with a delay angle α_P, which is less than $\pi/2$, while the negative group operates in the inverter region with a delay angle $\alpha_N = (180° - \alpha_P)$. Thus, the mean voltages of the two groups are equal, but the instantaneous values are quite different, and the voltage difference appears across the intergroup reactor and has the waveform shown in Fig. 11.16 (c). The circulating current flows through the two series-connected thyristor circuit and is limited only by the inductance of the intergroup reactor, assuming negligible circuit resistance. The circulating current is, therefore, proportional to the integral of the intergroup voltage and has the variation shown in Fig. 11.16(d). In the three-phase half wave system under consideration,

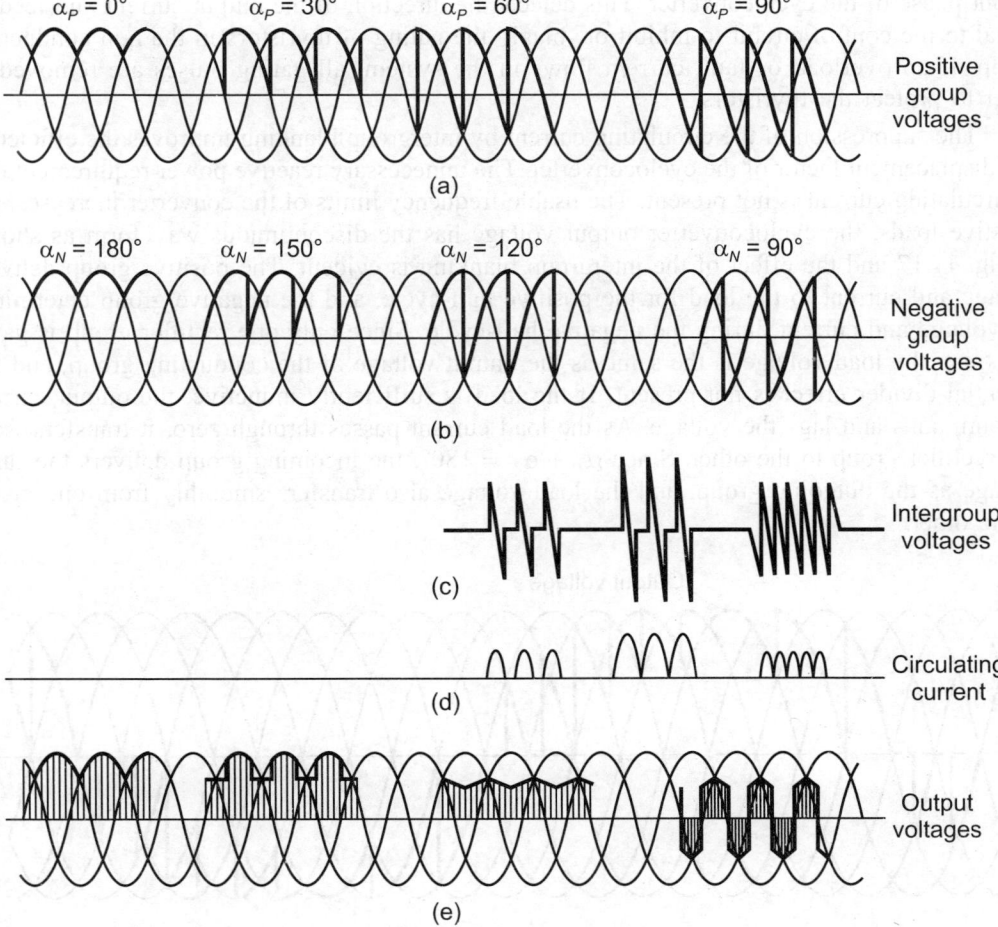

Fig. 11.16 Voltage waveforms across back to back connected converters

the maximum circulating current occurs when the firing angle α_P is 60°, and α_N is 120°. During the intervals, when the positive and negative thyristor groups have different instantaneous voltages, the intergroup reactor behaves as a potential divider, and the output voltage at the centre-tap is the average of the two group voltages, as shown in Fig. 11.16(e).

In the application of cycloconverter, the delay angles α_P and α_N are continuously modulated, and the output voltage waveform passes smoothly through each of the stages as shown in Fig. 11.16(e). The output waveform is considerably improved if a six-phase system is used. The intergroup voltages and circulating currents are then significantly reduced, and maximum circulating current occurs with a delay angle of 90°, corresponding to zero output voltage.

2. Suppression by intergroup blanking: Circulating currents can be completely avoided by inhibiting the firing pulses to idling converters. This is done when load current has its natural zero. The firing pulses to the converter are stopped. Control of converters requires additional zero sensing of current to determine the instant of blanking a converter. Modern control needs to be sophisticated if there is a possibility of discontinuous conduction. In modern control cycloconverter circuits, the circulating current is usually suppressed by blocking all thyristors in the rectifier group, which is not delivering load current. This is achieved by removing the firing pulses for the appropriate periods, and the intergroup reactors of Fig. 11.13(b) can then be reduced in size or completely eliminated. A current-sensing device is incorporated in each output phase of the cycloconverter. This detects the direction of the output current and feeds a signal to the control circuit to inhibit or 'blank' the gating of thyristors in the non conducting group. If an overload or fault current flows in the system, all gating pulses are removed in order to protect the thyristors.

The suppression of the circulating current by intergroup blanking improves the efficiency and displacement factor of the cycloconverter. The unnecessary reactive power requirement due to circulating current is not present. The usable frequency limits of the converter increases. On resistive loads, the cycloconverter output voltage has the discontinuous waveform as shown in Fig. 11.17 and the effect of the intergroup blanking is evident. The positive group delivers voltage and current to the load for the positive half-cycle, and the negative group determines the voltage and current during the negative half-cycle. Since only one rectifier group is gated at a time, the load voltage is the same as the output voltage of the conducting group, and the potential-divider effect is not present. If the load is sufficiently inductive, the output current is continuous and lags the voltage. As the load current passes through zero, it transfers from one rectifier group to the other. Since $\alpha_P + \alpha_N = 180°$, the incoming group delivers the same voltage as the outgoing group, and the load voltage also transfers smoothly from one group to the other.

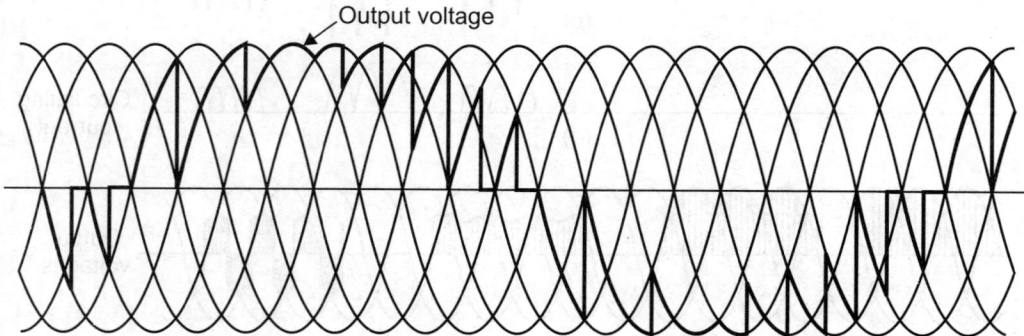

Fig. 11.17 The cycloconverter output voltage with discontinuous waveform

11.4 EFFECT OF SOURCE INDUCTANCE ON THE PERFORMANCE OF CYCLOCONVERTERS

Source inductance causes commutation delay. It delays the transfer of current from one thyristor to the other. It depends not only on the source inductance, but also on the converter firing angles as well as on the current to be commutated with high pulse number, there may be simultaneous commutations, the overlap affects the range of firing angles. The commutation overlap and hence source inductance modify the output voltage waveform and its distortion and amplitude of harmonies. The average value of voltage and displacement factor decreases and the commutation notches are produced in the output voltages. The rounding of the edges of input current waveform improves the harmonic currents of the input current.

11.5 COMPARISON OF THE CYCLOCONVERTER AND DC LINK CONVERTER

A cycloconverter is a frequency changer that converts AC power at one input frequency to output power at a different frequency with one stage conversion. Whereas DC link converter has two power controllers in cascade for conversion from input frequency to output frequency through the medium of high frequency. The cycloconverter and DC link converter, both deliver an alternating voltage at a frequency which is determined by the reference oscillator. In the cycloconverter, voltage control is also obtained by means of the control circuitry. A completely static supply is, therefore, available with a transient response in the millisecond range. This is also provided by DC link systems, which employ static voltage control techniques such as phase-shift control or pulse-width modulation.

The cycloconverter has some advantages and disadvantages over the DC link converter which are given below:

Cycloconverter	DC link Converter
1. In a cycloconverter, AC power at one frequency is converted directly to AC a lower frequency in a single conversion stage.	1. The DC link converter has two power controllers in cascade, and the full output power is converted twice.
2. The cycloconverter functions by means of phase commutation, and no auxiliary forced-commutating circuits are necessary. This results in a more compact power circuit and also eliminates the circuit losses associated with forced commutation.	2. DC link converter requires forced commutation for the inverter.
3. The cycloconverter is inherently capable of power transfer in either direction between source and load, and can, therefore, supply AC power to loads of any power factor. It is also capable of regenerative operation at full power over the complete speed range down to standstill. This feature is difficult to incorporate in static inverter drives, and hence, the cycloconverter is preferable for large reversing drives requiring rapid acceleration and deceleration. This type of application occurs principally in the metal rolling industry.	3. This feature is slightly difficult.

Cycloconverter	DC link Converter
4. In the cycloconverter, a commutation failure causes a short-circuit of AC supply, but if an individual thyristor fuse blows, a complete shut-down is not necessary. Since the cycloconverter continues to function with a somewhat distorted output waveform.	4. A balanced load is presented to the AC supply even with unbalanced output conditions.
5. Cycloconverter delivers a high-quality sinusoidal waveform at low output frequencies, since the low frequency wave is fabricated from a large number of segments of the supply waveform.	5. DC link, on the other hand, generates a stepped-wave voltage which may cause a jerky rotation of the AC motor at frequencies below 10 Hz. The distorted waveform also slightly accentuates the danger or system instability at low frequencies.
6. It requires large number of thyristors and its control circuitry is complicated.	6. DC link converter requires only twelve thyristors and its control circuit is less complicated.
7. At reduced output, power factor is poor.	7. In this power factor is good with diode rectifier.
8. Suitable for large power low speed reversing drives.	8. Suitable for high frequency.

SOLVED EXAMPLES

Example 11.1 A cycloconverter is made from two single-phase fully controlled bridges. The input to the bridges is 230 V, 50 Hz, single-phase AC. If the output frequency is one-fifth, the input frequency and the firing angle α is $\pi/4$, calculate the fundamental amplitude of the output. Assume a resistive load.

Solution Using the expression

$$v_{ols} = \frac{V_m}{m\pi} \sum_{n=1}^{m} \frac{m}{m-1}\left[\sin\frac{n\pi}{m} - \sin\left(\alpha - \alpha + \frac{(n-1)\pi}{m}\right)\right.$$

$$\left. + \frac{m}{m+1}\left\{ \sin\frac{n\pi}{m} + \sin\left(\alpha - \alpha + \frac{(n-1)\pi}{m}\right)\right\}\right]$$

$$v_{ols} = \frac{230 \times \sqrt{2}}{5\pi} \sum_{n=1}^{5} \frac{5}{5-1}\left[\sin\frac{n\pi}{5} - \sin\left(\frac{\pi}{4} - \frac{\pi}{4} + \frac{(n-1)\pi}{5}\right)\right.$$

$$\left. + \frac{5}{5+1}\left\{ \frac{\sin n\pi}{5} + \sin\left(\frac{\pi}{4} + \frac{\pi}{4} + \frac{(n-1)\pi}{5}\right)\right\}\right]$$

$$= \frac{\sqrt{2} \times 230}{5\pi}(6.24 + 4.75) = 228 \text{ V}$$

Similarly
$$v_{olc} = \frac{\sqrt{2} \times 230}{5\pi} \sum_{n=1}^{5} \left\{ \frac{5}{6} \left[\frac{\cos n\pi}{5} + \cos\left(\frac{\pi}{4} + \frac{\pi}{4} + \frac{(n-1)\pi}{5} \right) \right] \right.$$

$$\left. + \frac{5}{4} \left[\cos\frac{n\pi}{5} + \cos\left(\frac{\pi}{4} + \frac{\pi}{4} + \frac{(n-1)\pi}{5} \right) \right] \right\}$$

$$= \frac{\sqrt{2} \times 230}{5\pi}(-2.36 + 2.03) = -8.3 \text{ V}$$

Fundamental amplitude,
$$V_1 = \sqrt{v_{ols}^2 + v_{olc}^2}$$

$$= \sqrt{228^2 + 8.3^2} = 228.1 \text{ V}.$$

Example 11.2 A six pulse, blocked grouped cycloconverter is fed from a three-phase, 600 V (line), 20 Hz supply. The supply has an inductance of 1.146 mH/phase. If the cycloconverter is supplying a variable resistive load with a current of 28 A, estimate the peak and rms value of load voltage for firing angles of: (a) 0°, (b) 30°, and (c) 60°.

Solution Peak voltage is the mean voltage of the equivalent rectifier given by

$$V_{mean} = V_m \left(\frac{m}{\pi} \right) \sin\left(\frac{\pi}{m} \right) \cos\alpha - \frac{m\omega L}{2\pi} \cdot I_L$$

For this system
$$\frac{m\omega L}{2\pi} \cdot I_L = \frac{6 \times 2\pi \times 50 \times 1.146 \times 10^{-3} \times 28}{2\pi}$$

$$= 9.63 \text{ V}$$

The value of
$$\frac{m}{\pi} V_m \sin\left(\frac{\pi}{m} \right) = \frac{6}{\pi} \times \sqrt{2} \times 600 \sin 30° = 891.3 \text{ V}$$

(a) For $\alpha = 0°$, $\quad V_{mean} = 891.3 - 9.63 = 881.7 \text{ V} = V_{max}$ for cycloconverter

rms value of voltage of the cycloconverter
$$V_{rms} = \frac{V_{max}}{\sqrt{2}} = \frac{881.7}{\sqrt{2}} = 623.4 \text{ V}$$

(b) For $\alpha = 30°$, $\quad V_{mean} = 891.3 \cos 30° - 9.63 = 762.3 \text{ V}$

$$V_{rms} = \frac{762.3}{\sqrt{2}} = 539 \text{ V}$$

(c) For $\alpha = 60°$, $\quad V_{mean} = 891.3 \cos 60° - 9.63 = 436 \text{ V}$

$$V_{rms} = \frac{436}{\sqrt{2}} = 308.3 \text{ V}.$$

Example 11.3 A 3-phase to single-phase cycloconverter employs 3-pulse positive and negative group converters. Each converter is supplied from delta/star transformer with per phase turns ratio of 2:1. The supply voltage is 400 V, 50 Hz. The load has $R = 2\,\Omega$ and $X_L = 1.5\,\Omega$. In order to account for commutation overlap and thyristor turn off time, the firing angle in the inversion mode should not exceed 160°. Estimate: (a) fundamental rms output voltage, (b) rms output current, and (c) output power.

Solution (a) Given turns ratio $= 2:1$

    Per phase input voltage to transformer $= 400$ V

    Input voltage to converter/phase $= \dfrac{400}{2} = 200$ V

    Voltage reduction factor $r = \cos \alpha_{min} = \cos (180° - 160°)$
$$= \cos 20°$$

Expression for rms phase voltage delivered by cycloconverter,

$$V_{or} = r \left[V \left(\frac{m}{\pi} \right) \sin \left(\frac{\pi}{m} \right) \right]$$

(b) rms output current $\quad I_{or} = \dfrac{V_{or}}{Z} = \dfrac{155.4}{\sqrt{2^2 + 1.5^2}} \tan^{-1} \left(\dfrac{1.5}{2} \right)$

$$I_{or} = 62.17 \angle 36.8° \text{ A}$$

(c) Output power, $\quad P_{or} = I_{or}^2 R = (62.17)^2 \times 2 = 7731$ W.

Example 11.4 Three-phase to single-phase cycloconverter employs 6 pulse bridge converter. Each converter is fed from delta/star transformer with per phase turns ratio of 2:1 of supply voltage 400 V of 50 Hz. The RL load has $R_L = 2\,\Omega$ and $X_L = \omega_o L = 1.5\,\Omega$. The firing angle in the inversion mode should not exceed 160° in order to account for commutation overlap and thyristor turn off time. Compute: (a) fundamental rms output voltage, (b) rms output current, and (c) power output.

Solution (a) Given turns ratio $= 2:1$

    Per phase input voltage to converter

$$= \frac{400}{2} = 200 \text{ V}$$

    Input line voltage $= \sqrt{3} \times 200 = 346$ V

    Voltage reduction factor $\quad r = \cos (180° - 160°) = \cos 20°$

    For 6-pulse bridge converter,

$$m = 6$$

rms output voltage, $\quad V_{or} = r \left[V \left(\frac{m}{\pi} \right) \sin \left(\frac{\pi}{m} \right) \right]$

$$= \cos 20° \left[346 \left(\frac{6}{\pi} \right) \sin \left(\frac{\pi}{6} \right) \right] = 310.7 \text{ V}$$

(b) rms output current

$$= \frac{310.7}{\sqrt{2^2 + 1.5^2}} \angle - \tan^{-1} \left(\frac{1.5}{2} \right) \text{ A}$$

$$= 124.3 \angle - 36.8° \text{ A}$$

(c) rms output power P_o

$$= (124.3)^2 \times 2$$
$$= 30.92 \text{ kW.}$$

EXERCISES

Multiple Choice Questions

1. A cycloconverter is a device which:
   (A) converters AC into DC
   (B) converts DC into AC
   (C) converts AC of one frequency into AC of other frequency
   (D) none of these

2. Cycloconverter is concerned mostly:
   (A) with direct conversion of energy to a different frequency
   (B) with direct conversion of energy at the same frequency
   (C) with DC source only
   (D) none of these

3. Cycloconverter can be considered to be composed of two:
   (A) rectifiers connected back to back
   (B) inverters connected back to back
   (C) converters connected back to back
   (D) either of these

4. Whether the load is resistive, inductive, or capacitive, the firing of thyristors of a cycloconverter are:
   (A) commutated naturally
   (B) forced commutated
   (C) either of these
   (D) none of these

5. Cycloconverter is a:
   (A) frequency converter which has no intermediate DC state
   (B) device which converts AC to DC
   (C) device which converts DC to AC
   (D) none of these

6. Cycloconverter is capable of converting power to:
   (A) a lower frequency
   (B) a higher frequency
   (C) the same frequency
   (D) none of these

7. Cycloconverter has:
   (A) low input power factor
   (B) high input power factor
   (C) medium power factor
   (D) none of these

8. In a single stage conversion of cycloconverter, AC power at one frequency is converted directly:
   (A) to AC at low frequency
   (B) to AC at very high frequency
   (C) to DC only
   (D) none of these

9. The ruggedness and low weight of solid state cycloconverter makes it attractive for:
   (A) aircraft electrical system
   (B) battery operated vehicle
   (C) electric traction
   (D) all of these

10. The cycloconverter requires:
    (A) a large number of thyristors but its control circuit is very simple
    (B) a small number of thyristors and has complicated control circuit
    (C) a large number of thyristors and its control circuit is more complex
    (D) none of these

11. Soft start is a:
    (A) gradual application of power when switching on a load
    (B) flow of power in one direction
    (C) switching off
    (D) switching on

12. Cycloconverters are more efficient than DC link as they:
    (A) do not have any rotating part
    (B) use very less power
    (C) produce the output frequency only in one stage
    (D) none of these

13. A centre tapped transformer configuration of single-phase cycloconverter uses:
    (A) four thyristors
    (B) six thyristors
    (C) eight thyristors
    (D) eighteen thyristors

14. To obtain sinusoidal waveform at the output of a centre tapped transformer configuration of a single-phase cycloconverter, the firing angle is of value:
    - (A) 0°
    - (B) 45°
    - (C) 90°
    - (D) 180°

15. A single-phase bridge type cycloconverter uses:
    - (A) two thyristors
    - (B) four thyristors
    - (C) eight thyristors
    - (D) sixtyfour thyristors

16. A single-phase bridge type cycloconverter is used for producing an output frequency of:
    - (A) 12.5 Hz
    - (B) 25 Hz
    - (C) 50 Hz
    - (D) 75 Hz

17. Generally a single-phase centre tapped cycloconverter is used to produce an output frequency of:
    - (A) 12.5 Hz
    - (B) 50/3 Hz
    - (C) 25 Hz
    - (D) 75 Hz

18. Three-phase three pulse cycloconverter uses:
    - (A) six thyristors
    - (B) twelve thyristors
    - (C) eighteen thyristors
    - (D) none of these

19. The sum of firing angles of the positive group of thyristors and negative group of thyristors in case of three pulse phase cycloconverter is equal to:
    - (A) 90°
    - (B) 180°
    - (C) 360°
    - (D) all of these

20. Three-phase six pulse bridge cycloconverter uses:
    - (A) six thyristors
    - (B) twelve thyristors
    - (C) eighteen thyristors
    - (D) thirty six thyristors

21. Cycloconverters are mainly used in:
    - (A) speed control of DC shunt motors
    - (B) speed control of synchronous motors
    - (C) electric traction
    - (D) all of these

22. Which of the following statements is correct?
    - (A) Speed reversal of electric motors is not possible with the help of cycloconverter
    - (B) Regenerative braking cannot be achieved with the help of cycloconverter
    - (C) An input transformer is required for a bridge configuration of cycloconverter
    - (D) Minimum firing angle will produce the best possible waveform for a cycloconverter

23. From the point of view of low distortion the ratio of the output to input frequency of a cycloconverter should be:
    - (A) low
    - (B) high
    - (C) very high
    - (D) none of these

24. A cycloconverter is a:
    - (A) frequency changer from higher to lower frequency with one-stage conversion
    - (B) frequency changer from higher to lower frequency with two-stage conversion
    - (C) frequency changer from lower to high frequency with one-stage conversion
    - (D) either (A) or (B)

25. The cycloconverters require natural or forced commutation as under:
    - (A) natural commutation in both step-up and step-down cycloconverters
    - (B) forced commutation in both step-up and step-down cycloconverters
    - (C) forced commutation in step-up cyclo-converts
    - (D) forced commutation in step-down cyclo-converters

26. Three-phase to three-phase cycloconverters employing 18 SCRs and 36 SCRs have the same voltage and current ratings for their component thyristors. The ratio of VA rating of 36 SCR device to that of 18 SCR device is:
    - (A) $\dfrac{1}{2}$
    - (B) 1
    - (C) 2
    - (D) none of these

27. Three-phase to three-phase cycloconverter employing 18 thyristors and 36 thyristors have the same voltage and current rating for their components thyristors. The ratio of power handled by 36 thyristors to that handled by 18 thyristors is:
    - (A) 4
    - (B) 2
    - (C) 1
    - (D) none of these

28. The number of thyristors required for single-phase to single-phase cycloconverter of the mid-point type and for three-phase to three-phase 3-pulse type cycloconverter are respectively:

(A) 4, 6

(B) 4, 4

(C) 4, 18

(D) none of these

29. A three-phase to single-phase conversion device employs a 6-pulse bridge cycloconverter. For an input voltage of 200 V per phase, the fundamental rms value of output voltage is:

(A) $600/\pi$ V

(B) $300\sqrt{3}/\pi$ V

(C) $300/\pi$ V

(D) none of these

30. A three-phase to single-phase cycloconverter consists of positive and negative group of converters in this device, one of the two component converters would operate as a:

(A) rectifier, if the output voltage V_o and output current I_o have the same polarity

(B) inverter, if V_o and I_o have the same polarity

(C) rectifier, if V_o and I_o are of opposite polarity

(D) none of these

31. A three-phase to three-phase cycloconverter requires:

(A) 18 SCRs for 3-pulse device

(B) 18 SCRs for 6-pulse device

(C) 36 SCRs for 3-pulse device

(D) none of these

32. Which of the following statements are incorrect for cycloconverters?

(A) Step-down cycloconverter works on natural commutation

(B) Step-up cycloconverter requires forced commutation

(C) Load commutate cycloconverter works on line commutation

(D) Load commutated cycloconverter requires a generated emf in the load circuit

33. A single-phase voltage controller feeds an induction motor and a heater:

(A) in both the loads, fundamental and harmonics are useful

(B) in induction motor only fundamental and in heater only harmonics are useful

(C) in induction motor only fundamental and in heater harmonics as well as fundamental are useful

(D) none of these

34. Output frequency of a cycloconverter is significantly lower than the input frequency for:

(A) keeping commutation failure

(B) avoiding commutation failure

(C) reducing switching losses

(D) none of these

ANSWER KEY

1. (C)	**2.** (A)	**3.** (C)	**4.** (A)	**5.** (A)	**6.** (B)	**7.** (A)	**8.** (A)
9. (A)	**10.** (C)	**11.** (A)	**12.** (C)	**13.** (A)	**14.** (A)	**15.** (C)	**16.** (B)
17. (B)	**18.** (C)	**19.** (B)	**20.** (D)	**21.** (C)	**22.** (D)	**23.** (A)	**24.** (D)
25. (C)	**26.** (C)	**27.** (A)	**28.** (C)	**29.** (A)	**30.** (A)	**31.** (A)	**32.** (C)
33. (C)	**34.** (B)						

Review Questions

1. What is a cycloconverter? Discuss its principle.

2. Enumerate industrial applications of cycloconverter and discuss in detail.

3. Discuss why three-phase to single-phase cycloconverter requires positive and negative group phase controlled converters. How should the firing angles of the two converters be controlled?

4. Discuss the three-phase cycloconverter with waveforms.

5. What are the different applications of cycloconverters in the industry?

6. Discuss the advantages and disadvantages of cycloconverter over commutated inverter circuits.

7. Show that the fundamental rms value of per phase output voltage of low frequency for m pulse cycloconverter is given by $V_{or} = V\left(\dfrac{m}{\pi}\right)\sin\left(\dfrac{\pi}{m}\right)$.

Express V_{or} in terms of voltage reduction factor.

DC-to-AC Converters—Inverters

12.0 INTRODUCTION

Inverters are static circuits that convert power from a DC source to an AC source at a specified output voltage and frequency. The output voltage could be fixed or variable at a fixed or variable frequency. A variable output voltage can be obtained by varying the input DC voltage and maintaining the gain of the inverter constant. On the other hand, if the DC input voltage is fixed and it is not controllable, a variable output voltage, which is normally accomplished by pulse width modulation (PWM) control within the inverter. The inverter gain may be defined as the ratio of the AC output voltage to the DC input voltage. The output voltage waveform of ideal inverters are nonsinusoidal and contain certain harmonics. For low and medium power applications, square wave or quasi square wave voltages may be acceptable, and for high power applications, low distorted sinusoidal waveforms are required. With the availability of high-speed power semiconductor devices, the harmonic contents of the output voltage can be minimised or reduced significantly by switching.

12.1 INDUSTRIAL APPLICATIONS

Following are the important industrial applications of inverter, such as:

1. Variable speed AC motor drives
2. High voltage DC transmission lines
3. Induction heating
4. Aircraft power supplies
5. Standby power supplies
6. Uninterruptible power supplies.

   The input may be a battery fuel cell, solar cell, etc. Typical single-phase outputs are 220 V, at 50 Hz or 120 V at 60 Hz. For high power three-phase systems, output may be 220/380 V at 50 Hz or 115/200 V at 400 Hz.

12.2 TYPES OF INVERTER

Thyristor inverters may be classified according to the method of commutation, method of connections or according to the control systems used.

12.2.1 Inverters Based on the Method of Commutation

These inverters can be categorised into four types: (1) line commutated inverters, (2) load commutated inverters, (3) self-commutated inverters, and (4) forced commutated inverters.

1. **Line commutated inverters:** In the line commutated inverters, as the AC voltage available across the device passes through zero, the device is turned off. This process is also known as *natural commutation process*. It is possible only in case of AC circuits.
2. **Load commutated inverters:** It is similar to line commutation except that the commutating voltage is induced from the load circuit.
3. **Self-commutated inverters:** It relates to GTO, transistor, MOS devices where commutation is performed by gate or base drive signals.
4. **Forced commutated inverters:** In the forced commutated circuits some external means are needed to forcefully turn-off the device as the DC supply voltage does not pass through zero value. This process is known as *forced commutation process* and inverters based on this principle are called *forced commutated inverters*. Forced commutated inverter circuits can be further classified as:

   (a) parallel capacitor commutated inverter.
   (b) series commutated inverter.
   (c) impulse commutated inverter.

12.2.2 Inverters Based on the Method of Connections

According to the connections of the thyristors and the commutating components, the device can be classified mainly into three groups: (1) series inverters, (2) parallel inverters, and (3) the bridge inverters.

12.2.3 Inverters Based on the Control Source

Based on the control source, inverters can be classified into two types: (1) voltage source inverters (VSI), and (2) current source inverters (CSI). In the voltage source inverter, the input is DC voltage supply and inverter converts the input DC voltage into a square wave AC output voltage source as shown in Fig. 12.1 (a). In the current source inverter, the input is a DC current source and the inverter converts the input DC current into a square AC output current as shown in Fig. 12.1 (b).

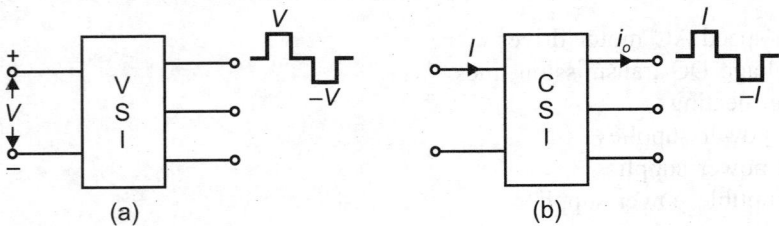

Fig. 12.1 Voltage source and control source inverters

12.3 FORCED COMMUTATED INVERTERS

In forced commutated circuit, some means are needed to forcefully turn-off the device and the inverter circuits based on this principles are known as *forced commutated inverter circuits*. These circuits independently provide an AC output of variable frequency and thus have much wider application. But as the device is to be turned off forcefully, it needs much more electronic circuits than the commutated.

Forced commutated inverters can be further classified into:

1. Parallel capacitor commutated inverters
2. Series commutated inverters
3. Impulse commutated inverters.

12.3.1 Parallel Capacitor Commutated Inverters

An alternating voltage to a load can be generated from a DC source by the use of centre-tapped transformer. Basically, by alternately switching the two thyristors, the DC source is connected in alternative senses to the two halves of the transformer primary, so inducing a squarewave voltage across the load in the transformer secondary. A thyristorised inverter consists of two parts: (1) the high power inverting elements, and (2) the low power oscillator.

The low power oscillator periodically triggers the thyristor in the inverting. The basic circuit is shown in Fig. 12.2(a) and its operation is explained below, using thyristors switching elements.

In the two thyristor circuits, capacitor is approximately charged to twice the DC supply voltage and is used alternately to turn-off the conducting thyristor. When thyristor Th1 is triggered on, DC source voltage V_s appears across half the transformer primary. A current causes an appropriate magnetic flux to rise in the top half of the transformer primary winding.

As this flux is common to both halves of the transformer winding, the source voltage V_s would be induced in the lower half and the commutating capacitor would have a voltage of $2V_s$ across it with the polarity shown. Subsequently, when Th2 is triggered on, the commutating capacitor applies a voltage of approximately $-2V_s$, to appear across Th1. When this reverse voltage is applied for a sufficient time across Th1, it will be turned off by the principle of parallel capacitor commutation. Thyristor Th2 will now be conducting and a voltage of $2V_s$ will appear across transformer primary and the commutating capacitor, but with a reverse polarity to that shown in Fig. 12.2(b). At the next trigger pulse Th1 will be again turned on and Th2 off. Thus, if trigger pulses are periodically applied to the alternate thyristors, an approximately rectangular voltage wave will be obtained at the transformer output terminals.

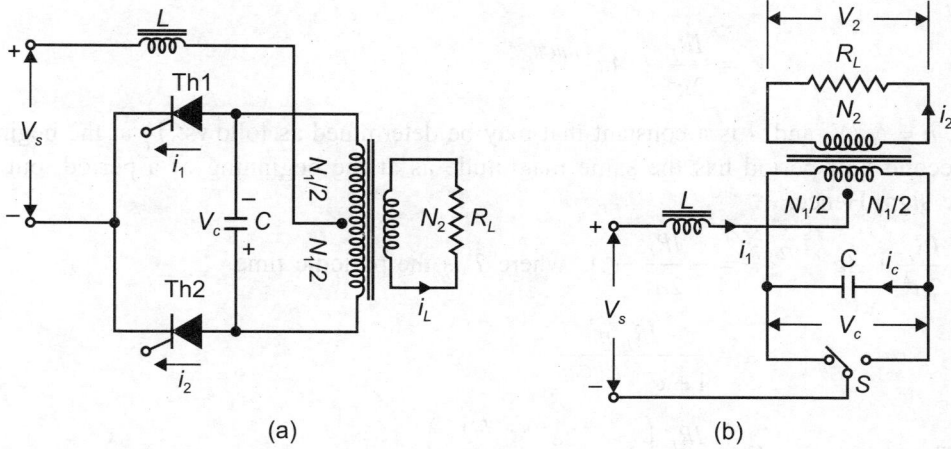

(a) (b)

Fig. 12.2 Single-phase parallel capacitor commutated inverter and analysis for circuit

Analysis of single-phase parallel capacitor commutated inverter: Assuming that Th1 is conducting, the switch has the position shown in Fig. 12.3(b) and the instantaneous conditions may be expressed by

$$V = V_s - V_{Th1} = L\frac{di_1}{dt} + \frac{V_c}{2} \qquad \qquad ...(12.1)$$

where, V_{Th1} = voltage drop across Th1

If a purely resistive load is assumed, then $V_2 = i_2 R_L$ and

$$V_c = i_2 R_L \frac{N_1}{N_2} \qquad \qquad ...(12.2)$$

where, N_2 is the secondary turns and N_1 is the primary turns.

Also, $i_2 N_2 = (i_1 - i_c) N_1/2 - i_c N_1/2$
$$= (i_1 - 2i_c) N_1/2 \qquad \qquad ...(12.3)$$

and $i_c = C\frac{dV_c}{dt} \qquad \qquad ...(12.4)$

Substituting the value of i_c in Equation (12.3), we get

$$i_2 N_2 = \left(i_1 - 2C\frac{dV_c}{dt} \right) N_1/2$$

and $i_2 = \left(i_1 - 2C\frac{dV_c}{dt} \right) \frac{N_1}{2N_2}$

In Equation (12.2), substituting the above value of i_2, we get

$$V_c = \left(i_1 - 2C\frac{dV_c}{dt} \right) \frac{R_L}{2} \left(\frac{N_1}{N_2} \right)^2 \qquad \qquad ...(12.5)$$

If the value of L is such that i_1 is substantially constant at a value I, then Equation (12.5) gives a first-order equation, i.e.

$$\frac{dV_c}{dt} + \frac{n^2}{CR_L}V_c = \frac{I}{2C}$$

and $V_c = \frac{IR_L}{2n^2} + Ae^{-n^2 t/CR_L}$

where, $n = N_2/N_1$ and A is a constant that may be determined as follows: V_c at the beginning of a second half period has the same magnitude as at the beginning of a period, but with reverse sign. Hence,

$$\frac{IR_L}{2n^2} + Ae^{-n^2 T/2/CR_L} = -\frac{IR_L}{2n^2} - A, \text{ where } T \text{ is the periodic time}$$

or $A = \dfrac{-IR_L/n^2}{1 + e^{-n^2 T/2/CR_L}}$

Thus, $V_c = \dfrac{IR_L}{2n^2} \left(1 - \dfrac{2e^{-(n^2 t/CR_L)}}{1 + e^{-n^2 T/2/CR_L}} \right)$

In the above equation $n = N_2/N_1$, but if N_1 is replaced by one-half the total primary turns N_1', then $IR/2n^2$ becomes $2IR/(N_2/N_1')^2$. This is the load resistance reflected across one-half of the primary winding and I is the current in this load when $C\,[(dV_c)/dt] = 0$.

Thus, $2IR_L/(N_2/N_1')^2 = 2V$

and
$$V_c = 2V \left(1 - \frac{2e^{-n^2 t/CR_L}}{1 + e^{-n^2\,T/2/CR_L}} \right) \qquad\qquad ...(12.6)$$

Equation (12.6) indicates that V_c increases exponentially towards a limiting voltage 2 V. Practical values of $(n^2\,T/2)/CR_L$ are generally such that

$$e^{-n^2\,T/2/CR_L} \ll 1$$

and hence $V_c = 2V\,(1 - 2e^{-n^2 t/CR_L})$ ...(12.7)

Design of commutating capacitor: The voltage across the commutating capacitor is given by

$$V_c = 2V\,(1 - 2e^{-n^2 t/CR_L}) \qquad\qquad ...(12.8)$$

when $V_c = 0$, let $t = t_0$, where $t_0 =$ turn-off time of the thyristor. From Equation (12.8)
$$0 = 2V - 4Ve^{-n^2 t_0/CR_L}$$

or $e^{-n^2 t_0/CR_L} = \dfrac{1}{2}$

$$\frac{n^2 t_0}{CR_L} = \log e^2 = 0.6931$$

Therefore, $t_0 = \dfrac{0.6931 \times CR_L}{n^2}$

The value of C, R_L, and n^2 must be chosen such that t_0 is greater than the turn-off time of thyristors.

Advantages:

1. It is the simplest forced commutation circuit.
2. A reasonably sinusoidal output can be obtained by an output filter.

Disadvantages:

1. The circuit is suitable only for fixed load.
2. This circuit is not suitable for large power for fixed value of L and C.
3. Voltage changes are not accommodated in the design.

12.3.2 Series Commutated Inverter

In this type of inverters, the commutating elements L and C are connected in series with the load constituting a series R–L–C resonant circuit. If the load is purely resistive then only it has resistance in the circuit. In case of inductive or capacitive load, its inductance or capacitance part is added to the commutating elements. Thyristors Th1 and Th2 are turned on alternately by applying trigger pulses to the gates to give a load current of required frequency. Figs 12.3 a) and (b) shows the circuit and waveforms of a simple series inverter. Let the initial voltage on capacitor C be V_c as shown. When thyristor Th1 is turned on, current flows for half a cycle $T/2$ through R–L–C series circuit. At point 'a' load current i is zero and thyristor Th1 will be turned off, also capacitor C will be charged to voltage V_c in reverse

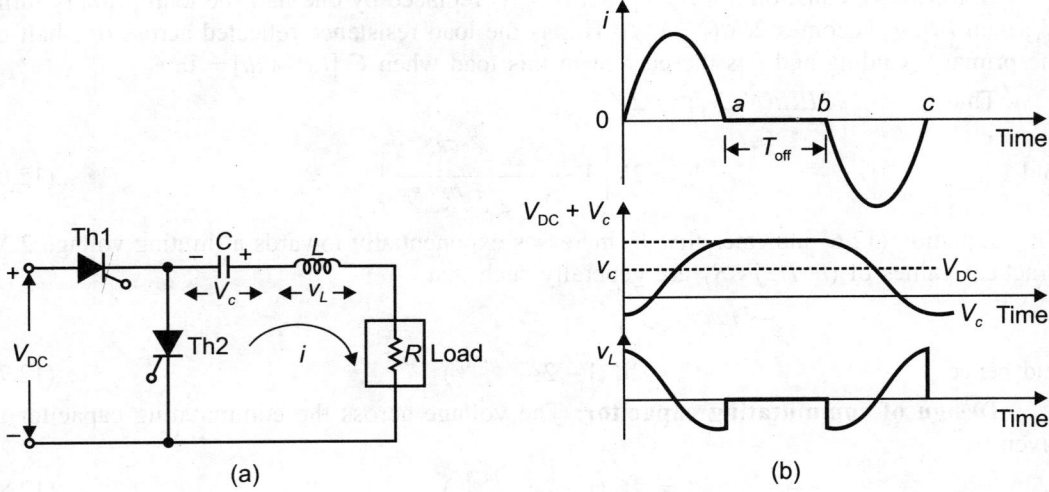

Fig. 12.3 Series commutated inverter circuited waveform

direction. Duration *ab* is the off period when the load is open circuited. So, the capacitor will retain voltage V_c.

When thyristor Th2 is fired at point *b* shown in Fig. 12.3(b), thyristor Th1 had already been turned off for a duration *ab* (T_{off}) more than the turn-off time it requires, then capacitor C will discharge through thyristor Th2 and underdamped R–L–C series circuit. Load current *i* will be in the opposite direction and again becomes zero at point *c*. Thyristor Th2 will then be turned off. A similar operation will occur when thyristor Th1 is turned on. The output frequency is given by: $f = \dfrac{1}{\dfrac{T}{2} + T_{off}}$ Hz, where T = time period of oscillations and T_{off} = time gap between turn-off of one thyristor and turn-on of the other thyristor. ...(12.9)

Thus, with the same LC components, variable frequency output can be obtained by changing the off time.

Mathematical analysis:

1. When thyhistor Th1 is conducting, assuming V_c is the initial voltage on the capacitor,

$$V_{DC} + V_c = Ri + L\frac{di}{dt} + \frac{1}{C}\int i\, dt \qquad \qquad ...(12.10)$$

Taking Laplace transform of Equation (12.10), we have

$$V_{DC}(s) = RI(s) + LsI(s) + \frac{1}{Cs}I(s) + Li(0) + v_c(0) \qquad ...(12.11)$$

where $i(0)$ is the initial value of current and $v_c(0)$ is the initial value of voltage across capacitance C, i.e. when thyristor Th1 is fired

$$v_c(0) = V_C$$

or $$V_{DC}(s) = RI(s) + LsI(s) + \frac{1}{C.s}I(s) + V_c$$

or $$I(s) = \frac{V_{DC}(s) - V_C}{Ls^2 + Rs + \dfrac{1}{C}} = \frac{[V_{DC}(s) - V_c]/L}{(s + R/2L)^2 + (1/C - R^2/4L^2)} \qquad ...(12.12)$$

Taking Laplace inverse of Equation (12.12), we have

$$i(t) = A \exp(-RT/2L) \sin \omega t \qquad \ldots(12.13)$$

where,

$$A = \frac{V_{DC} - V_C}{L\sqrt{1/LC - R^2/4L^2}} = \frac{V_{DC} - V_C}{\omega L}$$

and

$$\omega = \sqrt{\frac{1}{LC} - \frac{R^2}{4L^2}}$$

The voltage across capacitance C is given by

$$v_c(t) = V_{DC} - L\frac{di}{dt} - iR \qquad \ldots(12.14)$$

Putting the value of $i(t)$ from Equation (12.10) and after simplification, we have

$$v_c(t) = V_{DC} - (V_{DC} - V_C)\exp\left(-\frac{R}{2L}t\right) \times \left[\cos \omega t + \frac{\sin \omega t}{\sqrt{4L/R^2C - 1}}\right]$$

2. When thyristor Th2 is conducting and Th1 is off, then

from Equation (12.10), putting $V_{DC} = 0$, $V_C = v_c\left(\dfrac{T}{2}\right)$, and $t = \left(t - \dfrac{T}{2}\right)$, we have

$$i(t) = -\frac{v_c(T/2)}{\sqrt{1/LC - R^2/4L^2}} \exp[-R/2L(t - T/2)] \times \sin\sqrt{\frac{1}{LC} - \frac{R^2}{4L^2}}(t - T/2)$$

$$\ldots(12.15)$$

and similarly $\quad v_L = -v_c(T/2)\exp[-R/2L(t - T/2)]\cos\sqrt{\dfrac{1}{LC} - \dfrac{R^2}{4L^2}}(t - T/2)$

$$\times \frac{-\sin\dfrac{1}{LC} - \dfrac{R^2}{4L^2}(t - T/2)}{\sqrt{\dfrac{4L}{R^2C} - 1}} \qquad \ldots(12.16)$$

Output frequency $= \left[\dfrac{1}{\pi/\omega + T_{OFF}}\right] \qquad \ldots(12.17)$

and attenuation factor

$$= \exp\left(-\frac{R}{2L}\right) \cdot t \qquad \ldots(12.18)$$

Disadvantages:

1. Thyristor Th2 can only be fired when thyristor Th1 is off and vice-versa, otherwise, across DC supply, a dead short circuit will take place. This gives rise to distortion in the shape of the output waveform.
2. The maximum output frequency of the inverter can be obtained from this resonance of frequency of the resonant circuit.
3. The commutating elements must have a high rating because the load current is continuously flowing.

4. The ripple content is high as the power drawn from the supply is not continuous. It is not advisable to operate the series inverter at low frequencies because of high distortion in voltage waveform. A close sinusoidal output is obtained when X_L and X_C are appreciably larger than resistance R and resonant condition $X_L = X_C$ occurs. Then the inverter operating frequency is given by $\dfrac{1}{\sqrt{LC}}$.

12.3.3 Impulse Commutated Inverters

The thyristors in a voltage-fed inverter remain forward-biased by the DC supply voltage and are to be commutated by forced commutation. The popularly used (McMurray commutation technique) will be described here. In these circuits, it is assumed that the inverter is offering an inductive effect which imposes the most severe duty on the commutating current and inverse voltage is impressed across the device, helping to turn it off.

Simplifying assumptions made are as follows:

1. The load inductance is sufficiently large to maintain the load current constant at a value I_0 during the commutation interval. This is justified by the fact that the commutation process is completed in 10 to 100 microseconds.
2. The turn-on time of the thyristor is negligible. This is a reasonable assumption since the turn-on time is only 1 to 4 microseconds, and the rate of increase of current is determined by external circuit components with much longer time constants.
3. The brief pulse of reverse thyristor current at turn-off can be ignored.
4. The forward voltage drop of a conducting thyristor is negligible.

Two methods of impulse commutation will be considered as applied to inverters, namely: (1) McMurray inverter, and (2) McMurray–Bedford commutated inverter, which are discussed in detail.

12.3.3.1 *McMurray Inverters*

Figure 12.4 shows the circuit of McMurray inverter. Here every thyristor is commutated by means of an auxiliary thyristor. Thyristors Th1, Th2, Th3 and Th4 are the main thyristors and thyristors Th1A, Th2A, Th3A and Th4A are the four auxiliary thyristors. Thyristors Th1 and Th4 form the first pair of main thyristors and are fired simultaneously to produce the positive half of the AC output.

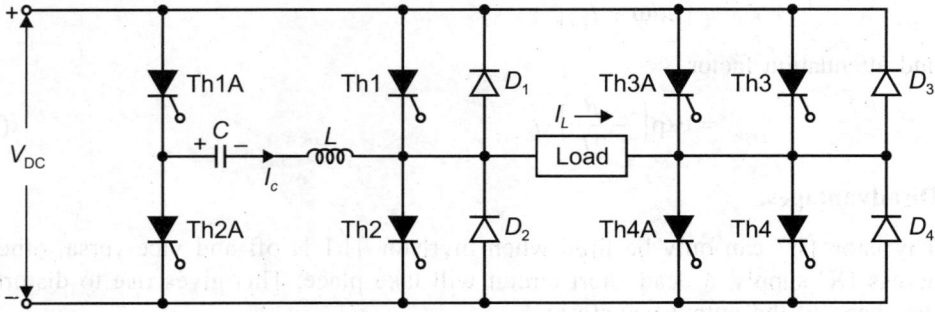

Fig. 12.4 The circuit of McMurray inverter

Thyristors Th3 and Th2 constitute the second pair and are fired simultaneously to give the negative half of the AC output. Let thyristors Th1 and Th4 are conducting, the path for the load current will be through the positive terminal of the main supply, Th1, load, Th4 and back to the negative terminal of the mains. At this instant, the right hand plate of capacitor C becomes positive. Now, for commutating Th1, the auxiliary thyristor Th1A has to be gated. The moment this happens, capacitor C will start discharging through L, feedback diode D_1, and Th1A. Thus, the discharge current I_C will try to neutralise the current flowing through Th1. As soon as the thyristor current is fully neutralised, thyristor Th1 will be turned off. When the capacitor is fully discharged it will start recharging with opposite polarities. When capacitor is fully charged with new polarity, thyristor Th1A will be automatically turned off and C will retain its charge. Now thyristors Th2 and Th3 can be gated. When thyristors Th2 and Th3 are conducting the load current I_L will be in the opposite direction following the path of positive terminal of the mains supply Th3, load, Th2 and back to the negative terminal of the mains.

To turn-off Th2, Th2A is gated. As soon as thyristor Th2A is gated, capacitor C starts discharging through the path Th2A, D_2 and L. This discharge current will again neutralise the current flowing through Th2 and turn it off. When the capacitor is fully discharged it will start recharging with the opposite polarities.

When the capacitor is fully charged, Th2A will be automatically turned off and capacitor C will retain its charge. This gives the original condition and thyristor Th1 can now again be gated for the next half-cycle. The cycle is repeated continuously, thus producing AC output voltage across the load. This circuit has been further improved by using complementary voltage commutation in the McMurray–Bedford circuit.

12.3.3.1.1 *Modified McMurray Inverter*

The modified version of the McMurray inverter is shown in Fig. 12.5. This is the auxiliary impulse-commutated (McMurray) circuit with the addition of two diodes (D_{1A} and D_{2A}), and a resistor (R) for return of capacitor overcharge to the supply. The operation and design of the power circuit is given in Fig. 12.5.

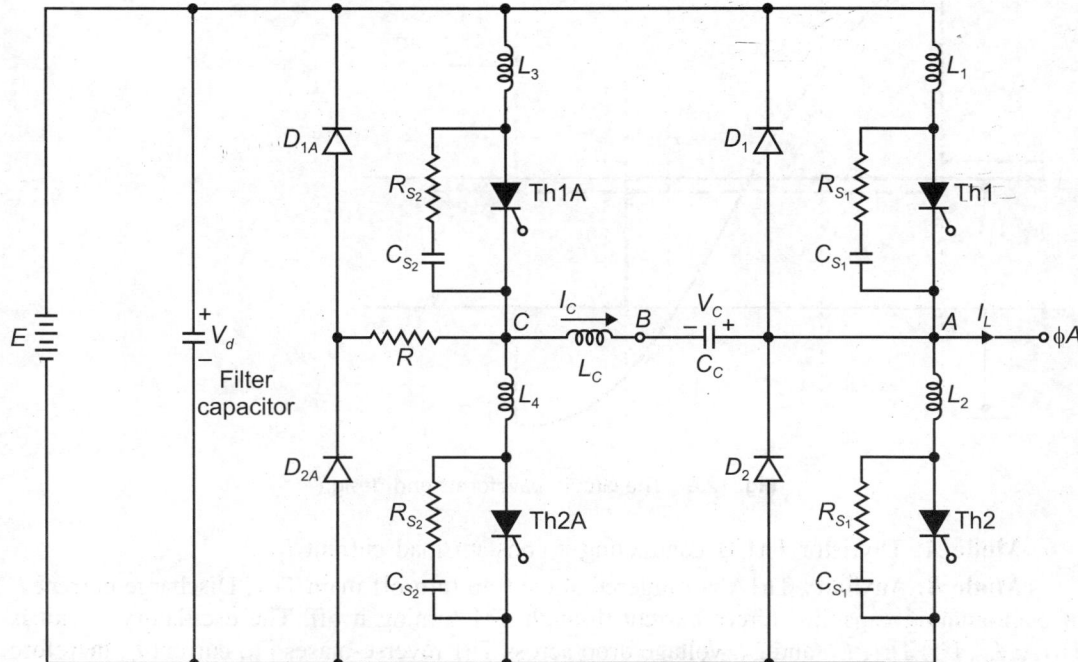

Fig. 12.5 The modified version of the McMurray inverter

A *LC* circuit is used to commutate both the thyristors. Additional thyristors are used for commutating purposes. Triggering of these thyristors initiates commutation and hence these are called *auxiliary thyristors*. The current transfer from the outgoing thyristor to the incoming one takes place in several stages. The commutating capacitor (C_c) is charged with the polarity as shown to a voltage V_c. To commutate main Th1, the auxiliary Th1A is turned on. This completes the *LC* circuit and an oscillation is initiated. As soon as commutating current I_c exceeds I_L, the excess current is flowing through the feedback diode D_1, thus providing a reverse-bias on Th1 sufficient to turn it off. At the end of the commutation process, the capacitor voltage V_c will be charged in the opposite direction. At this point Th2 will be carrying load current, and it can be turned off by triggering Th2A to initiate the commutation process.

12.3.3.1.2 *Operation of Modified McMurray Inverter Circuit with Lagging Load*

The circuit waveforms and design are explained assuming a lagging load in Fig. 12.6.

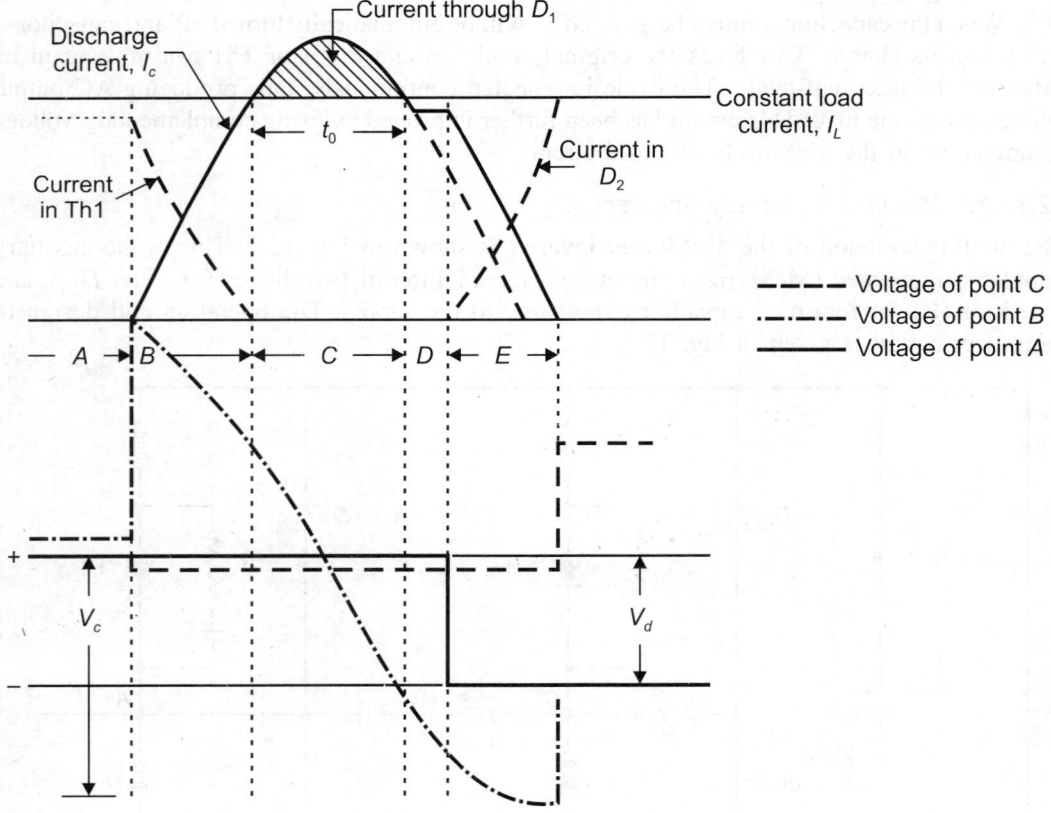

Fig. 12.6 The circuit waveforms and design

Mode *A*: Thyristor Th1 is conducting a constant load current I_L.

Mode *B*: Auxiliary Th1A is triggered at $t = 0$ to turn-off main Th1. Discharge current I_c of commutating capacitor forces current through Th1 turning it off. The oscillatory circuit is Th1A, L_C, C_C, Th1, L_1, and L_3. Voltage drop across Th1 reverse-biases D_1, current I_c, therefore flow only through Th1 and not through D_1 as shown in Fig. 12.6.

Mode C: After mode B, as the discharge current I_c exceeds the load current, excessive current flows through feedback diode D_1, thus reverse-biasing Th1 and turning it off. The discharge current reaches a maximum value when the capacitor voltage is zero, and then decreases as the capacitor charges in the reverse direction.

Mode D: After mode C, as discharge current I_c falls below load current I_L, the diode D_1 ceases to conduct. The load continues to draw a constant current through auxiliary Th1A, L_c and C_c. The commutating capacitor is charged in the reverse direction until it reaches supply voltage V_d.

Mode E: After mode D, the load current I_L continues to flow through L_c, until all the trapped magnetic field energy is converted to electrostatic energy in charging the capacitor. In the modified McMurray circuit, diodes D_{1A} and D_{2A} are across the auxiliary thyristors, the excess inductive energy is returned to the bus, thus reducing the commutating capacitor voltage. The overcharge across the commutating capacitor would appear as a forward voltage on the auxiliary thyristor, which requires a higher rated unit than would be normally used. Addition of diodes D_{1A}, D_{2A} and resistor R provide a path for this excess energy to be returned to the DC bus.

A full three-phase inverter will consists of three such identical units. A large filter capacitor shown at the input can sink the harmonics fed back to the source. The commutating capability of the McMurray circuit is a function of the load. As the load current increases, the capacitor voltage increases and hence the capability to commutate higher current. Further, the commutation losses are lower at lower values of load current.

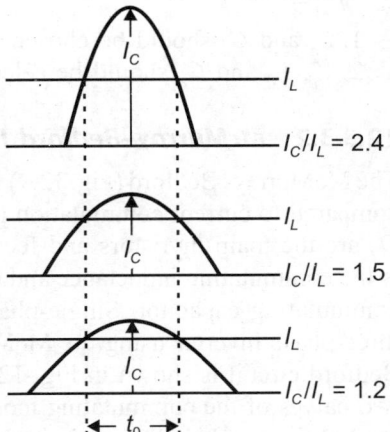

Fig. 12.7 The commutating current pulses of different magnitudes and widths

Design of McMurray inverter circuit: The commutating current pulses could be of various magnitudes and widths to successfully commutate thyristors in the McMurray inverter. Each current pulse would successfully commutate the thyristors if the turn-off time of the thyristors are below time t_0 shown in Fig. 12.7 observed that the best efficiency of the commutating circuit is achieved when the peak of commutating pulse current is 1.5 times the load current. Thus, the design criterion becomes $I_c/I_L = 1.5$.

For modes C and D of the modified McMurray inverter, the circuit parameters that come into play are L and C only. The commutating current pulse is shown in Fig. 12.8. This current pulse is obtained when thyristor Th1A is triggered to turn-off Th1. From Figs 12.5 and 12.8,

V_c = voltage across commutating capacitor, C_c

I_c = peak value of commutating current

C_c = commutating capacitor

L_c = commutating reactor

I_L = load current

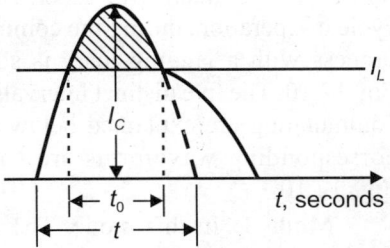

Fig. 12.8 The commutating current pulse

$\omega = 2\pi f$, where $f = \dfrac{1}{2t}$ = frequency of oscillation of the commutating pulse.

For maximum efficiency or minimum energy loss in the commutation circuits,
$$I_c = 1.5 I_L$$
For the commutation circuit:

$$I_c = \frac{V_c}{\omega L_c} = \frac{V_c}{2\pi f L_c} = \frac{V_c}{2\pi \cdot \dfrac{1}{2\pi\sqrt{L_c C_c}}} = V_c \sqrt{\frac{C_c}{L_c}} \qquad \text{...(12.19)}$$

$$f = \frac{2}{2\pi\sqrt{L_c C_c}} \qquad \text{...(12.20)}$$

Again selection of L_c and C_c are based on least stored energy for commutation. For $I_c/I_L = 1.5$, it can be shown that:

$$t \approx 1.68\sqrt{L_c C_c} \qquad \text{...(12.21)}$$

For successful turning-off of the main thyristors, the commutation circuit must satisfy the following:

1. L_c and C_c should be chosen such that t_0 = turn-off time of Th1
2. V_c, L_c and C_c should be selected such that $I_c = 1.5 I_L$.

12.3.3.2 *McMurray–Bedford Inverter*

The McMurray–Bedford (Fig. 12.9) works on the principle of complementary voltage commutation compared to current commutation principle as described before. The devices Th1, Th2 and D_1, D_2 are the main thyristors and feedback diodes. The tapped inductor in series with thyristors is the commutating inductance and C is the commutating capacitor. Single-phase of a three-phase inverter using the McMurray–Bedford circuit is shown in Fig. 12.9. The two halves of the commutating inductance are tightly coupled. The period of the AC output of the inverter is much greater than the turn-off time of the thyristors. The commutation process is initiated when the non conducting thyristor is triggered onto turn-off the conducting thyristor. One-half-cycle of operation, during the commutation process with a lagging load, is shown in Fig. 12.10. The five distinct intervals during commutating are explained below and the corresponding waveforms are shown in Fig. 12.10.

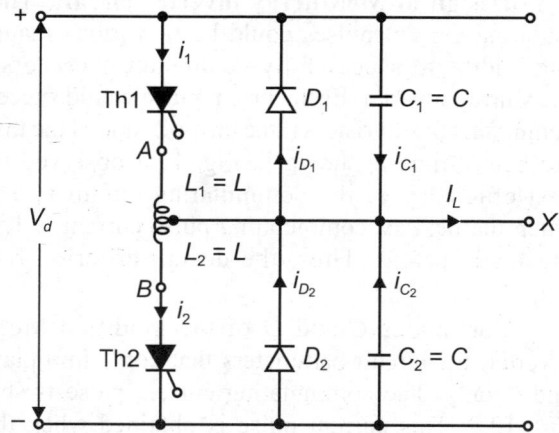

Fig. 12.9 The McMurray–Bedford inverter

Mode 1: In this mode, Th1 is conducting current to the load drawing power from the upper side of the DC supply. As the load current is almost constant with an inductive load, voltage drop across commutating inductance L is negligible. Hence, the terminal X of the load is at the positive of the DC supply voltage V_d. The capacitor C_1 is charged to zero voltage, and C_2 to V_d. The load current is equal to I_L.

Mode 2: Thyristor Th2 is triggered in this mode to turn-off Th1, at instant $t = 0$, point B potential falls to negative of the supply. Since the voltage across capacitors C_1 and C_2 cannot

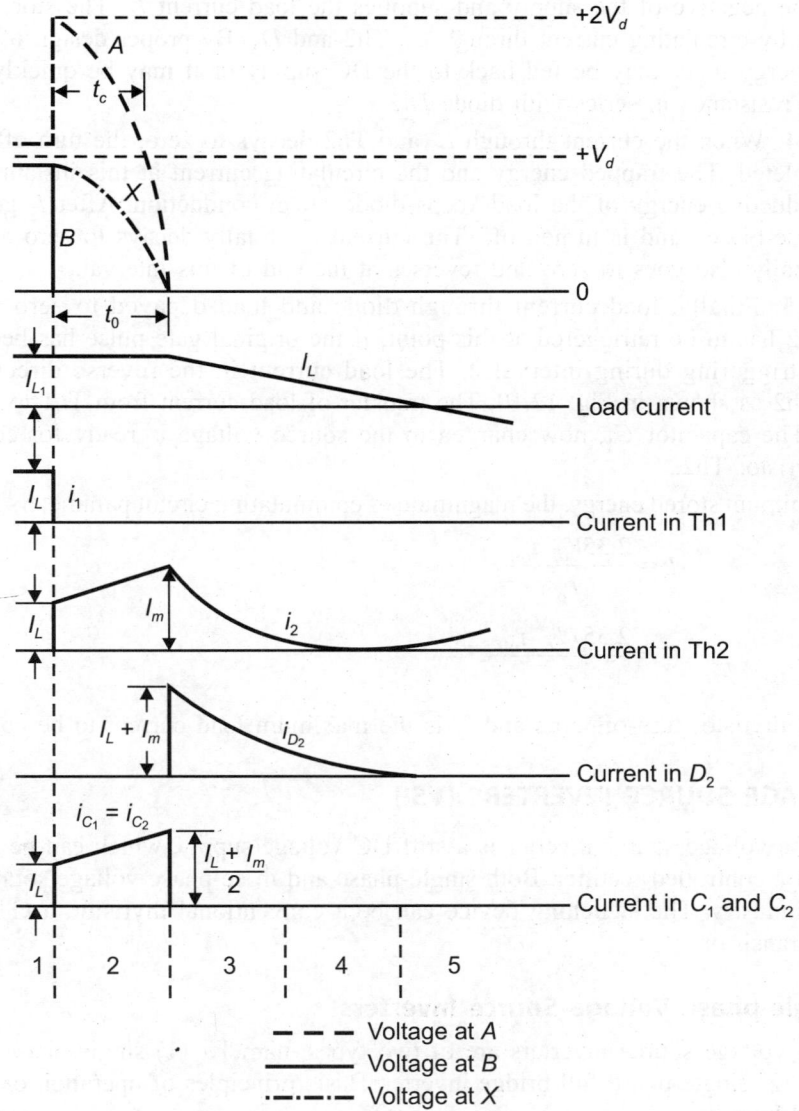

Fig. 12.10 The commutation process with a lagging load and corresponding waveforms

change instantaneously, a voltage V_d must appear across inductance L_2. The same voltage is induced across L_1 (V_d). Thus, the potential of point A is $2V_d$, which is across Th1 and turns it off. The load current I_L, which was flowing through Th1 and L_1, now transfers to winding L_2 and Th2 in order to maintain the energy stored in the inductance. The constant load current I_L is supplied by the capacitors C_1 and C_2. These initial conditions and waveforms are shown in Fig. 12.10. The waveform is part of a sinusoid of period $2\pi\sqrt{LC}$. The duration of the commutating interval t_0 is about 1/4 cycle. The time t_c available for the Th1 to turn-off is even shorter. During the commutating interval, the voltage across capacitor C_2 is also across L_2. As the voltage across capacitor C_2 discharges, the current i_2 increases from I_L to I_m and the voltage across C_2 goes to zero.

Mode 3: At $t = t_o$, capacitor C_1 is charged to supply voltage V_d and therefore no current can flow through C_1, i.e. $i_{c1} = 0$. The feedback diode D_2 starts to conduct and connects the point X to the negative of the supply and supplies the load current I_L. The stored energy in L_2 dissipates by circulating current through L_2, Th2 and D_2. By proper design of the circuit, the stored energy in L_2 may be fed back to the DC supply or it may be quickly dissipated by putting a resistance in series with diode D_2.

Mode 4: When the current through L_2 and Th2 decays to zero, the turn-off process of Th1 is completed. The trapped energy and the circulating current at this instant reduced to zero. The inductive energy of the load keeps diode D_2 in conduction. After i_2 goes to zero, Th2 is reverse-biased and is turned off. The current i_{D2} finally decays to zero and the load current I_L finally also goes to zero and reverses at the end of this interval.

Mode 5: Finally, load current through diode and load decayed to zero and is then reversed. Th2 has to be retriggered at this point, if the original gate pulse has been removed after initial triggering during interval 2. The load current in the reverse direction is now carried by Th2 as shown in Fig. 12.10. The transfer of load current from Th1 to Th2 is now completed. The capacitor C_1, now charged to the source voltage is ready for commutating the main thyristor Th2.

For minimum stored energy, the magnitude of commutating circuit parameters L and C are:

$$L = \frac{2.35 V_d \cdot t_{\text{off}}}{I_L}$$

and
$$C = \frac{2.35 \cdot I_L \cdot t_{\text{off}}}{V_d},$$

where, t_{off} is thyristor turn-off time and I_L is the maximum load current to be commutated.

12.4 VOLTAGE SOURCE INVERTERS (VSI)

The input of a voltage source inverter is a stiff DC voltage supply, which can be a battery or the output of a controlled rectifier. Both single-phase and three-phase voltage source inverters are used in industry. The switching device can be a conventional thyristor, a GTO thyristor or a power transistor.

12.4.1 Single-phase Voltage Source Inverters

Single-phase voltage source inverters are of two types, namely: (1) single-phase half-bridge inverter, and (2) single-phase full bridge inverter. Basic principles of operation of these types are discussed here.

12.4.1.1 *Single-phase Half-bridge Inverter*

The half-bridge configuration of the single-phase voltage source inverter is shown in Fig. 12.11 (a). The DC supply is centre-tapped. Thyristors Th1, Th2 act as ON/OFF solid state switches. Diodes D_1 and D_2 are known as *feedback diodes* because they can feedback load reactive energy.

During the positive half-cycle of the output voltage, thyristor Th1 is turned on, which makes $v_0 = +V/2$. During the negative half-cycle, thyristor Th2 is turned ON, which makes $v_0 = -V/2$. Figure 12.11 (b) shows the waveforms of gate pulses and output voltage v_0. Prior to turning on a thyristor, the other thyristor must be turned off, otherwise both the thyristors will conduct and short circuit the DC supply.

If the load is inductive, the current i_0 lags the output voltage v_0 as shown in Fig. 12.11 (c). Note that during $0 < t < T/2$, v_0 is positive, either thyristor Th1 or diode D_1 is conducting

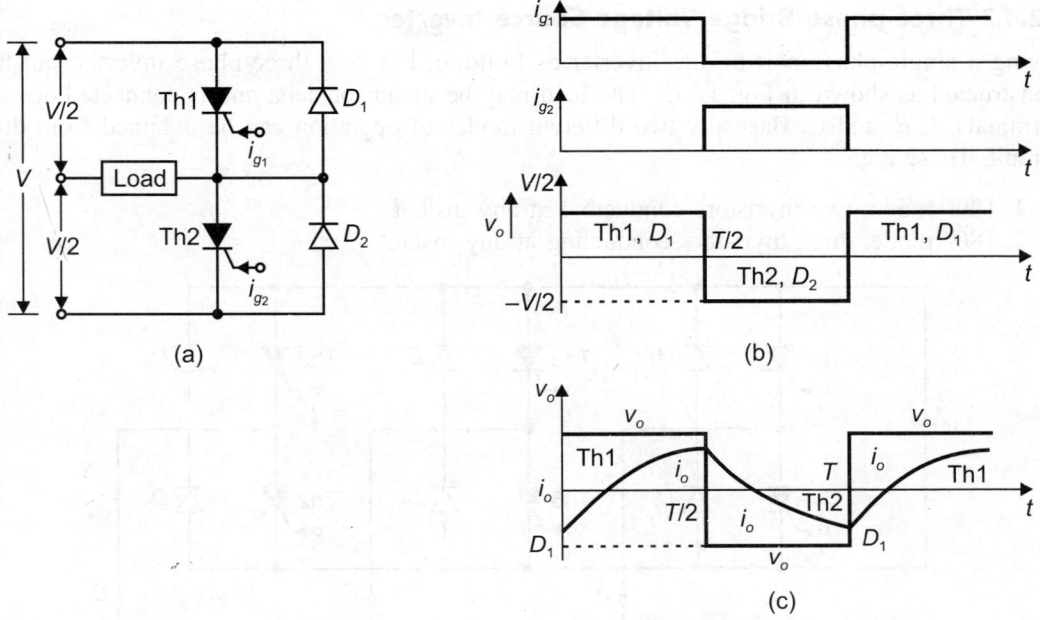

Fig. 12.11 The half-bridge configuration of single-phase voltage source inverter

during this interval. However, i_0 is negative during $0 < t < t_1$, therefore diode D_1 must be conducting during this interval. The load current i_0 is positive during $t_1 < t < T/2$ and therefore thyristor Th1 must be conducting during various intervals of time are shown in Fig. 12.11 (c). The feedback diodes conduct when the voltage and current are of opposite polarities.

12.4.1.2 *Single-phase Full-bridge Voltage Source Inverter*

The full-bridge configuration of a single-phase voltage source inverter is shown in Fig. 12.12 (a). Thyristors Th1 and Th2 are fired during the first half-cycle and thyristors Th3 and Th4 are fired during the second half-cycle of the output voltage. The output voltage is a square wave of amplitude V as shown in Fig. 12.12 (b). The frequency of the firing pulses decides the frequency of the inverter.

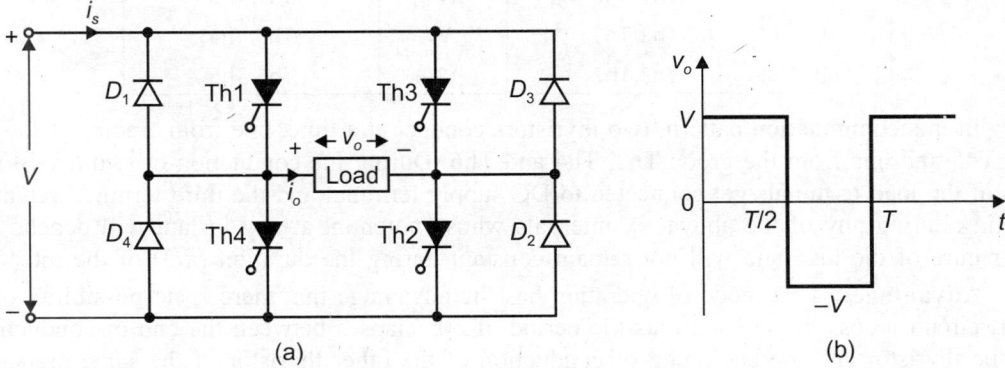

Fig. 12.12 The full-bridge configuration of a single-phase voltage source inverter

12.4.2 Three-phase Bridge Voltage Source Inverter

Using a single-phase half-bridge inverter as building block, a three-phase inverter can be constructed as shown in Fig. 12.13. The load may be in star or delta and is connected across terminals A, B, and C. Basically two different modes of operation can be obtained from this circuit. These are:

1. 120° mode; two thyristors conducting at any instant
2. 180° mode, three thyristors conducting at any instant

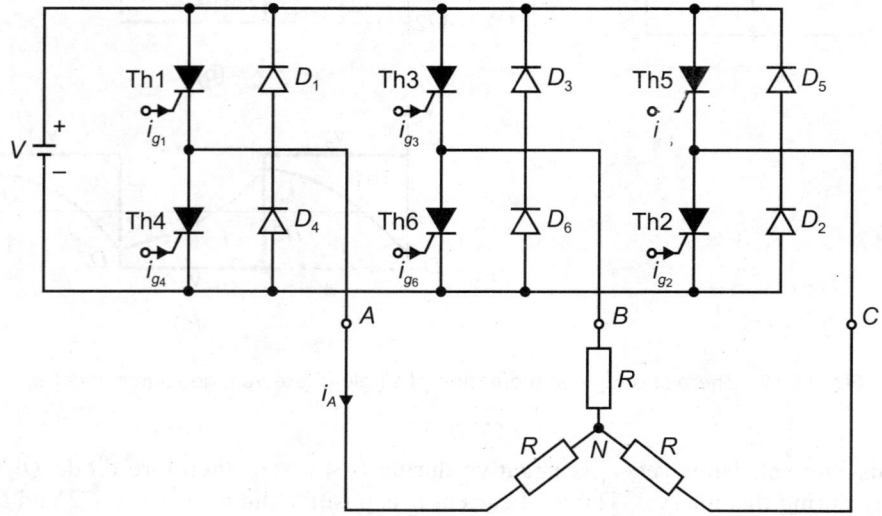

Fig. 12.13 A three-phase inverter with star connected load

Case I: 120° mode, two thyristors conducting at any instant.

Figure 12.14 shows various waveforms for the gating and commutation pattern in which each thyristor conducts for $2\pi/3$ radians. Six commutations per cycle are required in this case. One period of inverter operation has been divided into six intervals. In each interval, thyristors conduct according to the following pattern:

Interval	Thyristors conducting	Interval	Thyristors conducting
I	Th1, Th6	IV	Th4, Th3
II	Th2, Th1	V	Th5, Th4
III	Th3, Th2	VI	Th6, Th5

In this commutation pattern, two thyristors conduct at a time, one from among Th1, Th3 and Th5 and one from the group Th2, Th4 and Th6. During the conduction of two thyristors, two of the load terminals get connected to DC supply terminal and the third terminal remains floating during any of the above six intervals whose potential at any instant will depend on the nature of the load and will not remain constant during the duration $(\pi/3)$ of the interval.

Advantages: This mode of operation has the advantage that there is no possibility of a short circuit across the DC input as the period of 60° elapses between the end of conduction of one thyristor and the beginning of conduction of the other thyristor of the same branch.

Mathematical Analysis: Mathematical analysis of this inverter circuit for any type of load is complicated since the voltage of only two terminals is defined at any instant. However, for a balanced, three-phase star connected load, of Fig. 12.13, the output waveform of voltages will be seen in Fig. 12.14.

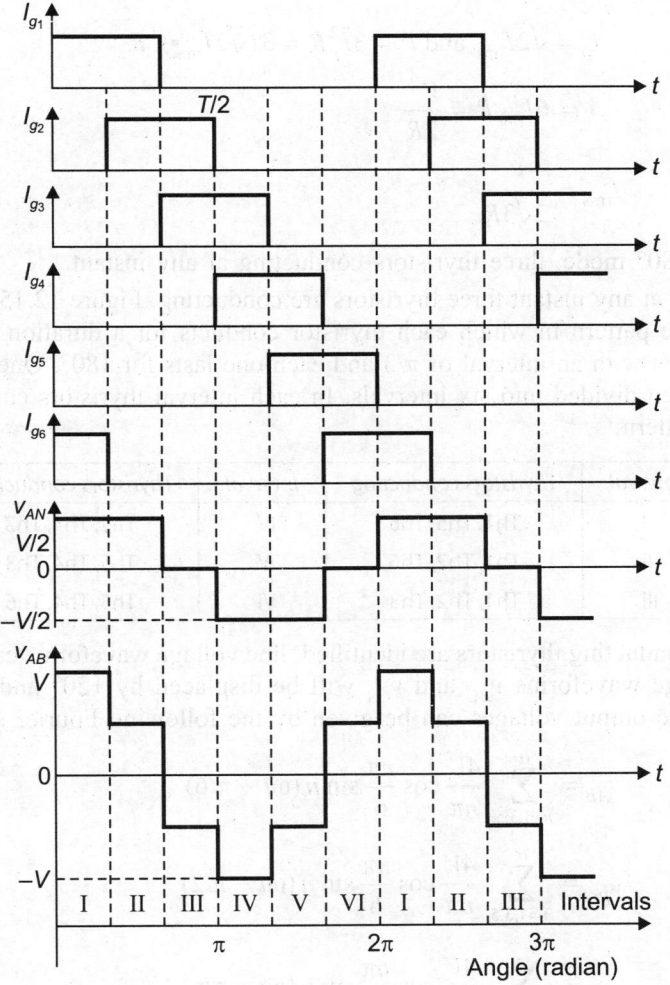

Fig. 12.14   The output waveform of voltages

The output power of the inverter,

$$P = \frac{3}{\pi} \int_0^{2\pi/3} \frac{(v_{AN})^2}{R} \, d(\omega t)$$

$$P = \frac{3}{\pi} \int_0^{2\pi/3} \left(\frac{V}{2}\right)^2 = \frac{1}{R} \, d\omega t = \frac{V^2}{2R} \text{ watts} \qquad ...(12.22)$$

where, v_{AN} is the phase voltage for a star connected resistance load connected across terminals *ABC*. Magnitude of $v_{AN} = V/2$ in the interval 0 to $2\pi/3$.

$$P_T = N V_{DRM} I_{rms} = 6 \times \frac{V}{2\sqrt{3}} \frac{V}{R} = \frac{\sqrt{3}V}{R} \qquad ...(12.23)$$

Utility factor $UF = \dfrac{P}{P_T} = \dfrac{V^2}{2R} \times \dfrac{R}{\sqrt{3}V^2} = \dfrac{1}{2\sqrt{3}} \approx 28.9\%$ ...(12.24)

Here $I_p = \sqrt{2}I_{rms}$ and $P = 3I_p^2 R = 3(\sqrt{2}I_{rms})^2 R$

$$= 6I_{rms}^2 R = \dfrac{V^2}{2R}$$ ...(12.25)

or $I_{rms} = \dfrac{V}{2\sqrt{3}R}.$ ...(12.26)

Case II: 180° mode, three thyristors conducting at any instant.

In this case at any instant three thyristors are conducting. Figure 12.15 shows the various waveform for the pattern in which each thyristor conducts for a duration of π. Gate signals I_{g1} to I_{g6} are given with an interval of π/3 and each one lasts for 180°. One period of inverter operation has been divided into six intervals. In each interval thyristors conduct according to the following pattern:

Interval	Thyristors conducting	Interval	Thyristors conducting
I	Th1, Th5, Th6	IV	Th2, Th4, Th3
II	Th1, Th2, Th6	V	Th5, Th4, Th3
III	Th1, Th2, Th3	VI	Th5, Th4, Th6

Once the conducting thyristors are identified, line voltage waveforms can be easily drawn. Other line voltage waveforms v_{BC} and v_{CA} will be displaced by 120° and will have similar waveforms. These output voltages can be given by the following Fourier series:

$$v_{AB} = \sum_{n=1,3,5}^{\alpha} \dfrac{4V}{n\pi} \cos \dfrac{n\pi}{6} \sin n\,(\omega t + \pi/6)$$ ...(12.27)

$$v_{BC} = \sum_{n=1,3,5}^{\alpha} \dfrac{4V}{n\pi} \cos \dfrac{n\pi}{6} \sin n\,(\omega t - \pi/2)$$ ...(12.28)

$$v_{CA} = \sum_{n=1,3,5}^{\infty} \dfrac{4V}{n\pi} \cos \dfrac{n\pi}{6} \sin n\,(\omega t - \pi/6)$$ ...(12.29)

If load is in delta, the phase and line currents can be calculated using the above equations. If load is in star and linear, superposition theorem can be used to obtain the phase currents and voltages. From Fig. 12.15, it can be seen that:

In interval I: $(0 < \omega t < \pi/3)$ phase voltage $v_{AN} = V/3$

In interval II: $(\pi/3 < \omega t < 2\pi/3)$ phase voltage, $v_{AN} = 2V/3$

In interval III: $(2\pi/3 < \omega t < \pi)$ phase voltage, $v_{AN} = V/3$.

Similarly, one can obtain other phase voltages v_{BN} and v_{CN} which will be lagging v_{AN} by 2π/3 and 4π/3 radian respectively.

The output power for a balanced star connected load is:

$$P = \dfrac{3}{\pi} \int_0^\pi \dfrac{(v_{AN})^2}{R} \, d\,(\omega t)$$

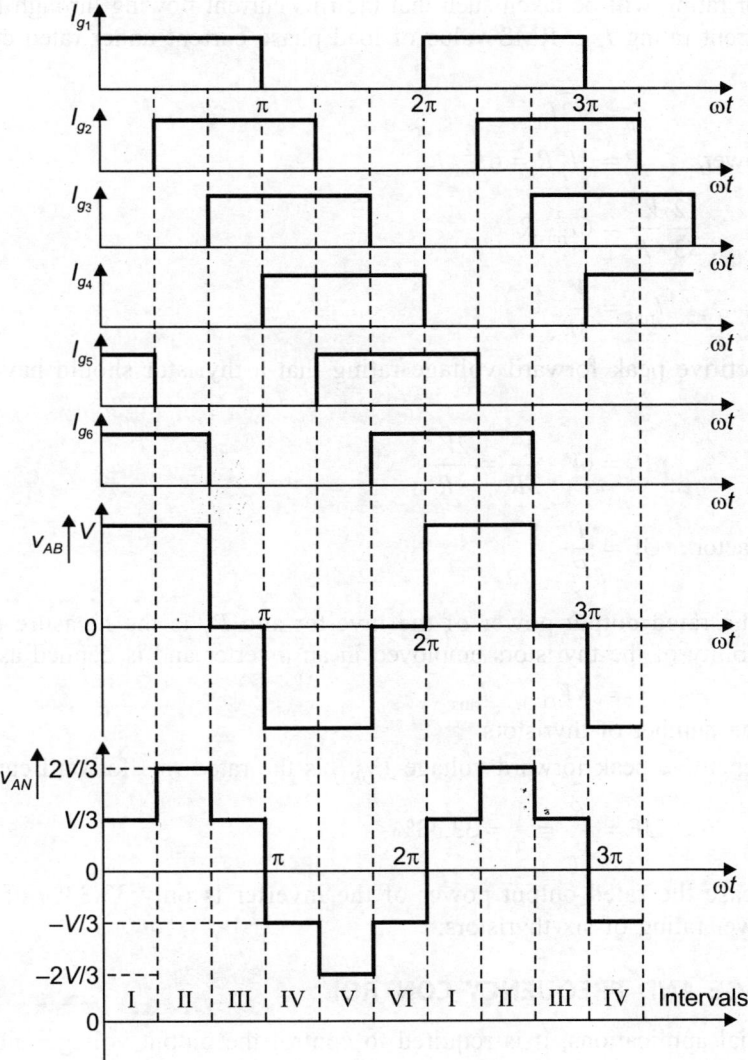

Fig. 12.15 The various waveforms for the pattern in which each thyristor conducts for a duration of π

using

$$v_{AN} = \frac{V}{3}, \quad \text{for } (0 < \omega t < \pi/3)$$

$$= \frac{2V}{3}, \quad \text{for } (\pi/3 < \omega t < 2\pi/3)$$

$$= \frac{V}{3}, \quad \text{for } (2\pi/3 < \omega t < \pi)$$

$$\therefore \qquad P = \frac{2V^2}{3R} \text{ watts} \qquad \qquad \qquad ...(12.30)$$

Thyristor rating will be taken such that the rms current flowing through it will be equal to its rms current rating I_{rms}. RMS value of load phase current under rated conditions, I_p is given by

$$I_p = \sqrt{2}I_{rms}$$

Load power, $\quad P = 3I_p^2 R = 6I_{rms}^2 R$

or

$$\frac{2}{3}\frac{V^2}{R} = 6I_{rms}^2 R$$

or

$$I_{rms} = \frac{V}{3R} \qquad\qquad ...(12.31)$$

The repetitive peak forward voltage rating that a thyristor should have for this load $V_{DRM} = V$.

Thus, $\qquad P_T = 6V \cdot \dfrac{V}{3R} = \dfrac{2V^2}{R} \qquad\qquad ...(12.32)$

Utility factor $\quad UF = \dfrac{P}{P_T} \qquad\qquad ...(12.33)$

where, P is the rated output power of the inverter and P_T is the measure of total power handling capability of the thyristors employed in an inverter and is defined as

$$P_T = NV_{DRM} \cdot I_{rms}$$

where, N is the number of thyristors.

I_{rms} is repetitive peak forward voltage V_{DRM} is the rated rms forward current

Thus, $\qquad UF = \dfrac{P}{P_T} = \dfrac{1}{3} = 33.33\% \qquad\qquad ...(12.34)$

In this case the rated output power of the inverter is only 33.33% of the combined maximum power rating of six thyristors.

12.5 VOLTAGE AND FREQUENCY CONTROL

In the industrial applications, it is required to control the output voltage and frequency of inverters for machine drive applications. The variation of voltage and frequency are required to cope with the variations of DC input voltage, for voltage regulation of inverters and for the constant volts/frequency control requirement. Most inverter applications require a means of voltage and frequency control. The three-phase inverter circuit changes DC input voltage to a three-phase variable frequency variable voltage output. The input DC voltage may be from a DC source or rectified AC voltage. Mostly, inverter applications require a variable voltage and frequency control. Voltage can be controlled by the *three* following methods:

1. Control of voltage supplied to the inverter
2. Control of voltage within the inverter
3. Control of voltage delivered by the inverter.

12.5.1 Control of Voltage Supplied to the Inverter

The control of voltage supplied to the inverter may be achieved in several ways. If an AC voltage is available, the commonly used method is the phase controlled rectifier. If the DC voltage is available then a chopper circuit is used to vary voltage supplied to the inverter.

Advantages: Voltage control method has the following advantages when the DC supply voltage to the inverter is to be controlled:

1. As the voltage is to be controlled, the inverter output voltage wave shape and its harmonic content are not significantly changed.
2. Voltage supplied to the inverter can be easily controlled.

Disadvantages: Disadvantages of voltage control scheme are as follows:

1. Power delivered by inverter is handled twice, once by the DC voltage control and once by inverter.
2. This scheme requires filtering in the DC circuit. This causes slower response time in a complete closed loop voltage regulated inverter power supply.
3. Generally there is excessive commutating voltage if the inverters are designed to reliably commutate the highest currents at a reduced DC voltage which produces increased circulating currents resulting in higher losses. This problem can be overcome by providing a fixed DC bus of constant commutating capacity.

12.5.2 Control of Voltage within the Inverter

The method of controlling the voltage within an inverter involves the use of pulse modulation technique. In this scheme, the output voltage is a pulse width modulated wave and the voltage is controlled by varying the duration of the output voltage pulses.

12.5.3 Control of Voltage Delivered by Inverter

The output voltage of the inverter may be controlled by using a transformer at the output. For wide frequency range this technique is not a practical solution.

12.6 CURRENT SOURCE INVERTER

A current fed inverter requires a stiff DC current source (ideally, infinite Theven impedance) at the input which is in contrast to the stiff voltage source desirable in a voltage fed inverter. A variable voltage source can be converted into a variable current source by connecting an inductance in series and controlling the voltage within a current control loop. The power semiconductors in a current fed inverters have to withstand reverse voltage, and therefore the devices such as GTOs, transistors and power MOS are not suitable. Applications of current fed inverters are:

1. Induction heating
2. Speed control of AC motors
3. Lagging VAR generation
4. Starting of synchronous motors in gas turbine.

Operation: Its operation and properties are markedly different from those of voltage source inverters. A block diagram and simplified schematic power circuit diagram are shown in Figs 12.16(a), (b). The filter inductor L_d is made very large increasing the source impedance to high value which keeps the supply current I_d practically constant. The load of the inverter is an induction motor which can be represented by a counter emf in series with equivalent leakage inductance. The power current, thus appears symmetrical about DC link. The DC current I_d is switched through the inverter thyristors so as to establish three-phase, six-stepped, symmetrical line current waves. Each thyristor conducts for 120° and at any instant one upper thyristor Th1 and one lower thyristor Th4 remain in conduction. The DC link

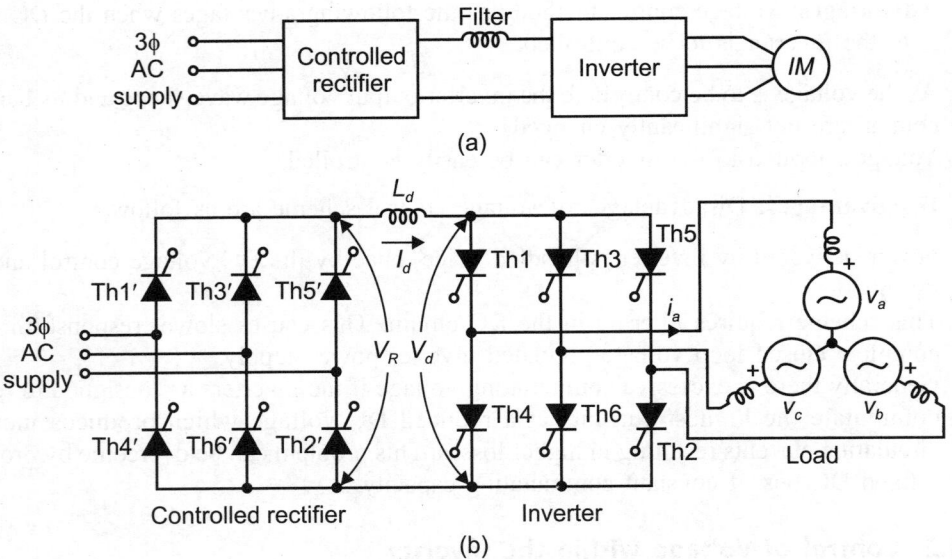

Fig. 12.16 A block diagram and simplified schematic power circuit diagram

current is considered harmonic free and the commutation effect is ignored. For variable speed drive applications, the inverter can be operated at a variable frequency with an adjustable magnitude of current I_d.

12.6.1 Inverter Operation Modes

The inverter firing angle α can be varied in the range 0 to 360° with respect to the counter emf wave, and the following four modes are:

Mode 1 ($0° \leq \alpha \leq 90°$) Load-commutated rectifier: In this mode, the commutation is performed by the load. Figure 12.17(a) shows the phase voltage and current waves for $\alpha = 45°$. When the incoming thyristor Th4 is fired, the outgoing thyristor Th2 will be impressed with the negative anode voltage v_{ca} and therefore will cause load commutation. The fundamental component of i_a will lag the voltage wave by 45°, and the corresponding phasor diagram is shown on the right side. The active power will flow from the load to the DC link, which will then be pumped back to the AC supply line by the rectifier operating in the line-commutated inverter mode.

The load will supply lagging VAR to the inverter, the leading VAR will be supplied to the load from the inverter. Such a condition can be met by a synchronous machine operating with over excitation. Therefore, this mode can be considered as a synchronous motor operating in a regenerative braking condition, as indicated in Fig. 12.17(e).

Mode 2 ($90° < \alpha \leq 180°$) Load-commutated inverter: This mode is explained in Fig. 12.17(b) for a typical angle $\alpha = 135°$. The outgoing thyristor Th2 is commutated by the load since v_{ca} is negative in this range. The active power flows to the load and the DC link voltage are positive. The load is required to operate at leading power factor as in the previous mode. This mode can therefore be considered as the motoring mode of a synchronous motor operating at over excitation.

Mode 3 ($180° \leq \alpha \leq 270°$) Force-commutated inverter: This mode is shown in Fig. 12.17(c). By dealing the inverter firing angle further beyond 180°, the advantage of load commutation is lost since the outgoing thyristor Th2 is impressed with a positive v_{ca} voltage.

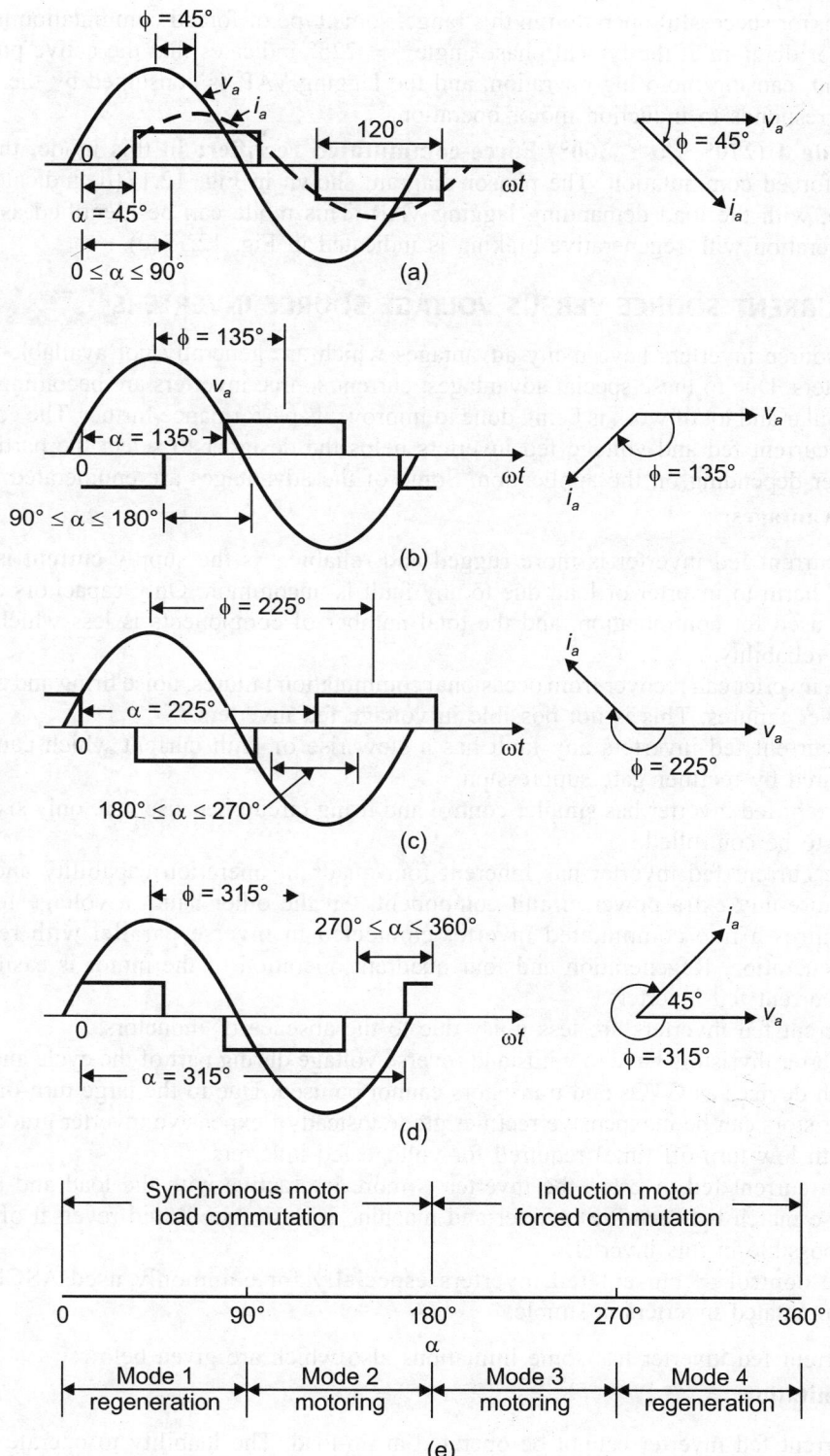

Fig. 12.17 Diagram of inverter operation mode

Therefore, for successful operation in this range, some type of forced commutation is required. The phasor diagram at the typical phase angle $\phi = 225°$ indicates that the active power flows to the load, causing motoring operation, and the lagging VAR is consumed by the load. This mode corresponds to induction motor operation.

Mode 4 (270° $\leq \alpha \leq$ 360°) Force-commutated rectifier: In this mode, the inverter requires forced commutation. The phasor diagram shown in Fig. 12.17(d) indicates rectifier operation, with the load demanding lagging VAR. This mode can be identified as induction motor operation with regenerative braking is indicated in Fig. 12.17(e).

12.7 CURRENT SOURCE VERSUS VOLTAGE SOURCE INVERTERS

Current source inverters have many advantages which are generally not available in voltage fed inverters. Due to these special advantages, current source inverters are becoming more and more popular and lot of work is being done to improve its performance further. The composition between current fed and voltage fed inverters helps the designer to select the particular type of inverter depending on the application. Some of the advantages are enumerated below:

Advantages:

1. A current fed inverter is more rugged and reliable. As the supply current is constant, any harm to inverter or load due to any fault is uncommon. Only capacitors and diodes are used for commutation, and the total number of components is less which increases the reliability.
2. The inverter can recover from occasional commutation failures, noise firing and momentary power failures. This is not possible in voltage fed inverters.
3. In current fed inverters any fault has a slow rise of fault current which can be easily cleared by rectifier gate suppression.
4. Current fed inverter has simpler control and firing circuit because here only six thyristors are to be controlled.
5. The current fed inverter has inherent four-quadrant operation capability and does not require any extra power circuit component. On the other hand, a voltage fed inverter requires a line commutated inverter connected in inverse parallel with rectifier for regeneration. Regeneration and four quadrant operation of the motor is easily possible in current fed inverters.
6. Current fed inverters are less noisy due to the absence of inductors.
7. Inverter thyristors have to withstand reverse voltage during part of the cycle and therefore such devices as GTOs and transistors cannot be used. Due to the large turn-off time, the thyristors can be inexpensive rectifier grade instead of expensive inverter grade thyristors (with low turn-off time) required for voltage fed inverters.
8. In a current fed inverter, the inverter is more interactive with the load and therefore a close match between the inverter and machine is desirable. Rapid reversal of the motor is possible in this inverter.
9. The control of current fed inverters especially for commonly used ASCI and load commutated inverters is simple.

Current fed inverter has some limitations also which are given below:

Limitations:

1. Current fed inverter cannot be operated at no-load. The inability to operate at no-load invalidates its operation in general-purpose power supply applications such as UPS systems.

2. Current fed inverters have a slow and sluggish response. The stability problem is more severe at light load and high frequency conditions. Stability problems are minimal in voltage fed inverters.

3. Its frequency range is lower.

4. It gives torque pulsation and harmonic heating problems in the motor which becomes highly objectionable at low frequency.

5. Large size of DC link inductor and commutating capacitors are required.

6. Multi machine load on a single inverter or multi inverter load on a single rectifier is very difficult with current fed inverters. In applications, where multi machine or multi inverter capability is required, a voltage fed inverter may prove very economical.

12.8 LOAD-COMMUTATED INVERTERS

In the preceding sections different modes of current-fed inverter with a counter emf type of load are discussed.

12.8.1 Single-phase Inverter

A single-phase bridge inverter with a passive R-L load is shown in Fig. 12.18. A capacitor of high value is connected across the load, so that resultant load has a leading power factor.

Thyristors Th1 and Th4 form one pair and Th3 and Th2 form the other pair. Thyristor pair Th1 and Th4 are gated to produce the positive half-cycle and the other pair Th2 and Th3 are gated to have the negative half-cycle of the AC output. Capacitor C is connected in parallel with R-L load which serves two purposes: (1) it improves the power factor, and (2) it acts as a commutating device in the circuit.

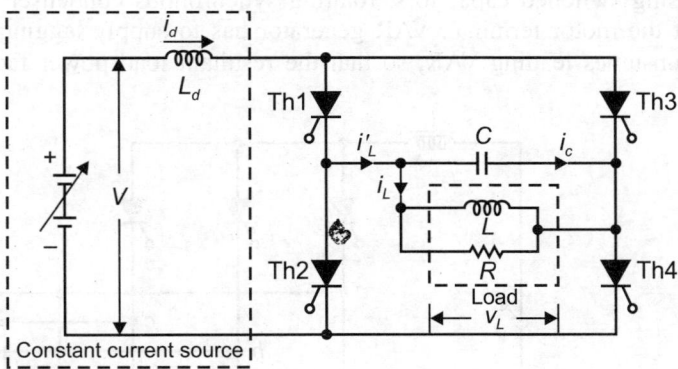

Fig. 12.18 A single-phase bridge inverter with a passive R-L load

When thyristors Th1 and Th4 conduct, capacitor C gets charged to the input voltage V with polarities shown. Current flow through load via path Th1, load and Th4. This constitute the positive half-cycle of the output. When the other pair of thyristors, Th3 and Th2 are gated, capacitor C provides positive polarity on the cathode of Th1 and negative polarity on the anode of thyristor Th4 which turn-off both thyristors Th1 and Th4. The current in the load passes in the reverse direction through the path Th3, load and Th2. This constitutes the negative half-cycle of the output. Now the capacitor gets charged with reversed polarities. In the next positive half-cycle when again thyristors Th1 and Th4 are turned on, capacitors reversed polarity provide reverse voltage to thyristor pair Th2 and Th3 and immediately turns them off. As a result AC output is obtained across the load.

The total load fundamental component of current is given by,

$$I_L' = \frac{2\sqrt{2}}{\pi} \cdot I_d$$

The active DC power supplied by source is consumed by load

$$VI_d = V_L^2/R_1 \text{ where, } R_1 = \frac{\omega^2 L^2}{R} \cdot$$

A typical application would be the supply to an induction motor. This circuit is mainly used for high frequency applications such as induction heating and is suitable for transfer of power from solar batteries.

12.8.2 Three-phase Inverter

The concept of load commutation can be extended to polyphase inverters. Figure 12.19 shows a three-phase bridge inverter with lagging power factor load where the load commutation is being achieved with a leading VAR load connected at the load terminal. For a variable load, a fixed capacitor bank can be connected at the terminal and the frequency of the inverter can be so chosen, so that load commutation occurs with a fixed advance angles. If the load is induction motor, the inverter frequency is given by the machine speed requirements. For this situation, the leading VAR requirement can be achieved by a variable VAR generator using switched capacitors, rotating synchronous condenser or cycloconverter type generator at the motor terminal. VAR generator has to supply lagging VAR demand of the load which consumes leading VAR, so that the resultant load power factor is leading.

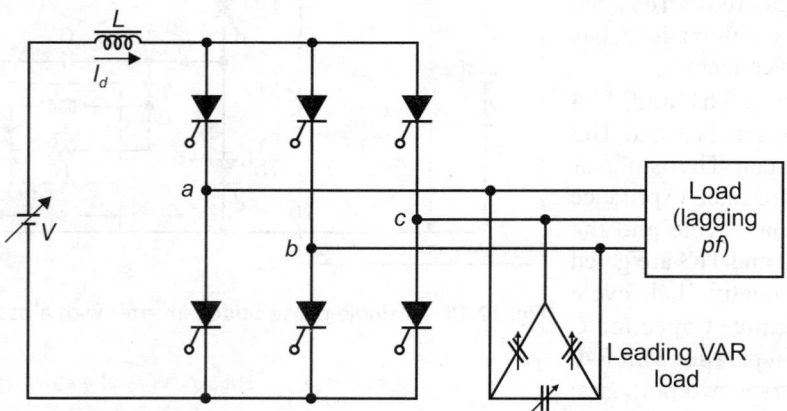

Fig. 12.19 A three-phase bridge inverter with lagging power factor load

12.9 FORCED-COMMUTATED INVERTERS

Current fed inverter requires forced commutation if it supplies a nonleading power factor load. Several types of forced-commutated inverters are:

1. Auto sequential commutated inverter (ASCI)
2. Individual auxiliary bridge commutated inverter (IABCI)
3. Third harmonic auxiliary commutated inverter (THACI)

12.9.1 Auto Sequential Commutated Inverter (ASCI)

This is most common type of force commutated current fed inverters used in practice. The auto sequential-commutated inverters can have various single-phase and polyphase circuit configurations. The three-phase bridge configuration is used exclusively in practice.

12.9.1.1 *Single-phase Bridge Auto Sequential Commutated Inverter*

A single-phase bridge inverter with auto sequential commutation is shown in Fig. 12.20. A constant DC current I_d is supplied from a variable voltage source. The current I_d is not affected by inverter load condition and therefore its ripple is neglected. The thyristor pairs Th1, Th2 and Th3, Th4 are alternately switched on to establish a nearly square wave current through the series R-L load. A commutating capacitor is connected across each of the upper and lower halves of the bridge. A diode is connected in series with each thyristor, which helps to isolate the capacitors from the load.

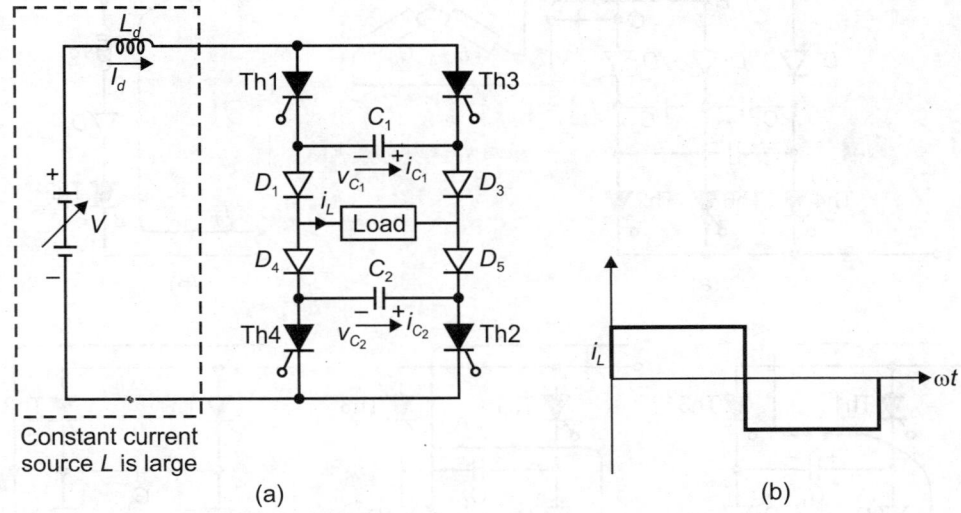

Fig. 12.20 A single-phase bridge inverter with auto sequential commutation

With the thyristors Th1 and Th2, on both capacitors are charged with their left hand plates positive. When thyristors Th3 and Th4 are fired, the capacitors are placed across thyristors Th1 and Th2 respectively, turning them off, current now flowing via Th3, C_1, D_1 load, D_2, C_2, Th4. The capacitor voltages will be reversed and in time, depending on the load voltage, diodes D_3 and D_4 will turn on, the supply current then being transferred over a time interval from D_1 to D_3 and D_4 to D_2, diodes D_1 and D_2 finally ceasing to conduct when the load current is completely reversed. The capacitor voltages will reverse ready for the next half-cycle.

The load current is square wave, ignoring the commutation period, with the load voltage typically being a sinewave in shape but containing voltage spikes at the commutating instants. A typical application would be the supply to an induction motor.

12.9.1.2 *Three-phase Bridge ASCI Inverter*

A three-phase bridge version of an autosequential commutated inverter is shown in Fig. 12.21. This circuit is popularly used in medium to large power induction motor drives. Since the induction motor operates at lagging power factor, the current fed inverters requires forced commutation.

Thyristor Th1 to Th6 are six thyristors, C_1 to C_6 are six commutating capacitors D_1 to D_6 diodes are used to prevent discharge of commutating capacitors through load. Any conducting thyristor is turned off by firing its adjacent thyristor. Hence, two thyristors are on together, one from each group, and it works in 120° mode. The commutation process is illustrated in Figs 12.21 (b) to (e). Figure 12.21 (b) shows the condition of inverter circuit with thyristors

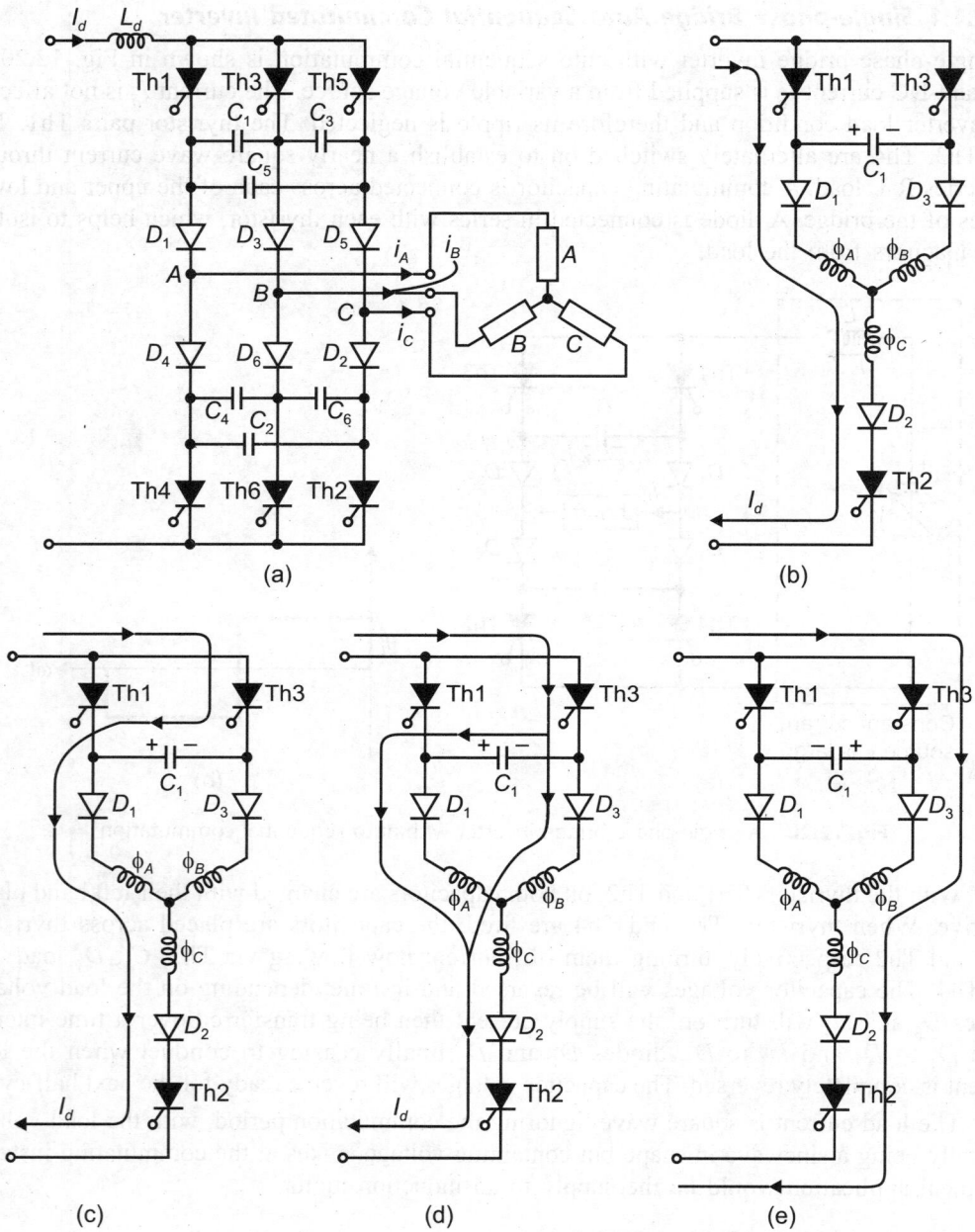

Fig. 12.21 Auto sequential commutated inverter circuit and commutation process

Th1 and Th2 are in ON condition, current path is shown as through Th1, D_1, ϕ_A, ϕ_C, D_2, Th2 and back to DC supply which charges capacitor C_1 as shown. Thyristor Th3 is turned on, capacitor C_1 applies a reverse voltage to thyristor Th1 through thyristor Th3. The load current is diverted through capacitor C_1 and thyristor Th3 as shown in Fig. 12.21 (c). During this period diode D_3 is reverse-biased. As current through C_1 (load current) is constant, capacitor C_1 is linearly charged in reverse direction. After some time diode D_3 is forward-biased and starts conducting as in Fig. 12.21 (d). Diode D_1 also conducts for some time. The load current is finally transferred to thyristor Th3, diode D_3 branch as shown in Fig. 12.21(e). Table 12.1 gives the commutation process after every firing.

Table 12.1 The commutation process after every firing

Thyristor turned ON	Thyristor commutated	Current path after commutation
Th4	Th2	$Th3 - D_3 - \phi_B - \phi_A - D_4 - Th4$
Th5	Th3	$Th5 - D_5 - \phi_C - \phi_A - D_4 - Th4$
Th6	Th4	$Th5 - D_5 - \phi_C - \phi_B - D_6 - Th6$
Th1	Th5	$Th1 - D_1 - \phi_A - \phi_B - D_6 - Th6$
Th2	Th6	$Th1 - D_1 - \phi_A - \phi_C - D_2 - Th2$
Th3	Th1	$Th3 - D_3 - \phi_B - \phi_C - D_2 - Th2$

The diodes cause the charge to be held on commutating capacitors. Without these diodes a capacitor would discharge through two phase loads.

The idealised waveforms of the input and output currents are shown in Fig. 12.22.

This circuit is widely used for pumps and fans. It has moderate cost and no extra circuitry is needed to feedback power into the mains. During commutation of the current, the motor winding forms a part of the turn-off circuit hence making it necessary for the inverter to be adapted to the motor. The major drawback of the current source inverter is that the impressed current makes it unsuitable for multi-motor drives. The power factor depends on the load and decreases sharply at low load and control is not suitable at no load.

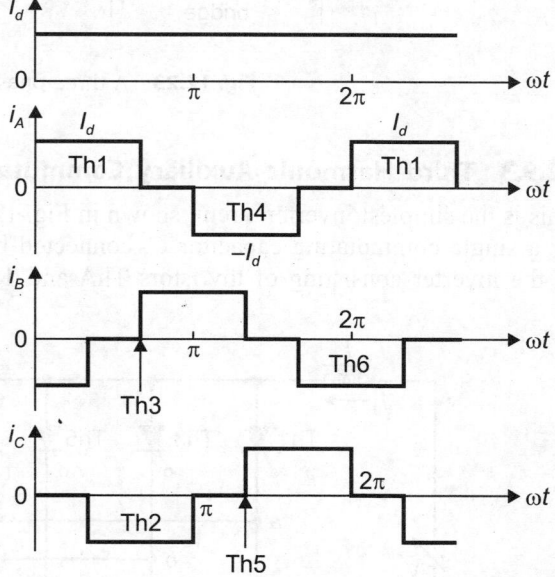

Fig. 12.22 The idealised waveforms of the input and output currents

12.9.2 Individual Auxiliary Bridge Commutated Inverter (IABCI)

A three-phase bridge inverter where forced commutation of thyristors is performed by an individual auxiliary bridge inverter is shown in Fig. 12.23.

In this circuit each thyristor in the main bridge has its corresponding commutating thyristor in the auxiliary bridge. There are three commutating capacitors C_1, C_2 and C_3, one for each phase. The commutating circuit has capacitors and auxiliary bridge. When a main thyristor is required to be commutated, its auxiliary thyristor and the incoming main thyristor are fired simultaneously. If the polarity of the capacitor voltage is correct, the outgoing thyristor turns off and the incoming thyristor conducts after the lapse of commutation time. At each commutation, the corresponding capacitor voltage is reversed, so that at 180° interval the opposite thyristor on the same lag can be commutated successfully. In this circuit, the auxiliary thyristor conducts during commutation only and the main thyristor conducts for nearly 120°. The duration of commutation widens causing much torque reduction at higher frequency. The combined rating of the main thyristor and the auxiliary thyristor becomes more than that ASCI. This inverter is definitely more expensive than that of ASCI inverter.

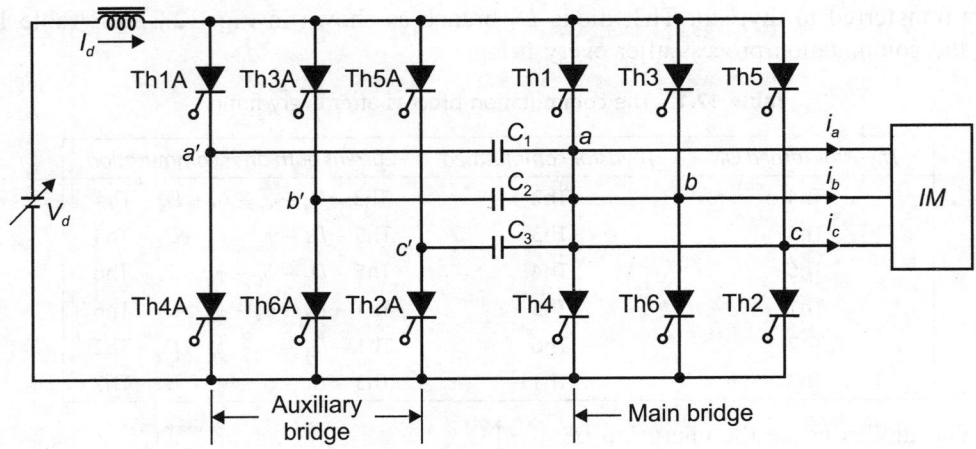

Fig. 12.23 A three-phase bridge inverter

12.9.3 Third Harmonic Auxiliary Commutated Inverter (THACI)

This is the simplest inverter circuit shown in Fig. 12.24, where force commutation is performed by a single commutating capacitor C, connected between machine neutral and the fourth leg of the inverter consisting of thyristors ThA and ThB.

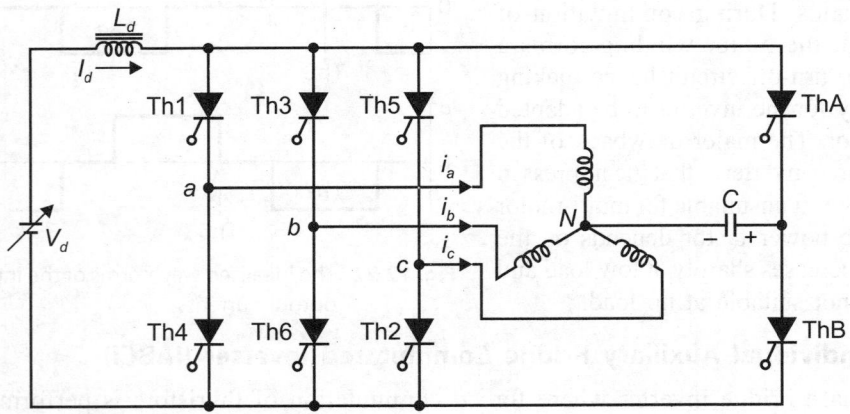

Fig. 12.24 The simplest inverter circuit

Firing of the auxiliary thyristor ThA commutates all the upper group of the main thyristors (Th1, Th3, Th5) and thyristor ThB commutates the main thyristors in the lower group (Th4, Th6, Th2). As the main thyristors in each group are commutated at 120° intervals, the auxiliary thyristors conduct alternately six times per cycle and the capacitor current i_c alternates three times per fundamental period. That is why this circuit is known as *third harmonic commutated inverter*.

This is rarely used for induction motor drive as torque is reduced substantially at high frequency. The circuit has been used successfully in starting synchronous motors and HVDC transmission for forced commutation of inverters.

12.10 CONTROL CIRCUIT FOR SQUARE WAVE INVERTER

Inverter control can be achieved by analog/digital hardware or by microcomputer software. In this section hardware control square wave inverter is briefly discussed. Figure 12.25 shows

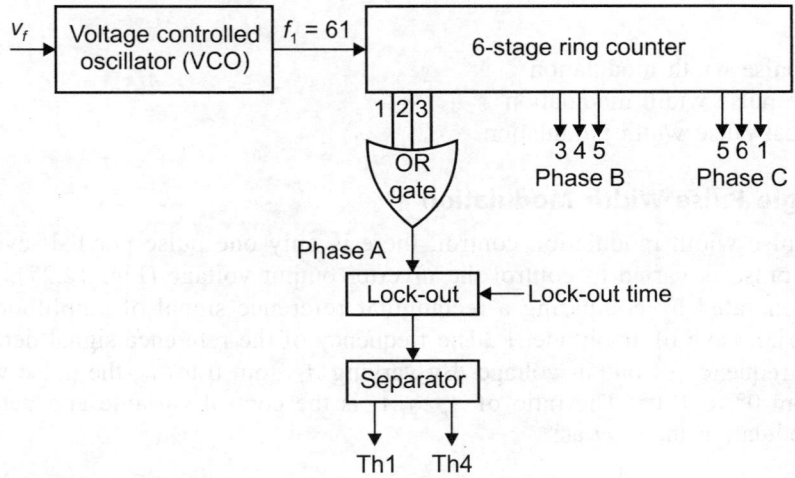

Fig. 12.25 The control block diagram of a six-stepped square wave inverter

the control block diagram of a six-stepped square wave inverter and Fig. 12.26 explains its operation. The inverter voltage and frequency are represented by the analog signals v_s and v_f, respectively. The signal v_s controls the gate firing angle of the phase-controlled rectifier in the front end to regulate the DC link voltage V_d. The signal v_f is converted to a pulse train of frequency f_1 through a voltage-controlled oscillator (VCO) which drives a six-state ring counter as shown in Fig. 12.25. The analog signal is modified such that $f_1 = 6f_s$, where f_s is the inverter fundamental frequency. For a constant and precision frequency requirement of the inverter, the frequency signal can be generated by a crystal oscillator. The adjacent three states of the ring counter are coupled through OR gates with overlapping of 60° to generate the three-phase square waves at the fundamental frequency f_s. These square waves are the basic logic drive signals of the respective phases of the inverter, where the upper devices should conduct in the positive half-cycle and the lower device should conduct in the negative half-cycle. However, at the transition of each half-cycle, a lock-out time interval is generated when the drive signal of both the upper and lower devices is inhibited. For a thyristor inverter, the commutation is initiated at the leading edge of the lock-out time so that at its trailing edge, the outgoing thyristor has completely turned off before gating on the incoming thyristor. The lock-out time can be generated by a one-shot timer. After the lock-out interval, the logic signals are separated to generate the respective logic drive signals shown in Fig. 12.26.

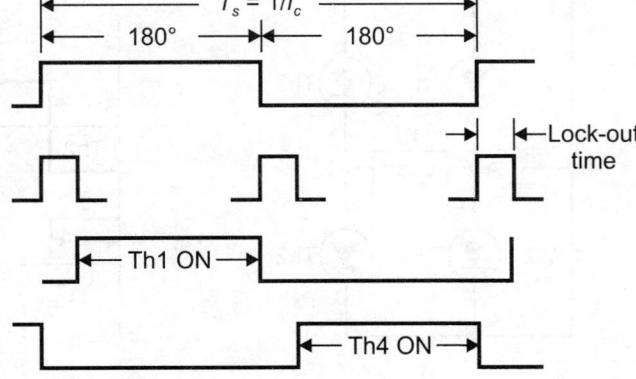

Fig. 12.26 The control block diagram of a six-stepped square wave inverter and its operation

12.11 PULSE WIDTH MODULATED INVERTERS

Out of the three methods of voltage control discussed above, second method, i.e. control of voltage within the inverter here is discussed in more detail. This method employs pulse width modulation techniques. The three most commonly used techniques for pulse width modulation are:

1. Single pulse width modulation
2. Multiple pulse width modulation
3. Sinusoidal pulse width modulation.

12.11.1 Single Pulse Width Modulation

In a single pulse width modulation control, there is only one pulse per half-cycle and the width of the pulse is varied to control the inverter output voltage (Fig. 12.27). The gating signals are generated by comparing a rectangular reference signal of amplitude A_r with a triangular carrier wave of amplitude A_c. The frequency of the reference signal determines the fundamental frequency of output voltage. By varying A_r from 0 to A_c, the pulse width δ can be varied from 0° to 180°. The ratio of A_r to A_c is the control variable and defined as the amplitude modulation index m as:

$$m = \frac{A_r}{A_c}$$

The rms output voltage is given by

$$V_0 = \left[\frac{2}{2\pi} \int_{(\pi-\delta)/2}^{(\pi+\delta)/2} V_s^2 d(\omega t) \right]^{1/2} = V_s \sqrt{\frac{\delta}{\pi}}$$

The Fourier series of voltage yields

$$v_0(t) = \sum_{n=1,3,5}^{\alpha} \frac{4V_s}{n\pi} \sin \frac{n\delta}{2} \sin n\omega t$$

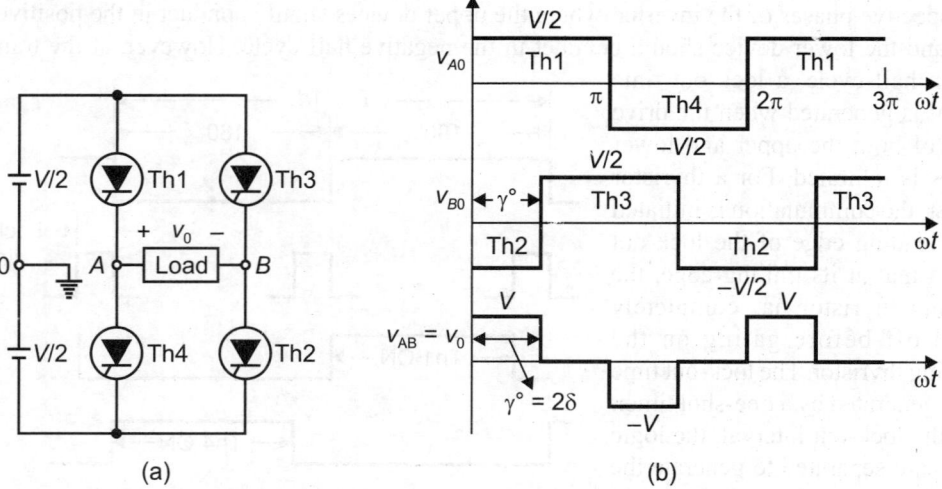

Fig. 12.27 Circuit of single pulse PWM and waveforms

12.11.2 Multiple Pulse Width Modulation

The harmonic content can be reduced by using several pulses in each half-cycle of output voltage. Natural extension of the single pulse modulation is the multipulse modulation in which several equidistant pulses per half-cycle are used. This permits a reduction in the harmonic content at low output voltages. The variation in the fundamental component of the output voltage is obtained by varying the pulse width δ. This type of modulation is achieved by comparing a DC control signal of variable magnitude 'A' with a triangular wave of magnitude A_m in a comparator as shown in Figs 12.28 (a) and (b). The pulse width is changed by varying the magnitude of DC control signal. The frequency of DC signal sets the output frequency f_o, and the triangular or carrier signal frequency f_c determines the number of pulse per half-cycle p. The modulation index controls the output voltage. This type of modulation is also known as *uniform pulse width modulation*. The number of pulses per half-cycle is given by

$$p = \frac{f_c}{2f_o} = \frac{m_f}{2} \qquad \qquad ...(12.35)$$

where, $m_f = f_c/f_o$ is defined as *frequency modulation ratio*.

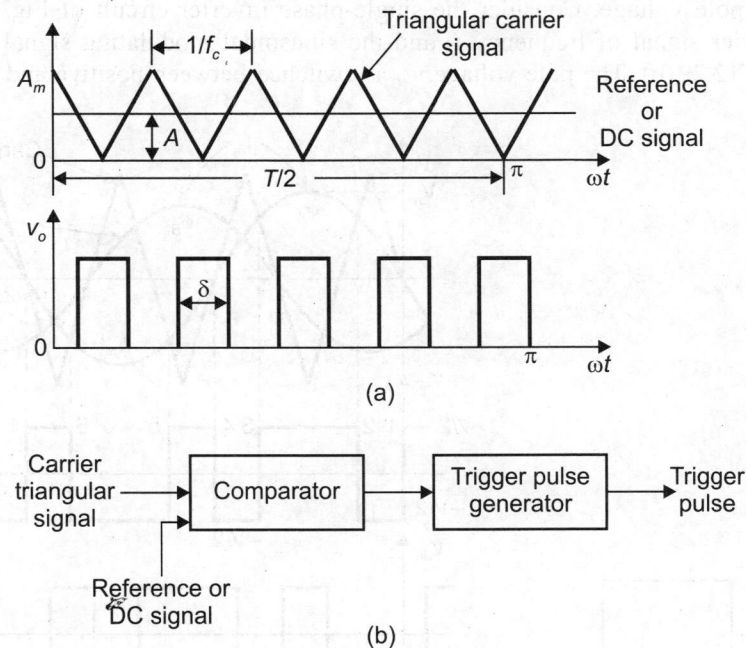

(a)

(b)

Fig. 12.28 Waveforms of multiple PWM

If δ is the width of each pulse, the rms output voltage can be given by:

$$V_0 = \left[\frac{2p}{2\pi} \int\limits_{(\pi/p-\delta)/2}^{(\pi/p+\delta)/2} V^2 d(\omega t) \right]^{1/2} = V \sqrt{\frac{\delta p}{\pi}} \qquad ...(12.36)$$

The general form of the Fourier series for the instantaneous output voltage is

$$v_0(t) = \sum_{n=1,3,5}^{\infty} B_n \sin n\omega t \qquad ...(12.37)$$

The coefficient B_n can be determined by considering a pair of pulses such that the positive pulse of duration δ starts at $\omega t = \alpha$ and the negative one of the same width starts at $\omega t = \pi + \alpha$. The effects of all the pulses can be combined together to obtain the effective output voltage. The method is particularly suitable for reducing harmonics at lower voltages. However, small higher order harmonics will predominate for large number of pulses and they have to be filtered out by load inductance.

12.11.3 Sinusoidal Pulse Width Modulation (SPWM)

The sinusoidal PWM technique is popular in industrial applications, and is reviewed extensively in this section.

It is noted that at lower values of pulse width, the inverter output voltage is rich in harmonic content. PWM inverters with multiple pulses in each half-cycle of the inverter output voltages can reduce the harmonic content. Various methods have been used to achieve this feature. One method popular in industrial applications is known as *sinusoidal PWM technique* which is described here.

In the sinusoidal PWM technique, a triangular carrier wave of frequency f_c and a modulating wave of frequency f_m, the same frequency as that of inverter output are used to modulate the pole voltage. Consider the single-phase inverter circuit of Fig. 12.29 (a). The triangular carrier signal of frequency f_c and the sinusoidal modulating signal ϕ_A and ϕ_B are shown in Fig. 12.29 (b). The pole voltage v_{A0} is switched between positive and negative buses

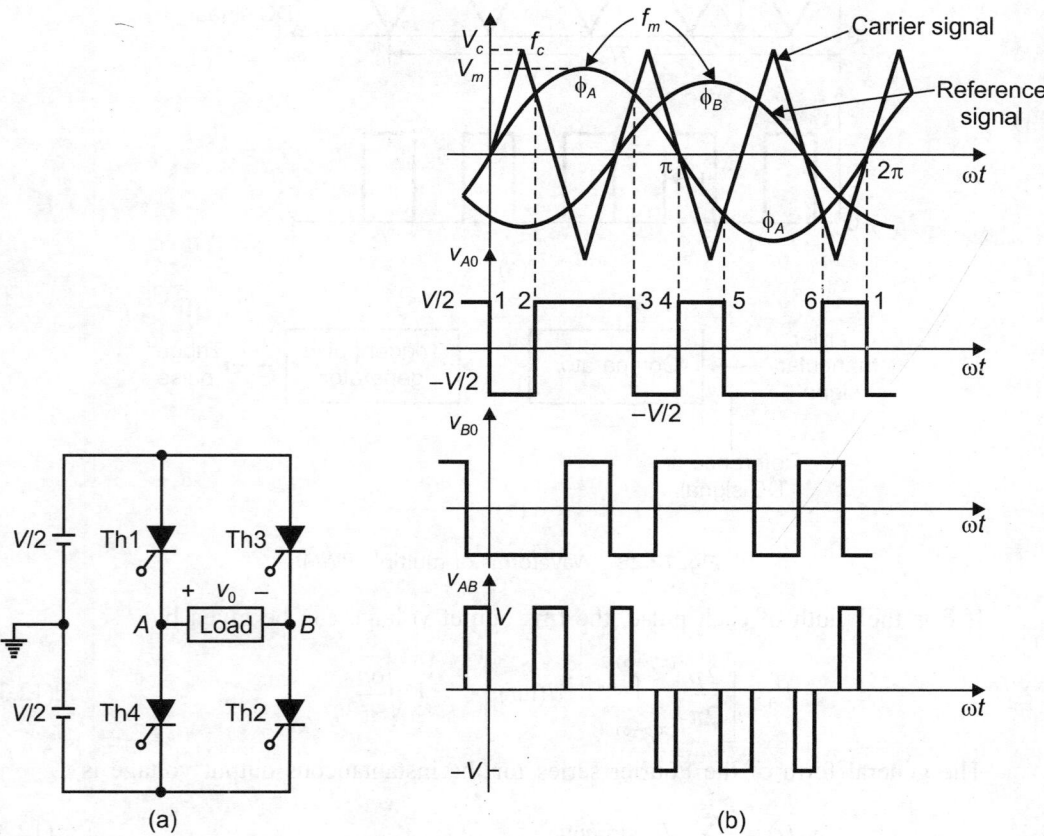

Fig. 12.29 Waveforms of sinusoidal pulse width modulation

at the intersections of the carrier signal and modulating signal ϕ_A. Similarly, the pole voltage v_{B0} is modulated by the carrier signal and the modulating signal ϕ_B. These pole voltages v_{A0}, v_{B0} and load voltages v_{AB} which is equals to $(v_{A0} - v_{B0})$ is shown in Fig. 12.29(b).

It is observed that in each half-cycle the pulses have different widths. The central pulse is wider than the side pulses.

The alternate pulse and notch widths of half-bridge inverter output are sinusoidally modulated and the waveform contains a fundamental component of which the frequency and amplitude can be varied by varying the frequency and voltage respectively of the modulating signal. A Fourier analysis of the output wave is given by:

$$v(t) = \frac{mV}{2} \sin(\omega_s t + \phi) + \text{Bessel function of harmonic terms} \qquad ...(12.38)$$

where, m is the modulation index which is the amplitude ratio of V_m/V_c; V_m and V_c are the peak values of modulating and carrier signal and ω_s is the fundamental frequency which is same as the modulating frequency, ϕ is the phase shift of output which depends on the position of the modulating signal. Ideally, m can vary between 0 and 1 to give a linear relationship between modulating and output voltage. Note that the carrier ratio determines the number of pulses in each half-cycle of the inverter. Output voltage and the modulation index determines the width of the pulses and hence the rms value of the inverter output voltage.

Sinusoidal pulse modulation is superior to the two described above, as this method uses a sinusoidal wave of peak amplitude 'V_m' and frequency 'f_m' and a triangular wave of peak value V_c and frequency f_c. Pulse width depends upon the position of the pulse as shown in Fig. 12.29(b). All lower order harmonics are eliminated for $V_m/V_c \leq 2$.

Disadvantages of PWM: There are two serious disadvantages of pulse width modulation method:

1. Large number of commutation are required in each cycle.
2. Magnitude of fundamental component is reduced.

12.12 HARMONIC REDUCTION

In the case of single-phase bridge inverters at the maximum output voltage condition, square wave voltage is delivered to the load. The square wave contains 33.3% third harmonic, 20% fifth harmonic and 14.3% seventh harmonic. In same applications harmonic current in load current should be less than 5% of the fundamental. One way of reducing the harmonic content to such a low value is to employ filter between the inverter and load. It is customary to use some following methods to eliminate low frequency harmonics so that the size of the filter can be reduced, losses can be minimised and the transient response is not affected.

12.12.1 Harmonic Elimination Using Different Transformer Connections

The output voltage of two inverters can be suitably combined by a transformer to provide combined voltage having harmonic content less than that of individual outputs. Figure 12.30(a) shows transformer connection adding two voltages v_1 and v_2 separated by a phase angle of $\pi/3$. It is observed from Fig. 12.30(b) that such a phase angle will eliminate third harmonic. This fact can be proved from mathematical analysis as given below:

$$v_1 = \sum_{n=1,3,5}^{\infty} b_n \sin n\omega t \qquad ...(12.39)$$

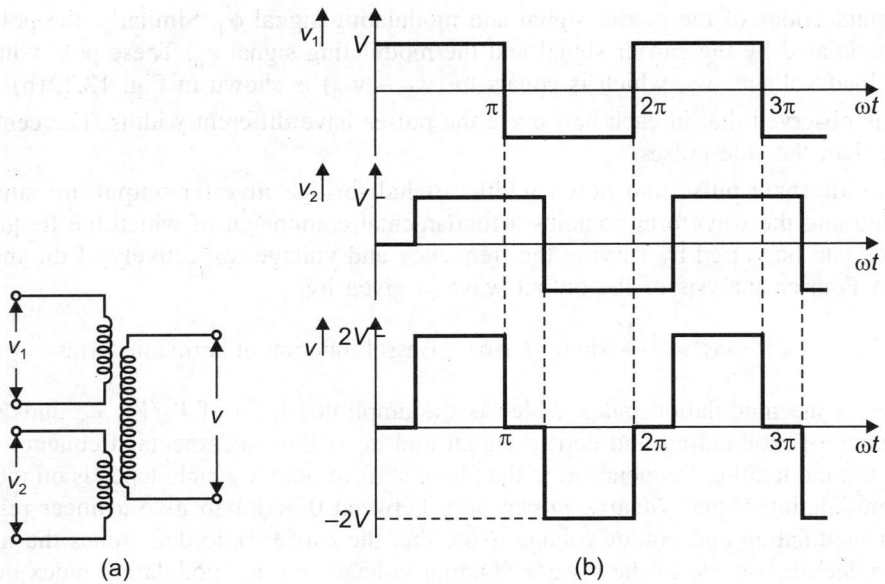

Fig. 12.30 Harmonic elimination

$$v_2 = \sum_{n=1,3,5}^{\alpha} b_n \sin n\,(\omega t - \pi/3)$$

$$v = v_1 + v_2 = \sum_{n=1,3,5}^{\alpha} b_n \sin n\omega t + \sum_{n=1,3,5}^{\alpha} b_n \sin n\,(\omega t - \pi/3)$$

$$= (b_1 \sin \omega t + b_3 \sin 3\omega t + ...) + [b_1 \sin (\omega t - \pi/3) + b_n \sin 3\,(\omega t - \pi/3) + ...]$$

$$= \sqrt{3}\,[b_1 \sin (\omega t - \pi/3) + b_5 \sin 5\,(\omega t + \pi/6) + ...] \qquad ...(12.40)$$

Thus, third harmonic has been eliminated completely but the output voltage of the inverter is $\left(\dfrac{\sqrt{3}}{2}\right)$ or 86.6% of the value which would have been obtained otherwise. If the phase angle of $\pi/5$ radian is used between v_1 and v_2 then it can be proved that 5th harmonic will be eliminated.

In practice both pulse width modulation and transformer connection methods are used together to reduce harmonics but this reduces the rating of the inverter considerably.

12.12.2 Harmonic Elimination Method

The undesirable harmonics of a square wave can be eliminated and the fundamental voltage component can be controlled by the harmonic elimination method. In this method, notches are created on the square wave at predetermined angles, as shown in Fig. 12.31. In the figure, half-cycle output is shown with quarter-wave symmetry. It can be shown that the four notch angles α_1, α_2, α_3 and α_4 can be controlled to eliminate three harmonic components and control the fundamental voltage. A larger number of harmonic components can be eliminated if the waveform can accommodate additional notch angles.

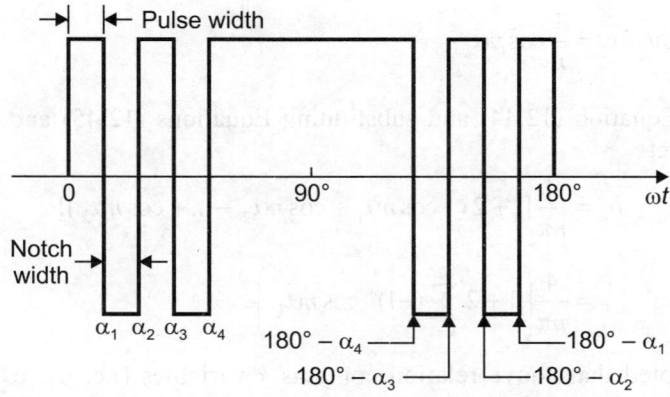

Fig. 12.31 Method of harmonic elimination

Mathematically, the general Fourier series of the wave can be given by:

$$v(t) = \sum_{n=1}^{\infty} (a_n \cos n\omega t + b_n \sin n\omega t) \qquad ...(12.41)$$

where,

$$a_n = \frac{1}{\pi} \int_0^{2\pi} v(t) \cos n\omega t \, d\omega t,$$

$$b_n = \frac{1}{n} \int_0^{2\pi} v(t) \sin n\omega t \, d\omega t$$

For quarter-cycle symmetry waveform, only the odd harmonics with sine components will exist. The coefficients are given by:

$$a_n = 0 \qquad ...(12.42)$$

$$b_n = \frac{4}{\pi} \int_0^{\pi/2} v(t) \sin n\omega t \, d\omega t \qquad ...(12.43)$$

Assuming that the wave has unit amplitude, i.e. $v(t) = \pm 1$, b_n can be expanded as

$$b_n = \frac{4}{\pi} \left[\int_0^{\alpha_1} (+1) \sin \omega t \, d\omega t + \int_{\alpha_1}^{\alpha_2} (-1) \sin n\omega t \, d\omega t + \int_{\alpha_2}^{\alpha_3} (+1) \sin n\omega t \, d\omega t + ... \right.$$

$$\left. + \int_{\alpha_{k-1}}^{\alpha_k} (-1)^{k-1} \sin n\omega t \, d\omega t + \int_{\alpha_k}^{\pi/2} (+1) \sin n\omega t \, d\omega t \right] \qquad ...(12.44)$$

Using the relation

$$\int_{\theta_1}^{\theta_2} \sin n\omega t \, d\omega t = \frac{1}{n} (\cos n\theta_1 - \cos n\theta_2)$$

The first and last terms are

$$\int_0^{\alpha_1} (+1) \sin n\omega t \, d\omega t = \frac{1}{n} (1 - \cos n\alpha_1) \qquad ...(12.45)$$

$$\int_{\alpha_k}^{\pi/2} (+1)\sin n\omega t \, d\omega t = \frac{1}{n}\cos n\alpha_k \qquad\qquad ...(12.46)$$

Integrating Equation (12.44) and substituting Equations (12.45) and (12.46) in Equation (12.44), we get

$$b_n = \frac{4}{n\pi}[1 + 2(-\cos n\alpha_1 + \cos n\alpha_2 - ... + \cos n\alpha_k)]$$

$$= \frac{4}{n\pi}\left(1 + 2\sum_{k=1}^{k}(-1)^k \cos n\alpha_k\right) \qquad\qquad ...(12.47)$$

It may be noted that above relation contains k variables (i.e. α_1, α_2, α_3, ..., α_k) and k number of simultaneous equations are required to solve their values. With k number of α angles, the fundamental voltage can be controlled and $(k - 1)$ harmonics can be eliminated.

12.13 CONTROL CIRCUIT FOR PWM INVERTER

The control block diagram of a PWM inverter is shown in Fig. 12.32, which consists of the following functional elements:

1. A voltage control oscillator converts the analog frequency signal v_f into a pulse train of frequency f_1.

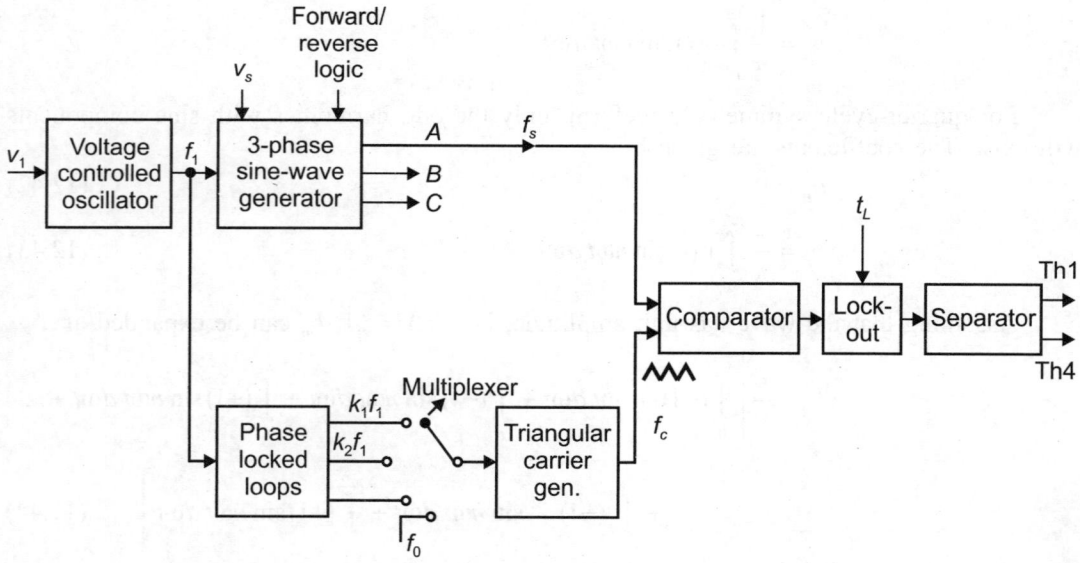

Fig. 12.32 The control block diagram of a PWM inverter

2. A three-phase sine-wave signal generator with variable amplitude and frequency as desired for the inverter output. The frequency f_s is a scaled-down value of f_1 and the amplitude is proportional to the analog voltage signal v_s. The phase sequence of the generator is reversible by a forward/reverse logic signal.
3. The phase-locked loops (PLL) which generate frequencies of different ratios.
4. A multiplexer, which selects the appropriate frequency channel of the PLL depending on the range of f_1.

5. A triangular carrier wave generator of fixed amplitude but variable frequency (f_c), which is synthesized from the multiplexer-selected frequency. The frequency f_c is a submultiple of the input frequency and the waveform can be phase synchronized with the sine signal wave.

6. An analog comparator for each phase, which compares the sine and triangular waves with a small hysteresis band.

7. The lock-out and separator circuit for each phase.

The elements of sine reference wave generator are shown in Fig. 12.33. The amplitudes of a unit sine wave at regular angular intervals are stored digitally in the form of a lookup table in a read-only memory (ROM). The frequency f_1 clocks the programmable up/down-counter, which generates the address of ROM in a normal UP counting mode. When the counter reaches the terminal count at the end of a cycle, it resets and a new cycle begins. The digital output from the ROM is fed to a multiplying digital-to-analog converter (DAC) to convert into an analog sine wave. The analog voltage v_s multiplies the DAC output to control the sine-wave amplitude.

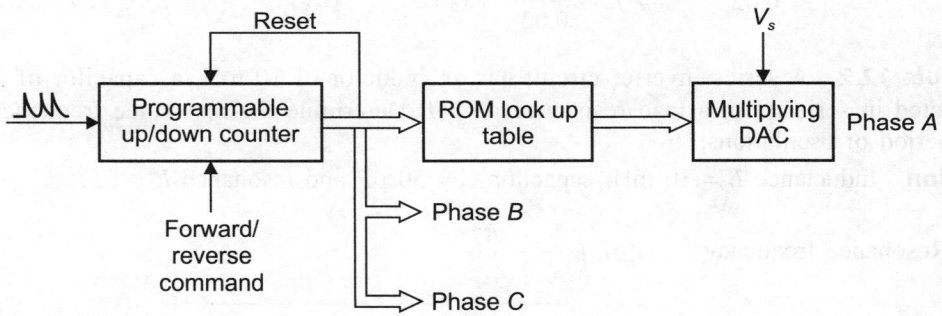

Fig. 12.33 The elements of sine reference Wave generator

The phases B and C have identical ROMs and DACs except that the look-up tables are mutually phase shifted by 120°. The phase sequence of the sine wave can be reversed by the forward/reverse command as shown. At the reverse command, the counter changes into a down-counting mode and generates the sine waves in a backward direction, which is the reversal of the phase sequence.

The triangular carrier wave is generated using the same principle as above except that it has only one phase, its magnitude is fixed, and it does not involve phase sequence reversal.

SOLVED EXAMPLES

Example 12.1 Calculate the time period of oscillation and resonance frequency of a series inverter with output frequency of 50 Hz. The time gap between turn-on of one thyristor and turn-off of the other thyristor is 10 msec.

Solution   Given $f = 50$ Hz, $T_{off} = 10$ msec

Output frequency of a series inverter,

$$f = \frac{1}{T_{off} + \pi/\omega}\,\text{Hz} = \left[\frac{1}{T_{off} + \pi/2\pi f}\right]$$

$$f = \frac{1}{T_{\text{off}} + 1/2f} = \left[\frac{1}{T_{\text{off}} + T/2}\right] \text{Hz}$$

where, T is the time period of oscillations.

$$50 = \left[\frac{1}{(T/2) + 10 \times 10^{-3}}\right]$$

or $\left[\dfrac{T}{2} + 10 \times 10^{-3}\right] = \dfrac{1}{50}$

or $T = 0.02$ sec

Time period of oscillation,

$$T = \frac{\pi}{\text{Resonance frequency} f_r}$$

or $0.02 = \dfrac{\pi}{f_r} \Rightarrow f_r = \dfrac{\pi}{0.02} = 157$ Hz.

Example 12.2 A series inverter circuit has an inductor of 10 mH, a capacitor of 47 μF connected in series with a load resistance of 5 Ω. Determine the resonance frequency and time period of oscillations.

Solution Inductance $L = 10$ mH, capacitor $C = 50$ μF and resonance $R = 5 \Omega$

Resonance frequency $f_r = \sqrt{\dfrac{1}{LC} - \dfrac{R^2}{4L^2}}$

$$= \sqrt{\frac{1}{10 \times 10^{-3} \times 50 \times 10^{-6}} - \frac{5^2}{4 \times (10 \times 10^{-3})^2}}$$

$$f_r = \sqrt{\frac{10^9}{500} - \frac{25}{4 \times 10^{-4}}} = \sqrt{2 \times 10^6 - 0.063 \times 10^6}$$

$$f_r = \sqrt{1.937 \times 10^6} = 1392 \text{ Hz}$$

Time period of oscillation,

$$T = \frac{\pi}{f_r} = \frac{3.14}{1392} = 2.25 \text{ msec.}$$

Example 12.3 Determine the output frequency of a series inverter having the following parameters: capacitance $C = 0.1$ μF, inductance $L = 10$ mH, $T_{\text{off}} = 0.2$ msec, load resistance $R_L = 0.4$ kΩ. Calculate the attenuation factor.

Solution Output frequency,

$$f = \frac{1}{T_{\text{off}} + \pi/\omega} = \frac{\omega}{\omega T_{\text{off}} + \pi}$$

$$\omega = \left[\frac{1}{LC} - \frac{R_L^2}{4L^2}\right]^{1/2} = \left[\frac{1}{10 \times 10^{-3} \times 0.1 \times 10^{-6}} - \frac{400^2}{4 \times (10 \times 10^{-3})^2}\right]^{1/2}$$

$$= 2.45 \times 10^4 \text{ rad/sec}$$

Hence, $f = \dfrac{2.45 \times 10^4}{2.45 \times 10^4 \times 0.2 \times 10^{-3} + \pi} = 3$ kHz

Attenuation factor $= e^{-(R/2L)t}$

$$= e^{-\left(\frac{400}{2 \times 10 \times 10^{-3}}\right) \times \frac{1}{3 \times 10^3}}$$

$$= e^{-(6.66)} = 1.28 \times 10^{-3}.$$

Example 12.4 Calculate the value of inductance in series with inverter circuit having the frequency of 5 kHz and a capacitance of 1 μF. If the inverter is operating under resonance condition.

Solution Inverter operating frequency,

$$f = \frac{1}{\sqrt{LC}}$$

or $\qquad L = \dfrac{1}{f^2 C} = \dfrac{1}{(5 \times 10^3)^2 \times 1 \times 10^{-6}}$ H $= 40$ mH.

Example 12.5 In the single-phase VSI bridge inverter, the load current is I_0
$$= 540 \sin(\omega t - 45°).$$

The DC supply voltage $V = 300$ V. Determine the average value of supply current and power from the DC supply. Also determine the power delivered to the load.

Solution Average value of supply current

$$i_s|_{\text{avg}} = I_s = \frac{1}{\pi} \int_0^\pi 540 \sin(\omega t - 45°)\, d\omega t$$

$$= 243.2 \text{ A}$$

Power $\qquad P_s = V \cdot I_s = 300 \times 243.2$
$$= 72.96 \text{ kW}$$

From the Fourier analysis of square wave, the rms value of fundamental output voltage:

$$V_{01} = \frac{4V}{\pi\sqrt{2}} = \frac{4 \times 300}{\pi\sqrt{2}} = 270.14 \text{ V}$$

$$P_{\text{out}} = V_{01} I_0 \cos \Phi$$

$$= 270.14 \times \frac{540}{\sqrt{2}} \cos 45°$$

$$= 72.94 \text{ kW}.$$

Example 12.6 A three-phase voltage sourced bridge inverter is supplied from 600 V source. For a star connected resistive load of 15 Ω/phase, find the rms value of current, load power and the thyristor current ratings for: (a) 120°, and (b) 180° conduction (Fig. E.12.6).

Solution

(a) **For 120° conduction:** As the load is star connected:

Load current $\qquad = \dfrac{600}{2 \times 15} = 20$ A

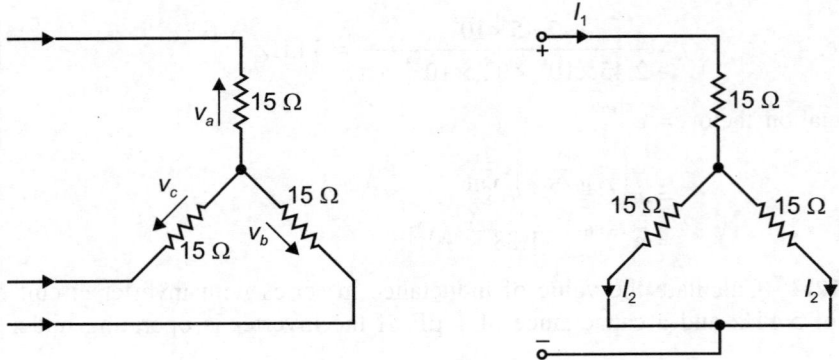

Fig. E.12.6 Circuit diagram of inverter

rms load current
$$= \left[\frac{1}{2\pi} \left(\int_0^{2\pi/3} 20^2 \, d\omega t + \int_\pi^{5\pi/6} 20^2 \, d\omega t \right) \right]^{1/2}$$

$$= [(20^2 + 20^2)/3]^{1/2}$$

$$= 16.33 \text{ A}$$

Load power $\qquad P = 16.33^2 \times 15 \times 3 = 12$ kW

Tyristor rms current $\qquad = (20^r/3)^{1/2} = 11.5$ A

Alternately: Using the equations derived

Thyristor rms current $\quad I_{rms} = \dfrac{V}{2\sqrt{3}R} = \dfrac{600}{2\sqrt{3} \times 15} = 11.5$ A

rms value of load current

$$I_p = \sqrt{2} I_{rms} = \sqrt{2} \times 11.5 = 16.3 \text{ A}$$

Load power $\qquad P = 3 I_p^2 R = 3 \times 16.3^2 \times 15 = 12$ kW.

(b) For 180° conduction: At any instant, load on the inverter

$$= 15 + \frac{15}{2} = 22.5\Omega$$

$$I_1 = \frac{V}{22.5} = \frac{600}{22.5} = 26.67 \text{ A}$$

$$I_2 = \frac{I_1}{2} = \frac{26.67}{2} = 13.33 \text{ A}$$

rms value of load current $= \left[\dfrac{1}{2\pi} \left\{ \int_0^{2\pi/3} 13.33^2 \, d\omega t + \int_{2\pi/3}^{4\pi/3} 26.67^2 \, d\omega t + \int_{4\pi/3}^{2\pi} 13.33^2 \, d\omega t \right\} \right]^{1/2}$

$$= \left[\frac{2 \times 13.33^2 + 26.67^2}{3} \right]^{1/2}$$

$$= 18.85 \text{ A}$$

Thyristors carry a current of 26.67 A for one sixth of a cycle and 13.33 for a half-cycle.

rms current in a thyristor,

$$\left(\frac{\text{rms load event}}{2}\right) = 13.33 \text{ A}$$

Load power $\qquad P = 3 \times 18.85^2 \times 15 = 15.99 \text{ kW}$

Alternately: Using the direct equation, we have thyristor current,

$$I_{\text{rms}} = \frac{V}{3R} = \frac{600}{3 \times 15} = 13.33 \text{ A}$$

rms value of load current $I_p = \sqrt{2}I_{\text{rms}} = \sqrt{2} \times 13.33 = 18.85 \text{ A}$

Load power $\qquad P = \dfrac{2V^2}{3R} = \dfrac{2 \times 600^2}{3 \times 15} = 160 \text{ kW}.$

Example 12.7 In McMurray inverter circuit shown in Fig. 12.5, calculate the values of L_c and C_c for the following conditions: (a) The maximum load current $I_L = 100 \text{ A}$. (b) Turn-off time t_{off} of Th1 is 40 μsec. (c) Input DC voltage $V_d = 300 \text{ V}$.

Solution For minimum energy loss in the commutation circuit, $I_c = 1.5I_L = 1.5 \times 100 = 150 \text{ A}$. Let us limit the capacitor voltage V_c to 450 volts.

$$I_c = 150 \text{ A} = V_c\sqrt{\frac{C_c}{L_c}} = 450\sqrt{\frac{C_c}{L_c}}$$

or $\qquad \sqrt{\dfrac{C_c}{L_c}} = \dfrac{150}{450} = \dfrac{1}{3}$ $\qquad\qquad$ ...(i)

Using the equation, $t_0 = 0.535t$ from Fig. 12.8. In the given problem $t_{\text{off}} = 40$ μsec. Taking 50% tolerance, $t_0 = 60$ μsec

$$t = \frac{60 \,\mu\text{sec}}{0.535} = 112 \,\mu\text{sec}$$

Using the equation $\qquad t = \pi\sqrt{L_c C_c}$

$$\sqrt{L_c C_c} = \frac{112 \times 10^{-6} \text{ sec}}{\pi} = 35.65 \times 10^{-6} \text{ sec} \qquad\qquad ...(ii)$$

From Equation (i) $\qquad \sqrt{\dfrac{L_c}{C_c}} = 3$ $\qquad\qquad$ ...(iii)

Solving Equations (ii) and (iii), we have

$$L_c = 107 \,\mu\text{H}$$
$$C_c = 11.88 \,\mu\text{F}.$$

Example 12.8 A single-phase load is supplied from a constant source inverter. If the load is 12 Ω resistive and DC voltage source is 120 V, determine suitable source inductance and turn-off capacitors at 20 Hz frequency operation. Assume a turn-off time of thyristors equal to 50 μsec.

Solution Figure E.12.8 refers to a single-phase constant source inverter.

Steady state load current $\quad = \dfrac{120}{12} = 10 \text{ A}$

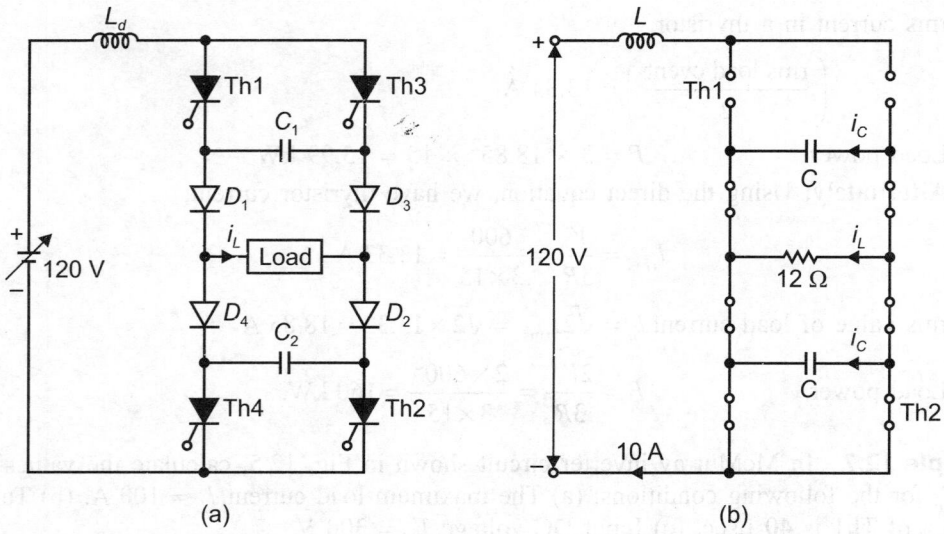

(a)   (b)

Fig. E12.8 Inverter connected to load

Let the load short circuit increase the load current by 1 A in one cycle then

$$\frac{di}{dt} = \frac{1}{1/20} = 20$$

$$L\frac{di}{dt} = 120 \text{ V} \quad \text{or} \quad L = \frac{120}{di/dt} = \frac{120}{20} = 6 \text{ H}$$

Just after turn-off thyristors Th1 and Th2, all the diodes are conducting as shown in Fig. E.12.8(b).

The voltage difference between source and load is taken across L by a very small rate of current change, small enough so that 10 A can be assumed constant.

Thus, the KCL and KVL equations for the circuit are:

$$2i_C + i_L = 10$$

and

$$12i_L = \frac{1}{C}\int i_C \, dt - 120 \text{ V}$$

Capacitor C is initially charged to -120 V. Taking the Laplace transform of above equations, we get

$$2i_C + i_L = \frac{10}{s}$$

and

$$12i_L = \frac{i_C}{s_C} - \frac{120}{s}$$

Taking the Laplace inverse, we get

$$i_L(t) = 10 - 20e^{-t/24C} \text{ A}$$

and

$$i_{c(t)} = 10e^{-t/24C} \text{ A}$$

The voltage across thyristor Th1 will become forward after the load voltage, i.e. $i_L = 0$.
Hence,

$$0 = 10 - 20^{-t/24C}$$

or
$$20e^{-t/24C} = 10$$
$$e^{-t/24C} = 1/2$$

Using turn-off time $\qquad = 50\ \mu sec$
$$C = 34\ \mu F.$$

Example 12.9 In the single-phase parallel capacitor commutated inverter circuit shown in Fig. 12.2 (a), the battery voltage $V_s = 240$ V, $\dfrac{N_1}{2} : N_2 = 1{:}1$. For a load resistance $R_L = 5\,\Omega$, write the expression for voltage on the capacitor and find the necessary value of C to obtain 40 µsec turn-off time on the thyristor. Assume inductor L is very large and the transformer is ideal. Sketch the waveforms of (a) capacitor voltage V_c, (b) voltage across Th1, (c) capacitor current i_c, and (d) current in Th2 (i_2). Show also on the sketch, triggering points in Th1 and Th2, in the above said circuit of Fig. 12.2 (a) w.r.t. time.

Solution The voltage across the capacitor is given by the equation
$$V_c = 2V\,(1 - 2e^{-n^2 t/CR_L})$$

Given: $\qquad \dfrac{N_2}{N_1/2} = 1,\ R_L = 5\,\Omega,\ t_0 = 40$ µsec

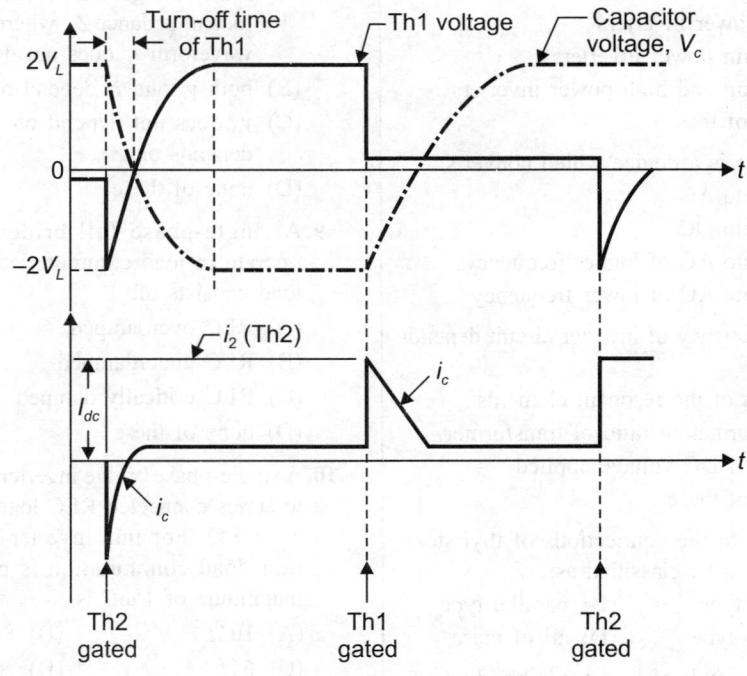

Fig. E.12.9

or
$$n = \frac{N_2}{N_1} = \frac{1}{2}$$

so
$$V_c = 2V\,(1 - 2e^{-t/4CR_L}).\ \text{This equation can be sketched in Fig. E.12.9}$$

or
$$t_0 = \frac{0.6931 CR_L}{n^2}, \text{ when } V_c = 0$$

Therefore,
$$C = \frac{40 \times 10^{-6}}{0.6931 \times 4 \times 5} = 2.88\,\mu F$$

The various waveforms of capacitor and thyristor voltage and currents over one cycle of operation are sketched in Fig. E.12.9.

EXERCISES

Multiple Choice Questions

1. The main part of an inverter is:
   (A) oscillator circuit
   (B) DC source
   (C) step up transformer
   (D) filter

2. Now-a-days mostly thyristors are used for developing:
   (A) low power inverters
   (B) medium power inverters
   (C) medium and high power inverters
   (D) none of these

3. An inverter is a device which converts:
   (A) DC into AC
   (B) AC into DC
   (C) AC into AC of higher frequency
   (D) AC into AC of lower frequency

4. Output frequency of inverter circuit depends on:
   (A) values of the resonant elements
   (B) transformation ratio of transformer
   (C) level of DC voltage applied
   (D) none of these

5. According to the connections of thyristors, inverters can be classified as:
   (A) series type   (B) parallel type
   (C) bridge type   (D) all of these

6. Oscillator circuit of inverter is build-up of:
   (A) resistor   (B) inductor
   (C) capacitor   (D) all of these

7. Which of the following statements is incorrect?
   (A) Bridge inverter does not use any transformer
   (B) Bridge inverters can be designed for single-phase as well as three-phase circuits

   (C) In McMurray inverter circuit the commutating pulse is obtained from the main power source
   (D) Current source inverters are mainly used for high frequency applications

8. In voltage source inverters:
   (A) load voltage waveform v_0 depends on load impedance Z_L, whereas, load current waveform i_L does not depend on Z_L
   (B) both v_0 and i_L depend on Z_L
   (C) v_0 does not depend on Z_L, whereas, i_L depends on Z_L
   (D) none of these

9. A single-phase full bridge inverter can operate in load commutation mode in case load consists of:
   (A) RLC overdamped
   (B) RLC underdamped
   (C) RLC critically damped
   (D) none of these

10. A single-phase bridge inverter delivers power to series connected RLC load with $R = 2\,\Omega$, $X_L = 8\,\Omega$. For this inverter-load combination, load commutation is possible, if the magnitude of $1/\omega C$ is:
    (A) $10\,\Omega$ (B) $8\,\Omega$
    (C) $6\,\Omega$ (D) zero

11. In single-pulse modulation of PWM inverters, third harmonic can be eliminated if pulse width is equal to:
    (A) $30°$
    (B) $60°$
    (C) $120°$
    (D) none of these

12. In single-pulse modulation of PWM inverters, fifth harmonic can be eliminated if pulse width is equal to:
(A) 30°   (B) 72°
(C) 90°   (D) 108°

13. In single-pulse modulation of PWM inverters, the pulse width is 120°. For an input voltage of 220 V DC, the rms value of output voltage is:
(A) 179.63 V   (B) 254.04 V
(C) 280 V   (D) none of these

14. So far as the harmonic content at the inverter output voltage is concerned, multiple pulse modulation will provide:
(A) less harmonics
(B) more harmonics
(C) perfect sinusoidal of fundamental frequency
(D) none of these

15. In single-pulse modulation used in PWM inverters, V is the input DC voltage. For eliminating third harmonic, the magnitude of rms value of fundamental component of output voltage and pulse width are respectively:
(A) $\dfrac{2\sqrt{2}}{\pi}V, 120°$ (B) $\dfrac{4V}{\pi}, 60°$
(C) $\dfrac{2\sqrt{2}}{\pi}V, 60°$ (D) $\dfrac{4V}{\pi}, 120°$

16. In multiple-pulse modulation used in PWM inverters, the amplitudes of reference square wave and triangular carrier wave are respectively 1 V and 2 V. For generating 5 pulses per half-cycle, the pulse width should be:
(A) 36°   (B) 24°
(C) 18°   (D) 6°

17. In multiple-pulse modulation used in PWM inverters, the amplitudes and frequency for triangular carrier and square reference signals are respectively 4 V, 6 kHz and 1 V, 1 kHz. The number of pulses per half-cycle and pulse width are respectively:
(A) 6, 90°   (B) 3, 45°
(C) 4, 60°   (D) 3, 40°

18. Which of the following statements is correct in connection with inverters?
(A) Voltage source inverter and current source inverter both require feedback diodes

(B) Only current source inverter requires feedback diodes
(C) GTOs can be used in current source inverter
(D) Only VSI requires feedback diodes

19. In a constant source inverter, if frequency of output voltage is f Hz, then frequency of voltage input to constant source inverter is:
(A) f (B) $2f$
(C) $3f$ (D) $4f$

20. In an inverter with fundamental output frequency of 50 Hz, if third harmonic is eliminated, then frequencies of other components in the output voltage wave in Hz, would be:
(A) 250, 350, 500, high frequencies
(B) 50, 50, 350, 550
(C) 50, 150, 250, 550
(D) none of these

21. A single-phase current source inverter has capacitor C as the load. For a constant source current, the voltage across the capacitor is:
(A) square wave
(B) triangular wave
(C) step function
(D) none of these

22. A single-phase full bridge VSI has inductor L as the load. For a constant source voltage, the current through the inductor is:
(A) square wave
(B) triangular wave
(C) sine wave
(D) none of these

23. Natural commutation is considered to a kind of:
(A) forced commutation
(B) line commutation
(C) commutation failure
(D) none of these

24. Which of the following is the limitation of current source inverter?
(A) Sluggish dynamic performance
(B) Stability problem
(C) Unreliable
(D) None of these

25. In three-phase inverter, leading VAR requirement can be met by a variable VAR generator using:
   (A) switched capacitor
   (B) rotating synchronous condenser
   (C) current fed inverter VAR generator
   (D) all of these

26. Which of the following classes is the most popular among forced commutated current fed inverters?
   (A) Autosequential commutated inverter
   (B) Auxiliary bridge commutated inverter
   (C) Third harmonic commutated inverter
   (D) None of these

ANSWER KEY

1. (A)	2. (C)	3. (A)	4. (A)	5. (D)	6. (D)	7. (C)	8. (C)
9. (B)	10. (A)	11. (C)	12. (B)	13. (A)	14. (A)	15. (A)	16. (C)
17. (B)	18. (D)	19. (B)	20. (B)	21. (B)	22. (B)	23. (B)	24. (A)
25. (D)	26. (A)						

Review Questions

1. What is an inverter? Give four applications of inverters.
2. What are line commutated inverters? How do they operate? Explain.
3. Explain the operation of Single-phase half-bridge inverters. What are its limitations?
4. In the parallel inverter circuit, an inductance is added in series with the DC source. What will happen if this inductance is shorted.
5. What is the purpose of connecting diodes in antiparallel with thyristors in inverter circuits?
6. What are the standard methods of controlling voltage of inverters?
7. Explain the working principle of forced commutated inverters.
8. Explain the working principle of bridge inverter employing voltage and current commutation process.
9. Discuss the principle of working of a three-phase bridge inverter circuit. Draw the line and phase voltage waveforms on the assumption of 120° conduction, resistive load in star connection.
10. What is the need of voltage control at the output of inverters? Describe and compare various methods for the control.
11. What is pulse width of modulation? List the various PWM techniques. How do these differ from each other?
12. Show that for a single pulse modulation, the output voltage can be expressed as:

$$v_0 = \sum_{n=1,3,5}^{\alpha} \frac{4V_s}{n\pi} \sin \frac{n\pi}{2} \sin nd \sin n\omega t, \text{ where } 2d \text{ is the pulse width.}$$

13. Why is it necessary to remove harmonics from the output of an inverter?
14. Explain with the help of a diagram working principle of McMurray inverter circuit.
15. Explain with the help of diagram working principle of McMurray–Bedford inverter circuit.
16. Explain in brief the working principle of voltage control inverters.
17. Explain in brief the working principle of current source inverters.
18. Compare the advantages and limitations of voltage source and current source inverters.
19. Explain the voltage control, techniques of inverters.
20. Distinguish between 120° and 180° mode operation of 3 phase inverters. Draw the line to neutral voltage waveform in each case for one complete cycle.

Thyristorised Controlled DC Motors

13.0 INTRODUCTION

Controlled rectifiers are used to provide the variable DC voltage to armature and field circuits in DC motor. In this chapter apart from other drives, separately excited motor will be focussed. An important factor to consider is the ripple content in the DC output of the rectifier. In a single-phase half-wave circuit, one pulse of current is produced every supply frequency, and this provides an output that is rich in harmonics, resulting in excessive heating and torque pulsations. Single-phase sources are used primarily for motors with horse power output of 5 HP or less, for their simplicity and economy that offsets the poor wave shape.

For larger horse power applications, the three-phase full wave bridge configuration is used, and this has 300 pulses per second, which improves the harmonics content. It is often found necessary to employ two parallel phase shifted three-phase full wave controlled rectifier circuits to provide lower ripple. Speed control is achieved by adjusting the thyristor's firing angles.

DC drives are widely used in applications requiring variable speed, frequent starting, braking and reversing. Important applications include rolling mills, paper mills, machine tools, electric traction, textile mills, excavators and cranes. FHP DC motors are widely used as servo motors for positioning and tracting applications. DC drives continue to be used because of their low critical cost and excellent drive performance.

Availability of high power thyristors brought about a great change in industrial control equipment and drive performance. The M-G set was used for a long time to obtain variable DC voltage. But now, this has been largely replaced by thyristor converters. Virtually all new variable speed drives use thyristor converters. The three basic methods are used to obtain the variable voltage from a fixed supply voltage by thyristor converters as illustrated in Fig. 13.1. These methods are:

1. Phase control
2. Chopper control
3. Integral cycle control.

In phase control and integral cycle control, the conversion from AC to DC is achieved by rectification. The supply AC voltage turns off the thyristor and no commutation is widely used. Integral cycle control has not been found satisfactory for speed control of motors. Phase controlled converters are simple and less expensive and are extensively used in industries. Power conversion efficiency in these converters is generally high, above 95%, because of

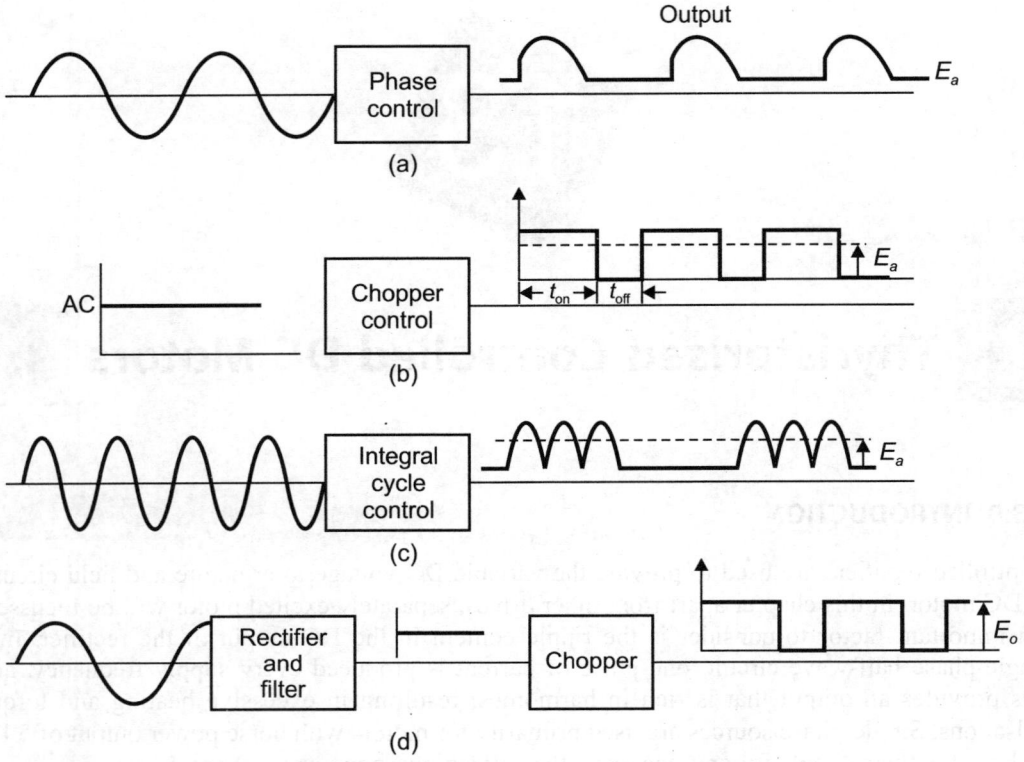

Fig. 13.1 Variable voltage from a fixed supply voltage by thyristor converters

relatively low losses in thyristors. If the supply is DC, chopper circuits are used in which auxiliary circuits are required to turn off the thyristors.

13.1 SPEED CONTROL OF DC MOTORS

The speed of a DC motor is given by:

$$N \propto \frac{E_b}{\Phi} = \frac{V - I_a R_a}{K\Phi} \qquad \qquad ...(13.1)$$

where, V is the supply voltage, I_a is the armature current, K is the motor speed constant and R_a is the armature resistance.

From the above equation it is clear that speed can be controlled by:

1. Armature voltage method
2. Flux control method, and
3. Armature resistance control method.

Armature voltage control is preferred because of high efficiency, good transient response and good speed regulation. But it can provide speed control only below base (rated) speed.

For speed above the rated speed, flux control method is employed. The maximum torque and power limitations of DC drives operating with armature voltage control and flux control are shown in Fig. 13.2. In armature voltage control at full field, torque $\propto I_a$ results in the maximum torque that the machine can deliver, has a constant value. In the field control method, at rated armature voltage, Power $P_m \propto I_a$ as $E_b \approx V$ = constant.

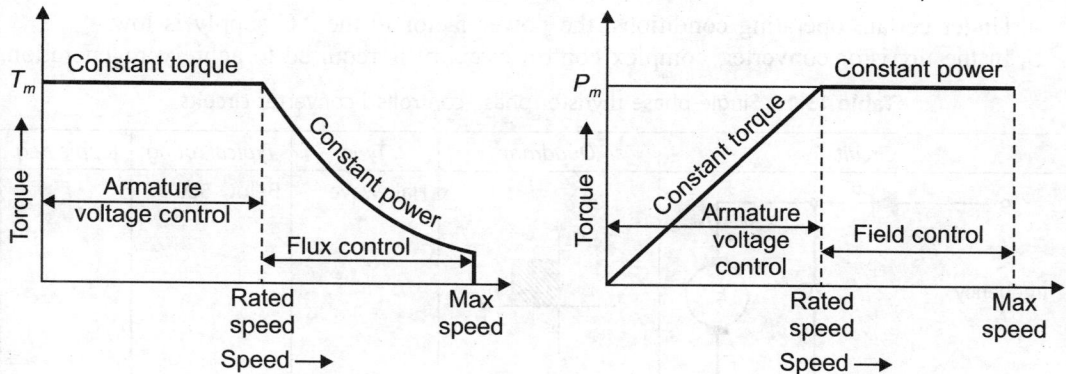

Fig. 13.2 The maximum torque and power limitations of DC drives operating with armature voltage control and flux control

Therefore, maximum power developed by the motor has a constant value. Thus, motor operation may be divided into two regions: (1) below the rated speed, and (2) above the rated speed. In the region below rated speed, with normal field flux and with a maximum value set on I_a, then equation of torque:

$$T = K_1 \, \Phi \, I_a \qquad \qquad \qquad ...(13.2)$$

indicates a constant maximum torque value. Above rated speed, when field Φ is reduced but I_a (rated) remains constant, the torque equation shows that permissible torque falls as speed increases. Thus, the horse power output given by

$$\mathrm{HP} = \frac{2\pi NT}{60} \qquad \qquad \qquad ...(13.3)$$

indicates that this is a region of constant permissible horse power output.

The terms constant torque and constant horse-power are occasionally used as descriptive to those two regions, not because operation is at these constant limits, but because maximum operation is fixed by them.

Advantages and Disadvantage of thyristorised drives: Thyristor drive has the following advantages and disadvantages:

Advantages:

1. The thyristor power module eliminates the electrical time lag of the generator field and armature. Time response is therefore faster, limited only by DC motor commutating ability and the inertia of the drive.
2. Basic operation is simple and reliable.
3. Operating efficiency is high, about 95%.
4. Small size, less weight and packaging flexibility result in reduced space.
5. Low initial cost.

Disadvantages:

1. Higher ripple content of the converter output adds to the motor heating and commutation problems.
2. The overload capability is lower than that of a comparable M.G. set.
3. Due to switching of thyristor distortion of the AC supply voltage and telephone, interference may be produced.

4. Under certain operating conditions, the power factor in the AC supply is low.
5. In the thyristor converter, complex control circuitry is required to achieve regeneration.

Table 13.1 Single-phase thyristor phase-controlled converter circuits

Circuit	Quadrant	Type	Typical rating	Ripple freq.
		Half wave	Below 500 W	f_s
		Semi converter	Upto 20 kW (75 kW in traction systems)	$2f_s$
		Full converter	Upto 20 kW (75 kW in traction systems)	$2f_s$
		Dual converter	Upto 20 kW	$2f_s$

13.2 PHASE CONTROLLED CONVERTERS

Each of the single-phase circuit configurations of Table 13.1 can be used to control the armature voltage and current of a separately excited DC motor. Phase controlled converters are classified as:

1. Single phase converters
2. Three-phase converters. The converters used for a particular application depends on:

   (a) the supply available whether single-phase or three-phase
   (b) rating of the device
   (c) amount of voltage ripple to be tolerated
   (d) reversible or nonreversible drive
   (e) need for regeneration, etc.

For the half-wave semi converter and full converter connections, the armature current is unidirectional whereas the dual converter permits the flow of armature current in either direction.

Semi converters are one quadrant converters which have one polarity of voltage and current at the DC terminals. Full converters are two quadrant converters in which voltage polarity can be reversed but current remains unidirectional because of unidirectional thyristors. Dual converters can operate in all the four quadrants. When a reversing switch is used at the DC terminals, semi converters can be operated in two quadrants and full converters in four quadrants.

Regeneration of power is possible with full converters. When regeneration is not required, semi converters are used for the sake of economy. A freewheeling diode D_{FW} is used to dissipate the stored energy in the motor inductance when the thyristor blocks. It provides protection against transients. In single-phase converters, the motor current may or may not be discontinuous but in three-phase converters, the current is mostly continuous.

13.2.1 Single Phase Separately Excited DC Motor Drives

In the circuit of Fig. 13.3, for a single-phase separately excited DC motor drive, the armature voltage is controlled by a semi converter or full converter and the field current is supplied from the AC supply through a diode bridge. The motor current cannot reverse due to thyristors in the converters. In semi converters the average DC output voltage E_a is always positive. Thus, power flow $E_a I_a$ is always from AC supply to DC load. In semi converters, freewheeling action (i.e. dissipation of armature inductance energy through the freewheeling path) takes place when thyristor blocks.

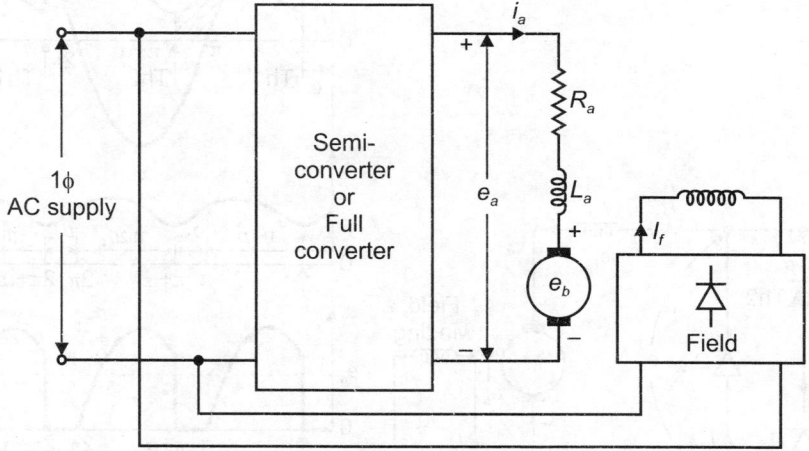

Fig. 13.3 Basic single-phase circuit for the speed control of a separately excited dc motor

Basic equations: While current flow is blocked in thyristors in the interval $0 \leq \omega t \leq \alpha$, no supply current flow and load current freewheels through D_{FW}. Each thyristor carries one of the pulses of supply current during each supply voltage wave. The armature circuit of the DC motor is represented by its back voltage e_b, armature resistance R_a and armature inductance L_a as shown in Fig. 13.3. Basic equations for back emf, torque and load voltages are given by:

Back emf voltage:	$e_b = K_a \Phi n$	
Average back voltage:	$E_b = K_a \Phi N$	...(13.4)
Developed torque:	$t = K_a \Phi i_a$	...(13.5)
Average developed torque:	$T_{av} = K_a \Phi I_a$	...(13.6)
Armature circuit voltage equation:	$e_a = R_a i_a + L_a \dfrac{di_a}{dt} + e_b$	...(13.7)

In terms of average values:
$$E_a = R_a I_a + E_b \qquad \text{...(13.8)}$$

The average speed:
$$N = \left(\frac{E_a}{K_a \Phi}\right) = \frac{E_b - I_a R_a}{K_a \Phi} \qquad \text{...(13.9)}$$

In phase controlled converters, the armature voltage e_a and armature current i_a change with time. The armature current i_a may be continuous or discontinuous. For both continuous and discontinuous current analysis is discussed here.

13.2.2 Analysis for Continuous Armature Current

Let us assume that the armature current is continuous over the whole range of operation. For semi converter and full converter system typical voltage and current waveforms are shown in Figs 13.4 and 13.5 respectively. The thyristors are symmetrically triggered. In the semi converter system, shown in Fig. 13.4, thyristor Th1 triggered at an angle α and thyristor Th2 at an angle $(\alpha + \pi)$ with respect to supply voltage v. In the full converter system shown in Fig. 13.5, thyristors Th1 and Th3 are simultaneously triggered at α and thyristors Th2 and Th4 are triggered at $(\pi + \alpha)$.

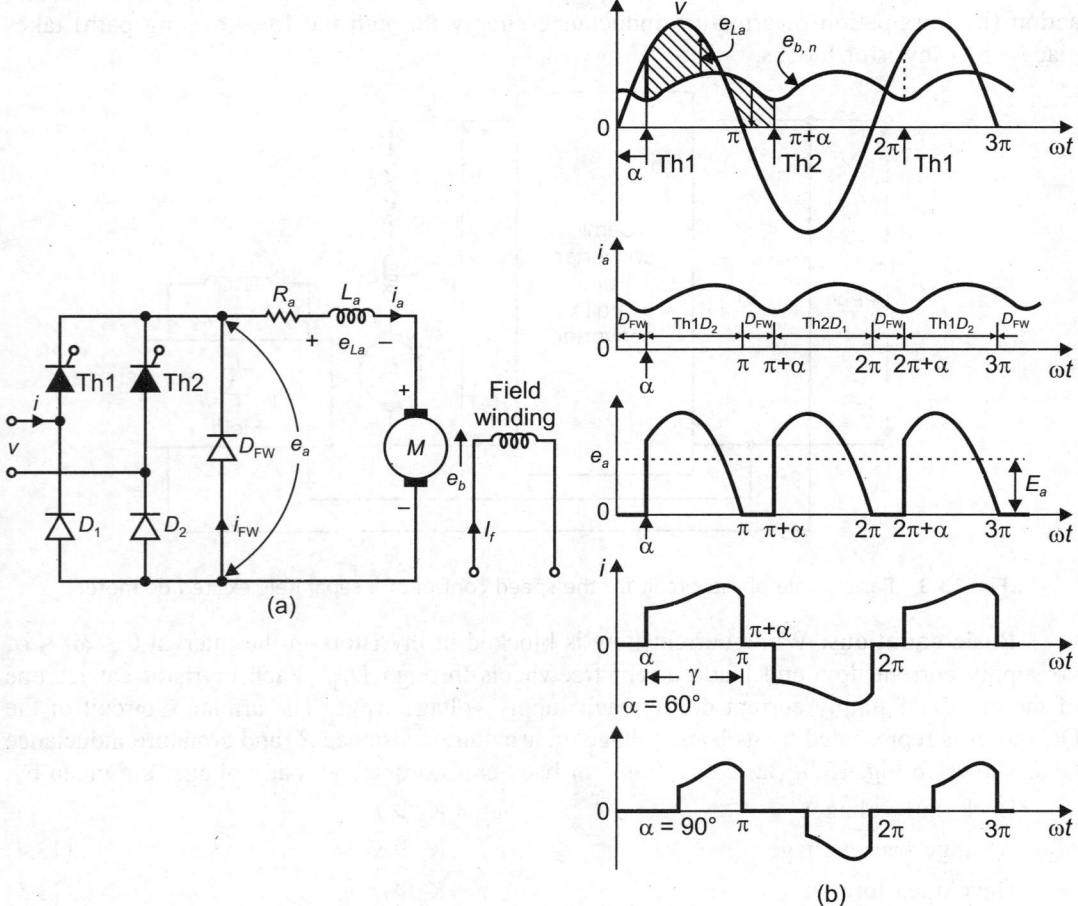

Fig. 13.4 Voltage and current waveforms for semi converter system

Consider the case in which the armature current is continuous, for which typical waveforms are shown in Fig. 13.4, when $\alpha = 60°$ and $90°$. In Fig. 13.4, the motor is connected to the input supply for the period $\alpha < \omega t < \pi$ through Th1 and diode D_2 and the motor terminal voltage e_a is the same as the supply input voltage v. Motor terminal voltage e_a tends to reverse as the input voltages change polarity beyond π which will forward bias the freewheeling diode D_{FW} and it will start conducting. The motor current i_a which was flowing from the supply through thyristor Th1, is transferred to diode D_{FW} (i.e. Th1 commutates). The motor terminals are shorted through the freewheeling diode during $\pi < \omega t < (\pi + \alpha)$, making e_a zero.

Energy from the supply is, therefore, delivered to the armature circuit when the thyristor conducts. This energy is partially stored in the inductance, partially stored in the kinetic energy of the moving system, and partially used to supply the mechanical load. During the freewheeling period, π to $\pi + \alpha$ energy is recovered from the inductance and is converted to mechanical form to supplement the kinetic energy in supplying the mechanical load. The freewheeling armature current continues to produce electromagnetic torque in the motor.

Single-phase full converter: The two quadrant, four thyristor, full converter circuit is applied to a separately-excited DC motor load as shown in Fig. 13.5, the motor 'M' is always connected to the supply through the thyristors. Thyristors Th1 and Th3 conduct during the interval $\alpha < \omega t < (\pi + \alpha)$ and connect motor to the supply. At $\pi + \alpha$, thyristors Th2 and Th4 are fired. Immediately the supply voltage appears across the thyristors Th1 and Th3, as a reverse-bias voltage and turns them off. The motor current i_a which was flowing from the supply through thyristors Th1 and Th3 is transferred to thyristors Th2 and Th4. During α to π, energy flows from the input supply to the motor. During π to $\pi + \alpha$, some of the motor system energy is feedback to the input supply.

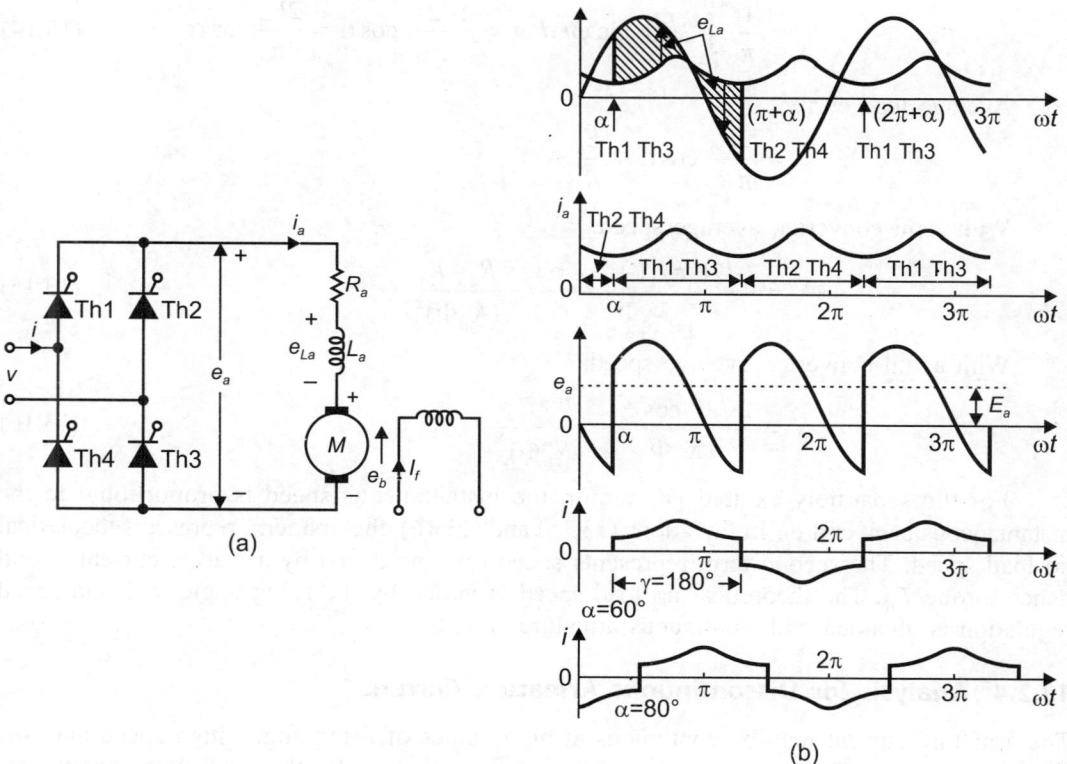

Fig. 13.5 Voltage and current waveforms of full converter applied to a separately-excited DC motor load

13.2.3 Torque Speed Characteristics

For a semi converter with freewheeling action, the load voltages are therefore:

$$e_a = v = R_a i_a + L_a \frac{di_a}{dt} + e_b; \ (\alpha < \omega t < \pi) \qquad \text{...(13.10)}$$

$$e_a = 0 = R_a i_a + L_a \frac{di_a}{dt} + e_b; \ (\pi < \omega t < \pi + \alpha) \qquad \text{...(13.11)}$$

The armature circuit equation for a full converter is

$$e_a = v = R_a i_a + L_a \frac{di_a}{dt} + e_b; \ (\alpha < \omega t < \pi + \alpha) \qquad \text{...(13.12)}$$

Let $\qquad v = \sqrt{2}V \sin \omega t$

The average motor terminal voltage with a semi converter is

$$E_a = \frac{1}{\pi} \int_\alpha^\pi \sqrt{2}V \sin \omega t \ d\omega t = \frac{\sqrt{2}V}{\pi} (1 + \cos \alpha) = \frac{V_m}{\pi} (1 + \cos \alpha) \qquad \text{...(13.13)}$$

Average load current,

$$I_a = \frac{E_a - E_b}{R_a} = \frac{V_m}{\pi R_a} (1 + \cos \alpha) - \frac{E_b}{R_a}$$

With a full converter

$$E_a = \frac{1}{\pi} \int_\alpha^{\pi+\alpha} \sqrt{2}V \sin \omega t \ d\omega t = \frac{2\sqrt{2}V}{\pi} \cos \alpha = \frac{2V_m}{\pi} \cos \alpha \qquad \text{...(13.14)}$$

Average load current

$$= \frac{2V_m}{\pi R_a} \cos \alpha - \frac{E_b}{R_a}$$

With semi converter, average speed,

$$N = \frac{(\sqrt{2}V/\pi)(1 + \cos \alpha)}{K_a \Phi} - \frac{R_a \cdot T_{av}}{(K_a \Phi)^2} \qquad \text{...(13.15)}$$

With a full converter, average speed,

$$N = \frac{2\sqrt{2}V \cos \alpha}{\pi K_a \Phi} - \frac{R \cdot T_{av}}{(K_a \Phi)^2} \qquad \text{...(13.16)}$$

For the separately excited DC motor, the instantaneous speed is proportional to the instantaneous back emf e_b. In Equations (13.15) and (13.16), the first term represents theoretical no load speed. The second term represents speed drop produced by armature current I_a and hence torque T_{av}. The theoretical no load speed is varied by the firing angle α. Good speed regulation is obtained with continuous armature current.

13.2.4 Analysis for Discontinuous Armature Current

The armature current may be continuous at high values of firing angle, high speed and low value of torque. In fact, the armature current is discontinuous for these operating conditions. If the armature current is discontinuous, no load speed will be higher and the speed regulation will be significantly poor in the region of discontinuous armature current.

The motor performance deteriorates and dynamic response becomes slow. It is, therefore, desirable to operate the DC motor in the continuous armature current mode. This can be achieved by having freewheeling action and using an external armature circuit choke to increase the time constant of the armature circuit.

Consider the condition where the armature current i falls to zero before the next pair of thyristors is switched on. The waveforms with semi converters and full converter with discontinuous armature current are shown in Figs 13.6 and 13.7 respectively. In Fig. 13.6, the motor is connected to the input supply for the period $\alpha < \omega t < \pi$ through thyristor Th1 and diode D_2. Beyond π, the motor terminal is shorted through the freewheeling diode D_{FW}. The armature current decays to zero at β before the thyristor Th2 is triggered at $\pi + \alpha$, thereby making the armature current discontinuous. During α to π, the motor terminal voltage e_a is the same as the supply voltage v. However, during π to β, the motor current freewheels through diode D_{FW} and so e_a is zero. In the interval $\beta < \omega t < (\pi + \alpha)$, the motor coasts and the motor terminal voltage e_a is the same as the back emf e_b. In Fig. 13.7, the motor is connected to the supply during $\alpha < \omega t < \beta$ and it coasts during $\beta < \omega t < \pi + \alpha$. As long as the motor is connected to the supply, its terminal voltage is same as the input supply voltage. In the discontinuous current mode, the calculations are cumbersome. A general approach valid for both continuous and discontinuous armature current is necessary.

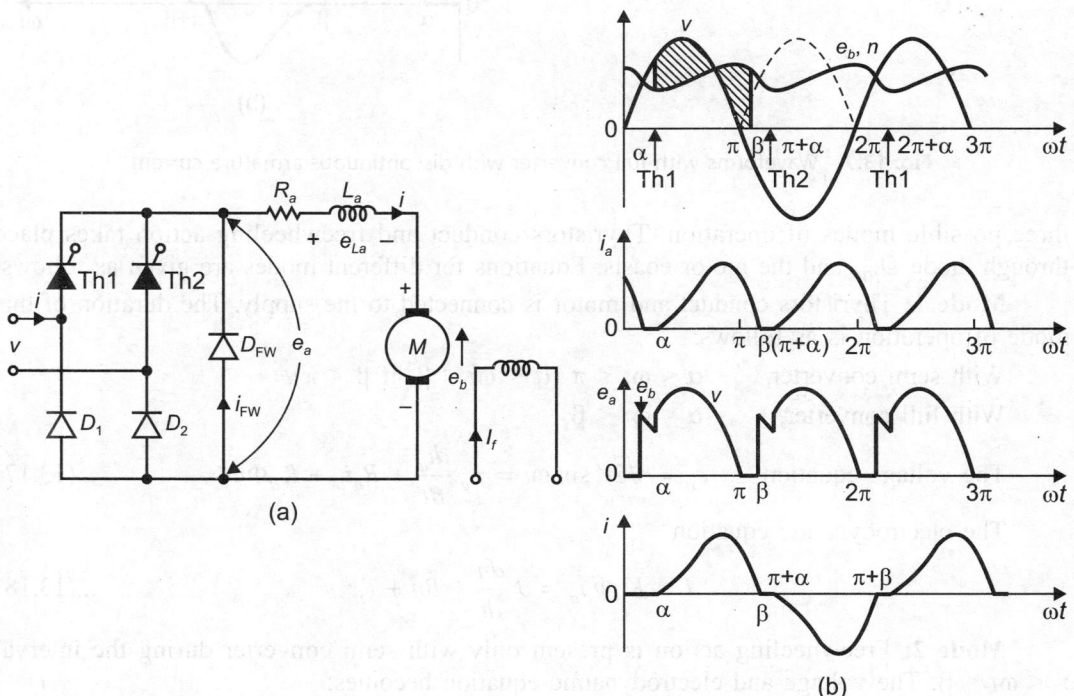

Fig. 13.6 Waveforms with semi converters with discontinuous armature current

13.2.5 General Analysis

Semi converter drives may exhibit discontinuous conduction. The power factor improves and ripple content decreases. The reactive power requirement is smaller even at low voltage ratios. In full converter drives, discontinuous conduction is present. The power factor of the converter is poor particularly at low voltages. The ripple content depends on pulse number. There are

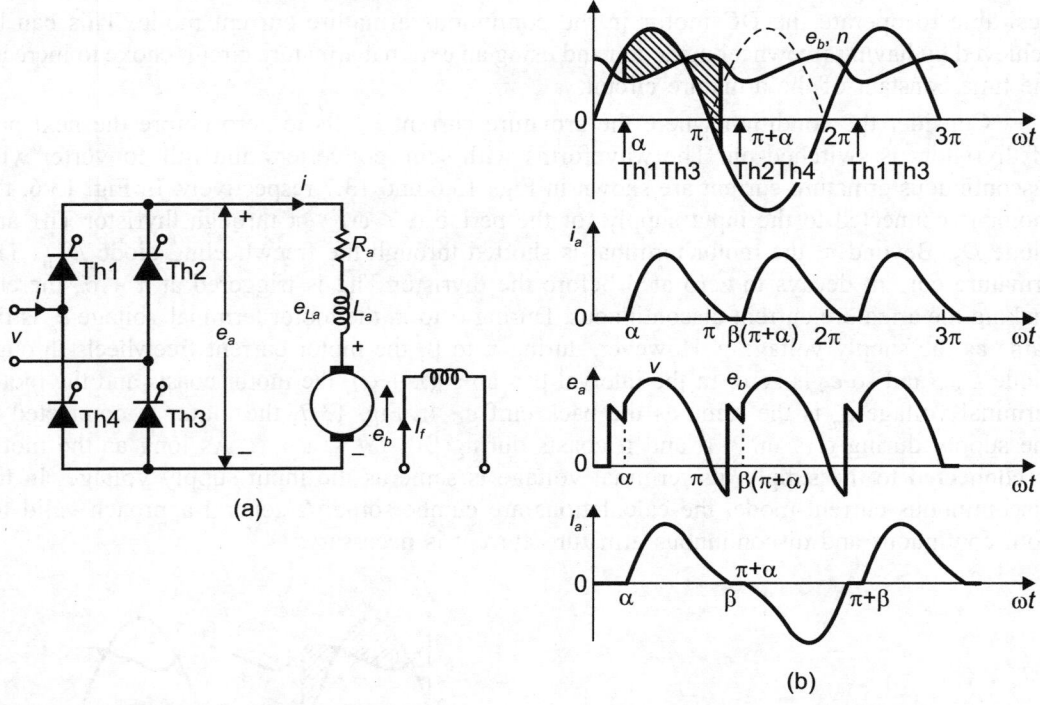

Fig. 13.7 Waveforms with full converter with discontinuous armature current

three possible modes of operation. Thyristors conduct and freewheeling action takes place through diode D_{FW} and the motor coasts. Equations for different modes are given as follows:

Mode 1: Thyristors conduct and motor is connected to the supply. The duration of this mode of operation is as follows:

With semi converter, $\quad \alpha < \omega t < \pi$ ($\alpha < \omega t < \beta$, if $\beta < \pi$).

With full converter, $\quad \alpha < \omega t < \beta$,

The voltage equation $\quad e_a = \sqrt{2}V \sin \omega t = L_a \dfrac{di_a}{dt} + R_a i_a + K_a \Phi n$ ...(13.17)

The electrodynamic equation

$$t = K_a \Phi i_a = J \frac{dn}{dt} + Bn + T_L \qquad \text{...(13.18)}$$

Mode 2: Freewheeling action is present only with semi converter during the interval $\pi < \omega t < \beta$. The voltage and electrodynamic equation becomes:

$$0 = R_a i_a + L_a \frac{di_a}{dt} + K_a \Phi n \qquad \text{...(13.19)}$$

$$K_a \Phi i_a = J \cdot \frac{dn}{dt} + Bn + T_L \qquad \text{...(13.20)}$$

If $\beta < \pi$, mode 2 will be absent.

Mode 3: Motor coasts during the interval

$$= \beta < \omega t < \pi + \alpha$$

The equations become $i_a = 0$.

$$= J \cdot \frac{dn}{dt} + Bn + T_L = 0 \qquad \text{...(13.21)}$$

Equations (13.17)–(13.21) can be solved by two methods, analytical and numerical, to determine the performance of the phase controlled DC drives.

13.2.6 Analytical Method

In analytical method, a steady state speed is assumed, and therefore only electrical equations have to be considered. From these equations it will be possible to obtain the expression for the motor current analytically. These equations will be used to compute the motor current and the steady state condition is noted by observing the variation of motor current at a particular instant over successive cycles. By this method, the steady state is reached after only a few cycles of computation. The performance parameters are determined after the steady state is reached.

If the current is discontinuous and at the instant of firing the thyristor, the supply voltage is less than the motor back emf 'e_b' then the thyristor will not turn on at α. It will start conducting when the input voltage v equals back voltage e_b. Let θ_s be the angle at which thyristor turns on.

Mode 1: Thyristor conducts
with full converter ($\theta_s < \omega t < \beta$);
with semi converter ($\theta_s < \omega t < \pi$)
($\theta_s < \omega t < \beta$ if $\beta < \pi$)

The voltage equation: $\sqrt{2}V \sin \omega t = L_a \dfrac{di_a}{dt} + R_a i_a + E_b$

Solution for i_a: $\qquad i_a = \dfrac{\sqrt{2}V}{Z} \sin(\omega t - \theta_a) - \dfrac{E_b}{R_a} + A_1 e^{-(R_a/L_a)t} \qquad \text{...(13.22)}$

where, A_1 is constant and $\qquad Z = \sqrt{R_a^2 + (\omega L_a)^2}$

and $\qquad\qquad\qquad\qquad \theta_a = \tan^{-1}(\omega L_a/R_a)$.

First term is the component of armature current due to the input supply voltage. The second term is due to back voltage E_b and third term is the transient component of current. From Equation (13.22).

Let the current i_a at $\omega t = \theta_s$ be I_{as}

$$I_{as} = \frac{\sqrt{2}V}{Z} \sin(\theta_s - \theta_a) - \frac{E_b}{R_a} + A_1 e^{-(R_a/\omega L_a)\theta_s} \qquad \text{...(13.23)}$$

where, $\qquad A_1 = \left[I_{as} - \dfrac{\sqrt{2}V}{Z} \sin(\theta_s - \theta_a) + \dfrac{E_b}{R_a} \right] e^{(R_a/\omega L_a)(\theta_s - \omega t)}$

Putting the value of constant A in Equation (13.22), we get

$$i_a = \frac{\sqrt{2}V}{Z} \left[\sin(\omega t - \theta_a) - \sin(\theta_s - \theta_a) e^{(R_a/\omega L_a)(\theta_s - \omega t)} \right]$$

$$+ \frac{E_b}{R_a} \left[e^{(R_a/\omega L_a)(\theta_s - \omega t)} - 1 \right] + I_{as} e^{(R_a/\omega L_a)(\theta_s - \omega t)} \qquad \text{...(13.24)}$$

Mode 2: Freewheeling action present with a semi converter, during the interval $\pi < \omega t < \beta$.

The voltage equation becomes

$$0 = \frac{L_a di_a}{dt} + R_a i_a + E_b \qquad \ldots(13.25)$$

Solution for i_a
$$i_a = A_2 e^{-(R_a/\omega L_a)\omega t} - \frac{E_b}{R_a} \qquad \ldots(13.26)$$

where, A_2 is a constant. The value of i_a at $\omega t = \pi$ can be computed from Equation (13.24). Let $i_a(\pi) = I_{a\pi}$, from Equation (13.26)

$$I_{a\pi} = A_2 e^{-(R_a/\omega L_a)\pi} - \frac{E_b}{R_a} \qquad \ldots(13.27)$$

or
$$A_2 = \left[I_{a\pi} + \frac{E_b}{R_a} \right] e^{(R_a/\omega L_a)\pi} \qquad \ldots(13.28)$$

Putting the value of A_2 in Equation (13.26), we get

$$i_a = I_{a\pi} e^{(R_a/\omega L_a)(\pi - \omega t)} + \frac{E_b}{R_a} [e^{(R_a/+\omega L_a)(\pi - \omega t)} - 1] \qquad \ldots(13.29)$$

Mode 2 will be absent if $\beta < \pi$. Mode 2 is absent with a full converter.

Mode 3: Motor coasts during the interval $\beta < \omega t < (\pi + \alpha)$

$$i_a = 0 \qquad \ldots(13.30)$$

Supply current:

Mode 1:

$$i = i_a \text{ with semi converter } \theta_s < \omega t < \pi \qquad \ldots(13.31)$$
$$\text{with full converter } \theta_s < \omega t < \beta$$

$$i = -i_a \text{ with semi converter } \pi + \theta_s < \omega t < 2\pi \qquad \ldots(13.32)$$
$$\text{with full converter } \pi + \theta_s < \omega t < \pi + \beta$$

From modes 2 and 3 $\qquad i = 0 \qquad \ldots(13.33)$

These performance parameters can be obtained either by analytical or a numerical method. In most motors, the mechanical time constant is large, and the time variation of speed in the steady state is insignificant. This justifies the assumption of constant steady state speed in the analytical method. If the current is discontinuous, the steady state is reached in one half cycle. If the current is continuous, computation over a few cycles may be necessary before the steady state is reached.

13.3 COMPARISON BETWEEN SEMI CONVERTER AND FULL CONVERTER

1. Supply power factor and displacement factor deteriorate with a decrease in speed in both semi converter and full converter. Semi converter systems have an edge over full converter systems and are preferred where inverter operation for regenerative braking is not required.
2. Harmonic content in the input supply current is substantially higher in a semi converter in the lower speed range. In the lower speed range, as the speed decreases, harmonic content increases in the semi converter system which remains essentially the same in full converter system.

3. Semi converter system produces less rms current and hence there is less heating of the motor. At rated torque and normalised speed full converter operation produces 1.44 times the heating produced by semi converter operation.

13.4 SINGLE-PHASE SERIES DC MOTOR DRIVES

Fractional horsepower drives are built to operate from single-phase line, because such power is available where such lower drives are needed, and because of lower cost of fewer thyristors. The single-phase drives extend from poorly regulated units with armature or tachometer feedback. In a series motor, the field circuit is connected in series with armature and the motor terminal voltage is controlled by a semi converter or full converter. Series motors are extensively used in many applications. They are particularly suited for applications that require high starting torque such as cranes, hoists, elevators, etc. Inherently, series motors can provide essentially constant power output, and particularly, suitable for traction drives.

Basic equations: The armature circuit resistance R_a and inductance L_a include the resistance and inductance of the series field winding. Back emf is

$$e_b = K_a \Phi n \qquad \qquad ...(13.34)$$

The flux Φ has two components. One component, say Φ_α, is produced by the armature current flowing through the series field winding. The other component, say Φ_{res}, is due to the residual magnetism. The latter is small and can be assumed constant.

$$\Phi = \Phi_\alpha + \Phi_{res} \qquad \qquad ...(13.35)$$

If magnetic linearity is assumed,

$$\Phi_\alpha = K_f i_a \qquad \qquad ...(13.36)$$

From Equations (13.34), (13.35) and (13.36)

$$e_b = K_a (K_f i_a + \Phi_{res}) n = K_a K_f i_a n + K_a \Phi_{res} n$$
$$= K_{af} i_a n + K_{res} n \qquad \qquad ...(13.37)$$

The back voltage due to residual magnetism is very small and is proportional to speed. The back voltage due to the flux produced by the armature current is the major voltage and it is present when both i_a and n are present. Average back voltage is

$$E_b = K_{af} N I_a + K_{res} N \qquad \qquad ...(13.38)$$

Developed torque, $\qquad \qquad t = K_a \Phi i_a \qquad \qquad ...(13.39)$

If the flux Φ_{res} is neglected, then

$$t \approx K_{af} \cdot i_a^2 \qquad \qquad ...(13.40)$$

Therefore, for either direction of current, torque is developed in the same direction. Hence, in the series motor, speed reversal can be achieved by reversing either the field winding or the armature terminals, but not both.

Average torque, $\qquad T_{av} = K_{af} i_a^2 |_{ave} = K_{af} I_{av}^2 \qquad \qquad ...(13.41)$

The armature circuit voltage equation is given by

$$e_a = R_a i_a + L_a \frac{di_a}{dt} + e_b \qquad \qquad ...(13.42)$$

In terms of average quantities,

$$E_a = R_a I_a + E_b \qquad \qquad ...(13.43)$$

A full wave drive circuit is usually built as shown in Fig. 13.8 (a). Using two diodes and two thyristors having a common cathode connection but a separate freewheeling diode is used. In small motors (L/R) ratio may be so small that a freewheeling diode will not

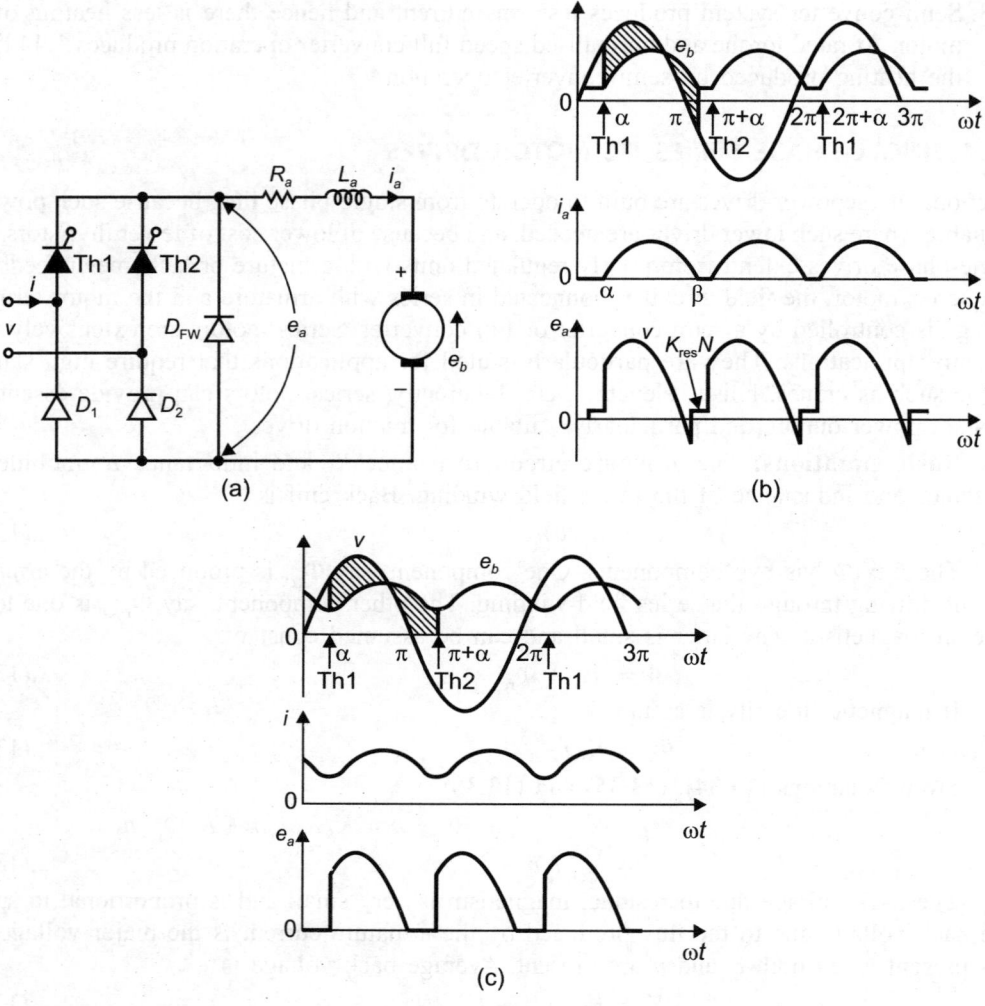

Fig. 13.8 Current and voltage waveforms for semi converter system

contribute any assistance to increase the conduction angle. However, it may provide same protection to the bridge if the line source is suddenly interrupted. The term discontinuous is applied to the conduction where the armature current reaches zero each half cycle before the next thyristor is fired. The term continuous is applied to the condition where the armature current never ceases, but continues to flow either through the thyristor or diodes. Current and voltage waveforms for semi converter and full converter series motor drives are shown in Figs 13.8 and 13.9 respectively. This leads discontinuous motor current. In series motor, the back voltage is proportional to motor current (speed is assumed constant and back voltage due to residual magnetism is very small and neglected). Therefore, back voltage decreases as the motor current decreases and so the motor current tends to be continuous. Only at high speed low current in the motor is likely to become discontinuous. Waveforms are drawn for both continuous and discontinuous currents. In the discontinuous current mode, the motor terminal voltage e_a is the same as the back voltage due to residual magnetism, which is very small.

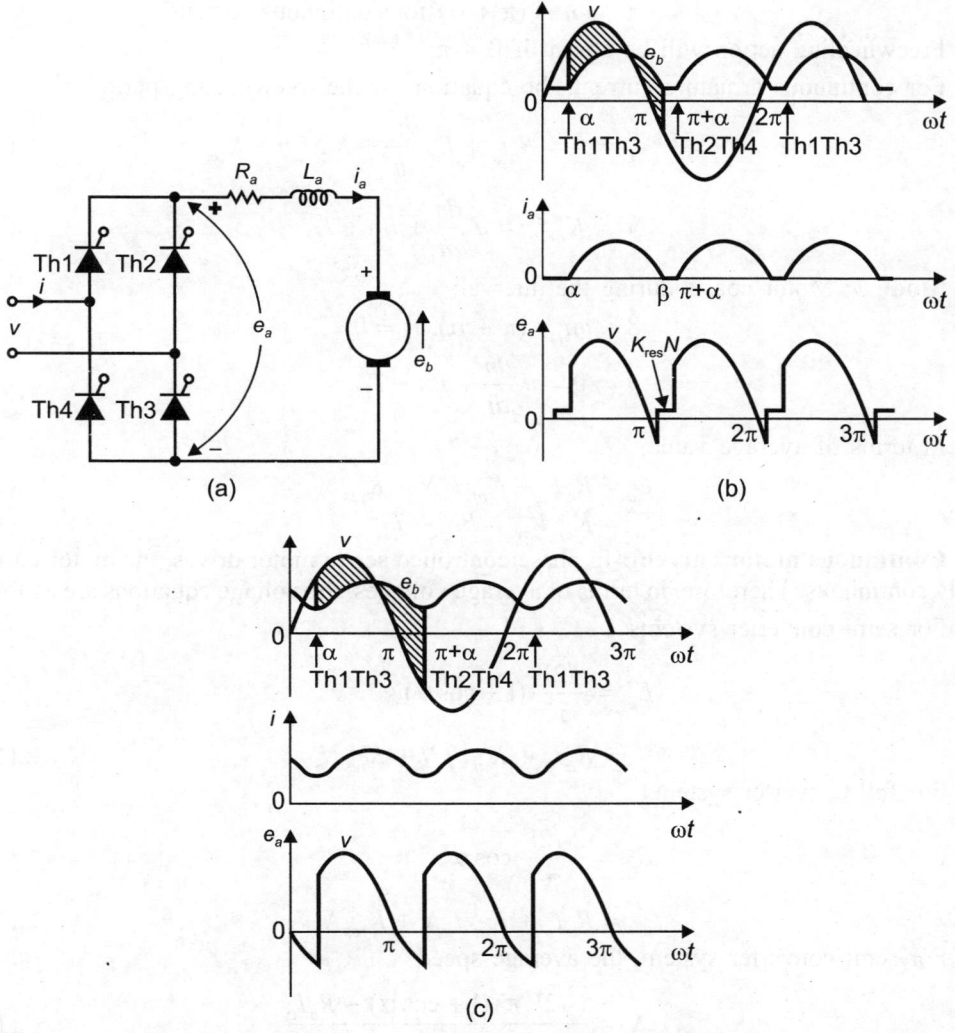

Fig. 13.9 Waveforms for full converter series motor drives

Operating modes: The system can operate in three modes.

Mode 1: Thyristor conducts during the interval:

$$\alpha < \omega t < \pi \text{ with semi converter and}$$
$$\alpha < \omega t < \beta \text{ with full converter.}$$

The armature-circuit equation for the conducting period

$$e_a = \sqrt{2}V \sin \omega t = R_a i_a + L_a \frac{di_a}{dt} + K_{af} i_a n + K_{res} n \qquad ...(13.44)$$

The electrodynamic equation

$$t = K_{af} i_a^2 = J \frac{dn}{dt} + Bn + T_L \qquad ...(13.45)$$

Mode 2: Freewheeling action in semi converter systems during the interval:

$$\pi < \omega t < \beta \text{ for discontinuous current and}$$

$$\pi < \omega t < (\pi + \alpha) \text{ for continuous current.}$$

Freewheeling action will be absent if $\beta < \pi$:

For continuous armature current, the equation for the freewheeling period

$$e_a = 0 = R_a i_a + L_a \frac{di_a}{dt} + K_{af} i_a n + K_{res} n \qquad ...(13.46)$$

$$t = K_{af} i_a^2 = J \frac{dn}{dt} + Bn + T_L \qquad ...(13.47)$$

Mode 3: Motor coasts during the interval:

$$\beta < \omega t < (\pi + \alpha), \ i_a = 0,$$

$$t = 0 = J \frac{dn}{dt} + Bn + T_L$$

In terms of average value,

$$E_a = R_a I_a + K_{af} I_a N + K_{res} N \qquad ...(13.48)$$
$$T_{av} = K_{af} I_{av}^2 = BN + T_L \qquad ...(13.49)$$

Continuous motor current: In phase controlled series motor drives, the motor current is mostly continuous. Therefore, in terms of average voltages, the voltage equations are as follows:

For semi converter systems

$$E_a = \frac{\sqrt{2}V}{\pi} (1 + \cos \alpha)$$

$$= R_a I_a + K_{af} I_a N + K_{res} N \qquad ...(13.50)$$

For full converter systems

$$E_a = \frac{2\sqrt{2}V}{\pi} \cos \alpha$$

$$= R_a I_a + K_{af} I_a N + K_{res} N \qquad ...(13.51)$$

For semi converter system, the average speed

$$N = \frac{(\sqrt{2}V/\pi)(1 + \cos \alpha) - R_a I_a}{K_{af} I_a + K_{res}} \qquad ...(13.52)$$

For full converter system $N = \dfrac{(2\sqrt{2}V/\pi)(\cos \alpha) - R_a I_a}{K_{af} I_a + K_{res}} \qquad ...(13.53)$

If the ripple in the motor current can be neglected, then $I_a \approx I_{av}$

Average torque, $\qquad T_{av} = K_{af} I_{av}^2 \approx K_{af} I_a^2$

For semi converter system

$$T_{av} = K_{af} \left[\frac{(\sqrt{2}V/\pi)(1 + \cos \alpha) - K_{res} N}{R_a + K_{af} N} \right]^2 \qquad ...(13.54)$$

For full converter system,

$$T_{av} = K_{af} \left[\frac{(2\sqrt{2}V/\pi)(\cos \alpha) - K_{res} N}{R_a + K_{af} N} \right]^2 \qquad ...(13.55)$$

13.5 THREE-PHASE DRIVES

Three-phase converters are extensively used in adjustable speed drives from about 10 HP upto several thousand horse power rating. Three-phase, half wave circuit shown in Table 13.2 is not greatly used because of the DC components inherent in its line currents. Mostly large horse

Table 13.2 Three-phase thyristor phase-controlled converter circuits

Circuit	Quadrant operation	Type	Typical rating	Ripple freq.
		Half wave	40 kW	$3f_s$
		Semi converter	10–100 kW	$3f_s$
		Full converter	75–120 kW	$6f_s$
		Dual converter	Upto 150–1500 kW	$6f_s$

power DC drives take power from three-phase sources. In these drives, the drive motor is controlled by three-phase controlled converters. The ripple frequency of the motor terminal voltage is higher than that of single-phase converters. The three-phase half wave converter is impractical for most purposes, because the supply currents contain DC components. Semi converters and full converters are most commonly used in practice. Dual converters are used in reversible drives having power ratings of several megawatts in steel industry and heavy applications.

Semi converter operation: The semi converter circuit includes a freewheel diode to assist in maintaining continuous load current. The semi converter circuit absorbs less reactive voltampere than the fully controlled converter. Figure 13.10 shows a three-phase semi converter drive circuit and wave-forms of voltages and currents. The diodes D_1, D_2, D_3 conduct during the intervals t_4 to t_6, t_6 to t_8 and t_2 to t_4, respectively. If the thyristors Th1, Th2 and Th3 were diodes, they would

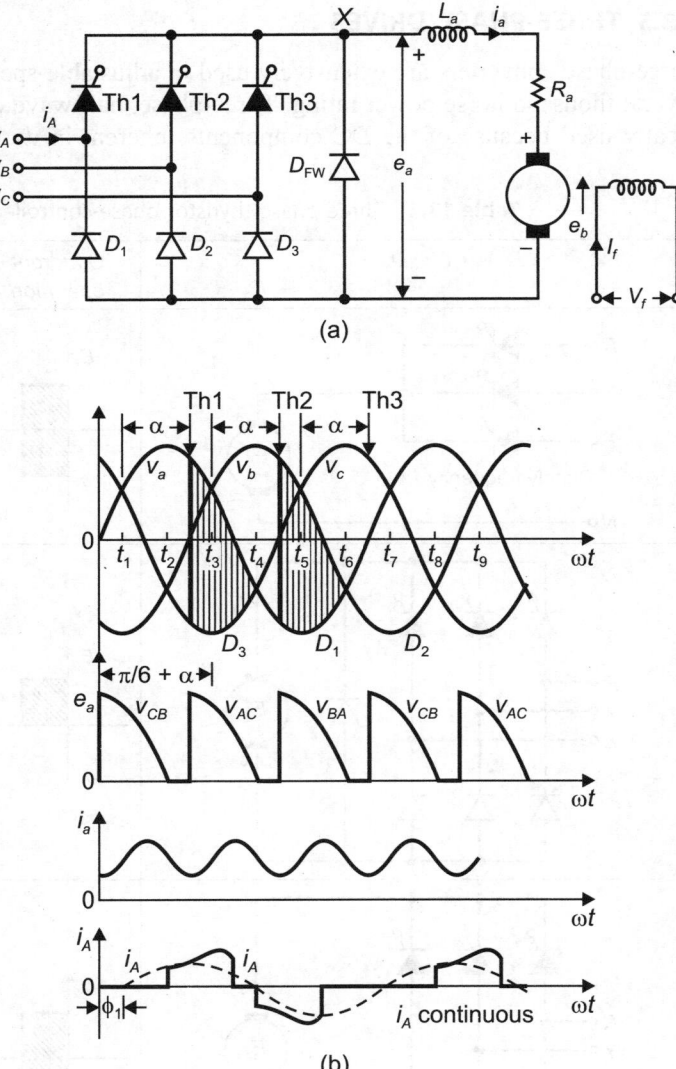

Fig. 13.10 Three-phase semi converter controlled dc motor system and waveforms

conduct during the intervals t_1 to t_3, t_3 to t_5 and t_5 to t_7 respectively. Therefore, the references for the triggering angles for thyristors Th1, Th2 and Th3 are the instants t_1, t_3 and t_5, respectively, i.e. these are the crossing points of the phase voltages v_A, v_B and v_C.

During the interval $(\pi/6 + \alpha) < \omega t < \omega t_4$, thyristor Th1 and diode D_3 conduct. Therefore, motor terminal X is connected to phase voltage v_A and the terminal Y is connected to phase voltage v_C. Thus, the motor terminal voltage during the period is $e_a = v_A - v_C = v_{AC}$. At ωt_4, e_a is zero, and from this time onward e_a tends to be negative. The freewheeling diode D_{FW} thus becomes forward-biased at ωt_4 and motor current flows through it until the next thyristor Th2 is turned on at $\left(\dfrac{\pi}{6} + \alpha + \dfrac{2\pi}{3}\right)$. In the absence of the freewheeling diode, freewheeling action would have taken place through thyristor Th1 and diode D_1. At large firing angles, the motor current can be continuous or discontinuous. When $\alpha = 0$, the average output voltage becomes identical to that of an uncontrolled three-phase bridge.

At large firing angles if the current demand is low and the speed is not low, the motor current may be discontinuous.

In terms of average voltages

$$E_a(\alpha) = I_a R_a + E_b = I_a R_a + K_a \Phi N \cdot$$

$$N = \frac{E_a(\alpha) - R_a I_a}{K_a \Phi} \qquad \qquad ...(13.56)$$

Let the supply phase voltages be as follows:

$$v_A = \sqrt{2}V \sin \omega t, \, v_B = \sqrt{2}V \sin\left(\omega t - \frac{2\pi}{3}\right), v_C = \sqrt{2}V \sin\left(\omega t + \frac{2\pi}{3}\right)$$

If the motor current is continuous,

$$E_a(\alpha) = \frac{3}{2\pi} \int_{\pi/6+\alpha}^{\pi/6+\alpha+2\pi/3} (v_A - v_C)\, d(\omega t) = \frac{3\sqrt{6}V}{2\pi}(1 + \cos \alpha) \qquad ...(13.57)$$

The displacement angle ϕ_1 increases as the firing angle increases. The input power factor will thus decrease as firing angle increases. Equation (13.57) is not valid for continuous current operation when $\alpha > 90°$, the average load voltage contribution of the upper half-bridge becomes negative. The result is that the overall load voltage E_a goes negative at some intervals during the cycle and freewheel diode D_{FW} conducts.

Full converter operation: A circuit diagram is given in Fig. 13.11, in which motor armature is represented by its equivalent with low armature inductance and large firing angle. The armature current may become discontinuous especially if the DC motor speed is high. If the motor armature circuit contains substantial series inductance and the firing angle is small then the armature current is likely to be continuous, even with large motor back emf. Three-phase converters are used in industrial applications upto 120 kW level, where two quadrant operation is required. Full converter drive circuit and the voltage and current waveforms are shown in Fig. 13.11. The instants of firing the thyristors are marked for $\alpha = 60°$ and the thyristors are fired at an interval of 60°, and the ripple in the motor terminal voltage is 6 pulses per cycle. As the thyristors are triggered at a faster rate, the motor current is mostly continuous.

At $\omega t = \pi/6 + \alpha$, thyristor Th1 turns on. Prior to this instant, thyristor Th6 was turned on. Therefore, during the interval $(\pi/6 + \alpha) < \omega t < (\pi/6 + \alpha + \pi/3)$, thyristors Th1 and Th5 conduct, and motor terminals are connected to phase A and phase B, making $e_a = v_{AB}$. At $\omega t = (\pi/6 + \alpha + \pi/3)$, thyristor Th2 is fired and immediately thyristor Th6 is reverse biased and turns off. The current from Th6 is transferred to Th2, and, therefore, the motor terminals are connected to phase A through thyristor Th1 and phase C through thyristor Th2 making $e_a = v_{AC}$. This process repeats after every 60° whenever a thyristor is fired. The motor terminal voltage can become negative. This is the inversion mode of operation at which regeneration can be made by reversing the motor voltage.

The average motor terminal voltage is

$$E_a(\alpha) = \frac{3}{\pi} \int_{\pi/6+\alpha}^{\pi/6+\alpha+2\pi/3} (v_A - v_B)\, d(\omega t)$$

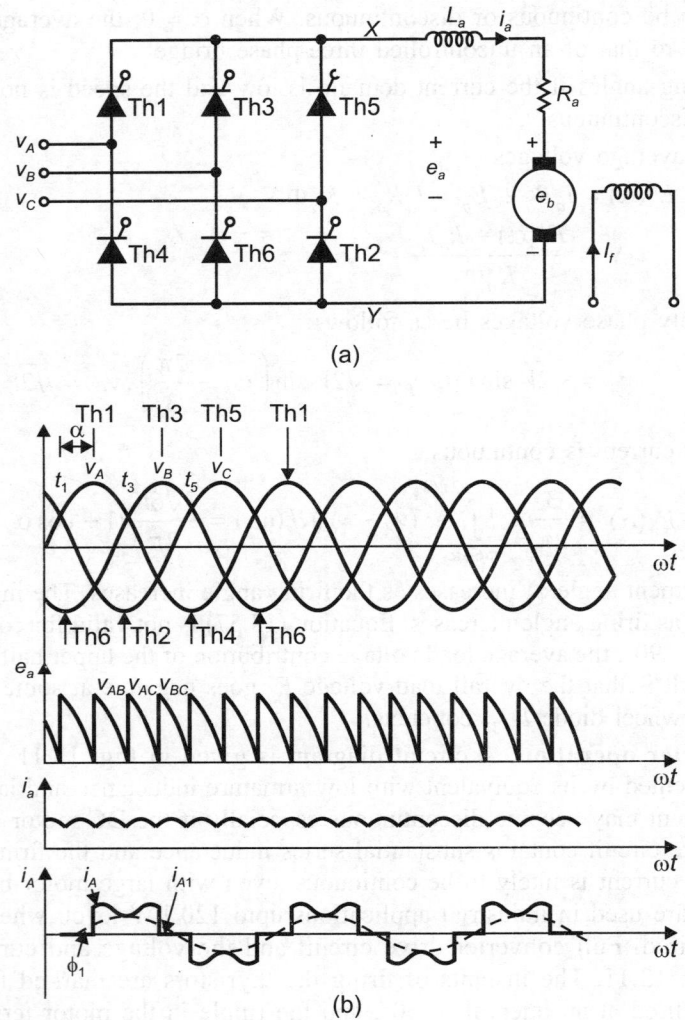

Fig. 13.11 Three-phase full converter drive system with waveforms for continuous motor current

$$= \frac{3\sqrt{6}V}{\pi} \cos \alpha \qquad \qquad ...(13.58)$$

The average speed is, $\quad N = \dfrac{E_a(\alpha) - R_a I_a}{K_a \Phi}$

In a separately excited motor,

$$T_{av} = K_a \Phi I_a$$

So, $\qquad \qquad N = \dfrac{E_a(\alpha)}{K_a \Phi} - \dfrac{R_a}{(K_a \Phi)^2} T_{av} \qquad \qquad ...(13.59)$

In Equation (13.59), the first term represents ideal no-load speed, which, therefore, depends on $E_a(\alpha)$. If the motor current is assumed continuous, motor terminal voltage depends only on the firing angle α. The second term represents the decrease in speed as the motor torque increases. Since the armature resistance R_a is small, the decrease in speed is small. In large motors, the motor current is likely to be continuous even at no-load condition as no-load current is not small. Therefore, three-phase drives provide better speed regulation and improved performance as compared to single-phase drives.

13.6 DUAL CONVERTERS

It is basically a four quadrant drive. Two modes of operation, i.e. circulating current and circulating current free operation are possible. In the former, circulating current forms a base load which makes the conduction continuous. The effects of discontinuous conduction are felt in the other mode. If two full converters are connected back to back, both the voltage and the current at DC terminal can be reversed and, therefore, the system will provide four quadrant operation. Such a system, shown in Fig. 13.12, is called *3-pulse dual converter*. This system is normally used for high power DC drives for speed control in both directions.

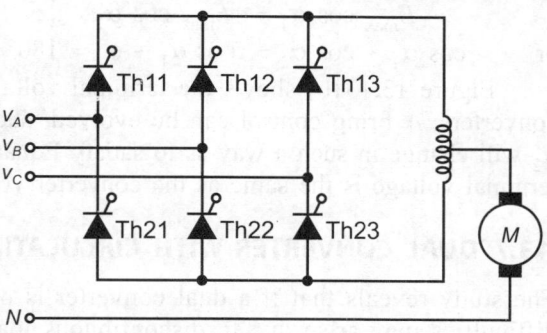

Fig. 13.12 Three pulse dual converter

Ideal dual converter: For ideal dual converter, there is no ripple at the DC output terminals. The magnitude of DC voltages varies as the cosine of firing angles of converters. The converters can be replaced as shown in Fig. 13.13 (a) by a DC voltage source in series with diodes that represent unidirectional current flow characteristic of converters. The magnitude of the DC voltage varies as the cosine of the firing angles of the converters. The firing angles of both converters are regulated by a control voltage E_c, so that their DC voltages are always exactly equal and of the same polarity. Thus, both converters produce the same terminal voltage, one converter operating as a rectifier and other operating as an inverter. From the equation

$$E_a(\alpha) = \frac{3\sqrt{6}}{\pi} V \cdot \cos \alpha,$$

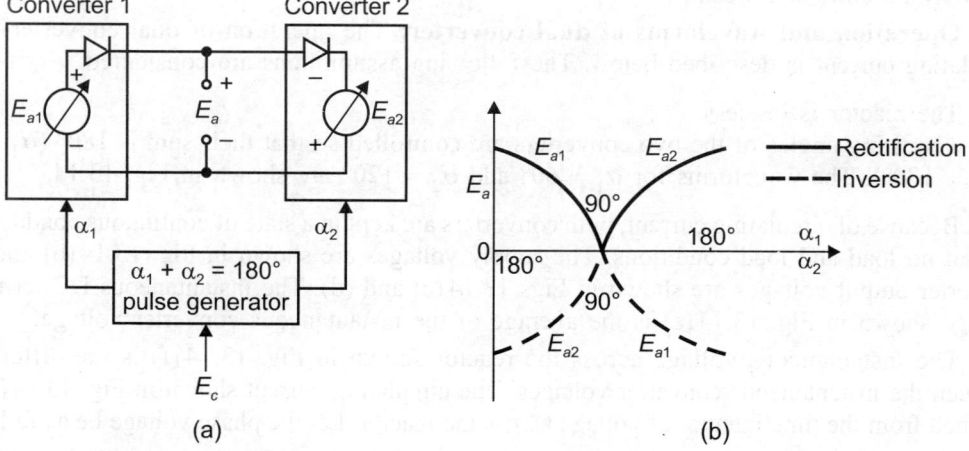

Fig. 13.13 Ideal dual converter with voltage waveform as a function of firing angle

$$E_{a1} = E_{\max} \cos \alpha_1 \qquad \qquad ...(13.60)$$

and
$$E_{a2} = E_{\max} \cos \alpha_2 \qquad \qquad ...(13.61)$$

In ideal dual converter,

$$E_a = E_{a1} = -E_{a2} \qquad \qquad ...(13.62)$$

From Equations (13.60), (13.61), and (13.62)

$$E_{\max} \cos \alpha_1 = -E_{\max} \cos \alpha_2$$

or $\quad \cos \alpha_1 + \cos \alpha_2 = 0$ or $\alpha_1 + \alpha_2 = 180° \qquad \qquad ...(13.63)$

Figure 13.13 (b) shows the terminal voltage as a function of firing angle for the two converters. A firing control can be evolved, such that as control signal E_c channels α_1 and α_2 will change in such a way as to satisfy Equation (13.63). In ideal dual converter, the load terminal voltage is the same as the converter voltages.

13.7 DUAL CONVERTER WITH CIRCULATING CURRENT

The study reveals that if a dual converter is operated without circulating current, several difficulties may arise due to discontinuous load current. These difficulties are avoided if circulating current is allowed to flow in the dual converter. In this scheme, triggering pulses are allowed to trigger thyristors of both the converters. The control voltage that generates these firing pulses satisfies the equation

$$\alpha_1 + \alpha_2 = 180°$$

Although both converters produce the same average voltage at their terminals, their instantaneous voltages are different which make the circulating current flow in the two converters. This circulating current is limited by a circulating current reactor. This scheme has the following advantages:

1. The time response is fast if the converters are in continuous conduction.
2. A natural mode is provided in the power circuit for the load current to flow in either direction at any time.
3. The circulating current keeps both converters in virtually continuous conduction over the whole control range, or independent of whether the external load current is continuous or discontinuous or, whether there is any load current at all.

However, the need for a reactor is a disadvantage. At high power levels, the size and cost may be quite significant.

Operation and waveforms of dual converter: The operation of dual converter with circulating current is described below. The following assumptions are considered.

1. The reactor is lossless.
2. The firing angles of the two converters are controlled so that their sum is 180° ($\alpha_1 + \alpha_2$ = 180°). The waveforms for $\alpha_1 = 60°$ and $\alpha_2 = 120°$ are shown in Fig. 13.14.

Because of circulating current, both converters are kept in a state of continuous conduction, both at no load and load conditions. The supply voltages are shown in Fig. 13.14 (b) and the converter output voltages are shown in Figs 13.14 (c) and (d). The instantaneous DC terminal voltage shown in Fig. 13.14 (e) is the average of the instantaneous converter voltage.

The instantaneous voltage across the reactor shown in Fig. 13.14 (f) is the difference between the instantaneous converter voltages. The circulating current shown in Fig. 13.14 (g) is obtained from the time integral of voltage across the reactor. Let the phase voltage be as follows:

$$v_A = \sqrt{2}V \sin \omega t, \, v_B = \sqrt{2}V \sin (\omega t - 2\pi/3), v_C = \sqrt{2}V \sin (\omega t + 2\pi/3) \quad ...(13.64)$$

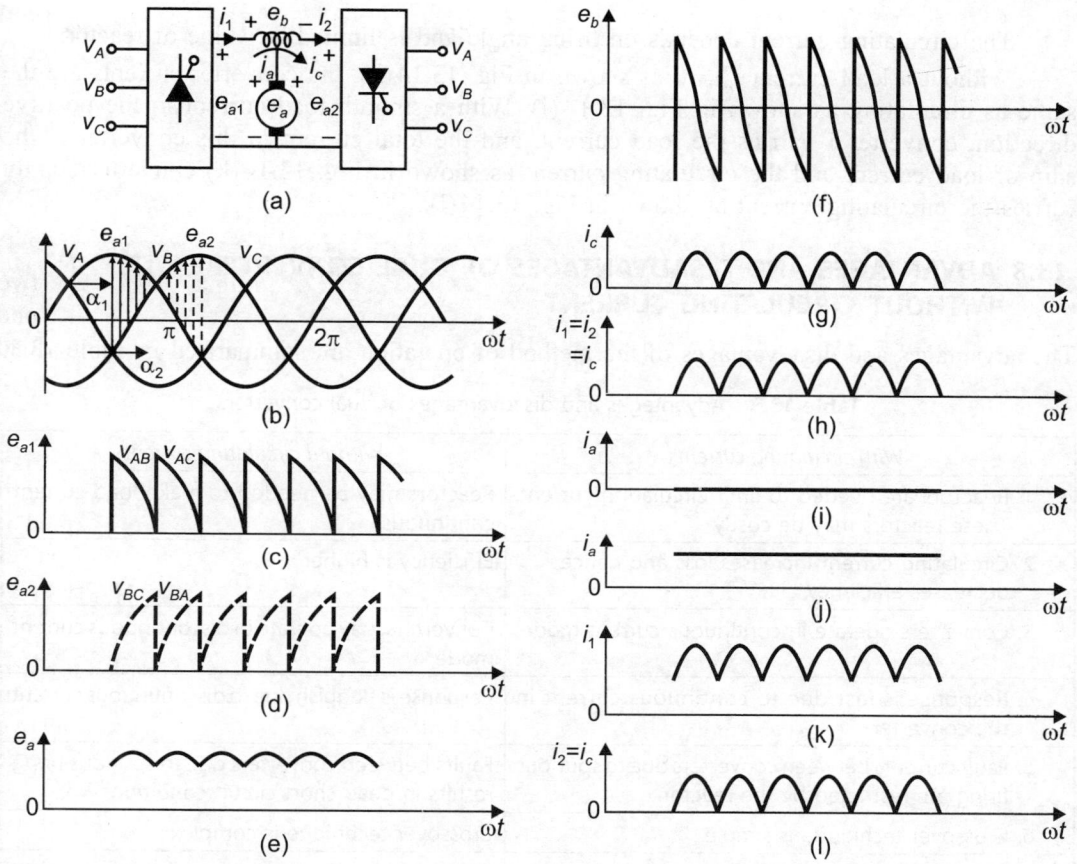

Fig. 13.14 Dual converter with circulating current

and
$$v_A + v_B + v_C = 0 \qquad \qquad ...(13.65)$$

During the interval
$$(\pi/6 + \alpha_1) < \omega t < (\pi/6 + \alpha_1 + \pi/3)$$

$$e_{a1} = v_A - v_B = v_{AB} \qquad \qquad ...(13.66)$$

$$e_{a2} = -(v_C - v_B) = v_B - v_C = v_{BC} \qquad \qquad ...(13.67)$$

$$e_r = e_{a1} - e_{a2} = v_A + v_C - 2v_B \qquad \qquad ...(13.68)$$

or
$$e_r = -3v_B$$

Circulating current is given by

$$i_c = \frac{1}{\omega L} \int_{\alpha_1 + \pi/6}^{\omega t} e_r \, d(\omega t)$$

$$= \frac{1}{\omega L} \int_{\alpha_1 + \pi/6}^{\omega t} -3\sqrt{2}V \sin(\omega t - 2\pi/3) \, d(\omega t)$$

$$= \frac{3\sqrt{2}V}{\omega L} [\cos(\omega t - 2\pi/3) - \cos(\alpha_1 - \pi/2)] \qquad \qquad ...(13.69)$$

The circulating current depends on firing angle and is limited by value of reactor.

Without a load current $i_a = 0$ as shown in Fig. 13.14(h), the converter currents are the same as circulating as shown in Fig. 13.14(i). With a smooth load current in the positive direction, converter 1 carries the load current, and the total current in this converter is the sum of load current and the circulating current as shown in Fig. 13.14(k) converter supply carries the circulating current as shown in Fig. 13.14(l).

13.8 ADVANTAGES AND DISADVANTAGES OF DUAL CONVERTERS WITH AND WITHOUT CIRCULATING CURRENT

The advantages and disadvantages of the method of operation are summarised in Table 13.3.

Table 13.3 Advantages and disadvantages of dual converters

With circulating current	Without circulating current
1. Reactors are needed to limit circulating current. These reactors may be costly	Reactors may be needed to make load current continuous
2. Circulating current increases loss and hence decreases efficiency	Efficiency is higher
3. Converters operate in continuous current mode	Converters may operate in discontinuous current mode
4. Response is fast due to continuous current in the converter	Response is sluggish due to discontinuous current
5. Fault currents between converters due to spurious firing are restricted by the reactor	Faults between converters due to spurious firing results in dead short circuit conditions
6. Crossover technique is simple	Crossover technique is complex
7. Converter loading is higher than the output load	Converter loading is the same as the output load
8. Linear transfer characteristics are obtained	Non linear transfer characteristics due to discontinuous currents are obtained.

13.9 POWER FACTOR, SUPPLY HARMONICS AND RIPPLE IN MOTOR CURRENT OF PHASE CONTROLLED DC DRIVES

Phase controlled DC drives have the following drawbacks:

1. **Low power factor:** When the motor is running at low speed, supply power factor is very poor because of low armature voltage.
2. **Distortion of supply:** Short circuit of lines cause short current pulses during commutation of thyristor which distorts the line voltages. This will produce radio frequency interference in communication equipment.
3. **Ripple in motor current:** Motor current has an AC ripple which superimposed on DC component. The ripple increases losses, derates motor and affects motor commutation. It causes discontinuous conduction which affects speed regulation.

13.10 CHOPPER CONTROLLED DC DRIVES

Chopper drives are widely used in traction applications and electric vehicles all over the world. A DC chopper is connected between a fixed voltage DC source and a DC motor to vary the armature voltage. A DC motor can be operated in one of the four quadrants by controlling the armature or field voltages. The possible control modes of DC chopper drives are:

1. Power control
2. Regenerative brake control and
3. Dynamic brake control.

Control of separately excited DC motors: Generally transistor choppers are preferred over thyristor, because they can be operated at such a higher frequency (2.5 to 10 kHz) than thyristors (upto 1 kHz). Because of lower voltage and current ratings of transistors use of transistor is restricted to 200 kW. For higher ratings, thyristor choppers are used. A transistor chopper controlled separately excited motor drive is shown in Fig. 13.15(a). A thyristor chopper is obtained when transistor is replaced by a thyristor with a forced commutation circuit. This is a one quadrant drive. Waveforms of motor terminal voltage v_a and armature current i_a for continuous conduction are shown in Fig. 13.15(b). Motor terminal voltage is V during on-period of the transistor $0 \leq t \leq t_{on}$. The operation is given by:

$$R_a i_a + L_a \frac{di_a}{dt} + E_b = V \quad (0 \leq t \leq t_{on}) \qquad ...(13.70)$$

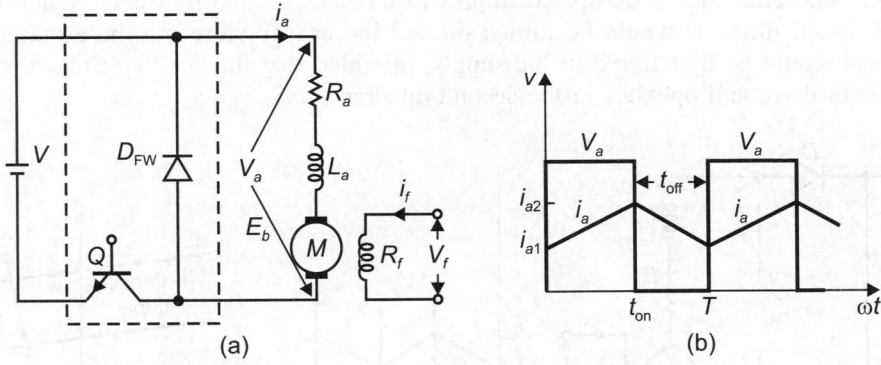

(a)   (b)

Fig. 13.15 Waveform of separately excited DC motor

In this interval, armature current increases from i_{a1} to i_{a2}. Since motor is connected to the source during this interval, it is called *duty interval*.

At $t = t_{on}$, transistor is turned off. Motor current freewheels through diode D_{FW} and motor terminal voltage is zero during interval $t_{on} \leq t \leq T$. Equation for this interval is given by

$$R_a i_a + L_a \frac{di_a}{dt} + E_b = 0, \quad (t_{on} \leq t \leq T) \qquad ...(13.71)$$

Motor current decreases from i_{a2} to i_{a1} during this interval.

Ratio of duty interval t_{ON} to chopper period T is called duty ratio or duty cycle (δ)

$$\delta = \frac{\text{duty interval}}{T} = \frac{t_{on}}{T}$$

From Fig. 13.15(b) $V_a = \frac{1}{T} \int_0^{t_{on}} V \cdot dt = \delta V = f \cdot t_{on} \cdot V \qquad ...(13.72)$

$$I_a = \frac{\delta V - E_b}{R_a} \text{ or } E_b = \delta V - I_a R_a$$

From the known equation of DC machines

Torque, $\qquad T = KI_a$ and $E_b = K\omega_m$

or $\qquad\qquad \omega_m = \dfrac{E_b}{K}$

$$\omega_m = \frac{\delta V - I_a R_a}{K}$$

$$= \frac{\delta V}{K} - \frac{R_a \cdot T}{K^2} \qquad\qquad ...(13.73)$$

13.11 REGENERATIVE BRAKING

Chopper for regenerative braking operation is shown in Fig. 13.16(a). Transistor Q is operated periodically with period T and on period t_{on}. Waveforms for motor voltage and armature current for continuous conduction are shown in Fig. 13.16(b). When transistor Q is on, current i_a increases from i_{a1} to i_{a2}. The mechanical energy converted into electrical by motor, now working as a generator, partly increases the stored magnetic energy in armature circuit inductance and remainder is dissipated in armature resistance and transistor. When transistor Q is turned off, diode D would be turned on and the energy stored in the armature circuit inductances would be transferred to the supply, provided that the supply is receptive. It is a one quadrant drive and operates in the second quadrant.

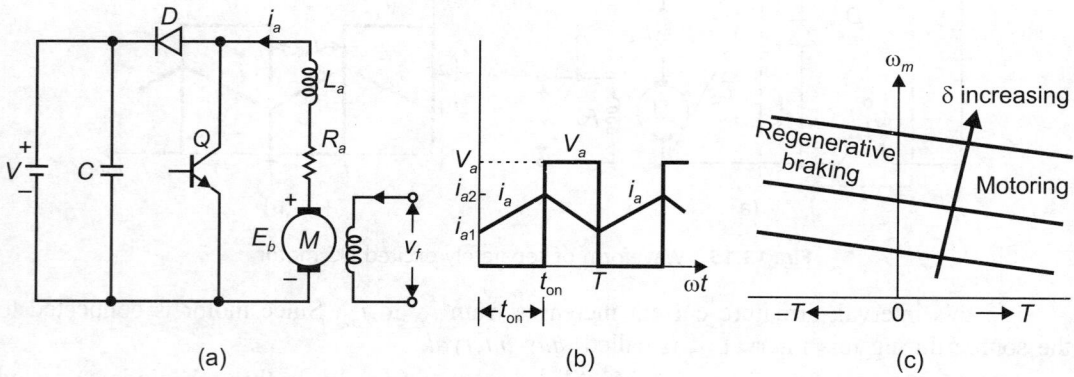

(a)	(b)	(c)

Fig. 13.16 Schematic of chopper for regenerative braking operation and waveforms

Average voltage across the transistor Q,

$$V_{avg} = \frac{1}{T} \int_{t_{on}}^{T} V \cdot dt = \delta V \qquad\qquad ...(13.74)$$

where duty ratio $\qquad \delta = \dfrac{\text{duty interval}}{T} = \dfrac{T - t_{on}}{T}$

$$I_a = \frac{E_b - \delta V}{R_a} \text{ or } E_b = \delta V + I_a R_a$$

Since armature current is reversed, torque,

$$T = -KI_a \text{ or } I_a = -T/K$$

From the equation $\qquad E_b = K\omega_m$

or $\qquad\qquad\qquad \omega_m = \dfrac{E_b}{K} = \dfrac{\delta V + I_a R_a}{K}$

$$= \dfrac{\delta V}{K} - \dfrac{T \cdot R_a}{K^2} \qquad\qquad ...(13.75)$$

Figure 13.16(c) shows speed torque characteristics.

A separately excited DC motor is stable in regenerative braking. The armature and field can be controlled independently to provide the required torque during starting. A chopper fed series and separately excited DC motors are both suitable for traction applications.

13.12 DYNAMIC BRAKING

In a dynamic braking, the energy is dissipated in a rheostat and it may not be a desirable feature. In mass rapid transit system, the energy may be used in heating the trains. This is a one quadrant drive and operates in the second quadrant. Figures 13.17(a) and (b) show dynamic braking and waveforms for voltage and current assuming that armature current i_a is continuous and ripple free. Energy E consumed by resistor R_b during a cycle of chopper operation is given by:

$$E = I_a^2 R_b (T - t_{on})$$

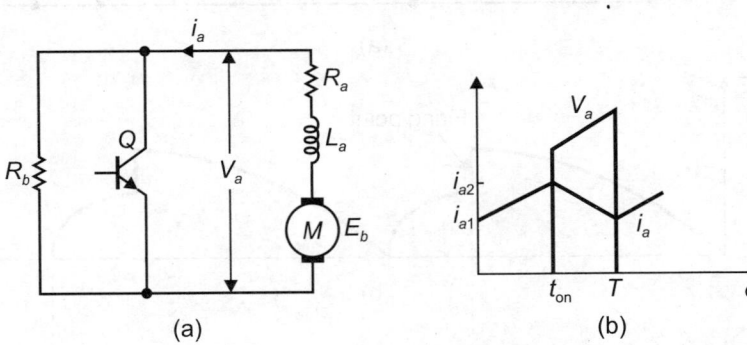

(a)   (b)

Fig. 13.17 Dynamic braking and waveforms

Average power consumed by R_b,

$$P = \dfrac{E}{T}$$

$$= I_a^2 R_b (1 - \delta) \qquad\qquad ...(13.76)$$

Effective value of R_b $\qquad = \dfrac{P}{I_a^2} = R_b (1 - \delta) \qquad\qquad ...(13.77)$

where $\qquad\qquad\qquad \delta = \dfrac{t_{on}}{T}$

From Equation (13.77), it is clear that the effective value of braking resistor can be changed steplessly from 0 to R_b as δ is controlled from 1 to 0. As the speed falls, δ can be increased steplessly to brake the motor at a constant maximum torque.

13.13 SPEED REGULATION BY DIFFERENT METHODS

There are various methods by which the speed of DC motors can be controlled using thyristors. Some of the methods which are frequently used are discussed below.

13.13.1 Speed Regulation by Armature Voltage Control

If the counter back emf is regulated then the motor speed must be constant except for the effect of armature reaction. The armature voltage differs from the counter emf E_b only by the amount of armature resistive drop $I_a R_a$, which is usually small, so that controlling the armature voltage will provide reasonably precise constant speed. This is the principle of basic circuit of Fig. 13.18(a).

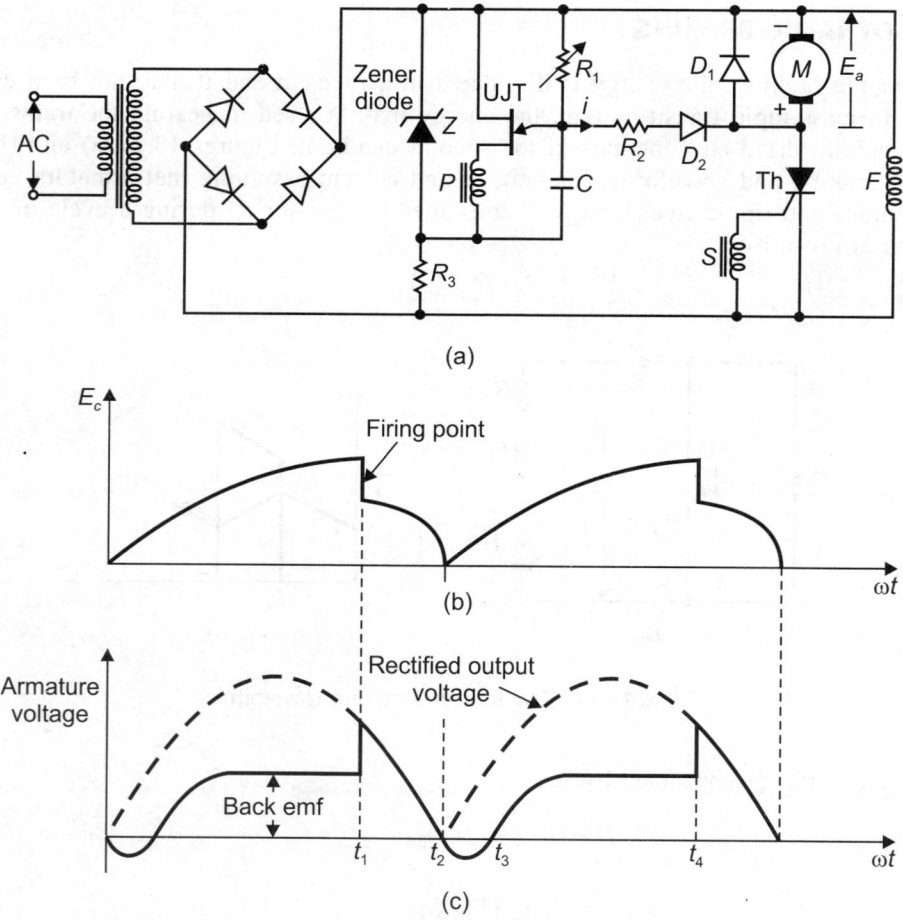

Fig. 13.18 Speed regulation by armature voltage control using UJT

Zener diode Z serves as a reference diode against which the sum of armature voltage E_a and voltage across the capacitor C is compared. Resistor R_1 sets the desired speed since it largely controls the charging rate of capacitor C. Diode D_2 prevents charging current from being supplied through the armature circuit unijunction transistor UJT fires whenever the voltage across capacitor C becomes equal to the Zener diode voltage multiplied by stand off ratio of UJT.

At time t_1, the thyristor is fired by UJT, a voltage is supplied to the armature until the end of rectified half cycle of the supply. The thyristor cuts off at voltage zero and the inductive energy stored in the armature decays by driving current through diode D_1 to time t_3. The drop across diode D_1 ensures that diode D_2 is blocked in that interval, and capacitor C charges through resistor R_1 as shown in Fig. 13.18(b). At time t_3, the inductive energy has dissipated and the armature voltage rises to the counter emf value E_a. Diode D_2 is now forward biased and the charging current is shunting from capacitor C through resistor R_2. The amount of this shunting action is dependent upon E_a, or the voltage across capacitor C is controlled so that its speed of rise is dependent upon the back emf. That is,

$$E_c + iR_2 + E_a = E_z \qquad \qquad ...(13.78)$$

and current D is subtracted from the charging current of capacitor C, is the balancing variable. Thus, the time of triggering the thyristor at t_4 is dependent upon and controlled by E_a or the speed of the motor.

Actually the time of triggering of the thyristor is also controlled by the motor output torque. With heavier mechanical load, the motor slows down, reducing E_a and drawing more current. The lowering of E_a reduces the shunting effect on capacitor C between t_3 and t_4 and allows C to charge faster and reach the firing condition earlier in the cycle. The heavier armature current, indicates increased load, means that more inductive energy is stored in the motor, and the decay period t_2 to t_3 is lengthened. But during this period, C is not shunted, and it charges at the maximum rate through R_1. Capacitor C discharges during the firing period t_1 to t_2 through diode D_2 and this thyristor thus resetting itself for speed and torque measurement in the next half cycle.

13.13.2 Speed Regulation by Armature Current Control

The torque varies as the input current in DC series motor, the current can be used to advance the thyristor firing angle to improve the speed regulation in a circuit suited for small motors, as shown in Fig. 13.19.

A simple version of solid state trigger circuit is shown in Fig. 13.19, the current uses a unijunction transistor UJT as the voltage sensitive trigger.

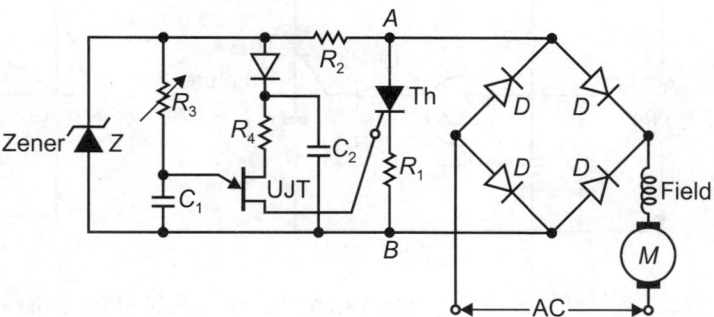

Fig. 13.19 Speed regulation using armature current control by UJT

The UJT breaks down when the emitter voltage to base 1 reaches a fraction of the voltage between the bases. The DC motor is connected to the AC supply using the bridge rectifier circuit. The bridge rectifier supplies rectified direct current to the control circuit. In turn, the thyristor Th in the control serves as a variable load on the DC output of the bridge. As the thyristor Th varies its loading effect by changing the firing angle, the current supplied to the motor is controlled. The bridge offers a low resistance in series with the motor, and the line voltage, less the drop across thyristor Th and resistor R_1, is applied to the motor. As the thyristor triggering time is delayed by manual adjustment of resistor R_3, the motor is slowed. As the control operates again to advance the triggering time, more current is supplied to the motor.

The Zener diode Z functions to clip the rectified voltage at a standard level and form the voltage which is applied to charging circuit $R_3 C_1$. Capacitor C_1 charges toward the voltage of

Zener diode Z. When it reaches the firing voltage of UJT, the unijunction transistor triggers the thyristor, capacitor C_1 then partially discharges through UJT and resistor R_1. However, while the thyristor is conducting, the voltage across A and B falls below the Z level, and the voltage applied to capacitor C_1 through resistors R_2 and R_3 depends upon the voltage at points A and B which is in itself dependent upon the motor current.

The larger the motor current during the conduction interval, the higher will be the voltage across capacitor C_1 when the thyristor ceases conduction at the end of the AC cycle. Capacitor C_1 then charges toward the Zener level and triggers UJT again. A high voltage across capacitor C_1 at the end of the thyristor conducting period means that it very quickly reaches the unijunction-transistor triggering level after the thyristor opens, and thus it fires the thyristor sooner in the next half cycle. It can be seen that the larger the motor current, the earlier the thyristor triggers, which supplies more current to maintain the motor speed at the increased torque, high value of resistor R_1 causes the speed to rise with increasing torque and lead to instability. The range of control is less than 180° and is of the order of 120°.

13.13.3 Speed Control of DC Series Motor

Speed control of DC series motor is achieved using a simple half wave circuit of Fig. 13.20(a). In this, thyristor Th is fired during the positive half cycle, which will supply a positive average voltage across the armature. The resistor R will control the rate of rise of voltage across the capacitor. When this voltage v_c becomes equal to armature voltage v_a plus the breakdown voltage V_{BR} of the diac, thyristor Th will get triggered.

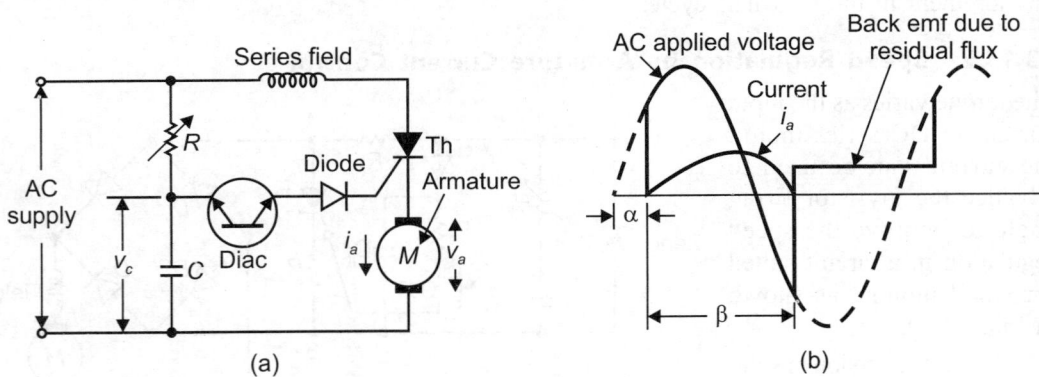

Fig. 13.20 Speed control of DC series motor

When thyristor Th is not conducting, voltage v_a across the armature will be due to residual flux. This is approximately proportional to speed. Thus, the firing angle α can be changed by varying resistance R, which in turn will change the speed of the motor. For a given value of R, let the speed fall below the set speed due to addition of load on the motor. Their voltage v_a across the armature will be low because of the reduction in speed. Therefore, in the next positive half cycle voltage v_a across the capacitor will be equal to $(v_a + V_{BR})$, much earlier than before, and the thyristor will get triggered. That is, the firing angle will be reduced and because of the increased average voltage across the armature, the motor speed will increase. Similarly, if the motor speed rises when the load is thrown off, the firing angle will automatically advance bringing down the speed of the DC motor.

13.13.4 Speed Control of Separately-excited DC Shunt Motor

Speed of separately excited DC motor can be controlled using a simple circuit of Fig. 13.21, where the counter emf in the armature is steady and circuit inductance is small, so that the current is discontinuous and does not spill into the next half cycle. The voltage and current waveforms are shown in Fig. 13.21 (b). The thyristor Th is fired at an angle α.

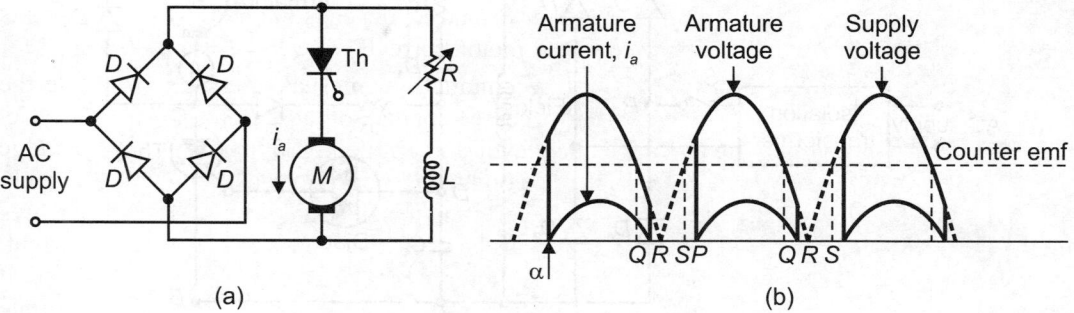

(a)   (b)

Fig. 13.21   Speed control of separately excited DC shunt motor

At point Q, the input voltage will be equal in magnitude to the counter emf. However, as the inductance of the armature is small, the current will go to zero, a little beyond Q and the thyristor Th will be turned off at R. Reverse bias will appear on the thyristor Th during the period from R to S. This period must be longer than the turn-off time of the thyristor Th. It is for this reason that such a circuit with large firing angles may not work properly as the counter emf will be low and the period from R to S will be shorter than it would be for smaller angles. If the current becomes zero beyond point S due to large armature inductance, then the thyristor Th will not turn off and control will be lost.

13.13.5   Voltage Dependent Resistor Method of Speed Regulation of DC Shunt Motor with Changes in Supply Voltage

The field and armature circuits will be effected by the increase in input supply voltage. In the circuit illustrated in Fig. 13.22 (a), the capacitor C_1 charges to the voltage across thyristor Th which is the voltage difference between the bridge rectifier output and the counter-emf across the armature. When the line voltage increases, it will lead to higher charging voltage and the motor will tend to speed up. But this tendency is partially counteracted by the fact that the increased voltage appears also across the field winding and leads to increase in field current and thus slows down the motor, as speed is inversely proportional to field flux.

At reduced speed, the effective line voltage compensation is provided by the circuit of Fig. 13.21 (b), which is almost identical with that of Fig. 13.21 (a) except for the addition of components R_3, R_4, C_2 and R_5. In this circuit, the component R_3 is a thyrite, which essentially is a voltage dependent resistor (VDR) or varistor. The resistance of varistor decreases when the voltage impressed across it increases and vice-versa. The characteristic curve of varistor has the characteristic that its resistance decreases exponentially with an increase in voltage and the current increases exponentially with an increase in voltage.

In the network consisting of components R_5, C_2 and R_3. The resistor R_5 is very small, capacitor C_2 effectively charges through R_3 toward the output voltage delivered by the bridge rectifier. The effect of an increased line voltage is to decrease exponentially the resistance of R_3, so that there is also an exponential increase in charging current of C_2. Because of the exponential increase in charging current, the rate of change of capacitor C_2 is relatively higher than the rate of increase in line voltage. The thyristor Th is turned ON when capacitor C_1 has charged to the breakover voltage of the diac. However, the thyristor Th is turned ON when capacitor C_1 has charged to the breakover of the diac plus the voltage across resistor

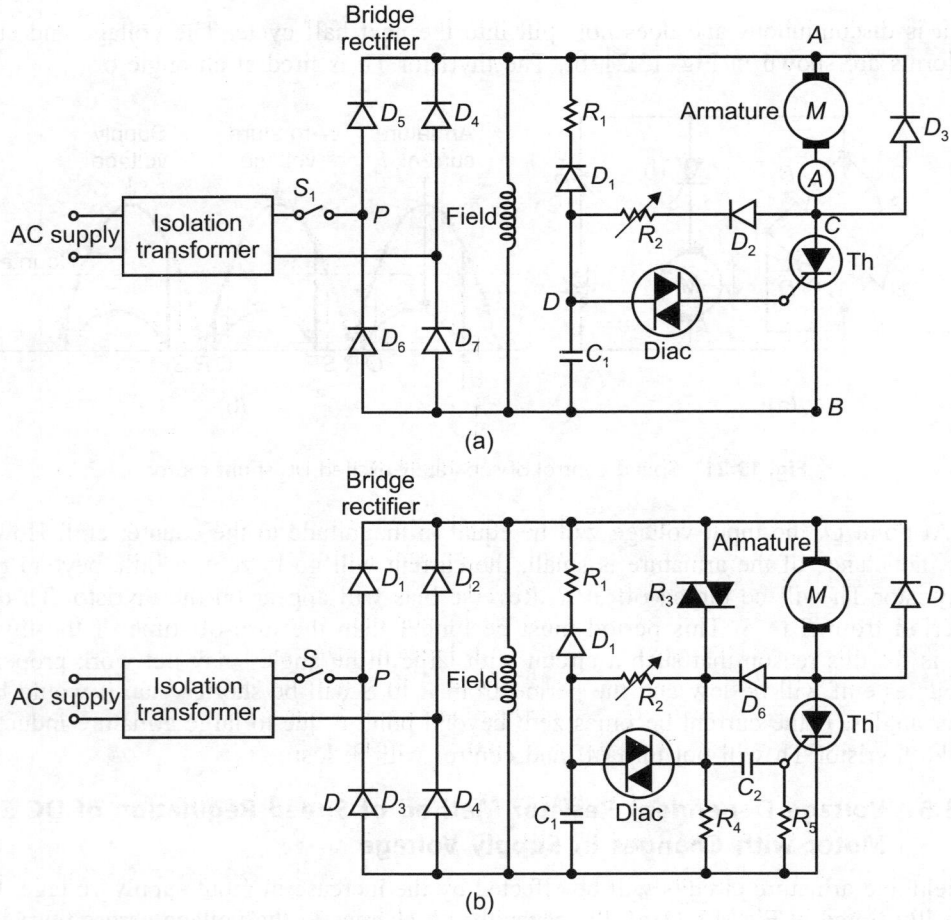

Fig. 13.22 Speed control of DC shunt motor using voltage dependent resistor

R_4. But the voltage across resistor R_4 is the same, effectively, as the voltage across capacitor C_2. Therefore, capacitor C_1 must reach the diac breakover voltage plus the voltage across capacitor C_2 before the thyristor Th is fired.

Let us study the effect of an increase in line voltage. The capacitor C_1, at the start of the half cycle, will charge at the faster rate, because it is charging toward a higher voltage. However, because the charge across capacitor C_2 is even greater than the line voltage, C_1 will have to reach a much higher voltage before the thyristor Th can be turned ON. Triggering time is, therefore, delayed reducing the power applied to the armature, hence offsetting the tendency of motor speed to increase with an increase in the line voltage.

13.13.6 Speed Regulation Using Saturable Reactor

In Fig. 13.23, a separately excited DC motor 'M' is fed through a centre tapped transformer whose speed is to be controlled. The voltage across resistor R_1 is the back emf, which is proportional to the actual speed of the motor. Part of this voltage is compared with a fixed reference voltage from resistor R_2. The error voltage is applied to the DC amplifier consisting of transistors T_1 and T_2. Depending upon the value of the error voltage, this amplifier controls the DC current applied to the saturable reactor.

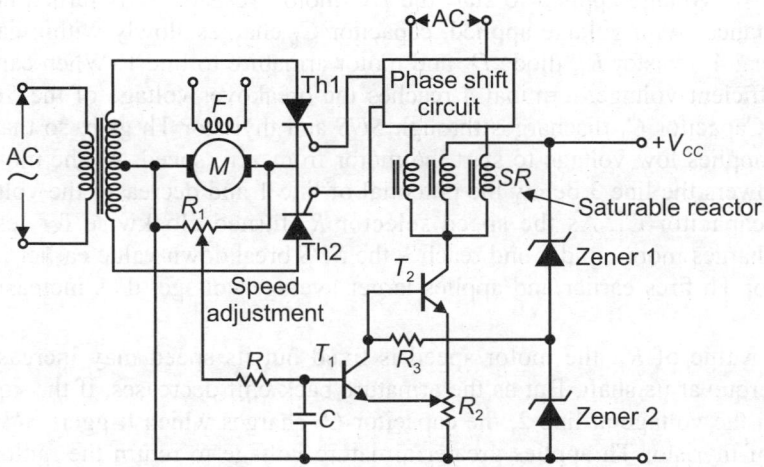

Fig. 13.23 Speed regulation of separately excited DC motor using saturable regulator

Let the speed of the motor is below the set value. Then, the error voltage will be less in such a way that transistor T_1 is less conducting. As a result, its collector current decreases and its collector voltage V_{c1} increases. This voltage is the input to transistor T_2, and therefore, its collector current I_{c2} will increase. This current passes through the DC winding of saturable reactor which gets more saturated. If current in saturable reactor increases, it becomes more saturated and its reactance decreases, so that firing angle decreases and thus the average current supplied by controlled rectifier will increase. As a result the speed of the motor increases.

The magnetic amplifier circuit has two features. First, the firing circuit can be made to be responsive to several signals by introducing them into separate control windings. Second, the pulse delivered to the gate has a width of $(180° - \alpha)$ so that the thyristor will refire during the interval if required to do so. However, the magnetic amplifier does introduce a time delay that complicates the compensation design of a feedback system.

13.13.7 Motor-speed Regulation by One Thyristor

The speed of 2.5 kW DC shunt motor may be regulated using the circuit of Fig. 13.24. The AC power is supplied to a bridge rectifier and supply steady shunt field current. The thyristor Th is fired in each half cycle, at a phase angle adjusted by R_1, thereby setting the motor speed. The thyristor Th and *SUS* are reset as each half wave of voltage drops to zero.

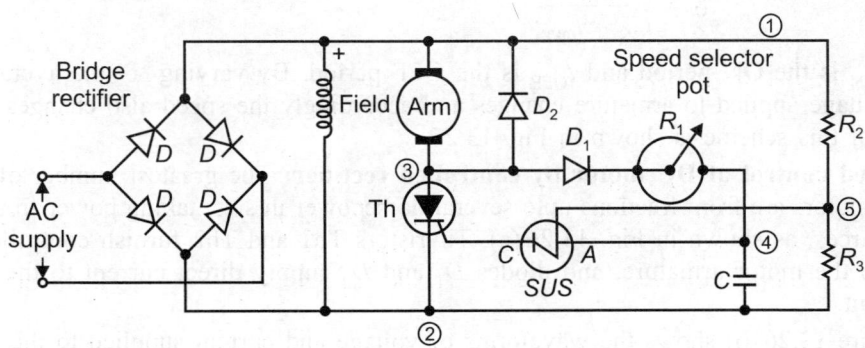

Fig. 13.24 Speed control of DC shunt motor using one thyristor

Prior to AC voltage applied to start the DC motor, resistor R_1 is turned counterwise to insert its resistance. With voltage applied, capacitor C_1 charges slowly within each half cycle through terminal 4, resistor R_1, diode D_1 and motor armature to line 1. When capacitor C_1 has charged to sufficient voltage, terminal 4 reaches the breakover voltage of the *SUS*, then *SUS* will conduct. Capacitor C_1 discharges through *SUS* and thyristor Th gate, so that thyristor Th fires late and applies low voltage to start the motor from zero speed. As the motor rotates, its counter emf lowers the line 3 below the potential of line 1 and decreases the voltage available for charging capacitor C_1. As the speed selector R_1 turned clockwise for less resistance, capacitor C_1 charges more rapidly and reaches the *SUS* breakdown value earlier in the AC half cycle. Thyristor Th fires earlier and applies larger average voltage, thus increasing the motor speed.

At fixed value of R_1, the motor speed is fixed but its speed may increase because of greater load torque at its shaft. But as the armature back emf decreases, if the voltage at point 3 is more than the voltage at line 2, the capacitor C_1 charges which triggers *SUS* sooner. The earlier firing of thyristor Th applies greater armature voltage to return the motor speed to its desired value. Thus, the speed is regulated to offset changes in load.

When the voltage at line 1 drops to zero between half cycles, the energy stored in the motor's inductance produces a positive voltage at line 3, that would force current to flow through thyristor Th to prevent its being reset. Therefore, diode D_2 is put so that this decreasing armature current can circulate through diode D_2, letting thyristor Th reset, ready to be fired in the next half cycle. The *SUS* acts as Zener diode and also a PNP transistor whose base connects to the *SUS* gate *G*.

13.13.8 DC Chopper Speed Control

This method of speed control has become very popular in the last few years mainly due to the advantage of solid state components which could be used as fast rate switching devices. In this method, the available AC supply is first rectified into DC using plain rectifiers and is then filtered by suitable loop reactor in the output. This DC is then fed to a chopper which allows the DC to pass through for a specific time interval. During the time, when a positive voltage is applied to the armature, the machine accelerates. As soon as the pulse is over, the declaration process starts and continues till the start of the next pulse.

If cycles are repeated continuously at a definite frequency and the elements of the cycle are maintained in a fixed relationship, the motor will run at definite average speed. The speed will change as soon as the relative values of ON and OFF periods are altered. Let δ be the duty ratio:

$$\delta = \frac{t_{ON}}{t_{ON} + t_{OFF}} \qquad \qquad ...(13.79)$$

where, t_{ON} is the ON period and t_{OFF} is the OFF period. By varying 'δ' the average value of DC voltage applied to armature changes and accordingly the speed also changes. A block diagram of this scheme is shown in Fig. 13.25.

Speed control of DC motor by controlled rectifier: The greatest number of rectifier supplied motors are from fractions upto several horsepower in size, taking power from single-phase sources, as shown in Fig. 13.26 (a). Thyristors Th1 and Th2 furnish controlled direct current to the motor armature, and diodes D_1 and D_2 supply direct current to the separate field circuit.

Figure 13.26 (b) shows the waveforms of voltage and current supplied to the armature with a firing angle θ_1, delayed so that conduction is discontinuous even though armature

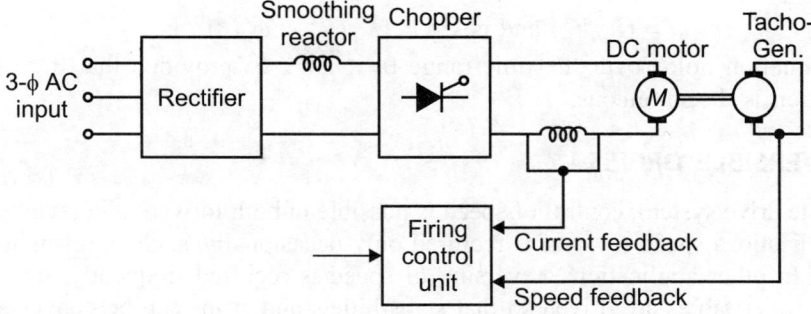

Fig. 13.25 Speed control of DC motor using chopper

inductance is present. Just before the instant of firing at θ_1, the current is zero and the armature voltage is equal to the counter emf E_b. At θ_1 rectifier thyristor Th1 fires, and the armature voltage jumps to v_1. The armature current then builds up slowly, owing to the armature inductance, and carries over to angle θ_2, when conduction ceases and the voltage jumps to E_b.

From the circuit, it is apparent that the net voltage available to build up current in the circuit is only $v_1 - E_b$ or $V_m \sin \omega t - E_b$. Only during the interval from θ_1 to θ_x at which $V_m \sin \omega t = E_b$ can be applied to produce a positive current in the circuit. The remainder of the conduction period from θ_x to θ_2 is due to the $L di/dt$ potential arising from the decreasing current in the armature inductance.

If θ_1 is advanced, the average voltage will increase and the speed will rise. If the torque load on the motor

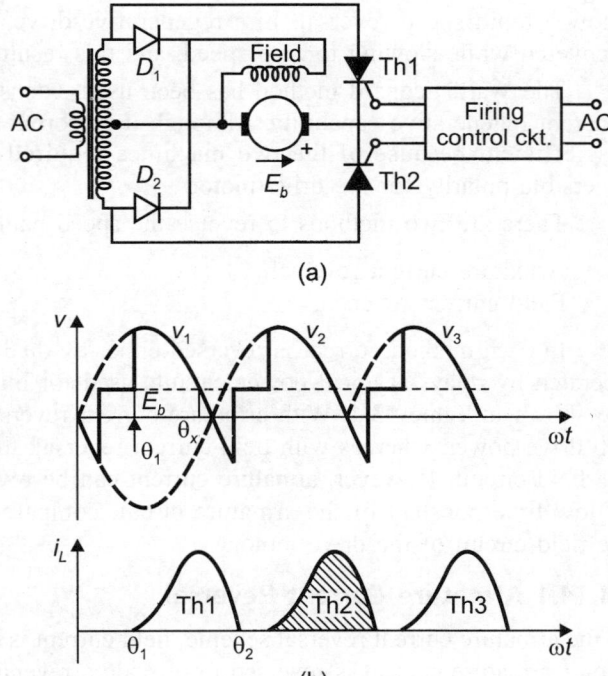

Fig. 13.26 Speed control of DC motor using controlled rectifier

rises, the armature current must increase, and this can happen only as a result of a reduced E_b value since the firing angle θ_1 is fixed. Since current can flow for only a short interval, the emf E_b must drop further than it would, if the motor were supplied continuously from a DC line. Since back emf E_b is proportional to speed, the result of the increase in load torque is an adverse effect on speed regulation.

The differential equation for conduction is

$$\frac{L di_L}{dt} + R_a i_L = V_m \sin \omega t - E_b$$

which has a solution

$$i_L = \left[\cos \phi \sin (\omega t - \phi) - a + \{a - \cos \phi \sin (\theta_1 - \phi)\}\, e^{-(R_a/\omega L)\,(\omega t - \theta_1)} \right]$$

where, $a = (E_b/V_m)$ and $\cos \phi = [R_d/(R_a^2 + \omega^2 L^2)^{1/2}]$

This equation holds over the time range $\theta_1 \leq \omega t \leq \theta_2$ provided that $\theta_1 > (\theta_2 - \pi)$ or that conduction is discontinuous.

13.14 REVERSIBLE DRIVES

In a reversible drive system, control of speed is possible in both forward and reverse directions. In some applications speed reversal is required only occasionally, such as reversing a subway car or train. In other applications, reversing of speed is required frequently, for example, in steel works, roller tables are driven so that stabs billets and strips can be conveyed from one stage of manufacture to the next. In hot strip mills, a reversible drive makes the hot strip pass through the mill back and forth several times until it is shaped into a desired form and length. In a drive system, where speed reversal is frequent, a regenerative drive is desired to allow a rapid speed reversal. In a regenerative drive, kinetic energy of the moving mass is recovered while slowing to zero speed, and this results in higher efficiency.

The Ward Leonard method has been used extensively as a reversible drive. It provides inherent regenerative capability. Although it performs satisfactorily, this system is costly and less efficient because of the two machines required to provide variable DC voltage with reversible polarity for the drive motor.

There are two methods to reverse the speed namely:

1. Armature current reversal
2. Field current reversal.

In each method current is reversed either by dual converter or by changeover contactors operated by relays. Typical power circuits used for high power reversing regenerative drives are shown in Table 13.4. With armature current reversal, the power circuit has to handle the full drive power, whereas with field current reversal, the power is controlled at a low level in the field circuit. However, armature current can be reversed faster than field current because of low time constant of the armature circuit compared to the much larger time constant of the field circuit of the drive motor.

13.14.1 Armature Current Reversal

In the armature current reversal scheme, field current is kept constant during the speed reversal. Motor armature current is reversed to provide a reversing torque, which, in turn, will reverse the speed. Armature current is reversed by either: (1) a converter and changeover contactors or (2) a dual converter.

The reversing drive system using changeover contactors is cheaper because only one converter is used. The dual converter is more expensive because two converters are used. If speed reversal is required frequently as in steel mills, the dual converter system is preferred, as contactor type system will require frequent maintenance and occasional replacement of contactors.

13.14.2 Field Current Reversal

To reverse motor speed, in the field current reversal scheme, the direction of armature current is unchanged and the field current is reversed. The power in the field circuit is much lower than the power in the armature circuit and is never more than a few kW even in large power drive system. Field current reversal is therefore cheaper compared to armature current reversal. The field current can be reversed either (1) by using a dual converter or (2) by a single converter with changeover contactors.

Table 13.4 Reversible drives

Type of reversal	Method of reversal	Power circuit	Features
Armature current reversal	Change over contactors		Good response, moving contactors require maintenance, armature zero current detection required for contactor changeover
	Dual converter		Fastest response, both converters are simultaneously fired. Losses in the two converters decrease efficiency
	Dual converter		Fast response, one converter to be fired at a time. Current detection and delay required to prevent firing of both converters together
Field current reversal	Dual converter		Cheaper
	Change over contactors		Response depends on field forcing

13.15 SELECTION OF THE DRIVES FOR SPEED REVERSAL

To select a drive system from these alternatives for use in a particular application, one must consider these factors:

1. Frequency of speed reversal
2. Capital cost
3. Response time
4. Efficiency
5. Maintenance
6. Reliability

For applications requiring frequent speed reversal, systems using dual converters are better suited because of rapid current reversal and minimum maintenance. Time response of current reversal is fast for dual converter in the armature circuit.

The field reversal method is less expensive because the power involved is at low level. The cost for a non regenerative drive has been considered as a reference. Note that selection of the controller significantly influences the capital cost.

Reversing methods using changeover contactors require considerable maintenance. A drive with better response time is generally more complex. To obtain faster response, additional components are necessary. They are normally either electronic or solid state in nature. Fortunately, the reliability of electronic equipment is very high.

13.16 25 kV AC TRACTION DRIVES USING SEMICONDUCTOR CONVERTER CONTROLLED DC MOTORS

Traditionally, the DC series motor has been used for traction drives. It has an ideal torque speed characteristic of high torque giving fast acceleration at low speeds, coupled with high speed motoring for cruising.

Earlier, electrified railway systems were fed by direct voltage either, via an overhead catenary, or by a third rail. For the high density short distance urban railway systems DC traction has been used. However, for long distances intercity railway electrification has been with single-phase alternating voltage fed from the public supply system at 50 Hz with voltage levels at 25 kV from the catenary.

An important consideration with the railway system is the harmonic current levels in the overhead catenary. The close and parallel proximity of the signalling cables to the railway lines mean that careful screening of these cables has to be arranged so as to avoid interference.

By the use of thyristors to control the traction motors in a railways locomotive, tappings on the transformer can be avoided, and a more accurate control of the motors is possible to utilize maximum wheel-to-rail adhesion without slipping. However, two main considerations with thyristors are to avoid harmonics and low power factor.

Figure 13.27 shows the basic layout of a typical locomotive arrangement. Four drive motors are shown, each being on a different axle. The armatures are connected in series-parallel arrangement fed from a unidirectional rectifier such that control is possible from zero to full voltage. The fields are supplied via fully controlled circuits, so that field weakening can be used to obtain high speeds and also to permit equal load sharing between the parallel groups. The armatures of the motors shown in Fig. 13.27 are fed from three series connected half controlled single-phase bridge rectifiers. For starting only bridge A is fired, firing being advanced as the speed and hence armature voltage builds up. When bridge A is full conducting

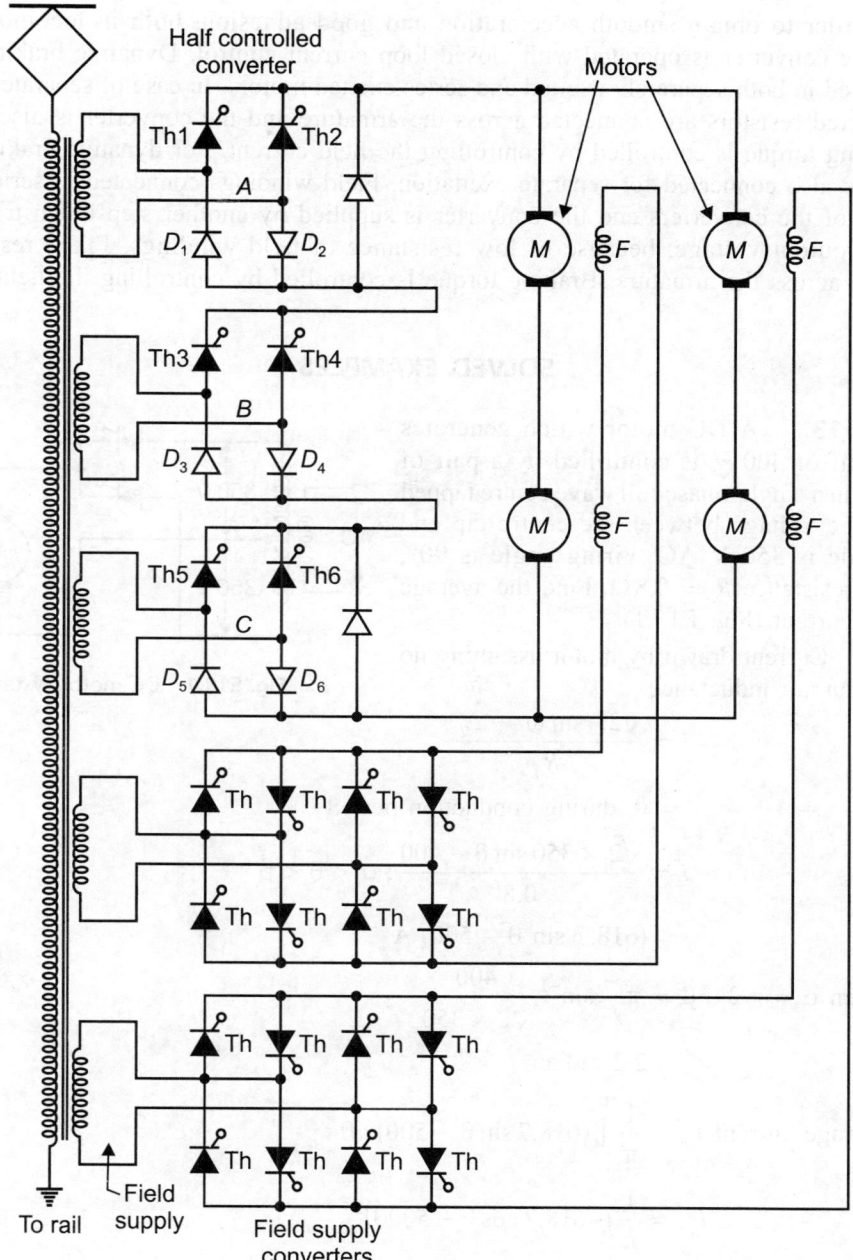

Fig. 13.27 Layout of typical locomotive arrangement

($\alpha = 0°$) then bridge B is brought into conduction until finally bridge C is fired, and its firing advanced until maximum armature voltage is reached. During the starting period the field currents are set to maximum. The purpose of using three bridges is to improve the power factor. In the above layout four separately excited motors are shown. The number of motors depends on ratings and they are connected in different combinations depending on application. Electrical multiple units (EMUs) in suburban traction employs four series motors connected in two parallel pairs with each pair having two series connected motors.

In order to obtain smooth acceleration and good adhesion, both in locomotives and EMUs, the converter is operated with closed loop current control. Dynamic braking can be incorporated in both separately excited and series excited motors. In case of separately excited motors, fixed resistors are connected across the armature and the converter is disconnected. The braking torque is controlled by controlling the field current. For dynamic braking, series motors are also connected for separate excitation. Field windings connected in series are fed from one of the converters and the converter is supplied by another step down transformer with low output voltage, because of low resistance of field windings. Fixed resistors are connected across the armature. Braking torque is controlled by controlling the field current.

SOLVED EXAMPLES

Example 13.1 A DC motor which generates a back emf of 400 V is controlled by a pair of thyristors in a single-phase full wave centre tapped circuit. The voltage between the centre tap and each anode is 350 V AC. Firing angle is 90°, armature resistance $R_a = 0.8\,\Omega$. Find the average armature current (Fig. E13.1).

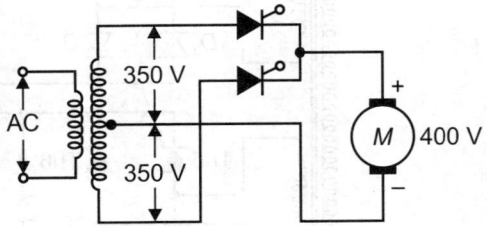

Fig. E13.1 DC motor system

Solution Current drawn by motor assuming no armature circuit inductance

$$i = \frac{\sqrt{2}V \sin \omega t - E_b}{R_a}$$

$$= 0, \text{ during conduction period}$$

$$i = \frac{\sqrt{2} \times 350 \sin \theta - 400}{0.8}; \alpha < \theta < \beta$$

$$= (618.7 \sin \theta - 500) \text{ A}$$

Given $\alpha = \pi/2$, $\beta = \pi - \sin^{-1} \dfrac{400}{\sqrt{2} \times 350}$

$$= 2.2 \text{ radian}$$

Average current $I_{av} = \dfrac{1}{\pi} \displaystyle\int_{\alpha}^{\beta} (618.7 \sin \theta - 500) \, d\theta$

$$I_{av} = \frac{1}{\pi} [-618.7 \cos \theta - 500\theta]_{\pi/2}^{2.2}$$

$$= \frac{1}{\pi} [364 - 500 \, (2.2 - \pi/2)]$$

$$I_{av} = 15.76 \text{ A}.$$

Example 13.2 A 15 HP, 220 V, 2000 rpm separately excited DC motor controls a load requiring a torque of $T_L = 45$ Nm at a speed of 1200 rpm. The field circuit resistance is $R_f = 147\,\Omega$, armature circuit resistance $R_a = 0.25\,\Omega$, voltage constant of the motor $K_a = 0.7032$ V/A rad/sec. Field voltage $V_f = 220$ V. Viscous friction and no load losses are negligible. Armature current may be assumed continuous and ripple free. Determine: (i) back emf, (ii) required armature voltage V_a, and (iii) rated armature current of the motor.

Solution

$$R_a = 0.25\,\Omega,\ R_f = 147\,\Omega,\ V_f = 220\text{ V}$$

Motor voltage constant $K_a = 0.7032$ V/A rad/sec

Load torque, $\quad T_d = T_L = 45$ Nm, $\omega = \dfrac{2\pi \times 1200}{60} = 125.66$ rad/sec

$$I_f = \frac{V_f}{R_f} = \frac{220}{147} = 1.497\text{ A}$$

Armature current, $\quad I_a = \dfrac{T_d}{K_a \Phi} = \dfrac{T_d}{K_a I_f} = \dfrac{45}{0.7032 \times 1.497} = 42.75$ A

(i) Average back emf, $E_b = K_a \Phi_n$ $n = 0.7032 \times 1.497 \times 125.66 = 132.28$ V

(ii) Armature voltage, $\quad E_a = E_b + I_a R_a$

$$= 132.28 + 0.25 \times 42.75 = 142.97\text{ V}$$

(iii) Since $\quad\quad\quad$ 1 HP = 746 W

$$I_{rated} = \frac{15 \times 746}{220} = 50.87\text{ A}.$$

Example 13.3 The speed of a 15 HP, 1400 rpm, separately excited DC motor is controlled by a single-phase full converter, the rated current of the motor is 40 A and armature resistance is $0.3\,\Omega$ and supply voltage is 260 V. Motor voltage constant $(K_a \Phi) = 0.182$ V/rpm. Calculate: (i) motor torque, (ii) speed of the motor, and (iii) power factor of supply. Assuming that there is sufficient inductance in the armature so that the motor current is continuous and ripple free

Solution Motor voltage constant

$$K_a \Phi = 0.182\text{ V/rpm}$$

$$= \frac{0.182 \times 60}{2\pi}\text{ V} \cdot \text{sec/radian} = \frac{1.74\text{ V sec}}{\text{rad}}$$

(i) Average torque, $\quad T_{av} = K_a \Phi I_a = 1.74 \times 40 = 69.6$ Nm

For full converter, $\quad E_a = \dfrac{2\sqrt{2}V \cos \alpha}{\pi}$

$$= \frac{2\sqrt{2} \times 260}{\pi} \cos 30° = 202.82\text{ V}$$

$$E_b = E_a - I_a R_a = 202.82 - (40 \times 0.3)$$

$$= 190.82\text{ V}$$

(ii) Average speed, $\quad N = \dfrac{E_a - I_a R_a}{K_a \Phi} = \dfrac{190.82}{0.182}$

$$= 1048.46 \approx 1048\text{ rpm}$$

(iii) Supply volt-ampere $\quad U = VI$

I is the rms supply current $= 40$ A

$$U = 260 \times 40 = 10400\text{ VA}$$

Neglecting the losses in the converter, power from the supply is

$$P_s = E_a I_a = 202.82 \times 40 = 8112.8\text{ W}$$

Supply power factor PF $\quad = \dfrac{P_s}{U} = \dfrac{8112.8}{10400} = 0.78.$

Example 13.4 The speed of a 15 HP, 1200 rpm, separately excited DC motor is controlled by a single-phase converter, the rated current of the motor is 38 A and armature resistance is $0.3\,\Omega$ and supply voltage is 260 V. The polarity of the motor back emf E_b is reversed by reversing the field excitation. Calculate: (i) the firing angle to keep the motor current at its rated value, and (ii) the power fed back to the supply.

Solution

(i)
$$E_a = \frac{2\sqrt{2}}{\pi} V \cos \alpha = \frac{2 \times \sqrt{2} \times 260 \cos 30°}{\pi} = 202.82 \text{ V}$$

The back emf at the time of reversing the polarity
$$E_b = E_a - I_a R_a = 202.82 - (38 \times 0.3) = 191.42 \text{ V}$$

Average armature voltage is
$$E_a = E_b + I_a R_a = -191.42 + (38 \times 0.3) = -180.02 \text{ V}$$

As
$$E_a = \frac{2\sqrt{2}V \cos \alpha}{\pi}$$

$$-180.02 = \frac{2\sqrt{2} \times 260}{\pi} \cos \alpha$$

or $\cos \alpha = 140.2°$

(ii) Power from the DC machine is
$$P_b = E_b I_a = 191.42 \times 38 = 7273.96 \text{ W}$$

Power lost to armature resistance is
$$P_a = I_a^2 R_a = 38^2 \times 0.3 = 433.2 \text{ W}$$

Power fed back to AC supply,
$$P_s = P_b - P_a$$
$$= (7273.96 - 433.2)$$
$$= 6840.76 \text{ W} = 6.84 \text{ kW}$$

or $P_s = E_a I_a = 180.02 \times 38 = 6840.76 \text{ W}.$

Example 13.5 A single-phase full converter is used to control the speed of a 7.5 kW, 220 V, 1800 rpm separately excited DC motor. The armature resistance is $0.25\,\Omega$ and the rated armature current is 20 A. The AC voltage is 250 V. The motor voltage constant is $K_a \Phi = 0.0278$ V/rad/min. For a firing angle $\alpha = 30°$ and rated motor current, determine: (i) the speed of the motor, and (ii) motor torque.

Assume continuous and ripple free armature current.

Solution For full converter, the voltage across the armature,
$$E_a = \frac{2\sqrt{2}V}{\pi} \cos \alpha = 2 \times \frac{\sqrt{2} \times 250}{\pi} \cos 30° = 194 \text{ V}$$

The back emf $\quad E_b = E_a - I_a R_a = 195 - (20 \times 0.25) = 190 \text{ V}$

(i) Angular speed $\quad \omega = \dfrac{E_a - I_a R_a}{K_a \Phi} = \dfrac{190}{0.0278} = 6835$ radian/min
$$= 113.9 \text{ rad/sec}$$

(ii) Average developed torque
$$T_{av} = K_a \Phi I_a = 0.0278 \times 60 \times 20$$
$$= 33.36 \text{ Nm}.$$

Example 13.6 The speed of a 15 HP, 480 V, 1200 rpm separately excited DC motor is controlled by a single-phase full wave bridge circuit. The rated motor armature current is 35 A, and the armature resistance is $0.15\,\Omega$. The AC supply voltage is 480 V. The motor back emf constant is $K_b\Phi = 0.45$ V/rpm. The motor current is continuous and ripple free. A rectifier operation, for a firing angle $\alpha = 60°$ and rated armature current, calculate: (i) torque, (ii) speed, and (iii) supply power factor.

Solution

(i)
$$K_b\Phi = 0.45 \text{ V/rpm}$$
$$= \frac{0.45 \times 60}{2\pi} \text{ V rad/sec} = 4.297 \text{ V rad/sec}$$
$$T = K_b\Phi I_a = 4.297 \times 35 = 150.4 \text{ Nm}$$

(ii)
$$E_a = \frac{2\sqrt{2}V}{\pi}\cos\alpha = \frac{2\sqrt{2} \times 480}{\pi}\cos 60° = 216.1 \text{ V}$$

$$E_b = E_a - I_aR_a = 216.1 - (35 \times 0.15) = 210.85 \text{ V}$$

Speed
$$N = \frac{E_b}{K_b\Phi} = \frac{210.85}{0.45} = 468.6 \text{ rpm}$$

(iii)
$$I_a = 35 \text{ A}$$
Input volt amp.
$$= VI = 480 \times 35 = 16800 \text{ VA}$$
Real power
$$= E_aI_a = 216.1 \times 35 = 7563.5 \text{ watts}$$
Power factor
$$= 7563.5/16800 = 0.45.$$

Example 13.7 In the above Example 13.6, in the inverter-operation, if the motor back emf polarity is reversed by reversing the field excitation, calculate: (i) the firing angle to keep the motor current at its rated value, and (ii) The amount of power fed back to the supply.

Solution

(i) At the instant of polarity reversal, the back emf is
$$E_b = 210.85 \text{ V}$$

For the generator equation,
$$E_a = E_b + I_a R_a = -210.85 + (35 \times 0.15)$$
$$E_a = -205.6 \text{ V}$$

Also
$$E_a = \frac{2 \times \sqrt{2} \times 480}{\pi}\cos\alpha = -205.6 \text{ V}$$

$$\cos\alpha = 0.47576 \text{ or } \alpha = 118.4°$$

(ii) Power from the DC machine as a generator is
$$P_g = 210.85 \times 35 = 7379.75 \text{ watts}$$

Power loss in the armature resistance
$$= 35^2 \times 0.15 = 183.75 \text{ watts}$$

Power fed back to the supply
$$= 7379.75 - 183.75 = 7196 \text{ watts}$$

Power fed back to the supply may be calculated as
$$E_aI_a = 205.6 \times 35 = 7186 \text{ watts}$$
$$= 7.186 \text{ kW}.$$

Example 13.8 A separately excited DC motor is supplied via a half-controlled single-phase bridge rectifier. The supply is 240 V, the thyristors are fired at 110° and the armature current continues for 50° beyond the voltage zero. Determine the motor speed at a torque of 1.8 Nm, given the motor torque characteristics as 2.0 Nm/A and armature resistance is 6 Ω. Neglect all rectifier losses.

Solution The voltage is supplied to motor between 110° to 180°. Zero voltage for 50° and motor back emf E_b from 50° to 110° (i.e. a period of 60°)

$$E_a = \frac{1}{\pi}\left[\int_{110°}^{180°} 240\sqrt{2}\sin\omega t\, d\omega t + E_b\pi\cdot\frac{60°}{180°}\right]$$

$$E_a = 71.1 + 0.333E_b$$

Average current $\qquad I_a = \dfrac{1.8}{1.0} = 1.8$ A

$$E_b = E_a - I_a R_a$$
$$= 71.1 + 0.333E_b - (1.8 \times 6)$$

Hence, $\qquad\qquad E_b = 90.43$ V

From the relation of gross mechanical power,

$$TN = E_b I_a$$

or $\qquad\qquad T/I_a = \dfrac{E_b}{N}$, therefore voltage characteristic is 1.0 V/rad/sec

Speed, $\qquad\qquad N = \dfrac{90.43}{2} = 45.22$ rad/sec.

Example 13.9 The speed of a 15 HP, 230 V, 900 rpm series motor is controlled by a single-phase semi converter. The combined field and armature circuit resistance is 0.15 Ω. Motor constants are $K_{af} = 0.03$ Nm/amp^2 and $K_{res} = 0.075$ V rad/sec. The supply voltage is 260 V. Assuming continuous and ripple-free motor current. For a firing angle α of 30° and speed $N = 900$ rpm, calculate: (i) motor torque, (ii) motor current, and (iii) input power.

Solution

(i) Speed, $\qquad\qquad N = 900$ rpm $= \dfrac{900 \times 2\pi}{60} = 94.2$ rad/sec

Motor torque $\qquad T = K_{af}\left[\dfrac{(\sqrt{2}V/\pi)(1+\cos\alpha) - K_{res}N}{(R_a + K_{af}N)}\right]^2$

$$= 0.03\left[\frac{(\sqrt{2}\times 260/\pi)(1+\cos 30°) - (0.075 \times 94.2)}{0.15 + (0.03 \times 94.2)}\right]^2$$

$$= 0.03\left[\frac{218.48 - 7.07}{2.98}\right]^2 = 150.9 \text{ Nm}$$

(ii) $\qquad\qquad T = K_{af}I_a^2$

$$I_a = \left[\frac{T}{K_{af}}\right]^{1/2} = \left[\frac{150.9}{0.03}\right]^{1/2} = 70.94 \text{ A}$$

(iii) Motor terminal voltage $E_a = \dfrac{\sqrt{2}V}{\pi}(1 + \cos\alpha)$

$$= \dfrac{\sqrt{2} \times 260}{\pi}(1 + \cos 30°)$$

$$= 218.48 \text{ V}$$

If the losses are neglected.

Input power $\qquad P_s = E_a I_a = 218.48 \times 70.94 = 15500 \text{ W} = 15.5 \text{ kW}.$

Example 13.10   For the above example, if the converter is a full converter, calculate: (i) motor torque, (ii) motor current, and (iii) supply power factor.

Solution   (i) For full converter motor torque

$$T = K_{af}\left[\frac{(2\sqrt{2}V/\pi)\cos\alpha - K_{res}N}{R_a + K_{af}N}\right]^2$$

$$= 0.03\left[\frac{(2\sqrt{2} \times 260/\pi)\cos 30° - (0.075 \times 94.2)}{0.15 + (0.03 \times 94.2)}\right]^2$$

$$= 0.03\left[\frac{202.79 - 7.07}{2.98}\right]^2 = 129.41 \text{ Nm}$$

$$I_a = \left[\frac{T}{K_{af}}\right]^{1/2} = \left[\frac{129.41}{0.03}\right]^{1/2} = 65.68 \text{ A}$$

(ii) For full converter system

$$E_a = \frac{2\sqrt{2}V}{\pi}\cos\alpha = \frac{2\sqrt{2} \times 260}{\pi}\cos 30°$$

$$E_a = 202.79 \text{ V}$$

Input power, $\qquad P_s = E_a I_a = 202.79 \times 65.68 = 13318.8 \text{ W}$

Input volt-ampere $\qquad U = VI_a = 260 \times 65.68 = 17076.8 \text{ VA}$

Power factor $\qquad PF = \dfrac{P_s}{U} = \dfrac{13318.8}{17076.8} = 0.78.$

Example 13.11   The speed of a separately excited DC motor is controlled by a single-phase semi converter. The field current, which is controlled by semi converter, is set to maximum possible value. The AC supply voltage to the armature and field converters is single-phase, 210 V, 50 Hz. The armature resistance is $R_a = 0.25\,\Omega$. The field resistance $R_f = 147\,\Omega$ and the motor voltage constant is $K_t = 0.7032$ V/A rad/sec. The load torque is $T_L = 45$ Nm at 1000 rpm. The inductances of the armature and field circuits are sufficient enough to make the armature and field current continuous and ripple free. Determine: (a) field current I_f, and (b) delay angle of the converter in the armature circuit α_a.

Solution

(a) $\qquad\qquad\qquad\qquad V = 210 \text{ V},$

$$V_m = \sqrt{2}V = \sqrt{2} \times 210 = 296.52 \text{ V}$$

$$R_a = 0.25\,\Omega$$

$$R_f = 147 \,\Omega$$
$$T_d = T_L = 45 \text{ Nm}$$
$$K_t = 0.7032 \text{ V/A rad/sec}$$
$$\omega = \frac{1000 \times 2\pi}{60} = 104.72 \text{ rad/sec}$$

The maximum field voltage and current is obtained for a delay angle of $\alpha_f = 0$ as given by equation

$$V_f = \frac{V_m}{\pi}(1 + \cos\alpha_f) = \frac{2V_m}{\pi} = \frac{2 \times 296.52}{\pi} = 188.86 \text{ V}$$

The field current
$$I_f = \frac{V_f}{R_f} = \frac{188.86}{147} = 1.284 \text{ A}$$

(b) From the equation of torque
$$T_d = K_t \Phi I_a = K_t I_f \cdot I_a$$

or
$$I_a = \frac{T_d}{K_t \cdot I_f} = \frac{45}{0.7032 \times 1.284} = 49.838 \text{ A}$$

From the motor voltage equation
$$E_b = K_v \omega I_f$$
$$= 0.7032 \times 104.72 \times 1.284 = 94.55 \text{ V}$$

The armature voltage
$$E_a = E_b + I_a R_a$$
$$= 93.82 + 49.838 \times 0.25 = 106.28 \text{ V}$$

From the relation
$$E_a = \frac{\sqrt{2}V}{\pi}(1 + \cos\alpha_a)$$

$$106.28 = \frac{296.52}{\pi}(1 + \cos\alpha_a)$$

or
$$\alpha_a = 82.79°.$$

Example 13.12 The speed of a separately excited DC motor is controlled by a single-phase full wave converter. The field circuit is controlled by a full converter and the field current is set to maximum possible value. The AC supply voltage to the armature and field converters is single-phase 440 V, 50 Hz. The armature resistance is 0.25 Ω and field resistance $R_f = 175 \,\Omega$ and the motor voltage constant is $K_v = 1.4$ V/A rad/sec. The armature current corresponding to the load demand is $I_a = 45$ A. If the delay angle of the armature converter is $\alpha_a = 60°$ and armature current is 45 A. Determine: (i) the torque developed by the motor T_d, and (ii) speed ω.

Solution

(i)
$$V_m = \sqrt{2}V = \sqrt{2} \times 440 = 622.25 \text{ V}$$
$$R_a = 0.25\,\Omega, \ R_f = 175\,\Omega$$
$$\alpha_a = 60°$$

and
$$K_v = 1.4 \text{ V/A rad/sec}$$

Maximum field voltage or current can be obtained for a delay angle of $\alpha_f = 0$

$$V_f = \frac{2V_m}{\pi} = \frac{2 \times 622.25}{\pi} = 396.14 \text{ V}$$

Field current $\qquad I_f = \dfrac{V_f}{R_f} = \dfrac{396.14}{175} = 2.26$ A

Developed torque $\qquad T_d = T_i = K_v I_f I_a$

$$= 1.4 \times 2.26 \times 45 = 142.5 \text{ Nm}$$

Armature voltage for full wave converter is:

$$E_a = \frac{2\sqrt{2}V}{\pi} \cos \alpha = \frac{2 \times \sqrt{2} \times 440 \cos 60°}{\pi} = 198.07 \text{ V}$$

The back emf $\qquad E_b = E_a - I_a R_a$

$$= 198.07 - 45 \times 0.25 = 186.79 \text{ V}$$

(ii) The speed is given by the equation

$$E_b = K_v I_f \cdot \omega$$

or $\qquad\qquad \omega = \dfrac{E_b}{K_v I_f} = \dfrac{186.82}{1.4 \times 2.26} = 59.05$ rad/sec

$$N = 565 \text{ rpm.}$$

Example 13.13 The speed of a 10 HP, 220 V, 1200 rpm, separately excited AC motor is controlled by a single-phase full converter. The rated armature current is 40 A. Armature resistance is $R_a = 0.25 \, \Omega$ and armature inductance $L_a = 10$ mH, the AC supply voltage is 265 V, motor voltage constant $K_a \Phi = 0.18$ V/rpm. For a firing angle of $\alpha = 30°$ and rated motor current, determine: (i) speed of the motor, (ii) motor torque, and (iii) power to the motor. Assume motor current is continuous and ripple free.

Solution

(i) Average terminal voltage $V_t = \dfrac{2\sqrt{2}V}{\pi} \cos \alpha$

$$= \frac{2\sqrt{2} \times 265 \times \cos 30°}{\pi} = 206.6 \text{ V}$$

Back emf, $\qquad E_b = V_t - I_a R_a = 206.6 - 40 \times 0.25$

$$= 196.6 \text{ V}$$

Speed $\qquad\qquad N = \dfrac{196.6}{0.18} = 1092$ rpm

(ii) $\qquad\qquad K_a \Phi = \dfrac{0.18 \times 60}{2\pi}$ V rad/sec $= 1.72$ V rad/sec

Torque, $\qquad\qquad T = 1.72 \times 40 = 68.75$ Nm

(iii) Power to the motor, $\qquad P = (i_a)_{\text{rms}}^2 R_a + E_a I_a$

Since current is ripple free

$$(i_a)_{\text{rms}} = (i_a)_{av} = I_a$$

$$P = I_a^2 R_a + E_a I_a = V_t I_a$$

$$= 206.6 \times 40 = 8264 \text{ W.}$$

Example 13.14 The speed of a 20 HP, 230 V, 1000 rpm DC series motor is controlled using a single-phase, (a) semi converter, and (b) full converter. The combination armature and field circuit resistance is $0.2\,\Omega$. Assume continuous and ripple free motor current and speed 1000 rpm and $K = 0.03$ Nm/amp^2. Determine for 250 V AC voltage, (i) motor current, and (ii) motor torque, for a firing angle of $\alpha = 30°$.

Solution

(i) *For semi converter*

Voltage across the armature,

$$E_a = \frac{\sqrt{2}V}{\pi}(1 + \cos \alpha)$$

$$= \frac{\sqrt{2} \times 250}{\pi}(1 + \cos 30°) = 210 \text{ V}$$

Angular speed

$$\omega = \frac{2\pi N}{60} = \frac{2\pi \times 1000}{60} = 104.66 \text{ rad/sec}$$

$$= \frac{(\sqrt{2}V/\pi)(1 + \cos \alpha) - I_a R_a}{K\Phi}$$

Assuming the armature current to be continuous and flux $\Phi \propto I_a$

$$\omega = \frac{(\sqrt{2}V/\pi)(1 + \cos \alpha) - I_a R_a}{KI_a}$$

$$K\omega I_a = (\sqrt{2}V/\pi)(1 + \cos \alpha) - I_a R_a$$

or $\quad I_a(K\omega + R_a) = (\sqrt{2}V/\pi)(1 + \cos \alpha)$

$$I_a = \frac{(\sqrt{2}V/\pi)(1 + \cos \alpha)}{(R_a + K\omega)}$$

Armature current

$$I_a = \frac{210}{0.2 + 0.03 \times 104.67} \text{ A}$$

$$= 62.87 \text{ A}$$

Motor torque

$$T_{av} = KI_a^2 = 0.03 \times 62.87^2$$

$$= 118.59 \text{ Nm}$$

(ii) *Full converter,*

Voltage across armature, $E_a = \left(\frac{2\sqrt{2}}{\pi}\right)V \cos \alpha$

$$= \left(\frac{2\sqrt{2}}{\pi}\right) \times 250 \cos 30° = 195 \text{ V}$$

Armature current

$$I_a = \frac{\left(\frac{2\sqrt{2}}{\pi}\right) \times V \cos 30°}{(R_a + K\omega)}$$

$$= \frac{195}{0.2 + 0.03 \times 104.67} = 58.39 \text{ A}$$

Therefore motor torque, $T_{av} = KI_a^2 = 0.03 \times 58.39^2$

$$= 102.25 \text{ Nm}.$$

Example 13.15 The DC series motor is powered by DC chopper from 600 V DC source. The armature resistance R_a and field resistances are $0.02\,\Omega$ and $0.03\,\Omega$. Back emf constant of the motor is $K_v = 15.27$ mV/A rad/sec. The average armature current $I_a = 250$ A. Armature current is continuous and ripple free. If the duty cycle of chopper is 60%, determine: (i) the input power from the source, (ii) equivalent input resistance of the chopper drive, (iii) motor speed, and (iv) developed torque.

Solution $\qquad\qquad\qquad\qquad V_s = 600$ V, $I_a = 250$ A, $\delta = 0.6$

(i) Total armature circuit resistance

$$R_m = R_a + R_f$$
$$= 0.02 + 0.03 = 0.05\,\Omega$$

Power supplied to motor, $P_o = V_a I_a = \delta V_s I_a$
$$= 0.6 \times 600 \times 250 = 90 \text{ kW}$$

Equivalent input resistance of the chopper drive.

(ii) $R_{eq} = \dfrac{V_s}{I_s} = \dfrac{V_s \cdot I}{I_a \delta} = \dfrac{600}{250 \times 0.6} = 4\,\Omega$

(iii) Average armature voltage $V_a = KV_s = 0.6 \times 600 = 360$ V

Back emf, $\qquad\qquad E_b = V_a - I_a R_a = 360 - 0.05 \times 250 = 347.5$ V

Motor speed $\qquad\qquad \omega = \dfrac{E_b}{I_a K_v} = \dfrac{347.5}{0.01527 \times 250} = 91.03$ rad/sec

$$= 869.3 \text{ rpm}$$

(iv) $T_d = K_v I_a^2 = 0.0152 \times 250^2 = 954.38$ Nm.

Example 13.16 A 230 V, 960 rpm and 200 A, separately excited DC motor has armature resistance of $0.02\,\Omega$. Motor is fed from chopper which provides both motoring and braking operations. Source has a voltage of 230 V. Assuming continuous conduction, calculate: (i) duty ratio of chopper for motoring operation at rated torque and 350 rpm, (ii) duty ratio of chopper for braking operation at rated torque and 350 rpm, and (iii) if the maximum duty ratio of chopper is limited to 0.95 and maximum permissible motor current is twice the rated, calculate maximum permissible motor speed obtainable without field weakening and power fed to the source.

Solution At rated operation, $E_b = V_a - I_a R_a$
$$= 230 - (200 \times 0.02) = 226 \text{ V}$$

(i) E_b at 350 rpm $\qquad = \dfrac{350}{960} \times 226 = 82.4$ V

Motor terminal voltage, $V_a = E_b + I_a R_a$
$$= 82.4 + (200 \times 0.02) = 86.4 \text{ V}$$

Duty cycle, $\qquad\qquad \delta = \dfrac{86.4}{230} = 0.376$

(ii) $\qquad\qquad V_a = E_b - I_a R_a = 82.4 - (200 \times 0.02) = 78.4$ V

$$\delta = \dfrac{78.4}{230} = 0.34$$

(iii) Maximum voltage available,
$$V_a = 0.95 \times 230 = 218.5 \text{ V}$$

$$E_b = V_a + I_a R_a = 218.5 + (200 \times 2 \times 0.02) = 226.5 \text{ V}$$

Maximum permissible motor speed,

$$N = \frac{226.5}{22.6} \times 960 = 962 \text{ rpm}$$

Assuming lossless chopper, power fed to the source,

$$V_a I_a = 218.5 \times 400 = 87.4 \text{ kW}.$$

Example 13.17 A 230 V, 960 rpm and 200 A separately excited DC motor has an armature resistance of $0.02\,\Omega$. Motor is operated in dynamic braking with chopper control with a braking resistance of $2\,\Omega$. Calculate duty ratio of chopper for a motor speed of 600 rpm and braking torque of twice the rated value.

Solution At speed of 960 rpm,

$$E_{b1} = V - I_a R_a = 230 - (200 \times 0.02) = 226 \text{ V}$$

From the equation $\quad \dfrac{E_{b2}}{E_{b1}} = \dfrac{N_2}{N_1}$

At 600 rpm, $\qquad E_{b2} = \dfrac{600}{960} \times 226 = 141.25 \text{ V}$

Effective value of braking resistance

$$R_{be} = \frac{E_b}{I_a} - R_a = \frac{141.25}{400} - 0.02$$

From the equation $\qquad R_{be} = R_b\,(1 - \delta) = 0.333\,\Omega$

or $\qquad\qquad (1 - \delta)\,R_b = 0.333$

or $\qquad\qquad \delta = 1 - \dfrac{0.333}{R_b} = 1 - \dfrac{0.333}{2} = 0.83\,\Omega$

Example 13.18 A 3-phase full converter is used to control the speed of a 150 HP, 600 V, 1000 rpm, separately excited DC motor. The converter is operated from a three-phase, 480 V, 60 Hz supply. The machine parameters are $R_a = 0.1\,\Omega$, $L_a = 5$ mH, $K\Phi = 0.3$ V/rpm. The rated armature current is 130 A. When the machine is operating in motoring operation draws a rated current and runs at 1500 rpm. Determine the firing angle and supply power factor, assuming that motor current is ripple free.

Solution Phase voltage, $\qquad V_p = \dfrac{V_L}{\sqrt{3}} = \dfrac{480}{\sqrt{3}} = 277 \text{ V}$

Given, $\qquad\qquad K\Phi = 0.3 \text{ V/rpm}$

$$E_b = K\Phi n = 0.3 \times 1500 = 450 \text{ V}$$
$$V_o = E_b + I_a R_a = 450 + 130 \times 0.1 = 463 \text{ V}$$

Average output voltage, $\quad V_o = \dfrac{3\sqrt{6}}{\pi} V_p \cos\alpha$

$$463 = \frac{3\sqrt{6} \times 277}{\pi} \cos\alpha$$

$$\alpha = 44.4°$$

Since the ripple is neglected the supply current is a square wave of magnitude 130 A and width 120°.

rms value of supply current

$$I_a = \left(\frac{1}{\pi} \times 130^2 \times \frac{2\pi}{3} \right)^{1/2} = \sqrt{2/3} \times 130 = 106.1 \text{ A}$$

Supply volt. amp are $\quad S = 3VI_A = 3 \times 277 \times 106.1 = 88169.1$ VA

Assuming no losses in the converter,

$$P_s = V_o I_o = 463 \times 130 = 60190 \text{ W}$$

Supply power factor, $\quad PF = \dfrac{P_s}{S} = \dfrac{60190}{88169.1} = 0.68.$

Example 13.19 The speed of a 25 HP, 300 V, 900 rpm separately excited DC motor is controlled by a three-phase full converter the field circuit is also controlled by a three-phase full converter. The AC input to the armature and field converter is three-phase star connected 208 V, 60 Hz. Armature resistance $R_a = 0.25\,\Omega$ and field circuit resistance $R_f = 145\,\Omega$ and motor voltage constant $K_v = 1.2$ V/A rad/sec. (i) If the field converter is operated at the maximum field current and the developed torque is $T_d = 116$ Nm at 900 rpm, determine the delay angle of armature converter α_a (ii) If the field circuit converter is set for maximum field current, developed torque $T_d = 116$ Nm and delay angle of armature converter $\alpha_a = 0$, determine the speed of the motor (iii) For the same load determined in part (ii), the delay angle of the field converter, if the speed has to be increased to 1800 rpm.

Solution

(i) $\qquad R_a = 0.25\,\Omega, \ R_f = 145\,\Omega, \ K_v = 1.2$ V/A rad/sec

$$V_L = 208 \text{ V, phase voltage, } V_p = \frac{208}{\sqrt{3}} = 120 \text{ V}$$

$$V_m = \sqrt{2} \times 120 = 169.7 \text{ V, } T_d = 116 \text{ Nm}$$

$$\omega = \frac{2\pi \times 900}{60} = 94.25 \text{ rad/sec}$$

For maximum field current at $\alpha_f = 0$

$$V_f = \frac{3\sqrt{3} \times 169.2}{\pi} = 280.7 \text{ V}$$

$$I_f = \frac{280.7}{145} = 1.936 \text{ A}$$

$$T_d = K_v \cdot I_f I_a \text{ or } I_a = \frac{T_d}{K_v I_f} = \frac{116}{1.2 \times 1.936} = 49.93 \text{ A}$$

$$E_b = K_v \cdot I_f \omega = 1.2 \times 1.936 \times 94.25 = 218.96 \text{ V}$$

$$E_a = E_b + I_a R_a = 218.96 + 49.93 \times 0.25 = 231.44 \text{ V}$$

$$E_a = 231.44 = \frac{3\sqrt{3} \times 169.7}{\pi} \cos \alpha_a$$

or $\qquad \alpha_a = 34.46°$

(ii) $\alpha_a = 0$, $\qquad E_a = \dfrac{3\sqrt{3} \times 169.7}{\pi} = 280.7$ V

$$E_b = 280.7 - 49.93 \times 0.25 = 268.22 \text{ V}$$

Speed $\qquad \omega = \dfrac{268.22}{1.2 \times 1.936} = 115.36 \text{ rad/sec} = 1101 \text{ rpm}$

(iii)
$$\omega = \frac{2\pi \times 1800}{60} = 185.5 \text{ rad/sec}$$

$$E_b = 268.22 = 1.2 \times 188.5 I_f$$

or
$$I_f = 1.186 \text{ A}$$

$$V_f = \frac{3\sqrt{3} \times 169.7}{\pi} \cos \alpha_f$$

Delay angle $\qquad \alpha_f = 52.20°.$

Example 13.20 The speed of a 10 HP, 600 V, 1800 rpm separately excited DC motor is controlled by a three-phase full converter, the converter is operated from three-phase, 480 V, 60 Hz supply, the rated armature current of motor is 16.5 A. The motor parameters are $R_a = 0.0874\,\Omega$, $L_a = 6.5$ mH and $K_a \Phi = 0.33$ V/rpm. The converter and AC supply are considered to be ideal. Calculate: (i) No load speeds at firing angle $\alpha = 0°$ and $\alpha = 30°$. Assuming that at no load, the armature current is 10% of rated current and is continuous. (ii) What will be the firing angle to obtain rated speed of 1800 rpm at rated motor current? Compute the supply power factor. (iii) Find the speed regulation.

Solution

(i) At no-load, for $\alpha = 0°$

Supply phase voltage, $\qquad V = \dfrac{V_L}{\sqrt{3}} = \dfrac{480}{\sqrt{3}} = 277 \text{ V}$

Motor terminal voltage, $E_a = \dfrac{3\sqrt{6}V}{\pi} \cos \alpha$

$$= \frac{3\sqrt{6} \times 277}{\pi} \cos 0° = 648 \text{ V}$$

$$E_b = E_a - I_a R_a = 648 - (16.5 \times 0.0874)$$

$$= 646.6 \text{ V}$$

No-load speed, $\qquad N_0 = \dfrac{E_b}{K_a \Phi} = \dfrac{646.6}{0.33} = 1960 \text{ rpm}$

For $\qquad \alpha = 30°$

$$E_a = \frac{3\sqrt{6} \times 277}{\pi} \cos 30° = 561.2 \text{ V}$$

$$E_b = E_a - I_a R_a = 561.2 - (16.5 \times 0.0874) = 559.8 \text{ V}$$

No-load speed, $\qquad N_0 = \dfrac{E_b}{K_a \Phi} = \dfrac{559.8}{0.33} = 1696 \text{ rpm}$

(ii) *Full load condition.*

Motor back emf E_b at 1800 rpm is,

$$E_b = K_a \Phi N = 0.33 \times 1800 = 594 \text{ V}$$

Motor terminal voltage at rated current is

$$E_a = E_b + I_a \cdot R_a = 594 + (165 \times 0.0874) = 608.4 \text{ V}$$

$$648 \cos \alpha = 608.4$$

$$\cos \alpha = \frac{608.4}{648} = 0.94$$

$$\therefore \qquad \alpha = 20.20°$$

The rms value of supply current,

$$I_A = \left(\frac{1}{\pi} \times 165^2 \times \frac{2\pi}{3} \right)^{1/2} = \sqrt{\frac{2}{3}} \times 165 = 134.64 \text{ A}$$

Supply volt-amp $= 3VI_A = 3 \times 277 \times 134.64 = 111885.8 \text{ VA}$

Assuming no losses in converter, power from the supply P_s is the same as power input to motor

$$P_s = E_a I_a = 608.4 \times 165 = 100386 \text{ W}$$

Supply power factor $= \dfrac{P_s}{U} = \dfrac{100386}{111885.8} = 0.9$

(iii) *Speed regulation*: At full load, the motor current is 165 A and speed is 1800 rpm. When the load is thrown off keeping the firing angle same at $\alpha = 20.20°$, motor current decreases to 10%, i.e. 16.5 A. So,

$$E_b = E_a - (I_a R_a) = 608.4 - (16.5 \times 0.0874) = 606.96 \text{ V}$$

and no-load speed, $\qquad N_0 = E_b/K_a \Phi$

$$= \frac{606.96}{0.33} = 1840 \text{ rpm}$$

% speed regulation $\qquad = \dfrac{1840 - 1800}{1800} \times 100 = 2.18\%.$

Example 13.21 The speed of a 15 HP, 300 V, 1800 rpm separated excited DC motor is controlled by a three-phase full converter drive. The field current is also controlled by a three-phase full converter and is set to maximum possible value. The AC input is star connected 208 V, 60 Hz supply. The armature resistance $R_a = 0.25\,\Omega$ and the field resistance is $R_f = 245\,\Omega$ and motor voltage constant $K_v = 1.2$ V/A rad/sec. The armature and field currents can be assumed to be continuous and ripple free. Determine: (i) delay angle of armature converter, if the motor supplies the rated power at the rated speed, (ii) no load speed if the delay angles are the same as in part (i) and the armature current at no load is 10% of the rated value, and (iii) speed regulation.

Solution $\qquad R_a = 0.25\,\Omega,$

$$R_f = 245\,\Omega$$

$$K_v = 1.2 \text{ V/A rad/sec,}$$

$$V = 208 \text{ V}$$

and $\qquad \omega = \dfrac{1800 \times 2\pi}{60} = 188.5 \text{ rad/sec}$

The phase voltage $\qquad V_p = \dfrac{V}{\sqrt{3}} = \dfrac{208}{\sqrt{3}} = 120 \text{ V}$

and $\qquad V_m = \sqrt{2} \times 120 = 169.7 \text{ V}$

$$I_{\text{rated}} = \dfrac{20 \times 746}{300} = 49.73 \text{ A}$$

Maximum possible field current, $\alpha_f = 0$ is given by

$$V_f = \frac{3\sqrt{3} \times 169.7}{\pi} = 280.7 \text{ V}$$

$$I_f = \frac{280.7}{245} = 1.146 \text{ A}$$

(i) $I_a = I_{rated} = 49.73 \text{ A}$

$E_b = K_v I_f \cdot \omega = 1.2 \times 1.146 \times 188.5 = 259.2 \text{ V}$

$E_a = E_b + I_a R_a = 259.2 + 49.73 \times 0.25 = 271.63 \text{ V}$

From equation $E_a = \dfrac{3\sqrt{3} \times \sqrt{2}V}{\pi} \cos \alpha$

$$271.63 = \frac{3\sqrt{3}V_m \cos \alpha}{\pi} = \frac{3\sqrt{3} \times 169.7}{\pi} \cos \alpha_a$$

or $\alpha_a = 14.59°$

(ii) $I_a = 10\% \text{ of } 49.73 = 4.973 \text{ A}$

and $E_b = E_a - I_a R_a = 271.63 - 0.25 \times 4.973 = 270.39 \text{ V}$

No-load speed $\omega_0 = \dfrac{E_b}{K_v \cdot I_f} = \dfrac{270.39}{1.2 \times 1.146} = 196.6 \text{ rad/sec}$

or $196.62 \times \dfrac{60}{2\pi} = 1877.58 \text{ rpm}$

(iii) Speed regulation $= \dfrac{\text{No load speed} - \text{full load speed}}{\text{Full load speed}} = \dfrac{1877.58 - 1800}{1800}$

$$= 0.043 = 4.3\%.$$

EXERCISES

Multiple Choice Questions

1. In a normal three thyristor phase rectifier, a thyristor can not be fired during:
   (A) first 10° of its anode voltage
   (B) first 30° of its anode voltage
   (C) first 50° of its anode voltage
   (D) first 80° of its anode voltage

2. Armature voltage of a DC motor can be controlled by means of:
   (A) cycloconverter
   (B) inverters
   (C) AC-DC converters
   (D) bridge rectifier circuit with fixed input

3. The speed of DC motor can be increased by:
   (A) increasing the thyristor firing angle in the armature circuit of a phase controlled converter

   (B) decreasing the thyristor firing angle in the armature circuit of a phase controlled converter
   (C) increasing the duty cycle of a chopper
   (D) none of these

4. A thyristor power converter is said to be in discontinuous conduction when:
   (A) the load current is zero even though the load voltage is present
   (B) both the load voltage and load current are zero simultaneously
   (C) when the load voltage is ripple free
   (D) none of these

5. A freewheeling diode in a phase controlled converter:

(A) increases the chances of discontinuous conduction in the load

(B) decreases the chances of discontinuous conduction in the load

(C) causes the chances of discontinuous conduction in the load

(D) none of these

6. The speed of DC shunt motor above normal speed can be controlled by:

(A) armature voltage control method

(B) flux control method

(C) both the methods

(D) none of these

7. A freewheeling diode in a phase controlled converter:

(A) causes smoothening of load current

(B) causes poor voltage regulation

(C) enables inverter operation

(D) none of these

8. A half-controlled converter:

(A) has poor voltage regulation

(B) has an improved power factor

(C) is capable of inversion also

(D) all of these

9. The ripple content of load current of a converter feeding RL load is decided by:

(A) load resistance only

(B) load inductance

(C) both load resistance and load inductance

(D) neither resistance nor inductance

10. For controlling the speed of DC motor of 150 HP rating, the following types of converters are normally used:

(A) single-phase full converters

(B) single-phase dual converters

(C) three-phase full converters

(D) three-phase dual converters

11. The overlap of the converter is responsible for:

(A) voltage regulation of the converter

(B) additional losses of the converter

(C) additional harmonics in the load current

(D) all of these

12. The advantage of the tachometer speed control method for DC motor is that, it senses:

(A) back emf   (B) armature current

(C) armature voltage (D) speed

13. An SCR is used to control the speed of a DC motor. At full speed, the motor is taking 1 A at 75 V. The maximum forward surge current rating and maximum forward breakover voltage rating respectively are of the order of:

(A) 3 A, 235 V   (B) 1 A, 300 V

(C) 5 A, 166 V   (D) 5 A, 200 V

14. A motor armature supplied through phase-controlled SCRs receives a smoother voltage shape at:

(A) high motor speed

(B) low motor speeds

(C) rated normal motor speeds

(D) none of these

15. In the speed control circuit shown below, if the speed of the motor decreases, the capacitor will charge (see figure below):

(A) slower

(B) faster

(C) independent of the speed

(D) to the supply voltage

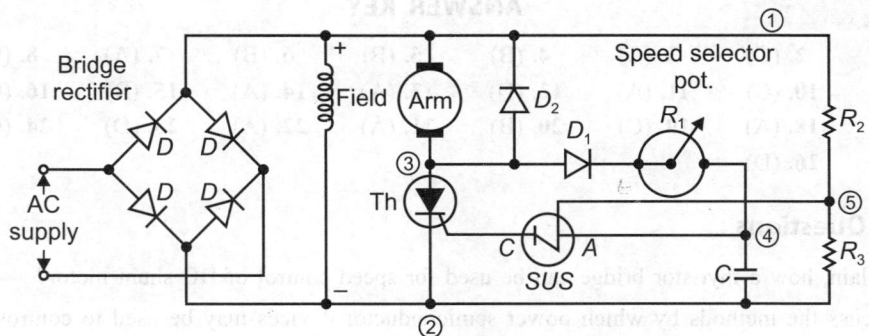

16. In the above circuit, if the speed of the motor decreases, the *SUS* will fire at:
    (A) higher the rated voltage
    (B) lower the rated voltage
    (C) the same voltage
    (D) voltage determined by the gate voltage of *SUS*

17. In the same circuit, if the speed of the motor decreases, the DC voltage across SCR will:
    (A) remains the same
    (B) decreases
    (C) increases
    (D) become more smooth

18. In the same circuit, if the motor speed decreases, the rms value of the voltage across SCR will:
    (A) decreases   (B) increases
    (C) remains the same (D) become zero

19. In the same circuit, the function of resistances $R_1 - R_2$ is to provide:
    (A) breakover voltage of *SUS*
    (B) anode voltage of *SUS*
    (C) reference voltage of *SUS*
    (D) none of these

20. In the same circuit, which of the following devices can not replace *SUS*?
    (A) Diac   (B) FET
    (C) UJT   (D) All of these

21. The function of diode D_2 is:
    (A) to dissipate $L\, di/dt$ of the armature
    (B) to rectify the armature current
    (C) to stop armature current to flow in the capacitor branch and thus charge it
    (D) to provide field current

22. If the speed of the motor decreases, the SCR will fire:
    (A) earlier
    (B) later
    (C) at the same angle
    (D) according to the anode voltage of *SUS*

23. The advantage of the tachometer speed control method for DC motors is that, it senses:
    (A) back emf
    (B) armature current
    (C) armature voltage
    (D) speed

24. A dual converter used for the speed control of DC motors, will have two bridges, they are:
    (A) two rectifiers
    (B) two inverters
    (C) one rectifier and one inverter
    (D) none of these

25. For DC motors a dual converter is used to obtain:
    (A) reversible speed control
    (B) regenerative braking
    (C) plugging
    (D) all of these

26. A DC chopper circuit controls the average voltage across the DC motor by:
    (A) controlling the input voltage
    (B) controlling the field current
    (C) controlling the line current
    (D) continuously switching ON and OFF the motor for fixed durations of t_{ON} and t_{OFF} respectively

ANSWER KEY

1. (B)	**2.** (C)	**3.** (C)	**4.** (B)	**5.** (B)	**6.** (B)	**7.** (A)	**8.** (B)
9. (C)	**10.** (C)	**11.** (A)	**12.** (D)	**13.** (A)	**14.** (A)	**15.** (B)	**16.** (C)
17. (C)	**18.** (A)	**19.** (C)	**20.** (B)	**21.** (A)	**22.** (A)	**23.** (D)	**24.** (C)
25. (D)	**26.** (D)						

Review Questions

1. Explain, how a thyristor bridge can be used for speed control of DC shunt motor?

2. Discuss the methods by which power semiconductor devices may be used to control and vary the speed of DC drives.

3. Discuss the speed control of a single-phase separately excited DC motor.

4. Draw the waveform of voltage and current for semi converter series motor drives.

5. Explain the working of AC chopper used for speed control of DC shunt motor.

6. Draw the waveforms for full converter series motor drives.

7. Discuss the speed regulation by armature current control.

8. What is dual converters? Discuss and explain where this method is used.

9. Discuss the speed regulation by armature voltage control.

10. How chopper is useful for controlling the speed of DC drives?

11. Explain the working of DC chopper used for regenerative braking of DC shunt motors.

12. Explain two factors on which the load voltage of a DC chopper circuit depends.

Thyristorised Controlled AC Motors

14.0 INTRODUCTION

The construction of the cage induction motor is relatively simple compared to other machines, but it does not lend itself-to speed adjustment, so readily as does the DC motor. The main use of power semiconductor device in AC motor control is centred on variable output frequency inverters.

In most of the applications requiring variable speed, DC motors are widely used. Main reason for the use of DC motors is its ability to provide a high torque at low speed and the wide speed range over which the speed can be varied. However, AC motors are widely used due to low cost and high reliability for fixed speed. Conventional methods of speed control of AC motors are either expensive or highly inefficient because of non availability of a variable frequency source as a commercial proposition. The most desirable feature of AC motors is the absence of commutators and brushes. Availability of thyristors and power transistors have allowed the development of variable speed induction motor drives.

The induction motor drives can be classified into two broad categories based on their applications:

1. **Adjustable-speed drives:** One important application of these drives is in process control by controlling the speed of fans, compressors, pumps and blowers, etc.
2. **Servo drives:** By means of sophisticated control, induction motors can be used as servo drives in computer peripherals, machine tools and robotics, etc.

Although variable speed induction motor drives are generally expensive than DC drives, they are used in a number of applications such as fans, blowers, mill run-out tables, cranes, conveyers, traction, etc. because of the advantages of induction motors.

14.1 SPEED CONTROL OF THREE-PHASE INDUCTION MOTORS

Three phase induction motors are of *two* types: (1) squirrel cage, and (2) wound rotor. In squirrel cage, the rotor consists of longitudinal conductor bars shorted by circular connectors at the two ends while in wound-rotor motor, the rotor has a balanced three-phase distributed winding having same poles as stator winding. However, in both, stator carries a three-phase balanced distributed winding. Induction motor always operates in the vicinity of synchronous speed. This indicates that in order to vary the speed of an induction motor one has to vary the synchronous speed which can be varied only by changing the value of supply frequency as the number of poles of the motor is fixed. Thus, variable frequency supply is one of the methods of achieving smooth speed control of induction motors.

For a three-phase induction motor, the voltage equation is expressed as:

$$E = 4.44 \Phi f N \text{ volts}$$

where, E is the supply voltage (watt)

Φ = air gap flux in webers

f = supply frequency (Hz)

N = number of turns

For constant torque operation below normal speed, Φ has to be kept constant. From above relation

$$\frac{E}{f} = 4.44 \Phi N \alpha \Phi \qquad \qquad ...(14.1)$$

That is to control the speed of an induction motor below its rated speed, not only the frequency but also the supply voltage has to be decreased proportionately to keep the (E/f) ratio constant.

In order to achieve constant horsepower operation above normal speed, the armature voltage has to be kept constant and flux has to be decreased. This can be achieved by keeping the supply voltage constant and increasing the value of supply frequency.

14.2 VARIOUS SCHEMES OF SPEED CONTROL OF INDUCTION MOTOR

The induction machine is most commonly used in adjustable-speed AC drive systems, for their low cost, less maintenance and simple construction. When operated at constant AC voltages, an induction motor operates at nearly constant speed. The speed of motors can be controlled from an AC or DC source. Different schemes for the speed control of induction motor are shown in Fig. 14.1.

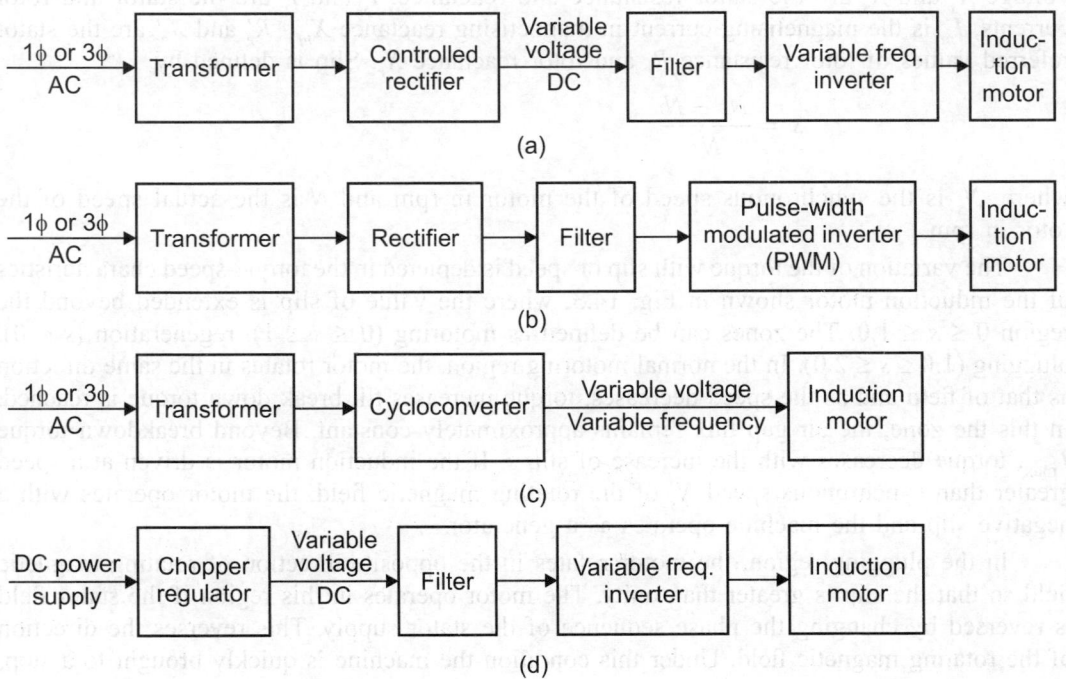

(a)

(b)

(c)

(d)

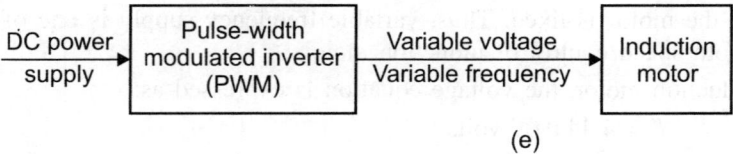

(e)

Fig. 14.1 Different schemes for speed control of induction motor

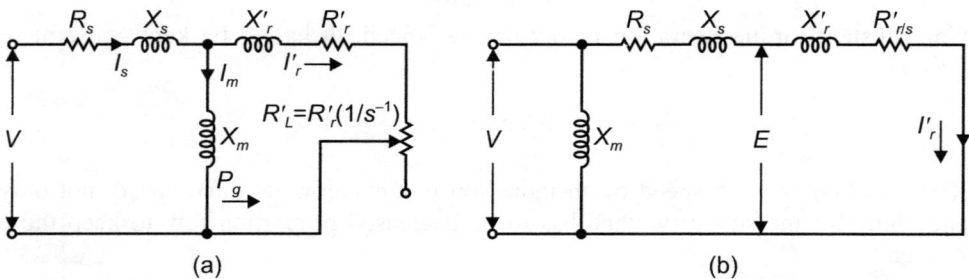

(a)   (b)

Fig. 14.2 Equivalent circuit of induction motor

14.3 CONSTANT FREQUENCY OPERATION OF INDUCTION MOTOR

When the induction motor is operated directly from constant voltage constant frequency AC line voltages, it operates at nearly constant speed. For a balanced set of stator voltages, the per phase equivalent circuit is shown in Fig. 14.2.

In the per phase equivalent circuit of a three-phase induction motor V is the stator supply voltage R_s and X_s are the stator resistance and reactance. I_s and I'_r are the stator and rotor currents, I_m is the magnetising current in magnetising reactance $X_m \cdot R'_r$ and X'_r are the stator referred values of rotor resistance R_r and rotor reactance X_r. Slip is defined by

$$s = \frac{N_s - N}{N_s}$$

where, N_s is the synchronous speed of the motor in rpm and N is the actual speed of the rotor in rpm.

The variation of the torque with slip or speed is depicted in the torque-speed characteristics of the induction motor shown in Fig. 14.3, where the value of slip is extended beyond the region $0 \leq s \leq 1.0$. The zones can be defined as motoring $(0 \leq s \leq 1)$, regeneration $(s < 0)$, plugging $(1.0 \leq s \leq 2.0)$. In the normal motoring region, the motor rotates in the same direction as that of field and as the speed decreases, torque increases till break down torque is reached. In this the zone, the air gap flux remains approximately constant. Beyond breakdown torque T_{max}, torque decreases with the increase of slip s. If the induction motor is driven at a speed greater than synchronous speed N_s of the rotating magnetic field, the motor operates with a negative slip and the machine operates as a generator.

In the plugging region, the motor rotates in the opposite direction of rotating magnetic field so that the slip is greater than unity. The motor operates in this region if the stator field is reversed by changing the phase sequence of the stator supply. This reverses the direction of the rotating magnetic field. Under this condition the machine is quickly brought to a stop,

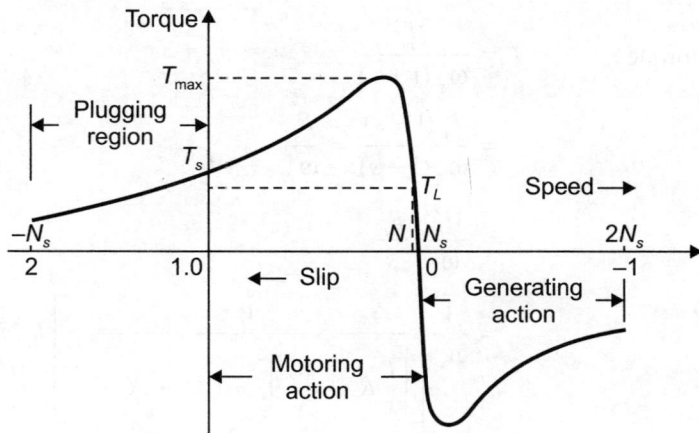

Fig. 14.3 Torque-speed characteristics

and if the supply is not disconnected, the motor starts to rotate in the opposite direction. This method of braking or rapidly reversing the induction motor is known as *plugging*. The energy due to the plugging brake is dissipated within the machine and therefore excessive machine heating must be taken care of.

14.4.1 Steady State Analysis at Constant Frequency

The important characteristics of induction motor like current, input and output power, losses, speed starting torque and maximum torque may be determined from the equivalent circuit.

From the equivalent circuit, the total power P_g transferred across the air gap for a three-phase motor is given by

$$P_g = 3I_r'^2 \frac{R_r'}{s} \qquad \text{...(14.2)}$$

From Fig. 14.2(b)
$$I_r' = \cfrac{V}{\left[\left(R_s + \cfrac{R_r'}{s} \right) + j(X_s + X_r') \right]} \qquad \text{...(14.3)}$$

Rotor copper loss
$$P_{Cu} = 3I_r'^2 R_r' \qquad \text{...(14.4)}$$

Electrical power converted into mechanical power,

$$P_m = P_g - P_{Cu}$$

$$= 3I_r'^2 \frac{R_r'}{s} - 3I_r'^2 R_r'$$

$$= 3I_r'^2 R_r' \left(\frac{1}{s} - 1 \right) = 3I_r'^2 R_r' \left(\frac{1-s}{s} \right)$$

$$= P_g (1 - s) \qquad \text{...(14.5)}$$

Torque developed by the motor,

$$T = P_m / \omega_r$$

where, ω_r is the rotor speed in rad/sec.

Developed torque $\qquad T = \dfrac{P_m}{\omega_s\,(1-s)}$

$$= \dfrac{P_g(1-s)}{\omega_s\,(1-s)} = \dfrac{P_g}{\omega_s}$$

$$= \dfrac{3I_r'^2}{\omega_s}\dfrac{R_r'}{s} \qquad\qquad\qquad\qquad ...(14.6)$$

$$= \dfrac{3}{\omega_s}\left[\dfrac{V^2}{\left[\left(R_s + \dfrac{R_r'}{s}\right)^2 + (X_s + X_r')^2\right]}\right]R_r'/s \ \text{Nm} \qquad ...(14.7)$$

If supply voltage V and frequency are constant; then the torque is a function of slip.

The motor output torque at the shaft is obtained by deducting friction windage and core loss torques from torque T.

Rotor copper loss $\qquad = sP_g = 3I_r'^2 R_r' \qquad\qquad\qquad\qquad ...(14.8)$

The term sP_g is known as *slip power* because it is proportional to slip for a given value of P_g.

14.4 VARIABLE FREQUENCY OPERATION OF INDUCTION MOTOR

For variable frequency operation of an induction motor, the constant voltage, constant frequency equivalent circuit of Fig. 14.2 is now replaced by a variable frequency equivalent circuit as shown in Fig. 14.4. If the stator frequency is increased beyond the rated value, the torque-speed curves, derived earlier can be plotted as shown in Fig. 14.5. The airgap flux and stator current decrease as the frequency increases and correspondingly,

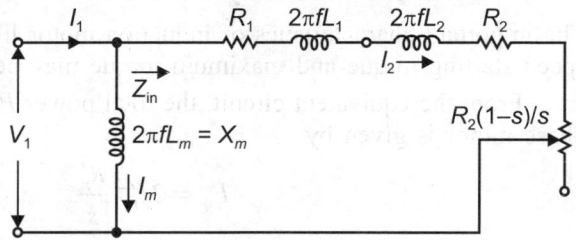

Fig. 14.4 Variable frequency equivalent circuit of induction motor

the maximum developed torque also decreases. As the frequency of stator voltage is varied, the magnetising reactance, stator reactance and rotor reactance change as a function of supply frequency. If the supply voltage is held constant and frequency is reduced to a lower value, the current in the motor would increase as the equivalent impedance of the equivalent circuit is a function of frequency. Hence, for variable frequency operation of induction motor, the voltage and frequency of the supply has to be changed in proper ratio to keep constant current in the motor. The power and torque developed in the motor is a function of current in the motor. The equivalent circuit parameters are all referred to the stator side. The total impedance of the equivalent circuit is given by

$$Z_{in} = [(R_1 + R_2/s) + j\,2\pi f\,(L_1 + L_2)]$$

The equivalent circuit parameters are as follows:

$$V_1 = \text{per phase voltage of supply}$$

R_1 and X_1 being stator resistance and reactance per phase

R_2 and X_2 are the rotor resistance and reactance per phase

X_m being the magnetising reactance

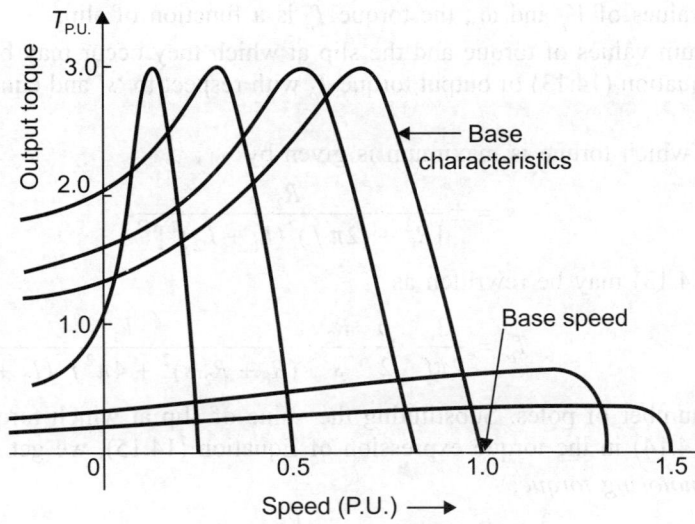

Fig. 14.5 Torque-speed curves of induction motor

I_1 is stator current and I_2 being the rotor current

$R_2 \left(\dfrac{1}{s} - 1 \right)$ is the load as a function of slip, f is the supply frequency in Hz.

For a frequency greater than the rated value, the airgap flux and stator current decreases and the maximum developed torque also decreases. For a frequency below the rated value, the air gap flux will saturate causing the stator current to rise to very high values. Thus, if w is to be reduced below the rated value, then the stator voltage should also be reduced in such a manner that the air gap flux is constant. In other words, the voltage applied to the motor should not be reduced proportional to the frequency at low frequencies. The volts/Hz must be higher at low frequencies than at higher frequencies.

Analysis of induction motor operation at variable frequency: Torque-speed characteristics of the induction motor are analysed at variable frequencies using the equivalent circuit of Fig. 14.4. The current I_2 flowing in the rotor of the motor is given by

$$I_2 = \frac{V_1}{\sqrt{(R_1 + R_2/s)^2 + 4\pi^2 f^2 (L_1 + L_2)^2}} \qquad \text{...(14.9)}$$

The power developed by the motor is given by

$$P_m = 3I_2^2 \frac{R_2(1 - s)}{s} \qquad \text{...(14.10)}$$

Assuming negligible losses in the machine,

$$P_m = P_{\text{output}} = P_o \qquad \text{...(14.11)}$$

Output torque of the motor is given by

$$T_o = \frac{P_o}{\omega_R} = \frac{3}{\omega_R} I_2^2 \frac{R_2(1 - s)}{s} \text{ Nm} \qquad \text{...(14.12)}$$

Actual speed of rotor, $\omega_R = (1 - s) \omega_s$. Therefore,

$$T_o = \frac{3}{\omega_s} \cdot \frac{R_2}{s} \frac{V_1^2}{[(R_1 + R_2/s)^2 + 4\pi^2 f^2 (L_1 + L_2)^2]} \qquad \text{...(14.13)}$$

For given values of V_1 and ω_s, the torque T_o is a function of slip.

The maximum values of torque and the slip at which they occur may be determined by differentiating Equation (14.13) of output torque T_o with respect to 's' and equating $dT_o/ds = 0$, and solving for s.

The slip at which torque is maximum is given by

$$s = \pm \frac{R_2}{[R_1^2 + (2\pi f)^2 (L_1 + L_2)^2]^{1/2}} \qquad ...(14.14)$$

Equation (14.13) may be rewritten as

$$T_o = \frac{3}{2\pi f} \cdot \frac{p}{2} \cdot \frac{R_2}{s} \cdot \frac{V_1^2}{(R_1 + R_2/s)^2 + 4\pi^2 f^2 (L_1 + L_2)^2} \qquad ...(14.15)$$

where, p is the number of poles. Substituting the value of slip at which torque is maximum from Equation (14.14) in the torque expression of Equation (14.15), we get:

1. *Maximum motoring torque,*

$$T_{m\,(max)} = \frac{3}{2\pi f} \cdot \frac{p}{4} \frac{V_1^2}{[R_1^2 + 4\pi^2 f^2 (L_1 + L_2)^2]^{1/2} + R_1} \;\text{Nm} \qquad ...(14.16)$$

2. *Maximum generating torque,*

$$T_{g\,(max)} = \frac{3}{2\pi f} \cdot \frac{p}{4} \frac{V_1^2}{[R_1^2 + 4\pi^2 f^2 (L_1 + L_2)^2]^{1/2} - R_1} \;\text{Nm} \qquad ...(14.17)$$

In the higher frequency operational range:

$$2\pi f (L_1 + L_2) \gg R_1$$

and therefore $\quad T_{m\,(max)} = T_{g\,(max)} = \dfrac{3p}{4} \dfrac{V_1^2}{(2\pi f)^2 (L_1 + L_2)} \;\text{Nm} \qquad ...(14.18)$

Equation (14.17) shows that if volts/Hz (V_1/f) is held constant, then the maximum torque or breakdown torque would remain constant.

At starting of the induction motor slip, $s = 1$. Using Equation (14.13), we get

$$T_{start} = \frac{3}{2\pi f} \cdot \frac{p}{2} \frac{R_2 V_1^2}{(R_1 + R_2)^2 + (2\pi f)^2 (L_1 + L_2)^2} \;\text{Nm} \qquad ...(14.19)$$

If the stator supply frequency, f is large

$$2\pi f (L_1 + L_2) \gg (R_1 + R_2)$$

From Equation (14.19), we get,

$$T_{start} = \frac{3p}{2} \cdot \frac{V_1^2 R_2}{(2\pi f)^3 (L_1 + L_2)^2} \qquad ...(14.20)$$

In the upper operational frequency range, if V_1/f is held constant, the starting torque would decrease with increasing frequency.

If the stator supply frequency, f is small.

$$2\pi f (L_1 + L_2) \ll (R_1 + R_2)$$

Hence, from Equation (14.19)

$$T_{start} = \frac{3p}{2} \frac{V_1^2 R_2}{2\pi f (R_1 + R_2)^2} \;\text{Nm} \qquad ...(14.21)$$

If V_1/f is held constant, when the operational frequency is in the lower range then the starring torque would increase with increasing frequency. Hence, it is seen from the above discussion, that there is an optimum value of frequency at which the starting torque is maximum. In case the stator supply frequency is increased to run the motor above rated speed, the voltage cannot be increased above the rated value, and hence the output torque of the motor would fall above rated speed.

14.5 OPERATION OF INDUCTION MOTOR FROM NON-SINUSOIDAL VOLTAGE SOURCE

In adjustable speed drives, the machines are fed by converters which contain harmonics at the output. These harmonics have the following harmful effects: (1) heating, and (2) torque pulsation. The time harmonics present in the applied voltage results in current at the harmonic frequencies. These harmonic currents increase the losses in the motor. The harmonic currents are responsible for producing speed and torque ripple in the motor. The output voltage of a three-phase inverter is supplied to the stator of the induction motor. The Fourier series expression for the voltage supplied to the motor is given by:

$$V_n = \sqrt{2}\,(V_1 \sin \omega t + V_5 \sin 5\omega t + V_7 \sin 7\omega t + ... V_n \sin n\omega t) \quad ...(14.22)$$

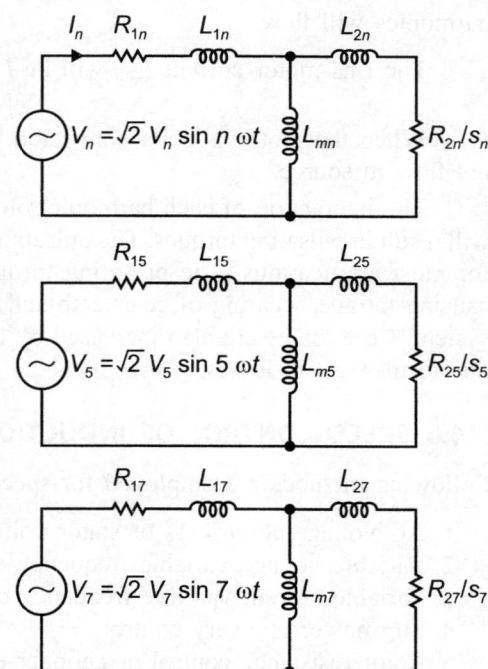

For each harmonic component, the machine can be approximately represented by a constant parameter linear equivalent circuit and the resultant current can be calculated by the superposition principle. The per phase equivalent circuit of Fig. 14.2 can be converted to a harmonic equivalent circuit as shown in Fig. 14.6. In this figure, n is the order of harmonic and s_n is the slip at the nth harmonic. The amplitude of the harmonic voltages decreases as order of the harmonics. The important harmonics are the fifth and the seventh. The equivalent circuits for the nth harmonic, the fifth, and the seventh harmonics are shown in Fig. 14.6.

The different regions of torque-speed curves of the motor for the fundamental, the fifth harmonic and the seventh harmonic are shown in Fig. 14.7. The motor operates at rated torque and rated speed depending on the fundamental frequency, the rotor appears stationary wrt harmonic field $s_n \cong 1.0$. Mathematically, the slip for the nth harmonic is given by

$$s_n = \frac{n\omega_s \mp \omega}{n\omega_s} \quad ...(14.23)$$

Fig. 14.6 Equivalent circuits for nth harmonic

where, ω_s = synchronous speed for the fundamental

ω = speed of rotor of motor

Negative sign is used for forward rotating fields and positive sign to those which produce backward rotating fields.

A harmonic component of the air gap flux induces rotor current at the same frequency and therefore torque is produced in the same direction as the rotating air gap flux. The 7th

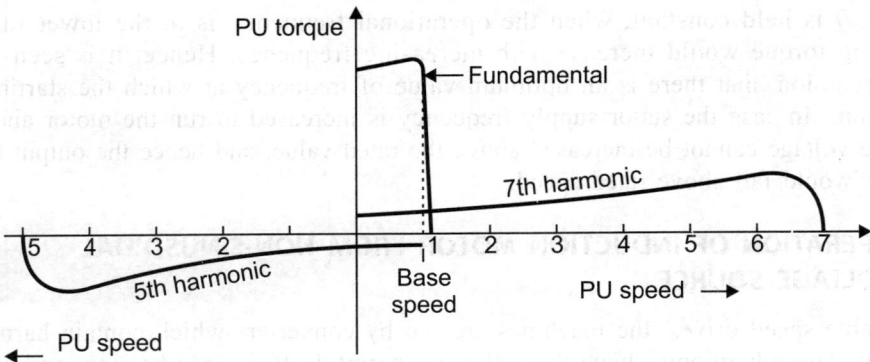

Fig. 14.7 Different regions of torque-speed curves of motor for fundamental, 5th and 7th harmonics

harmonic torque, adds to the fundamental torque, but the 5th harmonic torque opposes it. Hence, for the harmonics it is almost locked rotor condition, when the motor is running at its rated speed. Thus, the harmonic torques tend to cancel each other. Generally supply will have odd harmonics. When the stator is star connected, third harmonics and its multiple harmonics will flow.

The rms motor current I_{rms} will be $I_{rms}^2 = I_s^2 + \sum_{n=5,7,11...} I_n^2$

When the motor is delta connected, third harmonic current will flow in delta but will not flow in source.

The interaction of each harmonic rotor current with the air gap flux of another harmonic will result in pulsating torques. The pulsating torques are generally small and not very significant for most applications. The pulsating torque frequency may result in severe shaft vibration, causing fatigue, wearing of gear teeth and unsatisfactory performance in the feedback control system. Core losses are also increased by harmonics and hence the efficiency will be reduced due to increase in losses.

14.6 SPEED CONTROL OF INDUCTION MOTOR

Following methods are employed for speed control of induction motors:

1. AC voltage controllers or stator voltage control
2. Variable voltage variable frequency control
3. Variable current variable frequency control
4. Slip power recovery control
5. Rotor resistance control or chopper controller

14.6.1 AC Voltage Controller or Stator Voltage Control

By reducing stator voltage, speed of a high-slip induction motor can be reduced by an amount which is sufficient for the speed control of some fan and pump drives (Fig. 14.8). While torque is proportional to voltage squared, and current is proportional to voltage. Therefore, as voltage is reduced to reduce speed, for the same current, motor develops lower torque. Consequently, this method is used where torque requirement reduces with speed. The low-speed performance of induction motor is poor because the motor current is proportional to the applied voltage, whereas the electromagnetic torque varies approximately as the square of the voltage. Consequently, the torque per ampere is lower at reduced speeds, and large currents are required to develop appreciable torque.

Speed control is obtained by varying the firing angle of thyristors. These controllers are preferred over conventional variable resistance regulators. Saturable reactors have been used in the past to perform this function, but thyristor circuits now offer several advantages. (1) The thyristor unit is more compact, inspite of the heat sink requirements, and it weighs

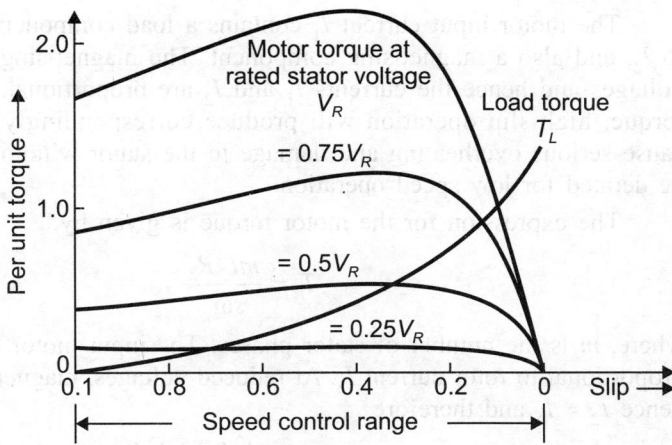

Fig. 14.8 Torque-speed performance

considerably less. (2) It has a higher efficiency, and the response time is a half cycle of the supply frequency, compared with a time lag of 0.1 second, or more, for a saturable reactor circuit. (3) Thyristor units from different manufacturers are interchangeable, whereas saturable reactor characteristics vary considerably.

Variable stator voltage at constant frequency for controlling the speed of induction motors, the circuit consists of two thyristors connected back-to-back, and triggered symmetrically at identical points in their anode-to-cathode voltage cycles. By using these circuits in the stator terminals, and controlling the conduction periods of the thyristors, the input voltage to the motor can be varied from zero to the full supply voltage. The motor is subjected to a chopped sine-wave voltage and the supply currents also have a high harmonic content, but satisfactory operation has been achieved with small and medium sized induction motors. For low rating machines, anti-paralleled thyristor pair in each phase can be replaced by a triac.

Principle of operation: Assuming the motor voltages and currents are balanced and sinusoidal, and also the rotor resistance is constant and independent of rotor frequency. If ω_s denotes the synchronous angular velocity and ω is the rotational angular velocity, the fractional slip is defined by $s = (\omega_s - \omega)/\omega_s$. The gross mechanical power output, including windage and friction losses $P_{mech} = T\omega$, where T is the internal motor torque. In accordance with standard theory, the total power input to the rotor, across the air gap from the stator, is $P_{ag} = T\omega_s$. The difference between the power input to the rotor and the mechanical power output is dissipated as heat losses in the rotor resistance. Thus, the rotor copper loss is given by

$$P_{Cu} = P_{ag} - P_{mech} \qquad \qquad ...(14.24)$$
$$= T(\omega_s - \omega) = sT\omega_s = sP_{ag}$$

A fraction s times the rotor input power is dissipated as rotor copper loss in the form of heat, and the remainder $(1 - s) P_{ag}$ is available as mechanical power output. Hence,

$$P_{mech} = (1 - s) P_{ag} \qquad \qquad ...(14.25)$$

Speed control using variable stator voltage at constant frequency is therefore inherently inefficient, since the rotor copper loss increases linearly with slip for a constant motor torque. Equation (14.25) shows that the rotor copper loss is proportional to the product of torque and slip, and the motor heat generated in a given duty cycle may be quickly approximated by evaluating the product of torque, slip, and time, for each portion of the duty cycle.

The motor input current I_1 contains a load component of current which is proportional to I_2, and also a magnetising component. The magnetising current may be neglected at low voltages and hence the currents I_1 and I_2 are proportional. The large rotor currents at high-torque, high-slip operation will produce correspondingly large stator currents which may cause serious overheating and damage to the stator windings. Consequently, the motor must be derated for low speed operation.

The expression for the motor torque is given by:

$$T = \frac{mI_2^2 R_2}{s\omega_s}$$

where, m is the number of stator phases. The input motor current I_1 comprises a component proportional to rotor current I_2. At reduced voltages, magnetising component is negligible and hence $I_2 \approx I_1$ and therefore

$$T \propto \frac{I_2^2 R_2}{s} \, \alpha \, \frac{I_1^2 R_2}{s} \qquad \qquad ...(14.26)$$

For stable operation to occur, the motor torque $T = T_L$ (the load torque)

For fan load, load torque $T_L \propto (\omega)^2 \propto (1 - s)^2$

For stable operation, $\qquad I_1 \propto I_2 \propto (1 - s) \, \sqrt{(s/R_2)} \qquad \qquad ...(14.27)$

This result indicates that the machine currents are inversely proportional to $\sqrt{R_2}$ and by differentiation, it is found that the currents have maximum values when $s = 1/3$, that is, at speed $\omega = 2/3\omega_s$.

For a constant-torque load, T_L is constant, and assuming motor and load torques are equal, gives the result

$$I_1 \propto I_2 \propto \sqrt{(s/R_2)} \qquad \qquad ...(14.28)$$

The motor input and rotor currents are again inversely proportional to $\sqrt{R_2}$, but they now increase slowly as the motor speed is reduced to zero, and are even larger in the plugging region where the slip is greater than a unity.

It is a fact that the rotor resistance is an important factor in determining the increase in motor current. If the normal full-load slip of a squirrel-cage motor is small, a very large increase in stator current will occur at low speeds when stator voltage control is employed. For satisfactory low-speed operation in a fan drive, the rotor should have a high resistance corresponding to a full-load slip of 10 to 14 percent. In a constant-torque drive, squirrel-cage rotors with variable-resistance, characteristics may also be used. The usual deep-bar or double-cage rotor is not very effective, and hence is not significant unless the slip exceeds about 0.5. An improved performance can be obtained by using a solid-iron rotor, or by placing a bar of permanent-magnet material, such as ALNICO, in the top of each rotor slot. The hysteresis loss in the ALNICO causes an increase in the effective rotor resistance, thereby increasing the torque per ampere at low speeds and giving the motor a better torque-speed performance.

Slip ring induction motors are commonly used in variable-voltage speed control methods. External rotor resistance is added to modify the torque-speed characteristic so that the torque per ampere at low speeds is increased. Motor heating is reduced, since part of the rotor circuit loss is now dissipated externally. Improvement in low speed performance may be obtained by connecting Alnico-cored saturable reactors, known as *saturistors* in the rotor circuit of the slip ring induction motor. To reduce the stator voltage when thyristor switching circuits are used either symmetrically or asymmetrically, the resulting voltage and current waveforms are highly distorted and the additional harmonic losses will further enhance the overheating problem at low speeds.

AC voltage controller or stator voltage controller for delta connected induction motor: Figure 14.9 shows two commonly used symmetrical three-phase AC voltage controller circuits for delta and star connected stators. For small size motors, triac may be used. The effective load voltage in three-phase AC circuit can be varied by a thyristor controller consisting of a pair of anti-parallel thyristors in a delta-connection as shown in Fig. 14.9(a) and in star connection shown in Fig. 14.9(b). Each pair of back-to-back thyristors controls the voltage delivered to one phase of the stator, and the phase voltage waveform consists of a series of sine-wave segments. A phase displacement of 120° is maintained between the sets of gating pulses delivered to each controller in order to produce a symmetrical reduction of the three-phase voltages. In the full-voltage condition, each thyristor receives a gating signal at the start of the positive half-cycle of its anode-to-cathode voltage. The series thyristor pair is then virtually a short-circuit and the stator phase receives a complete cycle of supply voltage. As the thyristor firing is delayed, the conduction period of each thyristor is shortened and the effective stator voltage is reduced. When the firing delay is 180°, the thyristor controller is open-circuited and the motor voltage and current are both zero. Using a switching device in the manner as a non-linear series impedance has the advantage that the power dissipation in the controller is considerably less than the load power, but the switching action also distorts the stator voltages. In delta connection third harmonic voltages produced by motor back emf causes circulating current through the windings which increases losses and thermal loading of the motor. Speed control is obtained by varying conduction period of thyristors. For low power rating motors anti-parallel thyristor pair can be replaced by a triac. AC voltage controllers are also used for soft start of motors.

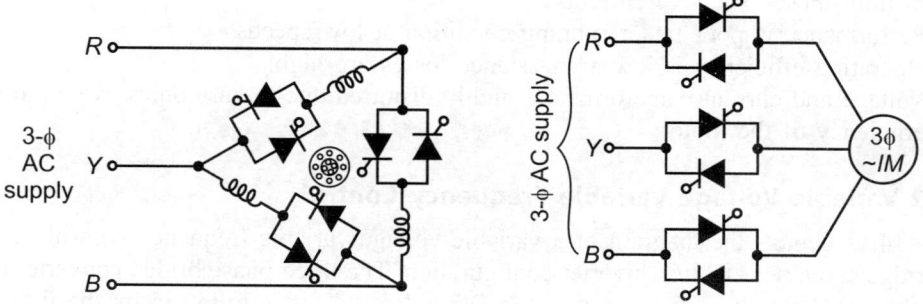

Fig. 14.9 AC voltage controller circuits for delta and star connected stators

The direction of rotation of a three-phase induction motor can only be reversed by changing the phase sequence of the applied voltages by introducing two additional thyristor controllers.

Figure 14.10 shows a commonly used closed loop system for variable speed AC drives, using stator voltage control technique. The DC tacho-generator develops a voltage proportional to the motor speed, and this is compared with the DC reference voltage representing the desired speed. The difference between the two signals is the error signal voltage, which controls the thyristor firing angles, and thereby changes the terminal voltage and motor speed so that the error signal is reduced. If the reference voltage is greater than the tacho-generator voltage, the conduction periods of the thyristors are increased. The increased stator voltage allows the development of an increased motor torque and hence the motor speed rises. If the tacho-generator voltage exceeds the reference voltage, the conduction angle of thyristor are reduced and the motor torque decreases, causing a reduction in shaft speed. When the motor speed is equal to the desired speed, the conduction angles are just sufficient to allow the development of a motor torque equal to the load torque. In a high-gain feedback system, the desired speed can be accurately maintained and there is no necessity for the motor to have a flat speed-torque characteristic, since the output speed is determined by the reference signal rather than the open-loop characteristic of the motor, stable operation may be obtained at any point of the induction motor speed-torque characteristic.

Merits of stator voltage control method:

1. The control circuitry is simple and more compact and it weighs less.
2. There is considerable savings in energy and thus it is an economical method as compared to other methods of speed control.
3. Its response time is quick.

Demerits:

1. Maximum torque available from the motor decreases with reduction in stator voltage.
2. At low speeds, motor currents are excessive and special arrangements should be provided to limit these excessive currents.
3. Performance is poor under running condition at low speeds.
4. Operating efficiency is low as resistance losses are high.
5. Voltage and current waveforms are highly distorted due to harmonics, which affects the efficiency of the motor.

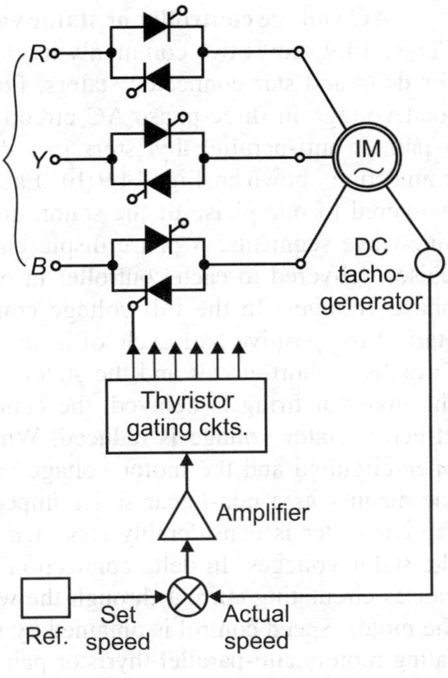

Fig. 14.10 Closed loop system for variable AC drive

14.6.2 Variable Voltage Variable Frequency Control

Figure 14.11 depicts the diagram of a variable voltage variable frequency control. It consists of a bridge converter DC link inverter configuration. The three-phase bridge converter converts three-phase AC supply voltage to variable DC voltage. This is followed by the filter circuit. The output of the filter is then fed to the input of the bridge inverter. The inverter produces a variable frequency variable voltage supply which is then used to control the input of AC motor. The circuit inverter contains six thyristors and six diodes. The firing circuits are not shown in

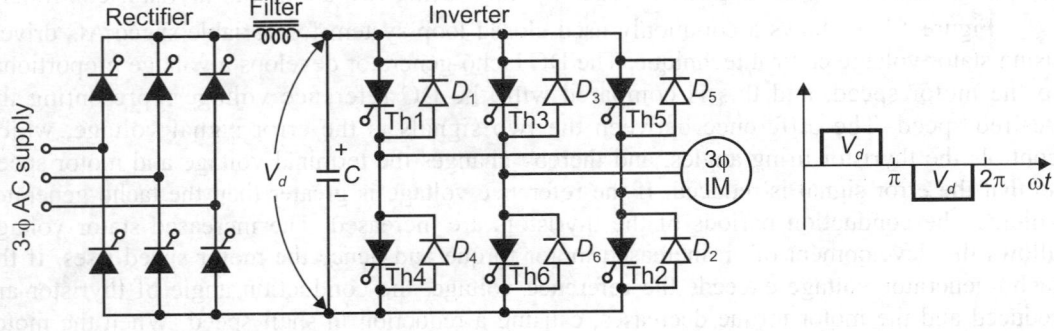

Fig. 14.11 Circuit of a variable voltage variable frequency control system and waveform

the figure. The firing of thyristor Th1 ensures the turning of thyristor Th4 since load current cannot reverse, the only path for this current is through diode D_1. The inverter generates a variable voltage variable frequency power supply to control the speed of the motor, to avoid saturation at low frequencies. The capacitor C, as shown in Fig. 14.11, supplies stiff voltage supply to the inverter and the inverter output voltage waves are not affected by nature of load.

The induced emf of stator phase is given by

$$V = 44.4 \Phi f\ V \text{ (volts)} \qquad \qquad ...(14.29)$$

where, V is the applied voltage, f, the supply frequency, Φ the air gap flux and N the number of turns per phase in the stator winding.

In order to achieve constant torque operation below rated speed, flux Φ has to be kept constant. The flux Φ can be kept constant, if the ratio (V/f) is kept constant. Thus, to control the speed of AC induction motor, below the rated value, not only frequency has to be decreased but also voltage has to be decreased in the same proportion such that (V/f) is constant.

The section between rectifier and inverter is known as *DC link*. It is possible to control the current in the DC link, which is a two stage frequency conversion device. AC is converted into DC using rectifier. The DC is inverted to AC of required frequency using a force commutated inverter. If the voltage of the DC link is to control the speed of the motor the inverter is called voltage source inverter. The voltage applied to the motor is held constant by using a capacitor. In this method at low speeds the motor operation is jerky. A voltage source inverter feeding a three-phase induction motor is shown in Fig. 14.11 along with voltage waveform. A controlled rectifier can be used, as it gives a fast response to any control demand but it suffers from the major disadvantage of all controlled rectifiers of lagging power factor on the AC supply.

14.6.3 Variable Current Variable Frequency Control

The controls shown in Fig. 14.12 function in a similar manner to those discussed in variable voltage variable frequency control. The inputs sets the inverter frequency and hence the speed, with the voltage adjusted to match the frequency. The torque is controlled during the acceleration periods. Figure 14.12 display the circuit diagram of variable current variable frequency control. A phase controlled rectifier generates variable DC voltage which is being converted to a direct current source by passing through a large inductor in series. The input current to the motor in such a case has a waveform shown in Fig. 14.12, which is a quasi wave. The inverter frequency is controlled so that the speed can be variable, with the voltage adjusted to match the frequency. The current is controlled during the acceleration periods.

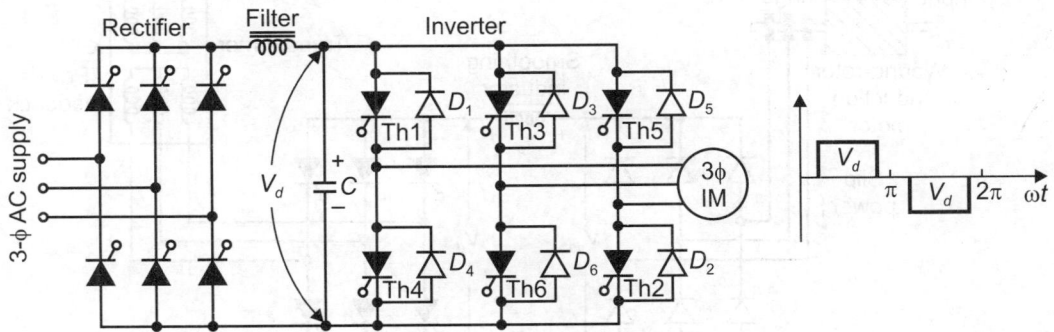

Fig. 14.12 Variable current variable frequency control circuit

The drive is very suitable for applications where violent changes in shaft torque must be avoided, because the very nature of the DC link inductor prevents the sudden current changes.

Another advantage of constant current inverter is that a fault such as a short circuit at motor terminals does not damage the inverter as the current remains constant. Here regeneration is possible without extra converter. The power circuit of the method is rugged and reliable. Besides, any sudden changes in current at the motor terminals due to short circuit do not damage the inverter because of the inductor.

14.6.4 Slip Power Recovery Control

The slip energy recovery system for speed control of a slip ring motor is shown in Fig. 14.13. It is possible to vary the motor speed efficiently by taking the power from the rotor. This method of speed control is applicable only with wound rotor induction motor and in addition to its high starting torque, it has the following of advantages:

1. variation of speed over a wide range below synchronous speed of motor,
2. simplicity of operation from mechanical as well as from automatic point of view,
3. low maintenance cost for mechanical and automatic controllers.

Besides the above advantages, it has few disadvantages:

1. low efficiency due to increased rotor resistance losses,
2. poor speed regulation.

Principle of slip power recovery control: In this method slip power is dissipated in the resistance and this reduces the efficiency of the motor at low speeds. This slip power can be recovered to the mains using inverter. The division of power in the rotor circuit has already been studied. The power delivered to the rotor across the air gap P_{ag} is divided between the mechanical power output P_{mech} and the rotor copper loss P_{Cu} and given by:

$$P_{Cu} = sP_{ag} \qquad \qquad ...(14.30)$$

and
$$P_{mech} = (1 - s) P_{ag} \qquad \qquad ...(14.31)$$

Also
$$P_{ag} = T\omega_s$$

where, T is the electromagnetic torque developed by the motor, and ω_s is the synchronous angular velocity.

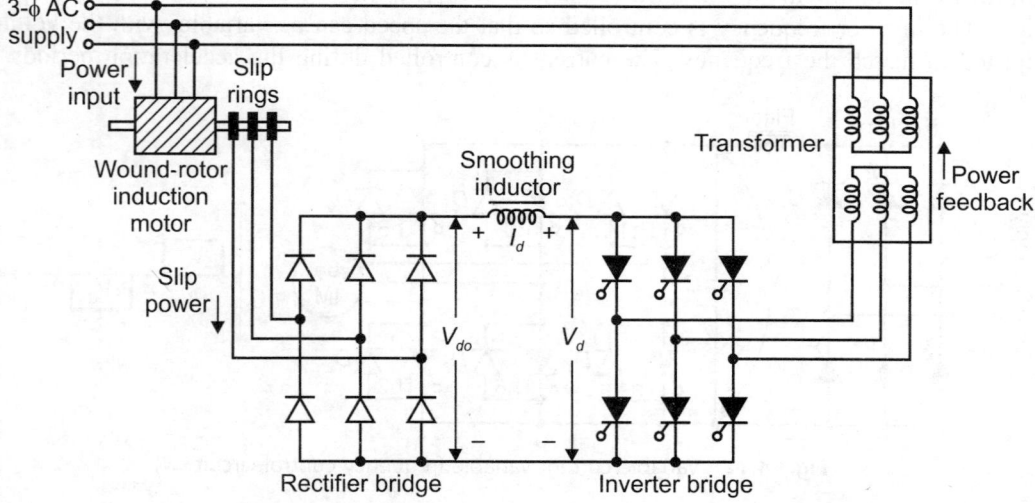

Fig. 14.13 Slip power recovery control circuit

The air gap flux of the machine is established by the stator supply and it remains practically constant if the stator drops and supply voltage fluctuations are neglected and hence the rotor copper loss is proportional to slip. Speed control of a wound-rotor motor by the introduction of external rotor resistance is, therefore, inherently inefficient. At half synchronous speed, the air gap power is divided equally between mechanical power output and rotor copper loss, giving an overall efficiency of less than 50 percent.

In general, at slip s, mechanical power is obtained from the air gap power with a per unit conversion efficiency of $(1 - s)$, and the overall motor efficiency is even lower than this. The drive system is not only efficient but the converter power rating is low because it has to handle only the slip power. This power rating becomes lowers for a more restricted speed range near the synchronous speed. The rotor resistance method of speed control is, therefore, uneconomical except for a very small sub-synchronous speed range. However, it is not essential that the slip power sP_{ag} be dissipated in resistance losses, as it can be removed from the rotor circuit and utilized externally, thereby improving the overall efficiency of the system. In these cascade connections, the slip power is either returned to the supply network or is used to drive an AC motor which is mechanically coupled to the same induction motor shaft.

Speed control of slip-ring induction motor for sub-synchronous speed (static Kramer drive): Instead of wasting the slip power in the rotor circuit resistance, it can be converted to line frequency and fed back to the line as shown in Fig. 14.13. This synchronous region speed control principle where the slip power is recovered back to the line through a converter cascade is known as a *static Kramer drive*. In this, control power is taken from the rotor at slip frequency to vary the motor speed efficiently. The voltage at slip rings is at slip frequency which is incompatible with stator frequency, therefore slip ring voltage is rectified by means of diode bridge into DC link. The DC link power is transferred back into the main supply via thyristor. In other words, the converter operates in the inverting mode. The rectifier and inverter are both phase commutated by alternating emfs appearing at the slip rings and supply bus bars respectively. The average counter emf of the inverter may be regarded as an injected emf opposing the rectified rotor voltage.

Mathematical analysis: Assuming commutation overlap negligible, the direct voltage output of the uncontrolled three-phase bridge rectifier is given from the theory of three-phase six-pulse bridge circuit.

$$V_{do} = 1.35V_r\, s \qquad \qquad ...(14.32)$$

where, V_r is the line-to voltage at standstill, and s is the fractional slip.

For a three-phase bridge inverter, the average counter emf, neglecting the stator and rotor voltage drops associated with the system, is given by

$$V_d = 1.35V_L \cos \alpha \qquad \qquad ...(14.33)$$

where, α is the inverter firing delay, which is in the range $90°$ and $180°$ and V_L is the AC line voltage.

On on-load, the motor torque is negligible and the rectified rotor current is almost zero. Since V_d and V_{do}, the two direct voltages of Equations (14.32) and (14.33) must balance in the ideal case.

Thus, $$1.35V_r S + 1.35V_L \cos \alpha = 0$$

or $$s = -(V_L/V_r) \cos \alpha = -a \cos \alpha = a\,|\cos \alpha| \qquad \qquad ...(14.34)$$

where, a is the effective stator-to-rotor turns ratio of the motor.

Maximum value of α is restricted to $165°$ for safe commutation of inverter thyristors. Slip can be controlled from 0 to $0.966a$, when α is changed from $90°$ to $165°$. By taking proper value of a required speed range can be obtained. At zero speed, voltage V_d is maximum which corresponds to $\alpha = 180°$ and at synchronous speed $V_d = 0$ when $\alpha = 90°$.

Speed control is, therefore, obtained by a simple variation of the inverter firing angle. If 'a' is unity, the no-load speed of the motor can be controlled from standstill to full speed.

In order to develop load torque, a rotor current I_2 is required, and the rectified rotor voltage must force current flow against the inverter back emf. As the induction motor is loaded, the speed falls slightly and the resulting increase in rectified voltage produces the necessary increase in rotor current.

If the rotor resistance is small, the fundamental rotor slip power, sP_{ag}, is approximately equal to the DC link power.

Thus, slip power $\quad\quad sP_{ag} = V_d I_d$

but $\quad\quad\quad\quad\quad\quad P_{ag} = T\omega_s$

and hence $\quad\quad\quad\quad T = \dfrac{P_{ag}}{\omega_s} = \dfrac{V_d I_d}{s\omega_s}$ ...(14.35)

If the speed drop on load is neglected, Equation (14.35) for the no-load slip can be substituted in this result. Substituting also for V_d from Equation (14.35) gives the torque expression

$$T = \frac{1.35 V_L I_d}{a\omega_s}$$

or $\quad\quad\quad\quad\quad\quad T_d \propto I_d$ ...(14.36)

Which indicates that the torque is proportional to the rectified rotor current I_d which in turn, is proportional to the difference between the rectified rotor voltage and the average counter emf of the inverter. The inverter emf is constant for a fixed firing angle, and hence the rotor slip increases linearly with load torque, having characteristic of a separately excited DC motor with armature-voltage control. In practice, the open-loop torque-speed characteristics have been shown in Fig. 14.14.

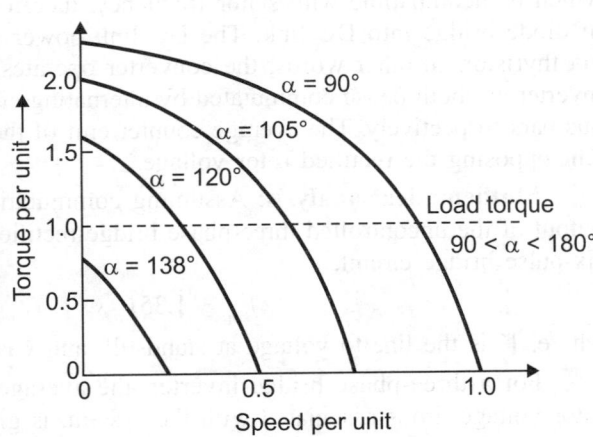

Fig. 14.14 Torque speed characteristics

Closed-loop control: The closed-loop control system shown in Fig. 14.15, a drop in speed automatically reducing the link voltage, hence allowing more current to circulate in the rotor. An inner loop limits the current in the converter to a safe level. The unidirectional power flow in the rotor circuit prohibits to a safe level. Figure 14.15 shows a block diagram of the control scheme. The inner control loop adjusts the current in the cascade circuit by variation of the thyristor firing angle, and this current determines the motor torque. The actual cascade current is measured by current transformers on the AC side of the inverter, and the desired current value is set by the outer speed loop, which measures the difference between the desired speed, as set on a potentiometer, and the actual tacho-generator value. Thus, a speed error produces a motor torque which reduces the error. Current limitation is readily incorporated by limiting the reference current value delivered by the speed-error amplifier. In this manner, the cascade current may be limited to 120 percent, say, of rated current, even under stalled conditions.

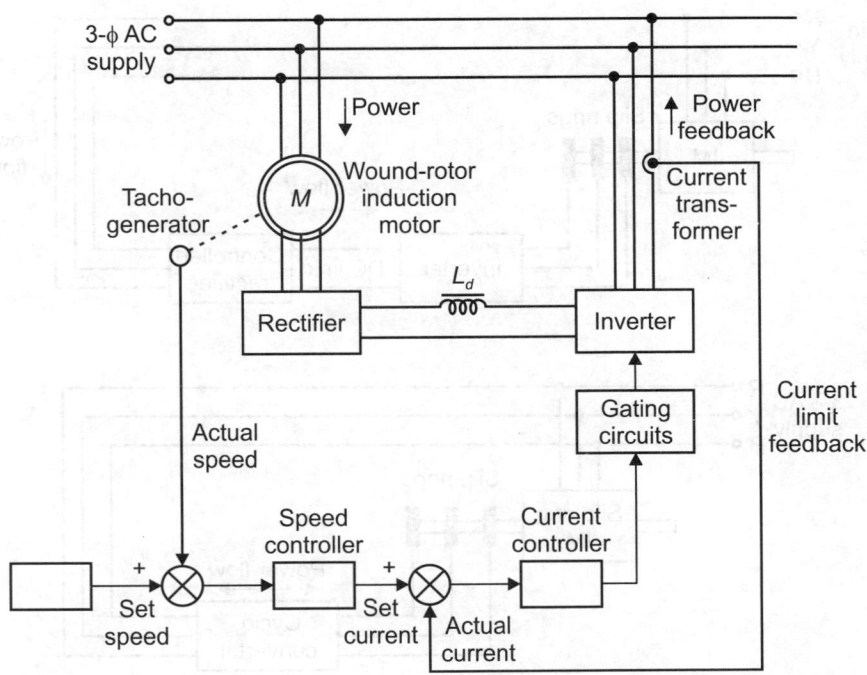

Fig. 14.15 Closed loop speed control circuit for slip-ring induction motor

If the desired speed greatly exceeds the measured value, the motor accelerates rapidly with maximum current and torque, until the desired speed is approached. The current and torque are then reduced automatically to the values required by the load. For a sudden reduction in demanded speed, the current is reduced to a low level, and the braking action of the load reduces the motor speed. The cascade drive control system is much simpler than any other variable-speed induction motor drive system in which the rotor slip is measured and controlled.

Speed control of slip-ring induction motor for super synchronous speed cascade: Super-synchronous speeds are possible with slip ring induction motor by injecting power into rotor circuit. Super-synchronous speeds may also be obtained by reversing the direction of power flow in the cascade circuit, thereby feeding additional power into the rotor. For this, to be possible, the arrangement of Fig. 14.16 can be used where the inverter can be self-commutated rotor voltage at speeds well away from synchronous speed. However, near to synchronous speed, the rotor voltage is low, and forced commutation must be employed in the inverter, which makes the scheme less attractive. The replacement of six diodes by six thyristors increases the converter cost appreciably, and also necessitates the introduction of a slip-frequency gating circuit. Difficulty is experienced near synchronism when the slip-frequency emfs are insufficient for phase commutation, and special connections or forced-commutation methods are necessary for the passage through synchronism. Thus, the provision of super-synchronous speed control unduly complicates the static converter cascade system and nullifies the advantages of simplicity and economy.

The dual-converter system in a static Kramer drive can be replaced by cycloconverter, as shown in Fig. 14.16 (b), the cycloconverter permits the slip power to flow in either direction. The slip-energy control scheme of Fig. 14.16 (b) seems ideally suited as an application of cycloconverter, being suited in the rotor circuit, provided the slip is less than (0.33) giving the cycloconverter frequency ratio above 3 to 1. Since the cycloconverter employs a large

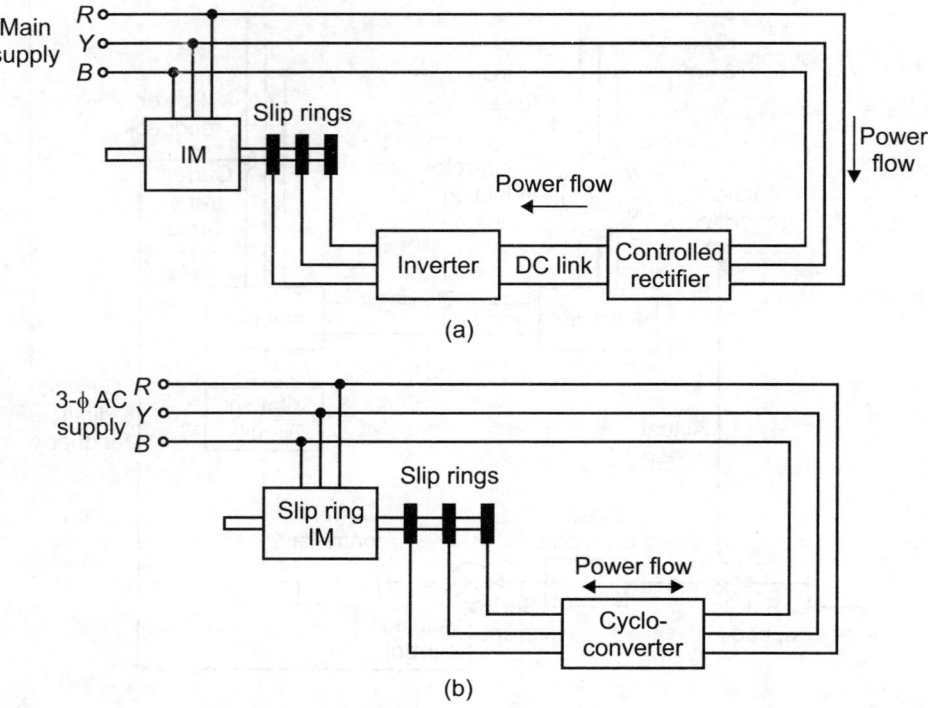

Fig. 14.16 Diagram for super synchronous cascade arrangement

number of thyristors, the drive is suitable for large capacity drives. The application of slip-ring induction motor is most attractive for pump and ventilation drives, which have a power requirement, i.e. approximately the cube of the shaft speed; thus only small speed changes are necessary to give quite large load power variation.

14.6.5 Induction Motor Control by Choppers, or Rotor Resistance

The speed of a wound rotor induction motor can be varied by varying the rotor resistance. The rotor resistance can be varied steplessly by using a simple chopper circuit as shown in Fig. 14.17(a). This method of speed control is very inefficient because slip energy is wasted

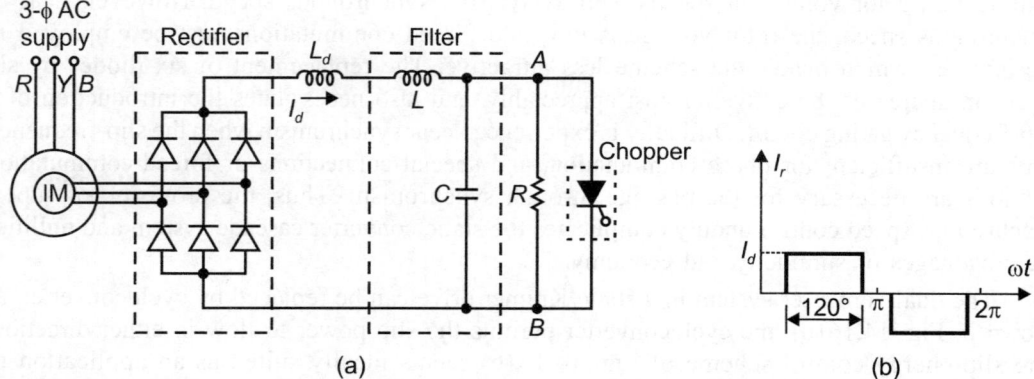

Fig. 14.17 Speed control of induction motor using chopper and its current waveform

in rotor circuit resistance. However, advantages are that high starting torque is available at low starting current and improved power factor is possible with wide range of speed control. The stator of the machine is directly connected to the line power supply, but in the rotor circuit the slip voltage is rectified to DC by bridge rectifier. The DC voltage is converted to a current source by connecting a large series inductor L_d and is fed to a chopper with an external shunt resistor R as shown.

The chopper periodically connects and disconnects the resistance R. When the chopper is off, the resistance is connected in the circuit and DC link current I_d flows through it. On the other hand, if the chopper is on, the resistance is short circuited and the current I_d is bypassed. The chopper operates with a duty cycle

$$\delta = \frac{t_{ON}}{T}$$

where, t_{ON} is the on time and T is the time period.

Equivalent resistance between terminals A and B can be given by

$$R_{AB} = (1 - \delta)\, R \qquad\qquad ...(14.37)$$

where, δ can be varied to vary the equivalent resistance.

Power consumed by resistance R_{AB} is

$$P_{AB} = I_d^2\, R_{AB} = I_d^2 \cdot R\,(1 - \delta) \qquad\qquad ...(14.38)$$

Rotor current waveform is shown in Fig. 14.17(b) when the ripple is neglected, the rms value of rotor current is

$$= \left[\frac{1}{\pi} \int_0^{2\pi/3} I_d^2\, d\omega t \right]^{1/2}$$

$$I_r = \sqrt{2/3} \cdot I_d \ \text{ or } I_d = \sqrt{\frac{3}{2}}\, I_r$$

So, power consumed by resistance per phase

$$= \frac{P_{AB}}{3}$$

$$= \frac{I_d^2}{3} R\,(1 - \delta) = \left(\sqrt{\frac{3}{2}} \right)^2 \frac{I_r^2}{3} \cdot R\,(1 - \delta)$$

$$P_{AB}/\text{phase} = \frac{I_r^2}{2} R\,(1 - \delta) = 0.5 I_r^2 R\,(1 - \delta)$$

This equation suggests that rotor circuit resistance per phase is increased by $0.5R\,(1 - \delta)$.

Thus, total rotor resistance per phase is

$$R_{rT} = R_r + 0.5R\,(1 - \delta) \qquad\qquad ...(14.39)$$

R_{rT} can be varied from R_r to $(R_r + 0.5R)$ as δ is changed from 1 to 0. Apart from stepless control, this method has an advantage of ensuring that rotor resistance remains balanced between three-phases for all operating points.

Hence, the rotor resistance can be varied steplessly by varying time t_{ON} and t_{OFF}. The following problems can be minimised by providing an additional LC filter. Use of simple circuit leads to following problems, which can be minimised by providing LC filter as shown in Figs 14.17(a) and (b).

1. Discontinuity in rotor current can occur, because the total inductance in the circuit is only due to leakage inductance of rotor.
2. For obtaining lower speed, higher values of R is to be used and this may lead to high voltage across thyristor, thereby requiring high PIV rating of thyristor Th.
3. Heating is produced in the rotor winding because rotor circuit is accompanied with high value of harmonics.

Closed-loop control: For satisfying the transient and steady state performance of induction motor, a closed loop control is normally employed. Low speed operation of a slip ring induction motor can be simply obtained by the introduction of external rotor resistance in the rotor to dissipate the slip power. Speed control is obtained by mechanical variation of the external resistance, and the drive is suitable for crane and hoist applications where sustained low-speed running is not required.

Now-a-days, a high-frequency thyristor chopper circuit allows the external rotor resistance to be varied statically and steplessly, and provides a low cost variable-speed drive with a good dynamic response. The schematic diagram of a closed loop induction motor drive circuit is shown in Fig. 14.18, in which the rotor slip power is rectified in a diode bridge rectifier and fed through a smoothing reactor L_d to a resistor R. A single thyristor in parallel with the resistor is switched on and off, at a frequency of about 1 kHz, by a chopper circuit. The chopper periodically connects and disconnects the resistance R. When the chopper is off, the resistance is connected in the circuit and the DC link current I_d flows through it on the other hand, if the chopper is on, the resistance is short circuited and the current I_d is bypassed. The ratio of on-time to off-time determines the effective value of rotor resistance and thus controls the motor speed by altering its torque-speed characteristic. By introducing a capacitor in series with the external resistor, it is possible to obtain a variation in the effective resistance from zero to infinity, thus permitting a wider range of speed control. This type of rotor circuit may also be used in combination with the stator-voltage-control circuits. When the load torque is small, speed control is obtained by variation of the stator voltage, and rotor resistance control is used in the high-torque range is bypassed.

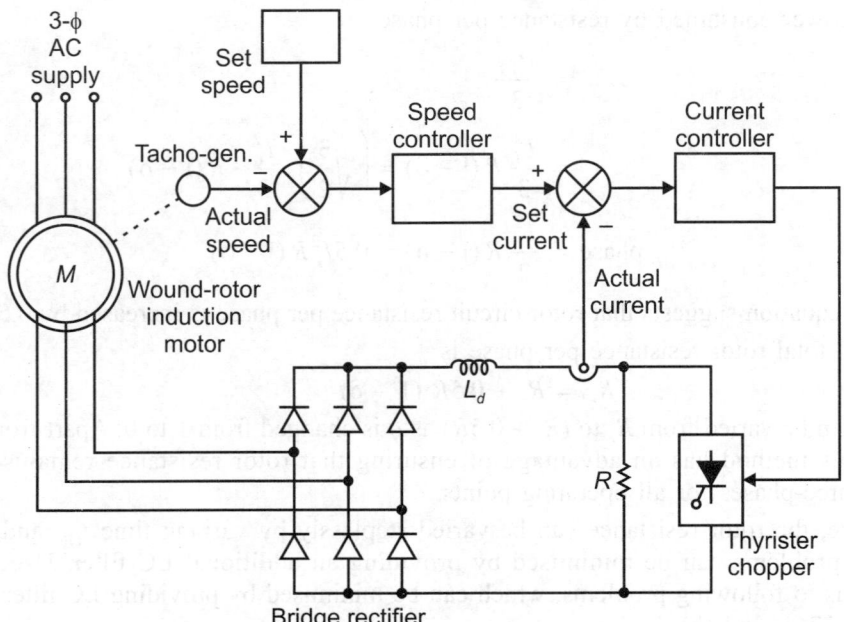

Fig. 14.18 The schematic diagram of closed loop induction motor drive circuit

14.7 STATIC SCHERBIUS DRIVE

If the inverter shown in Fig. 14.13 is a dual converter system replaced by a single-phase controlled line commutated cycloconverter of as shown in Fig. 14.19, then it is possible to have slip power flow in either direction and this scheme is called *static Scherbius drive*. This method has found applications in very large horse-power pump and blower type drives. The cycloconverter permits the slip power to flow in either direction, and therefore the speed of the machine can be controlled in both sub-synchronous and super-synchronous ranges with motoring and regeneration features. Various modes of operation shown in Fig. 14.20 assuming motor shaft torque as constant and the losses in the motor and cycloconverter are negligible can be explained as follows:

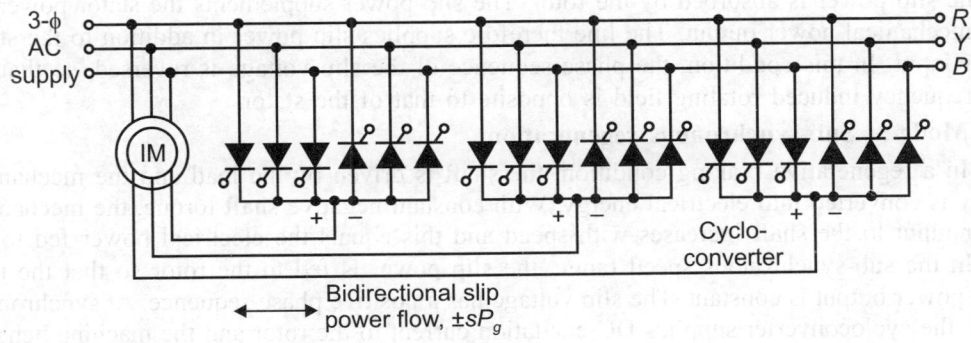

Fig. 14.19

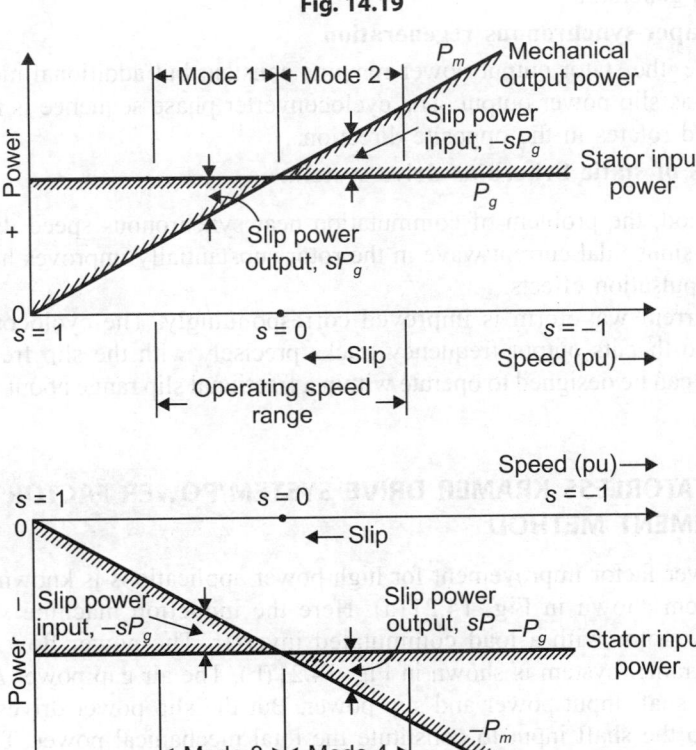

Fig. 14.20 Circuit of static Scherbius drive system

Mode 1: Sub-synchronous motoring

This mode is identical to that of the static Kramer system. The stator input or air gap power P_g remains constant and the slip power sP_g, which is proportional to the slip, is returned back to the line. Therefore, the line supplies the net mechanical power P_m consumed by the shaft. The slip frequency power in the rotor creates a rotating field in the same direction as in the stator and the rotor speed corresponds to the difference between these two frequencies. At slip $(s = 0)$, the cycloconverter supplies DC excitation to the rotor and the machine behaves like a synchronous motor.

Mode 2: Super-synchronous motoring

As the shaft speed increases beyond the synchronous speed, the slip becomes negative and the slip power is absorbed by the rotor. The slip power supplements the stator power for total mechanical power output. The line therefore supplies slip power in addition to the stator power input. In this condition, the phase sequence of the slip voltage is reversed, so that the slip-frequency induced rotating field is opposite to that of the stator.

Mode 3: Sub-synchronous regeneration

In a regenerative braking condition, the shaft is driven by the load and the mechanical energy is converted into electrical energy. With constant negative shaft torque, the mechanical power input to the shaft increases with speed and this equals the electrical power fed to the line. In the sub-synchronous speed range, the slip power is fed to the rotor so that the total stator power output is constant. The slip voltage has a positive phase sequence. At synchronous speed, the cycloconverter supplies DC excitation current to the rotor and the machine behaves as a synchronous generator.

Mode 4: Super-synchronous regeneration

In this mode, the stator output power remains constant but additional mechanical power input is reflected as slip power output. The cycloconverter phase sequence is now reversed so that the rotor field rotates in the opposite direction.

Advantages of static Scherbius drive system:

1. In this method, the problem of commutation near synchronous speed disappears.
2. The nearby sinusoidal current wave in the rotor substantially improves harmonic heating and torque pulsation effects.
3. The line current waveform is improved correspondingly. The cycloconverter is to be controlled so that its output frequency tracks precisely with the slip frequency.
4. This system can be designed to operate within a fractional slip range about the synchronous speed.

14.8 COMMUTATORLESS KRAMER DRIVE SYSTEM/POWER FACTOR IMPROVEMENT METHOD

A method for power factor improvement for high power applications is known as commutator less Kramer system shown in Fig. 14.21 (a). Here the induction machine shaft is coupled to a synchronous motor with a load commutated inverter. The power flow diagram in the commutatorless Kramer system is shown in Fig. 14.21 (b). The air gap power P_g flowing from stator is split into shaft input power and slip power. But the slip power drives a synchronous motor and adds to the shaft input to constitute the total mechanical power. The synchronous motor field is supplied from the line through a controlled rectifier. The speed and torque of the drive system are controlled by the inverter 'α' angle and field current, so that the load commutation of the inverter is possible at an optimum angle 'α' in a different speed. As a

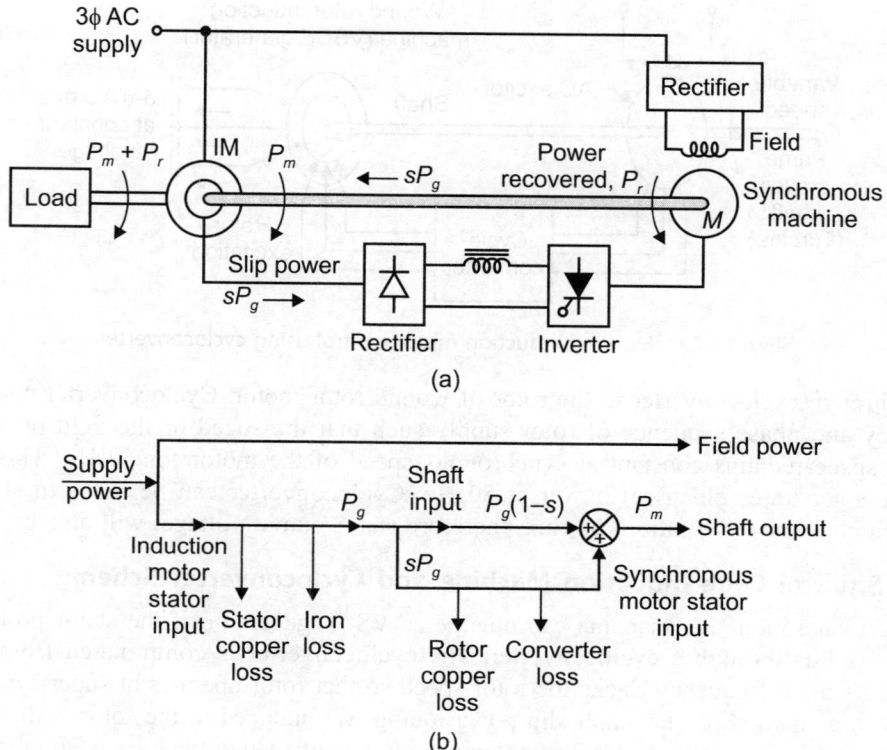

Fig. 14.21 Diagram of Kramer drive system

characteristic of the load commutated inverter drive, speed control is not possible at a low value because of insufficient counter emf. Besides having improved power factor, the system will operate reliably with momentary line voltage dip which will cause commutation failure in the static Kramer system.

The drive has a better power factor and lower harmonic content in the line current as compared to static Scherbius drive. In static Scherbius drives, reactive power and harmonics are associated with power fed back to the line. In static Kramer drive, since the power is not fed back to the line, problems associated with the feedback of power are also eliminated.

14.9 VARIABLE SPEED CONSTANT FREQUENCY GENERATION (VSCF)

Variable speed constant frequency generation (VSCF) involves generation of electrical power at fixed frequency and fixed voltage from a variable speed prime mover coupled to the generator shaft. This type of generator is used on aircraft and naval ships where speed of rotor varies with velocity and pressure of the wind. Both synchronous and induction machines can be used for (VSCF) generation. In generation, an adjustable speed AC drive system which has a regeneration capability can be used as a VSCF generator. Of several possible schemes two commonly used are as follows:

14.9.1 Slip Ring Induction Motor and Cycloconverter

In Fig. 14.22, rotor of the slip ring induction motor is coupled to the shaft of variable speed mover. An AC exciter is mounted on the same shaft and feeds its variable frequency

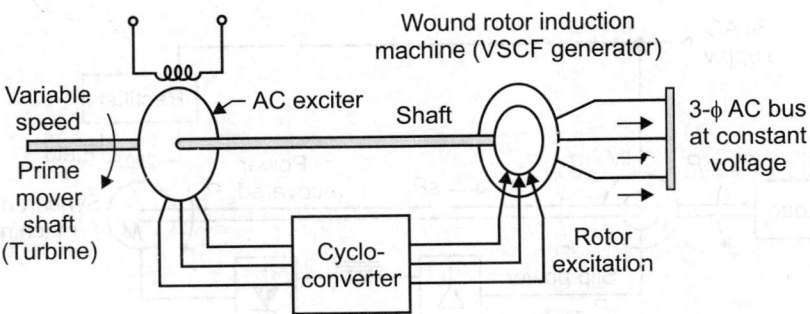

Fig. 14.22 Slip-ring induction motor control using cycloconverter

output through cycloconverter to the rotor of wound rotor motor. Cycloconverter controls the frequency and phase sequence of rotor supply such that the speed of the field produced by rotor in space remains constant at synchronous speed of the motor for 50 Hz. This ensures that stator generates electrical power at 50 Hz. Cycloconverter can be made to supply the rotor at a constant (V/f) ratio. The flux, therefore stator output voltage, will also be constant.

14.9.2 Squirrel Cage Induction Machine and Cycloconverter Scheme

A squirrel cage induction machine can operate as VSCF generator, if the stator power is fed to a 50 Hz bus through a cycloconverter. The cycloconverter is commutated from the line side and its input frequency tracks the rotor speed so that rotor operates at super-synchronous speed with a small slip. The small slip power output sP_g induced to the rotor is dissipated in the rotor circuit. The lagging VAR excitation requirement of the machine is supplied by the cycloconverter at stator frequency shown in Fig. 14.23.

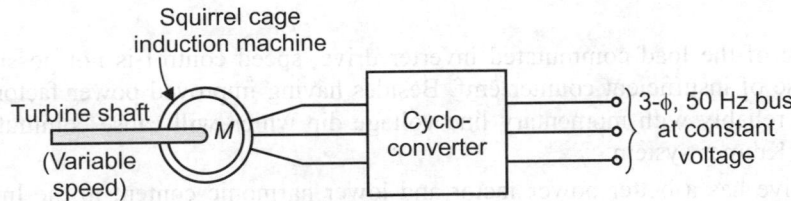

Fig. 14.23 Control circuit of squirrel cage IM using cycloconverter scheme

The cycloconverter can be replaced by a rectifier inverter system. The rectifier has to be force commutated so that motor lagging excitation VAR can be supplied. Cycloconverter also converts variable frequency and variable voltage power generated by the machine to power at constant voltage and frequency.

14.10 SPEED CONTROL OF SYNCHRONOUS MACHINES

The speed of a synchronous motor can be controlled by varying frequency of its source. The development of semiconductor variable frequency sources, such as inverters and cycloconverters has allowed their use in variable speed applications such as high power and high speed compressors, blowers, main line traction and servo drives, etc.

The speed control of synchronous machine may be achieved by voltage-fed inverters, current-fed inverters or cycloconverters. The control may be categorised into two groups.

1. Scalor control
2. Vector control

Scalar control: A synchronous machine drive system may have essentially two different modes of operation:

1. Open loop true synchronous motor mode
2. Self-control mode

Open loop true synchronous motor mode: In open loop true synchronous motor mode, the stator supply frequency is controlled from an independent oscillator. Frequency from its initial to the desired value is changed gradually so that the difference between synchronous speed and rotor speed is always small. This allows rotor speed to track the changes in synchronous speed. When the desired speed is reached, the rotor pulls into step after hunting oscillations.

Open loop true synchronous motor mode is an example of independent frequency control, shown in Fig. 14.24. This method of speed control is very popular in permanent magnet machine drives, where close speed tracking is essential in such applications as fiber spinning mill. In this method a number of machines are connected in parallel and are supplied by the same inverter, and the speed of the machine is related to the command frequency f, applied to the voltage source inverter through a delay circuit so that rotor speed is able to track the changes in frequency. A flux control block changes stator voltage with frequency (volt/Hz) to maintain a constant flux below rated speed and a constant terminal voltage above rated speed. The speed can be varied from zero to the full value by gradually varying the frequency. Beyond the base speed, the DC link voltage saturates and the machine enters into the field weakening constant power region, where the torque decreases with an increase in the frequency.

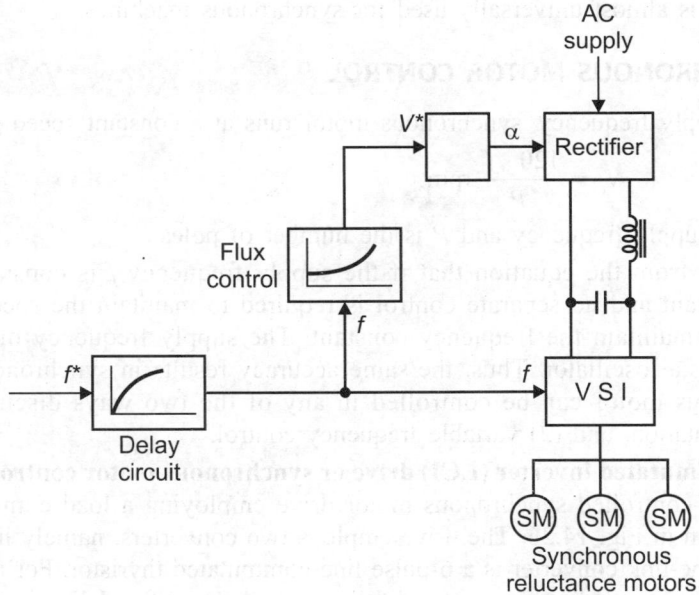

Fig. 14.24 Speed control of permanent magnet machine drive

Self-control mode: A synchronous machine in the self-control mode is supplied by a shaft position controlled electronic inverter. The self-control principle is illustrated in Fig. 14.25 by a current fed inverter. The machine rotor has a shaft position sensor and the sensor signal is processed. The self-control mode relates the inverter frequency uniquely with the machine speed. Figure 14.25 illustrates a current controlled brushless DC machine with separate field excitation. The self-controlled synchronous machine has the advantages that it cannot fall out

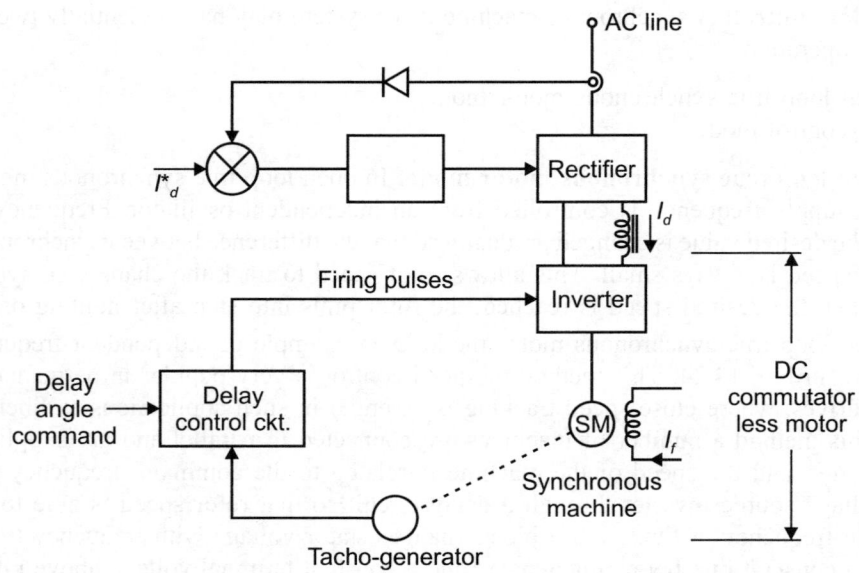

Fig. 14.25 Speed control of brushless DC machine

of step by steady state stability limit and rarely shows any transient stability problem. This type of control is almost universally used for synchronous machines.

14.11 SYNCHRONOUS MOTOR CONTROL

For a given supply frequency, synchronous motor runs at a constant speed given by

$$N_s = \frac{120 \cdot f}{P} \text{ rpm},$$

where f is the supply frequency and P is the number of poles.

It is clear from the equation that if the supply frequency f is constant, motor speed N_s is also constant and no separate control is required to maintain the speed constant, thus it is desired to maintain the frequency constant. The supply frequency may be accurately changed by crystal oscillator. Thus, the same accuracy results in synchronous motor speed. The synchronous motor can be controlled in any of the two ways discussed as follows: (1) load commutation, and (2) variable frequency control.

Load commutated inverter (LCI) drive or synchronous motor control with load angle sensing: A self-controlled synchronous motor drive employing a load commutated thyristor inverter is shown in Fig. 14.26. The drive employs two converters, namely link converter and load inverter. The link converter is a 6-pulse line-commutated thyristor. For firing angle range $0 \le \alpha \le 90°$, it works as a line-commutated fully controlled rectifier delivering E_{DC} and positive I_{DC} and for the range of firing angles $90 \le \alpha \le 180°$ it works as a line commutated inverter delivering negative E_{DC} and positive I_d. When synchronous motor operates on leading power factor by adjusting the excitation of synchronous motor this enables the inverter to operate as load commutated circuit. Load commutated circuits are also known as *self-controlled circuits*. System current source characteristics are given by reactor in the DC link. Firing pulses are supplied at the inverter at the instant determined by the rotor position sensor. This method fixes the position of stator voltage with respect to field vector, giving the characteristics of DC motor. This method can be operated on large speed range.

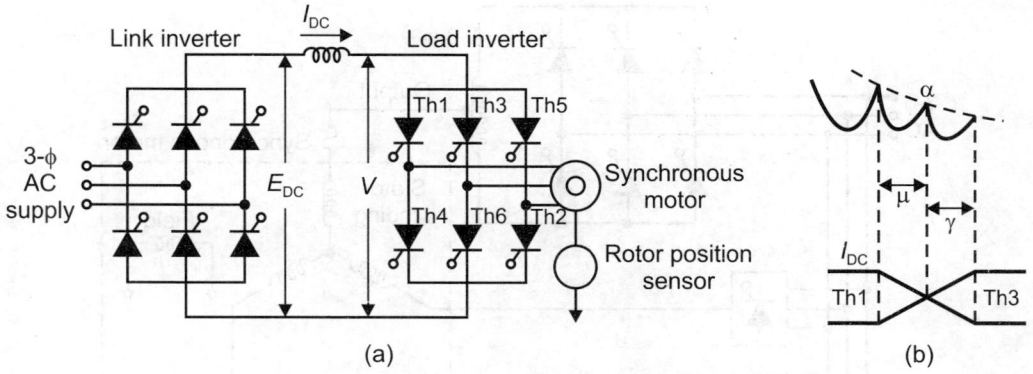

Fig. 14.26 Speed control of synchronous motor with load sensing device

For commutation from thyristor Th1 to thyristor Th3, thyristor Th3 is fired at instant 'α' as shown in Fig. 14.26(b). During the commutation period 'υ' current is transferred from thyristor Th1 to thyristor Th3. Commutation angle υ depends on the magnitudes of current and machine subtransient inductance. During period γ, counter emf from motor reverse biases thyristor Th1, hence turning it off. However, at speed below 10% of base speed counter emf from motor, is not sufficient to commutate device. This makes the machine difficult to start. Generally forced commutation is used under this condition. Forced commutation is commonly used for low speed commutation. Some of the important properties of LCI drives are as follows:

1. LCI is much simpler and has low losses compared to inverter used in CSI induction motor drives.
2. Use of synchronous motors in large power ratings results in overall drive efficiencies exceeding 95% at rated power.
3. There is no inrush current at starting.

Variable frequency control by cycloconverter: The speed of a synchronous machine is uniquely related to frequency, for very large horse power applications. Cycloconverters can be employed to control the speed of synchronous motors. Motor speed can be controlled by varying the frequency. For higher speeds, the machine is operated at a rated terminal voltage and variable frequency, and the pull out torque decreases with an increase in frequency. The principle and operation of cycloconverters was discussed in Chapter 13, where it was shown that a low-frequency supply can be directly synthesized from a higher frequency source by suitable switching of the cycloconverter elements. A major limitation of the cycloconverter is that its output frequency is limited to one-third of the input frequency, possibly slightly better for a higher-pulse configuration. If the input is 50 or 60 Hz, the maximum output frequency is around 20 Hz, the net result being that the cycloconverter application is limited to slow speed drives. If the power source is 400 Hz, then higher speeds are possible. Figure 14.27 shows that both voltage and frequency levels are directly controllable in the cycloconverter. However, although the cycloconverter is technically attractive for some applications, its use is severely limited on economic grounds compared to the inverter schemes.

The cycloconverter is expensive and the extensive electronic control circuitry is required. The system is inherently capable of braking by regeneration back into the AC source, thus four-quadrant operation is possible. Technically, cycloconverter use is limited by the low-output frequency range, and by the quite severe harmonic and power factor demands made on the AC supply system. A three-phase supply and a three-phase load arrangement is shown

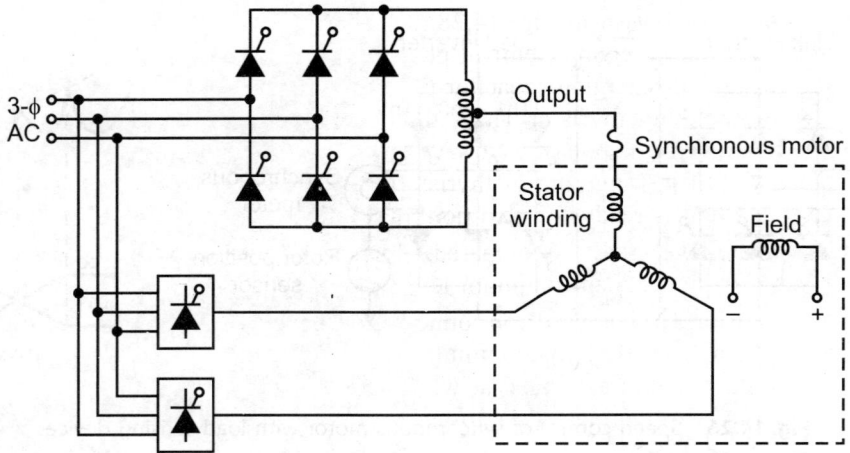

Fig. 14.27 Variable frequency control scheme

in Fig. 14.27. Only one phase of cycloconverter is shown in full. The other two phases are shown in blocks.

14.12 STARTING OF SYNCHRONOUS MOTOR

When operating with self-control, the starting current is low and starting torque is high, compared to direct online starting as an induction motor. Hence, self-control is employed for starting large synchronous machines in gas turbines and pumped storage power plants. Figure 14.28 shows forced commutation circuit for starting of synchronous motor. It consists of converter, DC link inductance and machine commutated inverter. Two auxiliary thyristors Th11 and Th22 are provided with capacitor C connected to the neutral of synchronous machine. Thus, the frequency of fundamental component of capacitive current will be three times the inverter frequency. Capacitor C is initially charged to required polarity by triggering one of the auxiliary thyristor and opposite branch main thyristor. The capacitor charges to twice the DC link voltage due to the presence of link inductance in charging circuit. Capacitor is charged

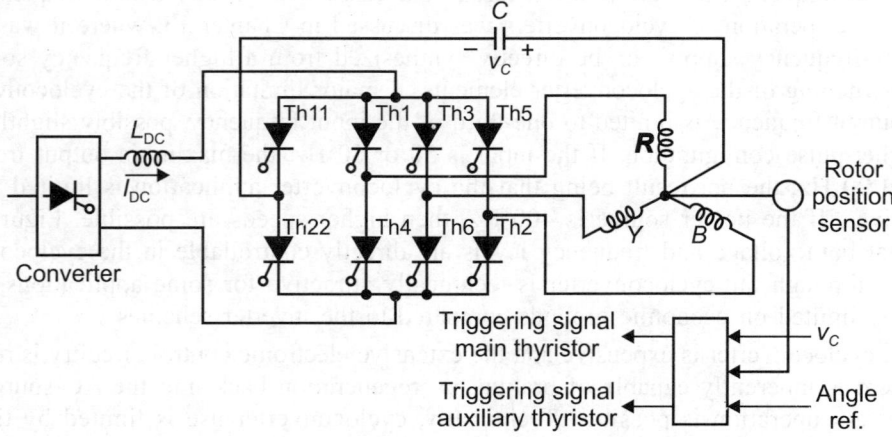

Fig. 14.28 Diagram of forced commutation circuit for starting of synchronous motor

previously to polarity as shown in Fig. 14.28. Now auxiliary thyristor Th22 is fired. As the voltage V_C is more than the counter emf in phase R, the current will start transferring from phase R. This also starts discharging capacitor C. After time T_1, complete current is transferred to capacitor C and auxiliary thyristor Th22, thereby turning off thyristor Th1. Capacitor C charges linearly and reverse its polarity. Now voltage across thyristor Th3 is of proper polarity to make it conducting. But triggering of thyristor Th3 is delayed till voltage V_C has attained sufficient high value and is ready for next commutation.

For this, voltage level of V_C is compared with desired level and triggering pulse is only permitted when V_C is higher than minimum level.

Auxiliary thyristor Th11 is used for commutation of thyristors Th2, Th4 and Th6 while auxiliary thyristor Th22 is used for commutation of thyristors Th1, Th3, Th5. When the drive has accelerated to sufficient speed at which transfer from forced commutation to load commutation is desired, triggering pulses to auxiliary thyristor Th11 and Th22 are blocked.

14.13 COMPARISON BETWEEN AC AND DC DRIVES

AC drives has many advantages over DC drives. Some of them are given below:

1. Simple and robust construction of AC motors as compared to DC motors.
2. Less maintenance and hence less running cost and more reliable.
3. As AC motor is cheaper than DC motor, but cost of static inverter is comparatively much higher than converter. Due to many other merits with AC drives, it may be preferable to use them even if they are marginally costlier as compared to DC drives.
4. As there is no commutation in AC motor, so no sparking problem.
5. Excellent synchronisation of multi-motor drives.
6. Synchronous reluctance motors run at constant synchronous speed, can be used for special applications.
7. In explosive and hostile atmospheric conditions, AC motors can perform better than DC drives.
8. With AC drives, stability is more severe than DC drives.
9. Control system associated with DC drives is simpler than AC drives. In AC drives, number of parameters such as stator voltage, current, frequency, slip, speed, etc., are to be controlled, whereas in DC drives only current, voltage and speed are to be controlled.
10. Rotation of AC motors can be reversed by changing any two phases of input supply through variation in firing sequence of devices.

14.14 CHOICE BETWEEN AC AND DC DRIVES

The merits and demerits of AC and DC drives have been discussed above. There are many advantages of AC drives, which make AC drives useful for different applications. AC drives are economical at relatively higher ratings, however, with the use of power transistor and gate turn off devices, it is expected they will be economical in lower rating also.

It is possible to use induction motor in a drive and control it to provide any desired output, such as constant speed, constant torque or braking.

In choosing a particular drive for a given application cost is certainly an important factor, but not the only criterion. Environmental consideration may demand no rubbing contacts, with the consequent risk of sparking; so forcing the use of AC motor. The supply authorities would look more favourably on a drive having a diode rectifier input, such as in some inverter, than the use of the phase-controlled thyristor rectifiers. The cooling of a motor depends on air turbulence set up within the motor by a fan on the shaft. If a motor is running below its rated speed, then the cooling will be less efficient, and hence either a reduced torque rating is specified, or the motor is force-cooled by external device. It is a fact that DC drives are well established at the present moment and AC drives control is more complex and it can be concluded that though DC drive has dark future, but still it has its own importance.

SOLVED EXAMPLES

Example 14.1 A three-phase four pole induction motor is operating on an input frequency of 75 Hz at 415 V and with slip of 4%. If the rotor resistance referred to stator is $1\,\Omega$, calculate: (i) torque developed by the motor, and (ii) rotor frequency.

Solution Torque,
$$T = \frac{sV^2}{R_r'}$$

Given
$$s = 0.04, \ R_r' = 10\,\Omega$$
$$T = \frac{0.04 \times 415^2}{1} = 6890 \ (\text{syn.}) \ \text{watt}$$

2. Rotor frequency $f' = sf = 0.04 \times 75 = 3$ Hz.

Example 14.2 A three-phase wound rotor induction motor is controlled by resistance controlled chopper as shown in Fig. E14.2. Resistance of $2\,\Omega$ is connected in the rotor circuit and a resistance of $4\,\Omega$ is additionally connected during chopper off time t_{OFF} of 4 msec. If the chopper frequency is 200 Hz and motor slip is 2%. Find the motor torque. Resistance values are referred to stator side.

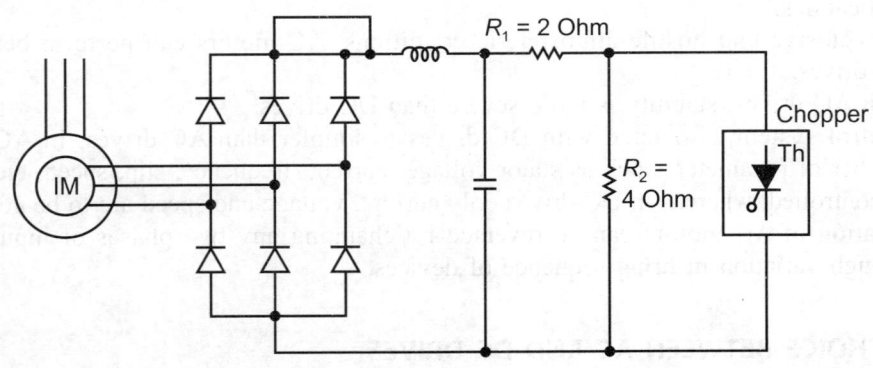

Fig. E14.2

Solution Torque of the motor $= V^2 s/R$
where, R is the motor resistance referred to stator in terms of R_1 and R_2

$$\therefore \qquad R = \frac{R_1 t_{\text{ON}} + (R_1 + R_2)\, t_{\text{OFF}}}{t_{\text{ON}} + t_{\text{OFF}}}$$

$$t_{\text{ON}} + t_{\text{OFF}} = T = \frac{1}{f} = \frac{1}{200} = 5 \text{ msec}$$

$$t_{\text{ON}} = 5 - t_{\text{OFF}} = 5 - 4 = 1 \text{ msec}$$

Thus,
$$R = \frac{2 \times 1 + (2 + 4)4}{4 + 1} = \frac{26}{5} \Omega$$

Motor torque
$$= \frac{V^2 s}{R} = \frac{(415)^2 \times 0.02}{26/5} = 662.3 \text{ (syn.) watt.}$$

Example 14.3 A 440 V, 50 Hz, 6 pole star connected slip ring induction motor has the following parameters $R_s = 0.5\,\Omega$, $R_r' = 0.4\,\Omega$, $X_r' = 1.2\,\Omega$, $X_m = 50\,\Omega$. Stator to rotor turns ratio is 3.5. Motor is controlled by static rotor resistance control. External resistance is chosen such that breakdown torque is produced at stand still for a duty ratio of zero. Calculate the external resistance. How duty ratio should be varied with speed so that motor accelerates at maximum torque?

Solution   Slip for maximum torque is

$$s_m = \frac{R_r'}{\sqrt{R_s^2 + (X_s + X_r')^2}}$$

With an external resistance whose referred equivalent value is R_e

$$s_m = \frac{R_e + R_r'}{[R_s^2 + (X_s + X_r')^2]} = \frac{R_e + 0.4}{[(0.5)^2 + (2.4)^2]^{1/2}}$$

or
$$R_e = 2.45 s_m - 0.4 \qquad \qquad \text{...(i)}$$

We know,
$$R_e = 0.5 R \, (1 - \delta) \, a^2$$

where a is the stator to rotor turns ratio.

Putting the value of $\qquad a = 3.5$
$$R_e = 0.5 R \, (1 - \delta) \, (3.5)^2$$
$$= 6.125 \, (1 - \delta) \, R \qquad \qquad \text{...(ii)}$$

Equating Equations (i) and (ii), we have
$$2.45 s_m - 0.4 = 6.125 \, (1 - \delta) \, R \qquad \qquad \text{...(iii)}$$

For maximum torque to occur at $s_m = 1$ (stand still) for $\delta = 0$
$$2.45 \times 1 - 0.4 = 6.125 \, (1 - 0) \, R$$

or
$$6.125 R = 2.05$$

or
$$R = \frac{2.05}{6.125} = 0.3347\,\Omega$$

Slip,
$$s_m = \frac{N_s - N}{n} \quad \text{and } N_s = \frac{120 \times 50}{6} = 1000 \text{ rpm}$$

Using Equation (iii)
$$2.45 s_m - 0.4 = 6.125 \, (1 - \delta) \, R$$

Putting the values of R and s_m, we have

$$2.45 \left(\frac{1000 - N}{1000} \right) - 0.4 = 6.125 \, (1 - \delta) \times 0.3347$$

After rearranging, we get $\delta = 1.195 \times 10^{-3} \, N$

It is clear from the above equation that for accelerating the motor at maximum torque, δ must change linearly with speed.

Example 14.4 A three-phase 5 hp, 440 V, 50 Hz, 4 pole slip ring induction motor controlled with variable frequency system is operating at a slip of 4%, calculate the chopper frequency for a pulse width of 2 msec, so as to obtain 150% torque at the same slip. Additional resistance introduced during off period is equal to rotor resistance.

Solution Torque
$$= \frac{V^2}{R} s\alpha \frac{1}{R}$$

where R is rotor resistance refer to stator.

If R_1 and R_2 are the rotor resistance before and after the introduction of control.

$$R_2 = \frac{1}{0.65} R_1 = 1.5 R_1$$

As
$$R_2 = \frac{R_1 t_{ON} + (R_1 + R_2) t_{OFF}}{t_{ON} + t_{OFF}}$$

$$1.5 R_1 = \frac{R_1 t_{ON} + 2R_1 t_{OFF}}{t_{ON} + t_{OFF}}$$

$$t_{ON} = t_{OFF} = 2 \text{ m sec or } T = 2 + 2 = 4 \text{ m sec}$$

or
$$f = \frac{1}{T} = 250 \text{ Hz.}$$

Example 14.5 An inverter supplies a 4 pole, 220 V, 50 Hz cage induction motor. Determine the approximate required output of the inverter for speeds, (i) 900 rpm, (ii) 1200 rpm, and (iii) 1500 rpm.

Solution Approximately, slip may be neglected, with the output inverter frequency related to synchronous speed.

As frequency,
$$f = \frac{(N \times P)}{120}$$

At each condition, voltage/frequency $= \frac{220}{50}$, hence the required inverter outputs are:

(i) Frequency $= \dfrac{900 \times 4}{120} = 30$ Hz, voltage $= \dfrac{220}{50} \times 30 = 132$ V

(ii) Frequency $= \dfrac{1200 \times 4}{120} = 40$ Hz, voltage $= \dfrac{220}{50} \times 40 = 175$ V

(iii) Frequency $= \dfrac{1500 \times 4}{120} = 60$ Hz, voltage $= \dfrac{220}{50} \times 60 = 264$ V.

Example 14.6 A 20 hp, 6 pole, 50 Hz, slip ring induction motor is controlled by a slip energy recovery method. Determine the angle of firing advance in the inverter at 600 rpm, if open circuit standstill slip ring voltage is 600 V and the inverter is connected to a 415 V, three-phase system. Neglect overlap and losses.

Solution Synchronous speed,
$$N_s = \frac{120 f}{p} = \frac{120 \times 50}{6} = 1000 \text{ rpm}$$

At 600 rpm, slip,
$$s = \frac{1000 - 600}{1000} = 0.4$$

Rotor voltage at 600 rpm

$$E_r = 600 \times 0.4 = 240 \text{ V}$$

Assuming three-phase converter bridge with $\alpha = 0$, $\gamma = 0$ and $p = 6$

The DC link voltage, $V_{DC} = \dfrac{6 \times 240 \times \sqrt{2}}{\pi} \sin \dfrac{\pi}{6} = 324 \text{ V}$

With $\gamma = 0$, $p = 6$:324 $= \dfrac{6 \times \sqrt{2}V}{\pi} \sin \dfrac{\pi}{6} \cos \beta$

$$= \dfrac{6 \times \sqrt{2} \times 415}{\pi} \sin \dfrac{\pi}{6} \cos \beta$$

Angle of firing advance $\beta = 54.8°$.

EXERCISES

Multiple Choice Questions

1. Thyristor switching circuits are used:
   (A) to reduce the stator voltage
   (B) to increase the stator voltage
   (C) to keep the stator voltage constant
   (D) none of these

2. Variable-speed drives using stator voltage control are normally:
   (A) open-loop system
   (B) closed loop system
   (C) both (A) and (B) are correct
   (D) none of these

3. For controlling the speed of a three-phase induction motor, the method generally used is the:
   (A) fixed voltage fixed frequency method
   (B) variable voltage variable frequency method
   (C) fixed voltage variable frequency method
   (D) none of these

4. Variable voltage fixed frequency supply can be obtained from:
   (A) three-phase cycloconverter
   (B) AC chopper
   (C) three-phase inverter
   (D) none of these

5. The slip power recovery method for the speed control of induction motor (slip ring):
   (A) increases the efficiency
   (B) decrease the efficiency
   (C) improves the power factor
   (D) none of these

6. In variable voltage variable frequency control, to achieve constant torque operation below base speed:
   (A) (V/f) has to be kept constant
   (B) flux has to be increased
   (C) flux has to be decreased
   (D) none of these

7. In AC motor control the ratio of voltage to frequency is maintained at constant value:
   (A) to make maximum use of magnetic circuit
   (B) to make minimum use of magnetic circuit
   (C) to maximise the current drawn from the supply to provide torque
   (D) none of these

8. Which of the following circuits can be used as a DC transformer?
   (A) controlled rectifier
   (B) inverter
   (C) magnetic amplifier
   (D) none of these

9. Power factor of synchronous motor can be made leading by adjusting its:
   (A) speed   (B) supply voltage
   (C) excitation   (D) supply frequency

10. It is recommended to employ UJT oscillator for gate triggering of thyristors for speed control of IM mainly because:
   (A) it is fairly simple
   (B) it is less expensive
   (C) it provides sharp rising pulses
   (D) none of these

ANSWER KEY

1. (A)   2. (B)   3. (B)   4. (C)   5. (A)   6. (A)   7. (A)   8. (B)
9. (C)   10. (C)

Review Questions

1. What are the different methods for controlling speed of induction motors?

2. Discuss the principle of stator voltage control of induction motor and how it is achieved?

3. Draw the circuit of stator voltage control of delta connected induction motor using close loop method.

4. Discuss the effects of harmonics in the system, and the possible means of reducing these harmonics.

5. Discuss Kramer method for speed control of induction motor.

6. Discuss the principal methods by using thyristor to control and vary the speed of both cage and slip-ring induction motors.

7. Describe the system of drive control in a DC link inverter fed cage induction motor.

8. How will you achieve speed less than synchronous speed in slip ring induction motors?

9. Discuss the method to attain the speed above synchronous speed of induction motor using thyristors.

10. How the speed of synchronous motor is controlled using SCRs?

Faults and Protection

15.0 INTRODUCTION

Due to reverse recovery process of semiconductor devices and switching actions in the presence of inductances overvoltages occur in the converter circuits. Due to short circuit fault conditions, an excess current flow through the devices. The heat produced by losses in the semiconductor devices must be dissipated sufficiently and effectively to operate the device within its upper temperature limit. The operation of the thyristor is greatly affected by temperature. The reliable operation of thyristor would require ensuring that the circuit conditions do not exceed the ratings of power devices, by providing protection against overvoltage, overcurrent and overheating. In practice, the power devices are protected from high dv/dt, di/dt by snubbers, thermal run away by heat sinks, transients and fault conditions by fuses, etc.

The power semiconductor device requires protection against excessive voltage, current, and certain rates of changes, so that the device is not damaged. In the earlier chapters, much stress has been given to the current and voltage levels experienced by the devices in given applications, but having chosen the appropriate device, no mention has been made of manner in which the device is protected against damaging circumstances.

15.1 CURRENT

Fault current is to be protected against the fault which provides a short circuit path to the supply source. The equation of current depends on the angle at which the fault current occurs in the alternating voltage cycle. The equation for i is

$$i = \frac{V_{max}}{Z} \sin(\omega t - \phi) + I e^{-t/\tau} \qquad ...(15.1)$$

where, Z is the supply impedance, ϕ the steady state current lag, I the transient component to satisfy initial condition and τ the time constant.

If the supply is strong, that it has a low source impedance, then the fault current will be of such a magnitude that the fault must be cleared well before the first peak. As the thermal mass of a semiconductor device is low, hence an over current will rapidly raise its temperature beyond the safe limit. Mechanical contactors frequently rely on clearing fault at the safe limit. Mechanical contactors frequently rely on clearing the fault at the first or a subsequent current zero, but this is far too slow for semiconductor device protection. Breaking the fault current before the first zero means the breaking phenomenon is similar for both ac and dc circuit breaking.

The device used for current protection is the fuse which should have the following characteristics which are discussed below:

1. It must carry continuously the device rated current.
2. Its thermal storage capacity must be less than that of the device being protected.
3. After breaking the current the fuse must be able to withstand any restriking voltage which appears across it.
4. The fuse voltage during arc must be high enough to force the current drawn and dissipate the circuit energy.

It may give adequate protection to place the fuses in the supply lines as shown in Fig. 15.1 particularly with a passive load.

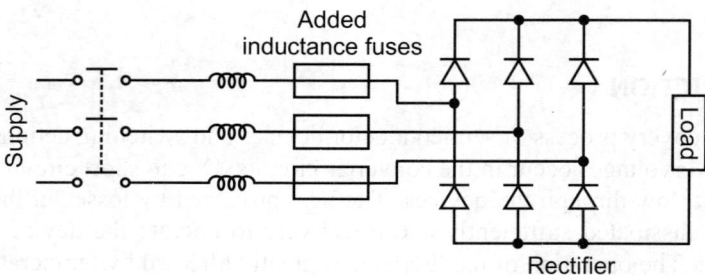

Fig. 15.1 The fuses in the supply lines

15.2 VOLTAGE

The peak voltage rating of a device used in a particular application must obviously be greater than the peak voltage, it experiences during the course of the circuit operation. However, the transients within the circuit may momentarily raise the voltage several thousand volts, and it is these voltages which must be prevented from appearing across the off-stage device.

The origin of the voltage transients having high dv/dt values may be from one of the three sources.

1. Supply source due to contactor switching, lighting or other supply surges.
2. Load source, say the voltage arising from the commutator arcing on a dc motor.
3. From within the convertor itself, due to switching of other devices, commutation oscillations or the fuse arc voltage when it is clearing a fault.

To protect against all three sources of voltage transients, it is necessary to protect each device individually.

A capacitor C across the thyristor (or diode) means that any high dv/dt appearing at the thyristor terminals will set-up an appropriate current (= $C\,dv/dt$) in the capacitor. The inductance in the circuit will reversely limit the magnitude of the current to the capacitor and hence limit dv/dt. The RC combination is often referred to as a snubber network and also serves to limit the induced voltage spike produced during reverse recovery of the thyristor storage charge.

15.3 SNUBBER CIRCUIT

To protect against dv/dt turn on of the thyristor, a snubber circuit, which is a R-C series circuit used across the thyristor.

The function of the snubber can be summarised as follows:

1. It protects the device from supply and load side voltage transients.
2. It reduces the switching losses of devices.
3. It reduces the magnitude of peak recovery voltage.
4. It reduces the off-state and reapplied dv/dt.

When the thyristor of Fig. 15.2 is fired, any charge in the capacitor will be discharged into thyristor, giving an excessively high di/dt, but this can be suitably limited by inclusion of the resistor. A diode D may be included to by-pass R for improved dv/dt protection. Typically the values of the protection components will be $C = 0.01$ to 1 μF, $R = 10$ to 1 kΩ and $L = 50$ to 100 μH. The exact values depend on the circuit voltage and stored capacity of the transient source.

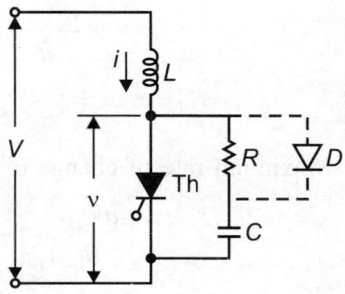

Fig. 15.2 Snubber circuit

15.3.1 Design of Snubber Circuits

It consists of a series combination of resistance R_S and capacitance C_S in parallel with the thyristor as shown in Fig. 15.3.

In practice, *a* dc voltage is switched across the thyristor, and the rate of change of voltage across the device must be below the dv/dt rating of the device, otherwise the device would start conducting even without the application of trigger pulse. Normally R_S, C_S and load circuit parameters form an underdamped circuit so that dv/dt is limited to acceptable values. The rate of change of voltage across thyristor may be calculated in the following *two* ways:

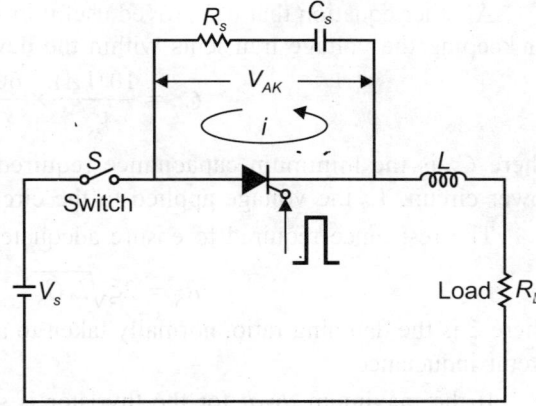

Fig. 15.3 Snubber circuit with series combination of $R_S C_S$

1. **When the switch is closed:** Assuming load resistance R_L is negligible, the voltage equation across the thyristor is as follows:

$$V_{AK} = V_S(1 - e^{-t/\tau}) \text{ (volts)} \qquad \qquad ...(15.2)$$

where

$$\tau = L/R_S$$

$$\therefore \qquad \frac{dV_{AK}}{dt} = \frac{R_S}{L} V_S \cdot e^{-t/\tau} \text{ (V/sec)}$$

At $t = 0$, rate of change of voltage is maximum

or

$$\left. \frac{dV_{AK}}{dt} \right|_{max} = \frac{R_S}{L} V_S \text{ (V/sec)} \qquad \qquad ...(15.3)$$

If
$$R_S = \sqrt{L/C_S}$$

then
$$\left.\frac{dV_{AK}}{dt}\right|_{max} = \frac{V_S}{\sqrt{LC_S}} \qquad ...(15.4)$$

2. When the switch is closed: The voltage oscillation across thyristor is given by

$$V_{AK} = V_m \sin \omega t \quad \text{where } \omega = \frac{1}{\sqrt{LC_S}}$$

∴
$$\frac{dV_{AK}}{dt} = \omega V_m \cos \omega t$$

$$= \frac{1}{\sqrt{LC_S}} \cdot V_m \cos \omega t$$

Maximum rate of change of voltage occurs at $t = 0$

$$\left.\frac{dV_{AK}}{dt}\right|_{max} = \frac{V_m}{\sqrt{LC_S}} \qquad ...(15.5)$$

Hence, by proper choice of L, C_S and R_S in the circuit dv/dt across the thyristor could be limited to an acceptable value, R_S also limits the discharge current through the thyristor when the thyristor is gated to turn it on.

Another equation that has proved useful in selecting the value of the capacitance required for keeping the voltage transients within the device rating is

$$C_S = \frac{10\,(VA)}{V_S^2} \times \frac{60}{f} \qquad ...(15.6)$$

where C_S is the minimum capacitance required, (VA) is the full load voltampere rating of power circuit, V_S the voltage applied to the circuit, f is the operating frequency.

The resistance required to ensure adequate damping is given by

$$R_S = 2\xi\sqrt{L/C_S} \qquad ...(15.7)$$

where ξ is the damping ratio, normally taken to about 0.65 and L is the effective commutating circuit inductance.

If the maximum dv/dt for the thyristor is specified, then the value of capacitance is

$$C = \frac{1}{2L}\left(\frac{0.564V_m}{dv/dt}\right)^2 \qquad ...(15.8)$$

where V_m is the peak input line to line voltage.

15.4 *dv/dt* PROTECTION

When a forward voltage is applied across the thyristor, the two outer junctions J_1 and J_3 are forward biased and the central junction J_2, is reverse biased as discussed in Chapter 5. A reverse biased junction p-n junction has the characteristic of a capacitor. Therefore, when a forward voltage is applied, a charging current flows:

$$i = \frac{C_j dV_{AK}}{dt} \qquad ...(15.9)$$

where C_j is the junction capacitor.

If the rate of application of the forward voltage is high, the charging current is so high that the thyristor will turn on without a gate pulse. This phenomenon is known as *dv/dt* turn on, will lead to improper operation in a circuit.

For proper operation in a circuit, the rate of rise of forward voltage (*dv/dt*) must be less than a specified limit which is typically in the range 20 to 500 V/μsec.

15.5 PROTECTION

Voltage Safety Factor: To protect the thyristor against voltage peaks, a safety factor V_s is introduced.

$$V_s = \frac{\text{Peak inverse voltage of the device}}{\text{Peak value of the applied voltage to the device}}$$

For line commutated converters, value of safety factor lies between 1.8 and 2.5 and for force commutated converters, its value can lie between 1.5 and 2.5. Generally higher values of safety factor is taken.

15.6 *di/dt* PROTECTION

When thyristor is forward biased and is triggered by a gate pulse, conduction of anode current starts in the immediate vicinity of the gate connection. The current thereafter spreads across the whole area of the junction. If the rate of rise of anode current is large, local hot spots will be created near the gate connection due to high current density. This localised heating may burn out the thyristor. Therefore, the rate of rise of the anode current (*di/dt*) at switching on must not exceed a specified limiting value. Normally a small inductor called a (*di/dt*) inductor is inserted in the anode circuit to limit (*di/dt*) of the anode current. A typical *di/dt* limit in thyristor is in the range of 20 to 500 A/μsec.

15.6.1 Thyristor *di/dt* Calculation

The maximum permissible rate of change of current through a device is dependent on the type of component being used. The manufacturer's data sheet specifies the maximum *di/dt* capability. The price of the device increases with *di/dt* capability. Through proper circuit design techniques, the rate of change of current through the thyristor must be kept to a minimum. This section calculates the *di/dt* through thyristor as a function of other circuit components and the applied voltage. The circuit is shown in Fig. 15.4.

Fig. 15.4 The circuit

The circuit voltage equation is given by

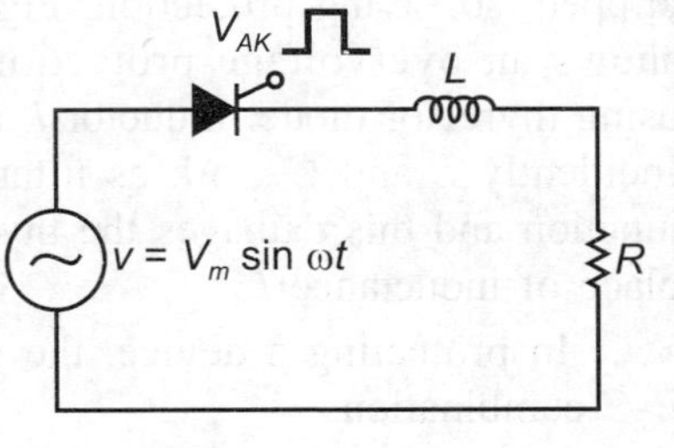

$$v = Ri + L \, di/dt \qquad \qquad ...(15.10)$$

Solving for current, $\qquad i = \frac{v}{R}(1 - e^{-t/\tau}) \text{ amp.}$

where $\qquad \qquad \tau = L/R$

$\therefore \qquad \qquad \dfrac{di}{dt} = \dfrac{v}{R} \cdot \dfrac{1}{\tau} \cdot e^{-t/\tau} \qquad \qquad ...(15.11)$

or $\qquad \qquad \dfrac{di}{dt} = \dfrac{v}{L} e^{-t/\tau} \qquad \qquad ...(15.12)$

The maximum di/dt is at $t = 0$,

or
$$\left.\frac{di}{dt}\right|_{\max} = \frac{v}{L} = \frac{V_m}{L} \text{ A/sec} \qquad ...(15.13)$$

It is seen from Equation (15.13) that the rate of change of current through the thyristor is dependent on the supply voltage and inductance in the current. The di/dt could be reduced by proper selection of inductance L in series with the thyristor.

15.7 OVERVOLTAGE PROTECTION

A thyristor may be subjected to overvoltage due to bad commutation, short circuit, transients due to switching operation, lightning stroke and so on. The thyristor is protected against these overvoltages by shunt connected non-linear resistance devices. These protective devices have falling resistance characteristics with increasing voltage. If a high voltage surge appears, the non-linear device produces a virtual short across the thyristor. Selenium thyristor diodes or Metal Oxide Varistors (MOV) are some of the common non-linear devices used for overvoltage protection.

15.7.1 Voltage Protection by Selenium Diodes and Metal Oxide Varistors

The selenium metal rectifier used extensively in the past for rectification may be used for protection. The metal to selenium junction has a low forward break overvoltage. The break over or clamping voltage is of the order of 72 V for the type used in protection. If such a device is placed in parallel with a power diode which has a higher peak reverse value, then any transient voltage will be clipped, so giving protection. Figure 15.5 shows an overvoltage protection circuit

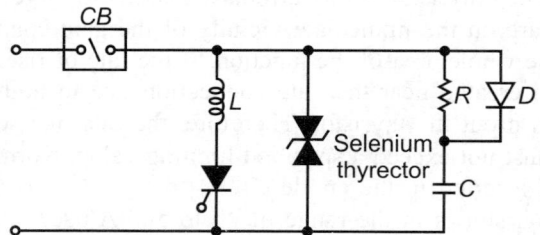

Fig. 15.5 An overvoltage protection circuit

using thyrector diode. Inductor L and capacitor C protect the thyristor against di/dt and dv/dt. Incidently L and C work as filter. A large di/dt will produce a hot spot temperature in the junction and this damages the thyristor. A saturable core reactor can be used for soft start in place of inductance L.

In protecting a device, the reliability of the selenium diode is not so good as that of R-C combination.

Varistors are non-linear variable impedance devices, consisting of oxide particles separated by an oxide film or insulation. As the applied voltage is increased, the film becomes conductive and the current flow is increased. The current is expressed as

$$I = kV^\alpha$$

where k is a constant and V is the applied voltage. The value of α varies between 30 and 40.

15.8 OVER CURRENT PROTECTION

The thermal capacity of semiconductor devices is small. Therefore continuous overloading, surge currents of long duration or large surge currents of short duration may increase the junction temperature beyond the permissible limit and destroy the device. Over current protection is provided by a circuit breaker and fast acting fuse in series with the thyristor.

The circuit breaker is used to protect the circuit breaker against continuous overloads or surge currents of long duration. The fast acting fuse is used to protect the thyristor against large surge currents of very short duration. The tripping time of the circuit breaker, fusing time of the fast acting fuse and the rating of the device have to be properly coordinated. The '$I^2 t$' rating of the fuse must be less than the corresponding value for the device provided by the manufacturer. The circuit breaker or the fuse must open the circuit before the thyristor can be damaged by a large current of any nature. The circuit breaker is used for the whole circuit and fuses may be used for individual devices.

15.8.1 Fuse

A fast-acting fuse can be used for protecting thyristors against large surge currents of very short duration, called *subcycle surge currents*. For reliable protection of the thyristor, the circuit breaker or the fuse must open the circuit before the thyristor suffers any permanent damage due to large currents.

The one-cycle surge rating $\hat{I}$ of thyristor is defined as the peak amplitude of the sinusoidal current, which the thyristor can carry for one half-cycle. The subcycle surge current rating may be obtained from

$$\hat{I}_{\text{subcycle}}(t) = \sqrt{\frac{\hat{I}^2 \times \dfrac{1}{100}}{t}} \qquad \qquad ...(15.14)$$

where $\hat{I}_{\text{subcycle}}$ is the peak amplitude of the sinusoid whose period is $2t$ and t is the duration of the surge in seconds. The fusing time for this circuit must be less than t to ensure reliable protection of the thyristor. To facilitate the right choice of the fuse, the manufacturers data include $I^2 t$ rating of the device. The fuse ratings give values of $I^2 t$ at different prospective fault currents and the corresponding peak let-through currents. From these data, the fusing time for a given subcycle surge current can be computed. Assuming that the fault current waveform is triangular, the fusing time t_c is given by

$$t_c = \frac{3I^2 t}{I_p^2} \qquad \qquad ...(15.15)$$

where I_p is the peak magnitude of the let-through fault current. The fuse will provide protection when the current is higher. For lower currents, the device will usually be protected by the circuit breaker.

With less-severe faults, the fusing time will be longer, so it is essential to ensure that at all fault levels, the fuse has an $I^2 t$ rating below that of the device being protected. With low level overloads, the fuse may not be adequate due to differing ambient temperatures, but here the slow-acting mechanical contactors would provide back-up protection.

It may give adequate protection to place the fuses in the supply lines, particularly with a passive load. With a line load, that is, a motor, it is possible for the load to provide the fault current, in which case individual thyristor can be fused as shown in Fig. 15.6(a). In general, the larger the equipment rating, the more tendency there is to fuse the device individually. Electromagnetic and explosive forces limit the maximum prospective current which a fuse can handle. It may be necessary to limit the rate of rise of the fault current, and hence avoid excessive stress on the fuse and device. The added inductance may be in the line, or in series with individual device and further to avoid excessive power losses, a saturable reactor can be employed as shown in Fig. 15.6(b). The added inductance will increase the overlap; a compromise is required between circuit performance and protection.

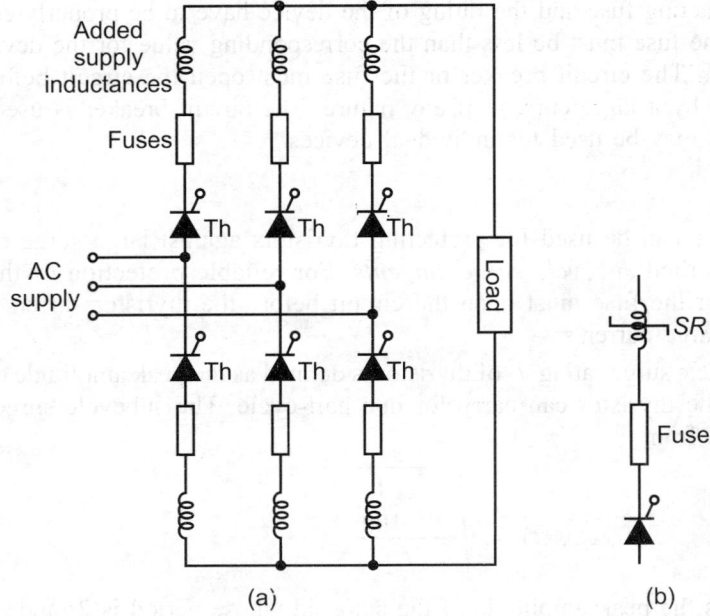

Fig. 15.6 A saturable reactor circuit

15.8.2 Electronic Crowbar Over Current Protection

An electronic crowbar protection provides rapid isolation of the power converter before any damage occurs.

Figure 15.7 illustrates the basic principle of electronic crowbar protection. A crowbar thyristor is connected across the input terminals. A current sensing resistor R detects the magnitude of converter current. If the value of converter current exceeds preset value, gate circuit provides the signal to crowbar thyristor and turns it on in a few microseconds. The input terminals are then short-circuited by crowbar thyristor and it shunts away the converter over current which depends on the source voltage and its impedance. After some time fuse interrupts the faults current which may be replaced by a circuit breaker if the thyristor has adequate surge current rating.

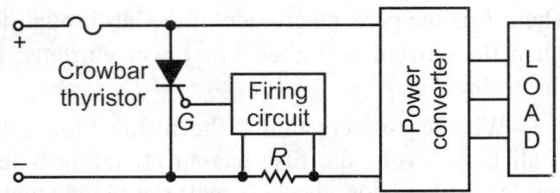

Fig. 15.7 Circuit diagram of electronic crowbar over current protection

15.8.3 Control Logic for Over Current Protection Circuit

A converter can be protected against high current by tripping the input circuit breaker. If the breaker trips frequently, the contacts in the breaker will deteriorate due to sparking. In addition, the breaker is not fast enough to protect semiconductor devices. Fast-acting semiconductor fuses are used to protect the devices. However, frequency blowing of the fuses is not economical. A simple over-current protection for the converter is achieved by blocking appropriate pulses whenever the current exceeds a preset limit. Figure 15.8(a) shows a circuit implementation of a control logic for over current protection. It consists mainly of *three* parts: (1) a voltage comparator C, (2) a flip-flop F-F with preset and clear, and (3) an AND gate A. The timing diagram to illustrate the operation of the circuit is shown in Fig. 15.8(b).

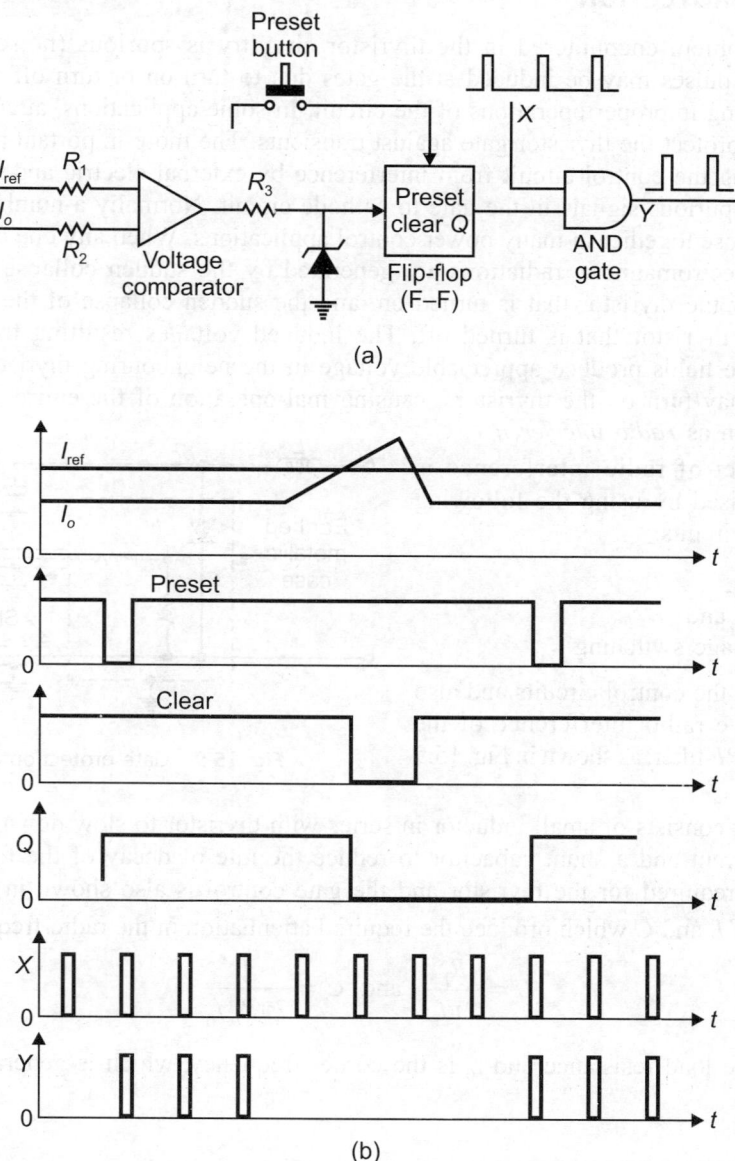

(a)

(b)

Fig. 15.8 Circuit implementation for over current protection

In Fig. 15.8(a), if the Q output of the flip-flop is high, the firing pulses X are transmitted through the AND gate as the Y output. When an over-current condition occurs, the Q output swings low and the AND gate is disabled so that the firing pulses X are blocked. I_{ref} represents the current limit, and I_o represents the converter or motor current. If I_o exceeds I_{ref}, output of the comparator C changes and the 'clear' input and Q output of the flip-flop are driven low. This disables the AND gate and blocks the firing pulses X. The AND gate remains disabled until the flip-flop is preset by pressing the present button. The Zener diode is used to limit the input voltage to the flip-flop.

15.9 GATE PROTECTION

A common problem encountered in the thyristor circuitry is spurious (noise) firing of the device, trigger pulses may be induced at the gates due to turn on or turn off a neighbouring thyristors causing improper operations of the circuit. In some applications, attention may have to be given to protect the thyristor gate against transients. The more important problem here is that of shielding the control circuit from interference by external electric and magnetic fields which induce spurious signals in the gate to cathode circuit. Normally a number of thyristors are mounted close together in many power control applications. When any one of the thyristors is triggered, electromagnetic radiations are generated by the sudden collapse of the electric field caused by the thyristor that is turned on, and the sudden collapse of the magnetic field caused by the thyristor that is turned off. The induced voltages resulting from the rate of change of these fields produce appreciable voltage in the neighbouring thyristor gate control circuits; this may turn on the thyristors, causing mal-operation of the entire control scheme which is known as *radio interference*.

The effect of radio interference is usually minimised by using the following elements and circuits:

1. Shielding,
2. *RF* filter, and
3. Zero voltage switching.

To shield the control circuits and also to minimise the radio interference of the thyristors, and *RF* filter, as shown in Fig. 15.9 may be used.

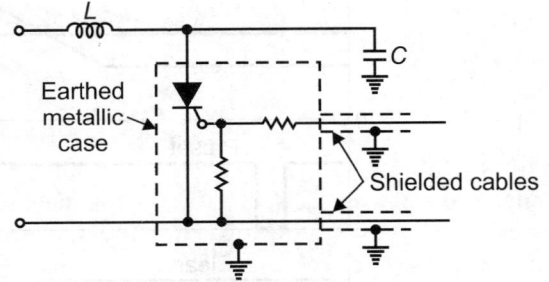

Fig. 15.9 Gate protection circuit

The filter consists of small inductor in series with thyristor to slow down the rate of rise of forward current and a shunt capacitor to reduce the rate of decay of the forward voltage. The shielding required for the thyristor and the gate control is also shown in the figure.

Values of L and C which produce the required attenuation in the radio frequency band are

$$L = \frac{R_L}{2\pi f_0} \quad \text{and} \quad C = \frac{1}{2\pi R_L f_0} \qquad \qquad ...(15.16)$$

where R_L is the load resistance and f_0 is the corner frequency, which is generally 50 kHz for the filter.

15.10 THYRISTOR MOUNTING

As the thyristor handles substantial amounts of power, it is subjected to high thermal stresses whenever it is conducting. These stresses are repetitive and result in internal mechanical forces. The thyristor has therefore to be properly braced to provide the required strength to withstand mechanical forces. The mounting of the thyristor must be so designed as to be able to carry away the internal heat and thereby limit the rise in junction temperature. For high power thyristors, the *PNPN* pellet is braced by two molybdenum plates and the anode is hard-soldered to an aluminium plate with a threaded stud. The thyristor is bolted to a heat sink through this stud. At normal power ratings, natural convection and conduction processes provide the required cooling. The heat sink may have fins and this is equipped with water cooling or forced air cooling. For effective thermal contact between the thyristor and the heat sink, spring washers are used with the bolt and the appropriate pressure applied while tightening the nut. A good electrical contact between the thyristor and the heat sink may not always be necessary. If electrical isolation is required, then mica or fibre glass disc type washers are fixed on the anode stud on either side of the heat sink as shown in Fig. 15.10. This results in effective thermal but very poor electrical contact between anode and the heat sink.

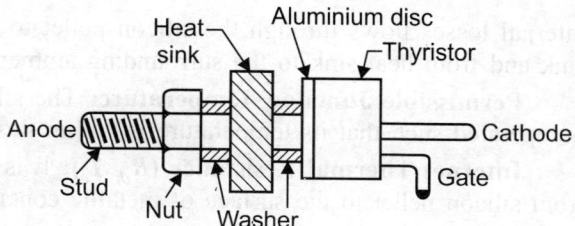

Fig. 15.10 Thyristor mounting circuit

15.11 HEAT SINKS

When a thyristor is conducting, different power losses occur in the thyristor namely:

1. Switching losses at turn on and turn off
2. Losses due to forward conduction
3. Losses due to leakage current during forward and reverse blocking
4. Gate trigging losses

At frequencies between zero and 400 Hz, the forward conduction loss is the major component. Switching losses become dominant at high operating frequencies. These electrical losses produce thermal heat which in turn increase the temperature of the device, and therefore must be dissipated from the junction region. These losses also increase with the increase of the rating. Heat produced due to losses may be dissipated by mounting the device on a heat sink. Therefore, thyristors are invariably mounted on heat sinks so that heat is conducted from the device to the heat sinks, and from there it is radiated and conducted to the atmosphere.

Heat sinks are made from metal with high thermal conductivity. Aluminium is the most commonly used metal. Heat dissipation takes place by connection from heat sinks, by increasing the surface area of the thyristor, by increasing the surface area of the thyristor by providing peripheral fins on the heat sink. Cooling by convection can be made more effective. Heat dissipation also takes place by radiation. Which are usually provided with black anodized finish to enhance the heat dissipation by radiation. For dissipating large losses in high power thyristors, water cooling is usually employed to get a compact size of the heat sink.

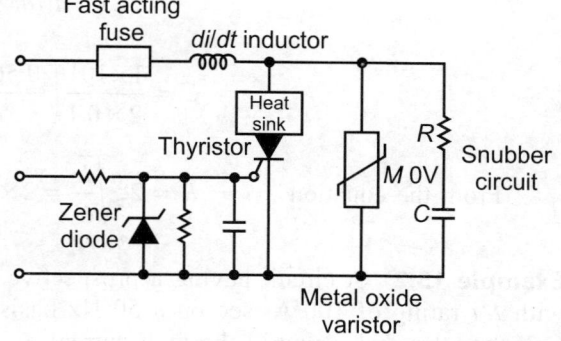

Fig. 15.11 Heat sink

Figure 15.11 shows the various components used to protect a thyristor against dv/dt, overvoltage, over current and gate protection, etc.

Thermal Resistance: Thermal energy flows from higher temperature region to a region of lower temperature. If the power loss P_{av} causes the temperature of two points to be at T_1

and T_2 °C $(T_1 > T_2)$ then thermal resistance $R_{th} = \dfrac{T_1 - T_2}{P_{av}}$ °C/W. The heat produced due to internal losses flows through the silicon pellet to thyristor container from container to heat sink and from heat sink to the surrounding ambient.

Permissible Junction Temperature: The silicon pellet of semiconductor device must be operated such that its temperature is within this value.

Internet Thermal Resistance (R_{th} T_c): It is the thermal resistance offered to the flow from silicon pellet to the surface of metallic container.

External Thermal Resistance: This is the thermal resistances offered to the flow from container to the heat sink (R_{ThCA}) R_{THCS} and from the heat sink to the cooling medium (T_{THSA}).

Total thermal resistance between junction and ambient,

$$R_{THJA} = R_{THJC} + R_{THCS} + R_{THSA} \qquad ...(15.17)$$

The difference in temperature between junction and ambient is:

$$T_j - T_A = P_{av} (R_{ThJC} + R_{ThCS} + R_{ThSA}) \qquad ...(15.18)$$

Heat Sink Specifications: The thyristor data sheet specifies maximum temperature, thermal resistances R_{THJC} and R_{THCS}. The manufactures of heat sinks provide catalogs in which sufficient data on heat sink is available in the form of curves for standard heat sinks of aluminium extrusions. These curves relate temperature difference between heat sink and ambient temperatures versus average power dissipation, P_{av} (watts).

SOLVED EXAMPLES

Example 15.1 Calculate the required parameters for a snubber network to provide reliable dv/dt protection to a thyristor used in a single-phase fully controlled bridge. Thyristor has maximum dv/dt capability of 40 V/μsec. The input line voltage has peak value 325 V and source inductance is 0.1 mH. Assume $\xi = 0.65$.

Solution Given; $\qquad L = 0.1$ mH, $V_m = 325$ V and $dv/dt = 40$ V/μsec

From the equation $\qquad C = \dfrac{1}{2L} \left(\dfrac{0.564 V_m}{dv/dt} \right)^2$

$$C = \dfrac{1 \times 10^3}{2 \times 0.1} \left(\dfrac{0.564 \times 325 \times 10^{-6}}{40} \right)^2 \approx 0.1\,\mu F$$

From the equation $\qquad R = 2\xi \sqrt{\dfrac{L}{C}} = 2 \times 0.65\sqrt{10^3} = 40.9$ Ohm.

Example 15.2 A circuit having a prospective fault current 1000 A is protected by a fuse with $I^2 t$ rating of 100 A^2-sec on a 50 Hz basis. The faulted circuit is opened in 5 m-sec. Calculate the peak value of the fault current.

Solution Assuming a triangular waveform for the fault current with a peak value of I_p, we have $t = 5$ m sec $= 5 \times 10^{-3}$ sec

$$t_c = 5 \times 10^{-3} = \dfrac{3 I^2 t}{(I_p)^2}$$

or
$$I_p^2 = 3 \times 100 \times \left(\frac{10^3}{5}\right) = 6 \times 10^4 \text{ amperes}$$

or
$$I_p = 244 \text{ A.}$$

Example 15.3 Determine: (a) the required snubber capacitance, (b) peak transient voltage, v_s; peak switching voltage ratio V_p/V_s, (c) damping resistance for a 5 kVA transformer having $120V_{rms}$ secondary voltage, switching frequency, $f_0 = 400$ Hz, circuit inductance of 100 μH at 400 Hz. The peak transient voltage is limited at 200 volts. Given ξ = 0.75.

Solution (a) Snubber capacitance,

$$C = \frac{100VA}{V_s^2} \cdot \frac{60}{f}$$

$$C = 10 \times \frac{5000}{120^2} \times \frac{60}{400} = 5 \, \mu F$$

(b) $\dfrac{V_p}{V_s} = \dfrac{200}{120\sqrt{2}} = 1.18$

(c) Damping resistance, $R = 2\xi\sqrt{L/C}$

$$= 2 \times 0.75 \sqrt{\frac{100}{0.5}} = 20 \text{ Ohm.}$$

Example 15.4 415 V, three-phase, 50 Hz bridge circuit is supplying to a 440 V, 136 A dc motor having armature resistance of 0.175 Ω and a time constant 30 m-sec. Design a voltage protection circuit and calculate energy dissipated per plate. Voltage capability of each plate is 30 V.

Solution Number of plates in each branch

$$= 400/30 \approx 15$$

Total number of plates $= 3 \times 15 = 45$

Peak armature current $= 136$ A

As time constant $\tau = \dfrac{L}{R} = 30$ msec

or Armature circuit inductance $= 0.175 \times 30 \times 10^{-3} = 5.25$ mH

Energy stored in armature circuit

$$= \frac{1}{2} LI_m^2 = \frac{1}{2} \times 5.25 \times 10^{-3} \times (136)^2 = 48.5 \text{ W}$$

Energy dissipated/plate $= \dfrac{48.5}{15} = 3.23$ W/sec.

Example 15.5 The thyristor in Fig. E15.5(a) is used to control power in a resistance R. The supply is 400 V and the maximum allowable di/dt and dv/dt for the thyristor are 50 A/μsec and 200 V/μsec, respectively. Determine the values of the di/dt inductance and the snubber circuit components R_s and C_s.

Solution The voltage across the capacitor cannot change instantaneously. Moreover, thyristor in its forward blocking state is equivalent to a high impedance. Therefore, immediately, after closing the switch S, the equivalent circuit is shown in Fig. E15.5(b). The voltage equation is

$$V = R_s i + L \frac{di}{dt} + Ri$$

$$\frac{di}{dt} = \frac{V - (R_s + R) i}{L}$$

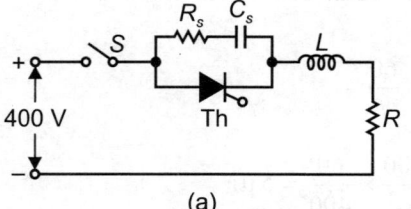 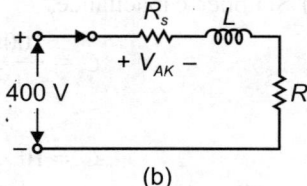

(a)   (b)

Fig. E15.5

The di/dt is maximum at $i = 0$. Therefore,

$$\left(\frac{di}{dt} \right)_{max} = \frac{V}{L} \qquad \qquad \text{...(i)}$$

Given $di/dt = 50$ A/μsec, $dv/dt = 200$ V/μsec

or

$$L = \frac{V}{\left(\dfrac{di}{dt} \right)_{max}} = \frac{400}{50} \, \mu H = 8 \, \mu H$$

Voltage across thyristor is $V_{AK} = R_S \cdot i.$ ...(ii)
Taking the derivative with respect to time, 't'

$$\frac{dV_{AK}}{dt} = R_s \cdot \frac{di}{dt}$$

$$\left(\frac{dV_{AK}}{dt} \right)_{max} = R_s \cdot \left(\frac{di}{dt} \right)_{max} \qquad \qquad \text{...(iii)}$$

From Equations (i) and (ii)

$$\left(\frac{dV_{AK}}{dt} \right)_{max} = R_s \cdot \left(\frac{V}{L} \right) \qquad \qquad \text{...(iv)}$$

or

$$R_s = \left(\frac{L}{V} \right) \left(\frac{dV_{AK}}{dt} \right)_{max} = \frac{8 \times 200}{400} = 4 \text{ Ohm}$$

When R_s is small, the energy lost in it will be high. Figure E15.5(a) shows that when supply is disconnected to the circuit, capacitor C_s will be charged to the supply voltage before the thyristor is triggered. Therefore, when the device is turned on, the capacitor C_s will provide a step increase in current through thyristor. If R_s is small, this current spike will be large. The value of R_s is thus normally greater than the limit dv/dt. The capacitor C_s is small, so that discharge through the thyristor when it is turned on, does not harm the device. Typical values of R_s and C_s are $10\,\Omega$ and $0.1\,\mu\text{F}$. If this value of R_s is used, the value of L has to increase to limit dv/dt to the specified value. From the equation

$$L = \frac{(VR_s)}{\left(\dfrac{dv}{dt}\right)_{\text{max}}} = \frac{400 \times 10}{200}\,\mu\text{H} = 20\,\mu\text{H}$$

This inductor value is not too high, and it is more than which is required for di/dt protection.

Example 15.6 For the circuit shown in Fig. E15.6 calculate: (i) the maximum values of di/dt, dv/dt for the thyristor; (ii) what are the rms values and average values of current ratings of thyristor for $\alpha = 90°$, neglecting the effect of inductance; and (iii) what is the voltage ratings of thyristor?

$e_s = 325 \sin 314t$

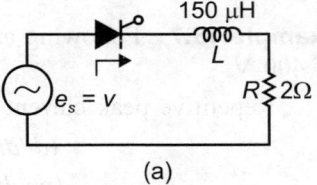

Fig. E15.6

Solution (i) Thyristor is turned on by applying a current pulse to the gate and there is a sudden increase of current on lowering of forward breakovervoltage. This rate of change of current di/dt should be within the thyristor rating and is controlled by external components like L and R.

$$e_s = v + Ri + L\,di/dt$$

Thus, current $i = \dfrac{v}{R}(1 - e^{-R/Lt})$

$$\frac{di}{dt} = \frac{v}{R} \cdot \frac{1}{\tau} \cdot e^{-t/\tau} \quad \text{where } \tau = L/R$$

$|di/dt|$ is maximum when $t = 0$

$$\left.\frac{di}{dt}\right|_{\text{max}} = \frac{v}{L} = \frac{V_m}{L}\ \text{A/sec}$$

$$= \frac{325}{150 \times 10^{-6}} = 2.17\ \text{A/}\mu\text{sec}$$

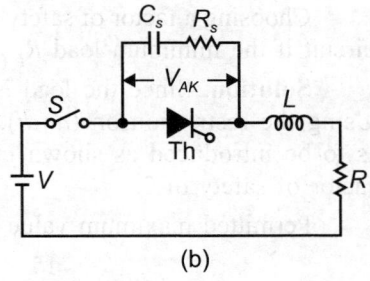

(a)

(b)

Fig. E15.6

dv/dt, the rate of change of application should not be exceeded, otherwise thyristor will switch on without gate pulse. R_s and C_s is called the *snubber circuit*.

When switch S is closed

$$V_{AK} = V(1 - e^{-t/\tau}) \quad \text{where } \tau = L/R_S$$

$$\frac{dV_{AK}}{dt} = \frac{R}{L}Ve^{-t/\tau}\ \text{(V/sec)}$$

$\dfrac{dV_{AK}}{dt}$ is maximum when $t = 0$

$$\left.\frac{dV_{AK}}{dt}\right|_{\text{max}} = \frac{VR_s}{L}\ \text{V/sec}$$

$$\left.\frac{dV_{AK}}{dt}\right|_{\text{max}} = \frac{325 \times 15}{150\,\mu\text{H}} = 32.5\ \text{V/}\mu\text{sec}$$

(ii)
$$I_{av} = \frac{V_m}{2\pi R}(1 + \cos\alpha)$$

$$= \frac{325}{2\pi \times 2}(1 + \cos 90°) = 25.86 \text{ A}$$

$$I_{rms} = \frac{V_m}{2R\sqrt{\pi}}\left(\pi - \alpha + \frac{\sin 2\alpha}{2}\right)^{1/2}$$

$$= \frac{325}{4\sqrt{\pi}}\left(\pi - \frac{\pi}{2} + \frac{\sin 180°}{2}\right)^{1/2}$$

$$= \frac{325}{4\sqrt{\pi}}\sqrt{\frac{\pi}{2}} = 57.45 \text{ A}$$

(iii) Voltage rating > (325) × 2 to 3 times
$$= 975 \text{ V}.$$

Example 15.7 Following are the specifications of a thyristor operating from a peak supply of 400 V.

Repetitive peak current $= I_{pk} = 200$ A

$(di/dt)_{max} = 15$ A/μsec

$(dv/dt)_{max} = 108$ V/μsec

Choosing a factor of safety of 2 for I_{pk}, $(di/dt)_{max}$ and $(dv/dt)_{max}$, design a suitable snubber circuit if the minimum load $R_{L\,(min)}$ be $10\,\Omega$.

Solution. Since the load is given to be resistive. Using the restriction on (di/dt), a series inductance L is to be introduced as shown in Fig. E15.7. Taking a factor of safety of 2.

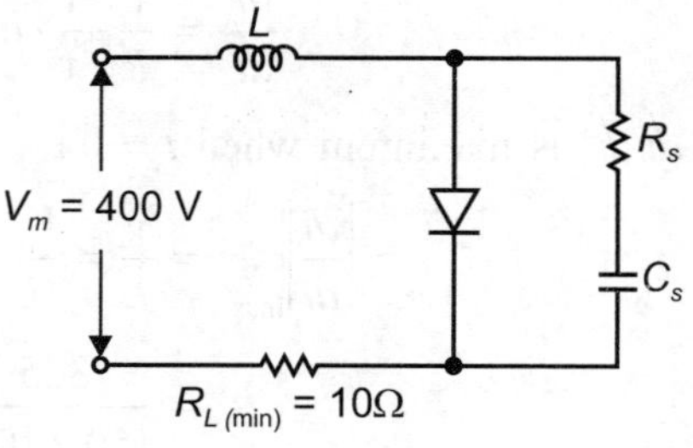

Permitted maximum value of

$$(di/dt) = \frac{15}{2} = 7.5 \text{ A/μsec}$$

Whence
$$L = \frac{V_m}{(di/dt)_{max}} = \frac{400}{7.5} = 53.23 \,\mu\text{H}$$

Fig. E15.7

This value of inductance will restrict the (di/dt) stress below the specified value under worse conditions.

For a peak value of voltage specified as 400 V and minimum load resistance as $10\,\Omega$, the peak current,

$$I_m = \frac{400}{10} = 40 \text{ A}$$

Assuming the thyristor to open instantaneously, entire load current shall tend to flow through snubber capacitor C_s, using the equation

$$C \cdot dv/dt = I$$

or
$$C = I_m/(dv/dt)_{max}$$

With a factor of safety 2,

$$\frac{dv}{dt} = \frac{108}{2} = 54 \text{ V/\mu sec}$$

or
$$C = \frac{40}{54 \text{ V/sec}} = 0.75 \,\mu\text{F}$$

The value of snubber resistance is such as to result in damping factor of 0.3 to 0.5. Choosing ξ as 0.4 the value of resistance,

$$R_S = 2\xi\sqrt{L/C}$$

$$R_S = 2 \times 0.4\sqrt{\frac{53.3 \times 10^{-6}}{0.74 \times 10^{-6}}} = 6.8 \,\Omega$$

Peak snubber discharge current

$$= 400/6.8 = 50.8 \text{ A}$$

When this is added to load current it becomes

$$= 58.8 + 40 = 98.8 \text{ A}$$

This current is below the peak repetitive current I_{pk} with a factor of safety of 2.

Example 15.8 R, L and C in an SCR circuit meant for protecting against dv/dt, di/dt of values $4\,\Omega$, $6\ \mu\text{H}$ and $6\ \mu\text{F}$, respectively. If the supply to the circuit is 300 V, calculate the maximum values of dv/dt and di/dt.

Solution di/dt protection is provided by the inductor. At the time of turn on, the applied voltage is shared by the inductor and the thyristor:

Rate of rise of current is decided by the inductance L:

$$L\,di_{dt} = V \text{ or } \frac{di}{dt} = \frac{V}{L} = \frac{300}{6 \times 10^{-6}} \text{ A/sec}$$

If i_s is the snubber current then rate of rise of device voltage is

$$dv/dt = \frac{d}{dt}(i_s R) + i_s/C$$

Resistance for adequae damping is

$$R_s = 2\xi\sqrt{L/C} \text{ or } 4 = 2\xi\sqrt{\frac{6 \times 10^{-6}}{6 \times 10^{-6}}} = \text{ or } \xi = 2$$

As the circuit is overlamped maximum dv/dt occurs at the instant of switching. Since $V_s = 300$ V. Peak snubber discharge current

$$= \frac{300}{4} = 75 \text{ A}$$

Then
$$I = 75 = C\,dv/dt$$

or
$$dv/dt = \frac{75}{C} = \frac{75}{6 \times 10^{-6}} \text{ V/sec} = 12.4 \text{ V/\mu sec.}$$

EXERCISES

Multiple Choice Questions

1. The device used for current protection is:
   (A) the fuse   (B) R-C network
   (C) snubber network (D) none of these

2. The power semi-conductor device requires protection against excessive:
   (A) voltage
   (B) current
   (C) certain rate of change
   (D) all of the above

3. A device has a critical di/dt rating which, if exceeded, can cause overheating:
   (A) during turn-on
   (B) during turn-off
   (C) both are true
   (D) none of these

4. The scheme which can be implemented if two thyristors turn on simultaneously, producing a short circuit on the supply, is known as:
   (A) protection by ringing
   (B) gate blocking
   (C) electronic crowbar
   (D) all the above

5. As soon as fault current is detected, it can be shunted away by turning on a parallel thyristor until the circuit breaker interrupts the fault current. This scheme is known as:
   (A) electronic crowbar
   (B) protection by ringing
   (C) gate blocking
   (D) none of these

6. Gates are protected against spurious (or noise) firing by using:
   (A) shield cables.
   (B) a Zener diode across the gate
   (C) a series resistance
   (D) all of the above

7. The gate can be protected against over-current by:
   (A) connecting a series resistance
   (B) a Zener diode across the gate
   (C) connecting a series inductor
   (D) none of these

8. The gate can be protected against over-voltage by:
   (A) a Zener diode across the gate
   (B) connecting a series resistance.
   (C) connecting a series inductor
   (D) heat sink

9. A suppressor is a device that responds to the rate of change of current or voltage to prevent:
   (A) a fall below a predetermined level
   (B) a rise above a predetermined level
   (C) overloading
   (D) none of these

10. Snubber network is R-C series network placed in parallel with a device to protect against:
    (A) over current transients
    (B) overvoltage transients
    (C) thermal losses
    (D) none of these

11. The thyristor can be safeguarded against overvoltages by:
    (A) thyristor diode
    (B) contactor
    (C) simple diode
    (D) all of these

12. Over voltages may be generated by:
    (A) switching of inductive loads
    (B) variations in the supply voltage
    (C) bad commutation
    (D) all of the above

13. Heat sink is a mass of metal that is added to a device for the purpose of:
    (A) absorbing heat
    (B) dissipating heat
    (C) absorbing and dissipating heat
    (D) none of these

14. Surge current rating of thyristor specifies the maximum:
    (A) repetitive current with sine wave
    (B) non-repetitive current with rectangular wave
    (C) non-repetitive current with sine wave
    (D) repetitive current with triangular wave

15. The di/dt rating of thyristor is specified for its:
    (A) decaying anode current
    (B) decaying gate current
    (C) rising gate current
    (D) rising anode current

16. For a thyristor, dv/dt protection is achieved through the use of:
    (A) RL in series with thyristor
    (B) RC across thyristor
    (C) L in series with thyristor
    (D) none of these

17. For thyristor di/dt protection is achieved through the use of:
    (A) R in series with thyristor
    (B) RL in series with thyristor
    (C) M in series with thyristor
    (D) none of these

18. The function of snubber circuit connected across a thyristor is to:
    (A) suppress dv/dt (B) increase dv/dt
    (C) decrease dv/dt (D) none of these

19. The object of connecting resistance and capacitance across gate circuit is to protect the thyristor gate against:
    (A) overvoltages   (B) dv/dt
    (C) noise signals   (D) none of these

ANSWER KEY

1. (D)	2. (A)	3. (A)	4. (A)	5. (A)	6. (A)	7. (A)	8. (A)
9. (B)	10. (B)	11. (A)	12. (D)	13. (C)	14. (C)	15. (D)	16. (B)
17. (C)	18. (A)	19. (C)					

Review Questions

1. What are the sources of overvoltages?

2. How the thyristors are protected against overvoltage?

3. What is snubber circuit and how it protects the thyristors?

4. Discuss, how the snubber network is designed?

5. Is it important to provide gate protection to thyristors?

6. Why heat sinks are provided in the practical circuits?

7. How the thyristors are protected against over current?

8. What is the need of protection of thyristors?

Industrial Applications Using Solid-State Devices

16.0 INTRODUCTION

The application of solid-state devices fall into one of two major areas, one is that which involves motor control and the other involves non-motor applications. With the advent of high power thyristors of increased reliability, the philosophy of industrial power control has changed drastically. In almost every process the old magnetic control methods have been replaced by more accurate, dependable, faster and economical thyristor control methods. Some of the industrial applications using solid-state devices, specially thyristors are dealt here to describe some additional industrial applications of power electronics such as residential applications, heating, welding and induction heating.

16.1 RESIDENTIAL APPLICATIONS

Residential homes and buildings require energy usage in good quantity. The residential applications include space heating and air conditioning, refrigeration and freezer, water heating, lighting, cooking, television, cloth washer and dryer, and many other miscellaneous applications.

The role of power electronics in residential applications is to provide energy conservation, reduced operating cost, increased safety, and greater comfort. Some of the dominant residential applications are discussed in this chapter.

Space heating and Air conditioning: Nearly 15 to 20% of electric energy is used for space heating and air conditioning. Heat pumps are now being used for this purpose. In conventional heat pump, the compressor operates essentially at a constant speed when the motor is running. The compressor output in this system is matched to the building heating or cooling load by cycling the compressor on or off.

A load proportional capacity modulated heat pump is shown in Fig. 16.1.

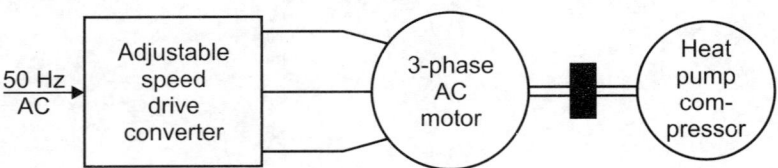

Fig. 16.1 Load proportional capacity modulated heat pump

The speed of the compressor motor and hence the compressor output is adjusted to match the building heating or cooling load, thus eliminating the on or off cycling of the compressor.

Either an induction motor drive or self-synchronous motor drive is used to adjust the compressor speed in proportion to the building load.

16.2 HEATING

A heating effect produced in the resistive load can be controlled to different power levels by use of triac connection shown in Fig. 16.2(a). For high power heating levels two thyristors in the inverse parallel connection can be used as shown in Fig. 16.2(b). To obtain common cathodes, and hence simplify the firing circuit, two diodes can be added as shown in Fig. 16.2(c).

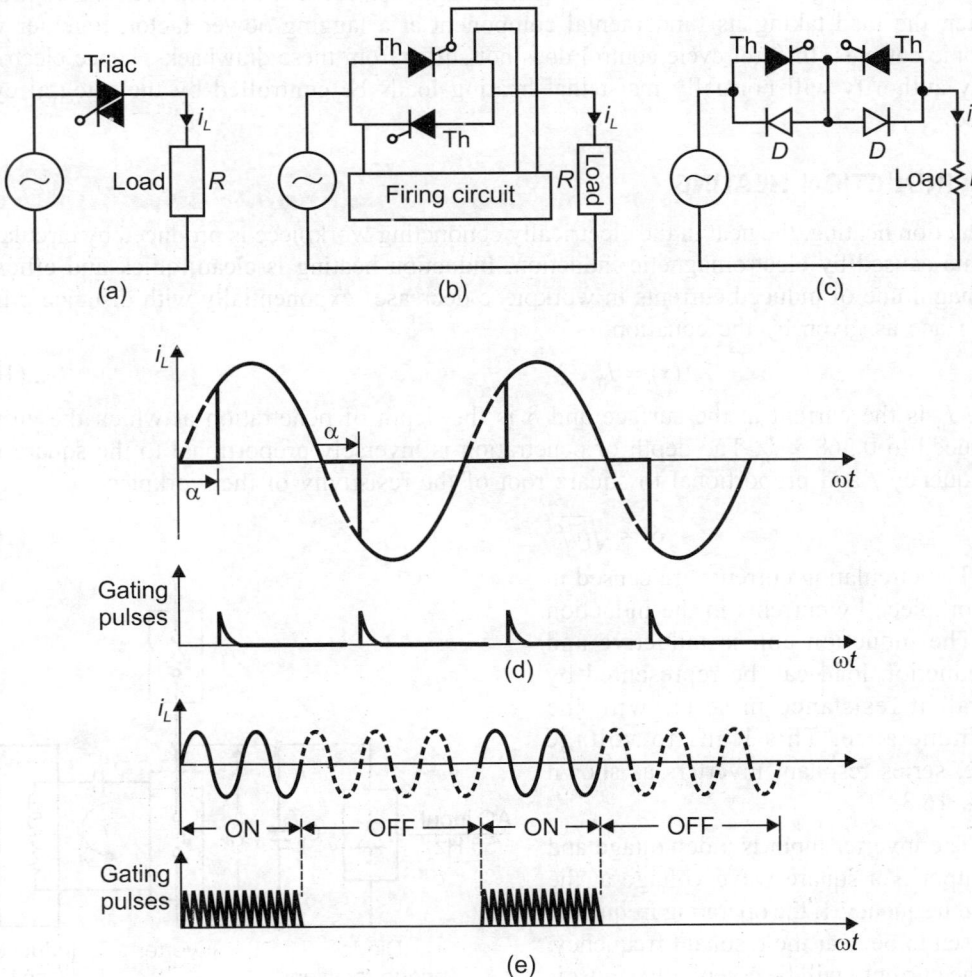

Fig. 16.2 Different power control circuits using phase angle control

Phase angle control can be used as shown in Fig. 16.2(d), where the start of each half cycle is delayed by an angle α. Assuming the load to be a pure resistive R and the supply to have maximum value of V_m, then

Power, $$P = \frac{V_{rms}^2}{R} = \frac{1}{\pi R}\int_{\alpha}^{\pi}(V_m \sin \omega t)^2 \omega t = \frac{V_m^2}{2\pi R}\left(\pi - \alpha + \frac{1}{2}\sin 2\alpha\right) \qquad ...(16.1)$$

The majority of heating loads have thermal time constants of several seconds or longer. In this case, little variation of the heater temperature will occur if control is achieved by allowing a number of cycles ON with a number of cycles OFF, as shown in Fig. 16.2 (e). This form of control is called *integral cycle control* or *burst firing* or *cycle syncopation*. The time of the ON-period consists of whole half-cycles. The power will be given by the equation,

Power, $$P = \frac{V_m^2}{2R} \times \frac{\text{ON time}}{(\text{ON + OFF})\text{ time}} \qquad ...(16.2)$$

While loading on the supply system, phase angle control will be similar to the controlled . rectifier, the load taking its fundamental component at a lagging power factor, together with harmonic currents. Integral cycle control does not suffer from these drawbacks, hence electricity supply authority will normally insist that heating loads be controlled by the integral cycle method.

16.3 INDUCTION HEATING

In induction heating, the heat in the electrically conducting workpiece is produced by circulating currents caused by electromagnetic induction. Induction heating is clean, quick and efficient. The magnitude of induced currents in workpiece decreases exponentially with distance x from the surface as given by the equation

$$I(x) = I_0 \, e^{-x/\delta} \qquad ...(16.3)$$

where I_0 is the current at the surface and δ is the depth of penetration at which the current is reduced to $0.368 \times I_0$. The depth of penetration is inversely proportional to the square root of frequency f and proportional to square root of the resistivity of the workpiece.

$$\delta \propto \sqrt{\rho/f} \qquad ...(16.4)$$

The circulating currents are caused in the workpiece by currents in the induction coil. The induction coil is inductive and the induction load can be represented by equivalent resistance in series with the coil inductance. This leads to voltage source, series resonant inverters as shown in Fig. 16.3.

The inverter input is a dc voltage and the output is a square wave voltage at the desired frequency. If the operating frequency is chosen to be near the resonant frequency, then the current i will be essentially sinusoidal due to impedance characteristic upto a few kilohertz, it is possible to use thyristors as switches in the inverter. The power to the load can be controlled by controlling the inverter frequency.

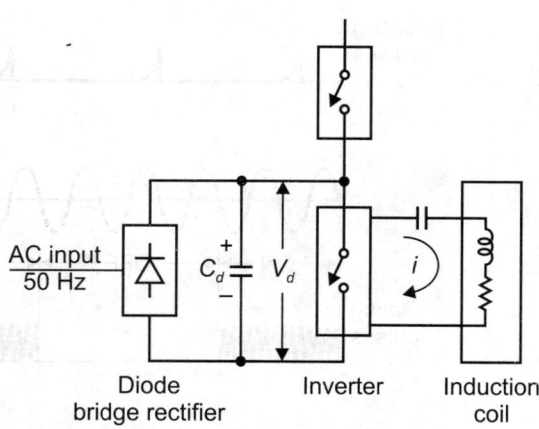

Fig. 16.3 Voltage source, inverter for induction heating

16.4 ELECTRIC WELDING

In electric arc welding, the melting energy is provided by establishing an arc between two electrodes, one of which is the metallic workpiece being welded.

It is desirable to have a very low ripple in the current once an arc is established. In all the welding applications, the electrical output needs to be electrically isolated from the input supply. This electrical isolation is provided by either a high frequency transformer or 50 Hz transformer.

In this process with a 50 Hz transformer, ac input voltage is first stepped down to a suitably low voltage. Then, it is converted to a controlled dc by means of the three schemes shown in Fig. 16.4. In Fig. 16.4(a) thyristor controlled bridge rectifier is used to control the output voltage and current. Second alternative shown in Fig. 16.4(b) uses a switch mode step down ac-dc converter to control the output. In scheme of Fig. 16.4(c), a diode bridge uncontrolled rectifier is used. The output of this diode rectifier is uncontrolled which is controlled by means of transistor series regulator to regulate the welder's output. All these techniques suffer from the losses in 50 Hz transformer, weight and size.

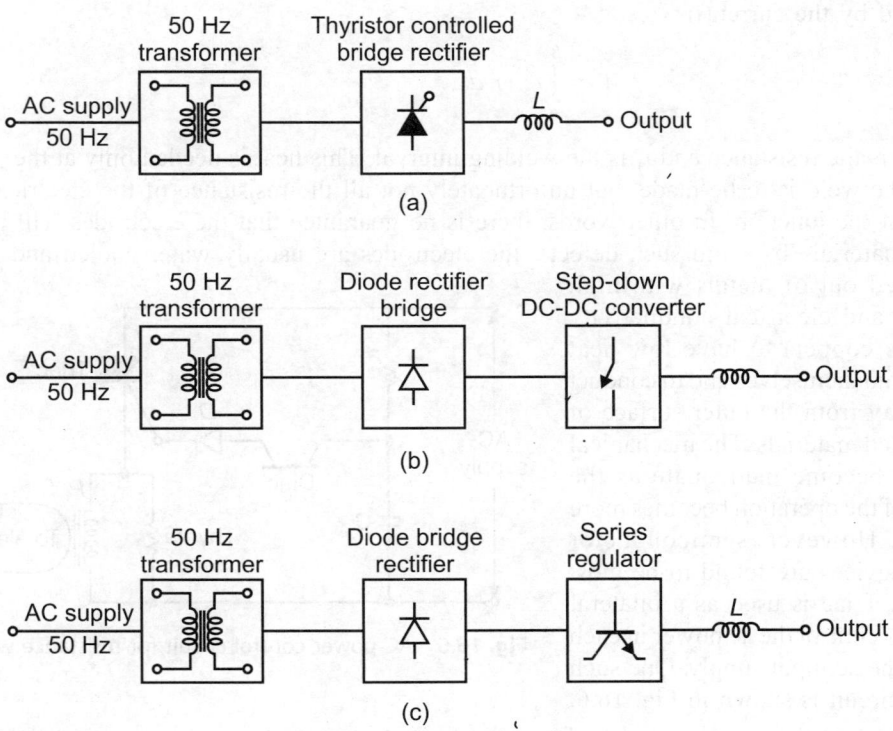

Fig. 16.4 Different circuits for electric welding

Figure 16.5 shows the block diagram of a switch mode welder, where the electrical isolation is provided by a high frequency transformer. A small inductance is needed at the output to limit the output current at high frequencies. The efficiency of such a welder is in the range of 85% to 90% with small weight and size as compared with welders employing 50 Hz transformer.

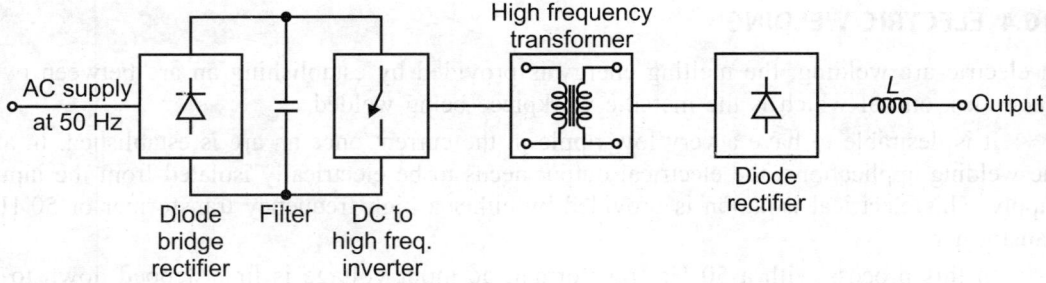

Fig. 16.5 Block diagram of switch mode welder

16.5 RESISTANCE WELDING

Resistance welding is the process of combining two or more pieces of metal together by passing heavy currents through the area of contact where heat is required. To achieve either controlled ac or short pulses of unidirectional currents are used. The total thermal energy produced by the current is

$$W = \int_{0}^{t_1} i^2 \cdot r \, dt$$

where r is the resistance and t_1 is the welding interval. This heat is needed only at the junction, where the weld is to be made, but unfortunately not all the resistance of the electrical circuit occurs at the junction. In other words, there is no guarantee that the electrodes will not weld to the material. To avoid such defects, the electrodes are usually water cooled and they are fashioned out of metals with high thermal and electrical conductivity (such as copper) to have low heat loss within themselves and to conduct heat away from the outer surface of the welded materials. The mechanical devices become inadequate as the timing of the operation becomes more critical. However, semiconductor power devices are found to be most suitable. Triac is used as a bilateral switch to control the ac power in each half of the ac input supply. One such simple circuit is shown in Fig. 16.6.

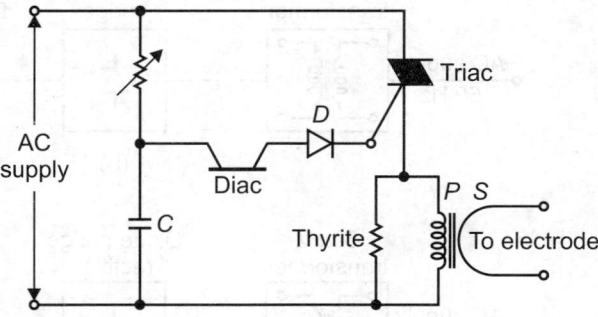

Fig. 16.6 AC power control circuit for resistance welding

External pressure is used to force the heated parts to be joined together through the regularisation of the grain structure. This produces a very homogeneous weld which has the same physical properties as that of the parent job metals. The welding equipment for resistance welding which accepts power supply from the mains voltage is made available across the primary of a welding transformer by means of triac. Thus, the triac connects the primary winding P of the welding transformer across the ac supply during the welding interval. The secondary winding is rated at several volts and currents that may reach hundreds of amperes for heavy welds. The rate of change of current in the primary of the welding transformer is so high that dangerously high transient voltages can appear across the transformer windings.

Therefore, it is often desirable to connect a thyrite resistor across the primary, as shown in Fig. 16.6. Thyrite has relatively high resistance at low applied voltages but the resistance falls as the voltage rises, and by proper choice of this, the occurrence of the voltages that would endanger the insulation of the transformer could be prevented. Firing circuit of triac consists of simple *R-C* network and a diac, which determines the instant when the triac will start conducting.

As soon as the voltage across the capacitor reaches the threshold voltage of the diac, it breaks down and a gate pulse is obtained to trigger the triac. By varying the resistance, the charging time of the capacitor to reach the desired threshold voltage is varied and thereby the amount of current can be controlled. In this scheme, power is controlled by controlling the conduction period of the triac in each half of the input ac wave.

16.6 SEAM WELDING

Seam welding is one of the techniques of resistance welding. In seam welding, two overlapping sheets of metal are welded together shown in Fig. 16.7(a) along a continuous line by rolling them between two wheel shaped electrodes and passing weld current for suitable intervals. Here, the seam is made up of overlapping spots. At slow speeds a weld current schedule of six cycles 'ON' and six cycles 'OFF' might be typical, but as the speed increases the duration of OFF-period is decreased. Sometimes the 'OFF'-period is made zero, when in each half cycle current makes a weld that overlaps the previous one. Steel barrels and refrigerator evaporator plates are among the objects manufactured by this process. For this type of welding, the circuit shown in Fig. 16.7 can be used for controlling the weld current and the speed of wheels is controlled by single motor through gear arrangement. The welding current and motor speed are controlled manually and independently. Usually a DC motor is used for driving the wheels, as these need large starting torque for controlling the speed of the dc motor. The speed of dc motor can be regulated by controlling either the field current or the armature voltage. For the application described here, the field is separately excited and the armature voltage is controlled. This constant torque mode of control is preferred because the full load torque of the motor remains the same, irrespective of the speed and the response of the motor is faster.

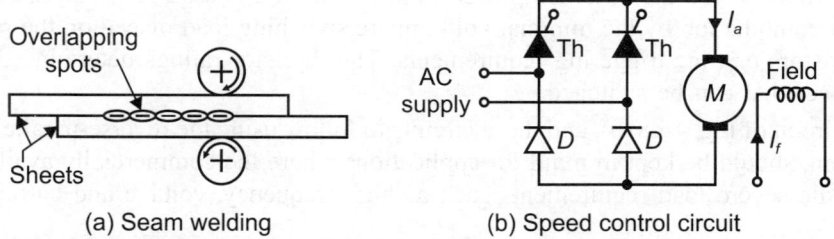

(a) Seam welding   (b) Speed control circuit

Fig. 16.7 Seam welding procedure

A simple speed control circuit called a half controlled circuit of symmetrical configuration is shown in Fig. 16.7(b) where both the thyristors are applied with the gating pulse in each of the half cycles. Since the cathodes of the two thyristors are at the same potential, but the thyristor which has the forward-bias at the instant of firing, will turn on. The free-wheeling action will take place through diode D_1 and thyristor Th1 or diode D_2 and thyristor Th2 when the input voltage polarity change. So the thyristor Th1 will not be turned off even though the input current and the output voltage become zero.

16.7 AC STATIC SWITCHES

The device is basically a switch automatically provides supply to the load. Change of resistance of the light dependent resistor (LDR) acts as an automatic switch. Since it does not have any moving parts, it called as static switch. In the circuit two thyristors are used for making and breaking ac circuits. The circuits of Fig. 16.8 provide high speed switching of ac power loads, and are ideal for applications with a high duty cycle. Any 'contact' resistance of the control device and load resistance should be just greater than the peak supply voltage divided by the peak gate current rating of the thyristor. If resistor R is made too low the load will result with consequent loss of load voltage and waveform distortion. The control device indicated can be either electrical or mechanical in nature.

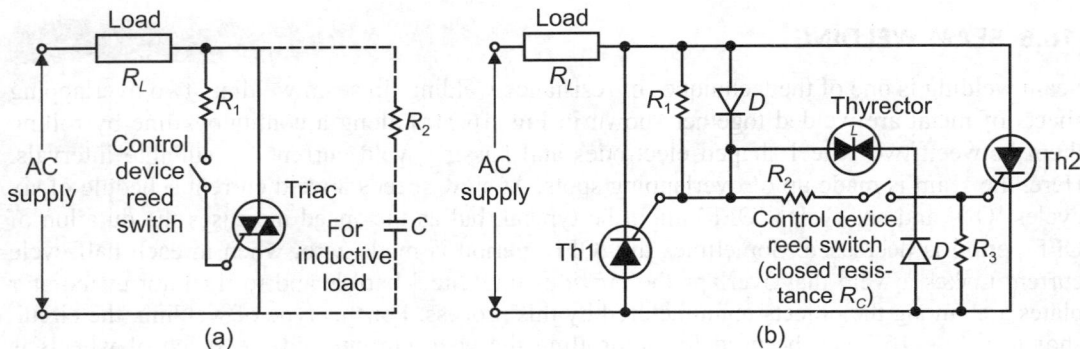

Fig. 16.8 High speed switches for AC power loads

Light dependent resistors and light activated semiconductors, photocouplers, magnetic cores, and magnetic reed switches are all suitable control elements. In particular, the use of the hermetically sealed reed switches as control elements in combination with thyristors and triacs offers many advantages. The reed switch can be actuated by passing ac or dc current through a small winding around it, or by the proximity of a small magnet. In either case complete electrical isolation exists between the control signal input, which may be derived from many sources, and the switched power output. Long life is assured by thyristors or triac/reed switch combination by the minimal volt-ampere switching load placed on the reed switch by the thyristors or triac triggering requirements. The thyristor ratings determine the amount of load power that can be switched.

The circuit of Fig. 16.8(b), and those circuits to follow using the reverse-parallel thyristors configuration, should be kept in mind for applications where the commercially available triacs cannot handle severe load requirements such as high frequency, voltage and current.

16.8 DC STATIC SWITCH

This circuit works in the opposite manner to that of a light-activated turn-off circuit. It may also be developed using the solid-state devices like diac, triac and LDR. Figure 16.9 illustrates a static thyristor for use in a dc circuit. When a low power signal is applied to the gate of thyristor Th1, this thyristor is triggered and voltage is applied to the load. The right hand plate of capacitor C charges positively with respect to the left hand plate through resistor R_1. When thyristor Th2 is triggered on, capacitor C is connected across thyristor Th1 so that this thyristor is momentarily reverse biased between anode and cathode. This reverse voltage turns thyristor Th1 off provided the gate signal is not applied simultaneously to both gates. The current through the load will decrease to zero in an exponential fashion as capacitor C becomes charged.

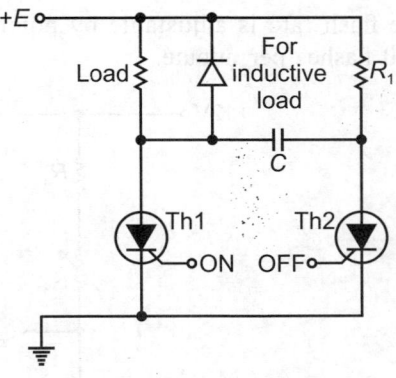

Thyristor Th1 should be selected so that the maximum load current is within its rating. Thyristor Th2 conduct only momentarily during the turn-off action, it can be smaller in rating than thyristor Th1. The minimum value of commutating capacitance C can be determined by the following equations:

For resistive load:

$$C \geq \frac{1.5 t_{OFF} I}{E} \mu F \qquad ...(16.6)$$

For inductive load:

$$C \geq \frac{t_{OFF} I}{E} \mu F \qquad ...(16.7)$$

Fig. 16.9 DC static switch

where
t_{OFF} = Turn-off time of thyristor in μ-seconds
I = Maximum load current in Ampere at time of commutation
E = Minimum dc supply voltage

The resistance of resistor R_1 should be ten to one hundred times less than the minimum effective value of the forward blocking resistance of Th2.

16.9 FLASHER CIRCUITS

The thyristor and the triac are ideally suited for this type of applications since they can function over a wide range of current and voltage with a much higher degree of liability than the commonly used electro-mechanical systems. The thyristor and the triac also offer an important advantage over power transistors in that they do not require excessive derating of current to handle the high inrush currents of incandescent lamps. The UJT, PUT and transistor make ideal trigger devices for the thyristor and triac in this type of the application since they permit an economical method for obtaining a wide frequency range and a high degree of frequency stability. Flasher circuits are used in traffic lights, aircraft beacons, navigational beacons and illuminated signs.

16.10 LOW POWER FLASHER

The circuit shown in Fig. 16.10 has been taken to illustrate the basic principles of a thyristor/ UJT flasher. Unijunction transistor, UJT, operates as a relaxation oscillator, delivering a train of trigger pulses to the two thyristor gates via resistor R_1. Assume thyristor Th2 is 'on' and the lamp is energized. When the next trigger pulse comes along it triggers thyristor Th1 and thyristor Th2 is turned off by the commutation pulse coupled to its anode via capacitor C_2. Because the commutation pulses have longer duration than the trigger pulses, thyristor Th2 cannot be retriggered inadvertently at this time. Thyristor Th2 is retriggered properly by the next trigger pulse from resistor R_2. 'Lock-up' (failure to flash caused by both thyristors being on together) is prevented by making thyristors Th1 turn off independent of the commutating capacitor. This is done by operating thyristors Th1 in a 'starved' mode, that is by making resistor R_2 so large that thyristor Th2 is unable to remain 'on' except to discharge capacitor C_2. During the remainder of the cycle thyristor Th1 is off, and capacitor C_2, therefore, is always able to develop commutating voltage for thyristor Th2. With the components shown, the flash rate is adjustable by potentiometer R_2 within the range 36 flashes per-minute to 160 flashes per minute.

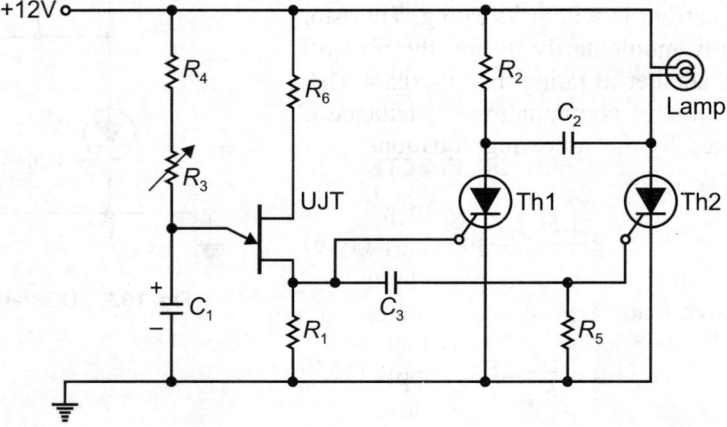

Fig. 16.10 Low power flasher circuit

16.11 AC FLASHER

The intensity of illumination of a lighting system can be controlled simply by the rheostatic method control. Solid-state methods of illumination control are more effective and efficient as compared to the rheostatic method of control. The power loss caused in a solid-state circuit is negligible and a smooth control of illumination is possible. Such circuits are widely used for automatic stage lighting. For heavy load requirements, the dc flashers have the disadvantage of requiring a large commutating capacitor. In such applications, the use of ac flashers ends up being more economical.

Figure 16.11 illustrates a power flip-flop flasher circuit that can handle two independent loads upto 2.0 kW each. Transformer Tr, diodes D_1 through D_4, resistor R_1 and capacitor C_1

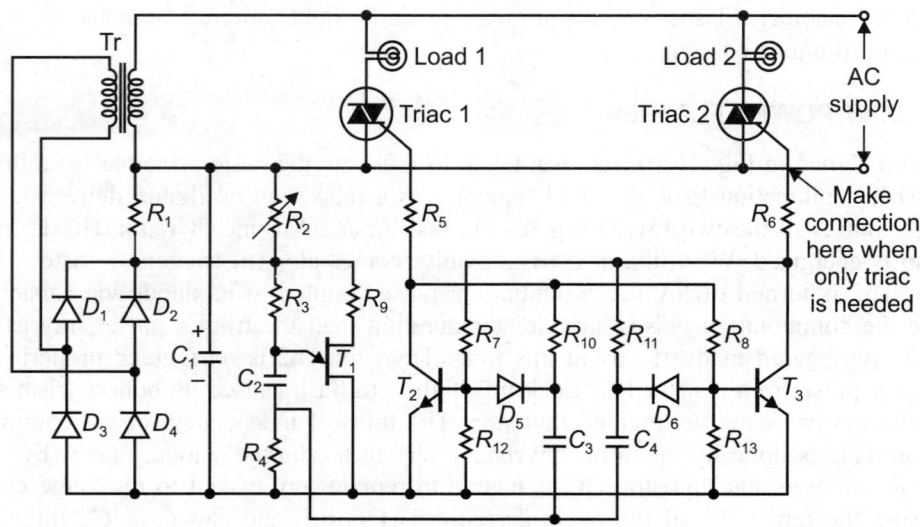

Fig. 16.11 AC flasher circuit

provide the dc supply to the free running unijunction oscillator and the transistor flip-flop of T_2, T_3. The interbase voltage for T_1 is taken directly from the positive side of the bridge rectifier in order to synchronize the free running unijunction oscillator with the supply frequency. The negative going output pulses developed across resistor R_4 trigger the transistors of the flip-flop which alternately turn the triacs on and off. The flashing rate is determined by the time constant of resistors R_2, R_3 and capacitor C_2.

16.12 HIGH SPEED SWITCH OR ELECTRONIC CROWBAR

'*Electronic crowbar*,' shown in Fig. 16.12 has proved very useful for protecting dc circuits against input line voltage transients and short circuit load conditions. If the dc supply exceeds the desired maximum value as determined by the setting of potentiometer R_1, the voltage at the emitter of UJT exceeds the peak point voltage causing UJT to trigger which in turn triggers the thyristor. The full supply voltage is then applied to the circuit breaker trip coil causing the circuit breaker to open the main dc supply bus. Besides increasing the speed of the circuit breaker action this circuit instantly loads down the dc bus, preventing the voltage on the load from rising until the circuit breaker has time to operate. The circuit also protects the load and the supply against short circuit conditions by monitoring the current through resistor R_2.

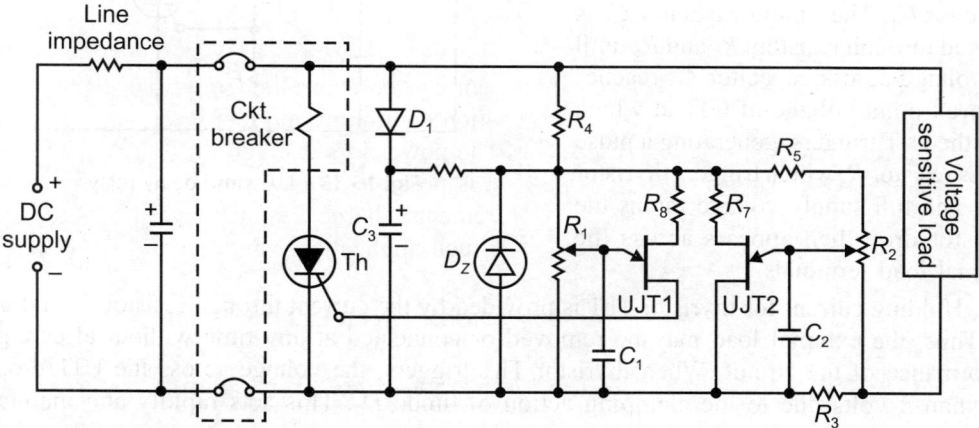

Fig. 16.12 Electronic crowbar circuit

When the voltage across resistor R_2 exceeds the desired maximum value as determined by the setting of potentiometer R_2 the voltage at the emitter of UJT 2 exceeds the peak point voltage, causing UJT 2 and the thyristor to trigger as before. Due to the stable firing voltage of the UJT the trip voltage across resistor R_3 can be very low, a value in the range from 100 millivolts to 500 millivolts being suitable for most applications. If only overvoltage protection is desired the circuit of Fig. 16.12 can be simplified by eliminating UJT 2 and its associated circuitry. Similarly, if only over current protection is desired UJT 1 and its associated circuitry can be eliminated.

In the circuit of Fig. 16.12 rectifier D_1 and capacitor C_3 are used to provide filtering against negative voltage transients which would otherwise result in false tripping of the circuit. The values of potentiometer R_1 and R_2 are chosen to have appropriate time constants with capacitor C_1 and C_2 so as to give the desired voltage-time response in the tripping action. The thyristor is ideal for this type of circuit because of its ability to switch on within a few microseconds after being triggered. This circuit is capable of carrying momentary currents as high as 2000 Amps for 2 milli-seconds without damaging the thyristor.

16.13 TIME DELAY CIRCUITS

The circuit diagram of thyristor-UJT operated timer circuit is shown in Fig. 16.13. It is basically a thyristor based switching circuit. UJT has been used as relaxation oscillator which serves as a triggering device for the thyristor. Time delay circuits are used frequently in industrial control, aircraft and missile systems to apply or remove power from a load at predetermined time after initiating signal is applied. Cascaded time delay circuits can be used to sequentially perform a series of timed operations.

UJT-Thyristor time delay relay: Figure 16.13 illustrates an extremely simple yet accurate and versatile solid-state time delay circuit. The operating current and voltage of the circuit depend only on the proper choice of the thyristor. Resistor R_5 and Zener diode D_1 provide a stable voltage supply for the UJT. Initially thyristor Th1 is off and there is no voltage applied to the load. Timing is initiated either by applying supply voltage to the circuit or by opening a shorting contact across capacitor C_1. The timing capacitor C_1 is charged through resistors R_1 and R_2 until the voltage across capacitor C_1 reaches the peak point voltage of UJT at which time the UJT triggers, generating a pulse across resistor R_4 which triggers thyristor Th1. The full supply voltage minus the thyristor drop then appears across the external load terminals.

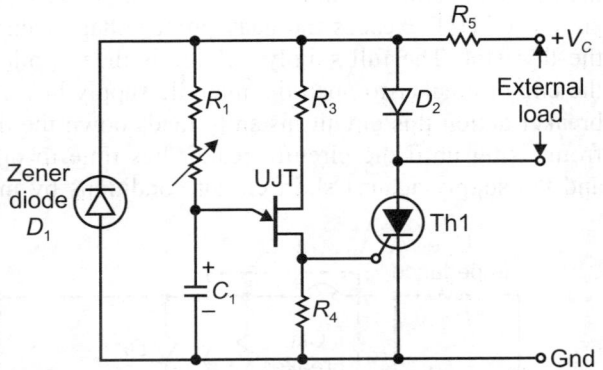

Fig. 16.13   UJT time delay relay

Holding current for thyristor Th1 is provided by the current through resistor R_5 and diode D_2. Thus, the external load may be removed or connected at any time without affecting the performance of the circuit. When thyristor Th1 triggers, the voltage across the UJT drops to less than 2 volts due to the clamping action of diode D_2. This acts rapidly and maintain a low voltage on capacitor C_1 so that the time interval is maintained with reasonable accuracy if the circuit is rapidly recycled. For the highest accuracy, however, additional means must be used rapidly and accurately, set the initial voltage on capacitor C_1 to zero at the beginning of the timing cycle. A pair of mechanical contacts connected across capacitor C_1 is ideal for this purpose.

The time delay of the circuit depends on the time constant $(R_1 + R_2) C_1$ and can be set to any desired value approximate choice of resistors R_1, R_2 and capacitor C_1. The resistor R_3 can serve as a temperature compensation for the circuit, increasing the value of resistor R_3 causes the time delay interval to have a more positive temperature coefficient. To reset the circuit in preparation for another timing cycle thyristor Th1 must be turned off either by momentarily shorting it with a switch contact or opening the dc supply.

16.14 FAN REGULATOR USING TRIAC

The speed of single phase motor can be controlled by controlling the rms value of input voltage. Figure 16.14 shows the circuit for a single phase motor using triac.

The capacitor C charges through resistor R, when the voltage across capacitor is equal to the breakdown voltage of the diac, a pulse is delivered to the triac, triggering it ON. The voltage will be applied to the load as soon as the triac becomes ON. If resistor R is low, the capacitor will charge to diac breakdown voltage in lesser time, i.e. the firing angle will be small. In this case the motor will run at high speed. If resistor R is high, firing angle will be more and speed of the motor will be low.

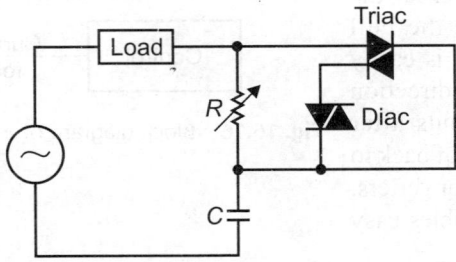

Fig. 16.14 Circuit of fan regulator

16.15 ELECTROCHEMICAL

Electrochemical plant requires low voltage direct current. This process includes electroplating, extraction and refining of metals by electrolysis, production of elements such as chlorine and hydrogen and electrochemical forming methods.

Electroplating is the process of depositing of metal on a workpiece immersed in a suitable electrolyte by making it the cathode and the material to be deposited is made the anode for which the Fig. 16.15 shows the process of electroplating. In electroplating current, densities are low with large spacing of the electrodes to ensure uniform plating. The amount of metal deposited is proportional to the quantity of current and time flow. The voltage required for electroplating vary from 5 V to 50 V, so normally half-wave rectifier circuits are used. Load voltage and current can be controlled by a regulating transformer and a diode rectifier or by the use of thyristors.

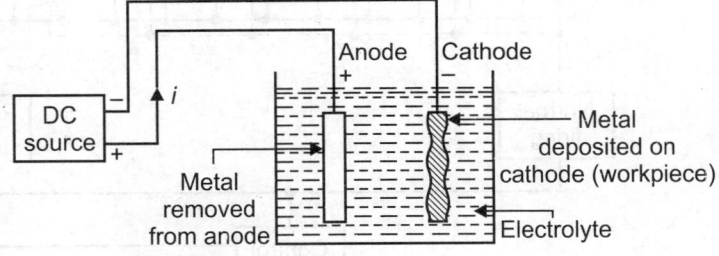

Fig. 16.15 Circuit of electroplating process

Metals such as aluminium and magnesium can be extracted by electrolysis which requires large quantities of electrical power. Aluminium is produced by the reduction of aluminium oxide by passing direct current through an electrolytic container containing the alumina dissolved in cryolite. The aluminium is precipitated at the carbon cathode sinking to the bottom of the container. Normally, each container requires 5 V but, when the alumina concentration drops, the voltage requirement can rise to 30 V to maintain the current level. Several containers are connected in series, so that typically the rectifier supplies 800 V, 70 kA to containers line.

A closed-loop feedback system, as shown in Fig. 16.16, is used so that the voltage variations in the load (pot line) can be compensated by change in the input voltage. A dc current transformer would sense the load current level, any reduction in the feedback signal automatically raising the input voltage, so bringing the load current back to its desired setting. Large scale production of chlorine, oxygen, caustic soda requires typically 50 kA, 50 V with half-wave connections for high efficiency.

In chemical industry, electrolytic processes such as copper refining produce a layer of hydrogen at the cathode which increases the cell voltage. This hydrogen gas is easily removed by reversing the direction of current for a few seconds after every few minutes. The use of back to back-connected thyristor converters, as shown in Fig. 16.17, enables easy periodic current reversal.

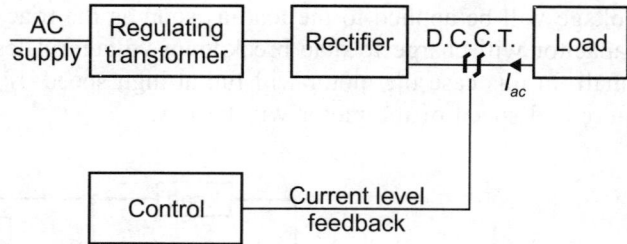

Fig. 16.16 Block diagram of a closed loop feedback system

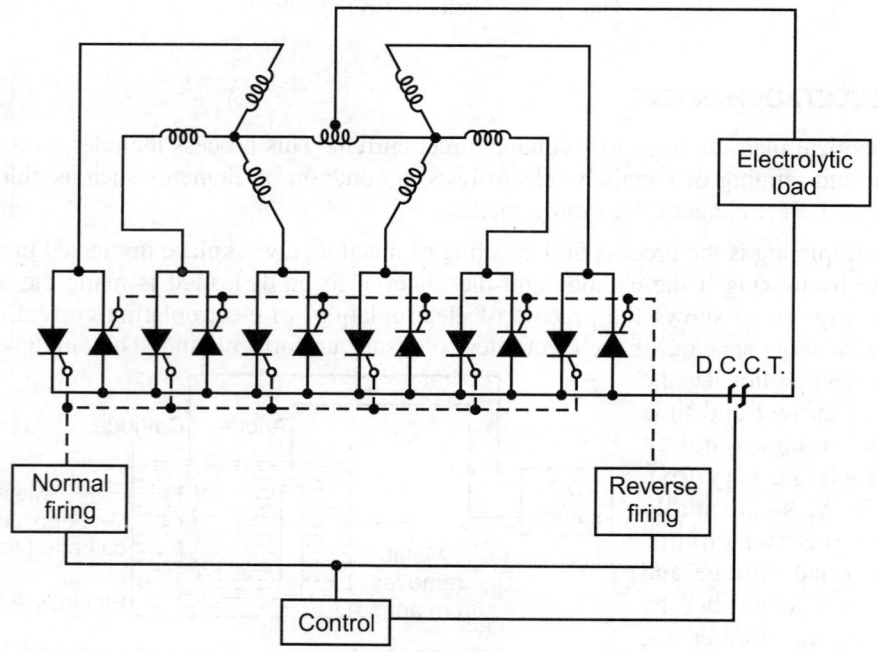

Fig. 16.17 Back to back connected thyristor converters circuit

16.16 MACHINING OR ELECTROCHEMICAL FORMING

Machining is the reverse of electroplating process. The principle is illustrated in Fig. 16.18, which shows how material is removed from the work (anode) and is carried away in the high velocity stream of electrolyte, thus avoiding plating on to the tool (cathode). Material is removed proportionally to the current density, to maintain the current. The gap is typically 0.2 mm, with a metal removal rate of 2 mm per minute. The process is used for hard materials, which is difficult to machine conventionally, for three dimensional shapes, or where the workpiece must not be distorted or stressed during machining.

In electrochemical forming equipment the input dc supply is less than 20 V, at currents of kilo ampere range. For this application, a diode rectifier fed from a regulated transformer would be used. Voltage regulation could be achieved by inverse-parallel thyristors in the supply lines to the transformer.

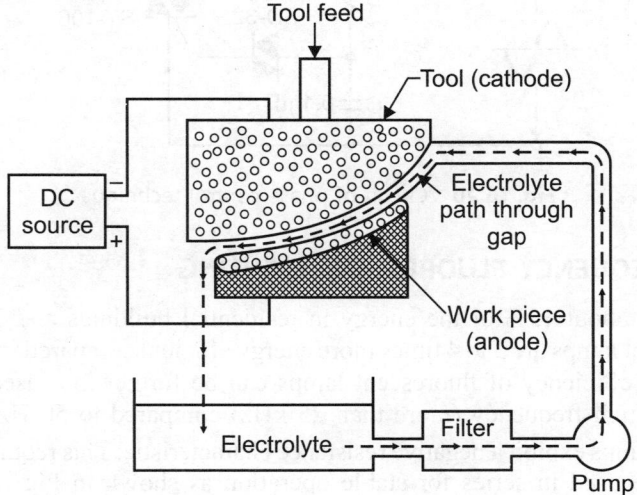

Fig. 16.18 Circuit of electrochemical forming

16.17 STATIC CIRCUIT BREAKER

An important application of SCR is as a switch. Since it does not have any moving parts, it is called as static conductor. Resistance R_1 is provided in the gate circuit to limit the gate current. Resistance R_2 protects the diode. Static ac circuit breaker is one in which the thyristors are used for making and breaking a circuit shown in Fig. 16.19. The trigger pulses are applied to the gates of thyristors through the control switch S when switch S is closed, thyristor Th1 will

fire at the beginning at the positive half-cycle. It will turn off when the current goes through the zero value. As soon as thyristor Th1 is turned off, thyristor Th2 will fire since the voltage polarity is already reversed and it gets the proper gate current. When any of the thyristor is triggered, the gate current will be negligible. To break the circuit, switch S is opened. As the current through the switch is small, opening the gate circuit has no problem. As no further gate signal will be applied to the thyristors when switch S is open, the thyristors will not be triggered and the load current will be zero.

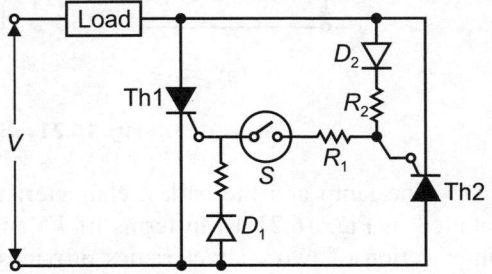

Fig. 16.19 Circuit of static circuit breaker

16.18 ILLUMINATION LEVEL OF INCANDESCENT LAMP

The illumination level of incandescent lamps can be controlled by this method using phase control technique.

A triac control circuit for lamp dimmers is shown in Fig. 16.20 using diac. A diac is gateless triac designed to breakdown at a low voltage when a charge on the capacitor C has attained the breakdown voltage, it will be discharged into the triac gate. Adjustment of resistance R determines the charging rate of capacitor C and hence the phase angle delay.

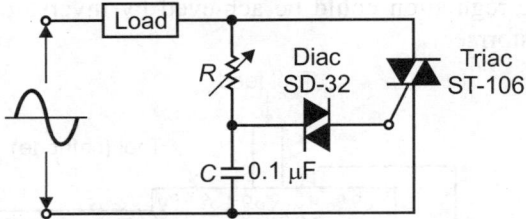

Fig. 16.20 Circuit of phase control technique

16.19 HIGH FREQUENCY FLUORESCENT LIGHTING

Lighting consumes about 15% of the energy in residential buildings and 30% in commercial buildings fluorescent lamps are 3 to 4 times more energy efficient compared with the incandescent lamps. The energy efficiency of fluorescent lamps can be further increased by 20 to 30% by operating them at high frequency (more than 25 kHz) compared to 50 Hz fluorescent lamps.

Fluorescent lamps exhibit a negative resistance characteristic. This requires that an inductive ballast or choke be used in series for stable operation as shown in Fig. 16.21 (a). Since the lamp impedance is essentially resistive, the three voltages in the circuit shown are related as

$$V^2_{ballast} + V^2_{lamp} = V^2_s$$

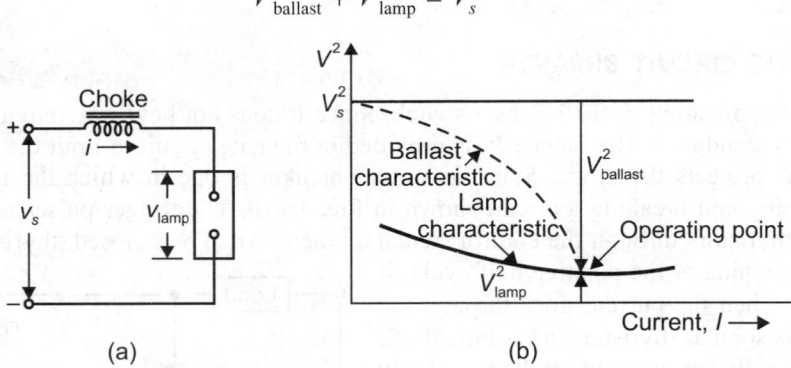

(a)   (b)

Fig. 16.21 High frequency lighting

The lamp and the ballast characteristics are plotted in Fig. 16.21 (b) in terms of V^2 and I the intersection of two characteristics provides stable operating point.

Figure 16.22 shows the circuit for 50 Hz rapid start system consisting of two tube lights (lamps) in series. At starting, the starting capacitor provides a shunt across lamp B and nearly all the input voltage appears across lamp A thus striking an arc. Once the arc discharge is established in lamp A, a high voltage appears across lamp B, which ignites an arc in lamp B. Then, the series combination of lamps A and B is in series with a power factor correction capacitor C_p, which is used to correct the power factor of operation.

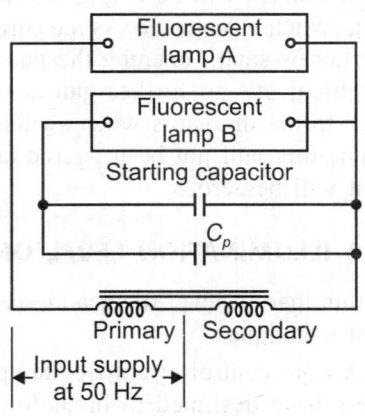

Fig. 16.22 Circuit for rapid start system

The high frequency fluorescent electronic lighting system is shown in Fig. 16.23 (a). The high frequency electronics choke or ballast converts 50 Hz ac input into high frequency output, usually in the range of 25 to 40 kHz.

The block diagram of high frequency electronics ballast shown in Fig. 16.23 (b) consists of diode rectifier bridge and a dc to high frequency ac inverter using class *E* resonant converter. An electromagnetic interference (EMI) filter is used before the rectifier bridge to suppress the conducted EMI. As the current drawn by the ballast from the supply will contain significant harmonics, hence the electronic ballast will operate at a poor power factor. This problem of harmonics can be remedied efficiently by an input current wave-shaping circuit. The electronic ballasts in general are more energy efficient compared to standard ballasts.

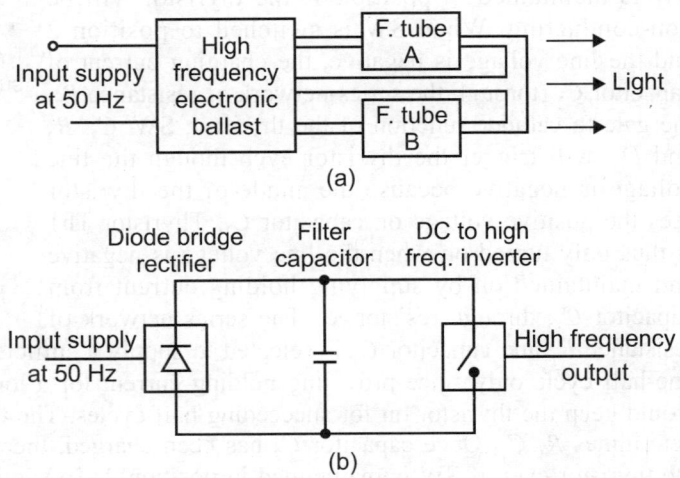

(a)

(b)

Fig. 16.23 Block diagram of high frequency electronics ballast

16.20 INDUCTION COOKING

In a standard electric or gas cooking range, a significant amount of heat escapes to the surroundings, thus resulting in poor thermal efficiency. This can be avoided by means of induction heating which is shown in Fig. 16.24 in a block diagram. The 50 Hz ac supply is converted to a high frequency ac of frequency 25 to 40 kHz which is supplied to an induction coil. This induces circulating currents in the metal pan on the top of induction coil thus directly heating the pan.

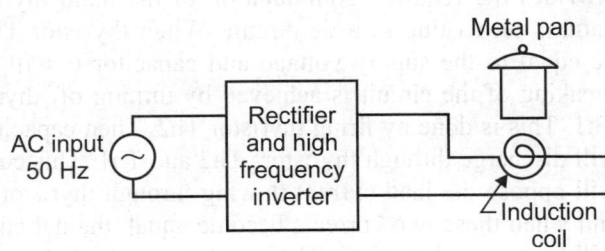

Fig. 16.24 Block diagram of induction heating

16.21 'ONE SHOT' THYRISTOR TRIGGER CIRCUIT

A circuit to trigger a thyristor for one complete half cycle of the ac supply is shown in Fig. 16.25. Triggering is initiated by closing push button switch SW, and the thyristor triggers always *near the beginning* of a positive half cycle, even though the switch may be closed randomly at any time during the two preceding half cycles. The thyristor will not trigger again until switch SW is opened and then reclosed. This type of logic is required for some test equipment supplies and for the solenoid drives of electrically operating stapling, guns, impulse hammers, etc., where load current must flow for one complete half cycle only.

During half cycles of ac line when the thyristor anode is positive, capacitor C_1 will be charged through the load, diode D_1 and resistance R_1. As long as switch SW is maintained in position 1, the thyristor will be non-conducting. When SW is switched to position 2 and the line voltage is negative, the charging current of capacitor C_2 (through the series-network of resistance R_3, the gate to cathode junction of the thyristor, SW, C_2, R_2 and D_3, will trigger the thyristor even though the line voltage is negative because the anode of the thyristor sees the positive voltage on capacitor C_1. Thyristor Th1 is thus only turned on when the line voltage is negative and maintained on by supplying holding current from capacitor C_1, through resistor R_1. The series network of

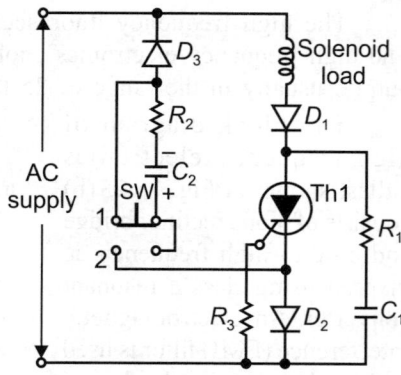

Fig. 16.25 Thyristor trigger circuit

resistance R_1 and capacitor C_1 is selected to supply a sufficient amount of holding current for one-half cycle only since providing holding current for a longer duration than one-half cycle would keep the thyristor on for succeeding half cycles. The thyristor holding current therefore determines $R_1 C_1$. Once capacitor C_2 has been charged, there will be no gate current through the thyristor even if SW is maintained in position 2. To trigger the thyristor again, SW would have to be flipped to position 1 in order to discharge capacitor C_2.

16.22 DC CIRCUIT BREAKER

Figure 16.26 shows a simple type of static dc circuit breaker in which the thyristors are used for making and breaking the dc circuits. Here thyristor Th1 is called the main thyristor and Th2 is the auxiliary thyristor. Thyristor Th2 has the function to turn off Th1 when required. In case of dc there is no natural zero to turn off. Therefore capacitor is used. Here, capacitor C provides the required commutation of the main thyristor, since the current does not have a natural zero value in a dc circuit. When thyristor Th1 is conducting, the load voltage will be equal to the supply voltage and capacitor C will get charged through resistance R_1. The breaking of the circuit is achieved by turning off thyristor Th1. This is done by firing thyristor Th2. Then capacitor C will discharge through thyristors Th2 and Th1. This current will oppose the load current flowing through thyristor Th1 and when these two currents become equal, the net current will be zero and thyristor Th1 will be turned off thereafter, capacitor will get charged through the load and during this time a reverse potential across thyristor Th1 will be applied. When capacitor C is fully charged, thyristor Th2 will be turned off because the current through the load is zero and the current through resistance R_1 is below the value of holding current of thyristor Th2.

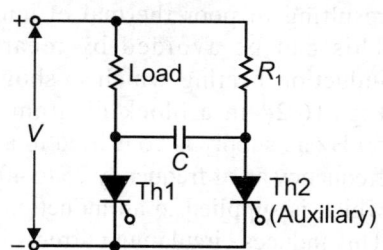

Fig. 16.26 DC circuit breaker

16.23 OVERVOLTAGE PROTECTION

Overvoltages caused by the bad regulation of supply voltage or by any switching action, will damage the equipments in the system and thus need protection. Thyristors can be used for protecting the equipments from overvoltages because of their switching action. The thyristor used for protection is connected in parallel with the load. Whenever the voltage exceeds a specified limit, the gate of the thyristor will get energised and trigger the thyristor. A large current will be drawn from the supply mains, which will reduce the overvoltage at the load. Since the applied voltage is ac, two thyristors are used; one for the positive half-cycle and the other for the negative half-cycle as shown in Fig. 16.27. Resistance R_1 limits the short circuit current when the thyristors are fired. This large current produces enough voltage drop in the source impedance, so that the terminal voltage is within safe limits. A Zener diode Z in series with resistances R_1 and R_2 constitutes a voltage sensing circuit. When the line voltage is above the specified limit, Zener diode Z will break. Then during the positive half-cycle, the gate of thyristor Th1 will get energised through R_1, D_1, R_1, D_2, and trigger it. In the negative half-cycle, the gate of thyristor Th2 will get energised through D_3, R_2, D_4, R_1, and thyristor Th2 will turn on if the overvoltage persists. As the load voltage returns to a safe value, the Zener diode Z will recover and the current through it will be very low so that the thyristors will not fire. Thus, the protection will be provided for the complete ac cycle.

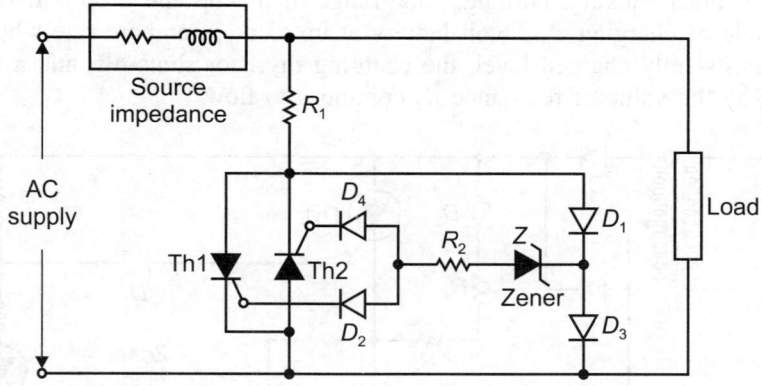

Fig. 16.27 Thyristorised circuit for overvoltage protection

16.24 SIMPLE BATTERY CHARGER

Figure 16.28(a) shows the diagram of battery charger using thyristor. Working of the circuit is as follows: A 230 V ac supply is step down to 15 V by centre tapped transformer. This ac voltage is rectified using a full wave diode rectifier. The Zener diode maintains point A at 15 V. When the battery is down, say about 12 V, line X will be 12 V. At point R, the voltage is a full wave rectified voltage as shown in Fig. 16.28(b). As the voltage at point B increases, voltage at point A also increases. At some point beyond P [see Fig. 16.28(b)], the thyristor Th

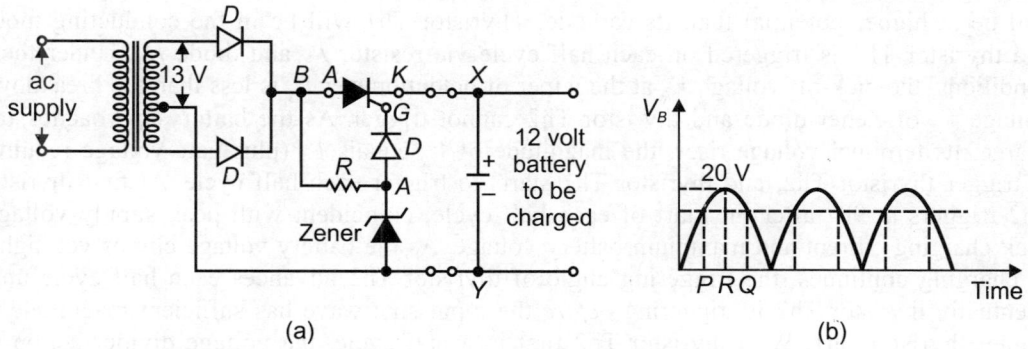

Fig. 16.28 Circuit of battery charger

is fired and the battery is connected to the rectifier. Once the voltage at point B falls below 12 V, the thyristor Th is turned off. Thus, the thyristor Th conducts and so charges the battery between P and Q. As the battery is charged, points P and Q move towards R and the charging period goes on reducing. When the battery is fully charged, say to about 14 V, the cathode of the thyristor Th will be at 14 V and the gate 14.3 V. V_{GK} of 0.3 would not be able to trigger the thyristor Th. Thus, the battery is cut-off from the charge.

16.25 BATTERY CHARGING REGULATOR

The battery charging circuit shown in Fig. 16.28 can be modified by making a provision for trickle charging arrangement. This can be incorporated by connecting another thyristor in the circuit, shown in Fig. 16.29. This circuit is an inexpensive means of utilizing the thyristor as a battery charging regulator, thus eliminating the problems inherent in electromechanical voltage relays-contact sticking, burning, wide range of pickup and drop out, wear, etc. This circuit is capable of charging a 12 volt battery at up to a six ampere rate when the battery voltage reaches its fully charged level, the charging thyristor shuts-off, and a trickle charge as determined by the value of resistance R_4 continues to flow.

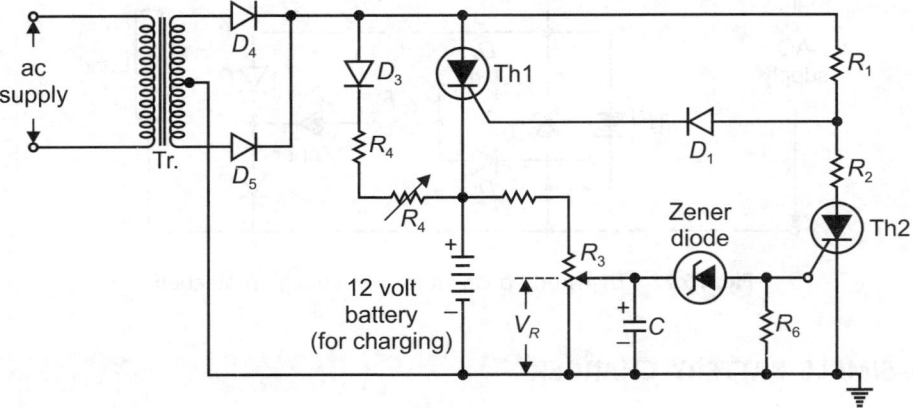

Fig. 16.29 Circuit of battery charging regulator

A single phase ac supply is stepped down with centre tapped transformer. This voltage is rectified to dc using D_4 and D_5 diodes. This rectified dc voltage is available across thyristor Th1 connected in series with 12 V battery which is to be charged. When the battery is in the discharged condition, its voltage will be low, and therefore, the anode of the thyristor Th1 will be at higher potential than its cathode. Thyristor Th1 will be in the conducting mode and thyristor Th1 is triggered on each half cycle via resistor R_1 and diode D_1. Under these conditions, the pick-off voltage V_R at the wiper of potentiometer R_3 is less than the breakdown voltage V_Z of Zener diode and thyristor Th2 cannot trigger. As the battery approaches full charge, its terminal voltage rises, the magnitude of V_R equals V_Z (plus gate voltage required to trigger thyristor Th2, and thyristor Th1 starts to trigger each half cycle. At first thyristor Th2 triggers at 90° after the start of each half-cycle, coincident with peak supply voltage, peak charging current and maximum battery voltage. As the battery voltage climbs yet higher as charging continues, the triggering angle of thyristor Th2 advances each half cycle until eventually thyristor Th2 is triggering *before* the input sine wave has sufficient magnitude to trigger thyristor Th1. With thyristor Th2 first in a half-cycle, the voltage divider action of resistors R_1 and R_2 keeps diode D_1 back-biased, and thyristor Th1 is unable to trigger. Heavy charging then ceases. Diode D_3 and resistor R_4 may be added, if desired, to trickle charge the battery during the normal 'off' periods. Heavy charging will recommence automatically when V_R drops below V_Z and thyristor Th2 stops triggering each cycle.

16.26 THYRISTOR CURRENT LIMITING CIRCUIT BREAKER

Circuit 16.30 illustrates that there are two thyristors Th1 and Th2, Thyristor Th1 connects the load across the dc supply and thyristor Th2 is used for disconnecting the load from the supply.

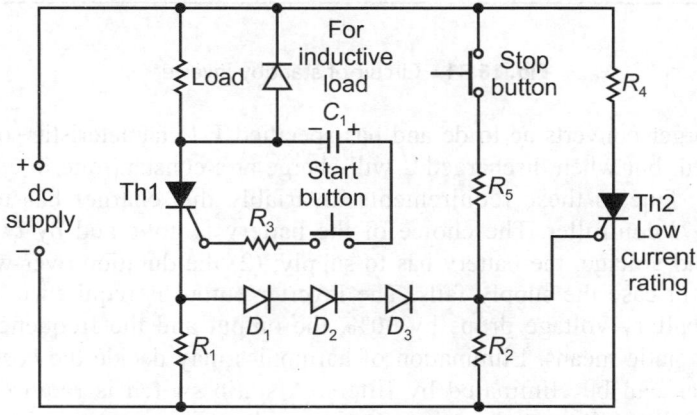

Fig. 16.30 Current limiting circuit breaker

Thyristorised current limiting circuit breaker of this type will provide protective functions very nicely. This circuit can be put in series with the dc output of a phase controlled rectifier or in series with the dc input to an inverter circuit. The circuit breaker is a parallel capacitor commutated flip-flop. When the 'start' button is momentarily pressed, thyristor Th1 starts to conduct and delivers power to the load, provided the load current is above the minimum holding current of thyristor Th1. Capacitor C_1 then charges to the load voltage through resistor R_4 as shown. When thyristor Th2 is trigged by closing of stop button, the positive terminal of capacitor C_1 is connected to the cathode of thyristor Th1, reversing the polarity across this thyristor and turning it off. This interrupts the flow of load current and opens the circuit. Thyristor Th2 will also be triggered by the voltage developed across resistor R_1 by load current. By adjusting the value of resistor R_1 and by proper number of diodes, the circuit can be made to trip out and interrupt overload of fault current at any predetermined level and saves the equipment from getting damaged. Thus, this device acts like a circuit breaker.

16.27 STANDBY INVERTERS/UNINTERRUPTIBLE POWER SUPPLIES (U.P.S.)

They are used to provide an emergency supply in the event of mains failure at main frequency. Basically it consists of:

1. battery charger, to maintain batteries charged and to supply current to the inverter
2. a battery, capable of supplying full load current for given time, say 10 minutes
3. an inverter which changes dc to ac of constant frequency and amplitude
4. a static contactor which connects the load from the inverter to the alternative ac supply.

Figure 16.31 includes an alternative direct link to the load from the ac supply. A change-over from the mains to inverter supply will demand that the inverter be synchronized with the mains to avoid waveform distortion.

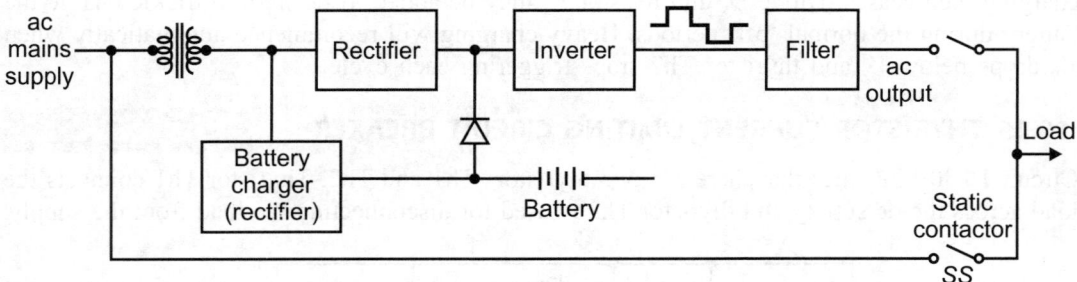

Fig. 16.31 Circuit of standby inverter

Battery charger converts ac to dc and has specified V/I characteristic so that the battery is not overcharged, but when discharged it will charge at a constant rate in addition to supply the load current. Due to these requirements invariably this charger has to be a thyristor converter with V/I controller. The choice of the battery is governed by two main factors: (1) the current and voltage, the battery has to supply; (2) the duration over which the battery has to take over in case the supply fails. The inverter output is required to be a sine wave. Even when the battery voltage drops by 30%, the output and the frequency is required to be within 1% by static means. Elimination of harmonics may decide the cost of the inverter. Higher harmonics can be eliminated by filters. A static switch is required to disconnect a faulty part and the switch to the other portion without interruption. Since the switch is required to operate within 10 m-sec, electromechanical contactors are unsuitable for this type of duty. A static switch is single pole switch and for change over a dual static contactor is required.

Three-phase four-wire supplies for standby inverters are constructed using three separate single-phase inverters feeding into the primaries of three-phase transformers, with one end of each secondary joined to form the neutral of the three-phase supply. In this connection of three-phase inverter, it is possible to supply unbalanced loads, while maintaining balance between the voltages.

Applications: Standby inverter (UPS) power supplies are needed for computers, data processors, data transmitters, microwave relay stations, nuclear reactor control, combustion and flame safeguard systems, communication links, etc. Due to the requirements of the industries, applications of UPS are increasing day by day to meet the demand.

16.28 STATIC EXCITATION SYSTEMS FOR ALTERNATORS

A long back, generally the excitation schemes utilise dc generators as exciters for the field of main alternators. These exciters are coupled to the same shaft of the main alternator. The field of the exciter is controlled by magnetic amplifier to form a feedback system. Although such systems have been working for long time, they suffer from slow speed of response and occupy a large amount of space by the side of the alternator. Besides, they need maintenance from time to time. Overall, such systems have poor reliability. Two static excitation schemes are available. They are: (1) brushless excitation system, and (2) thyristorised excitation system.

16.28.1 Brushless Excitation System

Large synchronous machines have their field windings on the rotor because less power for excitation is required through slip rings. If the brushless excitation is required, slip rings have to be eliminated. A typical system is shown in Fig. 16.32. An ac exciter is used in place of the conventional dc generator or dc supply, so that the leads can be taken through the hollow shaft. The output of the ac exciter is rectified by a rotating diode bridge and fed to the field winding of the main transformer. The excitation winding of the auxiliary alternator is on the stator and is fed from the thyristorised regulator. The feedback for the thyristor regulator is obtained from a set of *CTs* and *PTs*. If, any other stabilizing feedback is needed, it is possible to accommodate via the regulator. The scheme is shown in Fig. 16.32. The thyristor regulator consists of: (1) supply transformer, (2) thyristor bridge, (3) trigger circuit, (4) amplifier with reference potentiometer, and (5) field breaker with discharge resistor.

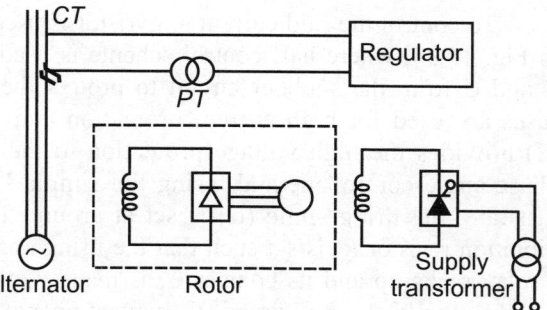

Fig. 16.32 Circuit of brushless exciter

16.28.2 Thyristorised Excitation System

If a change in load is rapid, the excitation control must give a fast response. This can be provided by a thyristor amplifier, which is shown in block diagram as phase controlled ac have commutated rectifier bridge in Fig. 16.33. The thyristor triggering signals are derived from load current via regulator. The excitation is kept constant for good power factor conditions until a certain load is reached and then rises in proportion to the load. Brushless excitation system is used for generators below 50 MVA capacity. Above this rating, thyristorised excitation system is used. This system is shown in Fig. 16.33. This system consists of: (1) supply transformer connected to the output of the alternator, (2) thyristor bridge, (3) trigger circuit, (4) amplifier with reference to potentiometer and feedback CT and PT set, (5) field breaker and discharge resistor. The available ceiling voltage is decided by the turn ratio of the supply transformer. The trigger angle of the thyristor bridge is controlled by the regulator with limit value control.

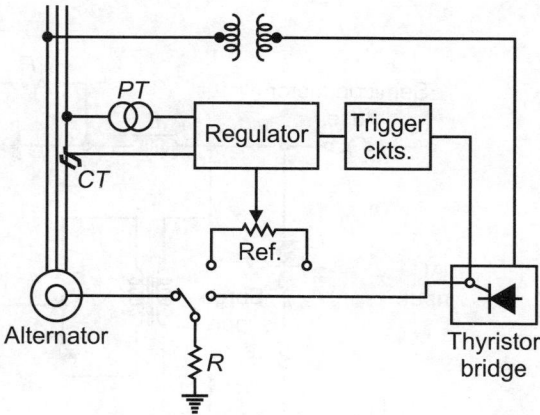

Fig. 16.33 Circuit of thyristorised excitation system

16.29 VIBROFEEDER CONTROL

With the increased importance attached to the industrial control, the need for accurate weighing and batching of various raw materials have also assumed more importance in the past few years. This has become possible mainly due to the improvement in the fields of electronic weighing, batching control with the introduction of microcomputers, etc. In a typical system it is often required to weigh in the desired amount of material, into a weigh hopper or a weighed skin by discharging the same from the fins through the vibrofeeders.

The vibration rate of the vibrofeeder is controlled by controlling the current in its coil. Usually, when the system starts filling the hopper or skin, the vibrofeeder is run at its fastest possible rate and when it reaches close to the desired weight the vibrofeeder is forced to run at a slower rate and stop when the desired weight is reached to reduce the error introduced due to the addition of the material in transit, once the vibrofeeder is stopped.

To control the said current a thyristor is used in series with the vibrofeeder coil, as shown in Fig. 16.34, where half control scheme is used. The inductor L is used for di/dt protection, R and C form the snubber circuit to protect the thyristor against large dv/dt, semiconductor fuses are used for high current protection and the high voltage back to back Zener diodes (Z) provides the high voltage protection to the thyristor. The power control is provided by phase-angle-control method, using the simple UJT circuit. For operating the vibrofeeder at fast rate, the firing angle (α) is set at around 10 degrees through resistor R_1, and for slow vibration resistor R_2 is set such that the firing angle is around 170 degrees. On start command, a relay picks up and its contact C_1 is made. On reaching the pre-set limit (below the desired weight), the 'fast-relay' drops (C_1 contact opens) and the 'slow-relay' picks up making contact C_2. When the desired weight is reached both the relays drop out and the contacts C_1 and C_2 are opened to stop power to the vibrofeeder. The vibrofeeder is provided only in the +ve half cycles and is turned off by natural commutation. A current sense circuit can be provided, to sense the actual current and compare the same with the maximum allowable current limit. Due to accidental short circuit of the coil or due to change in the gap between the coil and the vibrating plate, if the current increases beyond the set limit then the relay of the trip circuit picks up and opens out the contacts C. Thereby, the system can be operated always below the safe limit, without endangering the life of the coil and/or of the thyristor.

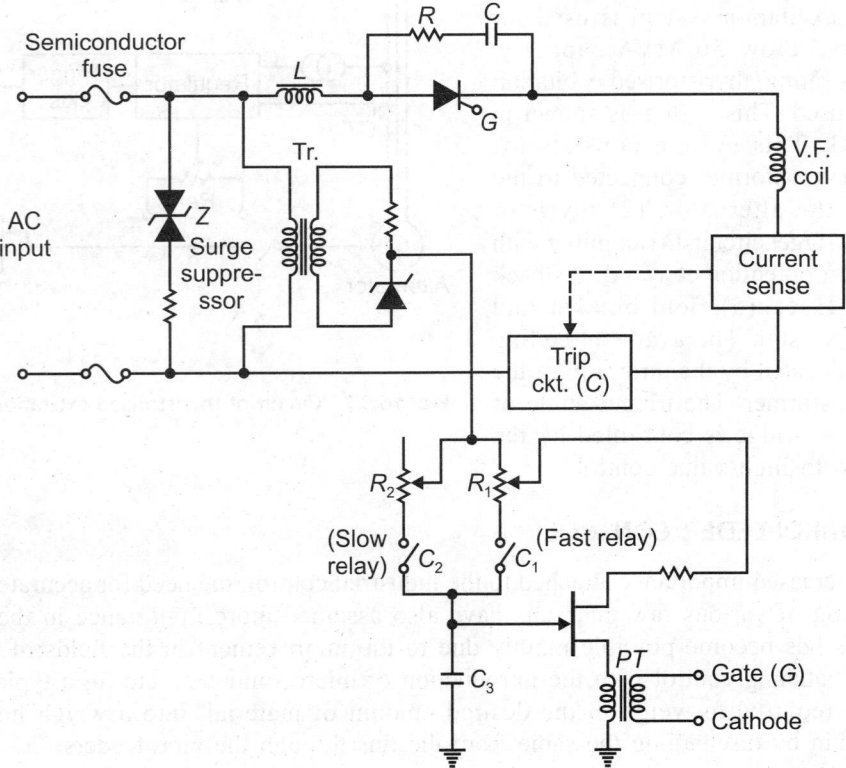

Fig. 16.34 Vibrofeeder control circuit

16.30 HIGH VOLTAGE DC TRANSMISSION

Power transmission through high voltage dc transmission line has many advantages over the ac transmission, because of the following advantages:

16.30.1 Advantages

1. Highly reliable.
2. More stable system as only thermal stability has to be considered and no ac stability problem due to oscillation.
3. Economical for the following reasons:

   (i) no ac skin effect, so for same rating conductor diameter is less.
   (ii) weight of conductor is less, so less number of transmission towers are needed.
   (iii) easy to manufacture cable as cable capacitance does not play any crucial role
   (iv) two conductors are needed instead of three conductors.

4. Flexibility. Independent control of power in two poles (+ve and −ve) with respect to earth is possible. There is no reactive power loss.
5. Intermediate switching stations are not needed.
6. Distributed capacitance and inductance play less important role.
7. Lightning can strike only one line, so fault can be cleared easily and transmission through other line can be continued.
8. Power of different frequencies after dc conversion can be transmitted.
9. Interference with the communication systems is less.
10. Energy flow can be controlled.

16.30.2 Disadvantages of HVDC transmission

1. Uneconomical for less than 800 km distance and for low power transmission.
2. Converters generate harmonics so costly filters are needed.
3. Power can not be tapped in the middle so again the inverter has to be used.

An application which requires strings of high voltage thyristors in the converters is that of high voltage direct current (HVDC) transmission. Transmission and distribution of electrical power is done using alternating current mainly due to economic reasons. However, due to recent advances in the development of high voltage thyristor converters and due to technical limitations of ac power transmission over long distances, high voltage dc link have been constructed using high voltage thyristor converters. Direct voltage transmission lines are more economical than alternating voltage transmission lines, but the device with which alternating voltages can be changed in level using transformers coupled with generator and motor device, makes the three-phase ac system the best overall both economically and technically.

In Fig. 16.35(a), the principle is to rectify the ac power to a voltage level of say ±200 kV, transmit power over a two cable line to a converter operating in the inverting mode, hence feeding power into the other ac system. Typically, each converter would be 12-pulse as shown in Fig. 16.35(b), the centre being earthed with one line at +200 kV and the other at −200 kV to earth for a ±20 kV system. The two ac systems linked by the HVDC transmission line must each be synchronous, that is, contain generators so as to maintain ac frequency constant. The two ac systems can be at different frequencies. Filters are used as shown in Fig. 16.35(a), to reduce the harmonics generated by the converters. Power flow can be in either direction, as each converter can operate in either the rectifying or the inverting mode.

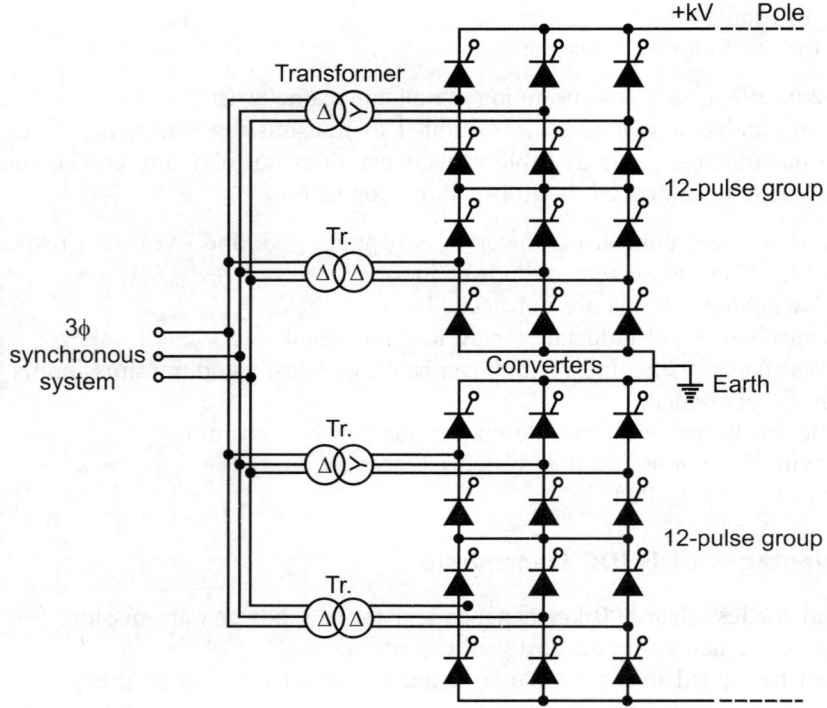

(a) Transmission and distribution

(b) Twelve pulse converter of electrical power

Fig. 16.35 HDVC transmission and distribution

16.31 THYRISTOR-SWITCHED CAPACITORS (TSC)

Figure 16.36 shows the basic arrangement where several capacitors can be connected to the supply voltage through a bidirectional switch consisting of back to back thyristors. Thyristor switched capacitor employs integral half cycle control where the capacitor is either fully in or out of the circuit.

The capacitor bank can be switched out by blocking the gate pulse to both the thyristors. The current flow stops at the instant of its zero crossing which also

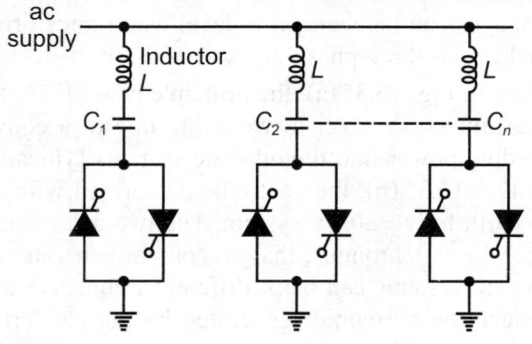

Fig. 16.36 Thyristorised switched capacitor circuit

corresponds to the capacitor voltage equal to the maximum ac supply voltage. The polarity of the capacitor voltage depends on the instant when the thyristor gate pulses are blocked. At switch on, thyristor must be gated at the proper instant of maximum ac voltage to avoid large over currents. Inductors are used here to limit over currents at switch on. By using a large number of thyristor switched small capacitor banks, it is possible to vary the reactive power in small steps.

16.32 THYRISTOR-CONTROLLED INDUCTORS (TCI)

Thyristor controlled inductors act as variable inductors where the inductive VARs supplied can be varied very quickly. This system may require either inductive or capacitive VARs depending on the system conditions. This requirement can be met by paralleling thyristor controlled inductors with a capacitor bank. The basic principle of thyristor controlled inductor can be understood by considering the per phase circuit of Fig. 16.37 where an inductor L is connected to the ac source through a bidirectional switch consisting of two back to back connected thyristors. If the resistive component of inductor is assumed to be negligibly small, the current through the inductor in steady state can be obtained as a function of the thyristor delay angle α. An equal value of α is used for both thyristors. The inductor current is not a pure sine-wave at $\alpha > 90°$. To prevent the third and multiples of third order harmonics, three-phase thyristor controlled inductors are connected in delta so that these harmonics circulate through inductors and do not enter ac system.

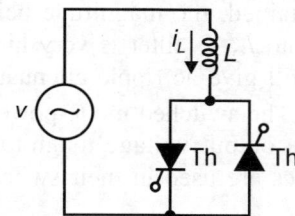

Fig. 16.37 Thyristorised controlled inductor circuit

16.33 WIND AND SMALL HYDRO-INTERCONNECTION

In the case of wind, power available varies with the cube of the wind velocity. For small hydro, the power available depends on pressure head and flow. For both wind and small hydro, to extract maximum amount of power it is desirable to let the turbine speed vary over a wide range to an optimum value dependent on the operating conditions. This would not be possible if a synchronous generator was directly connected to the supply system that dictated a constant speed. The induction generators connected to the supply system would allow the speed to vary only in a very narrow speed range. Therefore, allow the generator turbine speed to vary to optimise efficiency of power generation, the three-phase generator output is rectified into dc and then interfaced with the three-phase supply system by means of a switch mode converter. A block diagram is shown in Fig. 16.38, where a 50 Hz isolation transformer is included.

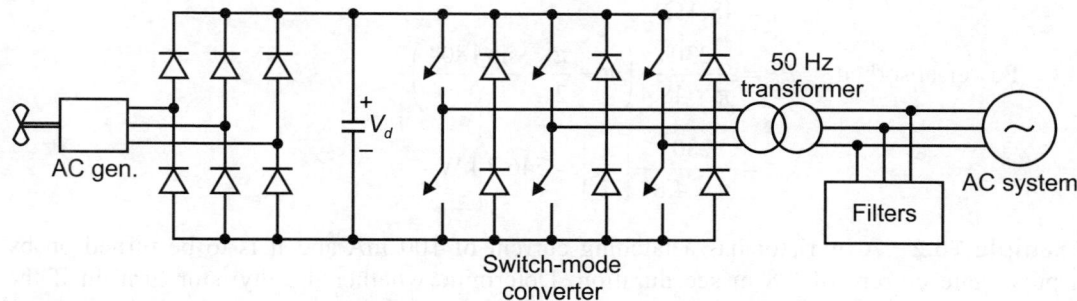

Fig. 16.38 A block diagram of wind and small hydro interconnection with isolation transformer

Because of medium power generation (a few tens of kWs) usually associated with the wind and small hydro generators, a three-phase supply interconnection is preferable.

16.34 SWITCHED MODE POWER SUPPLIES (SMPS)

In the low to medium power requirement (more than 50 W), a dc power supply is required which contains negligible ac ripple. For this application switched mode power supply circuits are used which is shown in Fig. 16.20.

The principle of the circuit shown is that a dc voltage is obtained from a rectifier being partially smoothed by the capacitor C_1. The switch is turned on and off rapidly to give a voltage which is chopped between voltage level and zero. This chopped voltage is fed into LC filter network which then smoothens the chopped waveform to give a level voltage to the load. By switching the rectifier output on and off very rapidly control of the output voltage is obtained, the magnitude being a function of the ON/OFF ratio. The ac ripple frequency fed into L_2 C_2 filter is very high, hence the values of L_2 and C_2 can be correspondingly low but still give ac ripple attenuation.

The switched mode power supply circuit is essentially a dc to dc converter with control of the output voltage magnitude. Like all power electronic equipment, the semi conductor devices are used in their switching mode in order to maximise efficiency.

SOLVED EXAMPLES

Example 16.1 A single phase 220 V, 1 kW, electric heater is connected to 230 V supply through TRIAC as shown in Fig. E.16.1. For a firing delay angle of 90°. Calculate the power absorbed in the heater element.

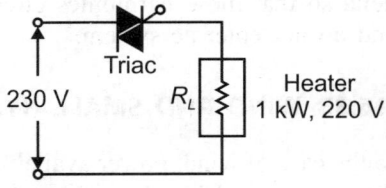

Fig. E16.1

Solution Triac is similar to antiparallel connected thyristors. Thus, the equation of power,

$$P_0 = \frac{V^2}{\pi R_L}\left(\pi - \alpha + \frac{\sin 2\alpha}{2}\right)$$

Given $\qquad V = 230$ V

$\qquad\qquad \alpha = 90°$

Power $\qquad P = \frac{V^2}{R_L}$ or $R_L = \frac{V^2}{P} = \frac{220^2}{1000}$

$\qquad\qquad = 48.4\,\Omega$

Power absorbed, $\quad P_0 = \frac{230^2}{\pi \times 48.4}\left(\pi - \frac{\pi}{2} + \frac{\sin 180°}{2}\right)$

$$= \frac{230^2}{\pi \times 4.81}\left[\frac{\pi}{2}\right] = 546.7 \text{ kW}.$$

Example 16.2 A thyristor has a latching current of 100 mA and it is to be turned on by a pulse gate current of 0.8 m sec duration. Determine whether the thyristor turn on if the inductance of the load is 1.8 H and the circuit operates from the main supply peak voltage 340 V.

Solution Time constant of the load must be at least ten times the width of the gate current pulse. Hence, the thyristor current increases more or less linearly.

The rate of change of current

$$\frac{dI}{dt} = \frac{V}{L}$$

$$= \frac{340}{1.8} = 189 \text{ A/sec}$$

In time 0.8 m sec the current has risen to

$$0.8 \times 189 = 151.2 \text{ mA}$$

Therefore, the thyristor will turn on.

Example 16.3 In integral cycling system supplies power to a 50 Ω load from 240 V main supply. Calculate the required $\left(\dfrac{t_{ON}}{t_{OFF}}\right)$ ratio. If the load power is to be: (i) 500 W, (ii) 750 W.

Solution
$$P = \frac{V^2}{R} = \frac{240^2}{50} = 1152 \text{ W}$$

(i) Given
$$P = 500 = 1152 \left[\frac{t_{ON}}{t_{ON} + t_{OFF}} \right]$$

or
$$\frac{1152}{500} = \frac{(t_{ON} + t_{OFF})}{t_{ON}} = \left(1 + \frac{t_{OFF}}{t_{ON}} \right)$$

$$2.304 = 1 + (t_{OFF}/t_{ON})$$

or
$$\frac{t_{ON}}{t_{OFF}} = \frac{1}{1.304} = 0.77$$

(ii)
$$750 = 1152 \left(\frac{t_{ON}}{t_{ON} + t_{OFF}} \right)$$

$$\frac{1152}{750} = (1 + t_{OFF}/t_{ON})$$

$$1.535 = 1 + \frac{t_{OFF}}{t_{ON}} ; \quad \frac{t_{OFF}}{t_{ON}} = 0.535.$$

Example 16.4 A HVDC system is rated 500 MW, ±250 kV and the converters are arranged. Find the rms current and peak reverse voltage of each thyristor.

Solution The direct current
$$= \frac{500 \times 10^6}{2 \times 250 \times 10^3} = 1000 \text{ A}$$

Each thyristor conducts 1000 A for one-third cycle, rms value of current,

$$I_{rms} = \frac{1000}{\sqrt{3}} = 577 \text{ A}$$

Each transformer feeds a six pulse group of mean voltage

$$= \frac{250}{2} = 125 \text{ V}$$

From the equation $\qquad V_{av} = \frac{3}{\pi} V_{L\,(max)}$

or $\qquad V_{L\,(max)} = \frac{V_{av} \times \pi}{3} = \frac{125 \times \pi}{3} = 131 \text{ kV.}$

Example 16.5 A resistance heating load is controlled from a single phase supply using a triac in the phase angle control mode. Determine the firing delay when the power is at: (i) 80%, (ii) 50%, and (iii) 30% of its maximum value.

Solution (i) Using the equation,

$$P = \frac{V_m^2}{2\pi R} \left(\pi - \alpha + \frac{1}{2} \sin 2\alpha \right)$$

When power is 100%, $\qquad \alpha = 0$

When power delivered 80%

$$P = 8 = \left(\pi - \alpha + \frac{1}{2} \sin 2\alpha \right)$$

which gives, $\qquad \alpha = 1.057 \text{ radian} = 60.5°$

  (ii) At half power $\qquad \alpha = 90°$

(iii) $\qquad \left(\pi - \alpha + \frac{1}{2} \sin 2\alpha \right) = 0.30$

$$\alpha = 1.896 \text{ radian} = 108.6°.$$

Example 16.6 A three-phase resistance load is controlled by three triacs from a 415 V supply. If the load is 15 kW, determine the required rating of the triacs. If thyristors were used instead of triacs, determine their rating.

Solution Rms line current $\qquad = \dfrac{15 \times 10^3}{\sqrt{3} \times 415} = 20.9 \text{ A}$

Hence, required triac rms rating

$$= 20.9 \text{ A}$$

In the off-state differences in the leakage current could result in the line voltage appearing across the triac. Hence, the required peak off-state voltage

$$= \sqrt{2} \times 415 = 586 \text{ V}$$

If the thyristors, were used, they would conduct for one half cycle of the semi-wave. So, the required rms current rating

$$= \frac{20.9}{\sqrt{2}} = 14.8 \text{ A}$$

Voltage rating will be the same as that of triac.

EXERCISES

Multiple Choice Questions

1. Uninterruptible supply is used in:
(A) computers
(B) communication links
(C) essential instrumentation
(D) all of the above

2. Integral cycle control does not suffer from the drawbacks given below:
(A) the load current takes its fundamental component at a lagging p.f. with harmonic currents.
(B) the load current takes its fundamental component at leading p.f.
(C) the load current takes its fundamental at u.p.f.
(D) none of these.

3. Electricity Supply Authorities normally insist that heating load be controlled by:
(A) a single thyristor with diode bridge
(B) phase angle control
(C) integral cycle control
(D) all of the above

4. The majority of heating loads have thermal-time constants of:
(A) zero second
(B) 0.2 second
(C) several seconds
(D) none of these

5. A heating load of resistance type can be controlled to different power levels by the use of:
(A) triac
(B) two thyristors in the inverse parallel connection
(C) connection with common cathodes
(D) all of the above

6. SCR can be used for protecting equipment from:
(A) under voltage
(B) overvoltages
(C) overloading
(D) none of these

7. In Thyristorised Excitation System, the ceiling voltage is decided by:
(A) triggering circuit
(B) turn ratio of supply transformer
(C) thyristor bridge
(D) all of the above

8. Static Excitation to alternators is provided by:
(A) brushless excitation system
(B) thyristorised excitation
(C) both (A) and (B)
(D) none of these

9. Thermal resistance is:
(A) the ratio of temperature difference to the heat power flow between the two interfaces
(B) the value of resistance at 0°C
(C) the value of resistance at room temperature
(D) none of these

10. AC load control in which regulation is by alternating continuous whole number of half cycles, on and off, is called:
(A) integral cycle
(B) inverse parallel connection
(C) phase angle control
(D) none of these

ANSWER KEY

1. (D)   **2.** (A)   **3.** (C)   **4.** (C)   **5.** (D)   **6.** (B)   **7.** (B)   **8.** (C)
9. (A)   **10.** (A)

Review Questions

1. What are the other industrial applications of thyristors?

2. Discuss how heating is achieved by thyristors?

3. What are different types of heating?

4. What is integral cycle control and how it is controlled?

5. Discuss the static excitation schemes available for alternators.

6. Discuss the use of thyristors in industry.

7. Discuss the relative merits of controlling a heating load by triacs by operating in: (i) phase angle delay, and (ii) integral cycle mode of control.

8. Discuss the use of inverters in industry for providing standby supplies. Describe the type of inverter which might be used for this purpose.

9. Discuss the application of power semiconductor devices to: (i) induction heating, (ii) electrochemical applications, and (iii) HVDC transmission.

10. Discuss the basic power supply current for a medium frequency induction heater.

11. Discuss the static excitation system for alternators.

12. How a thyristor can be used for the purpose of protection against overvoltage?

Regulated Power Supply

17.0 INTRODUCTION

An ordinary power supply consists of a rectifier and a filter circuit. With this arrangement, ac can be converted into dc. The dc output from an ordinary power supply remains constant so long as ac mains or load is unchanged. In many application, it is desired that dc output should remain constant irrespective of the changes in mains or the load under such condition an ordinary power supply cannot serve the purpose. Therefore, regulated power supply is used which keeps the dc output constant irrespective of the changes in ac mains or the load.

Limitations: An ordinary power supply has the following disadvantages:

1. The dc output voltage changes with a change in ac mains fluctuations.
2. The dc output voltage decreases considerably with load due to voltage drop in: (a) transformer, (b) rectifier and (c) filter. Due to these disadvantages power supply is being replaced by regulated power supply.

17.1 REGULATED POWER SUPPLY

Voltage regulators are a popular group of linear ICs. A voltage regulator receives input of a fairly constant dc voltage and supplies as output as somewhat lower value of dc voltage, which the regulator maintains fixed or regulated over a wide range of load current, or input voltage variation. Starting with an ac voltage supply, a steady dc voltage can be developed by rectifying the ac voltage, then filtering to a dc level, and finally regulating with voltage-regulator circuits, called a *regulated dc power supply*.

In many electronic applications, it is desired that dc voltage should remain constant irrespective of changes in ac supply or load. Under such conditions a regulated power supply is used which provides a fairly constant voltage. In this chapter various voltage regulating circuits are used to obtain regulated power supply.

A block diagram containing the parts of a typical power supply and the voltages at various points in the unit is shown in Fig. 17.1. The ac voltage, typically 220 V rms, is connected to a transformer which steps that voltage up or down to the level for the desired dc output. A diode rectifier then provides a half-wave or full-wave rectified voltage which is applied to a filter to smooth the varying signal. A simple capacitor filter is often sufficient to provide this smoothing action. The resulting dc voltage with some ripple or ac voltage variation is then provided as input to a regulator that provides as output a well-defined dc voltage level with extremely low ripple voltage over a range of load.

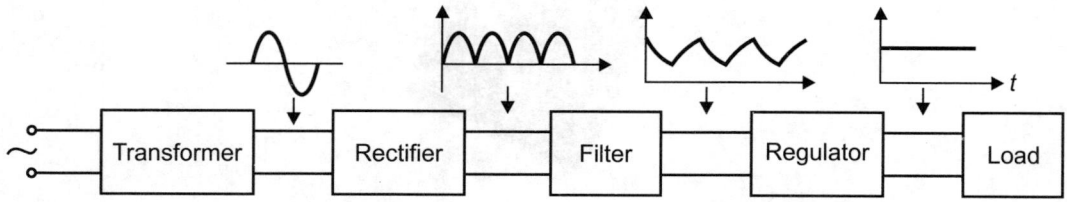

Fig. 17.1 Block diagram of power supply

17.2 DC AND AC POWER SUPPLIES

Generally electrical equipment require dc supply for operation. Close control and regulation are important factors to achieve the desired output. There are number of applications which require specially regulated dc supply such as electroplating, electric arc welding, electrostatic instruments in chemical industries.

The regulated dc power systems are of many types. The type of control scheme depends upon some important factors such as feasibility, efficiency, mechanical construction speed of response, fault protection and cost. Closer regulation requires a close loop control which may be classified into two groups namely: (1) AC regulation, (2) DC regulation.

In ac regulation scheme, the control element is incorporated on the ac side of the main rectifier and in the dc regulation scheme, the control element incorporated on the dc side of the main rectifier. In the ac regulation scheme, the common control elements are transductors, SCRs and regulating transformers.

The dc regulating elements generally are GTOs, high power bi-polar transistors and high power MOSFETs. The dc regulating system is superior to its ac regulating counterpart so far as transient behaviour is concerned.

In high power systems, the ac power is first converted to unregulated dc power by a simple diode circuit. The dc power is again converted to ac power by GTO or transistor invertor circuit at a much higher frequency compared to the power frequency.

The SCRs and GTOs can handle much larger power than power transistors and power MOSFETs but are inferiors to the latter so far as the gain-cut-off frequency is concerned. Moreover, the complexity of the control circuit of SCR or GTO system is much more severe compared to that of the transistor systems. Hence, transistorised dc regulators are used for low and medium power ranges. Transistors and FETs are generally connected in parallel to increase the power handling capacity.

17.3 CLASSIFICATION OF VOLTAGE REGULATOR

The voltage regulator can be classified into the following *two* categories:

    1. Linear regulator   2. Switch mode regulator

17.3.1 Parameters of Voltage Regulator

Some important terms used in regulator are discussed here:

    1. *Input Regulation*: It is defined as the change in output voltage expressed as a percentage of output voltage for a certain percentage change of input voltage.

2. *Output Regulation*: It is the percentage change in output voltage for a change in load current from one level to another.

3. *Output Resistance*: It is the resistance of the regulator looking from the output terminals under small signal condition.

4. *Output Voltage Change with Temperature*: It is the percentage change in output voltage for a change in temperature.

5. *Temperature Coefficient of Output Voltage* (αV_O): It is the ratio of the change in output voltage, expressed as a percentage of output voltage to the change in temperature.

$$\alpha V_O = \pm \left[\frac{V_0/T_2 - V_0/T_1}{V_0/25°C} \right] \times \frac{100}{T_2 - T_1} \%$$

6. *Output Voltage Change with Temperature:* It is the percentage change in the output voltage for change in temperature.

7. *Long Term Drift of Output Voltage*: It is the change in output voltage over a long period of time.

8. *Output Noise Voltage*: It is the rms noise voltage expressed as a percentage of output voltage with constant load and zero input ripple.

9. *Ripple Rejection*: It is the ratio of the peak-to-peak input ripple voltage to the peak-to-peak output ripple voltage.

10. *Current-Sense Voltage*: It is the voltage which is a function of load current and is used for the control of the current limiting circuit.

11. *Feedback Sense Voltage*: It is a voltage which is a function of the output and is used for feedback of the regulator.

12. *Reference Voltage*: It is a highly stable voltage that is compared with the feedback sense voltage to control the regulator.

13. *Bias or Quiescent Current*: It is the difference between the input and the output current.

14. *Standby Current*: It is the input current of the regulator with no output load and no reference voltage load.

15. *Dropout Voltage*: It is the low input-to-output differential voltage at which the circuit ceases to regulate against further reduction of input voltage.

16. *Short-Circuit Output Current*: It is the output current of the regulator when the output terminals are shorted.

17. *Peak Output Current*: It is the maximum current that can be drawn from the regulator current limiting circuits in the regulator.

17.3.2.1 *Linear Power Supplies*

Figure 17.2 (a) shows the schematic of a linear power supply. In order to provide electrical isolation between the input and the output and to deliver the output in the desired voltage range, a 50 Hz transformer is required. A transistor is connected in series that operates in its active region.

The control circuit in Fig. 17.2 (a) adjusts the transistor base current such that

$$V_0 = v_d - v_{CE} = V_{0\,ref}$$

The transistor in a linear supply acts as an adjustable resistor where the voltage difference $(v_d - V_0)$ between the input and desired output voltage appears across the transistor and causes power loss in it.

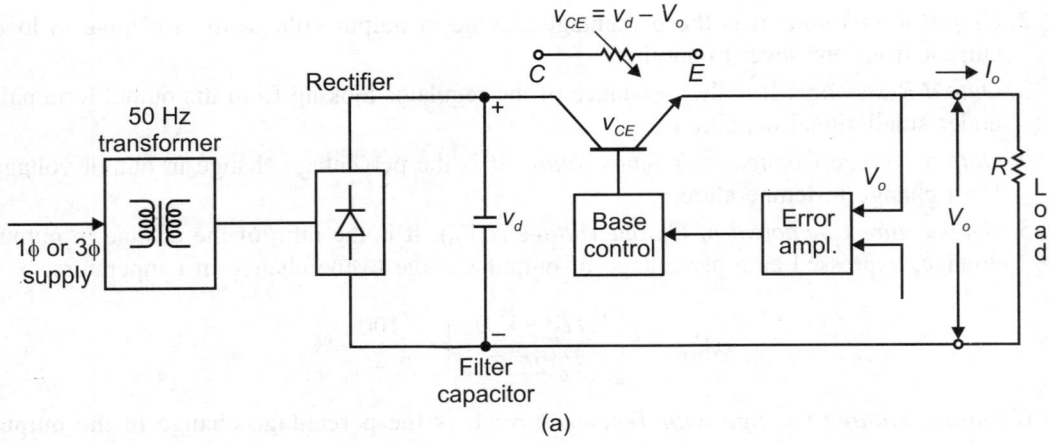

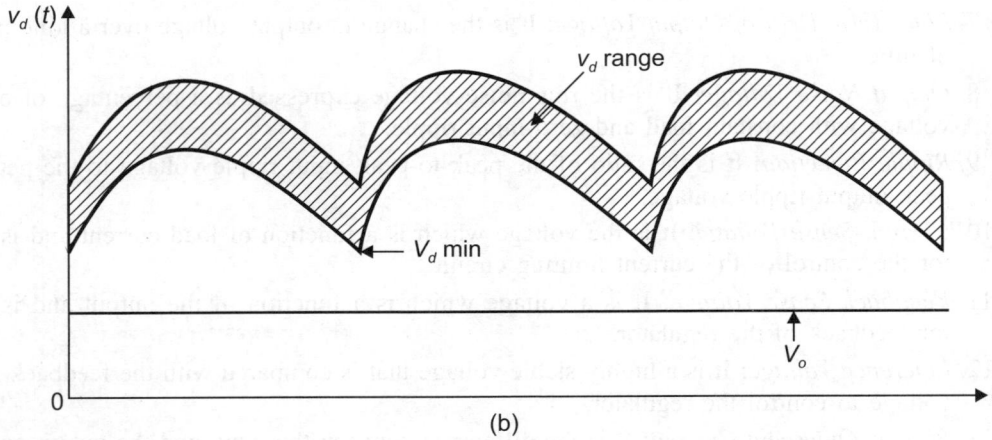

Fig. 17.2 (a) Schematic diagram of linear power supply (b) selection of transformer turns ratio

For a given range of 50 Hz ac input voltage, the rectified and filtered output $v_d(t)$ may be shown in Fig. 17.2 (b).

To minimise the transistor power losses, the transformer turns ratio should be carefully selected such that $V_{d\,min}$ in Fig. 17.2 (b) is greater than V_0 but does not exceed V_0 by a large margin.

Shortcomings of Linear Power Supply

1. A low frequency 50 Hz transformer is required. Such transformers are larger in size and less weight as compared to high frequency transformers.
2. The transistor operates in its active region incurring a significant amount of power loss, therefore the overall efficiencies of linear power supplies are generally in a range of 30 to 60%.

Advantages of Linear Power Supply

1. These power supplies utilize simple circuitry and therefore may cost less in small power ratings less than 25 watts.
2. These supplies do not produce large electromagnetic interference (EMI) with other equipment.

17.3.2 Linear Regulator

There are *two* types of linear regulator:

1. Shunt regulator
2. Series regulator

17.3.2.1 *Shunt Regulator with Zener Diode*

The simplest shunt regulator is the Zener Diode stabilizer circuit shown in Fig. 17.3. Since the power handling capacity of a Zener diode is small, it is used as a reference voltage source. Zener diode is a silicon device having a normal diode characteristic in the forward direction and a constant voltage property in the reverse direction with a reverse current above a certain value.

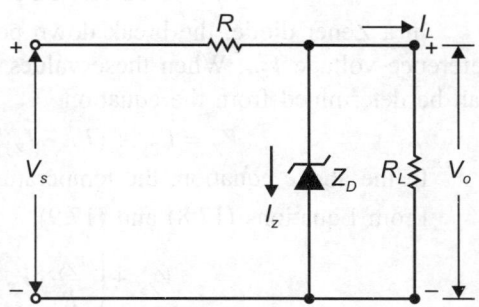

Fig. 17.3 Circuit of single stage Zener diode stabilizer

Basic equation relating to changes in the output of a stabilizing circuit is

$$dV_0 = \left(\frac{\partial V_0}{\partial V_s}\right)dV_S + \left(\frac{\partial V_0}{\partial I_0}\right)dI_0 + \left(\frac{\partial V_0}{\partial T}\right)\partial T... \qquad ...(17.1)$$

Here V_O is output voltage

V_s is unregulated input voltage

I_O is output current

and T is temperature

Stabilization Factor S;

$$S = \left(\frac{dV_o}{dV_S}\right) \text{ here } (I_O \text{ and } T \text{ are constant}) \qquad ...(17.2)$$

Ratio of percentage change in the output to the percentage change in the input voltage is termed as regulation change factor and denoted by S_F. Therefore,

$$S_F = \frac{V_s}{V_0}\left(\frac{\partial V_o}{\partial V_S}\right) \qquad (I_o \text{ and } T = \text{constant}) \qquad ...(17.3)$$

Output resistance R_o is defined as

$$R_o = -\left(\frac{\partial V_o}{\partial I_0}\right)\Omega \qquad (V_S \text{ and } T = \text{constant}) \ \Omega \qquad ...(17.4)$$

Temperature coefficient S_T is defined as:

$$S_T = \left(\frac{\partial V_o}{\partial T}\right)V/°C \qquad (V_S \text{ and } I_o \text{ are constant}) \qquad ...(17.5)$$

$$dV_o = SdV_S - R_o dI_o + S_\tau dT \qquad ...(17.6)$$

where $\qquad V_S = (I_Z + I_o) R + V_o \qquad ...(17.7)$

I_Z is the current through the Zener diode break down

R is series resistance including the source resistance

$$I_Z = \frac{V_S - V_o}{R} - I_o = \frac{V_S}{R} - V_O \left(\frac{1}{R} + \frac{1}{R_L} \right) \qquad \text{...(17.8)}$$

R_L = Load resistance

In a Zener diode, the break down point is not clearly defined, and instead, a specified reference voltage V_{ZS}. When these values are steady state values, V_Z at any other current I_Z can be determined from the equation

$$V_Z = V_{ZS} + (I_Z - I_{ZS}) r_{ZS} = V_o \qquad \text{...(17.9)}$$

In the above equation, the temperature effect is neglected.

From Equations (17.8) and (17.9)

$$V_0 = \frac{V_{ZS} + \left(\dfrac{r_{ZS}}{R} \cdot V_S \right) - I_{ZS} r_{ZS}}{1 + \dfrac{r_{ZS}}{R} + \dfrac{r_{ZS}}{R_L}} \qquad \text{...(17.10)}$$

Therefore, the stabilization factor is

$$S = \frac{r_{ZS}/R}{1 + \dfrac{r_{ZS}}{R} + \dfrac{r_{ZS}}{R_L}} = \frac{1}{1 + \left(\dfrac{R}{r_{ZS}} + \dfrac{R}{R_L} \right)}$$

The output resistance R_o is

$$R_o = \frac{R_{ZS} R}{r_{ZS} + R}$$

$$R > r_{ZS}, R_o \simeq r_{ZS}$$

Generally r_{ZS} is very small. For practical purpose $I_Z = I_{ZS}$, so

$$R = \frac{V_S - V_{ZS}}{I_Z + I_L}$$

For successful operation of the stabilizer, the following conditions are to be maintained.

$$I_{ZS\,(min)} + I_{L\,(max)} \leq \frac{V_{S\,(min)} - V_{Z\,(max)}}{R}$$

From which the maximum value of the series resistance R should be

$$R_{max} = \frac{V_{S\,(min)} - V_{Z\,(max)}}{I_{ZS\,(min)} + I_{L\,(max)}}$$

$V_{Z\,(max)}$ must be considered at $I_{Z\,(max)}$, which means that the equation for $I_{Z\,(max)}$ should be

$$I_{Z\,(max)} = I_{Z\,(min)} + I_{L\,(max)} - I_{L\,(min)}$$

Again the current flowing through the Zener diode is limited by the value of R. Therefore, the minimum value of R also becomes important because it dictates the power dissipation P_Z in the Zener diode which is

$$P_z = I_z V_z = \left[\frac{V_s - V_0}{R} - I_L \right] V_z$$

or
$$P_z = \left(\frac{V_s - V_z}{R} - I_L \right) V_z$$

because
$$V_0 = V_z$$

The maximum power dissipation on the Zener diode occurs when the load current is minimum, i.e.

$$P_{z\,(\text{max})} = \left[\frac{V_s - V_z}{R} - I_{L\,(\text{min})} \right] V_z$$

Considering the variation of input voltage

$$P_{z\,(\text{max})} = \left[\frac{V_{s\,(\text{max})} - V_{z\,(\text{max})}}{R} - I_{L\,(\text{min})} \right] V_{z\,(\text{max})} \qquad ...(17.11)$$

or
$$P_{z\,(\text{max})} = \left[\frac{V_{s\,(\text{max})} - V_{z\,(\text{min})}}{R} - I_{L\,(\text{min})} \right] V_{z\,(\text{min})} \qquad ...(17.12)$$

The rating of Zener diode should be choosen such that maximum dissipation capability of the device is greater than the value shown in above two Equations (17.11) and (17.12).

The minimum value of the series resistance R can be obtained from above two equations given by:

$$R_{\text{min}} \geq \frac{V_{s\,(\text{max})} - V_{z\,(\text{max})}}{\dfrac{P_{z\,(\text{max})}}{V_{z\,(\text{max})}} + I_{L\,(\text{min})}}$$

or
$$R_{\text{min}} \geq \frac{P_{s\,(\text{max})} - V_{z\,(\text{min})}}{\dfrac{P_{z\,(\text{min})}}{V_{z\,(\text{min})}} + I_{L\,(\text{min})}}$$

In this stabilizer, a reference voltage with $\pm 0.5\%$ stability can be easily achieved for input voltage variation of $\pm 10\%$ using a high stability series resistor R.

In order to estimate the range of input voltage which can be regulated, assume the minimum Zener current zero. If V_z and P_{max} are the Zener voltage and power rating of the diode, then $P_{\text{max}} = V_Z I_{Z\,\text{rated}}$.

The maximum Zener diode current is equal to

$$I_{Z\,(\text{max})} = 0.8 I_{Z\,\text{rated}}$$
and
$$I_{Z\,(\text{min})} = 0.2 I_{Z\,\text{rated}}$$
$$V_{i\,(\text{min})} = (I_L + I_{Z\,(\text{min})})\,R + V_Z \qquad ...(17.13)$$
and
$$V_{i\,(\text{max})} = (I_L + I_{Z\,(\text{max})})\,R + V_Z \qquad ...(17.14)$$

when there is no load, all the current will be taken by the Zener diode. From Equation (17.14)

$$\frac{V_{i\,(\text{max})} - V_Z}{R} < 0.8 I_{Z\,(\text{rated})}$$

or
$$R > \frac{V_{i\,(\text{max})} - V_Z}{0.8 I_{Z\,(\text{rated})}}$$

This gives the minimum series resistance

$$R_{\min} = \frac{V_{i\,(\max)} - V_Z}{0.8 I_{Z\,(\text{rated})}}$$

From Equation (17.13)

$$V_{i\,(\min)} - V_Z = (I_L + I_{Z\,(\min)})\,R$$

$$I_{Z\,(\min)} = \frac{V_{i\,(\min)} - V_Z}{R} - I_L$$

But

$$I_{Z\,(\min)} > 0.2 I_{Z\,(\text{rated})}$$

$$\frac{V_{i\,(\min)} - V_Z}{R} - I_L > 0.2 I_{Z\,(\text{rated})}$$

With an algebraic manipulation

$$R < \frac{V_{i\,(\min)} - V_Z}{0.2 I_{Z\,(\text{rated})} + I_L}$$

Which can be replaced by the equation

$$R_{\max} = \frac{V_{i\,(\min)} - V_Z}{0.2 I_{Z\,(\text{rated})} + I_L}$$

The diode maximum current rating must be such as to provide meaningful results. This requires that

$$R_{\max} > R_{\min}$$

$$\frac{V_{i\,(\min)} - V_Z}{0.2 I_{Z\,(\text{rated})} + I_L} > \frac{V_{i\,(\min)} - V_Z}{0.8 I_{Z\,(\text{rated})}}$$

$$\frac{V_{i\,(\min)} - V_Z}{V_{i\,(\max)} - V_Z} > \frac{0.2}{0.8} + \frac{I_L}{0.8 I_{Z\,(\text{rated})}}$$

$$\frac{0.8 V_{i\,(\min)} - V_Z}{V_{i\,(\max)} - V_Z} - 0.2 > \frac{I_L}{0.8 I_{Z\,(\text{rated})}} \qquad \qquad \text{...(17.15)}$$

In case of open-circuit, all the current will be taken by the diode. The current handling capability of the diode must satisfy the following inequality:

$$I_{Z\,(\max)} > I_L + I_{Z\,(\min)}$$

$$I_{Z\,(\max)} > 0.8 I_{Z\,(\text{rated})}$$

$$I_{Z\,(\min)} > 0.2 I_{Z\,(\text{rated})}$$

$$0.8 I_{Z\,(\text{rated})} > I_L + 0.2 I_{Z\,(\text{rated})}$$

or

$$0.6 I_{Z\,(\text{rated})} > I_L \qquad \qquad \text{...(17.16)}$$

Equations (17.15) and (17.16) set a lower limit on the current capability of Zener diode, the larger value supersede the lower one thus satisfying both inequalities simultaneously.

Advantage: Advantages of Zener diode over other voltage regulator is that it is small in size, light in weight, rugged and provides regulation over a wide current range.

Disadvantages

1. It is inefficient. As power is dissipated in the series resistance and the diode; thus the efficiency is poor.
2. The output voltage cannot be chosen at any value but depends on the breakdown voltage.

17.3.2.2 *Shunt Regulator Using Operational Amplifier*

A shunt regulator using an operational amplifier is shown in Fig. 17.4.

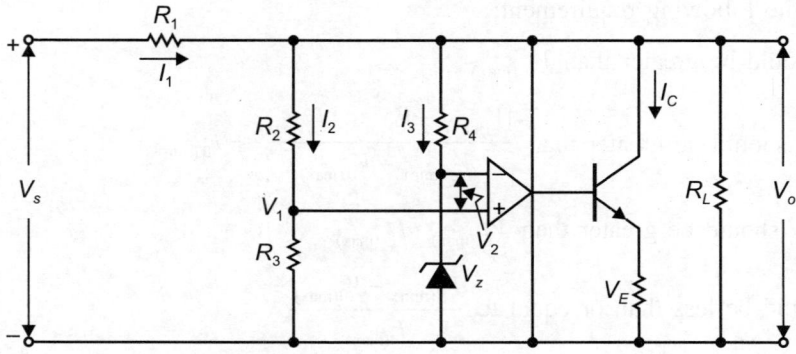

Fig. 17.4 Circuit of shunt regulator using-op-amp

In this circuit the inverting input of the op-amp is held at a fixed voltage with a Zener diode z_d the non-inverting input is fed from a potential divider $R_2 - R_3$. If the output voltage drops due to any reason V_1 decreases reducing the output voltage of the op-amp. This reduces the base bias of the transistor and its collector current decreases. The voltage drop across R_1 decreases proportionally, so as to increases output voltage to the initial value. The reverse sequence follows for any temporary rise in the voltage. The output voltage and the stabilization factor can be calculated from the Fig. 17.4.

$$V_1 = V_0 \left(\frac{R_3}{R_2 + R_3} \right)$$

The effective input of the op-amp V_2 is

$$V_2 = V_0 \left(\frac{R_3}{R_2 + R_3} \right) - V_{ZD}$$

Neglecting the base-emitter threshold voltage of the transistor, the transistor collector current I_c is obtained as

$$I_c = A \left(V_0 \frac{R_3}{R_2 + R_3} - V_{ZD} \right) h_{fe} \cdot \frac{1}{h_{ie}}$$

The stabilization factor is given by

$$S = \frac{1}{1 + R_1 \left(\dfrac{R_3}{R_2 + R_3} \right) A \cdot \dfrac{h_{fe}}{h_{ie}}}$$

$$\equiv \frac{1}{\dfrac{R_1 R_3}{R_2 + R_3} \cdot \dfrac{A h_{fe}}{h_{ie}}}$$

The regulator output is adjusted by the potiometric chain $R_2 - R_3$.

The shunt regulator is generally used in the case of low to medium voltages and medium current with relatively constant loads, the transistor must be capable of withstanding the full output voltage and full load current if the load varies from no load to full load. The transistor must fulfill the following requirement:

1. V_{CE} should be greater than $V_{0\,(max)}$

2. $I_{C\,(max)}$ should be greater than $\dfrac{[V_{(max)} - V_{0\,(min)}]\,I_{0\,(max)}}{V_{s\,(min)} - V_{0\,(max)}} - I_{0\,(min)}$

3. $P_{C\,(max)}$ should be greater than $V_{C\,(max)} \cdot I_{C\,(max)}$

4. R_1 should be less than or equal to $\dfrac{V_{S(min)} - V_{0(max)}}{I_{1(max)}}$

17.3.2.3 *Series Regulator*

The purpose of regulator is to convert a given dc or ac input voltage into a specific stabilized dc output voltage and maintain that voltage over a wide range of supply voltage and load conditions.

Series regulators are generally used in the case of variable voltages and variable load current. The building block of a series regulator is shown in Fig. 17.5.

1. A series control element to the change the input unregulated voltage to the desired output level over varying load conditions.
2. A reference source that provides a known stable voltage.
3. A sampling element to sample output voltage level.
4. A comparator that compares the output voltage sample to the reference voltage and creates an error signal.
5. An amplifier that amplifies the error signal to bring it to a level which can drive the series element.

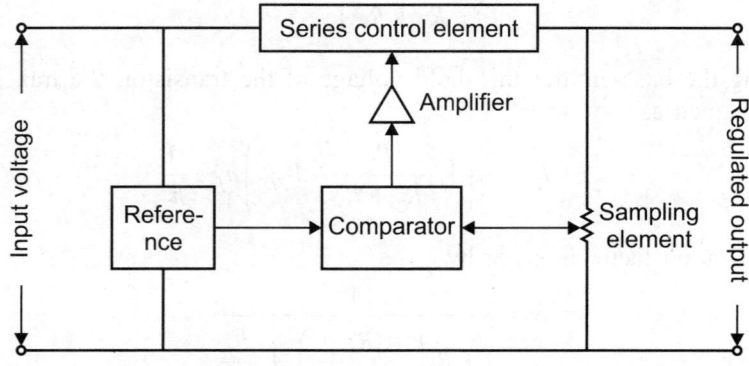

Fig. 17.5 Circuit of series regulator

The output voltage is regulated by modulating the series element, usually a transistor, which operates in the emitter follower mode. Any change in input voltage results in a change in the equivalent resistance of the series element. The product of this resistance and the load current creates a changing differential voltage that compensates for a changing input voltage. The reference element forms the foundation of all voltage regulators since the output voltage is either equal to or multiple of the reference. Any variation in the reference voltage will be interpreted as the output voltage error by the comparator and cause the output voltage to change accordingly. For good regulation, the reference voltage must be stable irrespective of all variations in the supply voltage and temperature.

The sampling element monitors the output voltage and changes it into a level equal to the reference voltage for a desired output voltage. Any variation in the output voltage causes the feedback voltage to change to some value greater or less than the reference voltage. This error voltage directs the series regulating element to respond appropriately to correct for the output voltage change to provide stabilized output.

17.3.2.4 *Simple Series Regulator with a Zener Diode*

The simplest series regulator is the emitter follower regulator shown in Fig. 17.6.

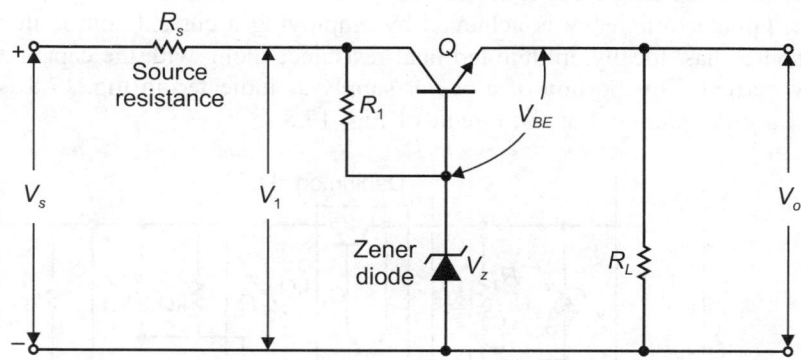

Fig. 17.6 Circuit of single emitter follower regulator

In this circuit the Zener diode maintains the base of the transistor at a constant potential with respect to the negative line. Naturally, if the input voltage changes, there is an equal change in the collector-to-base voltage to the transistor. Again, if there is any change in the output voltage, there is an equal change in the emitter-to-base voltage. These changes in potential tend to maintain this output voltage constant. A simple regulator has a number of limitations:

1. There is no freedom to vary the output.
2. The stability of Zener diode reference voltage is affected directly with input voltage variation and the stability of the output voltage is affected.
3. For the increased load, the current demand from Zener diode is large, and hence this diode should be of higher rating and the current gains of series transistor should be large.

The current gain of the series pass transistor is increased by compound connection transistor known as Darlington configuration as shown in Fig. 17.7.

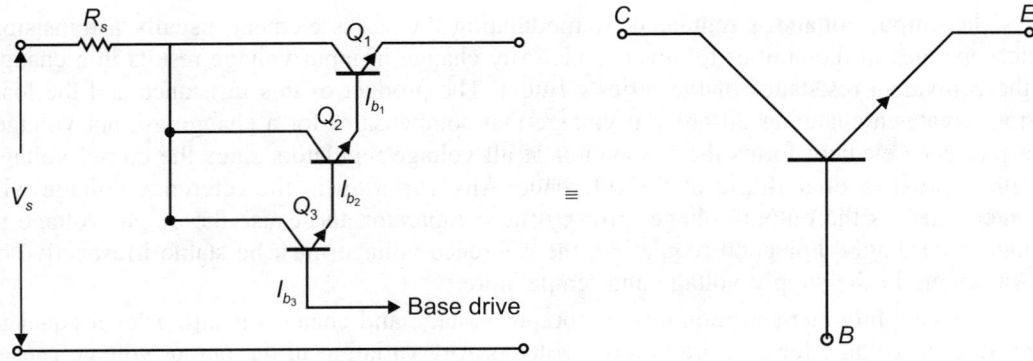

Fig. 17.7 Compound connection of transistors

17.4 COMPLETE VOLTAGE REGULATED POWER SUPPLY

A complete power supply employing a voltage regulator appears in Fig. 17.8. A *Darlington* circuit has replaced the single series transistor to increase the sensitivity of the regulator to change in V_L. Greater efficiency is achieved by employing a current source in place of R_3. The current source has, ideally, infinite terminal resistance along with the capability to supply the necessary current. This portion of a power supply as indicated in Fig. 17.8 is sometimes referred to as a *preregulator*. For the circuit of Fig. 17.8.

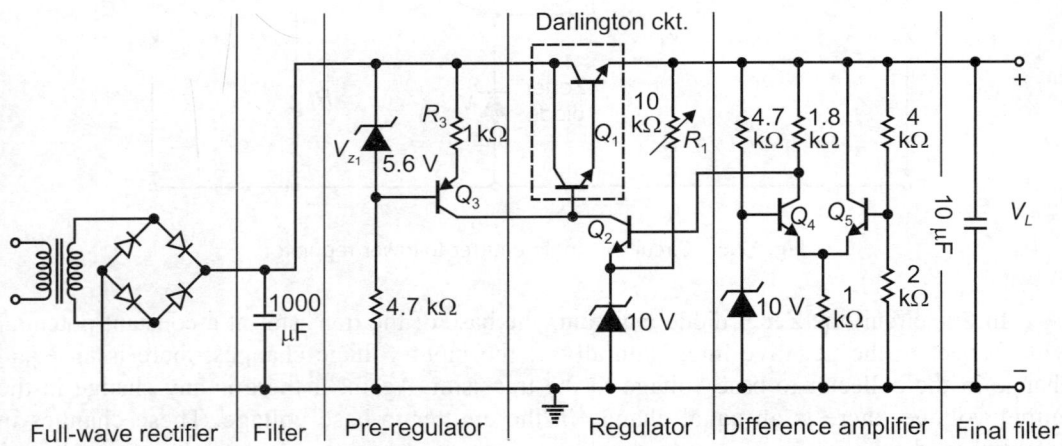

Fig. 17.8 Circuit voltage, regulated power supply

$$I_{\text{current source}} = I_{C_3} \cong \frac{V_{Z_1}}{R_3} \qquad \qquad ...(17.17)$$

To improve further the sensitivity of the regulator to changes in V_L a *difference* amplifier has been introduced, the output of which is fed to the control transistor. The unregulated input is a full-wave-rectified signal to be passed through a capacitive filter. The 10 μF capacitor at the output is to reduce the possibility of oscillations and further filter the supply voltage. The supply voltage V_L can be varied by changing R_1 while still maintaining regulation.

17.5 CURRENT REGULATOR

The analysis of current regulators will be limited to a brief discussion of the circuit of Fig. 17.9. A current regulator is designed to maintain a fixed current through a load for variations in terminal voltage. A decrease in $I_L = I_C$ due to a drop in V_L would result in a decrease in $I_E \cong I_C$ and, in turn, a drop in V_{RE}. The base-to-emitter potential is

$$V_{BE} = V_Z - V_{RE} \qquad ...(17.18)$$

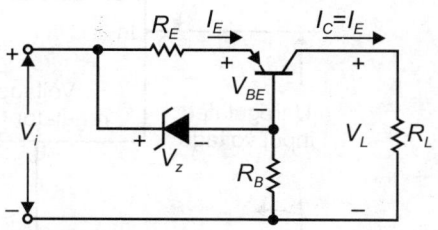

Fig. 17.9 Current regulator circuit

A decrease in V_{RE} will result in an increase in V_{BE} and the conductivity of the transistor, maintaining I_L at a fixed level.

17.6 IC VOLTAGE REGULATORS

This type of IC regulator provides very precise regulation of output voltage for both line and load variations. Voltage regulators comprise a class of widely used ICs. These units contain the circuitry for reference source, error amplifier, control device, and overload protection all in a single IC chip. Although the internal construction is somewhat different than that described for discrete voltage-regulator circuits, the external operation is quite same. Let us examine operation of (1) the popular 3-terminal fixed-voltage regulators (for both positive and negative voltages) (2) adjustable output voltage regulators and (3) precision voltage regulators.

A power supply can be built simply using a transformer connected to the ac supply to step the voltage to a desired level, then rectifying with a half- or full-wave circuit, filtering the voltage using a simple capacitor filter, and finally regulating the dc voltage using an IC voltage regulator.

A basic category of voltage regulators includes those used with only positive voltages, those used with only negative voltages, and those further classified as having fixed or adjustable output voltages. These regulators can be selected for operation with load currents from hundreds of milliamperes to tens of amperes corresponding to power ratings from milliwatts to tens of watts.

17.6.1 Three-Terminal Positive and Negative Voltage Regulators

Generally the three terminal regulators provide a fixed output voltage with no external components is schematically shown in Fig. 17.10. Most have interval current limiting and can provide fairly high load current upto 1.5 A, without an external pass transistor. Normally, this type of regulator is avoidable with a several fixed output voltage values. The fixed voltage regulator has an unregulated voltage, V_{in}, applied to one terminal, delivers a regulated output voltage, V_o, from a second terminal, with the third terminal connected to ground. For a particular IC unit, device specifications lists a voltage range over which the input voltage can vary to maintain the regulated output voltage, V_o, over a range of load current, I_o. An output-input voltage differential must be maintained for the IC to operate, which means that the varying input voltage must always be kept large enough to maintain a voltage drop across the IC to permit proper operation of the internal circuit. The device specifications also list the amount of output voltage change, V_o, resulting from changes in load current (load regulation) and also from changes in input voltage (line regulation).

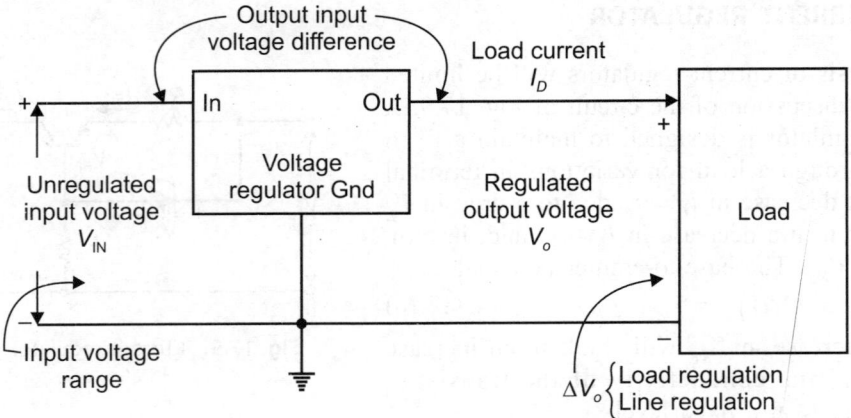

Fig. 17.10 Three terminal voltage regulator

The μA 7800 Regulators: These devices are examples of three terminal positive voltage regulators. A group of fixed-positive-voltage regulators is the series 78, which provide fixed voltages from 5 V to 24 V. Figure 17.11 (a) shows how many of these regulators are connected. A rectified and filtered unregulated dc voltage is the input, V_{in}, to pin 1 of the regulator IC. Capacitor connected from input or output to ground help to maintain the dc voltage and additionally to filter any high-frequency voltage variation. The output voltage from pin 2 is then available to connect to the load. Pin 3 is the IC circuit reference or ground. When selecting the desired fixed regulated output voltage, the two digits after the 78 prefix indicate the regulator output voltage.

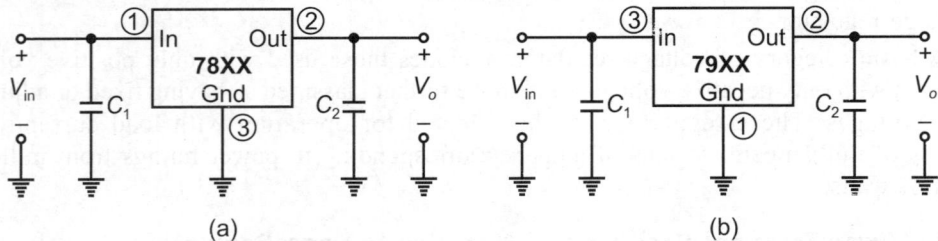

(a)   (b)

Fig. 17.11 Regulator using μA 7800 IC

Table 17.1 lists some typical data.

Table 17.1 Positive series 78XX voltage regulator ICs

IC Part number	Regulated positive voltage (V)	Minimum V_{in} (V)
7805	+5	7.3
7806	+6	8.35
7808	+8	10.5
7810	+10	12.5
7812	+12	14.6
7815	+15	17.7
7818	+18	21.0
7824	+24	27.1

Table 17.2 Fixed-negative voltage regulators in the 79XX series

IC Part number	Regulated output voltage (V)	Minimum V_{in} (V)
7905	–5	–7.3
7906	–6	–8.4
7908	–8	–10.5
7909	–9	–11.5
7912	–12	–14.6
7915	–15	–17.7
7918	–18	–20.8
7924	–24	–27.1

Negative Voltage Regulator: Negative voltage regulator ICs are available in the 79 series [Fig. 17.11 (b)], which provide a series of ICs similar to the 78 series but operating on negative voltages, providing a regulated negative output voltage. Table 17.2 lists the 79XX series of fixed-negative voltage regulators and their corresponding regulated voltages.

The LM340 5 V Regulator: This device is another example of three-terminal regulator. It is designed specifically for a +5 V output.

17.6.2 LM317 Adjustable Voltage Regulator

Voltage regulators are also available in circuit configurations that allow the user to set the output voltage to a desired regulated value. The LM317, for example, can be operated with output voltage regulated at any setting over the range of voltage from 1.2 V to 37 V. Figure 17.12 shows a typical connection using the LM317 IC which is an adjustable voltage regulator.

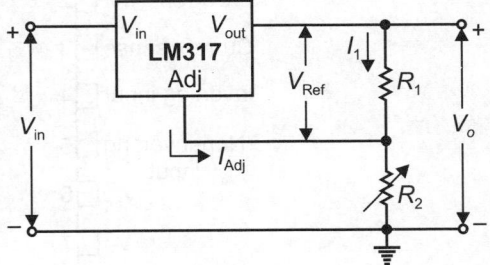

Fig. 17.12 Adjustable voltage regulator using LM317 IC

Selection of resistors R_1 and R_2 allow the setting of the output to any desired voltage over the adjustment range (1.2 to 37 V). The output voltage desired can be

$$V_o = V_{ref}\left(1 + \frac{R_2}{R_1}\right) + I_{adj}R_2$$

with typical values of $\qquad$ $V_{ref} = 1.25$ V and $I_{adj} = 100$ μA.

17.6.3 Precision Voltage Regulator

This type of IC regulators provide very precise regulation of the output voltage for both line and load variations. Generally these devices also provide for continuously adjustable output voltages within a specified range, current limiting and remote shutdown.

The μA 723 Regulator: This particular device exhibits the following list of features, and its packing configurations are shown in Fig. 17.13 (a), with an equivalent simplified schematic in part (b). This integrated circuit is mainly used as a voltage regulator and gives regulated output voltages ranges from 2 V to 37 V at a maximum current of 150 mA. The salient features of μA 723 are:

1. Output current limiting facilities

2. Adjustable output voltage
3. Better ripple rejection
4. Remote shut down control
5. Facilitates building of positive and negative voltage regulators
6. High current regulators possible with suitable series pass element.

Fold-Back Limiting in the μA 723: Fold-back current limiting is a method particularly in high-current regulators whereby the output current under overload conditions drops to a value well below the peak load current capability to present excessive power dissipation. As this device has internal circuitry for current limiting, only three resistors are required. A capacitor across the resistor is sometimes necessary for stability purposes (not shown).

The pin connections of μA 723 are shown in Fig. 17.13 (a).

Its circuit, shown in Fig. 17.13 (b), consists of temperature compensated voltage reference amplifier. The temperature is compensated by a Zener diode which has opposite nature of temperature co-efficient as compared to the base emitter diode of a transistor inside the IC. Constant current sources are employed in the voltage reference amplifier to keep the reference

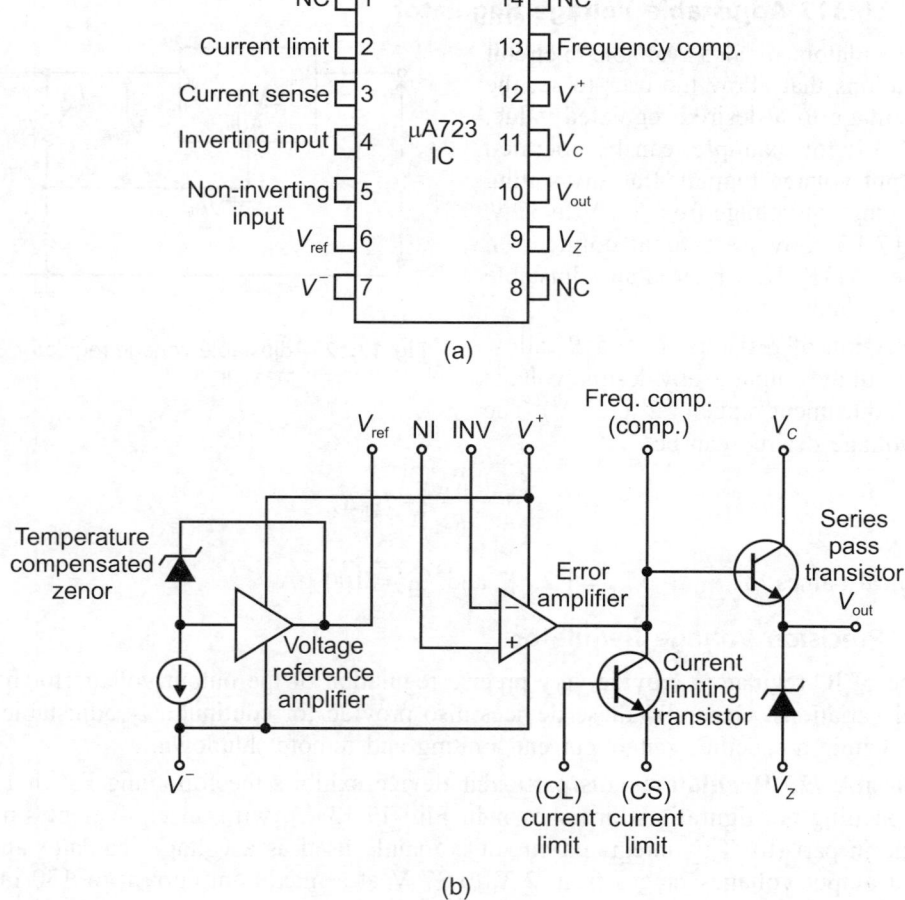

(a)

(b)

Fig. 17.13 Precision voltage regulator using μA 723 IC

voltage constant. These voltages are taken across the constant current fed Zener and base emitter in the IC and are constant irrespective of changes in the temperature, and the variations in main voltages.

The error amplifier is a differential amplifier which amplifies the difference between the constant reference voltage and the output of the IC. At the non-inverting input, part of reference voltage and at the inverting input part of IC output voltage is supplied. For better temperature stability, the differential amplifier is again fed by a constant current source.

The output of the differential amplifier is fed to the *series pass transistor* which conducts more or less and keeps the output voltage constant. The maximum current through the pass filter is 150 mA. The current limiting circuit functions as follows:

The terminals of the current limiting transistor, namely the base and emitter are brought outside. The part of the output of the IC is fed at the base of this transistor and the emitter is provided by a suitable current sense resistor connected in series with the load. When the current exceeds certain limit, the transistor is forward biased and divert the base drive of the series pass element, thereby limiting the output voltage of the IC and limit current.

17.7 PRACTICAL POWER SUPPLIES

A practical power supply can be built to convert the 120 V supply voltage into a desired regulated dc voltage with an IC three-terminal regulator. The standard circuit includes a transformer to step the voltage to a desired ac level, a diode rectifier to half-wave or full-wave, rectify the ac signal, and a capacitor filter to develop an unregulated dc voltage. The unregulated dc voltage is then connected as input to an IC voltage regulator, which provides the desired regulated output dc voltage. LED serves as a power-on indicator. The 7800 series three terminal regulator is available in a selection of fixed output voltages from 2.6 V to 24 V. Some systems may require several dc voltages, which can be achieved by distributing the rectified and filtered output to the required number of regulators.

17.8 SWITCH MODE REGULATOR

In switched mode power conversion, the controlling device is almost an ideal switch which is either closed or open. By controlling the ratio of the duration of time interval in the closed or open position often defined as the duty ratio, the power flow can be controlled in a very efficient way. Moreover, since the semiconductor power transistor switch operates always at saturated or cut-off mode except for a brief transition between these two states, the switching regulator can achieve good regulation despite large variations in input voltage and load changes and the efficiency is maintained high throughout.

The switching regulator regulates by varying the ON-OFF duty ratio of the power transistor switch. The switching frequency can also be made very high compared to the line supply frequency because of the advent of low loss power transistor particularly of the power MOSFET and devices like GTO, IGBT, SIT and SITH. As a result, the transformer and the filtering element used in the power supply can be made small, light weight, low in cost, high efficiency. The dc supply source applied to the switching regulator need not be well filtered thereby dispensing with expensive and large filter elements. Subsequently, the overall size and cost of the regulator diminish to a large extent.

The selection of switching frequency depends on many factors. The general trend of the designer is to use higher frequencies in order to simplify filtering and to increase the transistor efficiency, but in such a case, the transformer losses increase. The transformer is the main component responsible for size and weight of the regulator. The transformer size depends on the output power, the voltage level, the allowable temperature rise, efficiency and frequency.

It is seen that Cold Rolled Grain Oriented (CRGO) silicon iron core can be used upto a frequency of about 2 KHz and much higher frequencies can be applied to ferrite cores.

17.8.1 Advantages and Disadvantages of Switching Regulators

Advantages

Advantage of a switching regulator is its excellent load-transient properties. A step increase of load current causes a relatively small instantaneous change in the output voltage and the recovery time is of the order of a few hundred microseconds.

Disadvantage

Compared to the series or shunt regulator, the switching regulator has some disadvantages which preclude its use in some applications. The primary power source delivers current to the switching regulator in pulses, and for efficiency reasons they have short rise and fall times. In those applications where a significant series impedance appears between the supply and the switching regulator, the rapid change in current can generate sufficient noise. However, this problem can be minimized by reducing the series impedance increasing the switching time or by filtering the input to the regulator.

Another disadvantage of the switching regulator compared to the dissipative counterpart, is its response time to rapid changes in load current. The switching regulator will react to its new equilibrium only when the average inductor current reaches its new steady state value. In order to reduce this time, the values of the inductors can be made low or else the input output difference is made large.

The switching regulator has become increasingly popular in new equipments design not only in aerospace and defence applications but in Computers, Industrial process Control systems, Instrumentation and Communication.

17.9 SWITCH MODE REGULATORS

There are several schemes of switching regulators. The main purpose of the switching mode operation is to reduce the size of the isolating transformers and filter chokes and to increase the efficiency. Therefore, the voltage of the ac mains is first rectified and filtered to some extent. Then, this voltage is converted to a high-frequency ac voltage by an inverter. The output of the inverter is transformed by a high frequency transformer to the desired voltage, then rectified and filtered. The basic scheme of switch mode regulator is shows in Fig. 17.14.

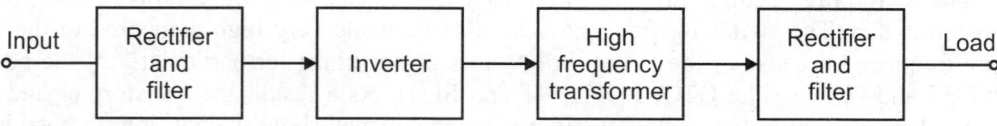

Fig. 17.14 Switch mode regulator scheme

The control may be introduced at any stage; the actual stage depends on the load requirements and the overall cost structure.

For low voltage multioutput supply, separate rectifier-filters and regulators are incorporated after high frequency transformer which has the required number of secondaries. Each output can be individually regulated by a series-pass regulator to provide a precise voltage regulation over wide variations of supply voltage and load current. Such a scheme would be costly because of the presence of a number of low voltage-high current series-pass regulators and, naturally, the overall efficiency is very low and may be to the extent of 25%.

In second scheme, regulators may be used to regulate the dc voltage input to the inverters as shown in Fig. 17.15.

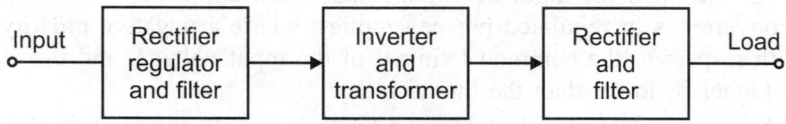

Fig. 17.15 Switching regulator with controller to regulate dc input to inverter

In this scheme phase controlled SCR circuits, series, shunt or switching regulators using high voltage, high power transistors can be used.

In other scheme, control is introduced in the inverter itself so as to provide a pulse modulated output. The scheme is shown in Fig. 17.16.

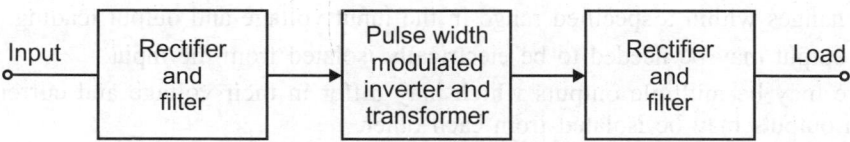

Fig. 17.16 Switching regulator with pulse width modulated control

Low output power single ended switching regulators are popular for low cost and good performance. The regulators are also called choppers. There are two types of such regulators as shown in Figs 17.17(a) and (b). In Fig. 17.17(a), the supply voltage is first transformed and converted to a suitable dc which is chopped to supply high frequency power pulses to an LC filter with a freewheel diode.

The duty cycle of the pulse is controlled to achieve the required regulation.

In the scheme shown in Fig. 17.17(b), chopping is performed at a high voltage dc obtained from direct rectification of the supply voltage. Here, a high frequency transformer is used for isolation and voltage transformation. The primary inductance of the transformer is the key to the operation of the filter capacitor to meet the load specification.

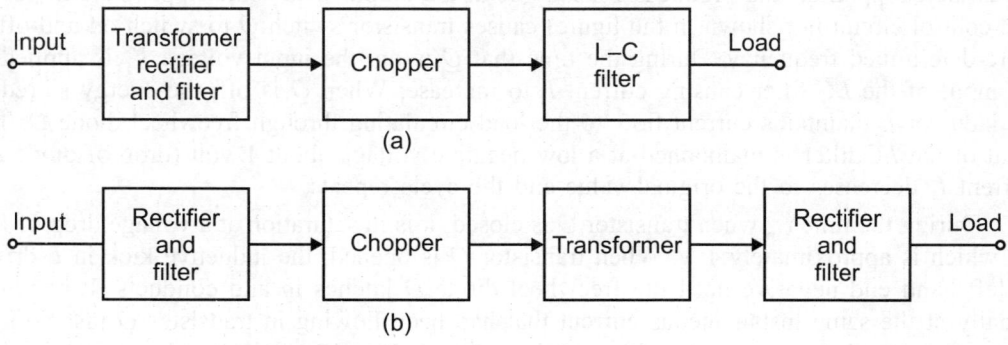

Fig. 17.17 Single ended switching regulator (a) isolation by power frequency transformer; and (b) isolation by high frequency transformer

Buck-Boost Regulator: The simplest converter topologies consist of a single switch, a single inductor and a single capacitor. These components may be arranged in different ways to form a buck, a boost or a buck-boost regulator. In a buck regulator, the output voltage is always less than the input dc voltage whereas in a boost regulator the output voltage is always greater than the input dc voltage. In buck boost regulator, the output voltage may be less, equal or greater than the input dc voltage. The main application of a step-down/up or Buck-boost converter is in regulated power supplies, where a negative polarity output may be desired with respect to the common terminal of the input voltage, and the output voltage can be either higher or lower than the input voltage.

A buck-boost converter can be obtained by the cascade connection of the two basic converters.

17.10 SWITCHING DC POWER SUPPLY

Most of analog and digital electronic circuits need regulated dc power supplies. They are designed to meet the following requirements.

1. **Regulated output:** The output voltage must be held constant within a specified limit for changes within a specified range in the input voltage and output loading.
2. The output may be needed to be electrically isolated from the input.
3. There may be multiple outputs which may differ in their voltage and current ratings. Such outputs may be isolated from each other.
4. The weight and size of regulated power supply reduces with improved efficiency.

17.10.1 Advantages of Switching Power Supplies

1. Advances in semiconductor technology have led to switching power supplies, which are smaller in size.
2. They are more efficient as compared to linear power supplies.
3. The cost of linear and switching power supplies depends on the power rating.

17.11 BASIC SWITCHING OF BUCK REGULATOR

The circuit diagram of the basic switching buck regulator is shown in Fig. 17.18.

In this circuit V_1 is the voltage applied at the input of the regulator. This voltage is chopped by a BJT Q to produce a square wave at the input to the low pass LC filter. The LC filter stores the chopped dc and produces a steady dc at the output with some ripple superimposed. The control circuit not shown in the figure) causes transistor switch Q to switch on and off at a pre-determined frequency. During the time that Q is on, the input voltage V_1 is applied to the input of the LC filter causing current I_1 to increase. When Q is off, the energy stored in the inductor L maintains current flow to the load circulating through freewheel diode D. The input of the LC filter is maintained at a low negative voltage about 1 volt (drop of diode D). Current I_1 decreases to the original value and the cycle repeats.

During the time t_{on} when transistor Q is closed, it is in saturation at a voltage drop of V_{CE} (sat) which is approximately 1 V. When transistor Q is opened, the inductive kick in L drives its left hand end negative until the freewheel diode D latches in and conducts. It conducts initially at the same instantaneous current that had been flowing in transistor Q just prior to its opening. At this instant, the voltage at the point A (Fig. 17.18) is V_1 which is about 1 V less than the rectified dc input voltage. After that, the voltage at A reduces to about 1 V below ground when Q is open due to the drop of freewheel diode D.

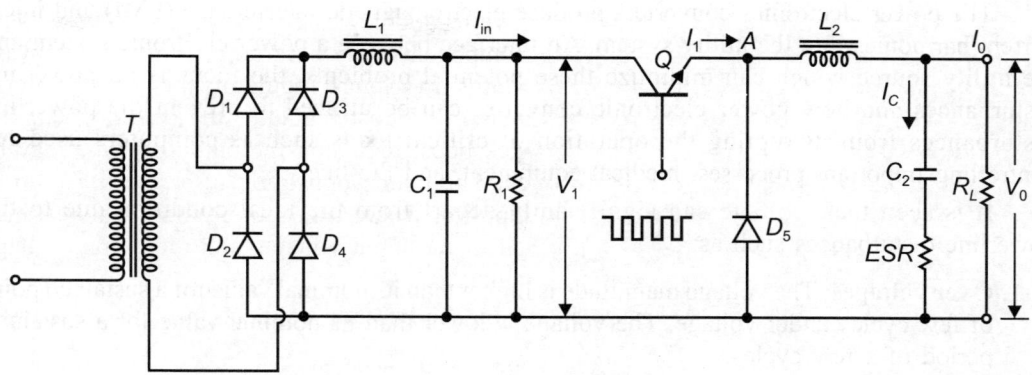

Fig. 17.18 Basic switching of buck regulator

17.11.1 Losses and Efficiency

The efficiency of a switching buck regulator depends on the factors given in the following equation

$$\text{Efficiency,} \quad \eta = \frac{P_{\text{out}}}{P_{\text{in}}} \times 100\% = \frac{V_0 I_0}{V_0 I_0 + \text{losses}}$$

The losses include the following:

1. dc loss in transistor, $P_Q = \left(V_{CE\,(\text{sat})} \cdot I_0 \cdot \dfrac{t_{\text{on}}}{T} \right) = \left(V_{CE\,(\text{sat})} \cdot I_0 \cdot \dfrac{V_0}{V_1} \right)$

2. switching loss in transistor   3. dc loss in diode
4. switching loss in diode   5. dc loss in inductor
6. ac losses in capacitor   7. control circuit losses

17.12 LINEAR VOLTAGE REGULATOR USING MOSFET

Many types of linear integrated circuit voltage regulators are available in monolithic and hybrid form. These devices are of low power and can handle load currents upto about 10 A. When the required power exceeds this limit, discrete design becomes essential. Power MOSFET when operated in the saturation region can be used in place of bipolar series pass transistor in linear regulators. MOSFET have several advantages over its bipolar counter part. The average gate drive current for a MOSFET is considerably less than the average base current of similarly rated bipolar transistor. A MOSFET can be operated at the rated breakdown voltage and rated drain current under certain pulsed condition. Consequently safe operating area (SOA) curve of a MOSFET is a fully rectangular whereas that of a bipolar transistor is not. MOSFETs are easy to operate in parallel if certain precautions are taken whereas bipolar transistor requires current equalizing emitter resistance for parallel operation. The design of gate control circuit is less complicated than that of the base control circuit because the power requirement is much less.

17.13 POWERLINE DISTURBANCES

There are many types of disturbances associated with the mains input. Power conditioner provide an effective way to protect sensitive electronic loads from these disturbances except for the power outages and frequency deviations.

The power electronics converters produce electromagnetic interference (EMI) and inject current harmonics into the utility system. An interface between a power electronic system and the utility source which can minimize these potential problems, the focus is on powerline disturbances and how power electronic converter can be utilized to prevent the powerline disturbances from disrupting the operation of critical loads such as computers used for controlling important processes, medical equipment, and the like.

It is seen that, voltage can significantly depart from the ideal condition due to the powerline disturbances such as:

1. **Over voltages:** The voltage magnitude is higher than its nominal value for a sustained point of few cycles under voltage. The voltage is lower than its nominal value for a sustained period of a few cycles.
2. **Under voltage:** The voltage is substantially lower than its nominal value for a few cycles.
3. **Output:** The utility system voltage collapses for a few cycles or more.
4. **Voltage spikes:** These are superimposed on the normal 50 Hz waveforms which occur occasionally (not on a repetitive basis). These can be either of a line-mode (differential-mode) or a common-mode type.
5. **Chopped voltage waveform:** This refers to a repetitive chopping of the voltage waveform and the associated ringing.
6. **Harmonics:** A distorted voltage waveform which contains harmonics voltage components at harmonic frequencies usually low-order multiples of the line frequency. These harmonics exist on a sustained basis.
7. **EMI:** This refers to high-frequency noise, which may be conducted on the powerline.

17.14 UNINTERRUPTIBLE POWER SUPPLIES (UPS)

UPS is the best solution to power conditioning for critical loads such as real-time processing computers, air route traffic control centres, industrial process control equipment, etc. They meet all the demand and specifications for these systems. The UPS are complex and costly. Basically UPS consists of the following equipments such as: (1) rectifier, (2) inverters, (3) battery, (4) filters.

For supplying very critical loads such as computers used for controlling important processes, some medical equipment, and the like, it may be necessary to use UPS. These provide protection against power outages as well as voltage regulation during powerline overvoltage and undervoltage conditions. They are also excellent in terms of suppressing incoming line transient and harmonic disturbances.

UPS in their block-diagram form are shown in Fig. 17.19. A rectifier is used for converting single-phase or three-phase ac input into dc, which supplies power to the inverter as well as to the battery bank to keep it charged.

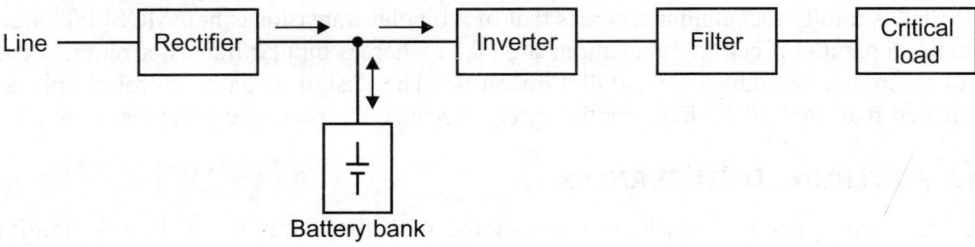

Fig. 17.19 Block diagram of uninterruptible power supply (UPS)

In the normal mode of operation, the power to the inverter is provided by the rectifier. In case of a line outage, power comes from the battery bank. The inverter produces either a single-phase or a three-phase sinusoidal waveform depending on the UPS. The output voltage of the inverter is filtered, prior to being applied to the load.

17.14.1 Rectifier

For supplying power to the inverter and for keeping the battery bank charged two rectifier arrangements are shown in Fig. 17.20. In a conventional arrangement shown in Fig. 17.20(a), a phase-controlled rectifier is used. It is also possible to use a diode rectifier bridge in cascade with a step-down dc-dc converter as shown in Fig. 17.20(b).

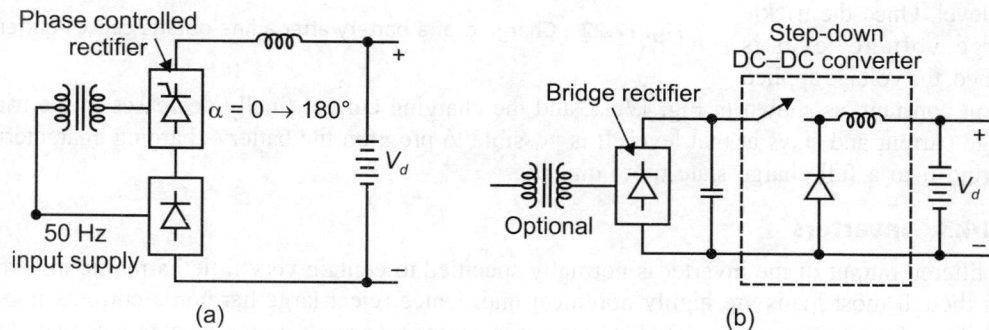

Fig. 17.20 Phase controlled rectifier

When an electrical isolation from the mains is required, it is possible to use a dc-dc converter with a high-frequency isolation transformer as shown in Fig. 17.21. The dc-dc converter with electrical isolation may be similar to the ones used in the switch-mode dc power supplies which may utilize resonant converter.

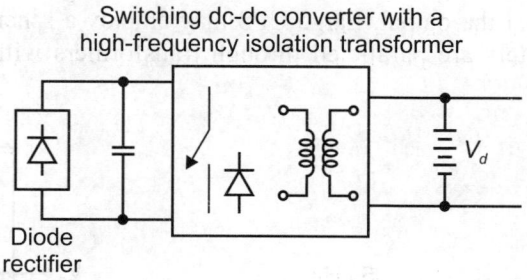

Fig. 17.21 Rectifier with high-frequency isolation transformer

17.14.2 Batteries

There are many different types of battery systems. But lead-acid batteries are commonly used for the UPS applications.

In the normal mode when the line voltage is present the battery is trickle charged to offset the slight self-discharge by the battery. This requires that a constant trickle charge voltage be applied across the battery and the battery continuously draws a small amount of current, thus maintaining itself in a fully charged state.

In the event of a line outage, the battery supplies the load. The battery voltage should not be allowed fall below the final discharge voltage level; otherwise the battery life is shorted. Typically, a 10 hour current is defined as the current in amperes that causes the fully charged battery to discharge in 10 hours to its final voltage level. Discharge currents in excess of the 10 hour current cause the final discharge voltage to be reached sooner than their magnitudes would suggest. Therefore, the higher discharge currents reduce the effective battery capacity.

Once the line voltage is restored, the battery bank in a UPS is brought to its fully charged state. This is accomplished by initially charging the battery at a constant charging current rate as shown in Fig. 17.22. This causes the battery terminal voltage to increase to its trickle charge voltage level. Once the trickle charge voltage level is reached, the voltage applied

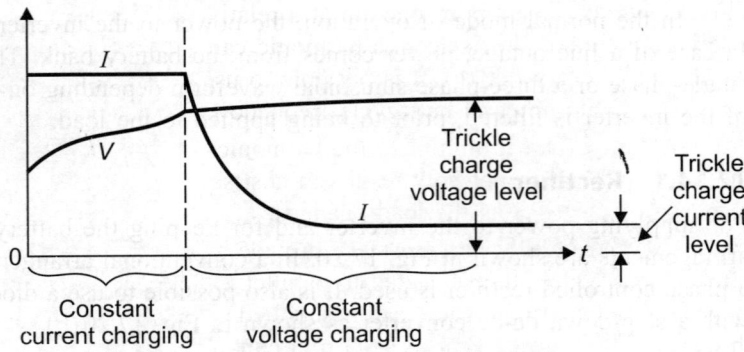

Fig. 17.22 Charging of a battery after a line-outage causes battery

is kept constant, as shown in Fig. 17.22, and the charging current finally decreases to the trickle charge current and stays at that level. It is possible to program the battery charging characteristic to bring it to a full charge state more quickly.

17.14.3 Inverters

The filtered output of the inverter is normally specified to contain very little harmonic distortion, even though most loads are highly nonlinear and, hence reject large harmonic currents into the UPS. Therefore, the inverter must allow almost instantaneous control over its output ac waveform.

Modern UPS normally use PWM dc-to-ac inverters with either a single-phase or three-phase ac output. A schematic is shown in Fig. 17.23 (a). An isolation transformer is generally used at the output. Large UPS may employ a scheme where the outputs of two or more such inverters are paralleled through transformers with phase shift, as shown in Fig. 17.23 (b).

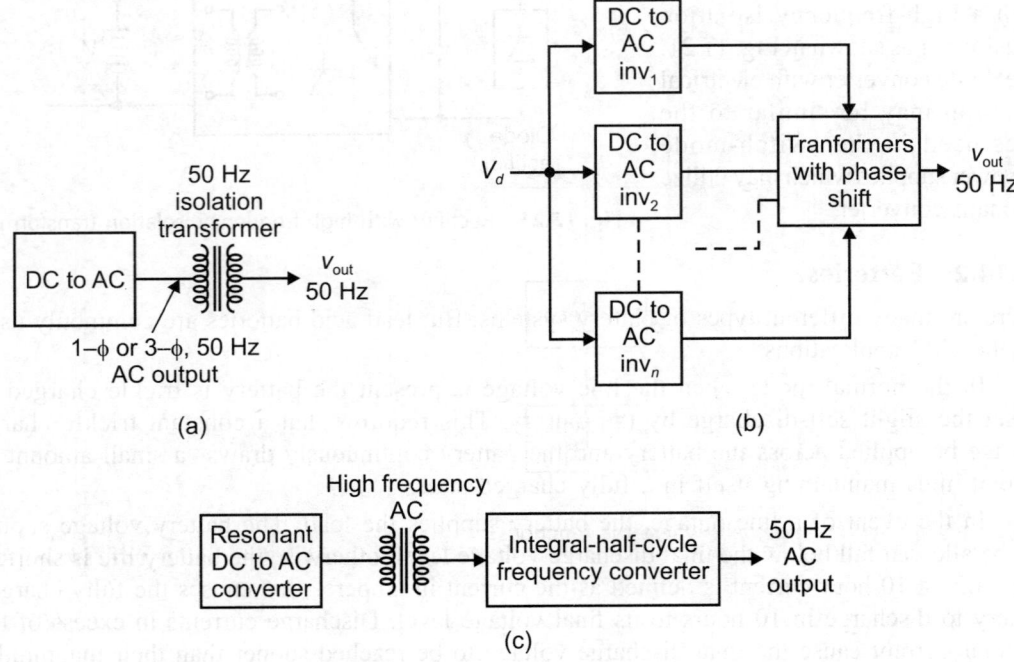

Fig. 17.23 Various inverter arrangements

This allows the inverters to operate at a relatively lower switching frequency, utilizing either a low-frequency PWM, selective harmonic cancellation, or a square-wave switching scheme. As shown in Fig. 17.23 (c), it is possible to use resonant converters, high frequency isolation transformers and integral half cycle frequency converter concepts.

It is important to minimize the harmonics content of the inverter output. This decreases the filter size, which not only results in cost savings but also results in an improved dynamic response of the UPS as the load changes.

Above a few kilowatts, most UPS provide power to several loads connected in parallel. As shown in Fig. 17.24, each load is supplied through a fuse. In the event of a short circuit in one of the loads, it is important for the UPS to blow that particular fuse and to keep on supplying the rest of the loads. Therefore, the current rating of the UPS under a sustained short-circuit condition should be sufficient to blow the fuse of the faulted load. In this respect, a rotating-type UPS with a large short-circuit current capacity is far superior to the power electronics type UPS.

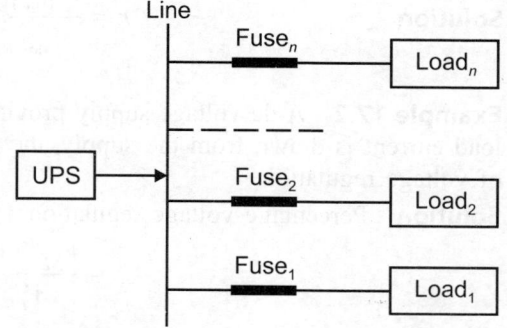

Fig. 17.24 A UPS supplying several load

An alternative scheme, where the functions of battery charging and the inverter are combined, is shown in Fig. 17.25. In the normal mode, the switching converter operates as a rectifier, charging the battery bank. In addition, it can draw inductive or capacitive currents from the mains, thus providing a fine regulation of the voltage supplied to the load. In case of a utility outage, the utility is isolated and the switching converter operates as an inverter, supplying power to the load from the battery bank. This arrangement is usually referred to as the 'standby power supply'.

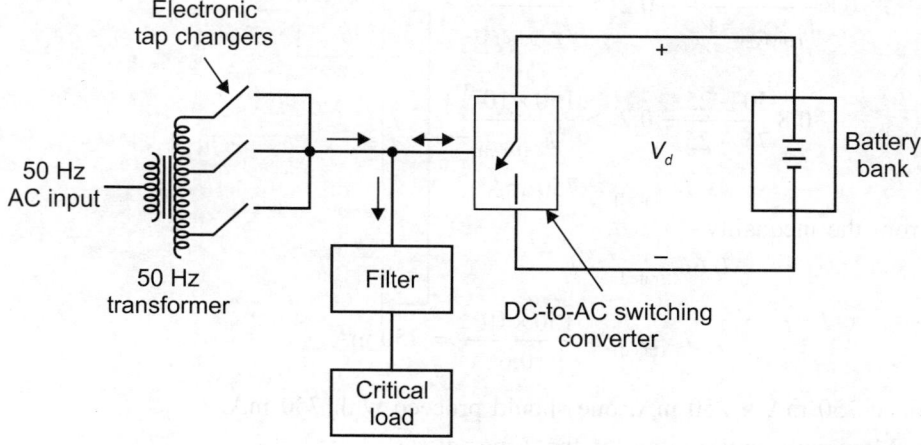

Fig. 17.25 UPS arrangement where the functions of battery charging and inverter are combined

SOLVED EXAMPLES

Example 17.1 Using a dc and ac voltmeter to measure the output signal from a filter circuit, a dc voltage of 25 V and an ac ripple voltage of 1.5 V rms are obtained. Calculate the ripple of the filter output.

Solution
$$r = \frac{V_{r\,(rms)}}{V_{dc}} \times 100\% = \frac{1.5\ V}{25\ V} \times 100 = 6\%$$

Example 17.2 A dc voltage supply provides 60 V when the output is unloaded. When full-load current is drawn from the supply, the output voltage drops to 56 V. Calculate the value of voltage regulation.

Solution Percentage voltage regulation
$$= \frac{V_{NL} - V_{FL}}{V_{FL}} \times 100\% = \frac{60\ V - 56\ V}{56\ V} \times 100\% = 7.14\%$$

If the value of full-load voltage is the same as the no-load voltage, the voltage regulation is 0%, which is the best.

Example 17.3 Design a suitable voltage regulator which will provide an output voltage of $V_Z = 25$ V dc to a load whose maximum load current will be $I_L = 150$ mA. The input voltage of the regulator is expected to vary from $V_{i\,(min)} = 50$ V to $V_{i\,(max)} = 75$ V dc. Find (a) the minimum power rating of the Zener diode, (b) the range of resistance R if a Zener diode is to be used which has a power rating $P_{max} = 40$ W.

Solution (a) From the inequality
$$0.8\frac{V_{i\,(min)} - V_Z}{V_{i\,(max)} - V_Z} - 0.2 > \frac{I_L}{I_{Z\,(rated)}}$$

$$\therefore \qquad 0.8\frac{50 - 25}{75 - 25} - 0.2 > \frac{150 \times 10^{-3}}{I_{Z\,(rated)}}$$

or
$$I_{Z\,(rated)} > 750\ mA$$

From the inequality
$$0.6I_{Z\,(rated)} > I_L$$

or
$$I_{Z\,(rated)} > \frac{150 \times 10^{-3}}{0.6} = 250\ mA$$

Since 750 mA > 250 mA, one should proceed with 750 mA.

∴ Maximum power rating of the Zener diode
$$= V_Z I_{Z\,(rated)} = 25 \times 750 \times 10^{-3} = 18.8\ W$$

(b)
$$P_{max} = V_Z I_{Z\,(rated)}$$

or
$$I_{Z\,(rated)} = \frac{P_{max}}{V_Z} = \frac{40}{25} = 1.6\ A$$

From the equation $\qquad R_{max} = \dfrac{V_{i(min)} - V_Z}{0.2 I_{Z(rated)} + I_L}$

$$R_{max} = \dfrac{50 - 25}{0.2 \times 1.6 + 0.15} = 53.19 \text{ Ohm}$$

and $\qquad\qquad R_{min} = \dfrac{V_{i(max)} - V_z}{0.8 I_{z(rated)}}$

$$R_{min} = \dfrac{75 - 25}{0.8 \times 1.6} = 39 \text{ Ohm}.$$

Example 17.4 Design the transistorised voltage regulater, employing a power transistor and Zener diode to supply a current of 1 A, at a constant voltage of 6 V. If the supply voltage is 12 V and it varies by symbol ±2 V. Given β = 50, V_{BE} = 0.5 V. Minimum Zener current is 8 mA.

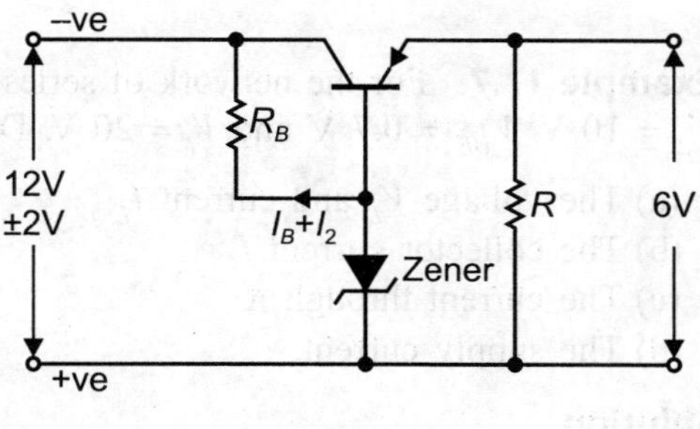

Fig. E17.1

Solution Collector current I_c = 1 A.

Base current, $\qquad I_b = \dfrac{I_c}{\beta} = \dfrac{1}{50} = 20 \text{ mA}$

Output voltage required is = 6 V

$$V_B = V_{BE} + V_Z$$
$$6 = 0.5 + V_Z$$
$$V_z = 6 - 0.5 = 5.5 \text{ V}$$

Hence, a Zener diode of 5.5 V must be used. Voltage drop across transistor (when supply voltage is 12 V) = 12 − 6 = 6 V.

Collector-base voltage drop (across R_B) = 12 − V_z

$$= 12 - 5.5 = 6.5 \text{ V}$$

$\therefore \qquad\qquad R_B = \dfrac{6.5}{I_B + I_Z} = \dfrac{6.5}{(20 + 8)\,\text{mA}} = 232 \text{ Ohms}.$

Example 17.5 Design a shunt regulator employing a single transistor is required to maintain 9 V at the output. If the maximum load is 150 mA, find the value of series resistance R when input voltage is 12 V ± 20% and V_{BE} = 0.2 V.

Solution $\qquad\qquad\qquad V_o = V_Z + V_{BE}$

$$V_o = 9 \text{ V}, V_{BE} = 0.2 \text{ V}$$
$$V_Z = V_o - V_{BE}$$
$$= 9 - 0.2 = 8.8 \text{ V}$$

Hence, a Zener of 8.8 V should be used.

Input voltage $\qquad\qquad\qquad = 12 \pm 20\%$

$$= 9.6 \text{ V to } 14.4 \text{ V}$$

Maximum voltage to be dropped across R

$$= 14.4 - 9 = 5.4 \text{ V}$$

Maximum current $= 150 \text{ mA} = \dfrac{15}{100} \text{ A}$

Value of resistance $R = \dfrac{15}{100} = 36 \text{ Ohm.}$

Example 17.6 A Zener diode is specified as having a breakdown voltage of 9.0 V with a maximum power dissipation of 360 mW. What is the maximum current the diode can handle?

Solution The maximum permissible current is

$$I_{Z\,max} = \frac{P}{V_Z} = \frac{360 \times 10^{-3}}{9.0} = 40 \text{ mA.}$$

Example 17.7 For the network of series regulator Fig. 17.6 if $R_L = 1$ k-ohm, $R = 5$ k-ohm, $V_Z = 10$ V, $V_{BE} = 0.7$ V and $V_i = 20$ V. Determine the following:

(a) The voltage V_L and current I_L
(b) The collector current I_C
(c) The current through R
(d) The supply current.

Solution

(a)
$$V_L = V_Z - V_{BE} = 10 - 0.7 = 9.3 \text{ V}$$
$$I_L = \frac{V_L}{R_L} = \frac{9.3}{1000} = 9.3 \text{ mA}$$

(b) $I_C \approx I_E = I_L = 9.3 \text{ mA}$

(c) $I_R = \dfrac{V_i - V_Z}{R} = \dfrac{(20 - 10)}{5000} = 2 \text{ mA}$

(d) Supply current, $I_S = I_C + I_R = 9.3 + 2 = 11.3 \text{ mA.}$

Example 17.8 For the network of current regulator Fig. 17.9, if $R_L = 4$ k-ohm, $R_S = 2$ k-ohm $V_Z = 10$ V, $V_{BE} = 0.7$ V and $V_i = 20$ V determine the following:

(a) The voltage V_L
(b) The current I_L
(c) Supply current through R_S
(d) The Zener current if $\beta = 50$.

Solution

(a) $V_L = V_Z + V_{BE} = 10 + 0.7 = 10.7 \text{ V}$

(b) $I_L = \dfrac{V_L}{R_L} = \dfrac{10.7 \text{ V}}{4000 \text{ Ohm}} = 2.675 \text{ mA}$

(c) $I_{RS} = \dfrac{V_i - V_L}{R_S} = \dfrac{(20 - 10.7)}{2000} = 4.65 \text{ mA}$

(d) $I_{RS} - I_L = 4.65 - 2.675 = 1.975 \text{ mA}$

$I_Z + \beta I_Z = I_Z(1 + \beta) = 1.975 \text{ mA}$

$$I_Z = \frac{1.975}{(1 + 50)} = 38.72 \,\mu A$$

$$I_C = 50 I_Z = 1.936 \text{ mA.}$$

Example 17.9 Determine the value of R_2 in the network of shunt regulator Fig. 17.4 to establish a load voltage of 20 V.

Solution

$$\frac{(20 - 10.7) \text{ V}}{1000} = \frac{10.7 \text{ V}}{R_2}$$

or

$$R_2 = \frac{10.7 \text{ V} \times 1000}{9.3 \text{ V}}$$

$$= 1151 \text{ Ohm.}$$

Example 17.10 For the network of current regulator Fig. 17.9, calculate V_L if $V_{BE} = 0.7$ V, $V_Z = 10$ V and $V_i = 20$ V. Resistor values are $R_E = 2$ k-ohm, $R_L = 5$ k-ohm and $R_B = 10$ k-ohm.

Solution

$$I_E = I_C = I_L = \frac{V_i - V_Z + V_{BE}}{R_E}$$

$$= \frac{20 - 10 + 0.7}{2000} = 5.35 \text{ mA}$$

$$V_L = I_L R_L = 5.35 \text{ mA} \times 5000 = 26.75 \text{ V.}$$

Example 17.11 What output voltage results in a circuit of LM317 Adjustable Voltage Regulator as shown in Fig. 17.12 with $R_1 = 240$ ohm and $R_2 = 3.3$ k-ohm?

Solution

$$V_0 = V_{ref} \left(1 + \frac{R_2}{R_1} \right) + I_{adj} \cdot R_2$$

$$= 1.25 \text{ V} \left(1 + \frac{3.3 \text{ k-ohm}}{240 \text{ Ohm}} \right) + 100 \,\mu A \,(3.3 \text{ k-ohm})$$

$$= 18.768 \text{ V.}$$

Example 17.12 (a) Analyze the operation of the +12 V voltage supply shown in Fig. E17.12 connected to a load drawing 400 mA, (b) determine the ripple across the filter capacitor of a voltage supply operating into a load that draws 250 mA. Assume $V_m = 25.5$ V.

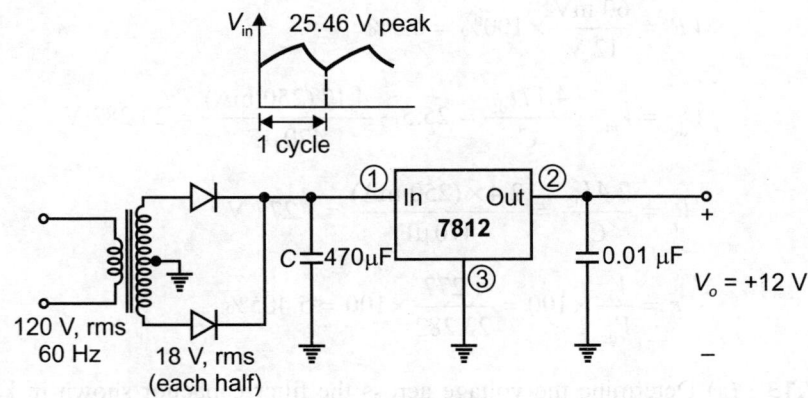

Fig. E17.12

Solution (a) The transformer steps down the line voltage from 120 V rms to a secondary voltage of 18 V rms across each transformer half. This results in a peak voltage across the transformer of

$$V_m = \sqrt{2}V_{\text{rms}} = \sqrt{2} \times 18 \text{ V} = 25.456 \text{ V}$$

The ripple voltage for $f = 60$ Hz is then

$$V_{r\,(\text{rms})} = \frac{V_m}{2\sqrt{3}fC} = \frac{2.4I_{\text{dc}}}{C} = \frac{2.4\,(400)}{470} = 2.043 \text{ V}$$

and the peak ripple voltage,

$$V_{r\,(\text{peak})} = \sqrt{3}V_{r\,(\text{rms})} = \sqrt{3}\,(2.043 \text{ V}) = 3.539 \text{ V}$$

The dc level of the voltage across the 470 μF capacitor C is,

$$V_{\text{dc}} = V_m - V_{r\,(\text{peak})} = 25.456 \text{ V} - 3.539 \text{ V} = 21.917 \text{ V}$$

The ripple factor of the filter capacitor when operating into a 400-mA load is

$$r = \frac{1}{\sqrt{2}\,(4f\,R_LC - 1)} = \frac{2.4I_{\text{dc}}}{CV_{\text{dc}}} \times 100\%$$

$$= \frac{2.4\,(400)}{(470)\,(21.917)} \times 100\%$$

$$\cong 9.3\%$$

The voltage across filter capacitor C has a ripple of about 9.3% and drops to a minimum voltage of

$$V_{\text{in (min)}} = V_m - 2V_{r\,(\text{peak})} = 25.456 \text{ V} - 2\,(3.\,539 \text{ V}) = 18.378 \text{ V}$$

Device specifications list V_{in} as required to maintain line regulation at 14.6 V. The lowest voltage being maintained across the capacitor is somewhat greater at 18.378 V.

Lowering the value of the filter capacitor or increasing the load current will result in greater ripple voltage and lower minimum voltage across the capacitor. As long as this minimum voltage remains above 14.6 V, the 7812 will maintain the output voltage regulated at +12 V.

Device specifications for the 7812 list the maximum voltage change as 60 mV. This means that the output voltage regulation will be less than

Percentage regulation,

$$VR = \frac{60 \text{ mV}}{12 \text{ V}} \times 100\% = 0.5\%$$

(b)
$$V_{\text{dc}} = V_m - \frac{4.17I_{\text{dc}}}{C} = 25.5 - \frac{4.17\,(250 \text{ mA})}{470\,\mu\text{F}} = 23.282 \text{ V}$$

$$V_r = \frac{2.4I_{\text{dc}}}{C} = \frac{2.4 \times (250 \text{ mA})}{470\,\mu\text{F}} = 1.277 \text{ V}$$

$$r = \frac{V_r}{V_{\text{dc}}} \times 100 = \frac{1.277}{23.282} \times 100 = 5.485\%.$$

Example 17.13 (a) Determine the voltage across the filter capacitor shown in Fig. E17.13, (b) Determine the ripple and dc voltage of positive 5 V power supply.

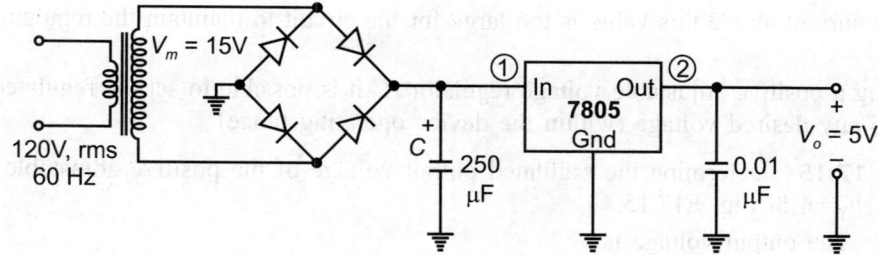

Fig. E17.13

Solution The specifications for the 7805 list an input of 7.3 V as the minimum allowable to maintain line regulation.

(a) At a load of I_{dc} = 200 mA, the ripple voltage is

$$V_{r\,(peak)} = \sqrt{3}V_{r\,(rms)} = \sqrt{3} \times \frac{(2.4)\,I_{dc}}{C} = \sqrt{3} \times \frac{2.4\,(200)}{(250)} = 3.326 \text{ V}$$

and the dc voltage across the 250 μF filter capacitor is

$$V_{dc} = V_m - V_{r\,(peak)} = 15 \text{ V} - 3.326 \text{ V} = 11.674 \text{ V}$$

The voltage across the filter capacitor will drop to a minimum value of

$$V_{in\,(min)} = V_m - 2V_{r\,(peak)} = 15 \text{ V} - 2\,(3.326 \text{ V}) = 8.348 \text{ V}$$

Since this is above the rated value of 7.3 V, the output will be maintained at the regulated +5 V.

(b) At a load of I_{dc} = 400 mA, the ripple voltage is

$$V_{r\,(peak)} = \sqrt{3} \cdot \frac{(2.4)\,(400)}{250} = 6.65 \text{ V}$$

around a dc voltage of

$$V_{dc} = 15 \text{ V} - 6.65 \text{ V} = 8.35 \text{ V}$$

which is above the rated 7.3 V lower level. However, the input swings around this dc level by 6.65 V peak, dropping during part of the cycle to

$$V_{in\,(min)} = 15 \text{ V} - 2\,(6.65 \text{ V}) = 1.7 \text{ V}$$

which is well below the minimum allowed input voltage of 7.3 V. Therefore, the output is not maintained at the regulated +5 V level over the entire input cycle. Regulation is maintained for load currents below 200 mA, but not at or above 400 mA.

Example 17.14 Determine the maximum value of load current at which regulation is maintained for the above circuit of Fig. E17.13.

Solution To maintain $\quad V_{in} \geq 7.3$ V

$$V_r\,(p\text{--}p) \leq V_m - V_{in\,(min)} = 15 \text{ V} - 7.3 \text{ V} = 7.7 \text{ V}$$

or

$$V_{r\,(rms)} = \frac{V_{r(p-p)}/2}{\sqrt{3}} = \frac{7.7V/2}{\sqrt{3}} = 2.2 \text{ V}$$

The value of I_{dc} (in mA)

$$I_{dc} = \frac{V_{r\,(rms)}C}{2.4} = \frac{(2.2)\,(250)}{2.4} = 229.2 \text{ mA}$$

Any current above this value is too large for the circuit to maintain the regulator output at +5 V.

Using a positive adjustable voltage regulator *IC* it is possible to set the regulated output voltage to any desired voltage (within the device operating range).

Example 17.15 Determine the regulated output voltage of the positive adjustable voltage regulator shown in Fig. E17.15.

Solution The output voltage is

$$V_o = 1.25 \text{ V} \left(1 + \frac{1.8 \text{ k}\Omega}{240\Omega} \right) + 100 \,\mu\text{A} \ (1.8 \text{ k}\Omega) \cong 10.8 \text{ V}$$

A check of the filter-capacitor voltage shows that an input-output voltage differential of 2 V can be maintained up to at least 200 mA of load current.

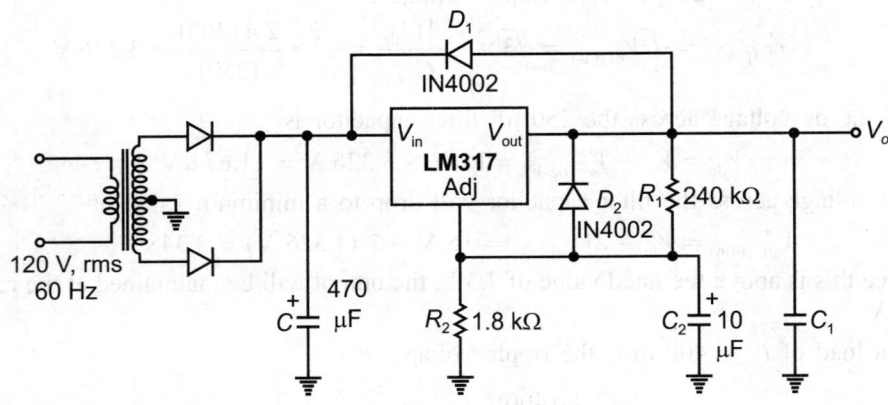

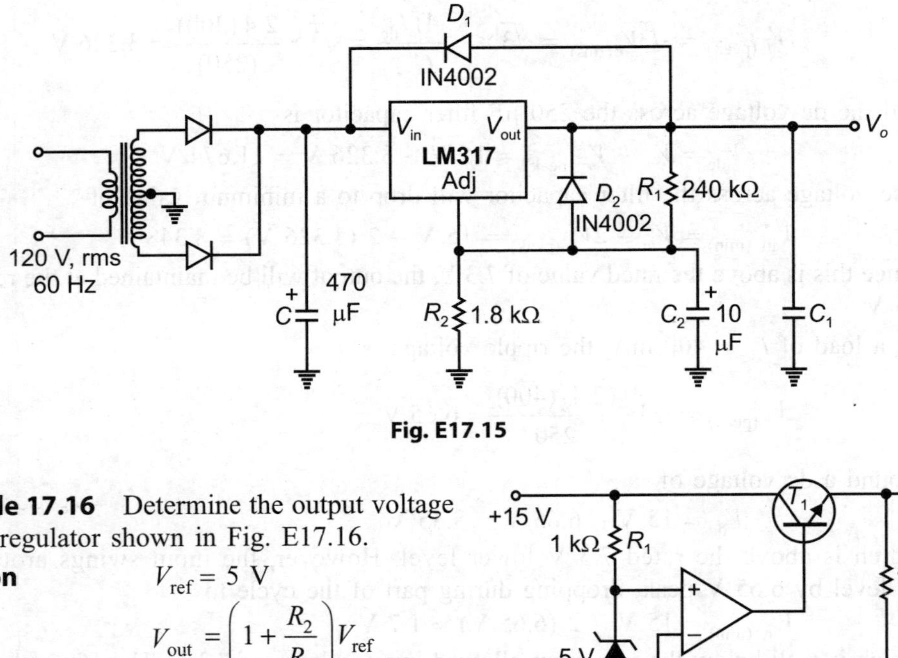

Fig. E17.15

Example 17.16 Determine the output voltage for the regulator shown in Fig. E17.16.

Solution

$$V_{\text{ref}} = 5 \text{ V}$$

$$V_{\text{out}} = \left(1 + \frac{R_2}{R_3} \right) V_{\text{ref}}$$

$$= \left(1 + \frac{10 \text{ k}\Omega}{10 \text{ k}\Omega} \right) 5 \text{ V}$$

$$= 10 \text{ V}.$$

Fig. E17.16

Example 17.17 Determine the output voltage for the µA 723 regulator in Fig. E17.17, if $V_{\text{ref}} = 1.6$ V. If the output is shorted, to what value the current is limited?

Solution

$$V_{\text{out}} = \left(1 + \frac{R_1}{R_2} \right) V_{\text{ref}} = \left(1 + \frac{8 \text{ k}\Omega}{4 \text{ k}\Omega} \right) 1.6 \text{ V}$$

$$= 4.8 \text{ V}$$

$$I_{L \text{ (max)}} = \frac{0.7 \text{ V}}{R_s}$$

$$= \frac{0.7}{580\Omega} = 1.2 \text{ mA.}$$

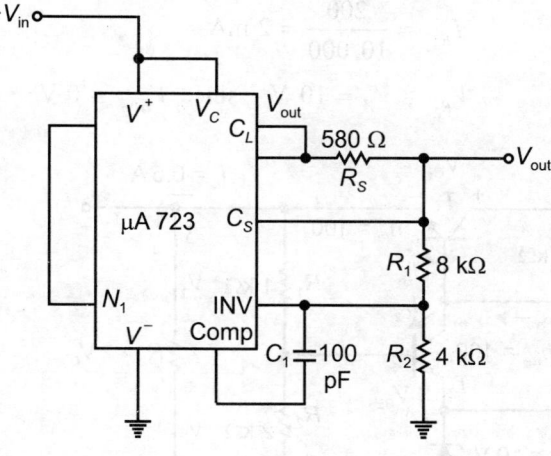

Fig. E17.17

Example 17.18 What output voltage results in a circuit shown in Fig. 17.12 with $R_1 = 240$ Ohm and $R_2 = 1.8$ k-ohm?

Solution

$$V_{\text{out}} = V_{\text{ref}}\left(1 + \frac{R_2}{R_1}\right) + I_{\text{adj}} \cdot R_2$$

$$= 1.25\left(1 + \frac{1.8 \text{ k-ohm}}{240 \text{ Ohm}}\right) + 100\,\mu\text{A} \ (1.8 \text{ k-ohm})$$

$$= 10.805 \text{ V.}$$

Example 17.19 A +5 V supply shown in Fig. E17.3 with $C = 330\ \mu\text{F}$ and load of 300 mA has what value of V_{min}? Will output be maintained at regulated +5 V level? Assume $V_m = 15$ V.

Solution

$$V_{r\,\text{(rms)}} = \frac{2.4\,(300)}{330} = 2.182 \text{ V}$$

$$V_{r\,(p-p)} = 2\,(2.182)\,\sqrt{3} = 7.558 \text{ V}$$

Since this is greater than $V_{\text{in (min)}}$ in Table 17.1 of 7.3 V, output regulation will be maintained.

Example 17.20 Determine the regulated output voltage using an LM317, shown in Fig. 17.12 of the text, with $R_1 = 240\Omega$ and $R_2 = 2.4$ kΩ.

Solution From the equation $V_o = V_{\text{ref}} = \left(1 + \frac{R_2}{R_1}\right) + I_{\text{adj}}R_2$

we have

$$V_o = 1.25 \text{ V}\left(1 + \frac{2.4 \text{ k}\Omega}{240\Omega}\right) + 100\,\mu\text{A} \ (2.4 \text{ k}\Omega)$$

$$= 13.75 \text{ V} + 0.24 \text{ V}$$

$$= 13.99 \text{ V.}$$

Example 17.21 Calculate various currents and voltages of the circuit of Fig. E17.21 (a) for the input shown. Using approximations $I_C \cong h_{fe} I_B$, $V_{BE} \cong 0$ V, and $I_C \cong I_E$.

Solution

$$V_{R4} = V_i - V_Z = 30 \text{ V} - 10 \text{ V} = 20 \text{ V}$$

and

$$I_{R_4} = \frac{200}{10,000} = 2 \text{ mA}$$

$$V_{R_2} \cong V_Z = 10 \text{ V} \quad \text{since } V_{BE_2} \cong 0 \text{ V}$$

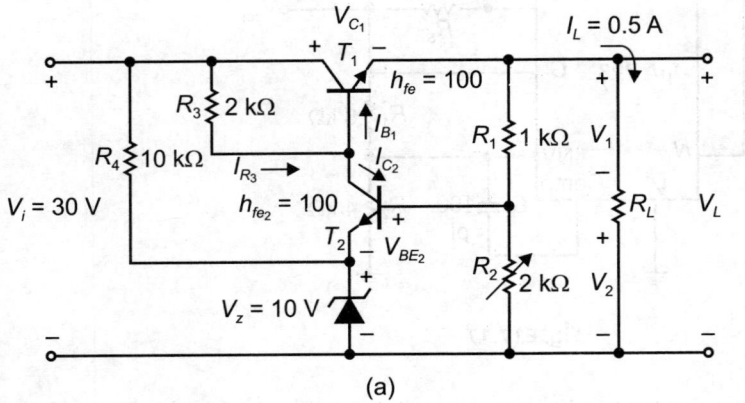 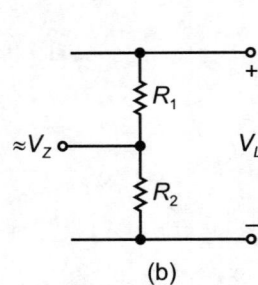

Fig. E17.21

and

$$I_{R_2} = \frac{10 \text{ V}}{2000} = 5 \text{ mA}$$

Assuming

$$I_{B_2} \ll I_{R_1}, I_{R_2}$$

Then

$$I_{R_1} = I_{R_2} = 5 \text{ mA}$$

and

$$V_L = 5 \text{ mA} \times 3 \text{ k}\Omega = 15 \text{ V}$$

with

$$V_{R_3} = V_i - V_L \ (V_{BE_1} \cong 0 \text{ V})$$
$$= 30 \text{ V} - 15 \text{ V} = 15 \text{ V}$$

and

$$I_{R_3} = \frac{15 \text{ V}}{2000} = 7.5 \text{ mA}$$

Similarly,

$$V_{C_1} = V_i - V_L = 30 \text{ V} - 15 \text{ V} = 15 \text{ V}$$

$$I_{E_1} \cong h_{fe} I_{B_1} = 100 I_{B_1}$$

and

$$I_{B_1} = \frac{I_{E_1}}{100} = \frac{(500 + 5) \text{ mA}}{100} = 5.05 \text{ mA}$$

$$I_{E_2} \cong I_{C_2} = I_{R_3} - I_{B_1}$$
$$= (7.5 - 5.05) \text{ mA} = 2.45 \text{ mA}$$

$$I_{B_2} \cong \frac{I_{C_2}}{100} = \frac{2.45 \text{ mA}}{100} = 24.5 \text{ μA}$$

(Certainly, $I_{B2} \ll I_{R2} \cdot I_{R2}$ as employed above is an excellent approximation). Finally,

$$I_Z = I_{R_4} + I_{E_2} = (2 + 2.45) \text{ mA} = 4.45 \text{ mA}$$

The fact that $I_{B_2} \ll I_{R_1}, I_{R_2}$ permits these of the circuit of Fig. E17.21 (b) to derive a rather useful equation for the circuit of Fig. E17.21 (a). Applying the voltage-divider rule results in

$$V_Z = \frac{R_2}{R_1 + R_2} V_L$$

or since V_Z is fixed,
$$V_L = V_Z \left(1 + \frac{R_1}{R_2} \right)$$

For the case above,
$$V_L = 10 \left(1 + \frac{1}{2} \right) = 15 \text{ V}$$

In Fig. E17.21 (a), R_2 is a variable resistor. Variations in this resistance will control V_L. The maximum voltage available is obviously 30 V (for $V_i = 30$ V) since at this point $V_{C1} = 0$ V (saturation). The minimum is 10 V attainable with either $R_1 = 0$ or $R_2 = \infty$.

EXERCISE

Multiple Choice Questions

1. Percentage change in the output voltage for a given change in input voltage is the:
   (A) line regulation
   (B) load regulation
   (C) percentage regulation
   (D) all of these

2. Percentage change in output voltage for a given change in load current is known as:
   (A) line regulation   (B) load regulation
   (C) both are correct   (D) none of these

3. The fundamental classes of voltage regulators are:
   (A) linear regulators
   (B) switching regulator
   (C) both (A) and (B) are correct
   (D) none of these

4. The basic components in a series regulator are:
   (A) control element   (B) reference voltage
   (C) error detector   (D) all of these

5. Basic configurations of switching regulators are:
   (A) step up   (B) step down
   (C) inverting   (D) all of these

6. A µA 7805 is a three terminal regulator with output of
   (A) 78 V   (B) 5 V
   (C) 80 V   (D) 12 V

7. The control element is a transistor in series with the load in case of:
   (A) series regulator
   (B) shunt regulator
   (C) series and shunt regulator both
   (D) none of these

8. The control element is Zener in parallel with the load in case of:
   (A) series regulator
   (B) shunt regulator
   (C) neither series nor shunt regulator
   (D) both (A) and (B) are correct

9. Precision voltage regulator belongs to
   (A) linear regulator
   (B) switching regulator
   (C) IC regulator
   (D) none of these

10. Three terminal regulator belongs to:
    (A) switching regulator
    (B) linear regulator
    (C) IC regulator
    (D) none of these

11. An ordinary power supply contains
    (A) rectifier only
    (B) filter only
    (C) rectifier and filter both
    (D) none of these

12. For a filter, no load voltage is 300 V and full load output voltage is 280 V. The regulation of the circuit is:
    (A) 9.6%   (B) 7.1%
    (C) 6.7%   (D) 4.2%

13. A typical value of filter capacitor for 50 Hz input is:
    (A) 100 pF   (B) 10 PF
    (C) 50 μF   (D) 10 μF

14. A typical value of induction filter for 50 Hz input is:
    (A) 10 H   (B) 100 H
    (C) 10 mH   (D) 100 μH

15. In a rectifier circuit, the load connected is of low value. For proper filter operation, it is required that:
    (A) a capacitor is to be included in the circuit
    (B) a bleeder resistance is to be placed in the circuit
    (C) an inductor filter is to be included in the circuit
    (D) all of these

16. The limitation of the voltage multiplying circuits is that:
    (A) the output has high ripple content
    (B) high output voltage is difficult to obtain
    (C) high output current is difficult to obtain
    (D) the size of the capacitors becomes very large

17. The function of bleeder resistance in filter circuit is:
    (A) to maintain minimum current necessary for optimum inductor filter operation
    (B) to work as voltage divider in order to provide variable output from the supply
    (C) to provide discharge path to capacitors so that output becomes zero when the circuit has been de-energised
    (D) all of these

18. A commercial power supply has voltage regulation of:
    (A) 1%   (B) 16%
    (C) 50%   (D) 100%

19. A Zener voltage regulator is used for:
    (A) small load current
    (B) heavy load current
    (C) no load current
    (D) none of these

20. An ideal regulated power supply should have:
    (A) 100% regulation
    (B) 50% regulation
    (C) 10% regulation
    (D) 0% regulation

21. An ordinary power supply has:
    (A) poor regulation
    (B) good regulation
    (C) very good regulation
    (D) none of these

ANSWER KEY

1. (A)	**2.** (B)	**3.** (C)	**4.** (D)	**5.** (D)	**6.** (B)	**7.** (A)	**8.** (B)
9. (C)	**10.** (C)	**11.** (C)	**12.** (B)	**13.** (C)	**14.** (A)	**15.** (C)	**16.** (C)
17. (D)	**18.** (A)	**19.** (A)	**20.** (D)	**21.** (A)			

Review Questions

1. Define line and load regulation.
2. What are the purposes of filters in rectifier circuits?
3. What is the purpose of regulator in power supply?
4. List the basic components in a series regulator.
5. How does the control element in shunt regulator differ from that in series regulator?
6. Name the advantages of a shunt regulator over a series type. Name a few disadvantages.
7. List three types of switching regulators.
8. Name two basic categories of IC voltage regulators.
9. Name the terminals of a three terminal regulator.
10. Label the functional blocks for voltage regulator.
11. Why is an unregulated power supply not good enough for some applications?

Industrial Drives Applications Using Microprocessor

18.0 INTRODUCTION

Low cost, low power consumption small physical size and versatility are responsible for the increasing applications of microprocessors in almost all branches of modern science and technology. Microprocessors based systems are suitable for dedicated applications in industrial instrumentations, etc. being very small and compact can form a part of this equipment to be controlled. They are capable of performing task of automatic control and measurement at a number of points using multiplexers and demultiplexers.

In recent years, a great deal of interest has been generated in the application of microprocessors in industrial drives. A microprocessor based-control system for motor drives promises several distinct advantages. Foremost among these is the flexibility. The control scheme is implemented in the software. Therefore, to change the control scheme in order to obtain a different drive characteristic or add a new control function, only the software needs to be modified, with nominal or no change in the hardware thus the controllers can be standardised. Microprocessor control can be completely digitized, with its sensitive to external influence thereby decreased. A microprocessor based control scheme requires fewer discrete components and less wiring, which will improve reliability. A microprocessor-based drive is expected to give high accuracy, better time response, and better speed regulation. The drive can be monitored from a supervisory control computer, because data is transmitted digitally at high speed.

As the price of microprocessors and their associated peripherals continues to fall, microprocessors-based control schemes will become cost competitive.

Applications of Microprocessors

The typical applications of microprocessors are as follows:

1. Microprocessor based control of firing circuit of thyristors.
2. Microprocessor based process controllers such as motor controllers (dc and ac) temperature monitoring, level controllers, controller for reversing drives, and pulse width controllers, etc.

3. Musical instruments such as synthesizers.
4. Medical instruments such as Stress Eco Cardiogram, ultrasound and blood analysers.
5. Instruments used in the defence such as RADAR and missiles, etc.
6. Computerised eye testing equipments.
7. High performance industrial drives such as rolling mills and paper mills.

In the following sections the basics of the microcomputer control in drive systems are discussed.

18.1 BASICS OF MICROPROCESSORS AND MICROCOMPUTERS

A microcomputer is a bus-oriented control unit using several LSI (large-scale integration) chips. Figure 18.1 shows the basic blocks in a microcomputer structure.

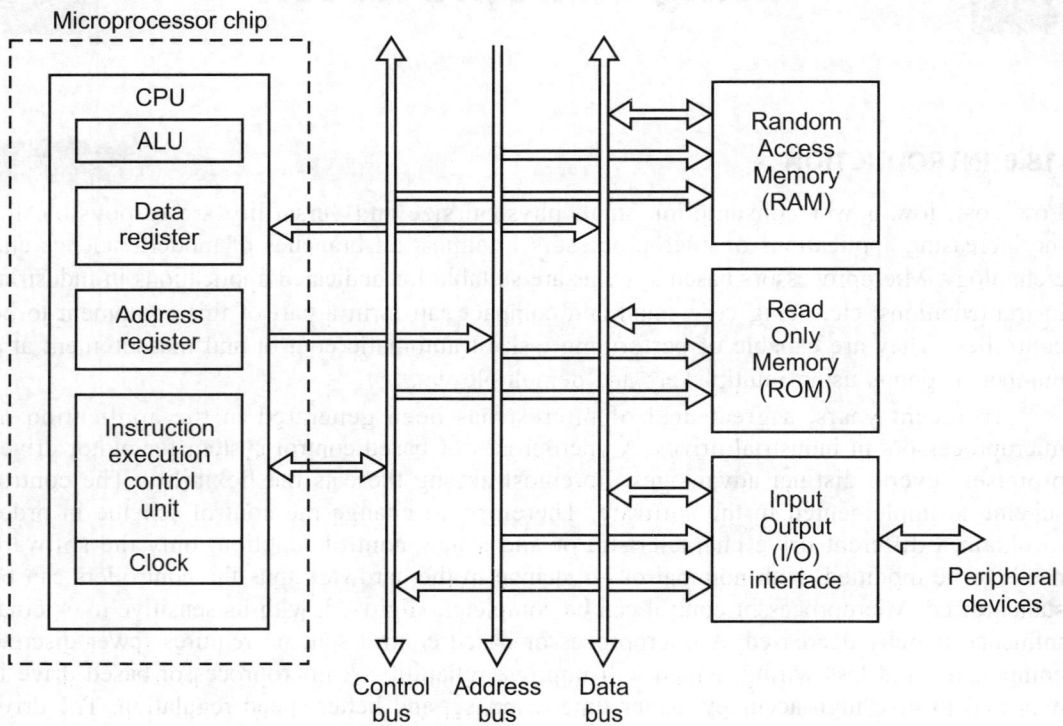

Fig. 18.1 Basic block diagram of a microcomputer

18.1.1 Microprocessor (μP)

The heart of the microcomputer is the microprocessor, which performs calculations and controls various functions. This central chip is also known as the central processing unit (CPU) or microprocessing unit (MPU). The microprocessor is the most important development in the electronics industry. This chip was introduced to meet the need for a universal large-scale integrated (LSI) circuit. Before the microprocessor, the LSI chip became more dedicated to a particular application.

The microprocessor has the ability to perform a wide variety of functions. It carries out an operation by executing as sequence of instructions called a software program that are stored in a memory connected to the microprocessor.

The various components within the microprocessor are shown in the simplified block diagram of a microprocessor (μP) chip in Fig. 18.1. The arithmetic logic unit (ALU) is the processing unit where all logical and arithmetic operations such as addition, subtraction, and bit manipulation are performed. The data registers are used for intermediate data manipulation and storage, thereby cutting down the number of transfers from the memory. They also receive and transfer data into and out of the CPU. The address registers are used for storing memory addresses and are used in conjunction with the data registers for data transfer from the memory of the input/output interface. The control unit controls and supervises the correct execution of instructions. The clock, or CPU timebase generator, is contained in the control unit. The clock frequency determines the basic operating speed of the microprocessor.

18.1.2 MEMORY ROM and RAM

The memory is a unit where instructions and data are stored. The memory contains a large number of locations or cells. The size of the memory is equal to the number of locations in the memory.

In the ROM (read only memory) instructions are stored permanently. They can only be read and cannot be destroyed. Most microcomputers are used for only one purpose. This dedicated use means that only a limited number of programs (i.e. sets of instructions) need to be stored in the memory. Therefore, most microcomputers store this program in a nonvolatile ROM, as the list of instructions never changes. The actual program is 'burned in' during manufacturing. Data constants, such as look-up tables, are also stored in ROMs.

In the RAM (random access memory) information can be written (i.e. stored) as well as read. Data or variable information can be stored in the RAM. The CPU, under program control, can read or alter the contents of a RAM location as desired.

A microcomputer system can have both ROMs and RAMs. Constants, look-up tables, and permanent programs can be stored in ROMs, while RAMs may contain problem variables and the portions of the program that may change in future.

18.1.3 Input-Output (I/O) Interface

The input-output interfaces permit communication between the CPU and the outside world. These interface circuits transfer data between the CPU and external devices. Apart from providing this link between the CPU and peripheral units, the interface circuits also convert the external data into a form usably by the microcomputer.

18.1.4 Data Bus, Address Bus, and Control Bus

Microcomputers are bus-oriented control units. Information is communicated among various parts of the system through the data bus, the address bus, and the control bus. In a microcomputer system using 8-bit microprocessors, there are typically 8 lines for the data bus, 16 lines for the address bus, and 6 lines for the control bus.

18.1.5 Program or Computer Software

The microprocessor can carry out various tasks only if it is provided with a set of instructions, called a *program*. The microprocessor, however, understands only binary codes and operates on binary instructions of 'ones' and 'zeros'. Writing a program in binary codes is very tedious and prone to error. A programmer, therefore, uses a convenient language to write the instructions. The most basic of the languages for a microprocessor is called *assembly language* in which symbolic (or mnemonic) codes are used. Every microprocessor has its own assembly language. In this symbolic coding system, an instruction is represented by a group of three or four letters chosen so as to suggest the function of the instruction. The symbolic coding of instructions cannot be understood by the microprocessor, which acts only on binary information. A program, called assembler program, converts the symbolic program or the source-program into a *binary program*, also called the object program, which the microprocessor can execute.

If the program is changeable and re-loadable, it is called *software*. The program that is not changeable by the user and is held in ROM is known as *firmware*. The electronics and circuitry of the microcomputer system are called *hardware*. The greatest advantage of microcomputer control of a drive system is the flexibility. The drive characteristics can be changed or new functions can be added in the future simply by changing the computer software (or firmware) with practically no change or minimal change to the hardware.

18.2 PRINCIPLE OF MICROPROCESSOR BASED SYSTEM

An instrumentation or process control system comprises a number of components which together perform measurement or control of one or more physical parameters. Functionally, the following three operations are performed:

1. Monitoring of one or more physical parameters and acquisition on relevant data.
2. Processing of the acquired data.
3. Controlling of the physical parameters by generating suitable control signals.

Most of the physical parameter of interest, i.e. speed, temperature, pressure level, firing angle are non-electrical in nature. The non-electrical quantities have to be converted into electrical signals with the help of devices such as transducers. Moreover, the electrical signals generated by the transducers are usually analog in form. Therefore, these analog electrical signals have to be converted into digital form. Extra hardware known as analog to digital converters (ADC) have to be used for this purpose. In many situations, process control variables are monitored by using opto-isolators, so that the noise and transients of an industrial environment do not disturb the processing elements.

On the other hand, analog signals have to be generated to control various physical parameters. The hardware which is used to convert digital data into equivalent analog form is known as the digital to analog converter (DAC). In many situations, relays have to be used to control process variables of high voltage or currents.

To realize an instrumentation, three major elements are required, namely:

1. Input devices
2. Signal processing devices
3. Output devices

The function of signal processing devices is performed by MPU and necessary input and output devices are to be interfaced to it. Block diagram of a typical process control system is shown in Fig. 18.2.

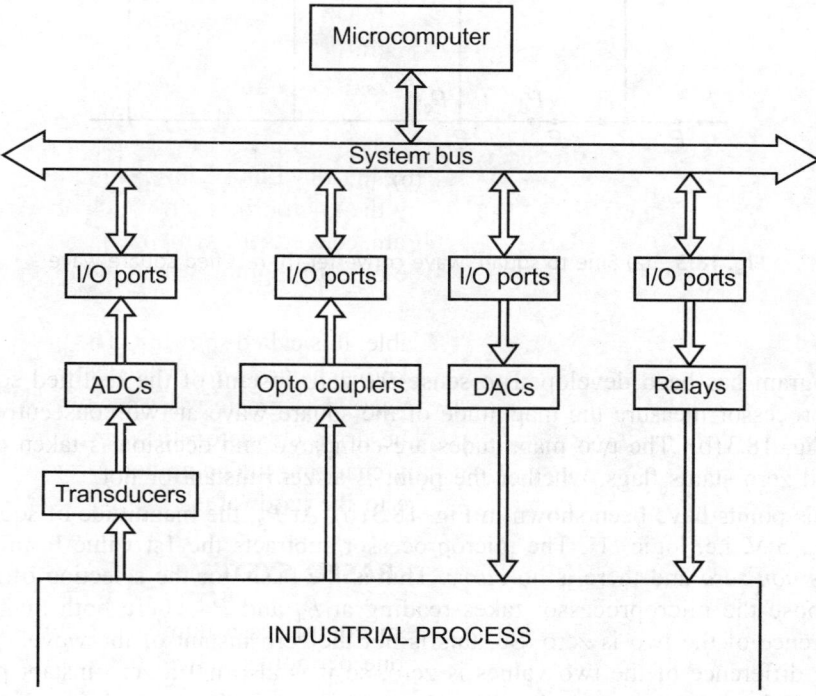

Fig. 18.2 Block diagram of process control system

18.3 MICROPROCESSOR BASED INDUSTRIAL DRIVES APPLICATIONS

18.3.1 Measurement of Electrical Quantities Like Frequency

To measure the frequency of a signal, the time period for half cycle is measured which is inversely proportional to the frequency. A sinusoidal signal is converted into square wave using a voltage comparator LM311 or operational amplifier LM747 as shown in Fig. 18.3 (a).

A diode is used to rectify the output signal. A potential divider is used to reduce the magnitude to 5 volts.

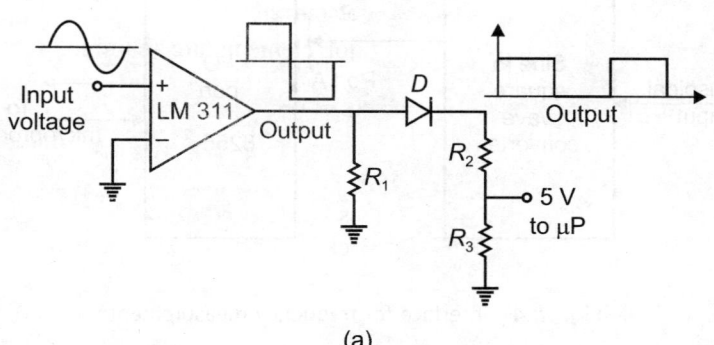

(a)

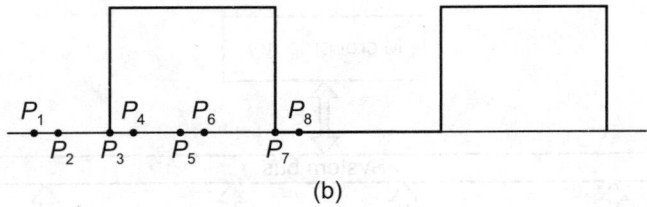

(b)

Fig. 18.3 (a) Sine to square wave converter (b) rectified square wave

A program has been developed to sense the zero instant of the rectified square wave. The microprocessor measure the magnitude of the square wave at two consecutive points as shown in Fig. 18.3(b). The two magnitudes are compared and decision is taken on the basis of carry and zero status flags, whether the point is at zero instant or not.

Various points have been shown in Fig. 18.3(b). At P_3, the magnitude of square wave is zero and P_4, 5 V, i.e. logic '1'. The microprocessor subtracts the 1st value from the 2nd, so the result is *non-zero* and there is no *carry*. This is the basis for the selection of zero instant point. Suppose the microprocessor takes reading at P_1 and P_2 where both magnitudes are zero. Difference of the two is zero. So, this is not the zero instant of the wave. At points P_5 and P_6, the difference of the two values is zero, so it is also not a zero instant point. At P_7 and P_8, the difference is non-zero but there is carry. So, it is the end point of the half-square wave.

As soon as the zero instant point is detected, the microprocessor initiates a register pair to count the number how many times the loop is executed. The microprocessor reads the magnitude of the square wave again and again and moves in the loop. It crosses the loop when the magnitude of the square wave becomes zero. Thus, the time for half cycle is measured. The count can be compared with the stored numbers in a look-up table and the frequency can be displayed. The count which is inversely proportional to the frequency of the input signal can be used for further processing and control as desired. An interfacing circuitry is shown in Fig. 18.4. The program flow chart is shown in Fig. 18.5. The port A is input. Control word is 98H.

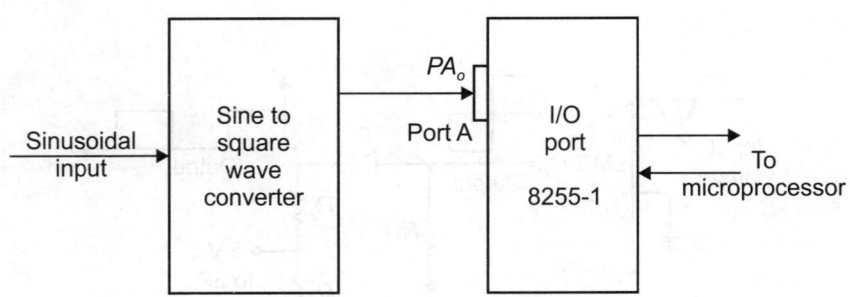

Fig. 18.4 Interface for frequency measurement

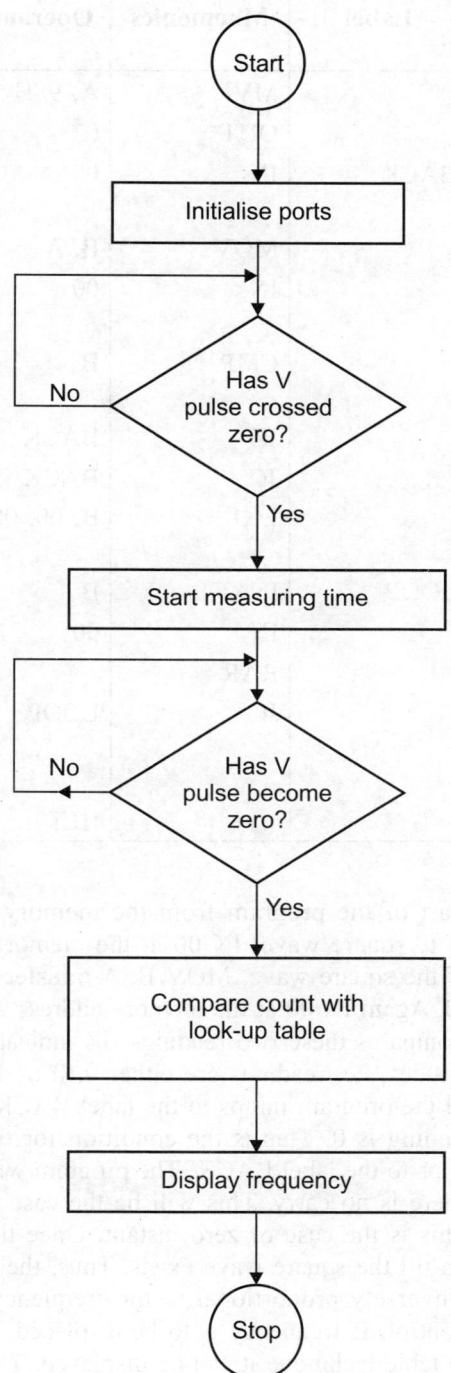

Fig. 18.5 Program of flow chart for frequency measurement

PROGRAM

Memory address	Machine codes	Label	Mnemonics	Operands	Comments
2000	3E, 98		MVI	A, 98H	Get control word.
2002	D3, 03		OUT	03	Initialise ports.
2004	DB, 00	BACK	IN	00	Read voltage pulse at port A.
2006	47		MOV	B, A	
2007	DB, 00		IN	00	Read voltage pulse again at port A.
2009	B8		CMP	B	Compare two readings.
200A	CA, 04, 20		JZ	BACK	
200D	DA, 04, 20		JC	BACK	
2010	01, 00, 00		LXI	B, 00, 00	Initialise B-C pair for counting.
2013	03	LOOP	INX	B	
2014	DB, 00		IN	00	Read voltage pulse.
2016	IF		RAR		
2017	DA, 13, 20		JC	LOOP	Check whether V has become zero. NO, go to LOOP.
201A	76		HLT		Stop.

In this program the part of the program from the memory address 2004 to 200F is to detect the zero instant of te square wave. IN 00 at the memory location 2004 is the 1st reading of the magnitude of the square wave. MOV B, A transfers the 1st reading from the accumulator to the register B. Again IN 00 at the memory address 2007 takes the 2nd reading of the magnitude. CMP B compares these two readings. JZ indicates the condition that both readings are of equal magnitudes, i.e. readings are either 0, 0 or 1, 1. Therefore, it is not a condition of zero instant and the program jumps to the label BACK. JC indicates that the 1st reading is 1 and the 2nd reading is 0. This is the condition for the end of rectified square wave. So, program again jumps to the label BACK. The program will move further only when the result is non-zero and there is no carry. This will be the case when the 1st reading is 0 and the 2nd reading is 1. This is the case of zero instant. Once the zero instant is detected the program moves in a loop till the square wave exists. Thus, the time period for half cycle is measured. The count is inversely proportional to the frequency. The count can be used for further processing and control. If frequency is to be displaced 7-segment displays can be interfaced and using loop-up table technique it can be displayed. The program given above is only for the frequency measurement. For its display, the program has to be extended.

Frequency measurement using SID line: The sinusoidal wave is converted into square wave using op. amp. The output is reduced to 5 V and applied to SID line of a microprocessor. The execution of RIM instruction reads the status of SID line and stores it in the 7th bit of the accumulator. It also reads status of interrupt MASKS and stores them in other bits of the accumulator. At present our interest is in the status of SID line, i.e. content of the 7th bit of the accumulator. A program to measure frequency is given below. Program from the memory location 2400 to 240D detects the rising edge of the square wave. Once it is detected H–L pair is initialised for counting. The program checks how long the wave remains high in one cycle. In other words, it measures time for half cycle because the wave remains high only for half cycle; for other half cycle it goes low. The time is given in terms of counts which is proportional to the frequency of the wave. The count is stored in H–L pair. Frequency can be displayed using look-up table or can be computed and then displayed. The counts in H–L pair can be used for control which depends on frequency.

18.3.2 Multiple Digit Display

Small number of 7-segment LED displays are interfaced to microprocessor through BCD to 7-segment decoder/driver and I/O ports. For example, to display 10 digits, 10 decoder/ driver and five 8-bit I/O ports are required. To reduce the number of interfacing components a technique known as multiplexing is used when large number of digits or alphanumeric characters are to be displayed.

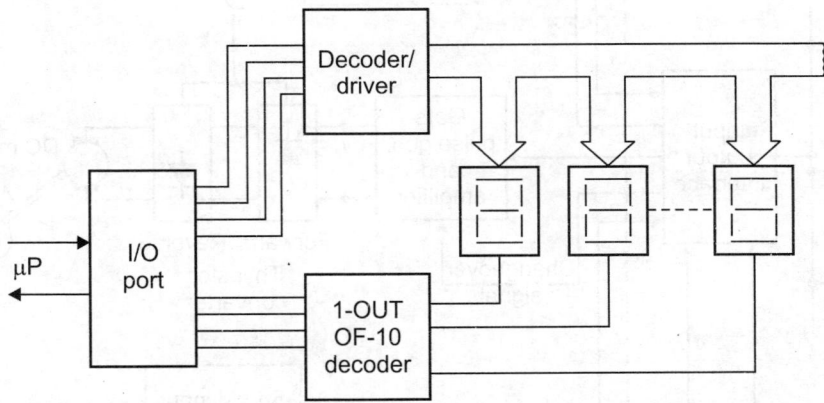

Fig. 18.6 Multiplexed 7-segment displays

A schematic diagram for multidigit display is shown in Fig. 18.6. To display a digit or character, the 7-segment code is sent to the decoder/driver through 4 lines of the output port. The ground point of the 7-segment display is connected to one of the output lines of a 1 out of 10 decoder, 4 input lines of which are connected to remaining 4 lines of the output ports. The microprocessor also sends a low signal to the ground terminal of the 7-segment display through out of 10 decoder. Each 7-segment LED is turned on and off in a sequence and the process is repeated continuously. In multiplexed technique only one digit is displayed at a time. Due to persistence of vision one can see the desired number on LEDs. A single decoder/ driver has been shown in the figure. The 8857 of National Semi-conductor can be used for the purpose. It can drive up to 10 7-segment display. It is logically equivalent to 7447 decoder/ driver but it produces higher current and does not require current limiting resistors. The 8857 are also designed for common cathode type LEDs.

The Intel 8279 has all necessary circuits to operate multidigit multiplexed LED display. This is used in a microprocessor-kit for multidigit display. It can control up to 16 LEDs to operate in multiplexed mode. In addition to multidigit display it also acts as an keyboard encoder for an 8 × 8 scanned keyboard.

18.3.3 Microprocessor Based Control of Reversing DC Drives

The work involves both hardware and software design. Hardware design is considered first followed by the software system design High-performance industrial drives, such as the reversing drive used for rolling mills, require high precision, fast response, and other complex control. The analog control scheme has been generally used in such drives with a fair degree of success. However, microcomputer control promises flexibility, improved performance, and economic viability. Figure 18.7 shows a simplified block diagram representation of a reversing dc drive using a microprocessor based control system.

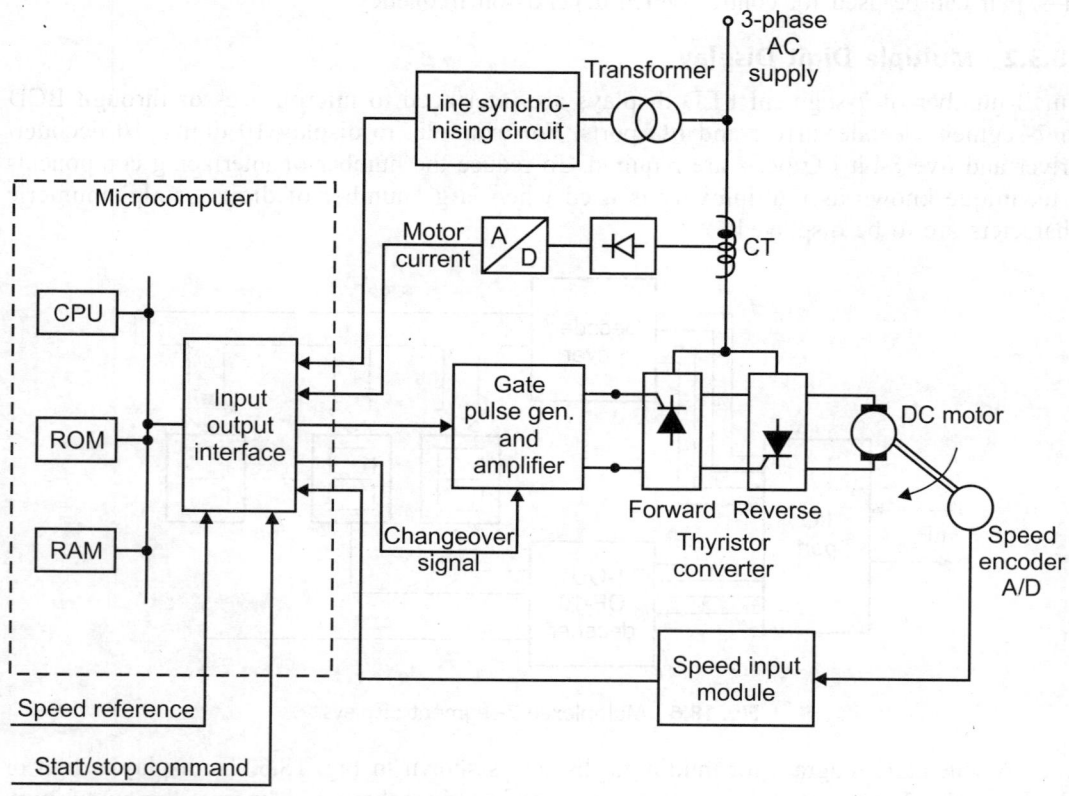

Fig. 18.7 Block diagram of microprocessor based system for reversing of DC drive

In the hardware system design, first consider the I/O devices needed for realising the system. A suitable transducer has to be used to convert speed information into equivalent analog electrical quantity. The speed information can be fed into the microcomputer using either a dc tachometer and an A/D (analog to digital) converter or by a digital tachometer and a digital counter. The motor current data is usually fed into the computer through a fast A/D converter. A synchronizing circuit interface is required so that the microprocessor can synchronise the generation of the firing pulse data with the supply line frequency. The gate pulse generator is shown as receiving a firing signal from the microprocessor. This generator could be incorporated as part of the microcomputer with gate pulse amplifiers as the only external elements. However, at present most microprocessors are not fast enough to allow the generation of gate pulses timed by software. Instead, data representing a required firing angle is fed to an external logic circuit, which generates the firing pulses in the proper sequence.

18.3.3.1 *Software Design*

Based on the hardware system design, the necessary software can be developed to implement the desired operation of speed monitoring and control. The software should perform the following operations. A set of instructions (i.e. a program) is stored in the memory, and those instructions are executed by the microprocessor for proper functioning of the drive. A typical program flow chart for this drive system is shown in Fig. 18.8. When the program starts, the first operation is to initialize the internal registers and the interface circuits. The microprocessor is now ready to receive the starting command. After receiving the start signal it begins executing the program in a continuous cyclic manner. The sequence of instructions allows the computer to process data for speed regulation, current regulation, and changeover reversal operation. A flow chart of these operations in the proper sequence is shown in Fig. 18.8.

The stability of the system depends partly on the operating speed of the microprocessor, but mainly on how the standard analog controller scheme (i.e. proportional or proportional-integral) is implemented in terms of soft-ware. Once the input-output hardware has been designed, a desired response can be obtained by simply changing the constants in the software representation of the system control equations.

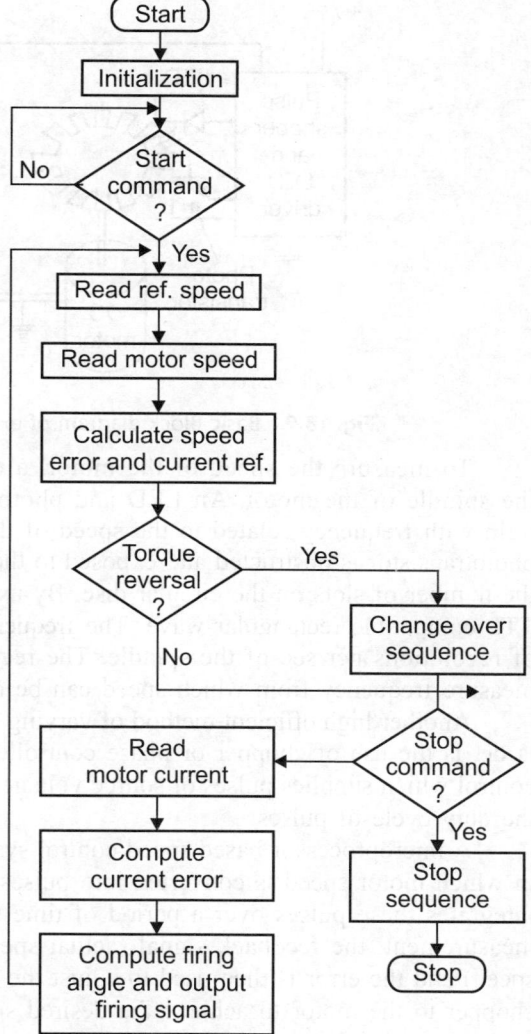

Fig. 18.8 Flow chart for reversing of DC drive

18.3.4 Microprocessor Based Speed Controller of DC Motor

In this controller, speed is maintained at specified value and displayed on LED, with a provision to set the speed using keyboard.

Hardware Design: The generalised equation for dc motor is

$$N = \frac{V_a - I_a R_a}{K\Phi}$$

Speed can be controlled by applying variable dc voltage across the armature coil of the motor thereby producing a variable armature current. The variable dc voltage is generated with the help of digital to analog converter (DAC). However, current amplifier is needed to drive the motor. The DAC is interfaced through I/O port. The desired speed is communicated through the keyboard and both the desired and measured speeds can be displayed on the LED display. The basic block diagram is shown in Fig. 18.9.

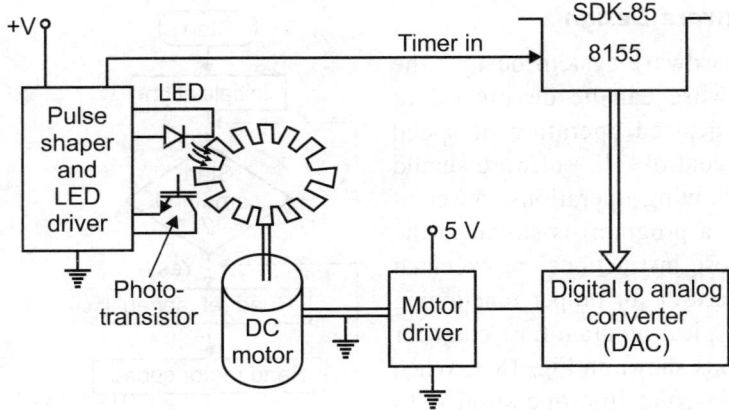

Fig. 18.9 Basic Block diagram of extra hardware for DC motor controller

To measure the speed of the motor, a circular disc with several slots is attached to the spindle of the motor. An LED and phototransistor assembly is used to generate pulse train with frequency related to the speed of the motor. For each rotation of the spindle, the phototransistor is obstructed and exposed to the light emitted by the LED. n times, where n is the number of slots on the circular disc. By using a suitable pulse shaper, it is converted into TTL compatible rectangular wave. The frequency of this wave is n times that of the number of revolutions per sec of the spindle. The rectangular wave is applied to a timer/counter to measure frequency from which speed can be determined.

Another high efficient method of varying the armature voltage, when the available supply is dc, is the use of chopper or phase controlled rectifier. A chopper is basically an ON-OFF control which supplies pulses of source voltage whose average voltage is controlled by varying the duty cycle of pulses.

A microprocessor based speed control system using dc chopper is depicted in Fig. 18.10 in which motor speed is converted into pulses by an optical transducer. The micro-processor integrates these pulses over a period of time to compute the actual rpm of the motor. After measurement, the feedback signal (actual speed) is compared with reference input (desired speed) and the error is then used to adjust the duty cycle of the voltage pulses applied by the chopper to the motor to achieve the desired speed. The output pulses of the tachometer in a given time give the speed of the shaft. These pulses are processed digitally to obtain required simplification. These are fed to microprocessor using proper interfaces. The measurement of speed, providing the speed error and control of chopper circuit are done by the microprocessor with suitable software design.

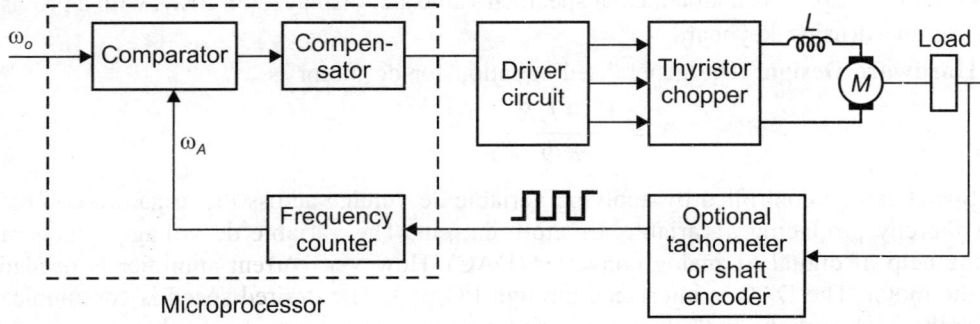

Fig. 18.10 Microprocessor based control system using dc chopper

18.3.4.1 *Software Design*

Software has to be designed to implement the following sequence of operations:

Step 1: Measure the motor speed, compare it with the desired speed to produce error.

Step 2: Compensator changes its output as a function of error.

Step 3: Duty cycle of the chopper is varied depending on the information transmitted from the compensator.

Step 4: Go to Step 1.

This cycle is implemented over and over to keep the motor speed at desired speed. A delay period could be included between steps 3 and 4 to allow the motor to respond to any change in the duty cycle made at the end of step 3. For accurate measurement of speed, the integration of pulses must be carried out over a fairly long interval of time.

Step 5: Convert the measured rpm from binary to BCD and output to the display unit.

Step 6: Go to Step 1.

18.3.5 Microprocessor Based Temperature Monitoring and Control System

This system can be used to control the temperature of water bath by controlling a heater ON or OFF. This application involves both hardware system and software design.

18.3.5.1 *Hardware System Design*

A suitable transducer has to be used to convert temperature into equivalent analog electrical quantity. The analog signal has to be converted into digital form by a suitable Analog to Digital Converter (ADC). A relay with necessary driver has to be used to switch the heater coil ON or OFF. To display temperature, the two digit segment 7-segment LED will be required. To set temperature, 8-bit switches can be used. All these I/O devices, ADC, relay, LED displays, and switches can be interfaced to the microcomputer through port lines. In EPROM of suitable size can be used to store the software and finally MPU will be necessary to act as the overall controller of the system. A block diagram of the hardware system is shown in Fig. 18.11.

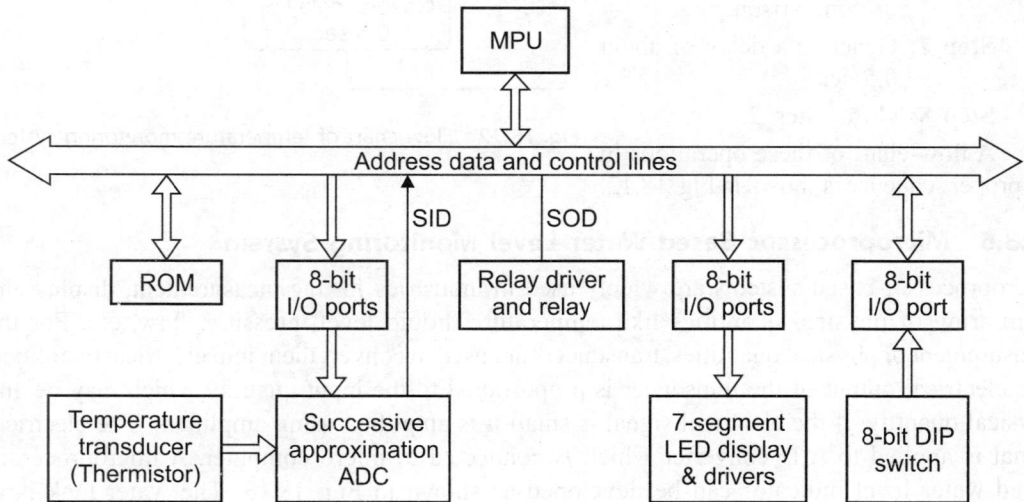

Fig. 18.11 Block diagram of microprocessor based temperature monitoring system

18.3.5.2 *Software Design*

Based on the hardware, the necessary software can be developed to implement the desired operation of temperature monitoring and control. These operations will be taken care in the software.

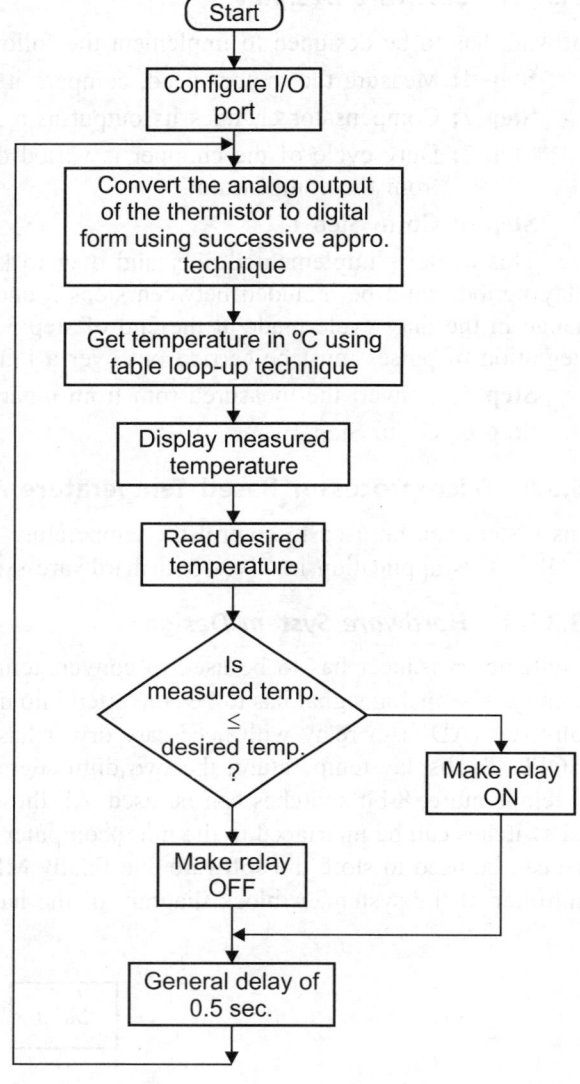

Fig. 18.12 Flow chart of temperature monitoring system

Step 1: Initialize the I/O ports of the system.

Step 2: Perform analog to digital conversion by successive approximation algorithm to convert the analog voltage developed across the thermistor into digital form.

Step 3: Get the temperature of the water bath using digital data.

Step 4: Display the measured temperature on the LED display.

Step 5: Read the desired temperature from the switch setting of the DIP switch.

Step 6: Compare the measured temperature with the desired temperature and make LOW or HIGH depending on the outcome of comparison.

Step 7: Generate a delay of about 0.5 sec.

Step 8: Go to Step 2.

A flow-chart of these operations in the proper sequence is shown in Fig. 18.12.

18.3.6 Microprocessor Based Water Level Monitoring System

Microprocessor-based systems are widely used in industries for the measurement, display and monitoring of physical quantities like temperature, liquid level, pressure, flow, etc. For the measurement of physical quantities, transducers are used to convert them into electrical quantities. The electrical output of the transducer is proportional to the input quantity which may be any physical quantity. If the electrical signal is small it is amplified using amplifiers. The electrical signal is applied to A/D converter which is connected to micro computer. A microprocessor based water level indicator can be developed as shown in Fig. 18.13. The water tank is at ground potential. Probes are connected to V_{dc} through resistors. When a probe is immersed in water it is at ground potential and its logic is 0. When it is not immersed in water, it is at

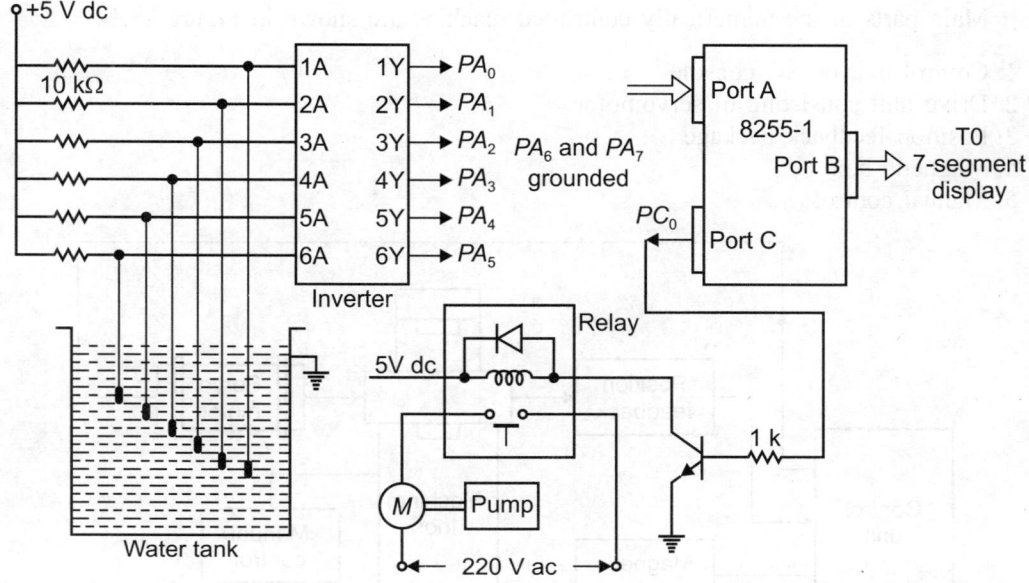

Fig. 18.13 Block diagram of microprocessor based water level monitoring system

5 V potential and its logic is 1. These probes are connected to an inverter which inverts the logic of the probes. After inversion the probe which is not immersed in water gives 0 logic and the probe immersed in water output logic 1. The outputs of the inverter are connected to Port A of 8255. In interface only six probes have been used. The 6 output points of the inverter are connected to 6 pins of Port A. The remaining 2 pins of Port A are grounded. The 2nd unit of 8255, has been used for this purpose. The microprocessor reads the binary logic corresponding to water level from the port, and displays it on 7-segment displays which are connected to Port B. The values of water levels at which probes are placed in the tank are stored in the memory in the form of look-up table. The microprocessor counts how many probes are immersed in water and determines the LSB of the memory address of look-up table. Suppose 1st, 2nd and 3rd probes are immersed in water. These probes are at logic 1. The microprocessor counts that 3 probes are immersed in water and it picks up the water level corresponding to memory location FD03 which is 30 cm. The values of water levels may be stored in metres or cms depending on the actual situation.

18.3.7 Computerized Numerically Controlled Machines (CNC Machine)

Machine tool may be classified as:

1. General Purpose
2. Automatic and
3. Numerically Controlled (N.C.)

Numerically controlled machines can be defined as machines controlled by number or a system in which actions are controlled by the direct insertions of numerical data at some point. The system must automatically interpret at least some portion of these data. In N.C. machine tools, the input information for controlling the machine tool motion is provided by means of magnetic tape or punched paper in a coded language.

Main parts of the numerically controlled machine are shown in Figure 18.14:

1. Control unit or NC console
2. Drive unit consisting of servomotor
3. Position feedback package
4. Magnetic box
5. Manual control

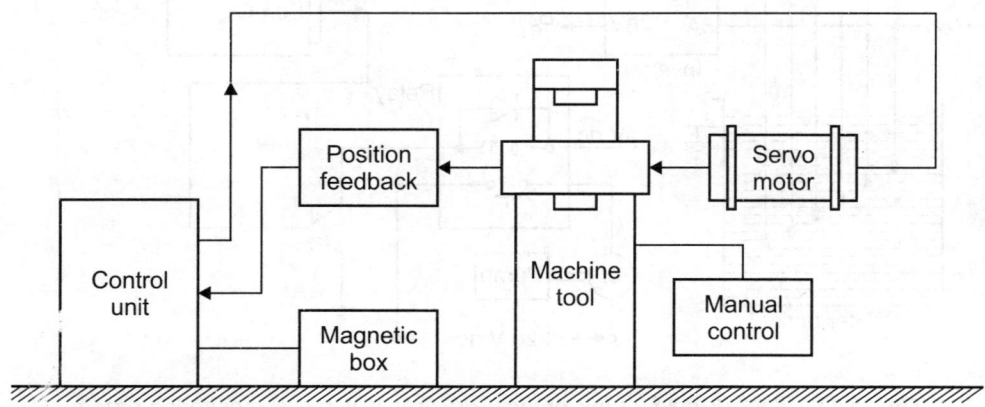

Fig. 18.14   Block diagram of CNC machine

In the control unit, the instruction for manufacturing a component are written in a coded language. These instructions are read and command signals are sent to the drive unit to control the motion of the job and also to the magnetic box to start and stop the motor selection of spindle speed, and actuation of tool etc. A feedback position transducer checks, whether the required length of travel is obtained to control the unit. Manual control helps the operator to perform some functions manually such as motor-start and stop, control of coolant supply, axes movement etc.

In computerised numerical control (CNC) for machine tools in which a microcomputer is included as an integral part of the control unit. Industrial robots were developed with CNC machines. In the closed loop, numerically controlled system of Fig. 18.15, the input to the control loop and the feedback signals may be a sequence of pulses, each pulse representing a unit e.g., 0.01 mm. The digital comparator correlates the two signals by means of digital to analog converter (DAC) outputs the position error which is used to drive dc servomotor. The feedback device is an encoder mounted on the other end of the lead screw.

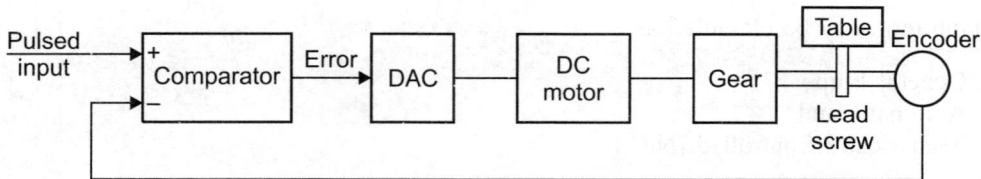

Fig. 18.15   Block diagram of closed loop of CNC machine

This is a device consisting of rotating disc divided into segments, which are alternately opaque and transparent. A photo cell and lamp are placed on the two sides of the disc. When the disc rotates, each change in light intensity provides a pulse. The rate of pulses per minute provided by encoder is proportional to the rpm of the lead screw.

18.3.8 Microprocessor Based Control of Firing Circuit of a Thyristor

A number of gate firing control schemes for phase-controlled converters have been described in previous chapters. Here typical scheme which is based on microprocess will be described. In this section, hardware and software are needed to implement the controller. The basic block diagram of the extra hardware is shown in Fig. 18.16 of a bridge-rectifier containing two thyristors and two diodes. Firing circuits of thyristors are to be controlled by microprocessor. Direct current (dc) flowing through the load can be controlled by controlling firing circuits of two thyristor of the bridge-rectifier. Bases of the transistors shown in the figure are connected to the ports of 8255. Emitter of the transistors is connected to a pulse transformer which is used to isolate microprocessor circuit from the bridge-rectifier circuit. Zero-cross detectors (not shown in the figure) are used to sense the positive going zero and negative going zero of the ac supply. A.P.T. is used to reduce ac supply to 3 volts. The input of zero-cross detectors is kept above 3 V_{ac}. The micro processor can sense the zero instant of the ac supply. Then a delay sub-routine is available in monitor's program is called. After the desired delay period a high pulse is sent to the gate of the thyristor. The transistor acts as a buffer which is selected to supply sufficient current to the gate needed to start firing. Again for the –ve half cycle of ac supply –ve going zero instant is sensed and delay subroutine is called. The firing is started by sending a high pulse by the microprocessor. The same delay is used for +ve as well as –ve half cycle. The process is repeated for every cycle of ac supply. This circuit can be used to control current in a load where such control is required. For example, to control temperature of an oven, the current flowing through the heating elements can be controlled by this method. For speed control of dc shunt motor current flowing through the armature can be controlled. The field circuit of the motor is supplied by a separate bridge-rectifier consisting of diodes which gives a fixed dc voltage. The gate firing control can be implemented by a single clip micro computer, which receives the converter voltage command from the feedback control loop and translates it into the firing angle control pulses for the thyristors.

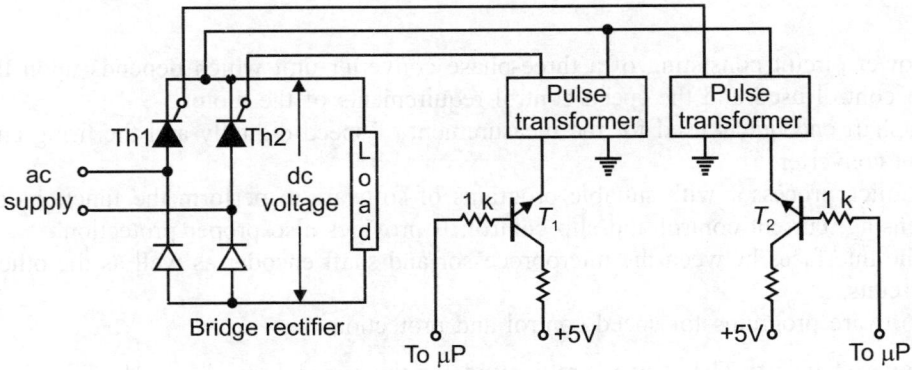

Fig. 18.16 Microprocessor based control of firing circuit of thyristor

18.3.9 Microprocessor Based Speed Control of Three-phase Induction Motor

The development of thyristors and thyristor power converters capable of providing a variable voltage variable frequency supply has made it possible to control the speed of a three-phase induction motor over a wide range. The induction motor is controlled to operate at constant flux to have good steady state and dynamic performances which is achieved by operating the inductor motor with constant V/f. However, at low frequencies the effects of stators resistance become appreciable and the operating flux decreases. The resistance drop has to be compensated at very low frequencies. This can be done automatically as a function of load by making it proportional to slip frequency. The voltage control as the frequency is varied can be achieved either external to or in the inverter feeding the induction motor. In the former case the rectifier firing angle is changed to have the voltage control. PWM principle can be used to control the voltage in the inverter.

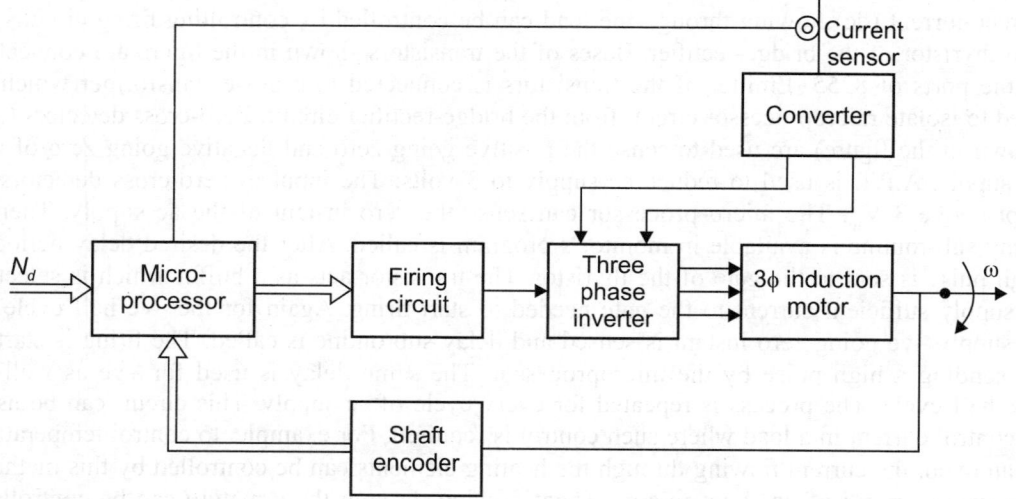

Fig. 18.17 Basic block diagram of a microprocessor speed control system

A microprocessor speed control system can be developed to perform the above functions. The basic block diagram of the system is shown in Fig. 18.17. The main constituents of the system are:

1. Power circuit consisting of a three-phase converter unit which depends upon the type of control used and the speed control requirements of the motor.
2. A shaft encoder is used for the measurement of speed digitally and the firing circuit of the converter.
3. A microprocessor with suitable programs of software to perform the functions of error sensing, current control and slip control. It provides also proper protection.
4. The interfaces between the microprocessor and shaft encoder as well as the other firing circuits.
5. Software programs for speed control and protection.

Power Circuit: The power circuit used for the speed control of induction motor is a converter which rectifies ac to dc using a rectifier. This dc is inverted to ac of desired frequency using VSI or CSI inverter. Both employ forced commutation. Typical circuit for an induction motor is shown in Fig. 18.18.

A shaft encoder is a disc having a definite number of slots at its periphery and mounted on the shaft of the motor. Using a light source and phototransistors a pulse train can be generated when the shaft rotates. These pulses measured in a definite amount of time give the value of the shaft speed. These pulses have to be shaped and amplified. They are fed to the microprocessor where they are counted using a pulse counting program. A proper interface circuitry should be used before they are fed to the microprocessor.

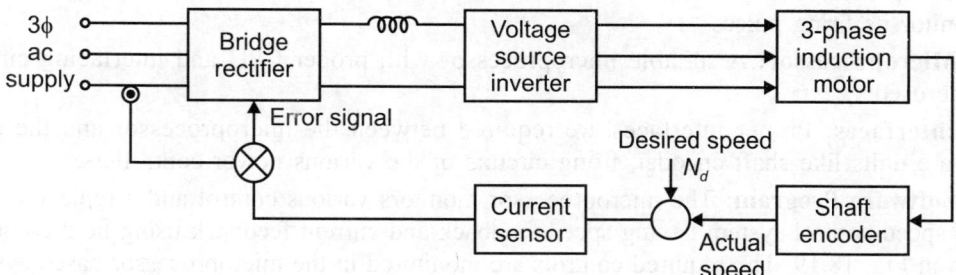

Fig. 18.18 Typical circuit for an induction motor

The inverter thyristors have to be fired by means of trigger pulses. These are provided by means of a firing circuit. This firing circuit is activated by the microprocessor programmed with suitable software. A proper interfacing is required here also for connecting the microprocessor to the firing circuit.

The microprocessor provides the various controls and serves as a control unit. A proper microprocessor with all the interfaces and required memories has to be selected.

The actual speed measured by means of a shaft encoder and speed measuring programme is compared with the desired speed which can be stored in the registers. Suitable programs produce a speed error using the above comparison. The frequency of the inverter is corrected using this error signal. The thyristors are turned ON and OFF on a particular sequence.

18.3.9.1 *Software Program*

These can be implemented by means of a suitably designed software program which may contain the following steps:

Step 1: Contains the necessary logic for choosing the mode by enabling proper interrupt.

Step 2: Necessary initialization.

Step 3: Speed counting and error generation routine.

Step 4: Frequency correction routine.

Step 5: Thyristor firing routine.

The microprocessor can be used to perform the protection also. It should be programmed to diagnose the fault and provide proper protection.

18.3.10 Microprocessor Based Speed Control of Synchronous Motors

A self-controlled current source inverter fed synchronous motor utilising machine voltage commutation is becoming very popular. A microprocessor based speed control system for a synchronous motors consists of:

1. Power circuit which is the main hardware.
2. Microprocessor with suitable programs of software.
3. Suitable interfaces with μP and shaft encoder and firing circuits.
4. Software design for speed controls and protection.

Power Circuit: The power circuit is a dc link converter. A combination of a three-phase bridge converter and current source inverter interconnected by means of smoothing inductance. Each of the converters has a firing circuit of its own. The firing pulses to these converters are issued by the microprocessor using the information of rotor position and dc link current. The optical shaft encoder used here gives the information of rotor position besides speed. The microprocessor may be used to control the power converter used in the field circuit and to monitor its firing pulses.

Microprocessor: A suitable microprocessor with proper CPU and interfacing circuits may be used.

Interfaces: Proper interfaces are required between the microprocessor and the other hardware units like shaft encoder, firing circuits of the various power controllers.

Software Program: The microprocessor monitors various control and a typical scheme of the speed control system having speed feedback and current feedback using field control is shown in Fig. 18.19. The required controls are monitored in the microprocessor based system. The various schemes for current regulation and speed regulation are executed by means of proper interrupt signals generated by the real time clocks. The motor speed is sensed by means of a shaft encoder is compared with the desired speed stored in the memory registers of the microprocessor. A speed error is developed and is processed by the speed controller which actuates the inner current loop by sending a link current demand. The saturation incorporated in the controller limits the current. The current regulating routine is also initiated by an interrupt signal. The reference value of the current is provided by the speed controller and this is compared with the actual value of current. The current error is processed to change the firing angle command to the rectifier. So, the software design takes care of providing proper interrupt signals to achieve desired speed.

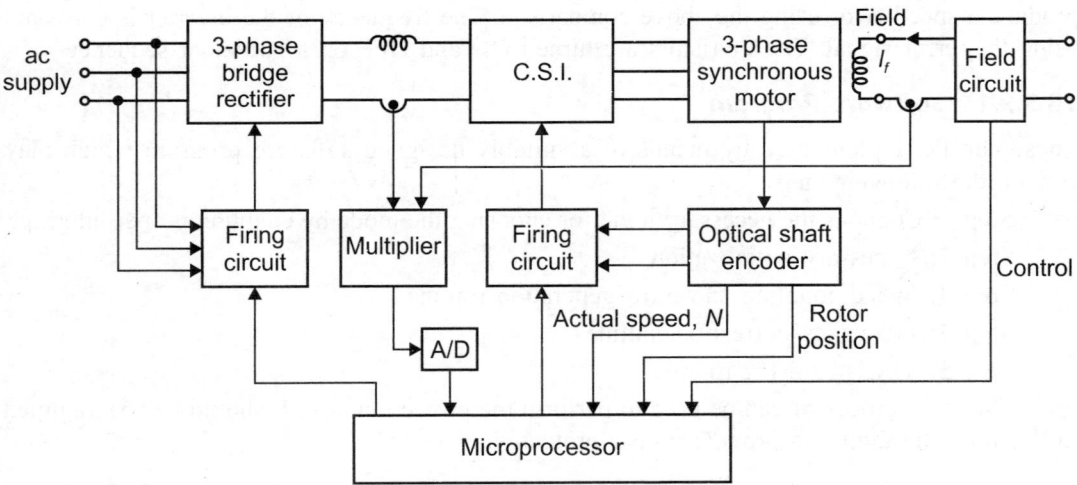

Fig. 18.19 A typical scheme of the speed control system having speed and current feedback

A microprocessor can provide independent control of these quantities to provide the desired features at all operating conditions. The microprocessor can be programmed so that the features of motoring and regeneration in either direction of rotation are available. All these tasks are implemented by suitably designed software for the microprocessor similar for dc machines and induction motor.

18.3.11 Microprocessor Based Process Control System

This process control system involves both hardware and software design. Hardware system design is considered first followed by software system design. In hardware system design,

I/O devices are needed for realising the system. Figure 18.20 shows a simple process control system in which the process parameters are converted to electrical signals and are transformed to digital form using A/D converters. These are interfaced with the rest of the system through input ports which are connected to a system bus.

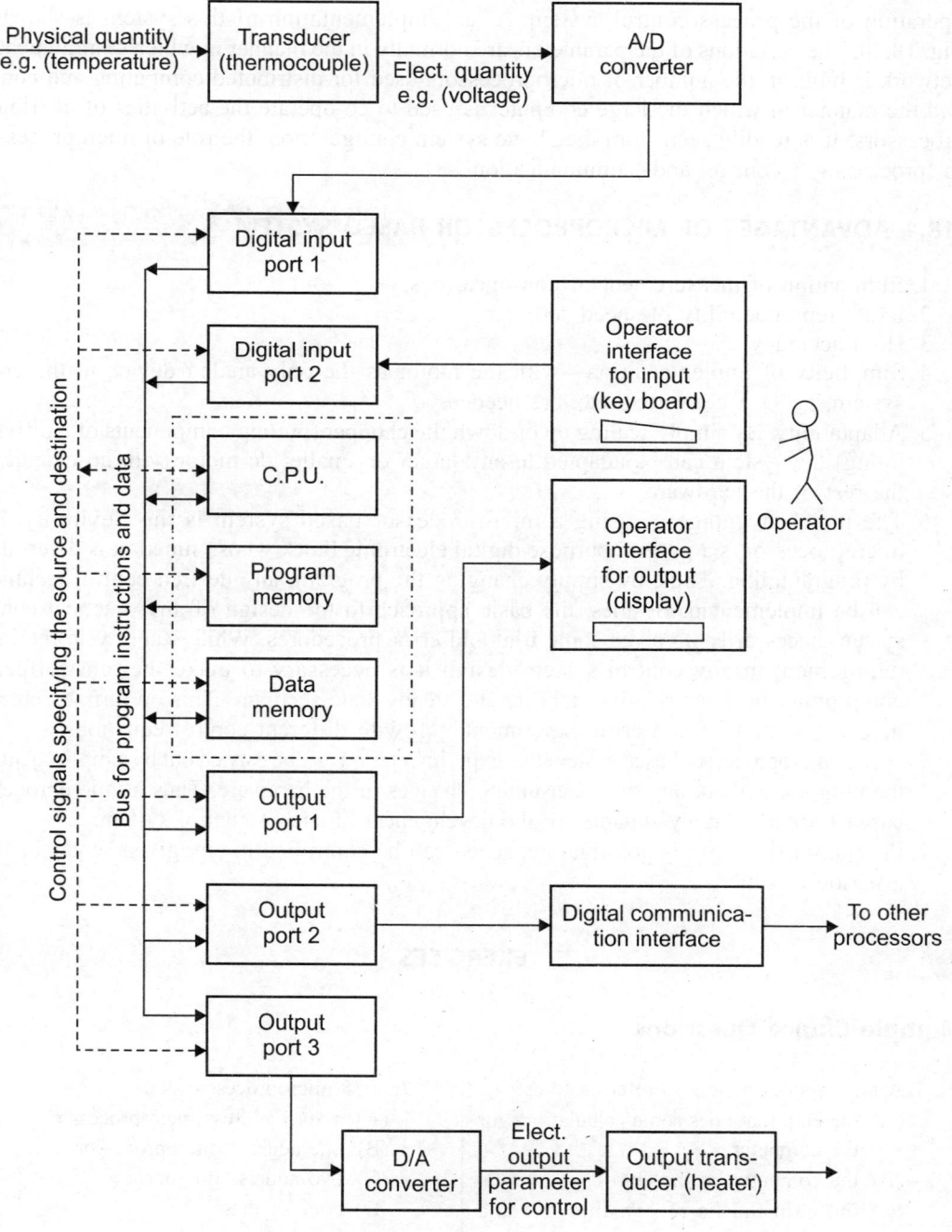

Fig. 18.20 Block diagram of process control system

The digital input port 1 as shown can be repeated for each process parameter. Another input port is also provided for keyboard entry. The microprocessor is interfaced to the system bus. One or more output ports are normally provided for displaying parameters for operator's feedback and data logging. Another output port normally provides communication with a large computer. One or more output ports provide signals to control the physical parameters.

Based on the hardware, the necessary software can be developed to implement the desired operation of the process control system. A real implementation of this system is shown in Fig. 18.20. The variations of the parameters are normally in the manner in which communication network is built, in the number of microprocessors used for distributed computing and control and the manner in which the large computer is used to co-operate the activities of distributed processors. It is readily seen from the above system configuration, the role of microprocessors in 'processing', 'control' and 'communication'.

18.4 ADVANTAGES OF MICROPROCESSOR BASED SYSTEM

1. Elimination of measurement of non-linearities.
2. Exact reproducibility of speed setting.
3. High accuracy.
4. Simplicity of implementation—with the motor as the only analog device in the entire system, no D/A converters etc. are needed.
5. Adaptability. By simply scaling up or down the chopper (putting components of a different rating) the system can be adapted to any larger or smaller dc motor with no changes in the rest of the hardware.
6. The main advantage of using a microprocessor based system is the flexibility. The microprocessor is a general purpose digital electronic block whose function is determined by programming. Thus, by simply changing the program, any desired control technique can be implemented. Besides, the basic approach to the design of any practical control system necessarily involves some trial and error procedures. While stability in the basic requirement in any control system design it is necessary to make the most effective compromise between relative stability and steady-state accuracy. This, in turn, essentially involves some trial and error experimentation with different control equations.

   In a microprocessor based system this experimentation can be carried out by simply changing the program without any time consuming changes in the hardware. Thus, a microprocessor based system is highly suitable for the development of such a control system.
7. Programmed control is possible, i.e. speed can be controlled in any given sequence over a period of time.

EXERCISES

Multiple Choice Questions

1. A microprocessor is also referred to as:
   (A) the chip that does some calculations for the computer
   (B) the computer on the chip
   (C) the chip that is responsible for data transfer
   (D) none of these

2. 8085 microprocessor is a:
   (A) a zero address microprocessor
   (B) one address microprocessor
   (C) two address microprocessor
   (D) none of these

3. As compared to 16-bit microprocessor, 8-bit microprocessor are limited in:

(A) speed

(B) directly addressable memory

(C) data handling capability

(D) all of the above

4. Microprocessors are programmed using:

(A) assembly language

(B) Pascal

(C) assembly language or some suitable high level language

(D) high level language such as C, C^{++}, Pascal

5. With reference to 8085, ANA R/M is:

(A) a logical instruction

(B) an arithmetic instruction

(C) data transfer instruction

(D) control instruction

6. The synchronization between microprocessor and memory is done by:

(A) ALE signal

(B) HOLD signal

(C) READY signal

(D) none of these

7. Two operands can be checked for equality using:

(A) OR-operation

(B) AND-operation

(C) EX-OR operation

(D) none of these

8. Microprocessor is called an n-bit micro-processor depending on:

(A) register's length

(B) size of internal data bus

(C) size of external data bus

(D) none of these

9. Programme counter is used to:

(A) store address of next instruction to be executed

(B) store temporary data to be used in arithmetic operation

(C) store the status of microprocessor

(D) none of these

10. Stack memory is used to:

(A) provide additional memory to the base memory

(B) save return addresses of sub routine

(C) save the status of the microprocessor

(D) none of these

ANSWER KEY

1. (B)   2. (B)   3. (D)   4. (C)   5. (A)   6. (C)   7. (C)   8. (B)
9. (A)   10. (B)

Review Questions

1. Show interface connections to measure and control temperature of several furnaces employing a microprocessor based system.

2. Show with a neat sketch and suitable interface a microprocessor-based scheme to measure, display and control water level.

3. Discuss how to measure, display and control speed of dc motor using microprocessor.

4. Show interface connections to control firing circuit of thyristors which are placed in a bridge rectifier.

5. Discuss a microprocessor based technique to measure resistance and reactance of an electric circuit.

6. Show interface connections to measure strain using a microprocessor based scheme. Discuss its operating principle.

MCQ for Semiconductor Devices and Commutation Techniques from Various Examinations

1. The thermal resistance between the body of a power semiconductor device and the ambient is expressed as:
   (A) voltage across the device divided by current through the device.
   (B) average power dissipated in the device divided by the square of the rms current in the device.
   (C) average power dissipated in the device divided by the temperature difference from body to ambient.
   (D) temperature difference from body to ambient divided by average power dissipated in the device.   **[Gate 1993]**

2. Figure show two thyristors, each rated 500 A (continuous) sharing a load current. Current through thyristor *y* is 120 A. The current through tyristor *x* will be nearly:

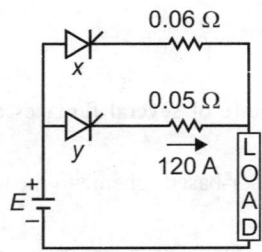

   (A) 10 A   (B) 25 A
   (C) 50 A   (D) 100 A
   [Gate 1995]

3. The Triac can be used in:
   (A) inverter
   (B) rectifier
   (C) multiquadrant chopper
   (D) ac voltage regulator   **[Gate 1996]**

4. Which semiconductor power device out of the following is not a current triggered device?

(A) Thyristor   (B) G.T.O.
(C) Triac   (D) MOSFET
          [Gate 1996]

5. Which of the following does not cause permanent damage of an SCR?
   (A) High current
   (B) High rate of rise of current
   (C) High temperature rise
   (D) High rate of rise of voltage **[Gate 1996]**

6. The MOSFET switch in its ON state may be considered equivalent to:
   (A) resistor   (B) inductor
   (C) capacitor   (D) battery
          [Gate 1998]

7. The uncontrolled electronic switch employed in power electronic converters is:
   (A) thyristor
   (B) bipolar junction transistor
   (C) diode
   (D) MOSFET   **[Gate 1998]**

8. In a commutation circuit employed to turn-off an SCR, satisfactory turn-off is obtained when:
   (A) circuit turn-off time < device turn-off time
   (B) circuit turn-off time > device turn-off time
   (C) circuit time constant > device turn-off time
   (D) circuit time constant < device turn-off time   **[Gate 1998]**

9. The latching current in the below circuit is 4 mA. The minimum width of the gate pulse required to turn on the thyristor is:

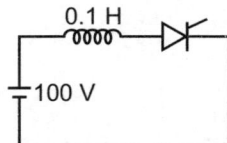

(A) 6 μs　　　　　(B) 4 μs
(C) 2 μs　　　　　(B) 1 μs **[IES 2001]**

10. Triac cannot be used in:
    (A) ac voltage regulators
    (B) cycloconverters
    (C) solid state type of switch
    (D) inverter　　　　　**[IES 2001]**

11. The snubber circuit is used in thyristor circuits for:
    (A) triggering　　　(B) *dv/dt* protection
    (C) *di/dt* protection　(D) phase shifting
    　　　　　　　　　　　　　[IES 2001]

12. It is preferable to use a train of pulse of high frequency for gate triggering of SCR in order to reduce:
    (A) *dv/dt* problem
    (B) *di/dt* problem
    (C) the size of the pulse transformer
    (D) the complexity of the firing circuit
    　　　　　　　　　　　　　[IES 2001]

13. Which one of the following is NOT the advantage of solid state switching of ac capacitors into ac supply over relay-based switching?
    (A) Low transients　(B) Low losses
    (C) Fast response　(D) Long life
    　　　　　　　　　　　　　[IES 2001]

14. The most suitable device for high frequency inversion in SMPS is:
    (A) BJT　　　　　(B) IGBT
    (C) MOSFET　　　(D) GTO **[IES 2001]**

15. When cathode of a thyristor is made more positive than its anode:
    (A) all the junctions are reverse biased
    (B) outer junctions are reverse biased and central one is forward biased
    (C) outer junctions are forward biased and central one is reverse biased
    (D) all the junctions are forward biased
    　　　　　　　　　　　　　[IES 2002]

16. The sharing of the voltages between thyristors operating in series is influenced by the:
    (A) *di/dt* capabilities
    (B) *dv/dt* capabilities
    (C) junction temperatures
    (D) static *V-I* characteristics and leakage currents　　　　　**[IES 2002]**

17. R-C snubber is used in parallel with the thyristor to:
    (A) reduce *dv/dt* across it
    (B) reduce *di/dt* through it
    (C) limit current through the thyristor
    (D) ensure is conduction after gate signal is removed　　　　　**[IES 2002]**

18. A thyristor controlled reactor is used to get:
    (A) variable resistance
    (B) variable capacitance
    (C) variable inductance
    (D) improved reactor power factor
    　　　　　　　　　　　　　[IES 2002]

19. In a switched-mode power supply (SMPS), after conversion of ac supply to a highly filtered dc voltage, a switching transistor is switched ON and OFF at a very high speed by a pulse width modulator (PWM) which generates very-high frequency square pulses. The frequency of the pulses is typically in the range of:
    (A) 100 Hz–200 Hz　(B) 500 Hz–1 kHz
    (C) 2 kHz–5 kHz　　(D) 20 kHz–50 kHz
    　　　　　　　　　　　　　[IES 2002]

20. Match List-I with List-II and select the correct answer:

List-I (Thyristors)		List-II (Symbols)
A.	Silicon-controlled rectifier (SCR)	1.
B.	Silicon-controlled switch (SCS)	2.
C.	Silicon-unilateral switch (SUS)	3.
D.	Light-activated SCR (LASCR)	4.

Codes:

	A	B	C	D
(A)	3	4	1	2
(B)	4	3	1	2
(C)	3	4	2	1
(D)	4	3	2	1

[IES 2003]

21. Turn-on and turn-off times of transistor depend on:
    (A) static characteristic
    (B) junction capacitances
    (C) current gain
    (D) none of the above **[IES 2003]**

22. Match List-I with List-II and select the correct answer using the codes given below the list:

List-I (Characteristic/Action)	List-II (Observations)
A. Turn-on time of thyristor	**1.** Depends on junction capacitances
B. Turn-on time of transistor	**2.** Is the sum of delay and rise times
C. Rate of rise of gate current in thyristor	**3.** Thyristor switches back to Off-state
D. If the gate pulse is removed, when the thyristor is carrying a current, less than latching current	**4.** Affect the delay time

Codes:

	A	B	C	D
(A)	2	1	4	3
(B)	1	2	3	4
(C)	2	1	3	4
(D)	1	2	4	3

[IES 2003]

23. Consider the following statement:

    Assertion (A): The critical rate of change of forward-voltage is the value of dv/dt, at which the device just goes into conduction without a gate pulse.

    Reason (R): Thyristors go to a state of conduction with the application of sharp rate of change of forward-voltage in the absence of gate pulse, even before the break-forward voltage limit is reached.

Of these statements:
(A) both (A) and (R) are true and (R) is the correct explanation of (A)
(B) both (A) and (R) are true but (R) is not the correct explanation of (A)
(C) (A) is true but (R) is false **[IES 2003]**

24. Figure shows a MOSFET with an integral body diode. It is employed as a power switching device in the ON and OFF states through appropriate control. The ON and OFF states of the switch are given on the V_{DS}-I_S plane by:

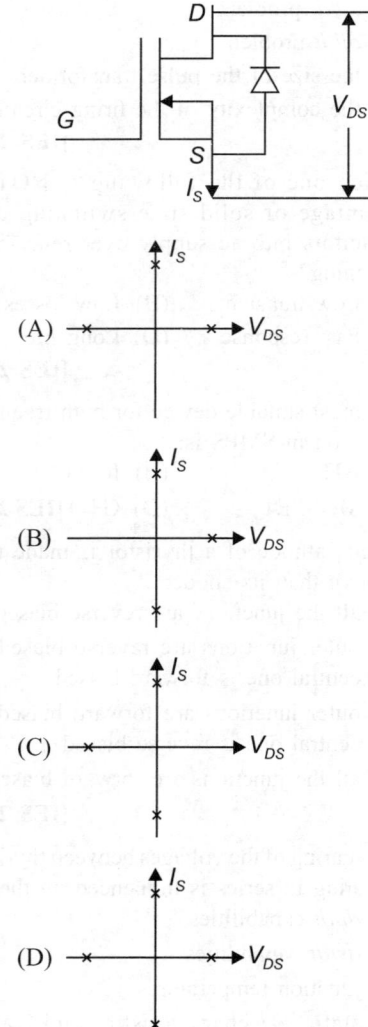

[Gate 2003]

25. Figure shows a thyristor with the standard terminations of anode (*A*), cathode (*K*), gate (*G*) and the different junctions named J1, J2 and J3. When the thyristor is turned on and conducting:

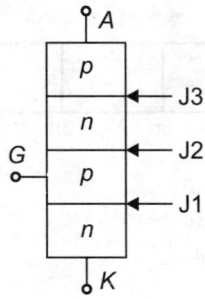

(A) J1 and J2 are forward biased and J3 is reversed biased

(B) J1 and J3 are forward biased and J2 is reverse biased

(C) J1 is forward biased and J2 and J3 are reverse biased

(D) J1, J2 and J3 are all forward biased

[Gate 2003]

26. Which one of the following statements is correct?

The turn-off times of converter grade SCRs are normally in the range of:

(A) 1 to 2 microseconds

(B) 50 to 200 microseconds

(C) 500 to 1000 microseconds

(D) 1 to 2 milliseconds **[IES 2004]**

27. Which one of the following statements is correct?

A triac is a:

(A) 2 terminal switch

(B) 2 terminal bilateral switch

(C) 3 terminal unilateral switch

(D) 3 terminal bidirectional switch

[IES 2004]

28. Which one of the following is the most suitable device for a DC-DC converter?

(A) BJT

(B) GTO

(C) MOSFET

(D) Thyristor **[IES 2004]**

29. Which one of the following is correct?

In a switched capacitor network for VAR compensation, the SCRs are commutated by:

(A) forced commutation

(B) resonant commutation

(C) natural commutation

(D) delayed commutation **[IES 2004]**

30. Which one of the following shows current fold back characteristics curve for an SCR controlled shunt regulated power supply?

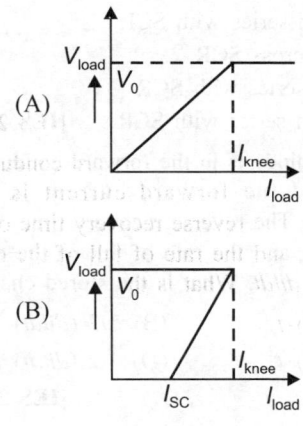

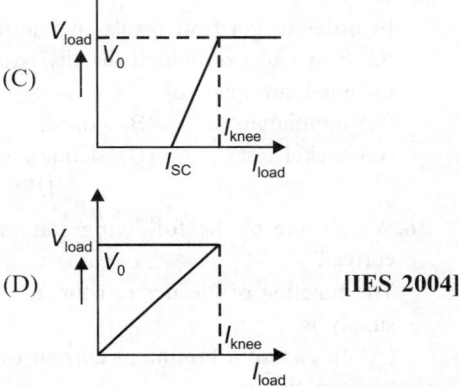

[IES 2004]

31. Which one of the following statement is correct?

In a transistor, the reverse saturation current I_{CO}:

(A) doubles for every 10°C rise in temperature

(B) doubles for every 1°C rise in temperature

(C) increases linearly with temperature

(D) decreases linearly with temperature

[IES 2004]

32. Which one of the following statement is correct?

In a thyristor, the holding current I_H is:

(A) more than the latching current I_L

(B) less than I_L

(C) equal to I_L

(D) equal to zero   **[IES 2004]**

33. Which one of the following statement is correct?

For an SCR, dv/dt protection is achieved through the use of:

(A) RL in series with SCR

(B) RC across SCR

(C) L in series with SCR

(D) RC in series with SCR   **[IES 2004]**

34. A power diode is in the forward conduction mode and the forward current is now decreased. The reverse recovery time of the diode is t_r and the rate of fall of the diode current is di/dt. What is the stored charge?

(A) $(di/dt) \cdot t_r$ (B) $1/2 \ (di/dt) \cdot t_r^2$

(C) $(di/dt) \cdot t_r^2$ (D) $1/2 \ (di/dt) \cdot t_r$

     [IES 2004]

35. Which one of the following statement is correct?

In order to get best results per unit cost, the heat sinks on which the thyristors are mounted, are made of:

(A) aluminium   (B) copper

(C) nickel   (D) stainless steel

     [IES 2004]

36. Which one of the following statement is correct?

The function of bleeder resistor in a power supply is:

(A) to ensure a minimum current drain in the circuit

(B) to increase the output dc voltage

(C) to increase the output current

(D) same as that of a load resistor

     [IES 2004]

37. Which one of the following is used as the main switching element in a switched mode power supply operating in 20 kHz to 100 kHz range?

(A) Thyristor   (B) MOSFET

(C) Triac   (D) UJT **[IES 2004]**

38. A MOSFET rated for 10 A, carries a periodic current as shown in figure. The ON state resistance of the MOSFET is $0.15 \, \Omega$. The average ON state loss in the MOSFET is:

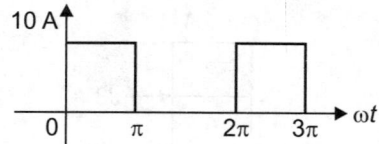

(A) 33.8 W

(B) 15.0 W

(C) 7.5 W

(D) 3.8 W

     [Gate 2004]

39. A gate turn-off (GTO) thyristor has capacity to:

(A) amplify the gate-current

(B) turn-off when positive current pulse is given at the gate

(C) turn-off when a gate-pulse is given at the gate even though it is reverse biased

(D) turn-off when a negative current pulse is given at the gate   **[IES 2005]**

40. Power electronic device with poor turn-off gain is:

(A) a symmetrical thyristor

(B) a conventional thyristor

(C) power bipolar junction transistor

(D) gate turn-off thyristor   **[IES 2005]**

41. Match List-I with List-II and select the correct answer using the code given below the lists:

List-I (Limiting factor)		List-II (Safe operating area portion)	
A.	The peak voltage limit	1.	PQ
B.	Secondary breakdown limit	2.	QR
C.	Power dissipation limit	3.	RS
D.	Peak current limit	4.	ST

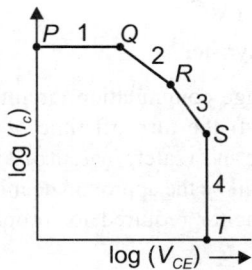

Codes:

	A	B	C	D
(A)	2	1	4	3
(B)	4	3	2	1
(C)	2	3	4	1
(D)	4	1	2	3

[IES 2005]

42. Carrier frequency gate drive is used for turn-on of a thyristor to reduce:

(A) di/dt

(B) turn-on time

(C) dv/dt

(D) size of pulse transformer [IES 2005]

43. Which one of the following statement is not correct?

(A) Power MOSFETs are so constructed as to avoid punch through

(B) In a power MOSFET, the channel length is relatively large and channel width is relatively small

(C) Power MOSFETs do not experience any minority carrier storage

(D) Power MOSFETs can be put in parallel to handle larger currents [IES 2005]

44. In a thyristor, di/dt protection is achieved by the use of:

(A) an inductance L in series with the thyristor

(B) a resistor in series with the thyristor

(C) RC in series with the thyristor

(D) RL in series with the thyristor [IES 2005]

45. Triacs cannot be used in AC voltage regulator for a:

(A) resistive load

(B) back emf load

(C) inductive load

(D) resistive inductive load [IES 2005]

46. The conduction loss versus device current characteristic of a power MOSFET is best approximated by:

(A) a parabola

(B) a straight line

(C) a rectangular hyperbola

(D) an exponentially decaying function

[Gate 2005]

47. The figure shows the voltage across a power semiconductor device and the current through the device during a switching transitions. Whether the transition is a turn ON transition or a turn OFF transition? What is the energy lost during the transition?

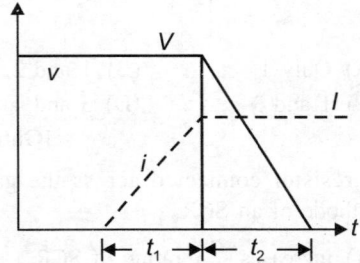

(A) Turn ON, $\dfrac{VI}{2}(t_1 + t_2)$

(B) Turn OFF, $VI(t_1 + t_2)$

(C) Turn ON, $VI(t_1 + t_2)$

(D) Turn OFF, $\dfrac{VI}{2}(t_1 + t_2)$ [Gate 2005]

48. An electronics switch S is required to block voltage of either polarity during its OFF state as shown in the figure (a). This switch is required to conduct in only one direction its ON states as shown in the figure (b).

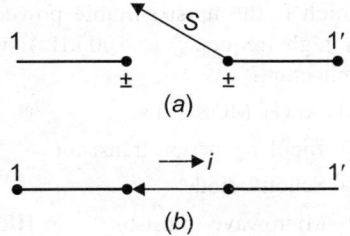

Which of the following are valid realizations of the switch S?

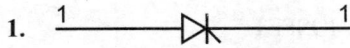

2.

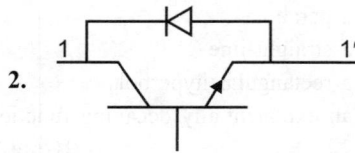

3.

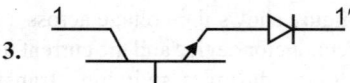

4.

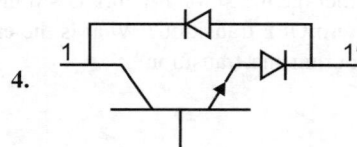

(A) Only 1   (B) 1 and 2

(C) 1 and 3   (D) 3 and 4

[Gate 2005]

49. A resistor connected across the gate and cathode of an SCR:

(A) increases $\dfrac{dv}{dt}$ rating of SCR

(B) increases holding current of SCR

(C) increases noise immunity of SCR

(D) increases turn-off time of SCR

[IES 2006]

50. Turn-on of a thyristor takes place when:

(A) anode to cathode voltage is positive

(B) anode to cathode voltage is negative

(C) there is a positive current pulse at the gate

(D) the anode to cathode voltage is positive and there is a positive current pulse at the gate   **[IES 2006]**

51. Which is the most suitable power device for high frequency (> 100 kHz) switching application?

(A) Power MOSFET

(B) Bipolar junction transistor

(C) Schotty diode

(D) Microwave transistor   **[IES 2006]**

52. Which of the following devices should be used as a switch in a low power switched mode power supply (SMPS)?

(A) GTO

(B) MOSFET

(C) TRIAC

(D) Thyristor   **[IES 2006]**

53. A voltage commutation circuit is shown in figure. If the turn off time of the SCRs is 50 μ sec and a safety margin of 2 is considered, what will be the approximate minimum value of capacitor required for proper commutation?

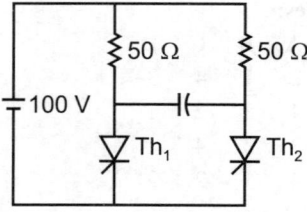

(A) 2.88 μF   (B) 1.44 μF

(C) 0.91 μF   (D) 0.72 μF

[Gate 2006]

54. An SCR having a turn ON time of 5 μ sec, latching current of 50 mA and holding current of 40 mA is triggered by a short duration pulse and is used in the circuit shown in figure. The minimum pulse width required to turn the SCR ON will be:

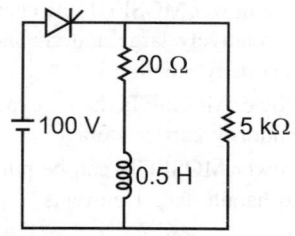

(A) 251 μ sec

(B) 150 μ sec

(C) 100 μ sec

(D) 5 μ sec   **[Gate 2006]**

55. Which one of the following is the main advantage of SMPS over linear power supply?

(A) No transformer is required

(B) Only one stage of conversion

(C) No filter is required

(D) Low power-dissipation   **[IES 2007]**

56. Match List-I with List-II and select the correct answer using the code given below the lists:

List-I (Type of device)		List-II (Characteristic/ application)
A. MOSFET	1.	Turn-off by negative gate pulse
B. GTO	2.	Bi-directional switching
C. UJT	3.	High speed switching
D. TRIAC	4.	Triggering circuit

Codes:

	A	B	C	D
(A)	3	1	4	2
(B)	3	1	2	4
(C)	1	2	3	4
(D)	1	2	4	3

[IES 2007]

57. Snubber circuits are used to protect thyristor from which of the following?

(A) High $\dfrac{di}{dt}$ and low $\dfrac{dv}{dt}$

(B) High $\dfrac{dv}{dt}$ and low $\dfrac{di}{dt}$

(C) Low $\dfrac{dv}{dt}$ and low $\dfrac{di}{dt}$

(D) High $\dfrac{dv}{dt}$ and high $\dfrac{di}{dt}$ **[IES 2007]**

58. Consider the following statement in respect of IGBT:

1. It combines the attributes of MOSFET and BJT.
2. It has low forward voltage drop.
3. Its switching speed is very much lower than that of MOSFET.
4. It has high input impedance.

Which of these statements are correct?

(A) 1, 2, 3 and 4   (B) 1, 2 and 4
(C) 1, 2 and 3   (D) 3 and 4

[IES 2007]

59. The circuit in the figure is a current commutated dc-dc chopper where, Th_M is the main SCR and Th_{AUX} is the auxiliary SCR. The load current is constant at 10 A. Th_M

is ON. Th_{AUX} is triggered at $t = 0$. Th_M is turned OFF between.

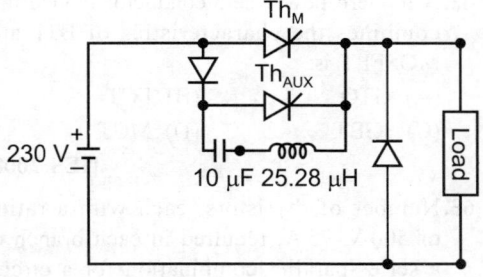

(A) $0\,\mu s < t \le 25\,\mu s$
(B) $25\,\mu s < t \le 50\,\mu s$
(C) $50\,\mu s < t \le 75\,\mu s$
(D) $75\,\mu s < t \le 100\,\mu s$ **[Gate 2007]**

60. In the circuit of adjacent figure, the diode connects the ac source to a pure inductance L.

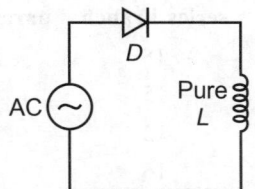

The diode conducts for:
(A) 90°   (B) 180°
(C) 270°   (D) 360°

[Gate 2007]

61. Match List-I with List-II and select the correct answer using the code given below the lists:

List-I (Device)		List-II (Monolithic construction)
A. Triac	1.	Two thyristors in anti parallel
B. Reverse conducting thyristor	2.	A thyristor and a diode in anti-parallel
C. Diac	3.	Two diodes in anti-parallel

Codes:

	A	B	C
(A)	1	2	3
(B)	3	2	1

(C) 2   3   1

(D) 3   1   2   **[IES 2008]**

62. A modern power semiconductor device that combines the characteristics of BJT and MOSFET is:

(A) GTO   (B) FCT

(C) IGBT   (D) MCT

[IES 2008]

63. Number of thyristors, each with a rating of 500 V, 75 A, required in each branch of a series-parallel combination for a circuit with a total voltage and current ratings of 7.5 kV and 1 kA respectively. If the device derating factor is 14%, then what is the number of thyristors in series and parallel branch respectively?

	No. of thyristors in series branch	No. of thyristors in parallel branch
(A)	18	16
(B)	15	14
(C)	12	12
(D)	16	18

[IES 2009]

64. The anode current through a conducting SCR is 10 A. If its gate current is made one-fourth, then what will be the anode current?

(A) 0 A   (B) 5 A

(C) 10 A   (D) 20 A **[IES 2009]**

65. In a power circuit of 3 kV, four thyristors each of rating 800 V are connected in series. What is the percentage series derating factor?

(A) 50   (B) 25

(C) 12.5   (D) 6.25 **[IES 2009]**

66. For an SCR, the gate cathode characteristic has a straight line slope of 140. For trigger source voltage of 20 V and allowable gate power dissipation of 0.5 Ω, what is the gate source resistance?

(A) 200 Ω (B) 255 Ω

(C) 195 Ω (D) 185 Ω

[IES 2009]

67. An SCR is rated for 650 V PIV. What is the voltage for which the device can be operated if the voltage safety factor is 2?

(A) $325V_{rms}$ (B) $230V_{rms}$

(C) $459V_{rms}$ (D) $650V_{rms}$

[IES 2009]

68. An SCR is considered to be a semicontrolled device because:

(A) it can be turned OFF but not ON with a gate pulse

(B) it conducts only during one half-cycle of an alternating current wave

(C) it can be turned ON but not OFF with a gate pulse

(D) it can be turned ON only during one half-cycle of an alternating voltage wave

[Gate 2009]

69. Match the switch arrangements on the top row to the steady-state *V-I* characteristics on the lower row. The steady-state operating points are shown by large black dots.

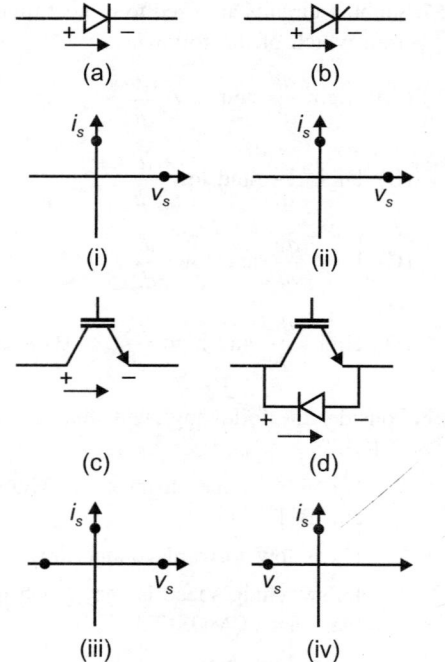

(A) (a)-(i), (b)-(ii), (c)-(iii), (d)-(iv)

(B) (a)-(ii), (b)-(iv), (c)-(i), (d)-(iii)

(C) (a)-(iv), (b)-(iii), (c)-(i), (d)-(ii)

(D) (a)-(iv), (b)-(iii), (c)-(ii), (d)-(i)

[Gate 2009]

70. In an LC series circuit connected to a dc supply of *E* volts via a thyristor when it turns-off, the voltage that appears across the thyristor is:

(A) +E
(B) +2E
(C) –E
(D) –2E **[IES 2010]**

71. Consider the following devices:
1. SCR
2. GTO
3. BJT
4. MOSFET
5. IGBT

Which of these devices do not belong to the family of transistors?
(A) 1 and 2 only
(B) 1, 2 and 3 only
(C) 2, 3 and 5 only
(D) 1, 2, 3, 4 and 5
[IES 2010]

72. An SCR is in conducting state, a reverse voltage is applied between anode and cathode, but it fails to turn-off. What could be the reason?
(A) Positive voltage is applied to the gate
(B) The reverse voltage is small
(C) The anode current is more than the holding current
(D) Turn-off time of SCR is large
[IES 2011]

73. A reverse conducting thyristor (RCT) normally replaces:
(A) a pair of anti-parallel thyristors in a circuit
(B) a combination of a thyristor and an anti-parallel diode in a circuit
(C) a thyristor in situation where it is not required to have reversed blocking capability at all
(D) conventional conversion grade thyristors having large turn-off time
[IES 2011]

74. A thyristor can be switched from a non-conducting state to a conducting state by applying:
1. Voltage more than forward break over voltage.
2. A voltage with high dv/dt.
3. Positive gate current with positive anode voltage.
4. Negative gate current with positive anode voltage.
(A) 1, 2, 3 and 4 are correct
(B) 1, 2 and 4 are correct
(C) 1, 2 and 3 are correct
(D) 2, 3 and 4 are correct
[IES 2011]

75. A structure obtained by lightly doped in drift region between the layers of a pn-junction a PIN diode is obtained. This structure is effective in:
(A) making the diode support large reverse blocking voltages
(B) making reverse recovery process slow
(C) making the diode have high on-state voltage drop
(D) reducing the voltage spike during turn-off due to stray inductance
[IES 2011]

76. Which one of the following statements is not correct for a MOSFET?
(A) Are easy to parallel for higher current
(B) Leakage current is relatively high
(C) Have more linear characteristic
(D) Overload and peak current handling capability are high
[IES 2011]

77. In a GTO, anode current begins to fall when the gate current:
(A) is negative peak at time t = 0
(B) is negative peak at t = storage period t
(C) just begins to become negative at t = 0
(D) just begins to become positive at t = 0
[IES 2011]

78. Consider the following statements:
1. A thyristor requires turn-off circuit while transistor does not.
2. The voltage drop of a thyristor is less than that of a transistor.
3. A thyristor requires a continuous gate current.
4. A transistor draws continuous base current.

Which of these statements are correct?
(A) 1, 2, 3 and 4
(B) 1 and 2
(C) 2 and 4
(D) 1 and 4
[IES 2011]

79. A field effect transistor with an anti-parallel body diode blocks:
(A) bidirectional voltage and passes unidirectional current
(B) bidirectional voltage and passes bidirectional current
(C) unidirectional voltage and passes unidirectional current
(D) unidirectional voltage and passes bidirectional current
[IES 2011]

80. Circuit turn-off time of an SCR is defined as the time:
    (A) taken by the SCR to turn off
    (B) required for the SCR current to become zero
    (C) for which the SCR is reverse biased by the commutation circuit
    (D) for which the SCR is reverse biased to reduce its current below the holding current **[Gate 2011]**

81. A voltage commuted chopper circuit, operated at 500 Hz, is shown below:

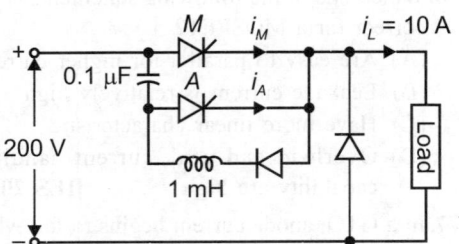

If the maximum value of load currents is 10 A, then the maximum current through the main (M) and auxiliary (A) thyristors will be:
    (A) $i_{M\max} = 12\,A$ and $i_{A\max} = 10\,A$
    (B) $i_{M\max} = 12\,A$ and $i_{A\max} = 2\,A$
    (C) $i_{M\max} = 10\,A$ and $i_{A\max} = 12\,A$
    (D) $i_{M\max} = 10\,A$ and $i_{A\max} = 8\,A$
    [Gate 2011]

82. Match List-I with List-II and select the correct answer using the code given below the lists:

| List-I | | List-II | |
(Device)		(Switching time)	
A.	TRIAC	1.	5–10 µs
B.	SCR	2.	100–400 µs
C.	MOSFET	3.	50–100 µs
D.	IGBT	4.	200–400 µs

Codes:

	A	B	C	D
(A)	4	3	2	1
(B)	1	2	3	4
(C)	4	2	3	1
(D)	1	3	2	4

[IES 2012]

83. The following is a unipolar device:
    (A) BJT (B) IGBT
    (C) GTO (D) MOSFET
    [IES 2012]

84. A thyristor has a PIV of 650 V. The voltage safety factor is 2. Then the voltage upto which the device can be operated is given by:
    (A) 1300 V (B) 650 V
    (C) 325 V (D) 230 V
    [IES 2012]

85. When a thyristor is in the forward blocking state, then:
    (A) all 3 junctions are reverse biased
    (B) anode and cathode junctions are forward biased but gate junction is reverse biased
    (C) anode junction is forward biased but other two are reverse biased
    (D) anode and gate junctions are forward biased but cathode is reverse biased
    [IES 2012]

86. An SCR triggered by a current pulse applied to the gate-cathode can be turned-off:
    (A) by applying a pulse to the cathode
    (B) by applying a pulse to the anode
    (C) by applying another pulse of opposite polarity to the gate-cathode
    (D) by reversing the polarity of the anode and cathode voltage **[IES 2012]**

87. In forward-bias portion of the thyristor's I-V characteristic, the number of stable operating regions is:
    (A) one (B) two
    (C) three (D) none of these
    [IES 2012]

88. A dc source of 100 volts supplies a purely inductive load of 0.1 H; the controller is an SCR in series with source and load. If the specified latching current is 100 mA, then the minimum width of the gating pulse to ensure turn-on of SCR would be:
    (A) 10 µs (B) 50 µs
    (C) 100 µs (D) 1 µs
    [IES 2012]

89. A single-phase two pulse converter feeds an R-L load with insufficient smoothing but the conduction is continuous. If the resistance of the load circuit is increased, then:
    (A) the ripple content of the load current will remain the same

(B) the ripple content of the load current will decrease

(C) the ripple content of the load current will increase

(D) there is possibility of discontinuous conduction due to an increase in the ripple content **[IES 2012]**

90. Among the following pairs, the one not correctly matched is:

(A) UJT—Intrinsic stand off ratio

(B) FET—Pinch-off voltage

(C) TRIAC—Breakdown voltage

(D) DIAC—Firing voltage **[IES 2012]**

91. The typical ratio of latching current to holding current in a 20 thyristor is:

(A) 5.0   (B) 2.0

(C) 1.0   (D) 0.5 **[Gate 2012]**

92. A thyristor has internal power dissipation of 40 W and is operated at an ambient temperature of 20°C. If thermal resistance is 1.6°C/W, the junction temperature is:

(A) 114°C   (B) 164°C

(C) 94°C   (D) 84°C

[IES 2013]

93. Consider the following statements regarding thyristor:

1. It conducts when forward biased and positive current flows through the gate.

2. It conducts when forward biased and negative current flows through the gate.

3. It commutates when reverse biased and negative current flows through the gate.

4. It commutates when the gate current is withdrawn.

Which of these statement(s) is/are correct?

(A) 1, 2 and 3

(B) 1 and 2 only

(C) 2 and 3 only

(D) 1 only

[IES 2013]

94. Which one of the following power semi-conductor device has bidirectional current capability?

(A) SCR   (B) MOSFET

(C) IGBT   (D) TRIAC

[IES 2014]

95. Consider the following statements:

SCR can be turned-on by:

1. Applying anode voltage at a sufficiently fast rate.

2. Applying sufficiently large anode voltage.

3. Increasing the temperature of SCR to sufficiently large value.

4. Applying sufficiently large gate curent.

Which of the above statements are correct?

(A) 1, 2 and 3

(B) 1, 3 and 4

(C) 1, 3 and 4

(D) 2, 3 and 4

[IES 2014]

96. Turn-on time of an SCR can be reduced by using a:

(A) rectangular pulse of high amplitude and narrow width

(B) rectangular pulse of low amplitude and wide with

(C) triangular pulse

(D) trapezoidal pulse **[IES 2014]**

97. Which of the following is the fastest switching device?

(A) JFET   (B) BJT

(C) MOSFET   (D) Triode

[IES 2014]

98. Which of the following does not cause damage of an SCR?

(A) High current

(B) High rate of rise of current

(C) High temperature rise

(D) High rate of rise of voltage

[IES 2014]

99. For the V-I characteristics of an SCR, which of the following statements are correct?

1. It will trigger when the applied voltage is more than the forward breakover voltage.

2. Holding current is greater than latching current.

3. When reverse biased, a small value of leakage current will flow.

4. It can be triggered without gate current.

(A) 1, 2 and 3   (B) 1, 3 and 4

(C) 1, 2 and 4   (D) 2, 3 and 4

[IES 2014]

100. Which of the following transistors is symmetrical in the same that emitter and collector or source and drain terminals can be interchanged?

(A) JFET

(B) MOSFET

(C) NPN transistor

(D) PNP transistor [IES 2014]

101. The snubber circuit used to shape the turn-on switching trajectory of thyristor and/or to limit *di/dt* during turn-on is:

(A) L-R snubber polarized

(B) R-C snubber polarized

(C) R-C snubber unpolarized

(D) L-R snubber unpolarized [IES 2014]

102. **Statement I:** The 'turn-on' and 'turn-off' time of a MOSFET is very small.

Statement II: The MOSFET is a majority carrier device.

Codes:

(A) Both statement I and II are individually true and statement II is the correct explanation of statement I.

(B) Both statement I and II are individually true but statement II is not the correct explanation of statement I.

(C) Statement I is true but statement II is false.

(D) Statement I is false but statement II is true. [IES 2014]

103. Figure shows four electronic switches (i), (ii), (iii) and (iv). Which of the switches can block voltages of either polarity (applied between terminals '*a*' and '*b*') when the active device is in the OFF state.

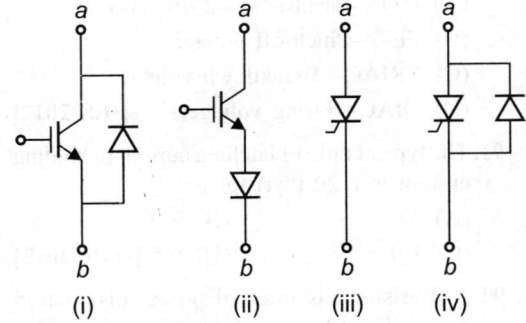

(i) (ii) (iii) (iv)

(A) (i), (ii) and (iii)

(B) (ii), (iii) and (iv)

(C) (ii) and (iii)

(D) (i) and (iv) [Gate 2014]

ANSWER KEY

1. (D)	**2.** (D)	**3.** (D)	**4.** (D)	**5.** (D)	**6.** (C)	**7.** (C)	**8.** (B)
9. (B)	**10.** (C)	**11.** (B)	**12.** (C)	**13.** (A)	**14.** (C)	**15.** (B)	**16.** (D)
17. (D)	**18.** (C)	**19.** (D)	**20.** (D)	**21.** (B)	**22.** (D)	**23.** (A)	**24.** (B)
25. (B)	**26.** (B)	**27.** (D)	**28.** (B)	**29.** (B)	**30.** (C)	**31.** (A)	**32.** (B)
33. (B)	**34.** (B)	**35.** (A)	**36.** (A)	**37.** (B)	**38.** (C)	**39.** (D)	**40.** (D)
41. (B)	**42.** (D)	**43.** (B)	**44.** (A)	**45.** (C)	**46.** (A)	**47.** (A)	**48.** (C)
49. (C)	**50.** (D)	**51.** (A)	**52.** (B)	**53.** (A)	**54.** (B)	**55.** (D)	**56.** (A)
57. (D)	**58.** (A)	**59.** (C)	**60.** (D)	**61.** (A)	**62.** (C)	**63.** (B)	**64.** (C)
65. (D)	**66.** (C)	**67.** (B)	**68.** (C)	**69.** (C)	**70.** (C)	**71.** (A)	**72.** (C)
73. (B)	**74.** (C)	**75.** (B)	**76.** (C)	**77.** (B)	**78.** (D)	**79.** (D)	**80.** (C)
81. (A)	**82.** (C)	**83.** (D)	**84.** (D)	**85.** (B)	**86.** (D)	**87.** (B)	**88.** (C)
89. (D)	**90.** (D)	**91.** (B)	**92.** (D)	**93.** (D)	**94.** (D)	**95.** (C)	**96.** (A)
97. (C)	**98.** (D)	**99.** (B)	**100.** (A)	**101.** (A)	**102.** (A)	**103.** (C)	

MCQ for Phase Controlled Rectifiers from Various Examinations

1. In the circuit shown in figure, L is large and the average value of 'i' is 100 A. The thyristor is gated in the half cycle of 'e' at a delay angle is equal to _____ where $e(t) = 200\sqrt{2} \sin 314t$.

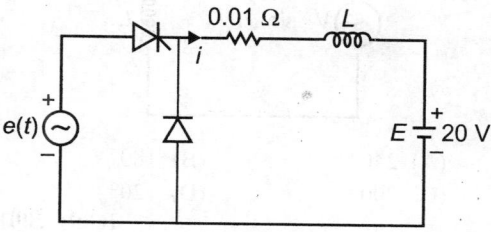

   (A) 2°   (B) 4°

   (C) 6°   (D) 8° **[Gate 1992]**

2. Referring to the figure, the type of load is:

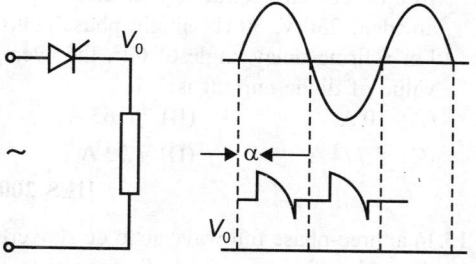

   (A) inductive load   (B) resistive load

   (C) dc motor   (D) capacitive load

                               [Gate 1994]

3. A single-phase diode bridge rectifier supplies a highly inductive load. The load current can be assumed to be ripple free. The ac supply side current waveforms will be:

   (A) sinusoidal   (B) constant dc

   (C) square   (D) triangular

                               [Gate 1995]

4. In a three-phase controlled bridge rectifier, with an increase of overlap angle the output dc voltage:

   (A) decreases

(B) increases

(C) does not change

(D) depends upon load inductance

                               [Gate 1996]

5. In a dual converter, the circulating current:

   (A) allows smooth reversal of load current, but increases the response time.

   (B) does not allow smooth reversal of load current, but reduces the response time.

   (C) allows smooth reversal of load current with improved speed of response

   (D) flows only if there is not interconnecting inductor.   **[Gate 1997]**

6. A three-phase, fully controlled, converter is feeding power into a dc load at a constant current of 150 A. The rms current through each thyristor of the converter is:

   (A) 50 A   (B) 100 A

   (C) $\dfrac{150\sqrt{2}}{\sqrt{3}}$ A   (D) $\dfrac{150}{\sqrt{3}}$ A

                               [Gate 1998]

7. When the firing angle α of a single-phase, fully controlled rectifier feeding a constant dc current into a load is 30°, the displacement power factor of the rectifier is:

   (A) 1   (B) 0.5

   (C) $\dfrac{1}{\sqrt{3}}$ (D) $\dfrac{\sqrt{3}}{2}$

                               [Gate 1998]

8. A thyristor based, three-phase, fully controlled converter feeds a dc load that draws a constant current. Then the input ac line current to the converter has:

   (A) an rms value equal to the dc load current

   (B) an average value equal to the dc load current

(C) a peak value equal to the dc load current

(D) a fundamental frequency component, whose rms value is equal to the dc load current **[Gate 2000]**

9. Match List-I with List-II and select the correct answer (α is the firing angle):

List-I (1-ϕ rectifier topology-feeding resistive load)		List-II (Average output voltage)
A.	Uncon-trolled-half wave	1.	$\dfrac{V_{peak}}{\pi}(1+\cos\alpha)$
B.	Controlled-half wave	2.	$\dfrac{2V_{peak}}{\pi}\cos\alpha$
C.	Controlled-full wave	3.	$\dfrac{V_{peak}}{\pi}$
D.	Semi-con-trolled	4.	$\dfrac{V_{peak}}{2\pi}(1+\cos\alpha)$

Codes:

	A	B	C	D
(A)	3	2	4	1
(B)	1	4	2	3
(C)	3	4	2	1
(D)	1	2	4	3

[IES 2001]

10. The total harmonic distortion (THD) of ac supply input current of rectifier is maximum for:

(A) single-phase diode rectifier with dc inductive filter

(B) three-phase diode rectifier with dc inductive filter

(C) three-phase thyristor rectifier with inductive filter

(D) single-phase diode rectifier with capacitive filter **[IES 2001]**

11. In a thyristor-controlled rectifier, the firing angle of thyristor is to be controlled in the range of:

(A) 0° to 90°   (B) 0° to 180°

(C) 90° to 180°   (D) 90° to 270°

[IES 2001]

12. AC to DC circulating current dual converters are operated with the following relationship between their triggering angles (α_1 and α_2).

(A) $\alpha_1 + \alpha_2 = 180°$ (B) $\alpha_1 + \alpha_2 = 360°$

(C) $\alpha_1 - \alpha_2 = 180°$ (D) $\alpha_1 + \alpha_2 = 90°$

[Gate 2001]

13. A half wave thyristor converter supplies a purely inductive load as shown in figure. If triggering angle of the thyristor is 120°, the extinction angle will be

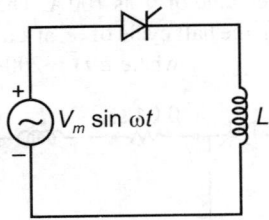

(A) 240°   (B) 180°

(C) 200°   (D) 120°

[Gate 2001]

14. A single-phase full-bridge converter with a free-wheeling diode feeds an inductive load. The load resistance is 15.53 Ω and it has a large inductance providing constant and ripple free dc current. Input to converter is from an ideal 230 V, 50 Hz single phase source. For a firing delay angle of 60°, the average value of diode current is:

(A) 10 A   (B) 8.165 A

(C) 5.774 A   (D) 3.33 A

[IES 2002]

15. In a three-phase full wave ac to dc converter, the ratio of output ripple-frequency to the supply-voltage frequency is:

(A) 2   (B) 3

(C) 6   (D) 12   **[IES 2002]**

16. A six-phase bridge-converter feeds a purely resistive load. The delay angle α is measured from the point of natural-commutation. The effective control of voltage can be obtained when α lies in the range:

(A) $0 \le \alpha \le 150°$ (B) $0 \le \alpha \le 120°$

(C) $0 \le \alpha \le 150°$ (D) $0 \le \alpha \le 180°$

[IES 2002]

17. When fed from a fully controlled rectifier, a dc motor, driving an active load, can operate in:

(A) forward motoring and reverse braking mode

(B) forward motoring and forward braking mode

(C) reverse motoring and reverse braking mode

(D) reverse motoring and forward braking mode **[IES 2002]**

18. A six pulse thyristor rectifier bridge is connected to a balanced 50 Hz three-phase ac source. Assuming that the dc output current of the rectifier is constant. The lowest frequency harmonics component of the ac source line current is:

(A) 100 Hz   (B) 150 Hz

(C) 250 Hz   (D) 300 Hz

[Gate 2002]

19. In the single-phase, diode bridge rectifier shown in figure, the load resistor is $R = 50\,\Omega$. The source voltage is $V = 200 \sin \omega t$, where $\omega = 2\pi \times 50$ radians per second. The power dissipated in the load resistor R is:

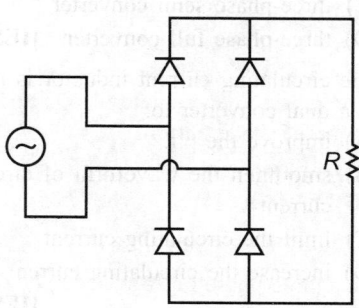

(A) $\dfrac{3200}{\pi}$ W   (B) $\dfrac{400}{\pi}$ W

(C) 400 W   (D) 800 W

[Gate 2002]

20. The characteristic features of discontinuous conduction compared to continuous conduction in a four-pulse, single-phase bridge converter are:

(A) larger average value of load voltage and larger ripple-content

(B) larger average value of load voltage and smaller ripple-content

(C) smaller average value of load voltage and smaller ripple-content

(D) smaller average value of load voltage and larger ripple-content **[IES 2003]**

21. If the rms source voltage is V volts, the minimum and maximum values of firing angles for a single-phase, half-wave controlled rectifier, supplying a load with a back emf of 40 volts are:

(A) $0°$ and $180°$

(B) $\alpha = \sin^{-1}(40/\sqrt{2}V)$ and $180°$

(C) $\alpha = \sin^{-1}(40/\sqrt{2}V)$ and
$[\pi - \sin^{-1}(40/\sqrt{2}V)]$

(D) $0°$ and $[\pi - \sin^{-1}(40/\sqrt{2}V)]$

[IES 2003]

22. A fully controlled natural commutated three-phase bridge rectifier is operating with a firing angle $\alpha = 30°$. The peak to peak voltage ripple expressed as a ratio of the peak output dc voltage at the output of the converter bridge is:

(A) 0.5   (B) $\sqrt{3}/2$

(C) $\left(1 - \dfrac{\sqrt{3}}{2}\right)$ (D) $\sqrt{3} - 1$

[Gate 2003]

23. A phase-controlled half-controlled single-phase converter is shown in figure. The control angle $\alpha = 30°$

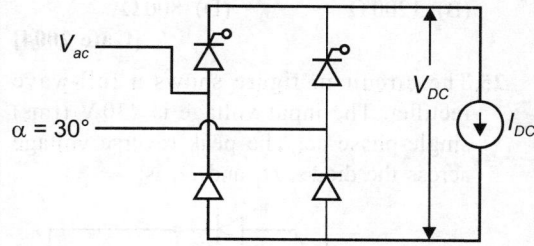

The output dc voltage wave shape will be as shown in:

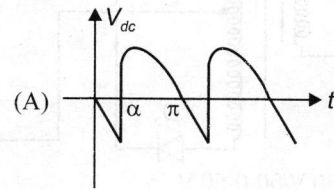

(A)

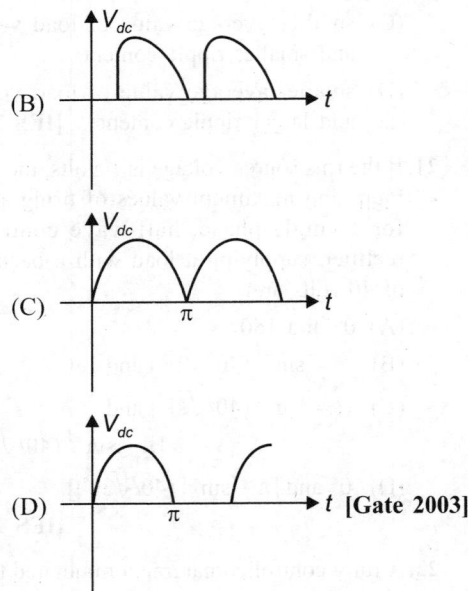

(B)

(C)

(D) [Gate 2003]

24. The triggering circuit of a thyristor is shown in figure. The thyristor requires a gate current of 10 mA, for guaranteed turn-on. The value of R required for the thyristor to turn-on reliably under all conditions of V_b variation is:

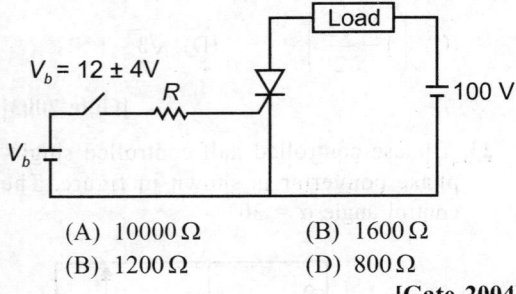

(A) $10000\,\Omega$
(B) $1600\,\Omega$
(B) $1200\,\Omega$
(D) $800\,\Omega$

[Gate 2004]

25. The circuit in figure shows a full-wave rectifier. The input voltage is 230 V (rms) single-phase ac. The peak reverse voltage across the diodes D_1 and D_2 is:

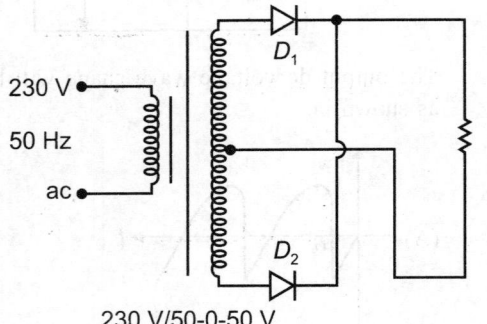

230 V/50-0-50 V

(A) $100\sqrt{2}$ V
(B) 100 V
(C) $50\sqrt{2}$ V
(D) 50 V

[Gate 2004]

26. The circuit in figure shows a three-phase half-wave rectifier. The source is a symmetrical, three-phase four-wire system. The line-to-line voltage of the source is 100 V. The supply frequency is 400 Hz. The ripple frequency at the output is:

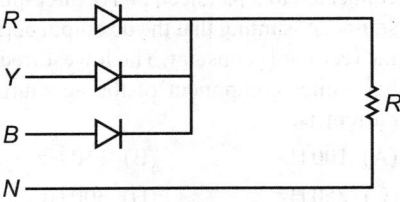

(A) 400 Hz
(B) 800 Hz
(C) 1200 Hz
(D) 2400 Hz

[Gate 2004]

27. A converter which can operate both in 3-pulse and 6-pulse modes is a:
(A) single-phase full converter
(B) three-phase half-wave converter
(C) three-phase semi converter
(D) three-phase full converter   [IES 2005]

28. The circulating current inductor is required in a dual converter to:
(A) improve the p.f.
(B) smoothen the waveform of circulating current
(C) limit the circulating current
(D) increase the circulating current
[IES 2005]

29. In a single-phase semi-converter with discontinuous conduction and extinction angle $\beta < \pi$, freewheeling action takes place for:
(A) α
(B) $\alpha - \beta$
(C) $\beta - \pi$
(D) zero degree
[IES 2005]

30. A three pulse converter is feeding a purely resistive load. What is the value of firing delay angle α, which dictates the boundary between continuous and discontinuous mode of current conduction?
(A) $\alpha = 0°$
(B) $\alpha = 30°$
(C) $\alpha = 60°$
(D) $\alpha = 150°$
[IES 2005]

31. A three-phase diode bridge rectifier is fed from a 400 V rms, 50 Hz, three-phase AC source. If the load is purely resistive, then peak instantaneous output voltage is equal to:

(A) 400 V   (B) $400\sqrt{2}$ V

(C) $400\sqrt{\dfrac{2}{3}}$ V   (D) $\dfrac{400}{\sqrt{3}}$ V

[Gate 2005]

32. Consider a phase-controlled converter shown in the figure. The thyristor is fired at an angle α in every positive half cycle of the input voltage. If the peak value of the instantaneous output voltage equals 230 V, the firing angle α is close to:

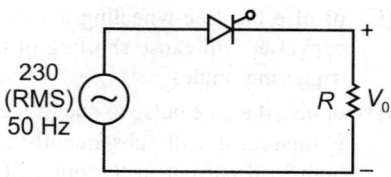

(A) 45°   (B) 135°

(C) 90°   (D) 83.6°

[Gate 2005]

33. In a dual converter, the circulating current:

(A) allows smooth reversal of load current, but increases the response time

(B) allows smooth reversal of load current with improved speed of response

(C) does not allow smooth reversal of load current, but reduces the response time

(D) flows if there is no interconnecting inductor   **[IES 2006]**

34. When the firing angle α of a single-phase fully controlled rectifier feeding constant dc current into the load is 30°, what is the displacement factor of the rectifier?

(A) 1   (B) 0.5

(C) $\sqrt{3}$ (D) $\dfrac{\sqrt{3}}{2}$

[IES 2006]

35. For the same voltage output, which one of the following has larger peak inverse voltage of the thyristor?

(A) Single-phase full wave centre tapped circuit

(B) Single-phase full wave bridge circuit

(C) Three-phase full wave bridge circuit

(D) Three-phase full wave centre tapped circuit   **[IES 2006]**

36. A fully controlled line commutated converter functions as an inverter when firing angle (α) is in the range:

(A) 0°–90°

(B) 90°–180°

(C) 90°–180° only when there is a suitable dc source in the load

(D) 90°–180° only when it supplies a back emf load   **[IES 2006]**

37. A single-phase half-wave uncontrolled converter circuit is shown in figure. A 2-winding transformer is used at the input for isolation. Assuming the load current to be constant and $v = V_m \sin \omega t$, the current waveform through diode D_2 will be:

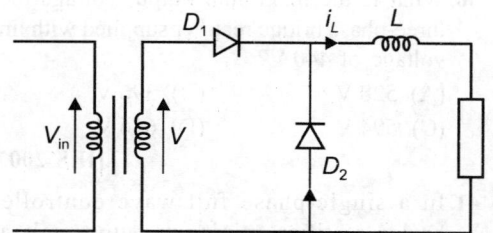

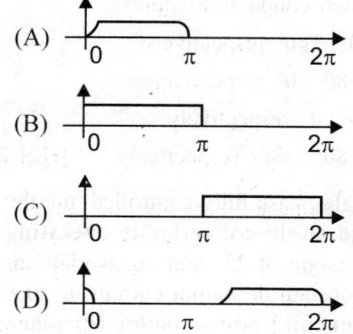

[Gate 2006]

38. A single-phase bridge converter is used to charge a battery of 200 V having an internal resistance of 2 Ω as shown in figure. The SCRs are triggered by a constant dc signal. If SCR 2 gets open circuited, what will be the average charging current?

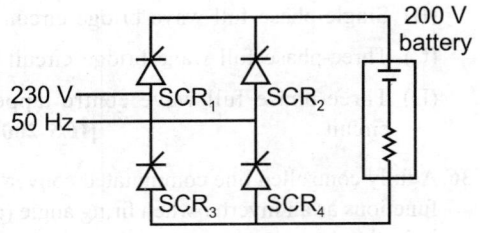

(A) 23.8 A   (B) 15 A

(C) 11.9 A   (D) 3.54 A

[Gate 2006]

39. Which one of the following is the correct statement?

In a two quadrant converter working in the 1st and 2nd quadrants:

(A) load current and load voltage are always positive

(B) load current is always negative

(C) load current can be positive or negative

(D) load current and load voltage are always negative   **[IES 2007]**

40. What is the maximum output voltage of a three-phase bridge rectifier supplied with line voltage of 440 V?

(A) 528 V   (B) 396 V

(C) 594 V   (D) 616 V

[IES 2007]

41. In a single-phase full wave controlled bridge rectifier, minimum output voltage and maximum output voltage are obtained at which conduction angles?

(A) 0°, 180° respectively

(B) 180°, 0° respectively

(C) 0°, 0° respectively

(D) 180°, 180° respectively   **[IES 2007]**

42. A single-phase fully controlled, the thyristor bridge ac-dc converter is operating at a firing angle of 25° and on overlap angle of 10° constant dc output current of 20 A. The fundamental power factor (displacement factor) at input ac mains is:

(A) 0.78   (B) 0.827

(C) 0.866   (D) 0.9 **[Gate 2007]**

43. A three-phase fully-controlled thyristor bridge converter is used as line commutated inverter to feed 50 kW power 420 V dc to a three-phase, 415 V (line), 50 Hz ac mains.

Consider dc link current to be constant. The rms current of the thyristor is:

(A) 119.05 A   (B) 79.37 A

(C) 68.73 A   (D) 39.68 A

[Gate 2007]

44. A single-phase full-wave half-controlled bridge converter feeds an inductive load. The two SCRs in the converter are connected to a common DC bus. The converter has to have a free-wheeling diode:

(A) because the converter inherently does not provide for free-wheeling

(B) because the converter does not provide for free-wheeling for high values of triggering angles

(C) or else the free-wheeling action of the converter will cause shorting of the AC triggering angles

(D) or else if a gate pulse to one of the SCRs is missed, it will subsequently cause a high load current in the other SCR.

[Gate 2007]

45. For a single-phase ac to dc controlled rectifier to operate in regenerative mode, which of the following conditions should be satisfied?

(A) Half-controlled bridge, $\alpha < 90°$, source of emf in load

(B) Half-controlled bridge, $\alpha > 90°$, source of emf in load

(C) Full-controlled bridge, $\alpha > 90°$, source of emf in load

(D) Full-controlled bridge, $\alpha < 90°$, source of emf in load   **[IES 2008]**

46. A half-controlled bridge converter is operating from an rms input voltage of 120 V. Neglecting the voltage drops, what are the mean load voltage at a firing delay angle of 0° and 180°, respectively?

(A) $\dfrac{120 \times 2\sqrt{2}}{\pi} V$ and 0

(B) 0 and $\dfrac{120 \times 2\sqrt{2}}{\pi} V$

(C) $\dfrac{120\sqrt{2}}{\pi} V$ and 0

(D) 0 and $\dfrac{120\sqrt{2}}{\pi} V$ **[IES 2008]**

47. A large dc motor is required to control the speed of blower from a three-phase ac source. What is the most suitable ac to dc converter?

(A) Three-phase fully controlled bridge converter

(B) Three-phase fully controlled bridge converter with freewheeling diode

(C) Three-phase half-controlled bridge converter

(D) A pair of three-phase converters in sequence control **[IES 2008]**

48. A single-phase half controlled converter shown in the figure feeding power to highly inductive load. The converter is operating at a firing angle of 60°.

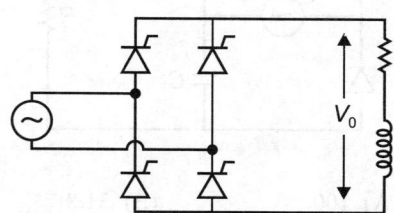

If the firing pulses are suddenly removed, the steady-state voltage (V_0) waveform of the converter will become:

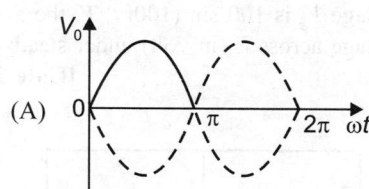

(A)

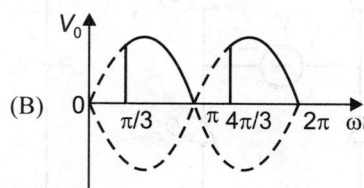

(B)

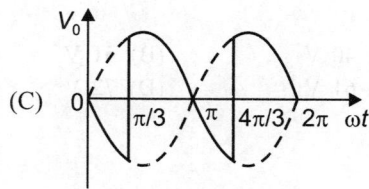

(C)

(D)

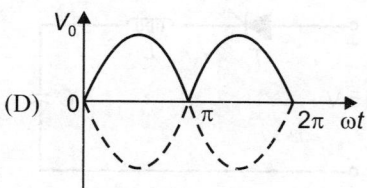

[Gate 2008]

49. The fully controlled thyristor converter in the figure is fed from a single-phase source. When the firing angle is 0°, the dc output voltage of the converter is 300 V. What will be the output voltage for a firing angle of 60°, assuming continuous conduction?

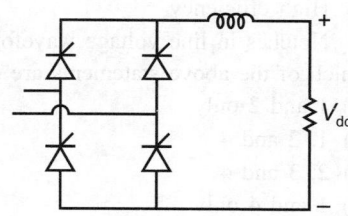

(A) 150 V (B) 210 V

(C) 300 V (D) 100π V

[Gate 2010]

50. A half-controlled single-phase bridge rectifier is supplying an R-L load. It is operating at a firing angle α and the load current is continuous. The fraction of cycle that the free-wheeling diode conduct is:

(A) $\dfrac{1}{2}$ (B) $\left(1 - \dfrac{\alpha}{\pi}\right)$

(C) $\dfrac{\alpha}{2\pi}$ (D) $\dfrac{\alpha}{\pi}$

[Gate 2012]

51. Thyristor T in the figure given on next page is initially off and is triggered with a single-pulse of width 10 μs. It is given that $L = \left(\dfrac{100}{\pi}\right)$ μH and $C = \left(\dfrac{100}{\pi}\right)$ μF. Assuming latching and holding currents of the thyristor are both zero and the initial charge on C is zero, T conducts for:

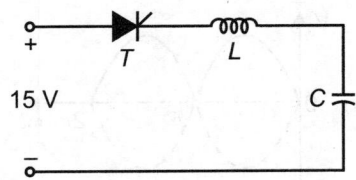

(A) $10\,\mu s$ (B) $50\,\mu s$
(C) $100\,\mu s$ (D) $200\,\mu s$

[Gate 2013]

52. Consider the following statements:

Phase controlled converters at small values of output voltage have:

1. Large harmonics in utility system
2. Poor power factor
3. High efficiency
4. Notches in line voltage waveform

Which of the above statements are correct?

(A) 1 and 2 only
(B) 1, 2 and 4
(C) 2, 3 and 4
(D) 1 and 4 only

[IES 2014]

53. In a three-phase controller bridge rectifier, the maximum conduction of each thyristor is:

(A) 60°   (B) 90°
(C) 120°   (D) 150°

[IES 2014]

54. A line commulated phase-controlled inverter is operating at its inverter limit. There can be a commutation failure if:

(A) the frequency decreases
(B) the voltage increases
(C) the frequency increases
(D) both voltage and frequency change such that v/f is constant   **[IES 2014]**

55. **Statement I:** For the same voltage output, the power factor of a single-phase semiconverter is better than a full converter.

Statement II: The single-phase semiconverter uses two diodes and two controlled switches.

Codes:

(A) both statement I and statement II are individually true and statement II is the correct explanation of statement I

(B) both statement I and statement II are individually true but statement II is not the correct explanation of statement I

(C) statement I is true but statement II is false

(D) statement I is false but statement II is true

56. In the following circuit, the input voltage V_{in} is $100\sin(100\pi t)$. For $100\pi RC = 50$, the average voltage across R (in volts) under stead-state is nearest to:

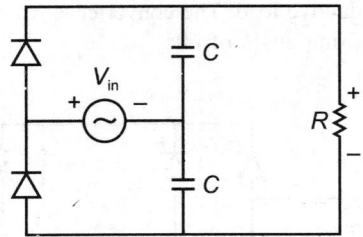

(A) 100   (B) 31.8
(C) 200   (D) 63.6

[Gate 2015]

57. In the given rectifier, the delay angle of the thyristor T_1 measured from the positive going zero crossing of V_s is 30°. If the input voltage V_s is $100\sin(100\pi t)\ V$, the average voltage across R (in volt) under steady-state is: **[Gate 2015]**

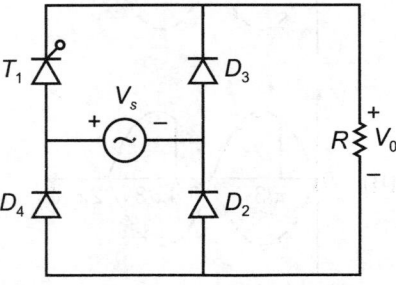

(A) 40 V   (B) 51 V
(C) 61 V   (D) 72 V

ANSWER KEY

1. (B)	**2.** (C)	**3.** (C)	**4.** (A)	**5.** (C)	**6.** (D)	**7.** (D)	**8.** (C)
9. (C)	**10.** (D)	**11.** (B)	**12.** (A)	**13.** (A)	**14.** (D)	**15.** (C)	**16.** (B)
17. (B)	**18.** (C)	**19.** (C)	**20.** (D)	**21.** (C)	**22.** (A)	**23.** (A)	**24.** (D)
25. (A)	**26.** (C)	**27.** (C)	**28.** (C)	**29.** (D)	**30.** (B)	**31.** (C)	**32.** (B)
33. (B)	**34.** (D)	**35.** (D)	**36.** (D)	**37.** (C)	**38.** (C)	**39.** (C)	**40.** (C)
41. (A)	**42.** (C)	**43.** (C)	**44.** (C)	**45.** (C)	**46.** (A)	**47.** (C)	**48.** (A)
49. (A)	**50.** (D)	**51.** (C)	**52.** (B)	**53.** (C)	**54.** (C)	**55.** (B)	**56.** (C)
57. (C)							

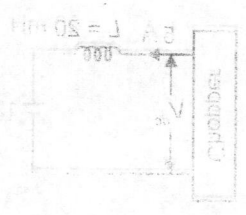

MCQ for Choppers from Various Examinations

1. A chopper operating at a fixed frequency is feeding an *R-L* load. As the duty ratio of the chopper is increased from 25% to 75%, the ripple in the load current:
   - (A) remains constant.
   - (B) decreases, reaches a minimum at 50% duty ratio and then increases.
   - (C) increase, reaches a maximum at 50% duty ratio and then decreases.
   - (D) keeps on increasing as the duty ratio is increased.   **[Gate 1993]**

2. A four quadrant chopper cannot be operated as:
   - (A) one quadrant chopper
   - (B) cycloconverter
   - (C) inverter
   - (D) bi-directional rectifier   **[IES 2001]**

3. A three-phase wound rotor induction motor is controlled by a chopper-controlled resistance in its rotor circuit. A resistance of $2\,\Omega$ is connected in the rotor circuit and a resistance of $4\,\Omega$ is additionally connected during OFF periods of the chopper. The OFF period of the chopper is 4 ms. The average resistance in the rotor circuit for the chopper frequency of 200 Hz is:
   - (A) $26/5\,\Omega$
   - (B) $24/5\,\Omega$
   - (C) $18/5\,\Omega$
   - (D) $16/5\,\Omega$

   [IES 2001]

4. For a step up dc-dc chopper with an input dc voltage of 220 volts, if the output voltage required is 330 volts and the non-conducting time of thyristor-chopper is $100\,\mu s$, the ON time of thyristor-chopper would be:
   - (A) $66.6\,\mu s$
   - (B) $100\,\mu s$
   - (C) $150\,\mu s$
   - (D) $200\,\mu s$

   [IES 2002]

5. An ideal chopper operating at a frequency of 500 Hz, supplies a load having resistance of $3\,\Omega$ and inductance of 9 mH from a 60 V battery. The mean value of the load voltage for on/off ratio of 4/1 (assuming that load is shunted by a perfect commutating diode and battery is loss-less) is:
   - (A) 240 V
   - (B) 48 V
   - (C) 15 V
   - (D) 4 V

   [IES 2003]

6. A boost-regulator has an input voltage of 5 V and the average output voltage of 15 V. The duty cycle is:
   - (A) 3/2
   - (B) 2/3
   - (C) 5/2
   - (D) 15/2 **[IES 2003]**

7. A chopper is employed to charge a battery as shown in figure. The charging current is 5 A. The duty ratio is 0.2. The chopper output voltage is also shown in figure. The peak to peak ripple current in the charging current is:

   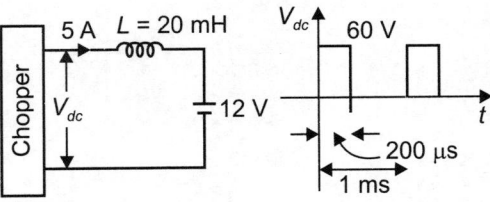

   - (A) 0.48 A
   - (B) 1.2 A
   - (C) 2.4 A
   - (D) 1 A

   [Gate 2003]

8. In the buck-boost converter, what is the maximum value of the switch utilization factor?
   - (A) 1.00
   - (B) 0.75
   - (C) 0.50
   - (D) 0.25 **[IES 2004]**

9. Figure shows a chopper operating from a 100 V dc input. The duty ratio of the main switch *S* is 0.8. The load is sufficiently inductive so that the load current is ripple

free. The average current through the diode *D* under steady-state is:

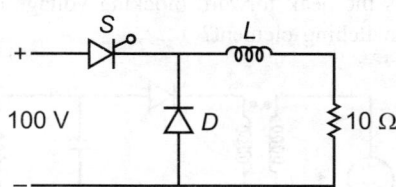

(A) 1.6 A   (B) 6.4 A
(C) 8.0 A   (D) 10.0 A

[Gate 2004]

10. The given figure shows a step-down chopper switched at 1 kHz with a duty ratio $D = 0.5$. The peak-peak ripple in the load current is close to:

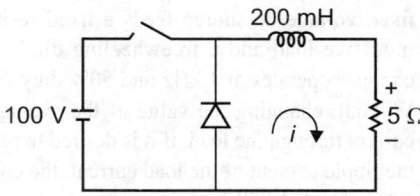

(A) 10 A
(B) 0.5 A
(C) 0.125 A
(D) 0.25 A

[Gate 2005]

Statement for common data 11 and 12.

A voltage commutated chopper operating at 1 kHz is used to control the speed of dc motor as shown in figure. The load current is assumed to be constant at 10 A.

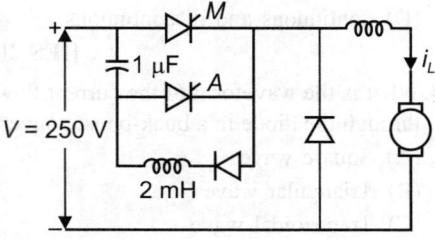

11. The minimum time in μ sec for which the SCR *M* should be ON is:
(A) 280 μs   (B) 140 μs
(C) 70 μs   (D) 0 μs

[Gate 2006]

12. The average output voltage of the chopper will be:
(a) 70 V   (B) 47.5 V
(C) 35 V   (D) 0 V

[Gate 2006]

13. The circuit shown in the below figure will work as which one of the following?

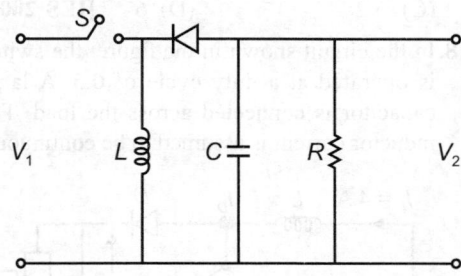

(A) Buck-Boost converter
(B) Buck converter
(C) Boost converter
(D) Dual converter   **[IES 2007]**

14. For a step-down dc chopper operating with discontinuous load current, what is the expression for the load voltage? (*K* is duty ratio of chopper)
(A) $V_0 = V_{dc} \times K$
(B) $V_0 = V_{dc}/K$
(C) $V_0 = V_{dc}/(1 - K)$
(D) $V_0 = V_{dc} (1 - K)$ **[IES 2008]**

15. An ideal chopper is operating at a frequency of 500 Hz from a 60 V battery input. It is supplying a load having 3 Ω resistance and 9 mH inductance. Assuming the load is shunted by a perfect commutating diode and assuming battery is loss less, what is the mean load current at an on/off ratio of 1/1?
(A) 10 A   (B) 15 A
(C) 20 A   (D) None of these

[IES 2008]

16. A two-quadrant dc to dc chopper can operate with which of the following load conditions?
1.   +ve voltage, +ve current
2.   −ve voltage, +ve current
3.   −ve voltage, −ve current
4.   +ve voltage, −ve current

Select the correct answer using the code given on next page.

(A) 1 only   (B) 1 and 2

(C) 1 and 4   (D) 3 and 4

[IES 2008]

17. A buck regulator has an input voltage of 12 V and the required output voltage is 5 V. What is the duty cycle of the regulator?

(A) 5/12   (B) 12/5

(C) 5/2   (D) 6   **[IES 2008]**

18. In the circuit shown in the figure, the switch is operated at a duty cycle of 0.5. A large capacitor is connected across the load. The inductor current is assumed to be continuous.

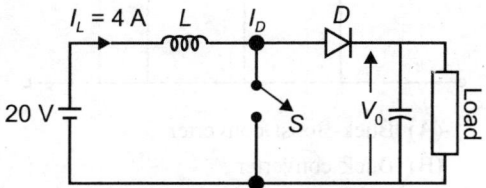

The average voltage across the load and the average current through the diode will respectively be:

(A) 10 V, 2 A   (B) 10 V, 8 A

(C) 40 V, 2 A   (D) 40 V, 8 A

[Gate 2008]

19. A DC chopper is used in regenerative braking mode of a dc series motor. The dc supply is 600 V, the duty cycle is 70%. The average value of armature current is 100 A. It is continuous and ripple free. What is the value of power feedback to the supply?

(A) 3 kW   (B) 9 kW

(C) 18 kW   (D) 35 kW

[IES 2009]

20. If a full wave fully controlled converter is modified as a full wave half controlled converter, what will be the maximum value of active power (P) and the maximum value of reactive power demand (Q)?

	P	Q
(A)	Double	Half
(B)	Unchanged	Unchanged
(C)	Half	Double
(D)	Unchanged	Half

[IES 2009]

21. For the isolated buck boost converter as shown in the circuit below, the output voltage is to be 35 V at a duty cycle of 30%. The DC input

is obtained from a front end rectifier without voltage doubling fed from a 115 V AC. What is the peak forward blocking voltage of the switching element?

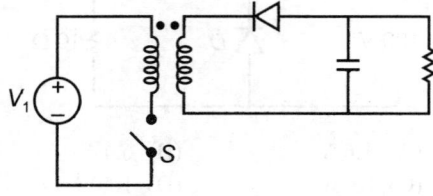

(A) 232.3 V

(B) 69.69 V

(C) 162.61 V

(D) 542 V   **[IES 2009]**

22. A dc to dc transistor chopper supplied from a fixed voltage dc source feeds a fixed resistive inductive load and a freewheeling diode. The chopper operates at 1 kHz and 50% duty cycle. Without, changing the value of the average dc current through the load, if it is desired to reduce the ripple content of the load current, the control action needed will be to:

(A) increase the chopper frequency keeping its duty cycle constant

(B) increase the chopper frequency and duty cycle in equal ratio

(C) decrease only the chopper frequency

(D) decrease only the duty cycle

[IES 2010]

23. In dc choppers, the waveforms for input and output voltages are respectively:

(A) discontinuous and continuous

(B) both continuous

(C) both discontinuous

(D) continuous and discontinuous

[IES 2011]

24. What is the waveform of the current flowing through the diode in a buck-boost converter?

(A) Square wave

(B) Triangular wave

(C) Trapezoidal wave

(D) Sinusoidal wave   **[IES 2011]**

25. In the circuit shown, an ideal switch S is operated at 100 kHz with a duty ratio of 50%. Given that Δi_c is 1.6 A peak-to-peak and I_0 is 5 A dc, the peak current in S is:

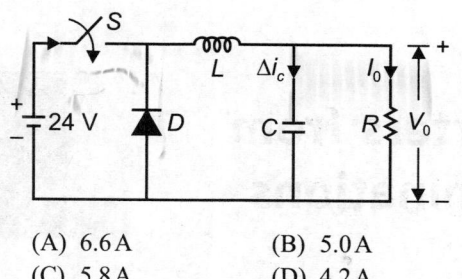

(A) 6.6 A   (B) 5.0 A

(C) 5.8 A   (D) 4.2 A

[Gate 2012]

26. **Statement I:** A forward dc-to-dc converter requires a minimum load at the output.

    Statement II: Without minimum load excess output voltage can be produced.

    (A) both statement I and II are individually true and statement II is the correct explanation of statement I

(B) both statement I and II are individually true but statement II is not the correct explanation of statement I

(C) statement I is true but statement II is false

(D) statement I is false but statement II is true   **[IES 2013]**

27. Which of the following regulator provide output voltage polarity reversal without a transformer?

    (A) Buck regulator

    (B) Boost regulator

    (C) Buck-Boost regulator

    (D) CUK regulator   **[IES 2014]**

ANSWER KEY

1. (C)	2. (B)	3. (A)	4. (D)	5. (B)	6. (B)	7. (A)	8. (B)
9. (A)	10. (C)	11. (B)	12. (B)	13. (A)	14. (A)	15. (A)	16. (C)
17. (A)	18. (C)	19. (C)	20. (B)	21. (A)	22. (A)	23. (D)	24. (C)
25. (C)	26. (A)	27. (C)					

MCQ for Inverters from Various Examinations

1. Triangular PWM control, when applied to a three-phase BJT based voltage source inverter, introduces:
   (A) low order harmonic voltages on the dc side.
   (B) very high order harmonic voltage on the dc side.
   (C) low order harmonic voltage on the ac side.
   (D) very high order harmonic voltage on the ac side.

   [Gate 2000]

2. Match List-I with List-II and select the correct answer:

List-I (Waveforms)

A.

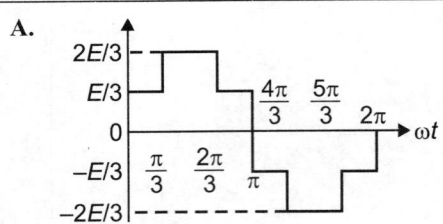

B.

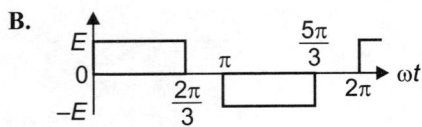

C.

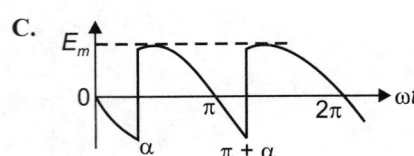

D.

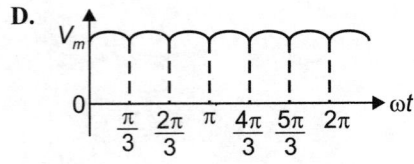

List-II (Descriptions)

1. Single-phase fully controlled ac to dc converter
2. Voltage commutated dc to dc chopper with input dc voltage E
3. Phase voltage of a three-phase inverter with 180° conduction and input dc voltage E
4. Line voltage of a three-phase inverter with 120° conduction and input dc voltage E
5. Three-phase diode bridge rectifier

Codes:

	A	B	C	D
(A)	3	4	1	5
(B)	5	1	4	2
(C)	3	1	4	5
(D)	5	4	1	2

[IES 2001]

3. In case of voltage source inverter, freewheeling can be needed for the load of:
   (A) inductive nature
   (B) capacitive nature
   (C) resistive nature
   (D) back emf nature   **[IES 2001]**

4. PWM switching is preferred in voltage source inverters for the purpose of:
   (A) controlling output voltage
   (B) output harmonics
   (C) reducing filter size
   (D) controlling output voltage, output harmonics and reducing filter size

   [IES 2001]

5. The operation of an inverter fed induction motor can be shifted from motoring to regenerative braking by:
   (A) reversing phase sequence

(B) reducing inverter voltage

(C) decreasing inverter frequency

(D) increasing inverter frequency

[IES 2002]

6. Compared to a single-phase half-bridge inverter, the output power of a single-phase full-bridge inverter is higher by a factor of:

(A) 12   (B) 8

(C) 4   (D) 2   **[IES 2002]**

7. Figure (a) shows an inverter circuit with a dc source voltage V_s. The semiconductor switches of the inverter are operated in such a manner that the pole voltage V_{10} and V_{20} are as shown in figure (b), what is the rms value of the pole-to-pole voltage V_{12}?

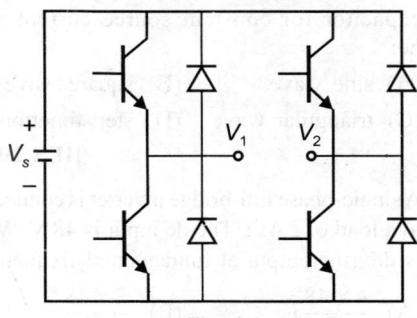

Fig. (a)

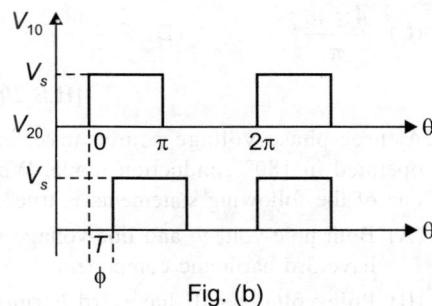

Fig. (b)

(A) $\dfrac{V_s\phi}{\pi/2}$ (B) $V_s\sqrt{\dfrac{\phi}{\pi}}$

(C) $V_s\sqrt{\dfrac{\phi}{2\pi}}$ (D) $\dfrac{V_s}{\pi}$ **[Gate 2002]**

8. In a self-controlled synchronous motor fed from a variable frequency inverter:

(A) the rotor poles invariably have damper windings

(B) there are stability problems

(C) the speed of the rotor decides stator frequency

(D) the frequency of the stator decides the rotor speed   **[IES 2003]**

9. A dc source is switched in steps to synthesize the three-phase output. The basic three-phase bridge inverter can be controlled. The angle through which each switch conducts, and at any instant, the number of switches conducting simultaneously are, respectively:

(A) 120° and 02   (B) 120° and 03

(C) 180° and 02   (D) 180° and 04

[IES 2003]

10. For a single-phase, full-bridge inverter supplying power to a highly inductive load as shown below, the correct sequence of operation of switches and didoes is:

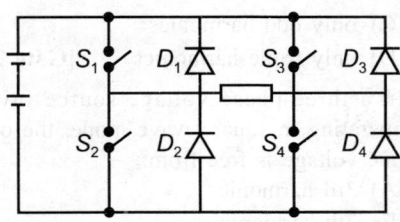

(A) $S_1\,S_4 - S_3\,S_2 - S_1\,S_4 - S_3\,S_2$

(B) $S_1\,S_2 - D_1\,D_2 - S_3\,S_4 - D_3\,D_4$

(C) $S_1\,D_3 - S_1\,S_4 - S_4\,D_2 - D_2\,D_3$

(D) $S_2\,D_4 - D_4\,D_1 - D_1\,S_3 - S_3\,S_2$

[IES 2003]

11. A single-phase, half-bridge inverter has input voltage of 48 V DC. Inverter is feeding a load of 2.4 Ω. The rms output voltage at fundamental frequency is:

(A) $\dfrac{2\times48}{\pi}V$ (B) $\dfrac{2\times48}{\sqrt{2}\pi}V$

(C) $\dfrac{\sqrt{2}\times48}{\pi}V$ (D) $\dfrac{2\times48}{2\sqrt{2}\pi}V$

[IES 2003]

12. A single-phase inverter has square wave output voltage. What is the percentage of the fifth harmonic component in relation to the fundamental component?

(A) 40%   (B) 30%

(C) 20%   (D) 10%

[IES 2004]

13. Which one of the following statements is correct?

A voltage source inverter is normally employed:

(A) when the source has low impedance and load has high reactance

(B) when the source has high impedance and load has low reactance

(C) when both the source and load have high values of impedance and reactance respectively

(D) when both the source and load have low values of impedance and reactance, respectively **[IES 2005]**

14. The output voltage waveform of a three-phase square-wave inverter contains:

(A) only even harmonics

(B) both odd and even harmonics

(C) only odd harmonics

(D) only triple harmonics **[Gate 2005]**

15. In a three-phase voltage source inverter operating in square wave mode, the output line voltage is free from:

(A) 3rd harmonic

(B) 7th harmonic

(C) 11th harmonic

(D) 13th harmonic

[IES 2006]

16. The pulse-width modulated inverter for the control of an ac motor is fed from which one of the following?

(A) Controlled rectifier

(B) Uncontrolled rectifier

(C) ac regulator

(D) Cycloconverter **[IES 2007]**

17. Consider the following statements:

1. Both voltage source inverter and current source inverter require feedback diodes.

2. Only current source inverter requires feedback diodes.

3. GTOs can not be used in a current source inverter.

4. Only voltage source inverter requires feedback diodes.

Which of these statements is/are correct?

(A) 1 only   (B) 2 and 3

(C) 3 and 4   (D) 4 only

[IES 2007]

18. A single-phase voltage source inverter is controlled in a single pulse-width modulated mode with a pulse width of 150° in each half cycle. Total harmonic distortion is defined as:

$$\text{THD} = \frac{\sqrt{V_{rms}^2 - V_1^2}}{V_1} \times 100$$

where V_1 is the rms value of the fundamental component of the output voltage. The THD of output ac voltage waveform is:

(A) 65.65%   (B) 48.42%

(C) 31.83%   (D) 30.49%

[Gate 2007]

19. A single-phase current source inverter is connected with capacitive load only the waveform of the output voltage across the capacitor for constant source current will be:

(A) sine wave   (B) square wave

(C) triangular wave   (D) step function

[IES 2008]

20. A single-phase full-bridge inverter is connected to a load of 2.4 Ω. The dc input is 48 V. What is the rms output at fundamental frequency?

(A) $\dfrac{4 \times 48}{\sqrt{2}\pi} V$ (B) $\dfrac{2 \times 48}{\sqrt{2}\pi} V$

(C) $\dfrac{4 \times 48}{\pi} V$ (D) $\dfrac{2 \times 48}{\pi} V$

[IES 2008]

21. A three-phase voltage source inverter is operated in 180° conduction mode. Which one of the following statements is true?

(A) Both pole-voltage and line-voltage will have 3rd harmonic components

(B) Pole-voltage will have 3rd harmonic component but line-voltage will be free from 3rd harmonic

(C) Line-voltage will have 3rd harmonic component but pole-voltage will be free from 3rd harmonic

(D) Both pole-voltage and line-voltage will be free from 3rd harmonic components
[Gate 2008]

22. A single-phase voltage source inverter is feeding a purely inductive load as shown in the figure.

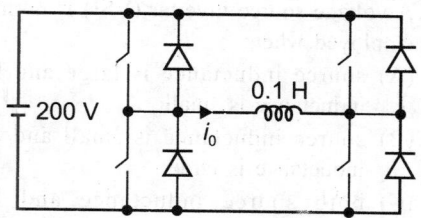

The inverter is operated at 50 Hz in 180° square wave mode. Assume that the load current does not have any dc component. The peak value of the inductor current i_0

(A) 6.37 A   (B) 10 A

(C) 20 A   (D) 40 A

                         [Gate 2008]

23. Consider the following statements, with respect to the power transistors used in inverters:
    1. Maximum collector-emitter voltage V_{CEO}.
    2. Maximum collector current.
    3. Maximum power dissipation.
    4. Maximum current gain at minimum load current.
    5. Maximum current gain at maximum load current.

    Which of these statements is/are correct?

    (A) 1 only   (B) 1, 2, 3 and 5

    (C) 2 and 3 only   (D) 2, 3 and 4

                             [IES 2009]

24. For elimination of 5th harmonics from the output of an inverter, what will be the position of pulse in a PWM inverter?

    (A) 72°   (B) 36°

    (C) 60°   (D) 90° **[IES 2009]**

25. In a single-phase VSI bridge inverter, the load current is $I_0 = 200 \sin(\omega t - 45°)$ mA. The dc supply voltage is 220 V. What is the power drawn from the supply?

    (A) 9.8 W   (B) 19.8 W

    (C) 27.25 W   (D) 34.03 W

                             [IES 2009]

26. What is the effect of blanking time on output voltage in PWM inverter?

    (A) Distortion in instantaneous voltage at current zero crossing

    (B) Low order space harmonics in output voltage

    (C) Distribution in instantaneous voltage at voltage zero crossing

    (D) High order time harmonics in output voltage   **[IES 2009]**

27. In single pulse modulation of PWM inverters, the pulse width is 120°. For an input voltage of 220 V dc, what is the rms value at the fundamental component of the output voltage?

    (A) 171.5 V   (B) 254.0 V

    (C) 127.0 V   (D) 89.81 V

                             [IES 2009]

28. A constant current source inverter supplies 20 A to a load resistance of 1 Ω to a load resistance change to 5 Ω, then the load current:

    (A) remains same at 20 A and the load voltage changes to 100 V

    (B) changes to 4 A from 20 A and the load voltage changes to 20 V

    (C) changes to 4 A from 20 A and the load voltage changes to 80 V

    (D) load voltage stay at 20 A and 20 V, respectively   **[IES 2010]**

29. The below figure shows an inverter circuit with a dc source voltage V_S. The semiconductor switches of the inverter are operated in such a way that the pole voltages of V_{10} and V_{20} are shown in the figure (b). What is the RMS value of the pole voltage V_{12}?

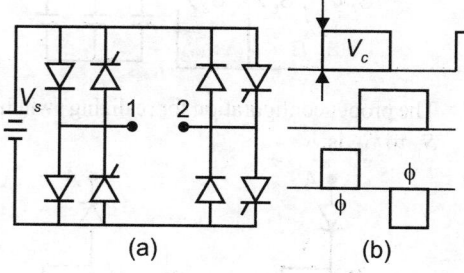

       (a)   (b)

(A) $\dfrac{V_S \phi}{\sqrt{2}\pi}$ (B) $V_S \sqrt{\dfrac{\phi}{\pi}}$

(C) $V_S \sqrt{\dfrac{\phi}{2\pi}}$ (D) $\dfrac{V_S}{\pi}$

                         [IES 2010]

30. **Assertion (A):** The L and C components of the communication circuit in McMurray inverter are chosen such, that the peak value of resonant current pulse during communication is sufficiently greater than the load current.

    Reason (R): A thyristor will successfully turn-off if the current is maintained below holding value for a time greater than the turn-off time of the device.

(A) both (A) and (R) are true and (R) is the correct explanation of (A)

(B) both (A) and (R) are true but (R) is NOT the correct explanation of (A)

(C) (A) is true but (R) is false

(D) (A) is false but (R) is true **[IES 2010]**

31. In a PWM inverter, f_0 and f are the frequencies in Hz for the carrier signal and reference signal respectively. Then the number of pulses per half cycle is:

(A) $N = f/f_0$ (B) $N = f/2f_0$

(C) $N = f_0/2f$ (D) $N = f_0/f$

[IES 2010]

32. A three-phase current source inverter used for the speed control of an induction motor is to be realized using MOSFET switches as shown below. Switches S_1 to S_6 are identical switches.

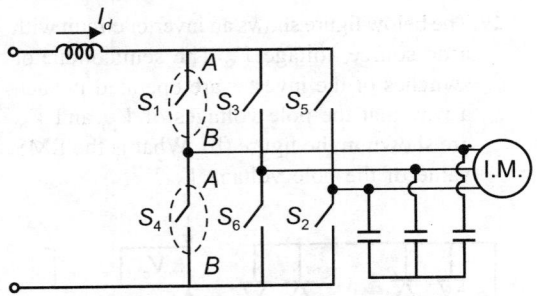

The proper configuration for realizing switches S_1 to S_6 is:

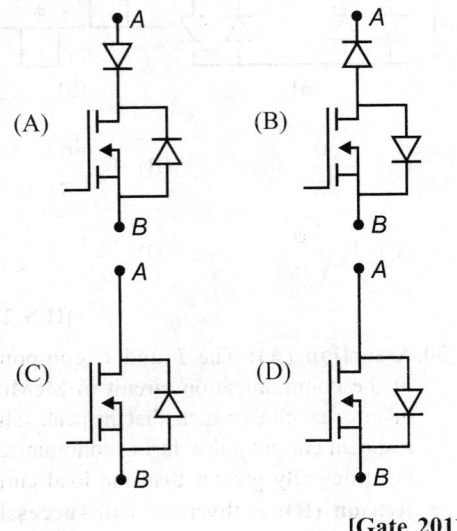

[Gate 2011]

33. A voltage source inverter (VSI) is normally employed when:

(A) source inductance is large and load inductance is small

(B) source inductance is small and load inductance is large

(C) both source inductance and load inductance are small

(D) both source inductance and load inductance are large **[IES 2012]**

34. A current source inverter is obtained by inserting a large:

(A) inductance in series with dc supply

(B) capacitance in parallel with dc supply

(C) inductance in parallel with dc supply

(D) capacitance in series with dc supply

[IES 2012]

35. **Statement I:** Multiple pulse width modulation is used to reduce the harmonic content in inverters.

Statement II: The high order harmonics can be easily filtered using passive filters.

(A) both statement I and II are individually true and statement II is the correct explanation of statement I

(B) both statement I and II are individually true but statement II is not the correct explanation of statement I

(C) statement I is true but statement II is false

(D) statement I is false but statement II is true **[IES 2012]**

36. A PWM switching scheme is used with a three-phase inverter to:

(A) reduce the total harmonic distortion with modest filtering

(B) minimize the load on the DC side

(C) increase the life of the batteries

(D) reduce low order harmonics and increase high order harmonics **[IES 2013]**

37. **Statement I:** The output current of a current source inverter remains constant irrespective of load.

Statement II: The load voltage in CSI depends on the load impedance.

(A) both statement I and II are individually true and statement II is the correct explanation of statement I

(B) both statement I and II are individually true but statement II is not the correct explanation of statement I

(C) statement I is true but statement II is false

(D) statement I is false but statement II is true. **[IES 2013]**

38. A single-phase, voltage source, square wave inverter feeds a pure inductive load. The waveform of the current will be:

(A) sinusoidal

(B) rectangular

(C) trapezoidal

(D) triangular **[IES 2014]**

39. What should be the frequency modulation ration (m_f) for a three-phase inverter, if the m_fth harmonic and its odd multiples are to be suppressed in the line-to-line voltages?

(A) m_f should be odd

(B) m_f should be even

(C) m_f should be an odd multiple of 3

(D) m_f should be an even multiple of 3

[IES 2014]

ANSWER KEY

1. (D)	2. (A)	3. (A)	4. (D)	5. (C)	6. (C)	7. (B)	8. (D)
9. (B)	10. (A)	11. (B)	12. (C)	13. (A)	14. (C)	15. (A)	16. (B)
17. (C)	18. (C)	19. (C)	20. (A)	21. (D)	22. (B)	23. (B)	24. (A)
25. (B)	26. (D)	27. (A)	28. (A)	29. (B)	30. (A)	31. (C)	32. (A)
33. (B)	34. (A)	35. (B)	36. (A)	37. (A)	38. (D)	39. (C)	

MCQ for Electrical Drives from Various Examinations

1. A three-phase semiconverter feeds the armature of separately excited dc motor, supplying a non-zero torque, for steady-state operation, the motor armature current is found to drop to zero at certain instances of time. At such instances, the voltage assumes a value that is:
   (A) equal is the instantaneous value of the ac phase voltage
   (B) equal to the instantaneous value of the motor back emf
   (C) arbitrary
   (D) zero   **[Gate 2001]**

2. A single-phase half-controlled rectifier is driving a separately excited dc motor. The dc motor has a back emf constant of 0.25 V/rpm. The armature current is 5 A without any ripple. The armature resistance is 2 Ω. The converter is working from a 230 V, single-phase ac source with a firing angle of 30°. Under this operating condition, the speed of the motor will be:
   (A) 339 rpm   (B) 359 rpm
   (C) 366 rpm   (D) 386 rpm
              [Gate 2004]

3. A variable speed drive rated for 1500 rpm, 40 Nm is reversing under no load. Figure shows the reversing torque and the speed during the transient. The moment of inertia of the drive is:

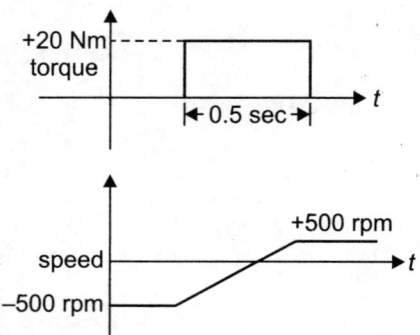

4. A motor armature supplied through phase controlled SCRs receives a smoother voltage shape at:
   (A) high motor speed
   (B) low motor speed
   (C) rated motor speed
   (D) none of these   **[IES 2005]**

5. An electric motor, developing a starting torque of 15 Nm, starts with a load torque of 7 Nm on its shaft. If the acceleration at start is 2 rad/sec², the moment of inertia of the systems must be (neglecting viscous and Coulomb friction)
   (A) 0.25 kg m²   (B) 0.25 Nm²
   (C) 4 kg m²   (D) 4 Nm²
              [Gate 2005]

6. A solar cell of 350 V is feeding power to an ac supply of 440 V, 50 Hz through a three-phase fully controlled bridge converter. A large inductance is connected in the dc circuit to maintain the dc current at 20 A. If the solar cell resistance is 0.5 Ω, then each thyristor will be reverse biased for a period of:
   (A) 125°   (B) 120°
   (C) 60°   (D) 55° **[Gate 2006]**

7. Consider the following statements:
   Assertion (A): Power electronic converters are extensively used in adjustable speed drives.
   Reason (R): Power electronic converters do not produce harmonic distortion.
   Of these statements:
   (A) both (A) and (R) are true and (R) is the correct explanation of (A)
   (B) both (A) and (R) are true but (R) is not the correct explanation of (A)
   (C) (A) is true but (R) is false
   (D) (A) is false but (R) is true **[IES 2007]**

(A) 0.048 kg m²   (B) 0.064 kg m²
(C) 0.096 kg m²   (D) 0.128 kg m²
           [Gate 2004]

8. A three-phase, 440 V, 50 Hz ac mains fed thyristor bridge is feeding a 440 V dc, 15 kW, 1500 rpm separately excited dc motor with a ripple free continuous current in the dc link under all operating conditions, neglecting the losses, the power factor of the ac mains at half the rated speed is:

(A) 0.354   (B) 0.372
(C) 0.90   (D) 0.955

[Gate 2007]

9. A cycloconverter-fed induction motor drive is most suitable for which one of the following?
(A) Compressor drive
(B) Machine tool drive
(C) Paper mill drive
(D) Cement mill drive   **[IES 2008]**

10. A single-phase fully controlled converter bridge is used for electrical breaking of a separately excited dc motor. The dc motor load is respectively by an equivalent circuit as shown in the figure.

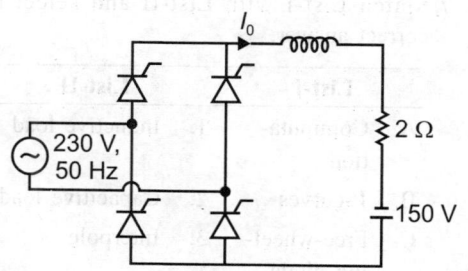

Assume that the load inductance is sufficient to ensure continuous and ripple free load current. The firing angle of the bridge for a load current of $I_0 = 10$ A will be:

(A) 44°   (B) 51°
(C) 129°   (D) 136°

[Gate 2008]

11. For low-speed high-power reversible operation, the most suitable drives are:
(A) voltage source inverter fed ac drives
(B) current source inverter fed ac drives
(C) dual converted fed dc drives
(D) cycloconverter fed ac drives

[IES 2011]

12. The separately excited dc motor in the figure below has a rated armature current of 20 A and a rated armature voltage of 150 V. An ideal chopper switching at 5 kHz is used to control the armature voltage. If $L_a = 0.1$ mH, $R_a = 1\,\Omega$, neglecting armature reaction, the duty ratio of the chopper to obtain 50% of the rated torque at the rated speed and the rated field current is:

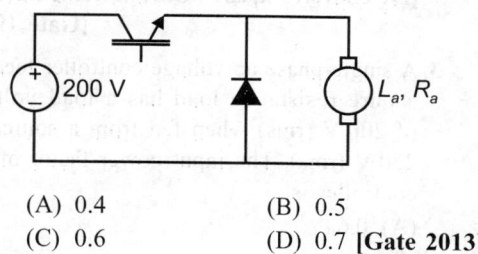

(A) 0.4   (B) 0.5
(C) 0.6   (D) 0.7 **[Gate 2013]**

ANSWER KEY

1. (B)	2. (A)	3. (A)	4. (A)	5. (C)	6. (D)	7. (C)	8. (A)
9. (B)	10. (C)	11. (D)	12. (D)				

MCQ for Miscellaneous from Various Examinations

1. If a diode is connected in anti-parallel with a thyristor, then:
   (A) both turn-off power loss and turn-off time decreases
   (B) turn-off power loss decreases but turn-off time increases
   (C) turn-off power loss increases, but turn-off time decreases
   (D) none of these   **[Gate 1997]**

2. Resonant converter's are basically used to:
   (A) generate large peaky voltage
   (B) reduce the switching losses
   (C) eliminate harmonics
   (D) convert a square wave into a sine wave
                                          [Gate 1999]

3. A single-phase ac voltage controller feeding a pure resistance load has a load voltage of 200 V (rms) when fed from a source of 250 V (rms). The input power factor of the controller is:
   (A) 0.64
   (B) 0.8
   (C) 0.894
   (D) difficult to estimate because of insufficiency of data   **[IES 2001]**

4. The most suitable solid-state converter for controlling the speed of the three-phase cage motor at 25 Hz is:
   (A) cycloconverter
   (B) current source inverter
   (C) voltage source inverter
   (D) load commutated inverter   **[IES 2001]**

5. The quality of output ac voltage of a cyclo-converter is improved with:
   (A) increase in output voltage at reduced frequency
   (B) increase in output voltage at increased frequency
   (C) decrease in output voltage at reduced frequency
   (D) decrease in output voltage at increased frequency   **[IES 2001]**

6. A cycloconverter is operating on a 50 Hz supply. The range of output frequency that can be obtained with acceptable quality, is:
   (A) 0–16 Hz
   (B) 0–32 Hz
   (C) 0–64 Hz
   (D) 0–128 Hz   **[IES 2001]**

7. Match List-I with List-II and select the correct answer:

	List-I		List-II
A.	Commutation	1.	Inductive load
B.	V-curves	2.	Capacitive load
C.	Free-wheeling diode	3.	Interpole
D.	Overlap	4.	Source inductance
		5.	Synchronous motor

Codes:

	A	B	C	D
(A)	3	5	1	4
(B)	2	4	3	5
(C)	3	4	1	5
(D)	2	5	3	4

                                    [IES 2002]

8. AC voltage regulators are widely used in:
   (A) traction drives
   (B) fan drives
   (C) synchronous motor drives
   (D) slip power recovery scheme of slip-ring induction motor   **[IES 2002]**

9. How many switches are used to construct a three-phase cycloconverter?

(A) 3   (B) 6

(C) 12   (D) 18   **[IES 2002]**

10. A three-phase cycloconverter is used to obtain a variable-frequency single-phase ac output. The single-phase ac load is 220 V, 60 A at a power factor of 0.6 lagging. The rms value of input voltage per phase required is:

(A) 376.2 V   (B) 311.12 V

(C) 266 V   (D) 220 V

                          [IES 2002]

11. Match List-I with List-II of conversion) and select the correct answer:

List-I (Converters)		List-II (Type of conversion)
A.	Controlled rectifier	1. Fixed DC to variable voltage and variable frequency AC
B.	Chopper	2. Fixed DC to variable DC
C.	Inverter	3. Fixed AC to variable DC
D.	Cyclocon-verter	4. Fixed AC to variable frequency AC

Codes:

	A	B	C	D
(A)	2	3	1	4
(B)	3	2	4	1
(C)	2	3	4	1
(D)	3	2	1	4

                          [IES 2003]

12. An AC voltage-regulator using back-to-back connected SCRs is feeding an RL load. The SCR firing angle $\alpha < \phi$ (ϕ is power factor angle of the load). If SCRs are fired using short-duration gate pulses, the output load-voltage waveform will be:

(A) symmetrical chopped ac voltage

(B) half-wave rectified

(C) full-wave rectified

(D) sinusoidal   **[IES 2003]**

13. An inter-group reactor is used in a single-phase cycloconverter circuit to:

(A) reduce current-ripples

(B) reduce voltage-ripples

(C) limit circulating current

(D) limit di/dt in the semiconductor switch

                          [IES 2003]

14. The triac circuit shown in figure controls the ac output power to the resistive load. The peak power dissipation in the load is:

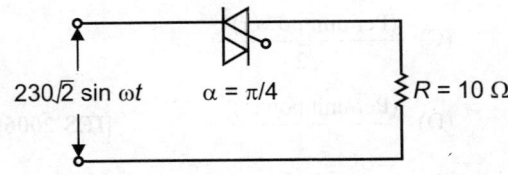

$$230\sqrt{2}\sin \omega t \qquad \alpha = \pi/4 \qquad R = 10\ \Omega$$

(A) 3968 W   (B) 5290 W

(C) 7935 W   (D) 10580 W

                          [Gate 2004]

15. For a single-phase ac voltage controller feeding a resistive load, what is the power factor?

(A) Unity for all values of firing angle

(B) $\left[\dfrac{1}{\pi}\left\{(\pi - \alpha) + \dfrac{1}{2}\sin 2\alpha\right\}\right]^{1/2}$

(C) $\left[\dfrac{1}{\pi}\left\{(\pi + \alpha) + \dfrac{1}{2}\sin 2\alpha\right\}\right]^{1/2}$

(D) $\left[\dfrac{1}{\pi}\left\{(\pi + \alpha) - \dfrac{1}{2}\sin 2\alpha\right\}\right]^{1/2}$

where α is firing angle measured from voltage zero.   **[IES 2005]**

16. Match List-I with List-II and select the correct answer using the code given below the lists:

List-I		List-II
A.	Voltage source inverter	1. Larger source inductance
B.	Current source inverter	2. Poor power factor
C.	Phase controlled ac to dc con-verter	3. Inverter limit
D.	Cycloconverter	4. Small source inductance

Codes:

	A	B	C	D
(A)	4	2	3	1
(B)	3	1	4	2
(C)	4	1	3	2
(D)	3	2	4	1	**[IES 2006]**

17. What is the power factor of a single-phase ac regulator feeding a resistive load?
(A) (Per unit power)^2
(B) (Per unit power)$^{1/2}$

(C) $\dfrac{\text{(Per unit power)}^2}{\sqrt{2}}$

(D) $\dfrac{\text{(Per unit power)}^{1/2}}{2}$ **[IES 2006]**

18. A single-phase ac voltage controller is controlling current in a purely inductive load. If the firing angle of the SCR is α, what will be the conduction angle of the SCR?
(A) π (B) $(\pi - \alpha)$
(C) $(2\pi - \alpha)$ (D) 2π **[IES 2007]**

19. What are the advantage of switching power supplies over linear power supplies?
1. The devices operate in linear/active region.
2. The devices operate as switches.
3. Power losses are less.
Select the correct answer using the code given below:
(A) 1 and 3   (B) 2 and 3
(C) 1 and 2   (D) 1, 2 and 3
[IES 2008]

20. Match List-I with List-II and select the correct answer using the code given below the lists:

	List-I		List-II
A.	Chopper controlled resistance in the rotor circuit of an induction motor	1.	Very low speed, high-power reversible drive
B.	Sub-synchro-nous converter-cascade in the rotor circuit of an induction motor	2.	Centrifuges in sugar industry
C.	3-phase ac voltage controller	3.	Blowers and compressors
D.	Cycloconverter	4.	Loads requiring good starting performance

	A	B	C	D
(A)	3	4	2	1
(B)	3	4	1	2
(C)	4	3	1	2
(D)	4	3	2	1	**[IES 2008]**

21. In the single-phase voltage controller circuit shown in the figure, for what range of triggering angle (α), the output voltage (V_0) is not controllable?

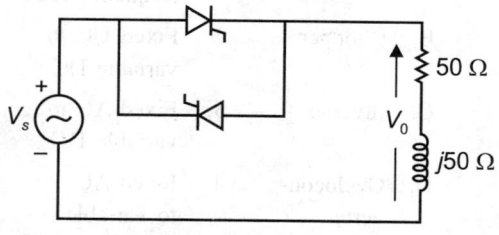

(A) $0° < \alpha < 45°$
(B) $45° < \alpha < 135°$
(C) $90° < \alpha < 180°$
(D) $135° < \alpha < 180°$

[Gate 2008]

22. In the ac regulator of Fig. 1, the supply voltage and gate currents waveforms are as in Fig. 2, what is the load voltage waveform for $R = 0$?

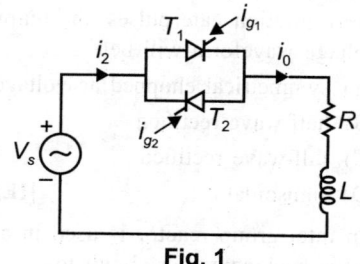

Fig. 1

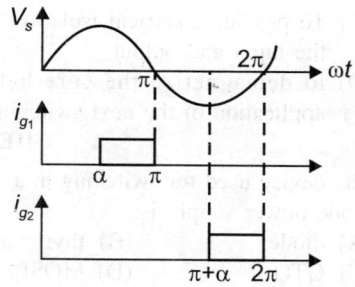

Fig. 2

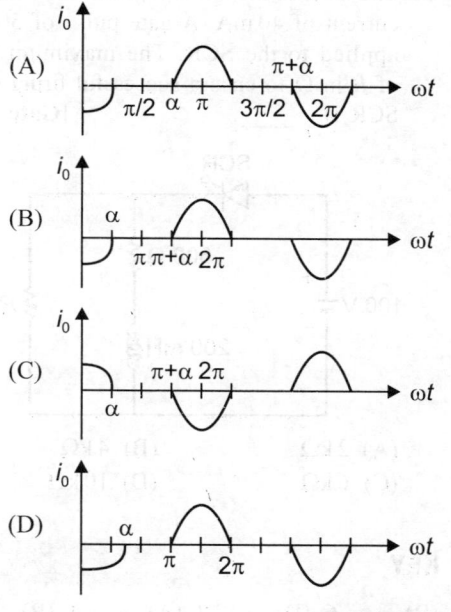

[IES 2009]

23. In push-pull type DC-DC converter the output voltage V_0 is given by:

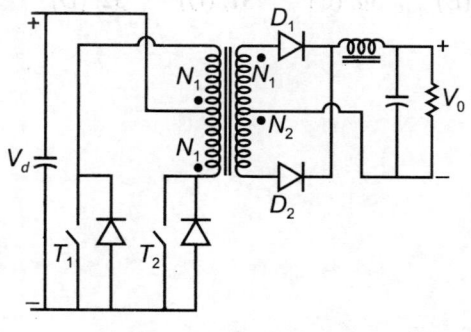

(A) $V_0 = 2\dfrac{N_2}{N_1} \cdot V_d \left(\dfrac{t_{ON}}{t_{ON} + t_{OFF}} \right)$

(B) $V_0 = \dfrac{N_2}{N_1} \cdot V_d \left(\dfrac{t_{ON}}{t_{ON} + t_{OFF}} \right)$

(C) $V_0 = 2\dfrac{N_2}{N_1} \cdot V_d \left(\dfrac{t_{ON}}{t_{OFF}} \right)$

(D) $V_0 = \dfrac{N_2}{N_1} \cdot V_d \left(\dfrac{t_{ON}}{t_{OFF}} \right)$ **[IES 2010]**

24. The power electronic converter shown in the figure has a single-pole double-throw switch. The pole P of the switch is connected alternately to throws A and B. The converter shown is a:

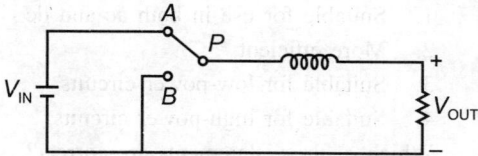

(A) step-down chopper (buck converter)

(B) half-wave rectifier

(C) step-up chopper (boost converter)

(D) full-wave rectifier **[Gate 2010]**

25. An advantage of a cycloconverter is:

(A) very good power factor

(B) requires few number of thyristors

(C) commutation failure does not short circuit the source

(D) load commutation is possible

[IES 2011]

26. An integral cycle ac voltage controller is feeding a purely resistive circuit from a single-phase ac voltage source. The current waveform consists alternately burst of N-complete cycle of conduction followed by M-complete of extinction. The rms value of the load voltage equals the rms value of supply voltage for:

(A) $N = M$ (B) $N = 0$

(C) $N = M = 0$ (D) $M = 0$

[IES 2011]

27. The maximum current through the battery will be:

(A) 14 A   (B) 40 A

(C) 80 A   (D) 94 A

[Gate 2011]

28. The kVA rating of the input transformer is:

(A) 53.2 kVA   (B) 46.0 kVA

(C) 22.6 kVA   (D) 19.6 kVA

[Gate 2011]

29. A single-phase ac regulator fed from 50 Hz supply feeds a load having 4 Ω resistance and 12.73 mH inductance. The control range of firing angle will be:

(A) 0° to 180°   (B) 45° to 180°

(C) 90° to 180°   (D) 0° to 45°

[IES 2012]

30. Consider the following statements:

Switched mode power supplies are preferred over the continuous types, because they are:

1. Suitable for use in both ac and dc
2. More efficient
3. Suitable for low-power circuits
4. Suitable for high-power circuits

Which of these statements are correct?

(A) 1 and 2   (B) 1 and 3

(C) 2 and 3   (D) 3 and 4

[IES 2013]

31. In a forward converter, a tertiary winding is used. What is the reason?

(A) to provide di/dt protection to the switching device.

(B) to provide dv/dt protections to the switching device.

(C) To provide electrical isolation between the input and output.

(D) to demagnetize the core before the application of the next switching pulse.

[IES 2014]

32. The device used for switching in a switched mode power supply is:

(A) diode   (B) thyristor

(C) GTO   (D) MOSFET

[IES 2014]

33. The SCR in the circuit shown has a latching current of 40 mA. A gate pulse of 50 μs is applied to the SCR. The maximum value of R in Ω to ensure successful firing of the SCR is _____.   **[Gate 2014]**

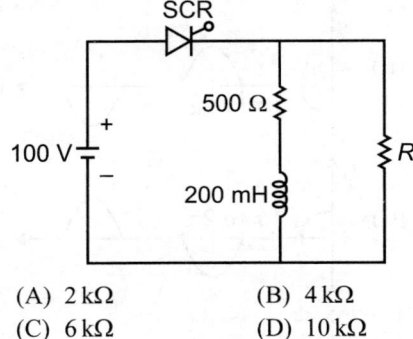

(A) 2 kΩ   (B) 4 kΩ

(C) 6 kΩ   (D) 10 kΩ

ANSWER KEY

1. (B)	**2.** (B)	**3.** (B)	**4.** (A)	**5.** (B)	**6.** (D)	**7.** (A)	**8.** (B)
9. (D)	**10.** (C)	**11.** (D)	**12.** (D)	**13.** (C)	**14.** (D)	**15.** (B)	**16.** (C)
17. (B)	**18.** (B)	**19.** (B)	**20.** (A)	**21.** (A)	**22.** (A)	**23.** (C)	**24.** (A)
25. (D)	**26.** (A)	**27.** (B)	**28.** (C)	**29.** (B)	**30.** (C)	**31.** (D)	**32.** (D)
33. (C)							

Index